Finite Element Mesh Generation

Finite Element Mesh Generation

DANIEL S.H. LO

CRC Press
Taylor & Francis Group
Boca Raton London New York

CRC Press is an imprint of the
Taylor & Francis Group, an **informa** business

A SPON PRESS BOOK

CRC Press
Taylor & Francis Group
6000 Broken Sound Parkway NW, Suite 300
Boca Raton, FL 33487-2742

First issued in paperback 2017

© 2015 by S.H. Lo
CRC Press is an imprint of Taylor & Francis Group, an Informa business

No claim to original U.S. Government works

ISBN-13: 978-0-415-69048-5 (hbk)
ISBN-13: 978-1-138-74924-5 (pbk)

Visit the Taylor & Francis Web site at
http://www.taylorandfrancis.com

and the CRC Press Web site at
http://www.crcpress.com

Contents

5   Mesh generation in three dimensions   239

Preface

Nowadays, the finite element method has diverse applications to problems in science and engineering ranging from simple two-dimensional static elasticity, non-linear large deformation analysis to three-dimensional fluid dynamic problems with shock waves. The prerequisite for a finite element analysis is a sound and valid finite element mesh, which can only be constructed efficiently by means of some well-devised and thoroughly tested computer algorithms. In contrast to the numerous textbooks, monographs, journal papers, etc., on the finite element method, comprehensive and concise accounts on mesh generation technologies seem to have been missing, except perhaps the book *Mesh Generation: Application to Finite Elements* written by P.J. Frey and P.L. George some 15 years ago. Anyway, finite element mesh generation has not been taken as a formal subject of teaching in universities, as it encompasses several disciplines including classical geometry, computational geometry and topology, finite element method, data structures and algorithms, computer programming and, to a certain extent, even computer graphics.

With the ever-improving performance of PCs, large-scale challenging engineering simulations and scientific computations by means of the finite element method are more accessible to daily design operations and even to research students. In line with this development, the mesh generation methodology is becoming increasingly recognised as a subject in its own right. As meshing technologies and their applications in new areas have developed pretty rapidly over the recent years, it is imperative to review and consolidate the progress in meshing technologies achieved thus far into a concise yet comprehensive text with a logical sequence as a valuable reference for laymen and experts alike.

Mesh generation over planar domains, curved surfaces and volumes with simplicial and non-simplicial elements on bounded and unbounded domains by means of a single processor or parallel processing will all be discussed in this text. Auxiliary techniques in facilitating finite element mesh generation will also be included to make the text self-contained and complete. From the geometrical and topological aspects and their associated operations and inter-relationships, each approach is vividly described and illustrated with examples. As the devil lies in the details, and the truth is also in the details, the basic concept along with every detail in the implementation of all the popular meshing techniques will be emphasised and elucidated with algorithms, flowcharts, pseudo-codes, illustrations and sample meshes.

The main theme (backbone) of the book is built on the presentation and formulation of various mesh generation methods in a logical natural sequence of meshing over two dimensions, curved surfaces and three dimensions. An introduction and the fundamentals in geometrical and topological computations have been added in the first two chapters to pave the way for a comfortable and enjoyable journey through the mesh generation algorithms developed for physical domains of different dimensions, geometries and characteristics. Equally important and indispensable in advanced applications to generate high-quality meshes for large-scale problems subject to difficult boundary constraints, mesh optimisation, parallel processing and auxiliary techniques will be discussed as well in the last three chapters to supplement and enhance the general meshing strategies described in the previous chapters. Innovative and unpublished materials could be found in various parts of this book; though they have not been thoroughly tested and verified, the preliminary results do look quite promising, and they would definitely inspire and stimulate new ideas for further improvements in mesh generation.

The content materials and writing style are not targeted to any particular group but rather to the general public who are interested in mesh generation technology and its developments. Sufficient details along with chapters of fundamentals and supplementary formulas and algorithms in the Appendix should allow even beginners in a self-learning mode to go through all the chapters without much difficulty. On the other hand, the book is concise and comprehensive enough to include all the popular mesh generation methods along with auxiliary techniques, new innovative materials and a long list of references that are of interest and value even to experts in the field of mesh generation.

Those who have little idea about mesh generation or even the finite element method can start reading from Chapter 1 down to the last chapter if they don't mind to spend time exploring everything about finite element mesh generation. Those who would just like to develop their own mesh generation computer programs can go directly to the relevant sections to consult the procedures and/or pseudo-codes for reference. From the abundant examples and tables of results in mesh quality and CPU time, experts will also find the various formulations and the corresponding algorithms useful as a reference and a possible source of comparison with their own. Although the book is not intended to be a textbook for undergraduates, the materials covered are broad and deep enough to support a one-semester university course. However, the best way to be familiarised with finite element mesh generation is to have first-hand experience in producing one's own version of mesh generation computer programs at least for the two classical methods, namely, the advancing-front approach and the Delaunay triangulation. To this end, computer listings of planar triangular mesh by means of the advancing-front approach and three-dimensional Delaunay triangulation of a set of spatial points are given in the Appendix. Finally, opinions and comments are most welcome to be sent to hreclsh@hku.hk.

S.H. Lo
Hong Kong

Acknowledgements

This work could not have been completed so smoothly without the help of many people from various places at different times to whom the author would like to express his deepest gratitude.

First of all, the author would like to thank his wife, Vivian, for her forbearance and care about their daughters Germaine and Maxine.

The Senior Research Fellowship awarded by the Croucher Foundation allowed the author to be relieved from his normal teaching and administrative duties for one year, and most of the background works were completed during this precious period of tranquility.

The author is indebted to Senior Editor Tony Moore of CRC Press and Spon Press (imprints of Taylor & Francis) for his kind invitation and encouragement to initiate this work as well as for his trust and patience throughout the entire course of writing.

Numerous research collaborators need special mentioning for their direct and indirect contributions. At the University of Hong Kong, the author ought to thank Prof. Y.K. Cheung for his sharing of the finite element method; Dr. C.K. Lee and T.S. Lau in surface meshing and adaptive refinement analysis – in particular, Dr. Lee of Nanyang Technological University has also provided examples of quadrilateral surface meshes; Dr. W.X. Wang in the surface intersection and packing of ellipses and spheres; and Prof. K.Y. Sze in the formulation of hybrid stress high-performance transition quadrilateral and hexahedral elements.

As for colleagues and friends at INRIA–Rocquencourt in France, the author would like to thank Prof. M. Bernadou, Prof. P.L. George and Prof. H. Borouchaki for their kind invitations for a number of Sabbatical visits; Prof. George's interesting work in Delaunay triangulation and boundary recovery; Prof. Borouchaki's inspiring discussion on anisotropic meshing and possibility of parallel Delaunay triangulation; and Prof. P. Laug in parametric surface meshing.

Special thanks go to Prof. J.F. Lau of Peking University for his work on Delaunay triangulation of non-uniformly distributed point sets and Prof. Z.Q. Guan of Dalian University of Technology for the collaborative work on boundary recovery and large displacement mesh optimisations. Finally, the collective works and efforts of all the researchers in the mesh generation community not only have made the subject so interesting and promising but also have been the most inspiring and stimulating in pushing the mesh generation technologies to a new frontier.

Chapter 1

Introduction

The introduction gives some ideas to those who know or don't know finite element mesh generation.

1.1 FINITE ELEMENT METHOD

In essence, the finite element method is a numerical technique that provides approximate solutions to the governing equations of a complicated system through a discretisation process. The system of interest can be either physical or mathematical. The domain of the system can be well defined or subject to continual changes (moving boundary problems such as transient-free surface water flow, large deformation problems, etc.). The boundary conditions can be well defined in terms of prescribed loads and displacements, or sometimes less well defined as in fluid–structure interactions or contact problems. The governing equations can be given in differential form or be expressed in terms of variation integrals.

Before an analysis is carried out, the entire system has to be divided into a number of individual subsystems or components whose behaviour is readily understood. The basic units of the discretised subsystems are called finite elements, which should neither overlap nor have gaps between each other. The finite elements used for a domain need not be of the same type, and the properties could also vary. Figure 1.1 shows how a smooth curved surface, as defined by function ϕ, is modelled by elements of various types. When three-node triangular elements are used, the ϕ surface is approximated by flat triangular facets, whereas the four-node and eight-node quadratic elements are able to represent warped and curved surfaces and can thus better approximate the actual function. Obviously, the approximation can also be improved by using more elements instead of increasing the order of the interpolation polynomial. This sketch illustrates the basic idea of the finite element method: piecewise approximation of a smooth function by means of simple polynomials, each of which is defined over a small region (element) and represented in terms of the values of the function at the element nodes.

1.2 WHAT IS FINITE ELEMENT MESH GENERATION?

A finite element mesh is a partition of a given domain into subdomains, which are called elements, such that every point of the domain is found in one of the elements. The entire domain has to be covered by the elements without overlapping, and the conditions of compatibility between finite elements on the boundary have to be satisfied as well. Two-dimensional domains can be discretised into triangular, quadrilateral or a mixture of triangular and

1

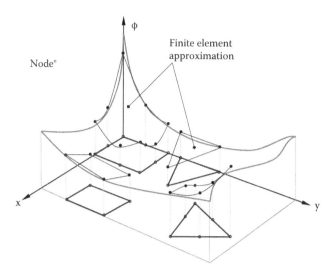

Figure 1.1 Smooth surface approximated by finite elements.

quadrilateral elements. Over three-dimensional domains, tetrahedral and hexahedral elements can be used; however, in some situations, wedges or pentahedral elements and pyramid elements could also be employed. As the topology of curved surfaces locally resembles that of a planar domain, similar to a two-dimensional problem, triangular and quadrilateral elements can be generated on surfaces. To reduce discretisation (numerical) error in a finite element analysis, the quality of the finite element meshes has to be optimised such that the element size is in compliance with the specified nodal spacing and that the shape of the elements ought to be as equilateral as possible. For conforming meshes, the boundary nodes of the finite element mesh have to lie on the boundary surface of the given domain, and for constrained meshes, apart from the geometrical requirements of a conforming mesh, additional topological requirements such as specified edges and faces have to be present in the mesh as well. Furthermore, higher-order curvilinear elements can also be employed to fit domains with curved boundaries to reduce discretisation error.

1.3 WHY FINITE ELEMENT MESH GENERATION?

Nowadays, the finite element method has tremendous applications to problems in science and engineering ranging from simple static elasticity, dynamic and transient, instability and damage mechanics analyses to more advanced applications including adaptive refinement analysis, large deformation non-linear analysis and fluid dynamic problems with shock lines. However, the accuracy of a finite element solution depends on the number of nodes in the mesh where they have been placed and the shape of the elements formed, and nowadays, a meaningful realistic engineering analysis may consist of thousands to millions of nodal points, which could be time-consuming and error-prone to be handled manually. The pre-requisite for a finite element analysis is a series of sound and valid finite element meshes, which can only be constructed efficiently by means of some theoretically sound and well-tested computer algorithms. What makes the finite element method stand out among other numerical techniques is its versatility, that is, the finite element method can be applied to domains of different dimensions and geometry subject to various boundary conditions, loading conditions and physics – static or dynamic, mechanical, thermal or coupled multi-field problems, etc. To meet all these requirements, we have to devise algorithms to generate rapidly finite element meshes of various characteristics on a planar domain, on curved surfaces and over three-dimensional volumes in a robust manner. Hybrid and mixed meshes, which are meshes consisting of various types of elements in different dimensions, may sometimes be required for certain problem types. Many mesh generation techniques, for instance, the Delaunay triangulation and the advancing-front technique (AFT), can also have applications in many other fields including data visualisation, terrain modelling, surface reconstruction, structural networking for arbitrary point sets, etc.

1.4 PROBLEM DEFINITION, SCOPE AND PHILOSOPHY: SCIENCE OR ART?

Owing to diverse applications for various disciplines under different situations, there are no formal universal rules as to how finite element mesh generation problems should be defined. However, domains represented by boundary specification are quite a common practice for meshing engineering objects, in which a planar domain is well defined by a series of boundary line segments, and three-dimensional volumes are bounded by triangular and/or quadrilateral facets without ambiguity. Other possibilities include volumes defined implicitly by a system of spatial points for which the boundary of the object can only be detected by means of some *in-or-out* inquiry mechanisms and meshing of a computational domain, which is large enough to contain the physical object or event under consideration. In summary, broadly speaking, there are three types of boundary settings for finite element mesh generation:

1. No boundary is defined, and just a large interior part extensive enough to cover the object or the event under consideration needs to be meshed, e.g. a background grid or the convex hull of a Delaunay triangulation, etc.
2. Geometrically conforming meshes, the boundary nodes of the mesh have to be on the boundary surface of the object.
3. Fully constrained meshes: apart from points, the boundary edges and faces of the mesh should all have a perfect match with those specified on the boundary surface of the object. As mesh generation is very sensitive to boundary requirements, even for the same physical domain, the mesh generation problem could be quite different subject to various boundary constraints, and very often, different mesh generation strategies have to be employed accordingly.

'Mesh generation: Art or science?' is a review paper written by Timothy J. Baker in 2005 in which no definite conclusion on whether the subject belongs to art or science has been given, except the comment 'Some of the advances were based on a sound theoretical understanding; many others were heuristic in nature, guided by an intuitive feel for what seemed like the right approach'. Mesh generation is a science in the sense that there are deterministic ways in producing certain mesh types, and there are systematic optimisation procedures in improving the quality of a finite element mesh; however, it is also an art in the sense that a solution may not exist, and there is freedom in choosing the element types, in using a different number and size of elements and in placing nodes at various positions to arrive at a solution. Boundary and internal constraints and optimal mesh quality further impose additional difficulties in the theoretical approach to the mesh generation problem, and what is the expected quality of the mesh satisfying all the boundary constraints being most likely an open question.

In view of the diverse possibilities in mesh generation, mesh generation using simplices and/or non-simplices on planar domain, on curved surfaces and over volumes bounded or unbounded will all be investigated and discussed in this text. According to Lohner (1997), there are only two basic ways to fill up a general bounded domain with elements: (i) filling the *empty*, i.e. an as-yet-unmeshed region, with elements, and (ii) modifying an existing mesh that is already covered with elements. However, there is perhaps a third way (iii) in which the mesh is refined, modified and stretched while its boundary is snapped onto the boundary of the object. A typical method for the first technique is the advancing-front approach (ADF); that for the second technique is the Delaunay triangulation; and examples for the third are the Meccano and grid/voxel methods. Finite element meshes can be broadly divided into two main types, namely, the structured mesh and the unstructured mesh. Structured meshes can be generated over smooth regular domains based on some deterministic procedures, whereas unstructured meshes are for complex irregular domains possibly with additional requirements such as element size variation and mesh directional properties, which in general can only be generated by means of some heuristic approaches.

1.5 GENERAL STRATEGIES, ROBUSTNESS, DIFFICULTIES AND METHODOLOGIES

As far as the existence of a solution is concerned, the most difficult problem or perhaps the only difficulty in mesh generation is the construction of a fully constrained finite element mesh for an arbitrary three-dimensional domain with irregular geometry and complicated boundary constraints. The difficulty is due to the fact that there exist polyhedra that can only be meshed with the introduction of interior points, the so-called Steiner points. As there is no systematic way to determine the number of Steiner points needed and their locations, we have to resort to heuristic means in an attempt to obtain a solution without degenerate elements. Since the first valid finite element mesh has special significance in indicating that the given domain is *meshable*, robustness is therefore of primary concern to a mesh generation algorithm, followed by mesh quality and speed of mesh generation. Moreover, based on the first finite element mesh, mesh quality can be further improved, and adaptive meshes with gradation in element size and directional characteristics can all be created by means of refinement and various mesh optimisation techniques. In general, simplicial meshes can have better adaptation to the more difficult boundary conditions and allow a progressive change in element size within the mesh, whereas quadrilateral and hexahedral meshes can be generated rapidly using mapping techniques over regular domains with simple boundaries. Popular mesh generation methods so far developed include Delaunay triangulation, ADF, Quadtree/Octree decomposition, Meccano transformation, refinement and coarsening, mapping and modification, optimisation by iterations, intersection and merging based on Boolean operations, etc.

1.6 MATHEMATICS

No doubt, in mesh generation, mathematics plays a vital role in providing values to various geometric quantities such as distance, angle, volume, mappings, shape measures and metric tensors in quantifying element shape and size, etc. However, mesh generation is more concerned with the number of nodes, where to place them and how they should be connected to form elements – topological operations in terms of nodal combinations for which there is no direct relationship with geometrical computations, though some estimations can be derived from the required element size and shape quality as additional constraints in mesh generation. In other words, for mesh generation, algorithms are as important as geometrical computations (Edelsbrunner 1987), except, of course, for the Voronoi tessellation or the duality of Delaunay triangulation, which is perhaps the only available interplay relationship between geometry and topology in the connection of a set of spatial points arbitrarily distributed in space. The minimum angle, which is a valid shape measure of triangular elements, is guaranteed in two-dimensional Delaunay triangulations. Based on Delaunay triangulation, some bound on the smallest interior angle can be established for two-dimensional triangular meshes conforming to a given boundary of line segments. Yet, there is no analogous valid shape measure for tetrahedral elements, which is guaranteed in three-dimensional Delaunay triangulations, and as a result, Delaunay triangulations may not be the most appropriate for numerical computations. Nevertheless, in mesh generation, it is really a crucial matter to have the first valid mesh, which can always be enhanced, modified and optimised through various transformations to turn it into a mesh apt for different purposes.

1.7 HISTORICAL DEVELOPMENT

The research on finite element mesh generation was formally started perhaps as early as the beginning of the 1970s (Mackerle 2001), and a comprehensive review of the finite element mesh generation schemes developed before 1980 was presented by Thacker (1980). In line with the advance of the finite element method, the irregular computational grid became increasingly popular for two reasons: (i) they allow points to be situated on curved boundaries of irregularly shaped domains and (ii) they allow points to be distributed at the interior of the domain with variable nodal spacing.

Co-ordinate transformation was an early attempt to map a regular reference domain onto a geometrically irregular computational physical domain with a possibility of smooth transition in element size. The finite difference method could also be applied to computational grids constructed based on co-ordinate transformation. The grids could be smoothed such that each interior point ought to be at the position determined by the average of the co-ordinates of its neighbours. In terms of mechanical analogy, the optimal grid should correspond to the equilibrium configuration of a system of springs between grid points. This idea of putting a node at the centroid of the surrounding polygon is in line with Laplace smoothing widely used in mesh optimisation up to these days. The spring analogy for the minimisation of energy has diverse applications nowadays in r-refinement (Li et al. 2001; Mosler and Ortiz 2007) and relocation of nodes by large displacements (Lin et al. 2014).

Finite element interpolation as a means of mesh generation was presented by Zienkiewicz and Phillips (1971) in which a curved domain is represented by a super-element, which could be further divided into smaller elements following the element reference co-ordinates. The blending function interpolation developed for local refinements to minimise the energy of the system is related to the r-refinement procedure that we are using today. Decomposition into simpler subregions, which is so intuitive as a means of mesh generation, was developed in the early days for the generation of structured meshes. Removing points from a fine grid generated by co-ordinate transformation and mapping a uniformly spaced zigzag boundary onto a curvilinear grid were two ideas to generate meshes of non-uniform element sizes.

Before Delaunay triangulation became widely used, finite element meshes were constructed by joining points randomly generated using heuristic connection rules. The drag method proposed by Park and Washam (1979) was perhaps the predecessor of the more sophisticated extrude and sweep methods that we are still using for mesh generation. The importance of gradation meshes was duly recognised, and various mesh generation methods based on Poisson's equation with a source term, mapping and removal of points and generation of random points with different densities were developed. Moreover, a touch on the three-dimensional problems primarily by the mapping techniques has also received quite some attention.

On the other hand, there was also substantial progress in many auxiliary techniques associated with the finite element mesh generation, namely, the node renumbering schemes for the reduction of matrix profile in the resolution of a system of linear equations (Cuthill 1972; Collins 1973; Akhras and Dhatt 1976; Lai et al. 1996; Lai 1998; Esposito et al. 1998; Kaveh and Bondarabady 2002; Fujisawa et al. 2003; Lim et al. 2006, 2007; Boutora et al. 2007; Wang et al. 2012), how boundary points and a desired point density for different regions are prescribed, the data input formats for mesh generation, etc. Data input for finite element mesh generation in batch mode and interactive mode was developed, and the latter, after years of evolution, can now be regarded as a proper model building CAD system. Sparked off by the review of Thacker (1980), unstructured mesh generation thrived in the early 1980s mainly driven by the development of the three popular unstructured mesh generation schemes, namely, the Delaunay triangulation, AFT and Octree decomposition.

The theoretical basis of Delaunay triangulation was established a long time ago by Dirichlet (1850), Voronoi (1908) and Delaunay (1934), and an efficient and robust construction algorithm by point insertion was only developed in 1981 respectively by Bowyer and Watson. However, Cavendish (1974), Lawson (1977) and Cavendish et al. (1985) were among the earliest to employ the method formally for 2D and 3D finite element mesh generation. Delaunay triangulation will only give the convex hull of the given point set, and for finite element mesh generation, geometrical and topological constraints on the boundary have to be enforced. Conforming and fully constrained Delaunay triangulations were studied, respectively, by Baker (1989b), Chew (1989) and George et al. (1990, 1991). Mesh generation over curved surfaces by means of parametric co-ordinates and anisotropic metric tensor to specify the size and shape of the elements was presented by Borouchaki and George (1996). Generation of anisotropic meshes in three dimensions by Delaunay triangulation coupled with AFT was proposed by Frey et al. (1998). Delaunay triangulation algorithms by parallel processing were developed by Blelloch et al. (1999), Chrisochoides and Nave (2003) and Lo (2012a,b), and algorithms for Delaunay triangulation of highly non-uniform distribution of large point sets were put forward by Lo (2013a).

The essence of AFT is not where mesh generation is started, whether it is from the boundary or radiating from an interior point, but the partition of the problem domain into a meshed zone and an unmeshed zone clearly delineated by the generation front, which is the common moving boundary between the zones. While the meshed and unmeshed parts can take any flexible arbitrary shape and form, and each of which may consist of several disconnected pieces, the frontal process allows us to focus on element generation at the front, which is one dimension less than the problem domain, and to pay no more attention to the meshed zones in which the mesh has already been generated. Mesh generation over arbitrary planar domains by AFT was presented by Lo (1985); in three dimensions by Lohner and Parikh (1988), Peraire et al. (1988) and Lo (1991b,c) and over surfaces by Lo (1989a), Lau and Lo (1996) and Lee (1999). Apart from direct mesh generation of simplicial elements on 2D and 3D surfaces, the advancing-front (ADF) concept can also be applied to many mesh-related operations such as the generation of quadrilateral meshes (Zhu et al. 1991a; Lee and Lo 1994; Owen et al. 1999), hexahedral elements (Blacker and Stephenson 1991; Owen and Saigal 2000), combined Delaunay–ADF approach (Borouchaki et al. 2000a), surface intersection (Lo 1995), ellipse and sphere packing (Lo and Wang 2005c,d) and merging of tetrahedral and hexahedral meshes (Lo 2012c, 2013c).

Octree decomposition (Yerry and Shephard 1984) as a method for finite element mesh generation was the direct extension of the Quadtree (Yerry and Shephard 1983) in two to three dimensions. Based on the nodal space requirements and the boundary characteristics, the enclosing space of an object to be meshed is recursively subdivided following the one-level refinement restriction. Mesh generation is achieved by snapping (projecting) points on the domain boundary and proper connection of points to form hexahedral and tetrahedral elements. By means of standard templates or the *marching cube* method, Octree partition of space is especially attractive for mesh generation of objects bounded by smooth surfaces analytically defined or implicitly defined by a system of spatial points. Grid/voxel methods in conjunction with the Octree partition found tremendous applications in meshing biomedical objects or domains into tetrahedral and hexahedral meshes (Viceconti et al. 1998, 2004; Zannoni et al. 1998; Smith et al. 2000; Ferrant et al. 2001; Prakash and Ethier 2001; Lapeer and Prager 2001; Samani et al. 2001; Verdonschot et al. 2001; Lacroix and Prendergast 2002; Antiga et al. 2003; Chabanas et al. 2003; Horgan and Gilchrist 2003; Kwok et al. 2003; Taddei et al. 2003, 2004; Fernandez et al. 2004; Tawhai et al. 2004; Wang et al. 2005, 2007b; Ramos and Simoes 2006; Zhang et al. 2006; Johnson et al. 2009).

Refinement and coarsening (de-refinement) are generally regarded as mesh modification procedures to meet the requirement of nodal-spacing functions and/or as a means to comply with the boundary constraints. Indeed, mesh refinement, by its own, is also a powerful mesh generation tool such as the recursive subdivision of a regular domain into smaller elements of similar type. Strictly speaking, Delaunay triangulation by point insertion is a form of mesh refinement in which the mesh is modified by the introduction of a newly inserted point, and more elements are created by proper connections with the inserted node. The merits of a refinement process are its robustness and speed in which, for each refinement, a valid finite element mesh is always maintained, and the results are only accepted if the refined mesh is superior to the original mesh before refinement. As refinement can usually be carried out by some local operations, therefore, it is fast with linear time complexity and lends itself to easy parallelisation. Coarsening can reduce the data points of a mesh yet maintain the main features of the underlying surface. In adaptive refinement analysis of fluid dynamic problems to capture the shock waves, de-refinement coupled with a proper relocation of nodes can reduce the data set and provide the framework for the generation of highly anisotropic meshes (McMorris and Kallinderis 1997; Alauzet and Frey 2005; Loseille and Alauzet 2009).

1.8 SO FAR ACHIEVED AND WHAT LIES AHEAD

Since planar domains are flat and Euclidean such that solutions exist for general arbitrary boundary constraints, fully constrained high-quality anisotropic meshes can be generated even for domains subject to the most difficult boundary conditions. By the parametric mapping method based on mesh generation on planar domains, complex surfaces can be divided into patches, over each of which mesh of different characteristics can be generated systematically with element size and shape in compliance with the metric tensor defined in terms of the surface curvatures. There are still two difficulties in mesh generation over three dimensions: (i) shape measure is not coherent with Delaunay triangulation such that flat degenerate tetrahedra will be generated in a Delaunay triangulation and such that the volume of an element can be arbitrarily small; and (ii) there is no systematic way in producing a fully constrained finite element mesh without degenerate element(s) for a general polyhedron. As both fully constrained and geometrically conforming 3D finite element meshes can have tremendous applications for different problem types, research will be continued to improve the quality of the finite elements on the boundary especially for those at some critical locations. Mesh parallelisation for the ADF, Delaunay triangulation and boundary handling (recovery) will also be interesting research topics as the scale and complexity of practical scientific computations and engineering problems are ever increasing.

1.9 TOPICS DISCUSSED IN THE CHAPTERS

Following this chapter, Chapter 2 presents the fundamentals in finite element mesh generation. Notations, symbols and abbreviations used in this text will first be listed out in Section 2.2, and terminologies and data structures pertinent to the finite element mesh generation are elaborated in Section 2.3. Geometrical operations and formulas are given in Section 2.4, whereas various topological operations and algorithms are provided in Section 2.5. Popular data-sorting methods such as *bubble sort*, *insertion sort*, *quick sort* and *bin sort* are all described and compared in Section 2.6. Background grids, namely, regular/irregular grids, Quadtree/Octree grids and kd-tree partitions as effective means to speed up searching and matching of various geometrical quantities are discussed with examples in Section 2.7.

Methods for finite element mesh generation are formally introduced in Chapter 3 in which various 2D mesh generation algorithms are presented. Following an introduction in Section 3.1, structured and unstructured meshes on planar domain are described, respectively, in Sections 3.2 and 3.3. Meshing by Quadtree decomposition is discussed in Section 3.4, and Delaunay triangulation over 2D domain is presented in Section 3.5. The ADF approach and its extension to combine with Delaunay triangulation will be explored, respectively, in Sections 3.6 and 3.7. As the Quadtree method may not be very effective in handling irregular boundaries, an enhanced scheme coupled with advancing-front technique (AFT) is proposed in Section 3.8, and finally, generation of quadrilateral meshes on a planar domain is presented with details and examples in Section 3.9.

Mesh generation on curved surfaces is discussed in Chapter 4. The parametric mapping method, surface curvatures and metric tensor specifications and mesh generation by the Delaunay–ADF scheme are described in Section 4.2. Mesh generation by packing of ellipse following an anisotropic curved surface metric and direct mesh generation on analytical curved surfaces are presented, respectively, in Sections 4.3 and 4.4. Mesh generation by means of a mesh-merging process through surface intersections is introduced in Section 4.5, and a brief account on the generation of quadrilateral meshes by schematic merging of triangles is given in Section 4.6.

Finite element mesh generation over three dimensions will be explored in Chapter 5. A detailed algorithm of Delaunay triangulation by a point inserted in 3D is described in Section 5.2. Boundary recovery procedures to achieve fully constrained Delaunay triangulations are discussed in Section 5.3, whereas boundary-protection techniques for geometry-conforming meshes are presented in Section 5.4. Classical ADF approach along with programming details are given in Section 5.5, and its extension to Delaunay–ADF meshing in 3D can be found in Section 5.6. Similar to ellipse packing in 2D, sphere packing in 3D as a means of generating tetrahedral meshes of variable element sizes is discussed in Section 5.7. The chapter ends with the introduction of various methods in Section 5.8 for the generation of structured and unstructured hexahedral meshes.

Chapter 6 is about mesh optimisation in which geometrical and topological operations for the enhancement of 2D and 3D finite element meshes are presented. In order to have an objective view apart from aesthetic judgements, various shape measures for simplices are discussed in Section 6.2. Mesh optimisation by means of shifting of nodes and topological operations such as face/edge swaps are described, respectively, in Sections 6.3 and 6.4.

Mesh generation by means of concurrent parallel processing is explored in Chapter 7. Before the development of any parallel meshing algorithms, the fundamentals and strategies for efficient parallel processing are discussed. An algorithm with detailed explanation for parallel Delaunay triangulation in 2D is given in Section 7.3. The 2D parallel algorithm, which turns out to be generic across dimensions, can be easily extended to three dimensions, as elucidated in Section 7.4. Another approach for parallel meshing by the method of domain partition is introduced in Section 7.5, in which a simple algorithm for the decomposition of general curved surfaces based on a given geometrical criterion is also presented.

Chapter 8 consists of all the auxiliary mesh generation techniques not yet covered in the previous chapters. Surface verification and preparation are perhaps mandatory for a large complicated object bounded by discretised surfaces. The topological consistency, geometrical tolerances and volume bounded by surfaces will all be evaluated and rectified if necessary by the procedures described in Section 8.1. For highly non-uniform point distributions, point insertion by means of a regular grid may not be the most efficient. To this end, a multi-grid insertion algorithm is proposed in Sections 8.2 and 8.3 for the 2D and 3D triangulation of non-uniformly distributed points, respectively. As stated early on in this introductory chapter, meshing by refinement is reliable as well as efficient. Mesh generation and adaptation by edge refinement are discussed in Section 8.4. As a related application of mesh refinement, meshing volumes bounded by analytical curved surfaces is described in Section 8.5. Merging of tetrahedral and hexahedral meshes through mesh intersection and local remeshing of well-defined tiny regions are presented, respectively, in Sections 8.6 and 8.7. The generation of curvilinear finite elements by means of a generic p1 mesh subdivision and optimisation is explored in Section 8.8, and finally, a concise account on the adaptive mesh generation using an example of the 3D elasticity is given in Section 8.9.

Some useful mathematical formulas and expressions related to finite element mesh generation are given in Appendices A1 to A11. Two FORTRAN computer programs on 2D ADF meshing and 3D Delaunay triangulation are provided in Appendices A12 and A13. A profuse list of a couple of hundreds of bibliography is included, which can be useful reference materials for various mesh generation problems at hand. The index list that follows will be helpful in providing a quick reference page to a particular author or item under consideration.

Chapter 2

Fundamentals

The fundamentals allow you to go through the other chapters with clarity and comfort.

2.1 INTRODUCTION

The basic concepts, notations, terminologies, geometrical and topological operations, sorting methods and background grids for mesh generation (MG) will all be presented in this chapter to pave the way for a formal discussion of finite element (FE) MG in the following chapters. The notations, symbols and abbreviations commonly employed in MG are given in Section 2.2. Although the symbols and abbreviations used are more or less those usually adopted in the MG community, however, the notations are quite unique in a way that the counter and index are employed to specify systematically a particular node, neighbour, edge or face of an FE. Terminologies related to MG and data structures for an FE mesh are presented in Section 2.3. For terminologies, instead of a formal mathematical definition, their characteristics are highlighted; in particular, those features intimately related to MG will be discussed in detail. Again, data structures are not those generally encountered in computer science but rather the data format and arrangement to specify an MG problem, i.e. how nodes of common FEs in 2D and 3D are labelled, how to store and address an FE mesh in a computer, etc.

Geometrical operations, namely, those for computing the distance between two simplices of the same or different dimensions, i.e. point-to-line segment and line segment to triangular facet, etc., frequently required in MG along with other useful formulas (for instance, normal at a point, solid angles, determination of intersection points, etc.), are given in Section 2.4. In MG, topological operations are equally if not more important than geometrical operations, as enquiries such as how many elements are connected to a particular node and how to list all the nodes or segments on the boundary of a mesh are always faced. As the number of nodes and the number of elements in a mesh can be very large, efficient topology computation algorithms are crucial to a robust MG scheme. Accordingly, algorithms for the common topological operations including elements connected to a node, adjacency relationship of a mesh, etc., in the form of detailed pseudo-code readily translated into C++ or FORTRAN programs are given in Section 2.5.

The ability to sort a large amount of data in an efficient and reliable manner is always a great asset in numerical computations. In mesh refinement based on the bisection of the longest edge, the edges in a mesh have to be sorted repeatedly such that the longest edge is always bisected in the refinement process, and very often, an MG process has to be carried out following a sequence according to the size of the elements, etc. Common sorting methods are introduced in Section 2.6; their performance on large data sets is compared, and the pseudo-codes of the sorting algorithms are also given. Perhaps the background grid is the most important technique in drastically reducing the CPU time for a large-scale meshing problem. Regular and irregular grids, Quadtree, Octree and kd-tree partitions are explored in Section 2.7, along with their constructions and performance, which will also be studied in great detail.

2.2 NOTATIONS, SYMBOLS AND ABBREVIATIONS

Notations, symbols and abbreviations will usually be explained when they first occur in the text. However, the most frequently used notations, symbols and abbreviations are listed below for easy reference.

2.2.1 Notations

Counter:	$i, j, k, l, m, n \in \mathbb{N}$
Index:	$a, b, c, d, e \in \{1, 2, 3, 4, 5, 6, 7, 8\}$
Co-ordinates:	x, y, z
Triangle:	$\Delta_i, i = 1, N_\Delta$
	$\Delta_i^a = a^{th}$ neighbour of Δ_i
	$V_i^a = a^{th}$ vertex of Δ_i
	$C_i =$ Circumscribing circle of Δ_i
Quadrilateral:	$Q_i, i = 1, N_Q$
	$Q_i^a = a^{th}$ neighbour of Q_i
	$V_i^a = a^{th}$ vertex of Q_i
Tetrahedron:	$T_i, i = 1, N_T$
	$T_i^a = a^{th}$ neighbour of T_i
	$V_i^a = a^{th}$ vertex of T_i
	$F_i^a = a^{th}$ face of T_i
	$S_i =$ Circumscribing sphere of T_i
Edge:	$E_i^a = a^{th}$ edge of Element i
Face:	$F_k =$ Facet of the triangulation, $k = 1, N_F$; identified with a tetrahedron i and an index a, such that $F_k = F_i^a$
$d(P, AB)$	Distance between point P and line segment AB
$d(P, ABC)$	Distance between point P and triangular facet ABC
$d(AB, CD)$	Distance between two line segments AB and CD
$d(PQ, ABC)$	Distance between line segment PQ and triangular facet ABC
$1, n$	$\equiv 1, 2, ..., n$
Italic	*Words carry special meaning*

2.2.2 Symbols

Vectors and matrices are usually presented, respectively, in boldface and block letters, and symbols with special meanings or appearing often in the text are listed as follows:

Ω	Problem domain
	Solid angle
$\partial\Omega$	Domain boundary
Γ	Boundary surface, Generation front
$\alpha, \beta, \gamma, \eta$	Shape factors
ρ	Node spacing function

	Radius of curvature
δ	Coefficient of conformity, distance measure
λ, μ	Lamé constants
	Principal stretches of metric tensor
	Parametric co-ordinates along a line segment
ξ, η, ζ	Reference element co-ordinates
L_1, L_2, L_3, L_4	Area/volume (barycentre) co-ordinates
H_a	Interpolation functions at the ath node
M	Metric tensor (matrix)
ε, σ	Strain tensor and stress tensor
ε	Tolerance
κ	Curvature
θ	Angle measure
NINT(x)	Nearest integer of real number x
INT(x)	Integral part of real number x
‖·‖	Euclidean norm
∇	Gradient operator
u, v	Vectors
a · b	Scalar product of vectors **a** and **b**
a × b	Cross product of vectors **a** and **b**
A:B	Scalar product of second order tensors A and B
M^T	Transpose of matrix M
ℝ	Set of real numbers
𝕀	Set of integers
ℕ	Set of natural numbers
ℍ	Hexahedral mesh
ℙ	Point set; pentahedral mesh
𝕋	Triangulation; triangular mesh; tetrahedral mesh
𝕐	Pyramid element mesh

2.2.3 Abbreviations

2D, 2d	Two-dimensional, two dimensions
3D, 3d	Three-dimensional, three dimensions
ADF	Advancing front
AFT	Advancing-front technique
BASE	The base triangle or tetrahedron in Delaunay point insertion
BRep	Boundary representation
CAD	Computer-aided design
CORE	The insertion cavity (polyhedron) in Delaunay triangulation
DT	Delaunay triangulation
FE	Finite element
FEM	Finite element method
hex	Hexahedron, hexahedra, hexahedral
LSF	Least square fit
MG	Mesh generation
NURB	Non-uniform rational B-spline(s)
PDE	Partial differential equation
quad	Quadrilateral
SPR	Super-convergent patch recovery

2.3 TERMINOLOGIES AND DATA STRUCTURES

The terminologies commonly encountered in MG are briefly described in this section, and their characteristics related to MG will be highlighted. Data structure is about the general forms and patterns for presenting the geometrical quantities of a mesh and how they are stored in a computer efficiently in some compact arrangements.

2.3.1 Triangulation

Let $\mathbb{P}$ be a finite set of points in $\mathbb{R}^d$ ($d = 2, 3$); the convex hull of $\mathbb{P}$ defines a domain Ω in $\mathbb{R}^d$. T is a simplex, which is a triangle in 2D and a tetrahedron in 3D; a triangulation $\mathbb{T} = \{T_i, i = 1, n\}$ of $\mathbb{P}$ is a set of n simplices satisfying the following conditions:

   i. The set of vertices in the simplices of $\mathbb{T}$ is exactly given by $\mathbb{P}$.
   ii. $\bigcup_{i=1,n} T_i = \Omega$.
   iii. The interior of $T_i \in \mathbb{T}$ is non-empty.
   iv. $T_i \cap T_j$ ($i \neq j$) = $\varnothing$, a point, an edge, or a face.

2.3.2 Delaunay triangulation

The triangulation $\mathbb{T}$ of a point set $\mathbb{P}$ is a Delaunay triangulation (DT) if the open circumscribing disc (circle/sphere) of every simplex T_i in $\mathbb{T}$ does not contain any point of $\mathbb{P}$. The properties of DT will be discussed in Chapters 3, 4 and 5, in particular, its role in MG on 2D, over surfaces and for 3D objects.

2.3.3 Constrained triangulation

In FE MG, it is necessary to distinguish two types of constrained triangulations of an object. The first type is called the semi-constrained (geometrically constrained) or conforming triangulation, which is a triangulation of the object with edges not cutting across the boundary of the object. The second type is called the fully constrained (topologically constrained) or simply constrained triangulation in which a specified set of sub-simplices ought to be found in the triangulation. Usually the discretised boundary of the object can be taken as the set of sub-simplices along with any other internal constraints as necessary.

2.3.4 Mesh and FE mesh

A mesh of an object or a domain is a partition of the interior, including the boundary, into simple polygons in 2D and polyhedra in 3D such that the domain is covered up, and no two distinct elements of a mesh overlap (except sharing a common node, edge or face). The non-empty criterion has not been emphasised as degenerate elements in a mesh are also acceptable as an intermediate step towards a valid mesh. Although conforming polygonal and polyhedral FEs exist (Sukumar and Tabarraei 2004; Rashid and Selimotic 2006; Sohn et al. 2012), simple polygons in 2D refer to triangles or quadrilaterals, and simple polyhedra in 3D refer to tetrahedron, hexahedron, pentahedron and pyramid commonly used in MG. The basic differences between a triangulation and a mesh are that a triangulation refers more to a set of points, whereas a mesh refers to an object or a domain of interest, and a triangulation contains only simplices, whereas a mesh may have simplices as well as non-simplices. A conforming triangulation is a triangular mesh of an object bounded by smooth

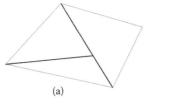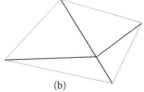

Figure 2.1 **FE mesh compatibility: (a) incompatible mesh; (b) compatible mesh.**

surfaces, and a constrained triangulation can be a mesh of an object with discretised boundary surfaces.

Only the geometrical and topological aspects, and not the functional aspects, of an FE mesh will be emphasised, and in this sense, there is not much difference between a mesh and an FE mesh. However, apart from the geometrical aspects, FE meshes are meshes to support numerical computations, which are stricter in the following aspects:

i. FE meshes have to be compatible, i.e. an edge in 2D and a face in 3D can only be shared by two elements except for the specially designed elements, which could still converge without full compatibility, as shown in Figure 2.1.
ii. The shape qualities of the elements in an FE mesh have to be optimised to reduce discretisation error; in particular, inverted and degenerate elements are not allowed.
iii. The size and shape of the elements have to comply with the specified node spacing function or metric.
iv. The node numbering and orientation have to be consistent.
v. Some nodal points, edges and faces have to be generated at specified positions.

In the sequel, a mesh simply means an FE mesh, and hence, the above five requirements will be enforced, and the quality of the elements will be optimised as much as possible.

2.3.5 Structured and unstructured meshes

A mesh is called structured if its connectivity is pre-determined, repeating periodically with a fixed pattern. In 2D, we have structured rectangular meshes and triangular meshes generated with the aid of a regular grid, and in 3D, regular hexahedral meshes or tetrahedral meshes derived from regular hexahedral meshes by a uniform pattern of subdivision of each hexahedron into five or six tetrahedra are structured meshes. As for unstructured meshes, the pattern of connectivity is not periodic, and the number of elements connected to a node varies and is unpredictable; in other words, the node element connection is quite random and copes with the geometry of the domain and other constraints over the entire mesh, and the size of the element may also vary quite significantly across the mesh.

2.3.6 Mixed and hybrid meshes

A mixed mesh is a mesh consisting of elements of various types. For instance, a 2D mixed mesh may contain triangular and quadrilateral elements, and a 3D mixed mesh may have tetrahedral, hexahedral and pentahedral elements in the mesh. On the other hand, a hybrid mesh is a mesh consisting of elements of different spatial dimensions; for example, a mesh of tetrahedral and triangular elements in space is a hybrid mesh (Lie et al. 2001).

2.3.7 Discretised manifold

In MG, a surface mesh is called manifold or more precisely discretised manifold if each edge is shared by exactly two triangular facets or connected to only one triangle on the boundary for open surfaces. The boundary of a mesh is a discretised manifold.

2.3.8 Control space

A control space for a domain to be meshed is a geometrical supporting structure covering the entire domain by means of a regular grid, Quadtree, Octree or kd-tree partition of space, a triangulation or a mesh generated earlier. A control space can facilitate MG in providing information such as the required element size at a point, the neighbouring geometrical quantities or whether a point is inside or outside the domain.

2.3.9 Adaptive mesh

In FE analysis, a solution error depends on the quality of the FE mesh, and this error will be reduced by putting elements of smaller size to areas of large discretisation error. An adaptive mesh is an FE mesh for which the size and shape of the elements ought to be in compliance with a specified node spacing function derived from the error of the FE solution.

2.3.10 Data structure

Data structures such as array, list, stack, queue, pointer and class, commonly employed in computer science, are not going to be discussed, but rather the data structure/format of various geometrical quantities in an FE mesh. An FE is defined by its type and the associated nodal point list in a pre-determined order according to the element type. For instance, a triangular element is defined by three nodal points $\{P_1, P_2, P_3\}$ in the counter-clockwise order on a planar domain, as shown in Figure 2.2. Similarly, a quadrilateral on a 2D plane is defined by a list of four nodal points $\{P_1, P_2, P_3, P_4\}$ also in the counter-clockwise order, as shown in Figure 2.3. On a surface in 3D space, the order of node numbers or orientation of

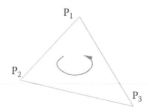

Figure 2.2 **A triangular element.**

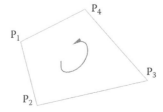

Figure 2.3 **A quadrilateral element.**

the triangular and quadrilateral elements has to be consistent such that their unit normals are pointing in the same direction, i.e. either towards the domain interior or outwards to the exterior of the domain according to the convention adopted.

2.3.10.1 Nodal points

The spatial position of a point is specified by its co-ordinates. Cartesian co-ordinates are the most common, and the nodal points of an FE mesh are given by

$\{x(i), y(i), z(i), i = 1, N\}$, where N = number of nodes in the mesh.

2.3.10.2 Boundary of a planar domain

In MG, a domain to be meshed is often defined by its boundary. A planar region is bounded by closed loops. There is only one exterior loop, but there can be as many interior loops as the number of openings within the domain, as shown in Figure 2.4. Boundary loops can be defined analytically or represented by a list of N directed line segments, $\mathbb{L} = \{A_i B_i, i = 1, N\}$, in which A_i and B_i are, respectively, the beginning node and the ending node of the ith line segment, as shown in Figure 2.4. Line segments of the exterior boundary loop will be entered in a counter-clockwise direction, whereas line segments of interior loops will follow a clockwise direction.

2.3.10.3 Boundary of a 3D domain

A 3D region is bounded by closed surfaces. There is only one exterior surface, but there can be as many interior surfaces as the number of openings within the domain, as shown in Figure 2.5. Boundary surfaces can be defined analytically or represented by a set of N oriented triangular facets, $\mathbb{S} = \{\Delta_i, i = 1, N\}$, as shown in Figure 2.5. The triangular facets on the interior or the exterior boundary surfaces have to be oriented in a systematic manner such that the normal to the surface always points towards the object or the domain to be meshed. It is remarked that in the generation of unstructured hexahedral or hex-dominated meshes, the domain boundary surfaces are sometimes discretised into quadrilaterals.

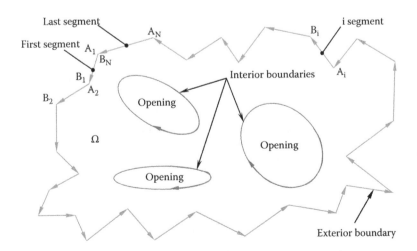

Figure 2.4 **Boundary of a planar domain.**

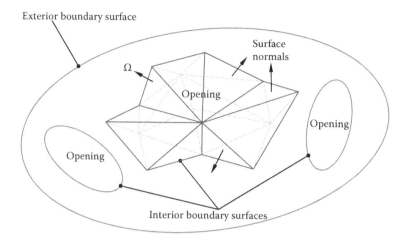

Figure 2.5 **Boundary of a 3D domain.**

2.3.10.4 Node labelling of FEs

A corner point of a geometrical object such as a triangle or a cube is called a vertex, and nodes of an FE are at the vertices; sometimes additional nodes for higher-order elements can be placed at the mid-point of an edge, at a face or even at the centre of an element. However, in linear p1 FE meshes, the nodes and vertices of an element are identical in number and in their positions, and hence, when there is no confusion, either one can be used to refer to a corner point of an element.

i. **Triangular and quadrilateral elements**
   The nodes of a triangular or quadrilateral element on 2D domains are labelled in a counter-clockwise manner, as shown in Figures 2.2 and 2.3, and on a spatial surface, the nodes of a triangular or quadrilateral element are labelled according to the orientation of the surface. A triangular mesh $\mathbb{T}$ or a quadrilateral mesh $\mathbb{Q}$ of N_E elements can be stored in a linear array:

$$\mathbb{T} = \left\{ V_i^a, a = 1, 2, 3 \text{ and } i = 1, N_E \right\}, \quad \mathbb{Q} = \left\{ V_i^a, a = 1, 2, 3, 4 \text{ and } i = 1, N_E \right\}$$

ii. **Tetrahedral, hexahedral, pentahedral and pyramid elements**
   Nodes on a face of the tetrahedral element or on the quadrilateral base of the pyramid element are labelled following the right-hand grip rule pointing towards the apex, as shown in Figure 2.6a and 2.6d. Nodes of a hexahedral or a pentahedral element are labelled in two layers in an order following the right-hand grip rule pointing towards the second layer, as shown in Figure 2.6b and 2.6c. A tetrahedral mesh $\mathbb{T}$, a hexahedral mesh $\mathbb{H}$, a pentahedral mesh $\mathbb{P}$ or a pyramid mesh $\mathbb{Y}$ of N_E elements can be stored sequentially in a linear array, i.e.

$$\mathbb{T} = \left\{ V_i^a, a = 1, 4 \text{ and } i = 1, N_E \right\}, \quad \mathbb{H} = \left\{ V_i^a, a = 1, 8 \text{ and } i = 1, N_E \right\}$$

$$\mathbb{P} = \left\{ V_i^a, a = 1, 6 \text{ and } i = 1, N_E \right\}, \quad \mathbb{Y} = \left\{ V_i^a, a = 1, 5 \text{ and } i = 1, N_E \right\}$$

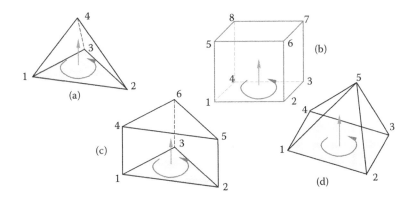

Figure 2.6 **Node labeling of 3D FE elements: (a) tetrahedral element; (b) hexahedral element; (c) penta-hedral element; (d) pyramid element.**

For a mixed mesh consisting of elements of various types, an additional pointer array has to be used to mark the starting point of each 3D solid element. Alternatively, a linear array of length $8N_E$ can also be reserved for a mixed mesh of N_E elements in such a way that each element is allocated eight nodal spaces irrespective of its type, and the excessive space can be filled up with zeros. At the expense of slightly more memory, there is an additional flexibility in deleting and adding elements to a mesh in the second storage scheme.

2.4 GEOMETRICAL OPERATIONS AND FORMULAS

2.4.1 Distance from a point P to a line segment AB, d(P, AB)

As shown in Figure 2.7, the unit vector along direction AB is given by

$$u = \frac{AB}{\|AB\|}$$

Compute $\alpha = AP \cdot u$

$$d = \|PC\| = \|AP - \alpha u\|$$

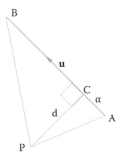

Figure 2.7 **Distance between point P and segment AB.**

If $\alpha < 0$, d = ||PA||
If $\alpha > $ ||AB||, d = ||PB||

2.4.2 Distance from a point P to a triangular facet ABC, d(P, ABC)

As shown in Figure 2.8, the unit normal to triangular facet ABC is given by

$$\mathbf{n} = \frac{\mathbf{v}}{\|\mathbf{v}\|} \quad \text{where } \mathbf{v} = \text{AB} \times \text{AC}$$

Compute $\alpha = \text{AP} \cdot \mathbf{n}$.
Let Q be the projection of P onto triangle ABC,

$$\text{AQ} = \text{AP} - \alpha\mathbf{n}$$

Let a, b and c be the area co-ordinates of Q with respect to points A, B and C, respectively.

If $a \geq 0$, $b \geq 0$ and $c \geq 0$, Q is inside triangle ABC, distance of P from facet ABC, $d = \alpha$.
If $a < 0$, Q is outside triangle ABC opposite to A, d = d(P, BC).
If $b < 0$, Q is outside triangle ABC opposite to B, d = d(P, CA).
If $c < 0$, Q is outside triangle ABC opposite to C, d = d(P, AB).
If $a < 0$ and $b < 0$, d = ||CP||; if $b < 0$ and $c < 0$, d = ||AP||; and if $c < 0$ and $a < 0$, d = ||BP||.

2.4.3 Distance between line segments in space, d(AB, CD)

Let AB and PQ be the two line segments in space and C and R be, respectively, the points on the line segments AB and PQ with parameterisation s and t, as shown in Figure 2.9,

$$\text{C} = \text{A} + s\text{AB} \text{ and } \text{R} = \text{P} + t\text{PQ} \quad s, t \in [0,1]$$

$$\text{CR} = (\text{P} + t\text{PQ}) - (\text{A} + s\text{AB}) = \text{AP} + t\text{PQ} - s\text{AB}$$

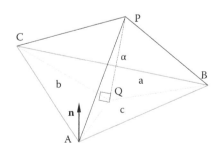

Figure 2.8 Distance between point P and facet ABC.

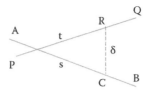

Figure 2.9 Distance between two line segments.

To find the shortest distance, the partial derivative of CR w.r.t. s and t has to be taken:

$$\frac{\partial}{\partial s}\|CR\|^2 = \frac{\partial}{\partial s}(AP + tPQ - sAB)^2 = 2(sAB^2 - AB \cdot AP - tAB \cdot PQ) = 0$$

$$\frac{\partial}{\partial t}\|CR\|^2 = \frac{\partial}{\partial t}(AP + tPQ - sAB)^2 = 2(PQ \cdot AP + tPQ^2 - sAB \cdot PQ) = 0$$

Let $a = AB \cdot AB$, $b = PQ \cdot PQ$ and $c = AB \cdot PQ$; we have

$$\begin{bmatrix} a & -c \\ -c & b \end{bmatrix}\begin{bmatrix} s \\ t \end{bmatrix} = \begin{bmatrix} e \\ -f \end{bmatrix} \Rightarrow \begin{bmatrix} s \\ t \end{bmatrix} = \frac{1}{d}\begin{bmatrix} b & c \\ c & a \end{bmatrix}\begin{bmatrix} e \\ -f \end{bmatrix} = \frac{1}{d}\begin{bmatrix} be - cf \\ ce - af \end{bmatrix}$$

with $d = ab - c^2$, $e = AB \cdot AP$ and $f = PQ \cdot AP$.

From s and t, compute the distance $\delta = \|AP + tPQ - sAB\|$.

If $s < 0$, δ = distance from point A to line segment PQ.
If $s > 1$, δ = distance from point B to line segment PQ.
If $t < 0$, δ = distance from point P to line segment AB.
If $t > 1$, δ = distance from point Q to line segment AB.
In case $d = 0$, the line segments are parallel, and the shortest distance δ is given by

$$\delta = \min (\delta_1, \delta_2)$$

where δ_1 is the distance from point A to PQ, and δ_2 is the distance from point B to PQ.

2.4.4 Intersection between two line segments on a plane

There are at least four methods for the determination of the intersection between two line segments P_1P_2 and P_3P_4 on a plane.

2.4.4.1 Analytical method

There is an intersection if points P_3 and P_4 are situated on the opposite sides of segment P_1P_2, and points P_1 and P_2 are situated on the opposite sides of segment P_3P_4, as shown in Figure 2.10.

$$d_1d_2 \leq 0 \text{ and } d_3d_4 \leq 0$$

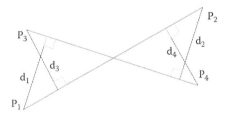

Figure 2.10 Intersection between two line segments.

where d_1, d_2, d_3 and d_4 are signed distances from a line given by

$$d_1 = (y_4 - y_3)x_1 + (x_3 - x_4)y_1 + c_2$$
$$d_2 = (y_4 - y_3)x_2 + (x_3 - x_4)y_2 + c_2$$
$$d_3 = (y_2 - y_1)x_3 + (x_1 - x_2)y_3 + c_1$$
$$d_4 = (y_2 - y_1)x_4 + (x_1 - x_2)y_4 + c_1$$

where $c_1 = y_1x_2 - y_2x_1$ and $c_2 = y_3x_4 - y_4x_3$.

The advantage of this method is that there is no intersection if either condition $d_1d_2 \leq 0$ or $d_3d_4 \leq 0$ is not satisfied, and in the process, the distance of a point to a line is also determined, which may be of use to judge whether a point is too close to a line or not.

2.4.4.2 Vectorial method

As shown in Figure 2.11, there is intersection if

$$(P_3P_1 \times P_3P_2) \cdot (P_4P_1 \times P_4P_2) \leq 0 \text{ and } (P_1P_3 \times P_1P_4) \cdot (P_2P_3 \times P_2P_4) \leq 0$$

2.4.4.3 Parametric method

As shown in Figure 2.12, the parametric representation of line segment P_1P_2 is given by

$$P = (x, y) = \lambda(x_2, y_2) + (1 - \lambda)(x_1, y_1)$$

Along line segment P_3P_4, the intersection point P is given by

$$P = (x, y) = \mu(x_4, y_4) + (1 - \mu)(x_3, y_3)$$

Solve for λ and μ, and there is an intersection if both λ and μ lie between 0 and 1.

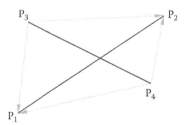

Figure 2.11 Vectorial method.

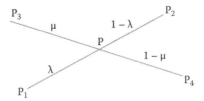

Figure 2.12 **Parametric method.**

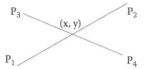

Figure 2.13 **Max–min method.**

2.4.4.4 The max–min method

The intersection point P(x, y) between line P_1P_2 and P_3P_4 is first calculated, and then it is checked if this point is on both segments. The intersection point is given by

$$x = [(x_2 - x_1)(x_4y_3 - x_3y_4) - (x_4 - x_3)(x_2y_1 - x_1y_2)]/D$$
$$y = [(y_2 - y_1)(x_4y_3 - x_3y_4) - (y_4 - y_3)(x_2y_1 - x_1y_2)]/D$$
$$D = (x_4 - x_3)(y_2 - y_1) - (x_2 - x_1)(y_4 - y_3)$$

As shown in Figure 2.13, there is intersection if

$$(x - x_1)(x - x_2) \le 0, \; (y - y_1)(y - y_2) \le 0$$

and

$$(x - x_3)(x - x_4) \le 0, \; (y - y_3)(y - y_4) \le 0$$

Remark: The analytical method and the vectorial method are more efficient for the detection of intersections where the point of intersection is not required, as there are fewer operations involved, and the division of zero can be avoided in case the two line segments are parallel. To enhance stability, before checking for intersections, it has to be ensured that the points P_1, P_2, P_3 and P_4 are all distinct; otherwise, it will be considered as a special case and will be dealt with accordingly.

2.4.5 Solid angle

Sometimes, the normal and the solid angle at a node on a triangulated surface have to be computed in order to detect any special features on the surface such as deep troughs, peaks, ridges and flat planer parts, as shown in Figure 2.14.

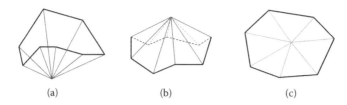

Figure 2.14 Surface features of meshed surfaces: (a) trough, $\Omega > 2\pi$; (b) peak, $\Omega < 2\pi$; (c) flat, $\Omega = 2\pi$.

Suppose that there are n triangular facets connecting to the node P under consideration. Let P_1, P_2, ..., P_n be the vertices connected to node P, such that the kth triangular facet connected to P is denoted by PP_kP_{k+1}, as shown in Figure 2.15. For each node P_k, k = 1, n, compute unit vector $\mathbf{u}_k$ in the direction PP_k, i.e.

$$\mathbf{u}_k = \frac{PP_k}{\|PP_k\|}$$

The unit normal to the kth triangular facet is given by

$$\mathbf{n}_k = \frac{\mathbf{u}_k \times \mathbf{u}_{k+1}}{\|\mathbf{u}_k \times \mathbf{u}_{k+1}\|}$$

The angle ϕ_k between the normals of adjacent triangular facets is given by

$$\phi_k = \cos^{-1}(\mathbf{n}_k \cdot \mathbf{n}_{k+1})$$

The solid angle Ω at node P can now be computed with the following formula:

$$\Omega = 2\pi - \sum_{k=1,n} \phi_k = 2\pi - \sum_{k=1,n} \cos^{-1}(\mathbf{n}_k \cdot \mathbf{n}_{k+1})$$

For planar surfaces, angles between normals are all equal to zero, and hence, Ω is equal to 2π at every point. For non-flat surfaces, ϕ_k can take a positive or negative value depending on the triple product $(\mathbf{n}_k \times \mathbf{n}_{k+1}) \cdot \mathbf{u}_{k+1}$.

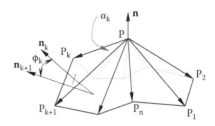

Figure 2.15 Solid angle and normal at a node.

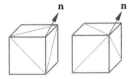

Figure 2.16 Normal at a corner of a cube.

2.4.6 Normal at a node

The normal to a plane is well defined by the cross product of any two non-parallel edges, and this definition has to be generalised to the normal at a point on a discretised surface. Based on the notations of Section 2.4.5, as shown in Figure 2.15, the angle P_kPP_{k+1} of the kth triangular facet is given by

$$\alpha_k = \cos^{-1}(\mathbf{u}_k \cdot \mathbf{u}_{k+1})$$

The normal at node P is computed by the weighted average of the surface normals:

$$N = \frac{\sum \alpha_k \mathbf{n}_k}{\sum \alpha_k} \text{ and unit normal, } \mathbf{n} = \frac{\mathbf{N}}{\|\mathbf{N}\|}$$

By this definition, the normal at a point depends only on the geometry of the surface but not on how it is discretised into elements. For instance, the surface normal at a corner of a cube is the same for the two surfaces meshed in two different patterns, as shown in Figure 2.16.

2.4.7 Intersection between a line segment and a triangular facet

The following are the procedures for the determination of the intersection between line segment Q_1Q_2 and triangle $P_1P_2P_3$ in space.

i. Compute the signed distance h_1 and h_2 of points Q_1 and Q_2 to the plane $P_1P_2P_3$. If $h_1h_2 < 0$, calculate $\xi = h_1/(h_1 - h_2)$. Intersection point Q on the plane is given by $Q = Q_1 + \xi(Q_2 - Q_1)$.
ii. Calculate the barycentre co-ordinates of Q, (L_1, L_2, L_3). If $L_1 \geq 0$, $L_2 \geq 0$, and $L_3 \geq 0$, Q is inside triangle $P_1P_2P_3$ and Q_1Q_2 intersects $P_1P_2P_3$, as shown in Figure 2.17.

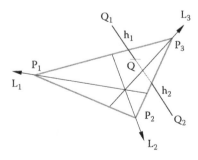

Figure 2.17 Intersection between a line segment and a triangular facet.

2.4.8 Distance between a line segment and a triangular facet in space, d(PQ, ABC)

The following are the steps in the determination of the distance between line segment Q_1Q_2 and triangle $P_1P_2P_3$ in space.

i. Compute the barycentre co-ordinates $(\lambda_1, \lambda_2, \lambda_3)$ and (μ_1, μ_2, μ_3) of Q_1 and Q_2 w.r.t. triangle $P_1P_2P_3$

$$\lambda_1 = e_1 \cdot P_2Q_1, \quad \lambda_2 = e_2 \cdot P_3Q_1, \quad \lambda_3 = e_3 \cdot P_1Q_1$$

$$\mu_1 = e_1 \cdot P_2Q_2, \quad \mu_2 = e_2 \cdot P_3Q_2, \quad \mu_3 = e_3 \cdot P_1Q_2$$

where e_1, e_2 and e_3 are unit vectors along the barycentre co-ordinate axes.

ii. As shown in Figure 2.18a, the minimum distance is given by the distance from an end point to the triangular facet if

$$\lambda_1 \geq 0, \lambda_2 \geq 0, \lambda_3 \geq 0 \text{ and } h_1(h_2 - h_1) \geq 0, \text{ then } d_{min} = h_1$$

On the other hand, if

$$\mu_1 \geq 0, \mu_2 \geq 0, \mu_3 \geq 0 \text{ and } h_2(h_1 - h_2) \geq 0, \text{ then } d_{min} = h_2$$

iii. Otherwise, the minimum distance is given by the distance between the line segment and an edge of the triangle, as shown in Figure 2.18b.

Q_1Q_2 is on the side P_2P_3 if $\lambda_1 \leq 0$ or $\mu_1 \leq 0$, $d_{min} = d(Q_1Q_2, P_2P_3)$

Q_1Q_2 is on the side P_3P_1 if $\lambda_2 \leq 0$ or $\mu_2 \leq 0$, $d_{min} = d(Q_1Q_2, P_3P_1)$

Q_1Q_2 is on the side P_1P_2 if $\lambda_3 \leq 0$ or $\mu_3 \leq 0$, $d_{min} = d(Q_1Q_2, P_1P_2)$

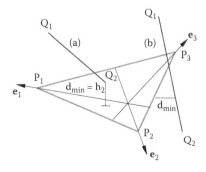

Figure 2.18 Distance between a line segment and a triangular facet: (a) nearest point inside triangle; (b) nearest point at an edge.

2.4.9 Dividing an edge into segments

In FE MG, it is often required to add more nodes to a line segment. The method employed to generate more nodes on an edge depends on the required node distribution and the data available. The minimum required information is the specified node spacing at the ending nodes of the given edge. In this case, for a smooth transition of element size, nodes are generated along the edge following a geometric progression.

2.4.9.1 Element size is specified at nodal points

Let ρ_1 and ρ_2 be the specified spacings at the ending nodes and d be the length of the edge.

a. $\rho_1 = \rho_2$
   The edge is to be divided following a uniform nodal distribution. Number of segment (division), $n = \mathrm{NINT}(d/\rho_1)$. Adjusted spacing between nodes, $s = d/n$.

b. $\rho_2 \neq \rho_1$
   Without loss of generality, assume $\rho_2 > \rho_1$. Suppose that the edge is to be divided into n segments with geometric progression ratio $r > 1$. The length of the first segment $s = \rho_1$ and we have for the last (n^{th}) segment $\rho_2 = \rho_1 r^{n-1}$. The length of the edge as a sum of all the n segments is given by

$$d = \frac{r^n - 1}{r - 1}\rho_1 = \frac{r\rho_2 - \rho_1}{r - 1} \Rightarrow r = \frac{d - \rho_1}{d - \rho_2}$$

Knowing the ratio r between adjacent segments, the number of segments n is given by

$$n = \mathrm{NINT}\left[\frac{\log\left(\dfrac{\rho_2}{\rho_1}\right)}{\log(r)}\right] + 1 \text{ and } r' = \left(\frac{\rho_2}{\rho_1}\right)^{\frac{1}{n-1}}$$

The geometric progression ratio r has to be adjusted as n is rounded up to a whole number, and the adjusted length of the first segment is given by

$$d = \frac{r'^n - 1}{r' - 1}s \Rightarrow s = \frac{r' - 1}{r'^n - 1}d$$

Remark: These formulas are applicable for edges with a length greater than a threshold value, i.e. $d > \dfrac{\rho_1^2 + \rho_1\rho_2 + \rho_2^2}{\rho_1 + \rho_2}$, and for edges of shorter lengths, no subdivision is needed.

Consider the division of the edges of a rectangle (100 × 50) into segments according to the element size specified at the corner points, as shown in Figure 2.19. Given the specified size of 2, 5, 16 and 8 at the corner points, the edge division formulas are applied to each edge in turn. Starting from the bottom edge following the counter-clockwise direction, the four edges are divided into 30, 5, 9 and 11 segments, respectively. It is noted that nodes are automatically clustered near the bottom-left corner where a smaller spacing of two units has been prescribed.

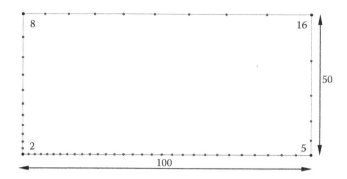

Figure 2.19 Dividing edges into segments.

2.4.9.2 *Element size is specified along the edge*

Apart from the nodal values specified at the ending nodes, if in addition the nodes have to be distributed along the edge following a specified spacing function, then a more sophisticated method has to be devised to generate nodes in compliance with the required node spacing. A set of new nodes has to be generated along the segments on the boundary **B**, such that the node distribution along the boundary segments complies with the given node spacing function ρ prior to MG:

$$\mathbf{B} = \{\mathcal{L}_i, i = 1, N_B\} \xrightarrow{\text{New nodes based on } \rho} \Gamma = \{L_i, i = 1, N_b\}$$

where N_B is the number of segments on the original domain boundary, and N_b is the number of segments after node generation along the segments in **B**.

The domain boundary **B** is first decomposed into straight lines composed of one or more boundary segments, each of which is bounded by two consecutive sharp corners (turns) found on boundary **B**. The task of boundary discretisation is to generate intermediate nodes along straight edges retrieved from **B** in compliance with the given node spacing function. A simple method assuming a linear node spacing variation between nodes is adopted for generating intermediate nodes on a straight line with n nodes and (n – 1) segments, as shown in Figure 2.20. In order to avoid any complicated node generation process involving integration and iteration (Frey 1987), an explicit formula that allows the position of new nodes to be calculated one by one along the straight line is employed.

With the exception of corner nodes and nodes where there is a sharp change of direction, nodes are all regenerated along each straight edge of **B** consisting of one or more line segments according to the given node spacing function. Depending on the given node spacing function, the number of segments N_b of the domain boundary after node regeneration may

Figure 2.20 Linear node spacing variation between nodes.

be more or less than the original number of segment N_B in **B** before node generation. As node spacing is sampled at each node point of **B**, for a more accurate node distribution, a uniform division can be first applied to all the line segments of **B**.

Let A be the node point just established and B be the next node to be generated, ρ_a and ρ_b being the node spacing at A and B, respectively. It is required that the average of the node spacing at A and B be equal to the distance between points A and B, i.e.

$$\|AB\| = \frac{1}{2}(\rho_a + \rho_b) \tag{2.1}$$

Two cases may arise as nodes are generated starting from one end of the straight line.

Case I: Points A and B are on the same line segment P_iP_{i+1}. Let ρ_i and ρ_{i+1} be the node spacing at P_i and P_{i+1}, respectively, r_i be the length of segment P_iP_{i+1}, and μ and λ be the parametric values of points A and B on segment P_iP_{i+1}, as shown in Figure 2.21. We have

$$\rho_b = \rho_i + \lambda(\rho_{i+1} - \rho_i), \quad \|AB\| = (\lambda - \mu)r_i = \frac{1}{2}(\rho_a + \rho_b)$$

$$\Rightarrow (\lambda - \mu)r_i = \frac{1}{2}[\rho_a + \rho_i + \lambda(\rho_{i+1} - \rho_i)] \Rightarrow \lambda = \frac{\rho_a + \rho_i + 2\mu r_i}{2r_i + \rho_i - \rho_{i+1}} \tag{2.2}$$

Case II: Points A and B are not on the same line segment. Referring to Figure 2.21, let B be on the line segment P_iP_{i+1} and ℓ be the distance between A and P_i. We have

$$\|AB\| = \ell + \lambda r_i = \frac{1}{2}(\rho_a + \rho_b)$$

$$\Rightarrow \lambda(2r_i + \rho_i - \rho_{i+1}) = \rho_a + \rho_i - 2\ell \Rightarrow \lambda = \frac{\rho_a + \rho_i - 2\ell}{2r_i + \rho_i - \rho_{i+1}} \tag{2.3}$$

To determine the position of a new node B, formula 2.2 is applied first to calculate λ. If λ is greater than 1, the next line segment $P_{i+1}P_{i+2}$ is included, and formula 2.3 with $\ell = (1 - \mu)r_i$ will be used to calculate λ. More line segments could be included if λ is still greater than 1, and formula 2.3 with ℓ updated is used to calculate the value of λ on the last included line segment. The last node generated does not necessarily terminate at the end point P_n of the straight line under consideration. There is, in general, a residue r when the end of the straight line is approached, as shown in Figure 2.22.

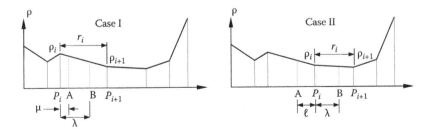

Figure 2.21 **Node generation along segments with linear variation** ρ.

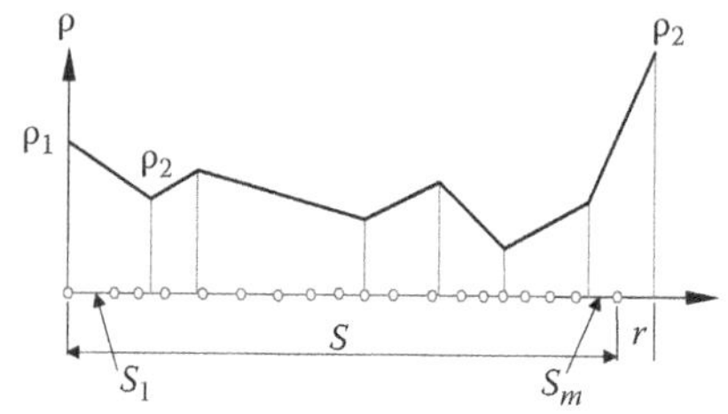

Figure 2.22 **Residue distributed to the segments generated.**

Let m be the number of line segments generated and S_i, $i = 1, m$, be the lengths of the line segments. Then the residue r can be distributed to the line segments already generated by

$$S_i^* = S_i\left(1 + \frac{r}{S}\right) \quad \text{where } S = \sum_{j=1}^{m} S_j$$

The positions (co-ordinates) of the nodes generated have to be revised using the adjusted node spacing S_i^*.

Remark: As node spacing can be specified explicitly by means of an analytical function or implicitly in terms of FE interpolation in adaptive meshing, node spacing is sampled, for both cases in exactly the same manner, at the nodal points of the boundary for node regeneration. For the first mesh, when there is no information about the node distribution, boundary line segments can be divided solely based on the domain geometry by specifying the appropriate element size at each corner node, and the boundary edges can be subdivided following a geometric progression, as described in Section 2.4.9.1.

2.4.10 γ Value of a tetrahedron cannot exceed the α value of its face

In the generation of fully constrained tetrahedral meshes over domains bounded by a given triangulated surface, one would like to ask, what is the expected quality of the tetrahedral meshes generated? The following derivation shows that γ_{min} is always smaller than or equal to α_{min}, where γ is the γ-quality of a tetrahedron, and α is the α-quality of a triangle.

As shown in Figure 2.23, γ value of tetrahedron ABCD is given by

$$\gamma = 72\sqrt{3}\,\frac{\text{volume}}{(\text{sum of edges squared})^{\frac{3}{2}}} = 72\sqrt{3}\,\frac{v}{s^{3/2}}$$

where v = volume of tetrahedron ABCD $= \dfrac{1}{3}$ Ah.

A = area of triangle ABC and

$$s = AB^2 + BC^2 + CA^2 + DA^2 + DB^2 + DC^2$$
$$= AB^2 + BC^2 + CA^2 + OA^2 + OB^2 + OC^2 + 3h^2$$

where O = 1/3 (A + B + C) is the centroid of triangle ABC.

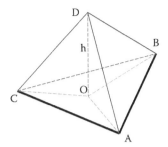

Figure 2.23 Tetrahedron ABCD.

$$\frac{\partial \gamma}{\partial h} = \frac{72\sqrt{3}}{s^3}\left(s^{1.5}\frac{\partial v}{\partial h} - v\frac{\partial s^{1.5}}{\partial h}\right) = \frac{72\sqrt{3}}{s^3}\left(s^{1.5}\frac{\partial v}{\partial h} - 1.5s^{0.5}v\frac{\partial s}{\partial h}\right) = \frac{72\sqrt{3}}{s^{2.5}}\left(s\frac{\partial v}{\partial h} - 1.5v\frac{\partial s}{\partial h}\right)$$

$$\frac{\partial \gamma}{\partial h} = 0 \Rightarrow \frac{72\sqrt{3}}{s^{2.5}}\left(s\frac{\partial v}{\partial h} - 1.5v\frac{\partial s}{\partial h}\right) = 0 \text{ or } s\frac{\partial v}{\partial h} - 1.5v\frac{\partial s}{\partial h} = 0 \Rightarrow \frac{1}{3}As = 1.5\left(\frac{1}{3}Ah\right)(6h)$$

or

$$9h^2 = s = AB^2 + BC^2 + CA^2 + OA^2 + OB^2 + OC^2 + 3h^2$$
$$\Rightarrow 6h^2 = AB^2 + BC^2 + CA^2 + OA^2 + OB^2 + OC^2$$

However,

$$OA^2 + OB^2 + OC^2 = \frac{1}{3}(AB^2 + BC^2 + CA^2)$$

Hence,

$$6h^2 = \frac{4}{3}(AB^2 + BC^2 + CA^2) \text{ or } h^2 = \frac{2}{9}(AB^2 + BC^2 + CA^2) \tag{2.4}$$

$$\gamma = 72\sqrt{3}\frac{v}{s^{3/2}} = \frac{72\sqrt{3}v}{\left(\frac{9}{2}h^2 + \frac{19}{32}h^2 + 3h^2\right)^{3/2}} = 8\sqrt{3}\frac{\frac{1}{3}Ah}{3h^3} = \frac{8\sqrt{3}A}{9h^2} = \frac{4\sqrt{3}A}{AB^2 + BC^2 + CA^2} = \alpha$$

Thus, γ attains the maximum value of α at the centroid O of triangle ABC with a height h given by Equation 2.4. Exercise: γ attains the highest value at the centroid of triangle ABC.

2.4.11 Determine whether a point is inside or outside of the problem domain

There are many ways to determine if a point is inside or outside a bounded domain in 2D or 3D. Since the mesh domain is bounded by oriented boundary segments on 2D and triangular facets on 3D, a simple yet robust method to determine its position relative to the problem domain can be devised by checking its orientation relative to the nearest boundary edge/face.

2.4.11.1 Two-dimensional domain

Let Ω be a 2D region bounded by boundary segments $\mathbb{B} = \{A_i B_i, i = 1, n\}$. Given a point P, the boundary segment nearest to P is given by

$$AB = \min_{i=1,n} d(P, A_i B_i), \quad A_i B_i \in \mathbb{B}$$

Calculate the signed area of triangle ABP. If the area of triangle ABP is greater than zero, P is inside Ω; if the area is less than zero, P is outside Ω; and if the area is equal to zero, P is on the boundary, as shown in Figure 2.24a. In case there are two nearest segments (P is nearest to a boundary node), choose the one that is not intersected by the other nearest edge.

2.4.11.2 Three-dimensional domain

Let Ω be a 3D region bounded by boundary faces $\mathbb{B} = \{A_i B_i C_i, i = 1, n\}$. Given a point P, the boundary face nearest to P is given by

$$ABC = \min_{i=1,n} d(P, A_i B_i C_i), \quad A_i B_i C_i \in \mathbb{B}$$

Calculate the signed volume of tetrahedron ABCP. If the volume of tetrahedron ABCP is greater than zero, P is inside Ω; if the volume is less than zero, P is outside Ω; and if the volume is equal to zero, P is on the domain boundary, as shown in Figure 2.24b. In case there are more than one nearest faces (P is nearest to a boundary node), choose the one that is

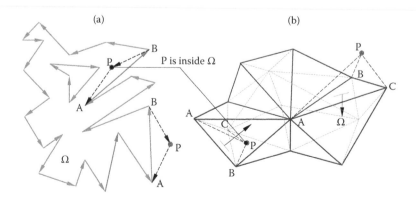

Figure 2.24 Check whether a point is inside or outside a bounded region: (a) 2D domain; (b) 3D domain.

not intersected by the other nearest facets. For a large mesh with a lot of boundary faces, a background grid can be used to speed up the searching process.

2.5 TOPOLOGICAL OPERATIONS AND ALGORITHMS

2.5.1 Find the neighbouring elements of a triangular mesh

Let $\mathbb{T} = \{T_i . i = 1, N_e\}$ be the set of triangular elements, with vertices V_i^j (j = 1,2,3) for the jth vertex of element i. We are going to find T_i^j, the jth neighbour (opposite to the jth vertex) of element i, and $T_i^j = 0$ if there is no neighbour on a boundary edge. The concept of the algorithm is based on the fact that two elements are neighbours if they share a common edge, and an edge is uniquely defined by the two ending nodes, as shown in Figure 2.25. The first part of the neighbour searching algorithm is to establish pointer array P in which P_k stores the number of nodes greater than and connected to a particular node k. Setting $P_1 = 1$, a cumulative count of nodal connections is obtained by adding $P_k + P_{k-1}$ to P_k such that nodes greater than k and connected to it are in the range P_{k-1} to $P_k - 1$ inclusive. With the aid of array Q, the second double loops i and j are to find any match of edge nodes with k being the smaller node of an edge. When the smaller node k is identified on an edge, searching along vector Q starting from $P_k - 1$ downwards, (i) if there is no match for the other node connected to k, register the larger node in Q and set a neighbouring element equal to 0; else (ii) establish the mutual neighbouring relationship between the current element and a previously registered element with the same edge nodes. The last part $\{L = m - 1;$ While $(Q_L \neq 0)$ $L = L - 1; Q_m = Q_{L+1}; Q_{L+1} = 0\}$ is to delete the matched edge, which will not affect the normal function of the algorithm but makes it run faster for large systems.

Algorithm Neighbour_T3 (Np, Ne, V, T, P, Q)

```
// Np and Ne are respectively the number of nodes and elements in the mesh.
// Input: V = Vertices of the elements
// Output: T = Neighbouring triangles
// Working arrays: P, Q
Initialize linear array P: P₁ = 1, {Pₖ = 0, k = 2,Nₚ}
        Loop: i = 1,Ne
                Loop: j = 1,3
                j1 = Vᵢᵐᵒᵈ⁽ʲ,³⁾⁺¹ ; j2 = Vᵢᵐᵒᵈ⁽ʲ⁺¹,³⁾⁺¹ ; k = min(j1, j2)
                Pₖ = Pₖ +1
                End loop j
        End loop i
        Loop: k=2,Np
```

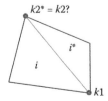

Figure 2.25 Adjacent elements *i* and *i** share common edge *k*1–*k*2.

```
                    Pk = Pk + Pk-1
                    End loop k
                    n = PNp
                    Set {Qk = 0, k = 1,n}
                    Loop: i = 1,Ne
                    Loop: j = 1,3
                    j1= Vimod(j,3)+1; j2= Vimod(j+1,3)+1 ; k1 = min(j1, j2); k2 = max(j1, j2); m = Pk1
1                   m = m - 1
                    If (Qm = 0) then
                    Qm = 3(i - 1) + j; Tij=0
                    else
                    k = Qm ; i* = (k + 2)/3; j* = k - 3(i* - 1)
                    j1*= Vi*mod(j*,3)+1; j2*= Vi*mod(j*+1,3)+1
                    k2* = max(j1*, j2*)
                    If (k2* ≠ k2) go to 1
                    Tij=i*; Ti*j*=i
                    L = m - 1
                    While (QL ≠ 0) L = L - 1
                    Qm = QL+1; QL+1 = 0
                    End if
                    End loop j
                    End loop i
```

2.5.2 Find the neighbouring elements of a tetrahedral mesh

Let $\mathbb{T} = \{T_i. i = 1, N_e\}$ be the set of tetrahedral elements, with vertices V_i^j ($j = 1,2,3,4$) for the jth vertex of element i. We are going to find T_i^j, the jth neighbour (opposite to the jth vertex) of element i, and $T_i^j = 0$ if there is no neighbour on a boundary triangular face. The concept of the algorithm is based on the fact that two elements are neighbours if they share a common face, and a triangular face is uniquely defined by the three nodes on the face. The neighbour searching algorithm for tetrahedral mesh is very similar to that for triangular meshes, except that, in the matching process, $3 - 1 = 2$ more nodes have to be checked for the matching faces connected to the same minimum node ($k1$) of the face, as shown in Figure 2.26. For the sake of clarity, two working arrays P and Q have been used in the *pseudo-code* of the neighbour searching algorithms. However, in the actual implementation, the two working arrays P and Q are merged into one such that the connection information of array Q is also stored in P, i.e. $\{P_{NP+k} = Q_k, k = 1,n\}$. In this case, the size of array P has to be greater than $N_P + n$, where N_P is number of nodes, and n is the number of connections in the mesh. As the number of connections in an FE mesh is directly proportional to the number of nodes or elements, the complexity of the neighbour searching algorithm is basically linear with respect to the number of nodes/elements.

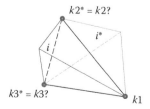

Figure 2.26 Adjacent elements *i* and *i** share a common face *k*1–*k*2–*k*3.

Algorithm Neighbour_T4 (Np, Ne, V, T, P, Q)

```
// Np and Ne are respectively the number of nodes and elements in the mesh.
// Input: V = Vertices of the elements
// Output: T = Neighbouring tetrahedra
// Working arrays: P, Q
// Index array I_ij = jth node of face i, I = {2,4,3; 1,3,4; 1,4,2; 1,2,3}
Initialize linear array P: P_1 = 1, {P_k = 0, k=2,N_p}
        Loop: i = 1,Ne
                Loop: j = 1,4
                a = I_{j1}; b = I_{j2}; c = I_{j3}
                j1 = V_i^a; j2 = V_i^b; j3 = V_i^c; k = min(j1, j2, j3)
                P_k = P_k + 1
                End loop j
        End loop i
        Loop: k = 2,Np
        P_k = P_k + P_{k-1}
        End loop k
        n = P_{Np}
        Set {Q_k = 0, k = 1,n}
        Loop: i = 1,Ne
        Loop: j = 1,4
        a = I_{j1}; b = I_{j2}; c = I_{j3}
        j1 = V_i^a; j2 = V_i^b; j3 = V_i^c
        k1 = min(j1, j2, j3); k3 = max(j1, j2, j3); k2 = j1+j2+j3-k1-k3
        m = P_{k1}
 1      m = m - 1
        If (Q_m = 0) then
        Q_m = 4(i-1) + j; T_i^j = 0
        else
        k = Q_m; i* = (k+3)/4; j* = k - 4(i*-1)
        a* = I_{j*1}; b* = I_{j*2}; c* = I_{j*3}
        j1* = V_{i*}^{a*}; j2* = V_{i*}^{b*}; j3* = V_{i*}^{c*}
        k3* = max(j1*, j2*, j3*); k2* = j1*+j2*+j3*-k1-k3*
        If (k2* ≠ k2 or k3* ≠ k3) go to 1
        T_i^j = i*; T_{i*}^{j*} = i
        L = m - 1
        While (Q_L ≠ 0) L = L - 1
        Q_m = Q_{L+1}; Q_{L+1} = 0
        End if
        End loop j
        End loop i
```

2.5.3 Find the elements connected to each node in a mesh

In finding the elements connected to each node in a mesh, in the first part of the algorithm, the number of elements connected to each node is determined and stored in array P. In the second part, the same two loops are executed again, but this time, knowing the number of elements connected to node k, the actual element i is stored in the space allocated to node k in P. It is noted that only one single linear array P has been used to mark off the starting point and the ending point of the elements connected to each node k, which are also stored in P according to the starting-point and ending-point positions. The algorithm can be applied to other element types, say, tetrahedral meshes or even meshes with mixed element types by modifying the j loop to go over all the nodes of each element in turn. The following

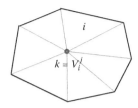

Figure 2.27 Triangular elements around node *k*.

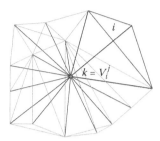

Figure 2.28 Tetrahedral elements around node *k*.

is the algorithm in finding all the elements connected to each node in a tetrahedral mesh. As shown in Figures 2.27 and 2.28, the element i containing node k (i.e. $k = V_i^j$ for j = 1,2,3 for triangular elements and j = 1,2,3,4 for tetrahedral elements) will be recorded in array P.

Algorithm Element_Node_T3 (Np, Ne, V, P)

```
// Find the elements connected to each node in a triangular mesh
// Np and Ne are respectively the number of nodes and elements in the mesh.
// Input: V = Vertices of the elements
// Output: Elements connected to node k: {Pᵢ, i = Pₖ+1, Pₖ₊₁}
Set zero to array P: {Pₖ = 0, k = 1,Nₚ+1}
      Loop: i = 1,Ne
              Loop: j = 1,3
              k = Vᵢʲ; Pₖ = Pₖ +1
              End loop j
      End loop i
      P₁ = P₁ + Np + 1
      Loop: k = 1,Np
      Pₖ₊₁ = Pₖ₊₁ + Pₖ
      End loop k
      Loop: i = 1,Ne
              Loop: j = 1,3
              k = Vᵢʲ; m = Pₖ; Pₘ = i; Pₖ = m - 1
              End loop j
      End loop i
```

Algorithm Element_Node_T4 (Np, Ne, V, P)

```
// Find the elements connected to each node in a tetrahedral mesh
// Np and Ne are respectively the number of nodes and elements in the
// mesh.
// Input: V = Vertices of the elements
```

```
// Output: Elements connected to node k: {Pᵢ, i = Pₖ + 1, Pₖ₊₁}
Set zero to array P: {Pₖ = 0, k = 1,Np + 1}
      Loop: i = 1,Ne
            Loop: j = 1,4
            k = Vᵢʲ; Pₖ = Pₖ + 1
            End loop j
      End loop i
      P₁ = P₁ + Np + 1
      Loop: k = 1,Np
      Pₖ₊₁ = Pₖ₊₁ + Pₖ
      End loop k
      Loop: i = 1,Ne
            Loop: j = 1,4
            k = Vᵢʲ; m = Pₖ; Pₘ = i; Pₖ = m - 1
            End loop j
      End loop i
```

2.5.4 Find the edges (unique line segments) of a triangular mesh

This simple algorithm is based on the idea that edge j1 – j2 shared by two elements following the orientation of the triangles will be that j1 < j2 in one element and vice versa, j2 < j1, in the other element, as shown in Figure 2.29. As j1 ≠ j2, the edge will only be counted once. To complete the set, all the boundary edges still have to be included, which are connected to only one element, and hence, the boundary condition $T_i^j = 0$ is also included in the *If* test.

Algorithm Edges_T3 (Np, Ne, V, T, E)

```
// Find the edges (unique line segments) of a triangular mesh
// Np and Ne are respectively the number of nodes and elements in the
// mesh.
// Input: V = Vertices of the elements, Neighbours of triangles T
// Output: Edges E: {Eᵢʲ, i = 1,n; j = 1,2}
      n = 0
      Loop: i = 1,Ne
            Loop: j = 1,3
            j1 = Vᵢ^mod(j,3)+1; j2 = Vᵢ^mod(j+1,3)+1
            If (j1 < j2 or Tᵢʲ = 0) then
            n = n + 1; Eₙ¹ = j1; Eₙ² = j2
            End if
            End loop j
      End loop i
```

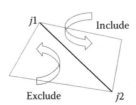

Figure 2.29 **Finding the edges of a triangular mesh.**

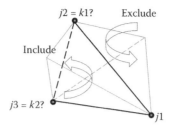

Figure 2.30 Finding the faces of a tetrahedral mesh.

2.5.5 Find the faces (unique triangular facets) of a tetrahedral mesh

The same idea for triangular mesh can be adopted, but it is slightly more complicated as the orientation of the faces with respect to a tetrahedron has to be considered. One way is to find out the minimum node and the maximum node of a triangular face, which will be counted if the order minimum node ($k1$) to maximum node ($k2$) is in line with the orientation of the tetrahedron under consideration, as shown in Figure 2.30.

Algorithm Faces_T4 (Np, Ne, V, T, F)

```
// Find the faces (unique triangular facets) of a tetrahedral mesh
// Np and Ne are respectively the number of nodes and elements in the
// mesh.
// Input: V = Vertices of the elements, Neighbours of tetrahedra T
// Output: Faces F: {Fⁱⱼ, i = 1,n; j = 1,3}
// Index array Iᵢⱼ = jᵗʰ node of face i, I = {2,4,3; 1,3,4; 1,4,2; 1,2,3}
n = 0
Loop: i = 1,Ne
      Loop: j = 1,4
      a = I_{j1}; b = I_{j2}; c = I_{j3}
      j1=Vᵢᵃ; j2=Vᵢᵇ; j3=Vᵢᶜ
      k1 = min(j1, j2, j3); k2 = max(j1, j2, j3)
      If ((k1 = j1 and k2 = j2) or (k1 = j2 and k2 = j3) or (k1 = j3 and
      k2 = j1) or Tᵢʲ = 0) then
      n = n + 1; Fₙ¹=j1; Fₙ²=j2; Fₙ³=j3
      End if
      End loop j
End loop i
```

2.5.6 Find the edges (unique line segments) of a tetrahedral mesh

As the number of edges incident to a point is a variable and unlimited similar to the case of a triangular mesh, the *Algorithm Neighbour_T3* can be modified to find out all the edges in a tetrahedral mesh. In scanning through the edges of an element, instead of the three edges of a triangle, all six edges of a tetrahedron have to be considered. Any edge of a tetrahedron ($n1 - n2$), if it is not already in array Q, is recorded as a new edge and registered in array Q; otherwise, it is just ignored, as shown in Figure 2.31.

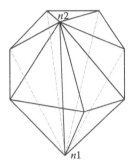

Figure 2.31 **Edge *n1–n2* is only recorded once.**

Algorithm Edges_T4 (Np, Ne, V, E, P, Q)

```
// Np and Ne are respectively the number of nodes and elements in the mesh.
// Input: V = Vertices of the elements
// Output: Edges E: {Eⱼᵢ, i = 1,n; j = 1,2}
// Working arrays: P, Q
Initialize linear array P: {Pₖ = 0, k = 1,Nₚ}
     n = 0
     Loop: i = 1,Ne
           Loop: j = 1,3
           Loop: k = j + 1,4
           k1=Vⱼᵢ; k2=Vᵏᵢ; m = min(k1, k2)
           Pₘ = Pₘ + 1
           End loop k
           End loop j
     End loop i
     Loop: k = 2,Np
     Pₖ = Pₖ + Pₖ₋₁
     End loop k
     Set {Qₖ = 0, k = 1, P_Np}
     Loop: i = 1,Ne
           Loop: j = 1,3
           Loop: k = j + 1,4
           k1=Vⱼᵢ; k2=Vᵏᵢ; n1 = min(k1, k2); n2 = max(k1,k2); m = P_n1 + 1
1    m = m - 1
     If (Qₘ = 0) then
     Qₘ = n2; n = n + 1; E¹ₙ=n1;E²ₙ=n2
     else
     if (Qₘ ≠ n2) go to 1
     End If
     End loop k
End loop j
End loop i
```

2.5.7 Retrieve the boundary (loop of line segments) of a triangular mesh

The boundary line segments (edges) of a triangular mesh can be easily recovered by considering solely the element adjacency relationship as given by the following algorithm.

Algorithm Boundary_T3 (Np, Ne, V, T, E)

```
// Find the boundary edges of a triangular mesh
// Np and Ne are respectively the number of nodes and elements in the mesh.
// Input: V = Vertices of the elements, Neighbours of triangles T
// Output: Boundary edges E: {Eⱼᵢ, i = 1,n; j = 1,2}
n = 0
Loop: i = 1,Ne
        Loop: j = 1,3
        If (Tⱼᵢ = 0) then
        n = n + 1; E¹ₙ = Vᵢᵐᵒᵈ⁽ʲ,³⁾⁺¹;  E²ₙ = Vᵢᵐᵒᵈ⁽ʲ⁺¹,³⁾⁺¹
        End If
        End loop j
End loop i
```

2.5.8 Retrieve the boundary (triangular facets) of a tetrahedral mesh

The boundary of a tetrahedral mesh can be determined purely by topological consideration. By definition, a boundary face is the triangular face of an element that is not to be shared with any other elements. Knowing the neighbours of each tetrahedral element, it won't be difficult to collect all those faces connected to only one single element, as shown in Figure 2.32. The following is a simple algorithm based on this idea.

Algorithm Boundary_T4 (Np, Ne, V, T, F)

```
// Find the boundary faces of a tetrahedral mesh
// Np and Ne are respectively the number of nodes and elements in the mesh.
// Input: V = Vertices of the elements, Neighbours of tetrahedra T
// Output: Boundary faces F: {Fⱼᵢ, i = 1,n; j = 1,3}
// Index array Iᵢⱼ = jᵗʰ node of face i, I = {2,4,3; 1,3,4; 1,4,2; 1,2,3}
n = 0
Loop: i = 1,Ne
        Loop: j = 1,4
        If (Tⱼᵢ = 0) then
        a = Iⱼ₁; b = Iⱼ₂; c = Iⱼ₃
        n = n + 1; F¹ₙ = Vᵢᵃ; F²ₙ = Vᵢᵇ; F³ₙ = Vᵢᶜ
        End If
        End loop j
End loop i
```

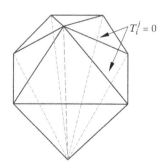

Figure 2.32 $T_i^j = 0$ on a boundary face.

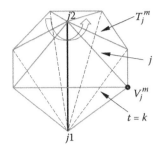

Figure 2.33 Tetrahedra around an edge, *coquille.*

2.5.9 Find the tetrahedral elements connected to an edge

In MG, it is often required to find the two elements sharing a common face, which is simply given by the adjacency relationship, or the ball of tetrahedra connected to a node, which is given by the algorithm **Element_Node_T4**. However, in mesh refinement, a node always has to be inserted at the mid-point of an edge so as to divide all tetrahedral elements connected to it. In order to do so, we have to first find out all the tetrahedra connected to a given edge, which is called a shell or *coquille* in French (George et al. 1991), as shown in Figure 2.33. An algorithm to find the tetrahedra connected to an edge is given as follows.

Algorithm T4_Edge (j1, j2, k, V, T, L)

```
// Find the tetrahedral elements connected to a given edge j1 - j2
// Input: k = tetrahedron containing j1 - j2
// Input: V = Vertices of the elements, T = Neighbours of triangles
// Output: List of tetrahedral elements, L = {Lᵢ, i = 1,n}
        n = 1
        L₁ = k
        Loop: i = 1,4
        if (Vᵏⁱ = j1 or Vᵏⁱ = j2 or Tᵏⁱ = 0) go to 1
        t = k
        j = Tᵏⁱ
2       n = n + 1
        Lₙ = j
        m = 1
        While (Vⱼᵐ = j1 or Vⱼᵐ = j2 or Tⱼᵐ = t) m = m + 1
        t = j
        j = Tⱼᵐ
        if (j = k) return
        if (j ≠ 0) go to 2
1       End loop i
```

2.5.10 Delete flagged elements from a tetrahedral mesh

In DT, tetrahedral elements to be deleted will first be flagged as they are still useful in the construction of the insertion cavity (CORE). It is also more efficient to allow elements to be formed within the capacity of the available memory and to only delete all the flagged elements in one go when the limit of the memory is to be exceeded. For a tetrahedral mesh of Ne elements with element node labels V, neighbours T and array F flagging the elements to be deleted, an algorithm to consolidate the tetrahedral mesh is given as follows.

Algorithm Compress_T4 (Ne, V, T, F)

```
// Consolidate a tetrahedral mesh
// Ne is the number of elements in the mesh.
// Input: V = Vertices of the elements, Neighbours of tetrahedra T,
// Flag array Fᵢ = 0 or 1, i = 1,Ne, tetrahedron i will be deleted if Fᵢ = 1.
// Output: A consolidated tetrahedral mesh {Ne*, V*, T*}
{       n = Ne
        Loop: i = 1,Ne
        If (Fᵢ ≠ 0)then
        n = n + 1
1       n = n - 1
        If (n < i) go to 2
        If (Fₙ ≠ 0) go to 1
        Loop: j = 1,4
        Vᵢʲ = Vₙʲ; k = Tₙʲ; Tᵢʲ = k ;
        If (k ≠ 0) find m such that Tₖᵐ = n and set Tₖᵐ = i
        End loop j
        Fᵢ = 0; Fₙ = 1; n = n - 1
        End If
        End loop i
2       Ne = n }
```

2.5.11 Find the tetrahedral elements within the boundary surface

For constrained boundary problems, following the boundary recovery process, all the tetrahedral elements within the closed boundary surface of triangular facets have to be identified. The same problem may be faced if the FE mesh and the boundary surface are constructed independent of each other. A simple algorithm based solely on adjacency relationship for a rapid retrieval of the elements within a bounded surface is given as follows.

Algorithm RBR3D (s, Ne, F, T, B, Z)

```
// Retrieve tetrahedral elements within a bounded surface
// s = seed tetrahedron inside the boundary surface, Ne = Number of
// elements in the mesh.
// Input: Fₖᵃ = aᵗʰ face of element k, Tₖᵃ = aᵗʰ neighbour of element k,
// Boundary surface B.
// Output: Zonal label Zᵢ = 0: outside or Zᵢ = 1: inside boundary surface,
// i = 1, Ne.
// Working array: A = List of active tetrahedra
{       n = 0; Set zero to array Z: (Zᵢ = 0, i = 1, Ne); Z_s = 1
        Loop: a = 1,4
        If (F_sᵃ ∉ B)then
        j = T_sᵃ ; n = n + 1; Aₙ = j; Z_j = 1
```

```
        End If
        End loop a
1       k = Aₙ;
        n = n - 1
        Loop: a = 1,4
        If (Fₖᵃ ∉ B) then
        j = Tₖᵃ
                if (Zⱼ = 0) then
                    n = n + 1; Aₙ = j; Zⱼ = 1
                End if
        End If
        End loop a
        If (n > 0) Go to 1
        Return }
```

2.6 SORTING

Data sorting or ordering is often needed to facilitate searching or to determine the sequence of processing following a certain criterion; for instance, mesh refinement by means of bisection of the longest edge requires a priority list of the edges for the order of subdivision or sorting of points in the construction of a k-dimensional tree (kd-tree), etc. Algorithms of various characteristics could be employed to rearrange an array of data following a specific ordering criterion.

2.6.1 Bubble sort

Bubble sort is one of the simplest sorting methods in which objects are rearranged by swapping adjacent items if they are in the wrong order. Operating from the top to the bottom, smaller items *bubble* up, and the largest item will sink to the bottom of the list in one pass. The same process is repeated with the remaining (n – 1) items, and in the second pass, the second largest item will sink to the (n – 1)th position. It is easy to see that the entire array of n items can be sorted in at most (n – 1) passes. However, for arrays already sorted at an intermediate stage, the process could stop earlier if there is no swap within a single pass. Let's consider the bubble sort of five numbers in an array, as shown in Figure 2.34a–c.

As shown in Figure 2.34a, the largest number 9 sinks to the bottom of the list in the first pass, the second largest number 7 sinks to the fourth position of the array in the second pass, as shown in Figure 2.34b, and the third largest number 4 sinks to the third position of the array in the third pass, as shown in Figure 2.34c. The numbers are now in good order, and the process terminates. For the sorting of an array of n items, in the first pass, (n – 1) comparisons and/or swaps are performed. In the second pass, (n – 2) comparisons and/or swaps are carried out, and the number of comparisons and/or swaps, N, in the worst case, is given by the triangular sum

$$N = (n-1) + (n-2) + \ldots + 2 + 1 = \frac{n(n-1)}{2}$$

Hence, it is an order $O(n^2)$ algorithm in the worst case and in the average case. However, the best performance is of order $O(n)$ when all the items are more or less in good order. The following is the *pseudo-code* for the bubble sort algorithm:

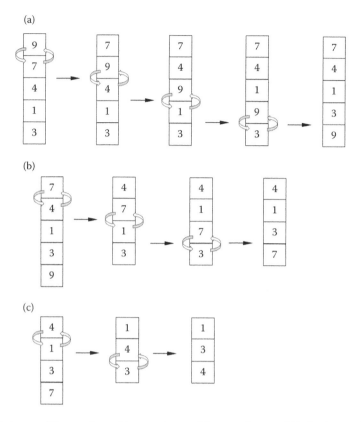

Figure 2.34 (a) First pass of five numbers, (b) second pass of four numbers and (c) third pass of three numbers.

Algorithm Bubble_Sort (n, A)

```
// A = Array to be sorted, A_i = Element in A with index i, n = Number of
// items
{      Loop: i = 1,n - 1
       k = 0
            Loop: j = 2,n - i + 1
                  If (A_{j-1} > A_j) then
                  Swap (A_{j-1} , A_j); k = 1
                  End if
            End loop j
       If (k = 0) return
       End loop i    }
```

2.6.2 Insertion sort

Insertion sort is another simple comparison sort in which the array is sorted by inserting one item at a time to the correct position of the current ordered list. Suppose the first k items have been sorted at an intermediate stage, the $(k + 1)^{th}$ item is inserted by finding from the list of ordered items the position of the first element, which is greater than or equal to the $(k + 1)^{th}$ item. Let's consider how the fourth number is inserted in the same array of five numbers [9, 7, 4, 1, 3] and how the fifth number is inserted, as shown in Figure 2.35.

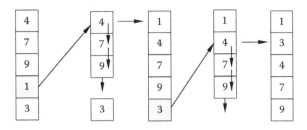

Figure 2.35 Inserting the fourth and fifth numbers.

In the insertion of the $(k + 1)^{th}$ item, k comparisons or number shiftings have to be carried out, and the number of operations for sorting n items is given by

$$N = 1 + 2 + \ldots + (n - 2) + (n - 1) = \frac{n(n-1)}{2}$$

Hence, insertion sort is also an order $O(n^2)$ algorithm. However, insertion sort has an average performance better than that of bubble sort as fewer swaps of elements are required in the insertion process. The following is the *pseudo-code* of the insertion sort algorithm.

Algorithm Insert_Sort (n, A)

```
// A = Array to be sorted, Aᵢ = Element in A with index i, n = Number of
// items
{       Loop: i = 2,n
        T = A_i
        j = 1
                While (A_j < T) j = j + 1
                Loop: k = i, j + 1, -1
                        A_k = A_{k-1}
                End loop k
        A_j = T
        End loop i     }
```

2.6.3 Quick sort

Quick sort is a *divide-and-conquer* algorithm. A list of n items is divided into two sub-lists by properly positioning a pivot taken out randomly from the original list. The sub-lists are further divided into shorter lists by introducing a pivot in each sub-list following exactly the same way as for the original list. Hence, a recursive procedure can be easily devised to deal with the lists and sub-lists in turn until there are only two items in the list or there are three items in the list with the pivot at the middle. In the ideal scenario, the pivots always cut the lists more or less into two equal halves. The level of subdivisions k reducing a list of n items down to a list of two or three items is given by

$$2^k = n \text{ or } k = \frac{\log(n)}{\log(2)}$$

At any level of subdivision, pivots can be inserted in all the lists by scanning through all the n items in the lists. Hence, the number of operations required to sort an array of n items can be estimated by

$$N \approx kn \approx n\log(n)$$

In the worst case, the pivots always cut at the head or the tail of a list such that the length of the sub-list is reducing at a very slow rate, resulting in an order $O(n^2)$ algorithm. However, the asymptotic performance of quick sort for random or evenly distributed data is of order $O(n\log(n))$, and thus, it is one of the most efficient and reliable sorting algorithms available.

Although the pivot can be chosen at random, it is more advantageous to pick the middle element as the pivot, as shown in Figure 2.36, for the sorting of the array [7, 9, 4, 1, 3, 6, 8].

As shown in Figure 2.36, the hatched elements serve as pivots, and their final positions are shaded. Once a pivot is selected, all the elements smaller than the pivot are placed to the left of the pivot by means of element swaps, and automatically, all the elements greater than or equal to the pivot will stay on the right of the pivot. When all the elements on the list are scanned through and the necessary swaps are done, the pivot will find its correct position on the original list. As shown in Figure 2.36, in the first subdivision, '1' is chosen as the pivot, and as it is the smallest number in the list, all the numbers are placed to the right or at the lower part of the list. As a result, '1' occupies the first position of the list, and it turns out to be the correct position on the list. In the next level of subdivision, '7' is chosen as the pivot; three numbers are smaller than '7', and they are put to the left or on the upper part of the list, and the result is that '7' is on the fourth position of the sub-list and the fifth position of the entire list if the head of the list is properly recorded, i.e. '6' is at the second position of the original list. At the third level of subdivision, two sub-lists are generated, each of which is assigned a pivot. '4' is the pivot for the first sub-list and '9' is the pivot for the second sub-list, and all the numbers find their correct positions on the original list after fixing the positions of the pivots by element swapping.

Although the quick sort can be easily implemented by means of recursive programming, the process could be speeded up by avoiding the recursive procedure using a loop over the sub-lists, provided that the bounds of the sub-lists are well kept track of. The following is the *pseudo-code* of the quick sort algorithm based on a loop over the sub-lists.

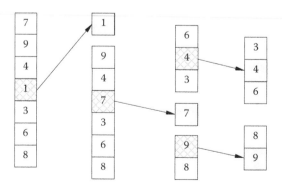

Figure 2.36 How lists are divided by pivots in Quicksort.

Algorithm QuickSort (n, A)

```
// H, T are integer arrays, H_L = Head of list L, T_L = Tail of list L
// I = Head of list, J = Tail of list, IP = Pivot position, L = Number of
// lists
{      L = 1; H_1 = 1; T_1 = n
       While (L > 0) {      I = H_L; J = T_L; L = L - 1
       Pivot (A, I, J, IP)
       If (I < IP - 1) then
       L = L + 1; H_L = I; T_L = IP - 1
       End if
       If (IP + 1 < J)
       L = L + 1; H_L = IP + 1; T_L = J
       End if }}
```

Procedure Pivot (A, I, J, IP)

```
// P = Pivot, IP = Position of pivot, I = Head of list, J = Tail of list,
// A_L = Element at index L
{      K =(I + J)/2 // Get pivot value at the middle of the list
       P = A_K; A_K = A_I; A_I = P; IP = I
       Loop: L = I + 1,J
       T = A_L
       If (T < P) then              // If (T < P or T = P and IP < K) then
       IP = IP + 1; A_L = A_IP; A_IP = T  // IP increases by 1 for each A_L smaller
                                    // than P
       End if
       End loop L
       A_I = A_IP; A_IP = P}
```

2.6.4 Bin sort

Bin sort is also known as bucket sort or address sort. In the previous sorting algorithms, items are sorted by means of comparisons and element swaps without specific reference to the actual values of the items, and as a result, the number of operations depends only on the positions of the data but not on the values of the data to be sorted. Bin sort employs an entirely different approach in which data are sorted based on their specific values and not on their order on the list. As the name of the sorting method suggests, a number of bins are prepared so that the items are spread into the bins according to their actual values, such that data of smaller values are put in the top bins, and data of larger values are put in the bottom bins in some linear interpolation manner. The number of bins used, N, depends on the available memory, and usually it is more efficient to use a large N that is equal to 2n or 3n, where n is the number of data to be sorted.

Let N be the number of bins available. Before a bin can be assigned to a particular value, we have to find out the range of the data. Suppose x_{min} and x_{max} are, respectively, the minimum and maximum values in the array; then the bin size, d, is given by

$$d = \frac{x_{max} - x_{min}}{N}$$

The ith element of array X will be assigned to bin k given by

$$k = INT\left(\frac{x_i - x_{min}}{d}\right) + 1$$

where $INT(x)$ = the largest integer smaller than or equal to x.

In case there are bins with more than one data, another round of bin sort can be applied to those bins until all the data points are separated in different bins. For evenly distributed data points, the data will be evenly sorted into the bins, and the bin sort is of linear complexity in this favourable situation. However, in the worst case, for each bin sort, all data points but one are put in one single bin, and it will take $n - 1$ sorts to separate all the points into different bins; the complexity of the algorithm is of order $O(n^2)$. Data values $x_k = k!$ will be such a difficult case for the bin sort; nevertheless, an ordinary computer cannot hold more than 200 data with such a wide range. For this simple analysis, it is clear that the efficiency of bin sort depends on data distribution. Devroye and Klincsek (1981) and Orenstein et al. (1983) showed that points distributed by functions of compact support and square integrable can be sorted in linear time by the recursive bin sort. In spite of its linear characteristics, whether bin sort would outperform quick sort depends on the values of data points as well as the initial order, as bin sort needs more space and overhead in finding the minimum and maximum values and in calculating the bin address for each data point. The procedure for bin sort is illustrated with an example of sorting eight data points, as shown in Figure 2.37.

The bin sort can be considerably speeded up using a loop structure instead of a recursive setting at the expense of using more storage. An array of n data points will first be sorted into n bins, and a pointer is used to mark the data points in each bin. Bin sort will be applied to all the bins with more than one data point until all the data points are separated into different bins. The following is the *pseudo-code* of the bin sort algorithm based on a loop structure.

$$k = MIN(INT(\tfrac{x_i - x_{min}}{d})+1, n), \quad n = 8, \ x_{min} = 1, \ x_{max} = 17, \ d = 2$$

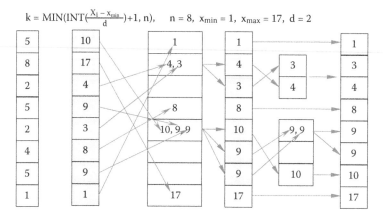

Figure 2.37 Bin sort of an array of n = 8 data points.

Algorithm BinSort (n, X)

```
// X = Array of data to be sorted, n = number of data points
// P, H, T are integer arrays, H_L = Head of bin L, T_L = Tail of bin L
// N = Number of bins with more than one data, points in bin i = {X_k :
// k = P_i + 1,P_{i+1}}
{      Bin (n, X, P)
       N = 0
       Loop i = 1,n
               If (P_{i+1} > P_i + 1) then
               N = N + 1; H_N = P_i + 1; T_N = P_{i+1}
               End if
       End loop i
       While (N > 0) {      J = H_N; K = T_N; N = N - 1; L = K - J + 1
               Bin (L, X(J), P)     // X_J in BinSort = X_1 in Bin
               Loop i = 1,L
                       If (P_{i+1} > P_i + 1) then
                       N = N + 1; H_N = P_i + J; T_N = P_{i+1} + J - 1
                       End if
               End loop i      }}
```

Procedure Bin (n, X, P)

```
// X_i = value of array X at index i, P = Integer array, data in bin
// k = {X_j, j = P_k + 1,P_{k+1}}
{      Set P to zero, P_i = 0, i = 1,n + 1
       x_min = X_1; x_max = x_min
       Loop i = 2,n
       y = X_i
       if (y < x_min) x_min = y
       if (y > x_max) x_max = y
       End loop i
       d = (x_max - x_min)/n
       if (d < tolerance) return  // tolerance = 10^{-99}

// Count the number of data points in each bin k, k = 1,n
       Loop i = 1,n
       k = INT((X_i - x_min)/d) + 1
       if (k > n) k = n
       P_k = P_k + 1
       End Loop i

// Compute P_k = Number of data points in bins 1 to k
       Loop k = 2,n
       P_k = P_k + P_{k-1}
       End loop k
       P_{n+1} = P_n

// Collect data points in each bin k, {Y_j, j = P_k + 1,P_{k+1}}
       Loop i = 1,n
       k = INT((X_i - x_min)/d) + 1
       if (k > n) k = n
       j = P_k; Y_j = X_i; P_k = j - 1
       End loop i
```

```
// Copy array Y back to array X
      Loop i = 1,n
      Xᵢ = Yᵢ
      End loop i    }
```

2.6.5 Comparison of the sorting methods

The sorting methods are to be tested with three sets of data of different characteristics. The first set consists of evenly distributed data, which are generated by the function

$$x_i = \sin (i + 0.23) \quad i = 1, n$$

The results are shown in Table 2.1 where *count* is the total number of data in all the lists sorted by quick sort or the total number of data in all the bins sorted by bin sort. Hence, *count* is a direct measure of complexity with respect to the number of data points n. *Time* is the CPU time of Intel® Core™ i7 CPU 870@2.93GHz running Visual Fortran in XP mode. The *count* and the CPU time taken by the four sorting methods for n = 10^2, 10^3, ..., 10^7 are plotted in Figure 2.38. As shown in Table 2.1, bubble sort and insertion sort are obviously $O(n^2)$ algorithms, as the CPU time increases by 100 times for each step of increment of n by 10 times. A direct comparison between the bubble sort and insertion sort reveals that the insertion sort is about twice more efficient than the bubble sort. As the $O(n^2)$ trend complexity is clear, the tests

Table 2.1 Counts and CPU time for sorting of data set I

n	Quick sort		Quick sort*		Bin sort		Bubble	Insert
	Count	Time(s)	Count	Time(s)	Count	Time(s)	Time(s)	Time(s)
100	681	9.77E-6	681	1.04E-5	148	1.31E-5	4.04E-5	1.9E-5
1000	11,799	1.42E-4	11,799	1.45E-4	2096	1.90E-4	3.6E-3	1.74E-3
10,000	165,812	0.00186	165,812	0.00197	21,767	0.00203	0.378	0.173
100,000	2,110,329	0.0196	2,110,329	0.0241	141,907	0.0133	37.6	17.5
1,000,000	25,852,531	0.284	25,852,531	0.292	1,622,397	0.21		
10,000,000	316,774,234	3.34	316,774,234	3.48	14,974,655	3.29		

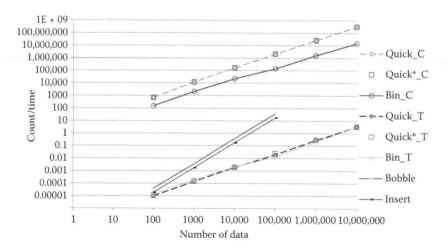

Figure 2.38 Data set I: x(i) = sin(i + 0.23).

for the bubble sort and insertion sort stop at n = 10^5. On the other hand, the quick sort and bin sort are far more efficient than the bubble sort and insertion sort, especially for large data sets. The nlog(n) complexity trend is quite obvious for the quick sort, as the *count* increases slightly more than 10 folds for each step of increment of n by 10 times. In spite of minor fluctuation for each increment of n, the bin sort follows more or less a linear trend. However, the overall performance of the quick sort and bin sort is quite similar for n up to 10^7.

The second set of data consists of packets of equal items, which are generated by the function

$$x_i = \sin (\mathrm{mod}(i,17) + 0.23) \quad i = 1, n$$

Only the quick sort and bin sort are tested, and the results are shown in Table 2.2. The *count* and the CPU time taken for this data set are plotted in Figure 2.39. As shown in Table 2.2, the *count* for the quick sort increases 100 times for each step of increment of n from 10^4 to 10^6. The $O(n^2)$ performance is the worst case for the quick sort due to the presence of packets of equal items. As for the bin sort, the *count* is absolutely linear, and the CPU time also follows a strictly linear trend, showing that the bin sort is very efficient with equal valued data. The fact that the bin sort has no difficulty with equal values is attributed to the program statement 'if (d < tolerance = 10^{-99}) return', and for an array of equal values, $x_{min} = x_{max}$ or $d = 0$, and as a result nothing will be done to this array.

Table 2.2 Counts and CPU time for sorting of data set 2

n	Quick sort		Quick sort*		Bin sort	
	Count	Time(s)	Count	Time(s)	Count	Time(s)
100	847	9.53E-6	711	1.03E-5	200	1.03E-5
1000	34,459	1.87E-4	10,259	1.34E-4	2000	9.90E-5
10,000	2,990,844	0.0108	138,376	0.00151	20,000	9.80E-4
100,000	294,591,139	0.987	1,879,235	0.0194	200,000	0.0096
1,000,000	29,417,088,191	96.7	21,759,630	0.221	2,000,000	0.1
10,000,000	2,941,235,588,192	9695	252,306,156	2.39	20,000,000	1.03

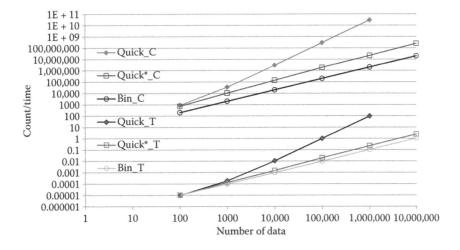

Figure 2.39 Data set 2: x(i) = sin(mod(i,17) + 0.23).

The weakness of quick sort on equal valued data can be easily rectified. The main difficulty is due to the improper placement of the pivot always at the first position for a list of equal valued data instead of at the middle. What has to be done is to replace the statement in *Procedure Pivot (A, I, J, IP)*, '*If (T < P) then*' by '*If (T < P or T = P and IP < K) then*'. In case of equal valued data, $T = P$, IP would not stop at the first position but would proceed to the Kth position right at the middle of the list. This modification has been implemented and tested, and the enhanced algorithm is denoted by quick sort*. As shown in Table 2.2, the quick sort* takes slightly more CPU time as compared to the quick sort to sort an array of small size, say, n = 100. For larger n = 10^3 onwards, the quick sort* takes much less CPU time than the quick sort, and as indicated by the *count*, the nlog(n) complexity trend is recovered for large n's. For evenly distributed data as shown in Table 2.1, the quick sort* only takes a few percent more CPU time compared to the quick sort.

The third set of data consists of widely spread and equal values of $\pm 10^{43}$, which are generated by the function

$$x_i = \exp\,(\text{mod}(i,100) + 0.1)\,\sin\,(\text{mod}(i,17) + 0.23)\quad i = 1, n$$

Again, only the quick sort, quick sort* and bin sort are tested, and the results are shown in Table 2.3. The *count* and the CPU times taken are plotted in Figure 2.40. As shown in Table 2.3, the bin sort takes slightly more CPU time as compared to the sorting of the set of

Table 2.3 Counts and CPU time for sorting of data set 3

n	Quick sort		Quick sort*		Bin sort	
	Count	Time(s)	Count	Time(s)	Count	Time(s)
100	763	1.02E-5	763	1.05E-5	998	9.20E-5
1000	12,413	1.43E-4	12,413	1.47E-4	8766	8.00E-4
10,000	183,760	1.93E-3	172,779	2.10E-3	76,384	6.12E-3
100,000	4,454,447	0.0283	1,891,357	0.0216	630,294	0.049
1,000,000	308,302,121	1.14	23,348,533	0.247	5,474,122	0.416
10,000,000	29,563,195,747	95.0	272,169,661	2.75	48,964,710	3.78

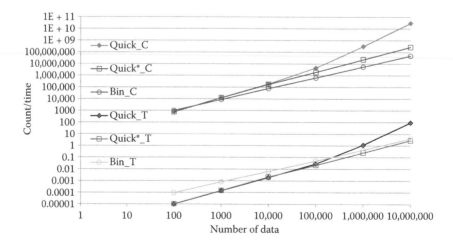

Figure 2.40 Data set 3: x(i) = exp(mod(i,100) + 0.1)sin.

evenly distributed data, showing that the performance of the bin sort is pretty consistent over data values in the practical range. However, the performance of the quick sort* is the best, indicating that the quick sort* is not sensitive to the values of the data set. The performance of the quick sort is rather disappointing for this data set, suggesting that the original version of the quick sort is quite sensitive to an even minor fraction of equal-valued data. From the three sets of data tested, the performance of the quick sort* is the most promising in terms of speed and memory requirement, yet a definite conclusion cannot be drawn unless more data sets are tested, and other programming techniques are tried out. However, one thing is clear that both the bin sort and quick sort* are efficient, and either one of them can be employed to sort data points up to $n = 10^8$ or more with confidence. Other sorting methods not mentioned in this section are also available; interested readers can take a look at the *merge sort* by Nardelli and Proietti (2006) and *sample sort* and *PE sort* by Chen (2006).

2.7 BACKGROUND GRID

In MG, a background grid is a partition of space to facilitate searching or to establish a neighbourhood relationship for various quantities (Shamos and Hoey 1991). Depending on the distribution of the points or objects under consideration and other specific requirements, grids of different characteristics can be constructed. By the use of a background grid, the space will be partitioned into cells, and the ideal scenario is to divide the space such that each cell will hold a roughly equal number of points. According to different point distributions, regular (uniform) grid, irregular (unequal spacing) grid, Quadtree/Octree, kd-tree, etc., have been devised.

2.7.1 Regular (uniform) grid (2D)

Regular (uniform) grid is a simple efficient scheme for uniformly distributed objects, for instance, points generated in a random process. Let $\{P_i = (x_i, y_i), i = 1, N\}$ be a set of N points on a 2D space. An algorithm for the construction of a regular grid is given as follows.

Determine the minimum and maximum values of the points in the x- and y-directions.

$$x_{min} = \min_{i=1,N} x_i, \quad x_{max} = \max_{i=1,N} x_i, \quad y_{min} = \min_{i=1,N} y_i, \quad y_{max} = \max_{i=1,N} y_i$$

Given the average number of points in a cell, n, calculate the size of a cell.

$$R_x = x_{max} - x_{min}, R_y = y_{max} - y_{min}, \alpha = R_x/R_y$$

The number of divisions in the x-direction is given by

$$N_x = \text{NINT}\left(\sqrt{N/(\alpha n)}\right)$$

The number of divisions in the y-direction is given by

$$N_y = \text{NINT} (\alpha N_x), \text{ where NINT(.)} = \text{nearest integer}$$

The grid spacing is given by

$$d_x = R_x/N_x, d_y = R_y/N_y$$

The number of cells is given by

$$N_c = N_x N_y$$

The following *pseudo-code* assigns nodes to each cell.

Algorithm Regular_Grid (N, X, Y, P)

```
// N = Number of points, X,Y = x-, y-co-ordinates, P = Pointer to record
// points in each cell
{     Set pointer array P to zero, P_i = 0, i = 1,N_c + 1
      Loop: i = 1,N // Determine the number of points in each cell
      I_x = (x_i - x_min)/d_x + 1; if (I_x > N_x) I_x = N_x
      I_y = (y_i - y_min)/d_y + 1; if (I_y > N_y) I_y = N_y
      k = (I_y - 1)N_x + I_x
      P_k = P_k + 1
      End loop i
      Loop: k = 1,N_c // Compute P_k = cumulative number of points up to cell k
      P_k+1 = P_K+1 + P_k
      End loop k
      Loop: i = 1,N // Assign points to each cell
      I_x = (x_i - x_min)/d_x + 1; if (I_x > N_x) I_x = N_x
      I_y = (y_i - y_min)/d_y + 1; if (I_y > N_y) I_y = N_y
      k = (I_y -1)N_x + I_x
      m = P_k; P_m = i; P_k = m-1
      End loop i    }
```

Points assigned to cell k are given by $\{P_m, m = P_k + 1, P_{k+1}\}$. The construction of a regular grid is simple and fast with a linear complexity as the operations in processing a point are independent of the number of points in the system. Given a point $p(x, y)$, the containing zone k is given exactly by the same formulas as in the point-allocation procedure.

$$k = (I_y - 1)N_x + I_x, \qquad I_x = [(x - x_{min})/d_x] + 1, I_y = [(y - y_{min})/d_y] + 1 \qquad [.] = \text{Integral part}$$

Let's consider the construction of a regular grid for 40 randomly generated points (slightly biased towards a corner), as shown in Figure 2.41, for which the co-ordinates of the points are given in Table 2.4.

$x_{min} = 0$, $y_{min} = 0$, $x_{max} = 100$, $y_{max} = 100$, $R_x = x_{max} - x_{min} = 100$, $R_y = y_{max} - y_{min} = 100$, $N_x = 3$, $N_y = 3$, $d_x = R_x/N_x = 100/3$, $d_y = R_y/N_y = 100/3$, $N_c = N_x N_y = 9$ and length of pointer array $P = N + N_c + 1 = 40 + 9 + 1 = 50$. Following the procedure to construct a regular grid for 40 randomly generated points, the resulting pointer array **P** is given in Table 2.5.

There are nine zones (cells) in the partition, and points in the first zone are given by

$$\{P_m, m = P_1 + 1, P_2\} = \{P_m, m = 11, 25\} = \{1, 7, 12, 13, 16, 21, 24, 25, 26, 27, 30, 32, 34, 36, 37\}$$

Points in zone 9 = $\{P_m, m = P_9 + 1, P_{10}\} = \{P_m, m = 49, 50\} = \{18, 28\}$

2.7.2 Regular (uniform) grid (3D)

A 3D uniform grid for a set of N points $\{P_i = (x_i, y_i, z_i), i = 1, N\}$ can be constructed following the same procedure of the 2D uniform grid. A regular grid of $N_x \times N_y \times N_z$ cells

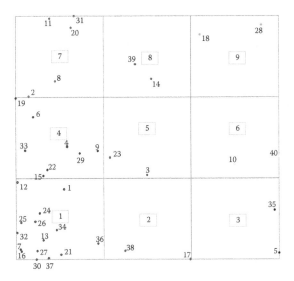

Figure 2.41 Regular grid of 9 zones for 40 randomly generated points.

Table 2.4 Co-ordinates of 40 randomly generated points

i	x_i	y_i	x_{10+i}	y_{10+i}	x_{20+i}	y_{20+i}	x_{30+i}	y_{30+i}
1	18.11	26.07	12.77	89.71	16.94	1.82	21.96	90.62
2	4.88	60.75	0.552	28.80	11.77	33.35	0.398	9.97
3	49.41	31.42	10.16	7.23	35.44	37.91	3.46	40.56
4	19.35	41.89	51.09	67.15	8.79	17.21	15.29	11.03
5	99.08	2.37	10.25	31.14	1.87	13.75	97.39	18.38
6	6.47	53.00	2.01	3.04	7.17	14.14	30.84	5.95
7	1.61	3.81	65.75	0.0386	7.91	3.23	12.22	0.507
8	14.85	66.41	69.45	83.72	92.51	87.27	41.08	3.25
9	30.75	40.41	0.005	60.12	24.03	39.45	44.98	72.67
10	82.47	35.36	20.77	86.27	7.56	0.00	99.04	38.28

Table 2.5 Pointer array P for a regular grid of 40 points

i	1	2	3	4	5	6	7	8	9	10
P_i	10	25	27	29	37	39	41	46	48	50
P_{10+i}	{1	7	12	13	16	21	24	25	26	27
P_{20+i}	30	32	34	36	37}	{17	38}	{5	35}	{4
P_{30+i}	6	9	15	19	22	29	33}	{3	23}	{10
P_{40+i}	40}	{2	8	11	20	31}	{14	39}	{18	28}

(boxes, zones, bins) subdivides uniformly the space of interest into a number of regions such that for any given point, the cell containing it can be rapidly determined without much calculation.

Compute $\alpha = \sqrt[3]{\dfrac{N}{nR_xR_yR_z}}$ where n = average number of points in a cell:

$R_x = x_{max} - x_{min}$, $R_y = y_{max} - y_{min}$, $R_z = z_{max} - z_{min}$,

The number of divisions along each direction is given by

$$N_x = NINT(\alpha R_x), N_y = NINT(\alpha R_y), N_z = NINT(\alpha R_z), \text{ number of cells, } N_c = N_x N_y N_z$$

Compute the size of a cell along each direction:

$$d_x = R_x/N_x, d_y = R_y/N_y, d_z = R_z/N_z$$

Determine cell k to which point $P_i = (x_i, y_i, z_i)$ belongs:

$$I_x = [(x_i - x_{min})/d_x] + 1, I_y = [(y_i - y_{min})/d_y] + 1, I_z = [(z_i - z_{min})/d_z] + 1$$

$$k = (I_z - 1)N_{xy} + (I_y - 1)N_x + I_x \quad \text{with } N_{xy} = N_x N_y \quad [x] = \text{Integral part of } x$$

Points assigned to cell k are given by $\{P_m, m = P_k + 1, P_{k+1}\}$. Whether it is a 2D grid or a 3D grid, by means of the pointer approach, the memory required for the grid construction is always a linear array **P** of size equal to $N + N_c + 1$. Given a point $p(x, y, z)$, the containing cell k is given by the same point-allocation formula,

$$k = (I_z - 1)N_{xy} + (I_y - 1)N_x + I_x,$$

where

$$I_x = [(x - x_{min})/d_x] + 1, \quad I_y = [(y - y_{min})/d_y] + 1, \quad I_z = [(z - z_{min})/d_z] + 1$$

2.7.3 Searching for general objects by means of a background grid

Very often, objects to be located are not points but some quantities of a finite size, such as a line segment, triangle, hexahedron, etc. Objects of finite size can be assigned not to a single cell but to a series of cells intersected by the object. For instance, we would like to determine cells that are cut across by a line segment in a regular 3D grid. Three methods are introduced to find the cells intersected by a line segment over a 3D grid.

2.7.3.1 Method 1: Search by neighbourhood

The method is often employed in computer graphics applications, and the procedure is given as follows. Consider line segment $P_1 P_2$.

i. Calculate the cell (I_x, I_y, I_z), which contains the point $P_1 = (a_1, b_1, c_1)$.

$$I_x = [a_1/d_x] + 1, \quad I_y = [b_1/d_y] + 1, \quad I_z = [c_1/d_z] + 1, \quad x_{min} = y_{min} = z_{min} = 0$$

ii. Following the direction of the arrow $P_1 P_2$, trace for the face that is being intersected, as shown in Figure 2.42, and go onto a neighbouring cell of (I_x, I_y, I_z), i.e.

$$I_x \mapsto I_x, I_x + 1 \text{ or } I_x - 1; \quad I_y \mapsto I_y, I_y + 1 \text{ or } I_y - 1; \quad I_z \mapsto I_z, I_z + 1 \text{ or } I_z - 1$$

Along a particular axis, throughout the journey, the trace follows a single direction, i.e. either always +1 or always –1.

iii. Line $P_1 P_2$ will be recorded in each cell visited, and the address of the cell is updated on the way. The tracing process continues until point $P_2 = (a_2, b_2, c_2)$ is reached.

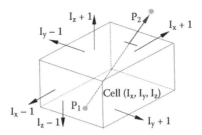

Figure 2.42 Method of ray tracing.

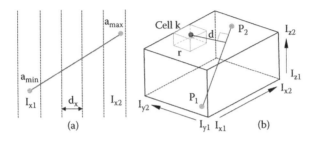

Figure 2.43 Method of checking distance: (a) min-max of a line segment; (b) distance from a cell.

2.7.3.2 Method 2: By checking the distance

i. Calculate the limits of the line segment as shown in Figure 2.43a.

$$a_{min} = min(a_1, a_2), b_{min} = min(b_1, b_2), c_{min} = min(c_1, c_2)$$

$$a_{max} = max(a_1, a_2), b_{max} = max(b_1, b_2), c_{max} = max(c_1, c_2)$$

Cells potentially intersected by line segment: $(I_{x1} \rightarrow I_{x2}) \times (I_{y1} \rightarrow I_{y2}) \times (I_{z1} \rightarrow I_{z2})$

where

$I_{x1} = [a_{min}/d_x] + 1, I_{x2} = [a_{max}/d_x] + 1, I_{y1} = [b_{min}/d_y] + 1, I_{y2} = [b_{max}/d_y] + 1, I_{z1} = [c_{min}/d_z] + 1, I_{z2} = [c_{max}/d_z] + 1.$

ii. Let r be the radius of the sphere containing cell k. Then cell k will not be intersected if $d > r$, where d is the projected distance of the centre of the sphere to line P_1P_2, as shown in Figure 2.43b.

2.7.3.3 Method 3: By elimination

A typical cell Ω can be defined by the intersection of half spaces bounded by hyper-planes passing through the six sides of the cell, as shown in Figure 2.44. Hence, we have

$$\Omega = \Omega_1 \cap \Omega_2 \cap \Omega_3 \cap \Omega_4 \cap \Omega_5 \cap \Omega_6$$

where Ω_i are the half spaces as bounded by the hyper-planes H_i along the sides of the cell.

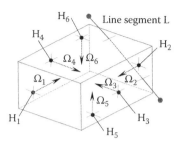

Figure 2.44 Cell bounded by hyper-planes H_1–H_6.

The intersection between cell Ω and a line segment L can be obtained by considering individual portions of the line segment cut off by the hyper-planes of the cell

$$L \cap \Omega = L \cap (\Omega_1 \cap \Omega_2 \cap \Omega_3 \cap \Omega_4 \cap \Omega_5 \cap \Omega_6)$$
$$= (L \cap \Omega_1) \cap (L \cap \Omega_2) \cap (L \cap \Omega_3) \cap (L \cap \Omega_4) \cap (L \cap \Omega_5) \cap (L \cap \Omega_6)$$
$$= L_1 \cap L_2 \cap L_3 \cap L_4 \cap L_5 \cap L_6$$

$L_i = L \cap \Omega_i$ = portion of L cut by H_i

If the original line segment L is mapped onto the interval [0,1], then intersection of line segments L_1, L_2, ..., L_6 can be easily done on this closed interval [0,1]. Whenever the resulting portion is reduced to an empty set during the process, there is no intersection between the cell and the line segment L.

2.7.4 Determine the cells intersected by a triangular facet

To find the intersection between two solid objects, it is necessary to consider the intersection of their boundary surfaces, which are usually discretised into triangular facets. A 3D background grid can be employed to speed up the searching of all potential triangular elements for intersection within the size range of a given triangular facet. The cells intersected by triangular facet $P_1 P_2 P_3$ can be found as follows.

    i. Calculate the limits of the triangular facet

$a_{min} = \min(a_1, a_2, a_3)$, $b_{min} = \min(b_1, b_2, b_3)$, $c_{min} = \min(c_1, c_2, c_3)$

$a_{max} = \max(a_1, a_2, a_3)$, $b_{max} = \max(b_1, b_2, b_3)$, $c_{max} = \max(c_1, c_2, c_3)$

    where

$P_1 = (a_1, b_1, c_1)$, $P_2 = (a_2, b_2, c_2)$, $P_3 = (a_3, b_3, c_3)$.

As shown in Figure 2.45, cells potentially intersected by triangle $P_1 P_2 P_3$ are given by

$(I_{x1} \rightarrow I_{x2}) \times (I_{y1} \rightarrow I_{y2}) \times (I_{z1} \rightarrow I_{z2})$

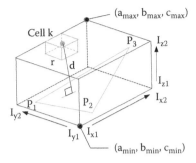

Figure 2.45 **Cells intersected by $P_1P_2P_3$.**

where

$$I_{x1} = [a_{min}/d_x] + 1, \; I_{x2} = [a_{max}/d_x] + 1,$$

$$I_{y1} = [b_{min}/d_y] + 1, \; I_{y2} = [b_{max}/d_y] + 1,$$

$$I_{z1} = [c_{min}/d_z] + 1, \; I_{z2} = [c_{max}/d_z] + 1.$$

ii. Cells too far away from the triangle are eliminated, i.e. $d > r$, where d is the distance from the centre of the cell to triangular facet $P_1P_2P_3$ (Section 2.4.2), and r is the radius of the enclosing sphere, as shown in Figure 2.45.

2.7.5 Irregular grid

Rectangular grids of a set of N points with irregular (unequal) spacing along the x- and y-direction can be constructed to cope with a non-uniform distribution of points. In an irregular grid, points are put to a cell according to its rank along the x- and y-axes rather than on their x- and y-values. To define irregular grid spacing along the x-direction, the points have to be sorted (ranked) in the x-direction, and similarly, points are also sorted along the y-direction to determine the grid spacing in the y-direction.

Let I_i and J_i be, respectively, the rank of point i along the x-direction and y-direction, i.e. I_1 is the point with the smallest x-value, I_N is the point with the largest x-value, etc. The following *pseudo-code* will arrange sorted points into an irregular grid of $N_c = N_xN_y$ cells. Given the average number of points in a cell, n, calculate the number of divisions N_x and N_y.

$$R_x = x_{max} - x_{min}, \quad R_y = y_{max} - y_{min}, \quad \alpha = R_x/R_y$$

The number of divisions in the x-direction is given by

$$N_x = NINT\left(\sqrt{N/(\alpha n)}\right)$$

The number of divisions in the y-direction is

$$N_y = NINT(\alpha N_x) \text{ where } NINT(.) = \text{earest integer}$$

The number of points in a strip along the x-axis is $N_1 = N/N_x$, and along the y-axis, $N_2 = N/N_y$.

Algorithm Irregular_Grid (N, I, J, P)

```
// N = Number of points, P = Pointer to record points in each cell
// I, J = Ranking of points in x- and y-direction
{       Set pointer array P to zero, P_i = 0, i = 1, N_c + 1
        Loop: i = 1,N // Determine the number of points in each cell
        I_x = I_i/N_1 + 1; if (I_x > N_x) I_x = N_x
        I_y = J_i/N_2 + 1; if (I_y > N_y) I_y = N_y
        k = (I_y - 1)N_x + I_x
        P_k = P_k + 1
        End loop i
        Loop: k = 1, N_c        // Compute P_k = cumulative number of points up
                                // to cell k
        P_{k+1} = P_{K+1} + P_k
        End loop k
        Loop: i = 1,N // Assign points to each cell
        I_x = I_i/N_1 + 1; if (I_x > N_x) I_x = N_x
        I_y = J_i/N_2 + 1; if (I_y > N_y) I_y = N_y
        k = (I_y - 1)N_x + I_x
        m = P_k; P_m = i; P_k = m - 1
        End loop i      }
```

The memory requirement of an irregular grid is the same as that of a regular grid, i.e. the length of pointer array **P** is given by $N + N_c + 1$. An irregular grid of the same set of 40 randomly generated points is constructed, as shown in Figure 2.46. The rectangle is divided into three vertical strips at $x_{19} = x_{min}$, x_{24}, x_{29}, $x_5 = x_{max}$, and there are exactly 13 points in each of the first two strips (between x_{19} and x_{24}, x_{24} and x_{29}) and 14 points in the third strip between x_{29} and x_5. Similarly, the rectangle is also divided into three horizontal strips at $y_{17} = y_{min}$, y_{34}, y_{29}, $y_{31} = y_{max}$, and there are exactly 13 points in each of the first two strips (between y_{17} and y_{34}, y_{34} and y_{29}) and 14 points in the third strip between y_{29} and y_{31}. The pointer array P for the irregular grid is shown in Table 2.6. The effort and the procedure in constructing an irregular grid are similar to those of a regular grid except that the points have to be sorted along the x- and y-directions to determine the positions in dividing up the

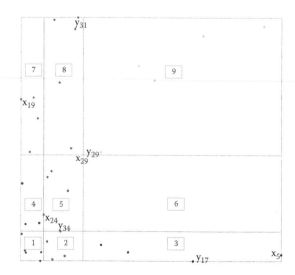

Figure 2.46 Irregular grid of nine zones for 40 randomly generated points.

Table 2.6 Pointer array *P* for an irregular grid of 40 points

i	1	2	3	4	5	6	7	8	9	10
P_i	10	15	18	22	25	30	35	39	44	50
P_{10+i}	{7	16	27	30	32}	{13	21	37}	{5	17
P_{20+i}	36	38}	{12	25	26}	{17	15	22	24	34}
P_{30+i}	{3	10	23	35	40}	{2	6	19	33}	{4
P_{40+i}	8	11	20	31}	{9	14	18	28	29	39}

rectangular domain into zones along the horizontal and vertical directions. However, as shown in this example, the points in each cell are much more evenly distributed: six points in cell 9; five points in cells 1, 5, 6 and 8; four points in cells 3 and 7; and three points in cells 2 and 4.

Analogous to regular grid construction, the points assigned to cell k are given by {P_m, m = $P_k + 1$, P_{k+1}}. Given a point p(x,y), the containing cell k is given exactly by the same formulas as in the point-allocation procedure: k = (I_y – 1)N_x + I_x, where x lies between grid lines I_x and $I_x + 1$, and y lies between grid lines I_y and $I_y + 1$.

An irregular grid over three dimensions can be constructed following the same procedure as a 2D irregular grid, except that points have to be sorted along the x-, y- and z-directions. Let I_i, J_i and K_i be, respectively, the ranking of the points along the x-, y- and z-directions; then the cell k to which point i belongs is given by

$$k = (I_z - 1)N_{xy} + (I_y - 1)N_x + I_x,$$

where

$$I_x = I_i/N_1 + 1, \text{ if } (I_x > N_x) \, I_x = N_x,$$

$$I_y = J_i/N_2 + 1, \text{ if } (I_y > N_y) \, I_y = N_y,$$

$$I_z = K_i/N_3 + 1, \text{ if } (I_z > N_z) \, I_z = N_z$$

In summary, irregular grids can effectively deal with moderately non-uniformly distributed points; however, as compared to regular grids, additional effort in the sorting of points is required.

2.7.6 Quadtree

To cope with the highly uneven distribution of points, cells can be refined locally until there are only a few points in each cell. Such a partition of space in 2D was named by Finkel and Bentley (1974) as Quadtree. The Quadtree partition of space has tremendous applications in computer graphics and MG (Bentley 1975; Bentley and Ottmann 1979). Efficient image and video coding scheme by means of the Quadtree was proposed by Sullivan and Baker (1994) to reduce the cost of image communication and storage. A Quadtree-based adaptive mesh refinement scheme using material force and residual error estimates was presented by Tabarraei and Sukumar (2005, 2007), and in modelling of discontinuous field by Tabarraei and Sukumar (2008). To construct a Quadtree subdivision for a set of N points on a 2D plane, first, a bounding rectangle containing all the points has to be defined

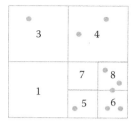

Figure 2.47 Cells containing more than four points are split.

(Holroyd and Mason 1990; Yiu et al. 1996). A maximum capacity of a cell can be set, such that the cells containing more points than the allowable limit will be split into four quadrants or child cells symmetrically along the x- and y-directions, as shown in Figure 2.47. In this example, not more than 4 points are allowed in each cell, and as a result, cell 2 is further subdivided into four cells, which are labelled as cells 5, 6, 7 and 8. The process can be repeated on all the subdivided cells until no cell contains more points than it is allowed to hold.

The subdivision can be done conveniently by means of a recursive procedure; however, a more efficient looping algorithm is presented in Section 2.7.7 in which points are input one by one until all the points are processed.

Algorithm Quadtree (N, X, Y, P, Q)

```
// N = Number of points, x_min, x_max, y_min, y_max = x, y limits of the points
// NP = Number of points allowed in a cell, NZ = Number of zones (cells)
// created
// P_k = Number of points in cell k, or the first child of cell k if P_k > NP
// Q = Array recording points allocated to the zones, (X_i, Y_i) =
// Co-ordinates of point i, i = 1,N
{      Set pointer array P to zero, P_i = 0, i = 1,N
       NZ = 4; xm =(x_min + x_max)/2; ym =(y_min + y_max)/2
       Loop: i = 1,N
       z = 1; u = X_i; v = Y_i
       x1 = x_min; x2 = x_max; y1 = y_min; y2 = y_max
       If (u > xm) then
       z = z + 1; x1 = xm; x2 = x_max
       Endif
       If (v > ym) then
       z = z + 2; y1 = ym; y2 = y_max
       Endif
       Zone (z, i, u, v, X, Y, x1, x2, y1, y2, NZ, P, Q)
       End loop i    }
```

Procedure Zone (z, i, u, v, X, Y, x1, x2, y1, y2, NZ, P, Q)

```
// Insert point i (u,v) to zone z (x1,x2,y1,y2)
{ 1:    n = P_z
        If (n < NP) then
        k = NP(z-1) + n + 1; Q_k = i; P_z = n+1; return
        End if
        if (n = NP) Split_Zone (z, i, u, v, X, Y, x1, x2, y1, y2, NZ, P,
        Q); Return
```

```
ym =(x1 + x2)/2; ym =(y1 + y2)/2
if (u > xm) n = n + 1; x1 = xm    else x2 = xm
if (v > ym) n = n + 2; y1 = ym    else y2 = ym
z = n
Go to 1 }    // Convention: if applies to the entire line
```

Procedure Split_Zone (z, i, u, v, X, Y, x1, x2, y1, y2, NZ, P, Q)

```
// Zone z is split into zones NZ + j, j = 1,4 and points in zone z are
// put in zones NZ + j
{       k = NP(z - 1)
1:      xm =(x1 + x2)/2; ym =(y1 + y2)/2; Pz = NZ + 1
        Assign (i, u, v, xm, ym, NZ, P, Q)
        Loop j = 1,NP
        L = Qk+j;
        Assign (L, XL, YL, xm, ym, NZ, P, Q)
        End Loop j
        NZ = NZ + 4
        Loop j = 1,4
        z = NZ + j - 4
                If (Pz > NP) then
                If (j > 2)    y1 = ym       else y2 = ym
                If (j = 2 or j = 4) x1 = xm      else x2 = xm
                Go to 1
                End if
        End Loop j    }
```

Procedure Assign (i, u, v, xm, ym, NZ, P, Q)

```
// Assign point i (u,v) to zone NZ + j, j = 1,2 3 or 4 according to its
// position relative to (xm,ym)
{       z = NZ + 1
        If (u > xm) z = z + 1
        If (v > ym) z = z + 2
        Pz = Pz + 1; k = NP(z-1)+ Pz; Qk = i         }
```

The algorithm runs as follows. The insertion point is first assigned to one of the four primary zones 1, 2, 3 or 4 according to its position relative to the central division lines. In the *Procedure Zone*, if the number of points in the current zone is less than NP, the insertion point is assigned to this zone; if the capacity of the zone (NP) is exceeded by the addition of the insertion point, then the current zone is split symmetrically into four cells, and further splitting of cells might be necessary if all the points are found in one of the subdivided cells. In case the cell is already split, the co-ordinates of the insertion point are checked against the central division lines to determine which child cell contains the insertion point. The *Procedure Zone* can be repeated until the point settles in a non-saturated cell.

The Quadtree division of the same set of 40 randomly generated points is shown in Figure 2.48, in which the rectangle containing all the points is partitioned into 28 zones. In this example, the maximum number of points allowed in each zone is set equal to 4, i.e. a zone will be split into four quadrants once the number of points in it exceeds 4. The pointer array for the Quadtree division of the points is given in Table 2.7.

A P_z value less than or equal to NP = 4 gives the number of points in the zone z, whereas if the P_z value is greater than NP, it represents the first child in the subdivision into four child

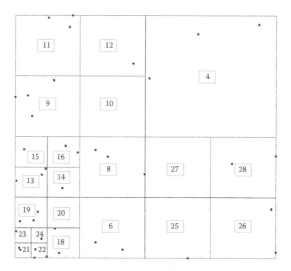

Figure 2.48 Quadtree division for 40 randomly generated points.

Table 2.7 Pointer array P for quadtree division of 40 points

i	1	2	3	4	5	6	7	8	9	10
P_i	5	25	9	3	17	2	13	3	4	0
P_{10+i}	3	1	3	1	1	2	21	2	3	0
P_{20+i}	2	3	1	1	1	2	0	2		

cells. Hence, $P_1 = 5$ means that cell 1 has been divided into four cells 5, 6, 7 and 8; $P_5 = 17$ shows that cell 5 has been divided into cells 17, 18, 19 and 20; $P_{17} = 21$ indicates that cell 17 has been divided into cells 21, 22, 23 and 24; and $P_{21} = 2$ indicates that there are two points in cell 21. There are three empty cells, namely, cells 10, 20 and 27, and there is only one cell containing 4 points, which is cell 9. For a zone with P_z smaller or equal to NP, the points allocated in zone z are recorded in array **Q** (length NZ × NP), given by

$$\{Q_{k+j}: j = 1, P_z; k = NP(z - 1)\}, z = 1, NZ.$$

Given a point p(x,y), the containing cell can be found following a similar procedure of point insertion, as shown by the algorithm *QTSearch*.

Algorithm QTsearch (P, x, y, z, x1, x2, y1, y2)

```
// Input: (x,y) = co-ordinates of the specified point
// Output: z = zone containing the given point, (x1, x2, y1, y2) =
// bounding rectangle of zone z
{       z = 0
        If (x < x_min or x > x_max) return
        If (y < y_min or y > y_max) return
        xm = (x_min + x_max)/2; ym = (y_min + y_max)/2
        z = 1; x1 = x_min; x2 = xm; y1 = y_min; y2 = ym
        If (x > xm) then
```

```
         z = z + 1; x1 = xm; x2 = x_max
         End if
         If (y > ym) then
         z = z + 2; y1 = ym; y2 = y_max
         End if
         Search_Zone (x, y, z, x1, x2, y1, y2, P)          }
```

Procedure Search_Zone (x, y, z, x1, x2, y1, y2, P)

```
{ 1:    n = P_z
        if (n ≤ NP) return
        xm = (x1 + x2)/2; ym = (y1+y2)/2
        if (x > xm) n = n + 1; x1 = xm    else x2 = xm
        if (y > ym) n = n + 2; y1 = ym    else y2 = ym
        z = n
        Go to 1        }
```

Quadtree partitioning into zones for a set of N points constructed by the point insertion algorithm has an average complexity of Nlog(N), as each point is assigned to a zone following a subdivision procedure such that the size of a cell is halved in each cell split, and it takes, on average, log(N) times of cell splits before it settles in a non-saturated cell. Of course, a point will go through more levels of subdivision if all the points but one are clustered in a small area. The performance of the insertion algorithm has been tested with randomly generated points from 10^3 points to 10^7 points, as shown in Table 2.8. A quasi-straight line is resulted in a plot of the CPU time against Nlog(N), as shown in Figure 2.49.

Table 2.8 **CPU time(s) for Quadtree, Octree and kd-trees**

N	$Nlog_{10}(N)$	Quadtree	Octree	2-d tree	3-d tree
1000	3000	0.000138	0.000123	0.000931	0.000928
10,000	40,000	0.00177	0.00163	0.01622	0.0165
100,000	500,000	0.0225	0.0215	0.28	0.259
1,000,000	6,000,000	0.35	0.311	4.426	4.316
10,000,000	70,000,000	5.77	4.41	74.03	70.27

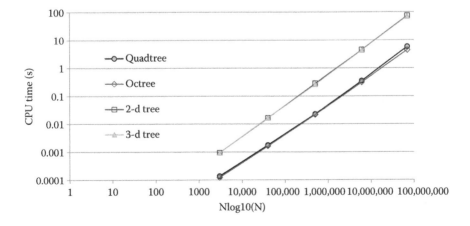

Figure 2.49 CPU time vs Nlog(N) for Quadtree, Octree and kd-trees.

2.7.7 Octree

The Quadtree scheme for the partition of space by local refinement can be readily extended to higher dimensions. In 3D, each cubic cell (zone) is subdivided into eight octants or eight child cells symmetrically about the three axes; such a construction is known as Octree. Similar to Quadtree, Octree subdivision is often applied to computer graphics, the generation of tetrahedral and hex meshes, surface construction from cloud points, detection of collision in space (Vemuri et al. 1998a,b), etc. Based on Octree for node positioning, Krysl and Ortiz (2001) proposed a variational Delaunay approach to the generation of tetrahedral FE meshes. In conjunction with an Octree scheme, Rassineux (1998) presented the generation and optimisation of tetrahedral meshes by AFT. With an Octree partition of space, Vemuri et al. (1998a) developed a fast collision detection scheme with applications to particle flows. Mesh size at a point is determined and stored by means of Octree (Quadros et al. 2004). Data-parallel Octrees for surface reconstruction was developed by Zhou et al. (2011). Automatic MG based on the Octree for cardiac electro-physiology problems was presented by Prassl et al. (2009).

The point insertion algorithm for the construction of Quadtree can also be applied to 3D to construct an Octree partition for a given set of spatial points. The following is the *pseudo-code* of Octree construction for a set of N points $\{(X_i, Y_i, Z_i), i = 1,N\}$ by inserting one point after another until all the points are processed.

Algorithm Octree (N, X, Y, P, Q)

```
// N = Number of points, xmin, xmax, ymin, ymax, zmin, zmax = x, y, z
// limits of the points
// NP = Number of points allowed in a cell, NZ = Number of zones (cells)
// created
// Pk = Number of points in cell k, or the first child of cell k if Pk > NP
// Q = Array recording points allocated to the zones
{     Set pointer array P to zero, Pi = 0, i = 1,N
      NZ = 8; xm = (xmin + xmax)/2; ym = (ymin + ymax)/2; ; zm = (zmin + zmax)/2
      Loop: i = 1,N
      z = 1; u = Xi; v = Yi; w = Zi
      x1 = xmin; x2 = xmax; y1 = ymin; y2 = ymax; z1 = zmin; z2 = zmax
      if (u > xm) z = z + 1;      x1 = xm;      x2 = xmax
      if (v > ym) z = z + 2;      y1 = ym;      y2 = ymax
      if (w > zm) z = z + 4;      z1 = zm;      z2 = zmax
      Zone (z, i, u, v, w, X, Y, Z, x1, x2, y1, y2, z1, z2, NZ, P, Q)
      End loop i    } // Note: if applies to the entire line
```

Procedure Zone (z, i, u, v, w, X, Y, Z, x1, x2, y1, y2, z1, z2, NZ, P, Q)

```
// Insert point i (u,v,w) to zone z (x1,x2,y1,y2,z1,z2)
{ 1:  n = Pz
      If (n < NP) then
      k = NP(z-1) + n + 1; Qk = i; Pz = n + 1; return
      End if
      If (n = NP) then
      Split_Zone (z, i, u, v, w, X, Y, Z, x1, x2, y1, y2, z1, z2, NZ, P, Q)
      Return
      End if
      ym = (x1 + x2)/2; ym = (y1 + y2)/2; zm = (z1 + z2)/2
```

```
if (u > xm) n = n + 1; x1 = xm   else x2 = xm
if (v > ym) n = n + 2; y1 = ym   else y2 = ym
if (w > zm) n = n + 4; z1 = zm   else z2 = zm
z = n
Go to 1        }
```

Procedure Split_Zone (z, i, u, v, w, X, Y, Z, x1, x2, y1, y2, z1, z2, NZ, P, Q)

```
// Zone z is split into zones NZ + j, j=1,8 and points in zone z are put
// in zones NZ + j
{      k = NP(z-1)
1:     xm = (x1 + x2)/2; ym = (y1 + y2)/2; zm = (z1 + z2)/2
       P_z = NZ+1
       Assign (i, u, v, w, xm, ym, zm, NZ, P, Q)
       Loop j = 1,NP
       L = Q_{k+j}
       Assign (L, X_L, Y_L, Z_L, xm, ym, zm, NZ, P, Q)
       End Loop j
       NZ = NZ + 8
       Loop j = 1,8
       z = NZ + j - 8
               If (P_z > NP) then
               if (mod(j,2)= 0)     x1 = xm else x2 = xm
               if (mod(j + 1,4)<2) y1 = ym else y2 = ym
               if (j > 4)    z1 = zm else z2 = zm
               Go to 1
               End if
       End Loop j    }
```

Procedure Assign (i, u, v, w, xm, ym, zm, NZ, P, Q)

```
// Assign point i (u,v,w) to zone NZ + j, j = 1,8 according to its
// position relative to (xm,ym,zm)
{       z = NZ + 1
        if (u > xm) z = z + 1
        if (v > ym) z = z + 2
        if (w > zm) z = z + 4
        P_z = P_z + 1
        k = NP(z-1)+ P_z; Q_k = i      }
```

An Octree partition of space for 100 randomly generated points (slightly biased towards a corner) is shown in Figure 2.50. The capacity of a cell (NP) was set equal to eight, so that a cell would be split into eight child cells symmetrically along the three axes once the number of points in it exceeded eight. Fifty-six cells were created in which some are empty cells, and some are mother cells divided into child cells. In Figure 2.50, points within a cell are marked with the same colour for easy identification. A spatial point within the cube is specified (marked as a hollow circle), and the containing cell (marked with purple colour) can be automatically located following the point insertion procedure, as shown in Figure 2.50. Similar to the Quadtree construction, the complexity of the Octree point insertion algorithm is expected to be Nlog(N) for a set of N spatial points as the average number of

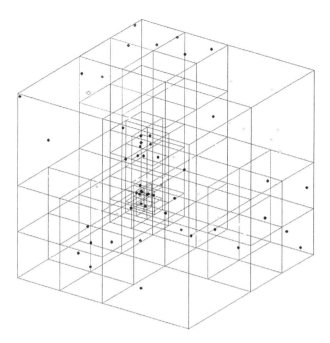

Figure 2.50 An Octree of 100 points.

subdivision for a given point is log(N). The insertion algorithm was tested with 10^3 to 10^7 randomly generated points (slightly biased towards a corner), and the results are shown in Table 2.8. A quasi-straight line is produced in a plot of the CPU time against Nlog(N), as shown in Figure 2.49. It is surprising to note that the construction of Octree takes slightly less CPU time compared to the construction of Quadtree for the same number of points. Although the number of regions is similar, the number of subdivisions might be quite different for each inserted point as Octree cells (NP = 8) are allowed to hold more points than Quadtree cells (NP = 4).

Not only is the point insertion algorithm for the construction of Quadtree and Octree simple and efficient but also the memory requirement is quite low. What we need are two vectors of size NZ and NP*NZ. $\{P_k, k = 1,NZ\}$ represents the number of points in cell k or the reference of the first child cell if P_k > NP. The points allocated to cell k are given by $\{Q_{m+j}: j = 1,P_k; m = NP(k - 1)\}$, k = 1,NZ. To further save memory by reducing the size of vector **Q**, the insertion algorithm can be run for the first time to establish the pointer vector **P** without recording points allocated to the cells. In the second run, points are recorded in vector **Q** in a compact manner without leaving any gap according to the number of points in each cell already computed and stored in vector **P**. This is a trade-off between the CPU time and the memory used; basically the program has to be run twice, but the memory reduced could be more than half especially for the Octree construction where there are quite a lot of empty cells.

In the construction of Quadtree or Octree, the cells are divided symmetrically along the co-ordinate axes without referencing to the given point set. As a result, the level of subdivision (depth of a tree) depends on the distribution of the points. The most favourable condition is that the points are evenly distributed. On the contrary, in the worst scenario, all the points but one are clustered within a tiny region, and it will take many subdivisions of cells to separate the points. The capacity of a zone (cell), NP, can only be set between 1 and 4 in *Algorithm Quadtree* and 1 and 8 in *Algorithm Octree*; this restriction could be removed by a simple modification of the programs. What has to be done is a proper separation of the information in P_k in order not to mix up with the number of points in a cell and the cell reference to the first child. *Algorithm OCTSearch* for the location of the containing zone of a specified point in an Octree partition is not provided in this chapter; readers are advised to produce their own version of *OCTSearch* following the idea presented in *Algorithm QTSearch* as an exercise.

2.7.8 Kd-tree

In the Quadtree and Octree construction, cells are split symmetrically in the spatial directions. Division of cells entirely based on geometry without referencing to the physical position of the points will sometimes require an excessive number of cell subdivisions to separate points that are closely clustered together. On the contrary, each cell of a kd-tree is split at the median of the points it contains, such that the spatial positions of the points are automatically taken into account, and the points are evenly distributed in the subdivided cells. Based on the idea of space partition, a kd-tree is a data structure in the form of a branching tree for organising points in a k-dimensional space. kd-trees provide a hierarchic spatial relationship between data points, which can be used to find nearest neighbours (Kanungo et al. 2002), to locate points within a zone and for other searching operations, such as construction and point insertion of DT (Devroye 1981, 1985; Lemaire and Moreau 2000; Devroye et al. 2004).

2.7.8.1 Construction of 2-d tree

A 2-d tree is generated by selecting a pivot point Q_1 from the given point set to separate the remaining points into two regions by passing a vertical line through the chosen point. Points with x-values smaller than that of Q_1 will be put to the region on the left, whereas points with x-values greater than that of Q_1 will be put to the region on the right, as shown in Figure 2.51, and points with equal x-values can be put on either side. In the second step, pivot points Q_2 and Q_3 are selected, respectively, from the regions to further divide them each into two sub-regions. This time, the regions are divided into smaller regions by horizontal lines passing points Q_2 and Q_3. Obviously, this subdivision process can be repeated

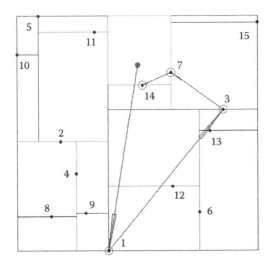

Figure 2.51 **2-d tree of 15 points.**

by selecting one point from each region as the pivot to create more sub-regions. As one point, the pivot, is taken away for each subdivision, for a set of N points, there are N divisions, and the space will be partitioned into N + 1 regions.

A balanced tree will be resulted if the median is selected as the pivot for each region subdivision, and in this case, each region will contain approximately equal number of points (differ at most by one point). For a balanced tree construction, the level of subdivision can be calculated as the number of points in a region is reduced by half in each subdivision, and all the leaf nodes (nodes at the bottom of the tree) are about the same distance from the root. To construct a balanced kd-tree for a set of N points, the level of subdivision, L, is given by

$$2^L \geq N \Rightarrow L \geq \log_2 (N) \Rightarrow L = INT (\log_2 (N)) + 1$$

Hence,

$$
\begin{aligned}
L &= 1, \quad N = 1 \\
L &= 2, \quad N = 2, 3 \\
L &= 3, \quad N = 4, 5, 6, 7 \\
L &= 4, \quad N = 8, 9, 10, 11, 12, 13, 14, 15
\end{aligned}
$$

as shown in Figure 2.51.

For a more symmetrical partition of space, the line of subdivision, vertical or horizontal, is rotated for each change of level of subdivision, i.e. L = 1, vertical; L = 2, horizontal; L = 3, vertical; and so forth.

2.7.8.1.1 Sequence of points

A kd-tree can be conveniently defined by proper sequencing of the given set of N points, $P = \{P_i, i = 1, N\}$. The following describes how a region P_i is divided into two sub-regions P_k and P_{k+1}. Find the median from P_i as a pivot along the x- or y-axis depending on the level of subdivision. Set Q_i = median, which divides P_i into P_k containing points on the left of P_i and P_{k+1} containing points on the right of P_i. As Q_i is the median in P_i, the number of points in P_k and the number of points in P_{k+1} are given by

$$
N_k = \begin{cases} \dfrac{N_i - 1}{2} & \text{if } N_i \text{ is odd} \\[2mm] \dfrac{N_i}{2} - 1 & \text{if } N_i \text{ is even} \end{cases}
\qquad
N_{k+1} = \begin{cases} \dfrac{N_i - 1}{2} & \text{if } N_i \text{ is odd} \\[2mm] \dfrac{N_i}{2} & \text{if } N_i \text{ is even} \end{cases}
$$

where N_i = number of points in P_i. The subdivision will be carried out following the order of construction of the regions. As one point, Q_i, will be taken away for the subdivision of region P_i, the sequence $\{Q_i, i = 1, N\}$ will be established after exactly N subdivisions.

Set $P_1 = P = \{P_i, i = 1, N_1\}$, where $N_1 = N$. P_1 will give rise to P_2 and P_3, and P_2 will generate P_4 and P_5, whereas P_3 will generate P_6 and P_7. In turn, P_4 will generate P_8 and P_9, and P_k will generate P_{2k} and P_{2k+1} and so on, as shown in Figure 2.52.

The following is the *pseudo-code* for the construction of 2-d tree by a proper sequencing of a given point set of N points, $\{(X_i, Y_i), i = 1, N\}$.

Level

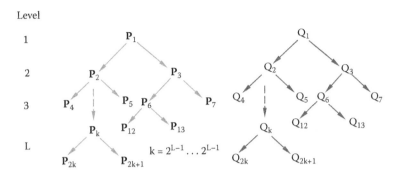

Figure 2.52 Sequence of sub-regions and pivots.

Algorithm 2d_tree (N, X, Y, Q)

```
// Construction of a balanced 2d-tree
// N = Number of points, X, Y = x, y co-ordinates of the points, Q =
// Sequence of 2d-tree
// Working arrays: P, L, R
{     Loop i = 1,N // Set initial values to P
      Pᵢ = i
      End loop i
      n = 1; L₁ = 1; R₁ = N; J2 = 0; k = 0; l = 0; level = INT(log(N +
0.5)/log(2.0))
      While (l < level) { l = l + 1; J1 = J2 + 1; J2 = n;
      Loop j = J1, J2
      left = Lⱼ; right = Rⱼ
      if (mod(l,2)= 1) Sort (left, right, P, X)        //Sort P according
                                                        //to X from left to
                                                        //right
      if (mod(l,2)= 0) Sort (left, right, P, Y)        //Sort P according
                                                        //to Y from left to
                                                        //right
      m =(left + right)/2         // m is the median
      k = k + 1; Qₖ = Pₘ   // median node assign to Qₖ
      n = n + 1; Lₙ = left; Rₙ = m - 1
      n = n + 1; Lₙ = m + 1; Rₙ = right
      End loop j     }
                J1 = J2 + 1; J2 = n;
                Loop j = J1, J2
                left = Lⱼ; right = Rⱼ
                k = k + 1; Qₖ = 0
                if (left = right) Qₖ = P_left
                End loop j     }
```

2.7.8.1.2 2-d tree diagram

Let's consider the relationship of the points in the sequence $\{Q_k, k = 1, N^+\}$ associated with region subdivisions, $P_1, P_2, ..., P_N$, where $N^+ = 2^L - 1 \geq N$. Q_1 is the root node, which has the child nodes Q_2 and Q_3, and in general, Q_k generates Q_{2k} and Q_{2k+1}. Hence, Q_{2k} and Q_{2k+1} are the children of Q_k, and Q_k is the father of Q_{2k} and Q_{2k+1}, as shown in Figure 2.52. On the other hand, $Q_{k/2}$ is the father node of Q_k, and in case $k/2 = 0$ or $k = 1$, Q_k is already at the top of the tree. Generalising the previous results, the grandfather (two levels up) of Q_k is given by $Q_{k/4}$.

Let's now consider the construction of the 2-d tree for 15 points, as shown in Figure 2.51. The median in the x-value of the given 15 points is determined and assigned to Q_1. The points sorted to the left of the median are put to region P_2, and the points sorted to the right of the median are put to region P_3. The process continues with the subdivision of P_2 and P_3 along the y-axis to give rise to points Q_2 and Q_3 and regions P_4, P_5, P_6 and P_7. Subdivision of regions P_4, P_5, P_6 and P_7 will generate points Q_4, Q_5, Q_6 and Q_7 and regions P_8, P_9, P_{10}, P_{11}, P_{12}, P_{13}, P_{14} and P_{15}. At this stage, there is exactly one point in each region P_8 to P_{15}, which is assigned to Q_8 to Q_{15}, and the construction of the 2-d tree is completed.

As shown in Figure 2.51, Q_1 on a vertical line generates Q_2 and Q_3 on horizontal lines, and Q_2 and Q_3 in turn generate Q_4, Q_5, Q_6 and Q_7 on vertical lines. Hence, the father and the son have a perpendicular relationship, and the grandfather and the grandchildren are on parallel lines. For instance, Q_1, Q_4 and Q_6 are on vertical lines, and Q_2, Q_8 and Q_{11} are on horizontal lines.

To search over a 2-d tree for the region that contains a specified point, we have to go down the 2-d tree from the root node all the way to the leaf node. Take for example the 15-point 2-d tree shown in Figure 2.51. The region that encloses a given point p, depicted in Figure 2.51 as a red solid dot, has to be located. p is on the right of Q_1, and p is bounded by a rectangle $[x_1, x_{max}, y_{min}, y_{max}]$, where x_1 is the x-co-ordinate of Q_1, x_{min} and x_{max} are, respectively, the minimum and maximum x-values of all the points, and y_{min} and y_{max} are, respectively, the minimum and maximum y-values of all the points. p is higher than Q_3, which divides the rectangle $[x_1, x_{max}, y_{min}, y_{max}]$, and the bounding rectangle shrinks to $[x_1, x_{max}, y_3, y_{max}]$. p is on the left of Q_7, which divides the rectangle $[x_1, x_{max}, y_3, y_{max}]$, and the bounding rectangle is reduced to $[x_1, x_7, y_3, y_{max}]$. Finally, the leaf node Q_{14} is reached, and the bounding rectangle is updated to $[x_1, x_7, y_{14}, y_{max}]$. It is noted that, in general, the leaf node on the bounding rectangle may not be the node nearest to the given node. A general algorithm searching for the nearest node in a kd-tree, and its complexity with respect to dimension k, is given by Chanzy et al. (2001). The following is a *pseudo-code* searching for the bounding rectangle of a specified point p(x,y).

Algorithm Search (x, y, X, Y, Q, k, x₁, y₁, x₂, y₂)

```
// Search over 2d-tree for leaf node k and bounding rectangle [x₁, y₁, x₂, y₂]
// (x, y) = input node, X, Y = x, y co-ordinates of the points, Q =
// Sequence of 2d-tree
{       x1 = x_min; y1 = y_min; x2 = x_max; y2 = y_max
        i = 1
1:      k = Q_i
        xk = X_k
        i = i + i
        If (x > xk) then
        i = i + 1
        if (xk > x1) x1 = xk
        else
        if (xk < x2) x2 = xk
        End if

        if (Q_i = 0) return
        k = Q_i; y_k = Y_k; i = i + i
        If (y > y_k) then
        i = i + 1;
        if (y_k > y1) y1 = y_k
```

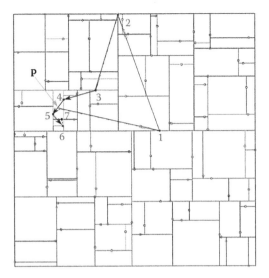

Figure 2.53 Squarish 2-d tree for 100 randomly generated points.

```
        else
        if (yk < y2) y2 = yk
        End if
        If (Qᵢ > 0) go to 1  }
```

In the partition of regions, if the regions are cut along their longest side instead of rotating through the axes, the pattern and the performance of the kd-tree are quite different. The behaviour of the kd-trees is much better as the points are more closely connected together, and they are called *squarish kd-trees* (Devroye et al. 2000). A *squarish* 2-d tree is constructed for 100 randomly generated points, as shown in Figure 2.53, in which the elongated rectangles are eliminated, and the cells are much closer to a square shape. As expected, the location of the bounding box for an arbitrarily specified point **p** takes $INT(\log_2(N = 100)) + 1 = 7$ steps. Starting from point 1 (root node), we go to points 2, 3, 4 and 5 as **p** is always on the left, then to point 6 as **p** is on the right of point 5 and finally to point 7 as **p** is above point 6.

2.7.8.2 Construction of 3-d tree

The concept and the procedure for the construction of a 2-d tree can be generalised naturally to higher dimensions. In 3D, $k = 3$, the space/regions are partitioned by a plane perpendicular to the x-, y- or z-axis in turn according to the level of subdivision. If Q_k is on the x-aligned plane, child nodes Q_{2k} and Q_{2k+1} are on the y-aligned plane, and the grandchild nodes Q_{4k}, Q_{4k+1}, Q_{4k+2} and Q_{4k+3} are on the z-aligned plane and so forth. The construction of a static kd-tree of n points takes $O(n\log^2(n))$ time if an $O(n\log(n))$ sort is employed to compute the median for each region subdivision. The complexity reduces to $O(n\log(n))$ if a linear median-finding algorithm is used (Lemaire and Moreau 2000; Devroye et al. 2004). The performance of *Algorithm kd_tree* was evaluated with 10^3 to 10^7 randomly generated points, as shown in Table 2.8. k-d construction takes substantially more time than Quadtree and Octree, as quick sort has to be employed to find the median for each cell subdivision. As shown in Figure 2.49, a straight line is resulted in a plot of CPU time against Nlog(N). It is

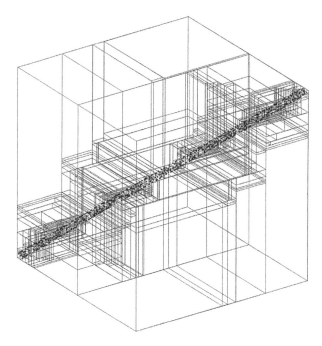

Figure 2.54 Squarish 3-d tree of 1000 points.

noted that the construction of a 2-d tree and a 3-d tree takes a roughly equal amount of CPU time, as the procedure is similar and the number of operations involved is exactly the same. A *squarish* 3-d tree of 1000 points distributed along a straight line is shown in Figure 2.54, in which small cells are found near the diagonal.

The following is the *pseudo-code* for the construction of a *squarish* 3-d tree by a proper sequencing of a given set of N points, $\{(X_i, Y_i, Z_i), i = 1, N\}$, for which additional arrays X1, X2, Y1, Y2, Z1 and Z2 are needed to store the corner points of the cells of spatial partition.

Algorithm 3d_tree (N, X, Y, Q, n, P, L, R)

```
// Construction of a squarish 3d-tree
// N = Number of points; X, Y, Z = x, y, z co-ordinates of the points; Q =
// Sequence of 3d-tree
// n = Number of cells; Points in cell k = {P_i, i = L_k, R_k}
{     Loop i = 1,N // Set initial values to P
      P_i = i
      End loop i
      X1_1 = X_min; X2_1 = X_max; Y1_1 = Y_min; Y2_1 = Y_max; Z1_1 = Z_min; Z2_1 = Z_max;
      n = 1; L_1 = 1; R_1 = N; J2 = 0; k = 0; l = 0; level = INT(log(N +
      0.5)/log(2.0))
      While (l < level) { l = l + 1; J1 = J2 + 1; J2 = n;
      Loop j = J1, J2
      left = L_j; right = R_j;      m = (left + right)/2  // m is the median
      U1 = X1_j; U2 = X2_j; V1 = Y1_j; V2 = Y2_j; W1 = Z1_j; W2 = Z2_j;
```

```
If (U2-U1>V2-V1 and U2-U1>W2-W1) then    //divide cell j in
                                         //x-direction
Sort (left, right, P, X)    //Sort points along x, assign corner
                            //points to cells n + 1 and n + 2
c = Pm; X1n+1 = U1; X2n+1 = Xc; Y1n+1 = V1; Y2n+1 = V2; Z1n+1 = W1; Z2n+1 = W2;
X1n+2 = Xc; X2n+2 = U2; Y1n+2 = V1; Y2n+2 = V2; Z1n+2 = W1; Z2n+2 = W2;
else
if (V2-V1>W2-W1) then        //divide cell j in y-direction
Sort (left, right, P, Y)    //Sort points along y, assign corner
                            //points to cells n + 1 and n + 2
c = Pm; X1n+1 = U1; X2n+1 = U2; Y1n+1 = V1; Y2n+1 = Yc; Z1n+1 = W1; Z2n+1 = W2;
X1n+2 = U1; X2n+2 =U2; Y1n+2 = Yc; Y2n+2 = V2; Z1n+2 = W1; Z2n+2 = W2;
else    //divide cell j in z-direction
Sort (left, right, P, Z)    //Sort points along z, assign corner
                            //points to cells n + 1 and n + 2
c = Pm; X1n+1 = U1; X2n+1 = U2; Y1n+1 = V1; Y2n+1 = V2; Z1n+1 = W1; Z2n+1 = Zc;
X1n+2 = U1; X2n+2 = U2; Y1n+2 = V1; Y2n+2 = V2; Z1n+2 = Zc; Z2n+2 = W2;
        endif
Endif
k = k + 1; Qk = c              // median node assign to Qk
n = n + 1; Ln = left; Rn = m - 1
n = n + 1; Ln = m + 1; Rn = right
End loop j    }
                J1 = J2 + 1; J2 = n;
                Loop j = J1, J2
                left = Lj; right = Rj
                k = k + 1; Qk = 0              // Qk = 0, for cells
                                              // containing no point
                if (left = right) Qk = Pleft // assign Qk for cells with
                                             // one point
                End loop j    }
```

Chapter 3

Mesh generation on planar domain

A plane is flat, Euclidean and everywhere visible, so it ought not to be very hard.

3.1 INTRODUCTION

Finite element (FE) mesh generation (MG) probably started its journey from the discretisation of a bounded region on a planar domain. Many MG algorithms, including the mapping method, Delaunay triangulation (DT), Quadtree decomposition, advancing-front approach, etc., had their ideas and development found first on two dimensions. This is a long chapter as all the basic concepts and ideas in FE MG will be fully explained and illustrated with examples. Following the classification in Section 2.3.5, the MG schemes developed on 2D domains can be classified into structured and unstructured meshing methods.

Transformation by FE interpolation and transfinite mapping, drag and sweeping methods are the common techniques employed at the very early stage in producing structured FE meshes, which are still useful nowadays to generate regular meshes. These methods can also be generalised quite naturally into higher dimensions following a similar procedure, and a short description of these methods can be found in Section 3.2. Section 3.3 presents unstructured meshing methods developed since the early 1970s; many MG schemes have been proposed among which the DT, AFT and MG using contour, coring method, Quadtree and mesh refinement have all been proved to be effective schemes for MG on 2D.

Quadtree decomposition as a spatial partition scheme takes up a unique place in the generation of quadrilateral elements in compliance with a node spacing function. In fact, the Quadtree method to be presented in Section 3.4 is apt in generating one-level refinement meshes for which efficient transition quadrilateral elements can be applied for adaptive refinement analysis. As the boundary treatment is the major weakness of Quadtree meshing, an enhanced version, in which triangular elements are generated by AFT in filling up the gap between the Quadtree mesh and the domain boundary, will be presented in Section 3.8.

The DT and the AFT are perhaps the most popular for the generation of unstructured meshes on planar domains, and a comprehensive account of these two methods will be given, respectively, in Sections 3.5 and 3.6. The Delaunay–AFT, which is the resulting scheme by combining these two popular methods, will also be presented in Section 3.7. Triangular and quadrilateral elements are the 2D FEs employed in FE MG. While it is straightforward to generate regular quadrilateral meshes by structured meshing, unstructured adaptive quadrilateral meshes can be generated through an indirect process by converting triangular meshes generated by the DT or AFT. Such an indirect approach for the generation of unstructured quadrilateral meshes is presented in Section 3.9.

3.2 STRUCTURED MESH ON PLANAR DOMAIN

Structured meshing techniques, which provide a straightforward way to generate meshes of regular patterns, were among the earliest approaches in MG. By a direct copy of the topology or how points are connected on a regular reference domain, the idea is intuitive to reproduce the same mesh pattern on irregular physical domains with possible curved boundaries. The only restriction for structured meshing is that the number of boundary edges of the physical domain and that of the reference domain are equal, so that corner points will map to corner points, and the edge of the reference domain will map to the corresponding edges of the physical domain connected to the same ending points. While the mesh connection is not a problem that is already established in the reference domain, there is an issue as to where to place the points in the physical domain.

Two approaches have been developed to allow points to be located in irregular domains. By the algebraic techniques (Gordon and Hall 1973a,b,c), points of the reference domain are directly mapped onto the physical domain through co-ordinate transformations, FE interpolation, transfinite mapping, etc. Another approach are solution-based methods (Thompson 1982; Hansen et al. 2005) in which partial differential equations (PDEs) are solved over the physical domain to generate contours where points can be located. It is interesting to point out that an FE mesh is required for a numerical solution of some physical problems governed probably by differential equations, yet for the purpose of MG, another set of differential equations have to be solved in advance. As a result, from the point of view of computation, the algebraic techniques are direct methods and are more efficient than the solution-based methods in which PDEs have to be solved.

The idea of structured meshing is intuitive, and regular mesh patterns can be rapidly generated over irregular physical domains as long as a convenient mapping can be identified. This method is particularly attractive to generate a large number of quadrilateral elements over regular domains where the quality and the speed of MG are guaranteed. However, this approach is rather restrictive for MG in a number of ways: (i) The number of divisions has to be equal on opposite sides; (ii) the elements may be distorted due to the mapping; (iii) a progressive change of element size is less flexible as element connections are fixed; and (iv) element distortion at the corner and in the interior of the physical domain cannot be predicted prior to the application of the mapping.

The structured meshing techniques are still very useful for their simplicity and reliability in daily use and in commercial software, which could also be combined with unstructured meshing methods to produce even better results. Mapping due to co-ordinate transformations can be found in standard mathematical texts, e.g. Cartesian–polar transformation described in Appendix A.11, and thus this will not be pursued further in this chapter. However, the FE interpolation technique and the transfinite mapping were quite popular in the early stage of MG before the development of the unstructured meshing methods. The FE interpolation for linear and quadratic quadrilateral and triangular domains will be given in Section 3.2.1, and the transfinite mapping of a general region bounded by four analytical curves will be described in Section 3.2.2.

3.2.1 FE interpolation

The domain is defined by means of the corner nodal points along with the corresponding FE interpolation functions. For instance, a quadrilateral domain with straight edges can be defined using the bilinear Lagrangian element Q4, as shown in Figure 3.1, such that a point **x** within the quadrilateral domain is given by

$$\mathbf{x} = H_1\mathbf{x}_1 + H_2\mathbf{x}_2 + H_3\mathbf{x}_3 + H_4\mathbf{x}_4$$

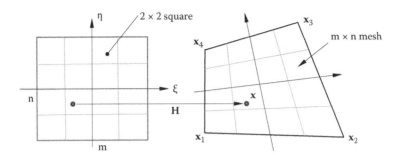

Figure 3.1 Quadrilateral mesh by finite element interpolation.

where $\mathbf{x}_i = (x_i, y_i)$, $i = 1,4$, are the nodal points at the corners, and $H_1 = H_1\,(\xi, \eta)\,\xi, \eta \in [-1, 1]$, are interpolation functions of the Q4 element, which are given by

$$H_1 = \frac{1}{4}(1-\xi)(1-\eta), \quad H_2 = \frac{1}{4}(1+\xi)(1-\eta),$$

$$H_3 = \frac{1}{4}(1+\xi)(1+\eta), \quad H_4 = \frac{1}{4}(1-\xi)(1+\eta)$$

The grid lines dividing the domain into smaller quadrilaterals can be constructed by setting a co-ordinate to constant and varying the other co-ordinate between –1 and 1. For example, a grid line along the ξ-co-ordinate can be generated by setting η = constant and varying ξ between –1 and 1. Similarly, to generate a grid line along the η-co-ordinate, simply set ξ = constant and vary η between –1 and 1. The intersection of these two sets of co-ordinate lines will produce an m × n mesh of quadrilateral elements, where m and n are, respectively, the number of subdivisions along the ξ and η directions, as shown in Figure 3.1.

A quadrilateral domain composed of boundary segments of parabolic curves can be represented by means of higher-order quadrilateral FEs, named by Zienkiewicz and Phillips (1971) as super-elements. In terms of the quadratic Serendipity element Q8, a curved boundary quadrilateral domain is given by $\mathbf{x} = \sum_{i=1}^{8} H_i\mathbf{x}_i$, where $\mathbf{x}_i$ are the nodal points at the corners and on the mid-sides of a curved edge, and $H_i = H_i(\xi, \eta)\,\xi, \eta \in [-1, 1]$ are interpolation functions of the Q8 element given by

$$H_1 = \frac{1}{4}(1-\xi)(1-\eta)(-\xi-\eta-1), \quad H_2 = \frac{1}{4}(1+\xi)(1-\eta)(\xi-\eta-1),$$

$$H_3 = \frac{1}{4}(1+\xi)(1+\eta)(\xi+\eta-1), \quad H_4 = \frac{1}{4}(1-\xi)(1+\eta)(\eta-\xi-1),$$

$$H_5 = \frac{1}{2}(1+\xi^2)(1-\eta), \quad H_6 = \frac{1}{2}(1-\eta^2)(1+\xi),$$

$$H_7 = \frac{1}{2}(1-\xi^2)(1+\eta), \quad H_8 = \frac{1}{2}(1-\eta^2)(1-\xi).$$

Again, the grid lines can be generated by setting one co-ordinate constant and varying the other co-ordinate between –1 and 1. For instance, to divide the original curved domain into 4 × 4 = 16 sub-domains, set ξ at –1, –0.5, 0, 0.5, 1, and vary η then set η at –1, –0.5, 0, 0.5,

1 and vary ξ to generate two sets of five contour lines. The intersection of these two sets of contour lines will produce the required sub-domains, as shown in Figure 3.2.

As for triangular domains, that is, regions bounded by three straight lines and/or curved edges, the triangular element interpolations have to be used. The interpolation functions for linear triangular element T3 are given by

$$H_1 = L_1, \quad H_2 = L_2, \quad H_3 = L_3,$$

where L_1, L_2 and L_3 are the area co-ordinates, as shown in Figure 3.3.

The interpolation functions for the quadratic triangular element T6 are given by

$$H_1 = L_1 (2L_1 - 1), \quad H_2 = L_2 (2L_2 - 1), \quad H_3 = L_3 (2L_3 - 1),$$

$$H_4 = 4L_2L_3, \quad H_5 = 4L_3L_1, \quad H_6 = 4L_1L_2$$

A point $\mathbf{p}$ in the reference domain will divide up the triangle into three zones by joining the point with the corners of the triangle, as shown in Figure 3.3. The area co-ordinates of point $\mathbf{p}$ are given by the ratios of the areas of the three zones so formed to the total area of the triangle.

$$L_1 = \frac{\Delta_1}{\Delta}, \quad L_2 = \frac{\Delta_2}{\Delta}, \quad L_3 = \frac{\Delta_3}{\Delta}, \quad \text{where } \Delta = \Delta_1 + \Delta_2 + \Delta_3$$

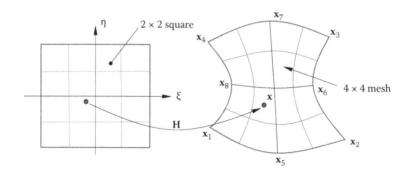

Figure 3.2 Q8 curved element divided into sub-elements.

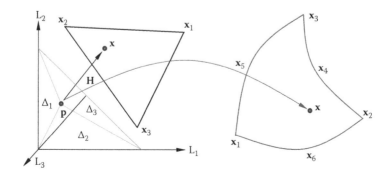

Figure 3.3 Triangular mesh by finite element interpolation.

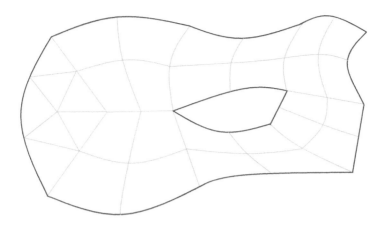

Figure 3.4 **Block decomposition of a multi-connected domain into super-elements.**

A point **p** in the reference domain, which is a triangle of convenient regular shape, say, a right-angled triangle, is mapped onto the physical real domain through the use of appropriate interpolation functions, as shown in Figure 3.3. For straight-edge triangles, we have

$$\mathbf{x} = H_1\mathbf{x}_1 + H_2\mathbf{x}_2 + H_3\mathbf{x}_3$$

and for six-node curved triangles with quadratic edges, we have

$$\mathbf{x} = H_1\mathbf{x}_1 + H_2\mathbf{x}_2 + H_3\mathbf{x}_3 + H_4\mathbf{x}_4 + H_5\mathbf{x}_5 + H_6\mathbf{x}_6$$

where $\mathbf{x}_i$ are the nodal points of the triangle under consideration. By means of the FE interpolation, a mesh of the reference domain can be reproduced on the actual physical domain by mapping properly all the points on the reference domain point by point onto the physical domain.

Based on the block decomposition (Bykat 1976; Sezer and Zeid 1991; Srinivasan et al. 1992; Egidi and Maponi 2008), a complex or multi-connected domain can be divided into a number of super-elements, each of which can be further subdivided into smaller elements of the same type, as shown in Figure 3.4.

3.2.2 Transfinite mapping

In the event that the curved boundary cannot be satisfactorily represented by FE polynomial interpolation functions, more general analytical functions have to be used. The transfinite mapping provides a means to map a unit square into a region bounded by four arbitrary analytical curves. Let $C_1(t)$, $C_2(t)$, $C_3(t)$ and $C_4(t)$, $t \in [0, 1]$, be the parametric bounding curves of a region, as shown in Figure 3.5. For instance, a straight line between two points A and B with parameter t is given by

$$C(t) = (1 - t)A + tB \quad t \in [0, 1]$$

and a circular arc from point A to point B is given by

$$C(t) = M + r(\cos(\phi), \sin(\phi))$$

where $\phi = (1 - t)\theta_a + t\theta_b$, mid-point of AB, $M = (A + B)/2$, $r = \|AB\|/2$ and θ_a and θ_b are, respectively, the angles of MA and MB making with the x-axis.

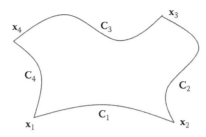

Figure 3.5 Region bounded by four arbitrarily shaped curves.

As a unit square can be readily discretised along the co-ordinate lines into a rectangular mesh of a specified element size, a mesh of the region can be generated by means of the following transfinite mapping function, which maps the unit square onto the region bounded by the curves C_1, C_2, C_3 and C_4.

$$\mathbf{x}(\xi, \eta) = (1 - \eta)\,\mathbf{C}_1(\xi) + \xi\mathbf{C}_2(\eta) + \eta\mathbf{C}_3(\xi) + (1 - \xi)\mathbf{C}_4(\eta) - [(1 - \xi)(1 - \eta)\mathbf{x}_1 + \xi(1 - \eta)\mathbf{x}_2$$
$$+\, \xi\eta\mathbf{x}_3 + (1 - \xi)\eta\mathbf{x}_4]\quad \xi, \eta \in [0, 1]$$

where $\mathbf{x}_1$, $\mathbf{x}_2$, $\mathbf{x}_3$ and $\mathbf{x}_4$ are the corner points of the region where two boundary curves meet. Prior to MG, one has to verify if the four curves meet at the corner points, i.e.

$$\mathbf{C}_1(0) = \mathbf{C}_4(0) = \mathbf{x}_1;\ \mathbf{C}_2(0) = \mathbf{C}_1(1) = \mathbf{x}_2;\ \mathbf{C}_3(1) = \mathbf{C}_2(1) = \mathbf{x}_3;\ \mathbf{C}_4(1) = \mathbf{C}_3(0) = \mathbf{x}_4$$

Figure 3.6 shows some regions meshed with the aid of transfinite mapping. Figure 3.6a is a mesh of a quadrilateral region bounded by straight edges, which could as well be meshed

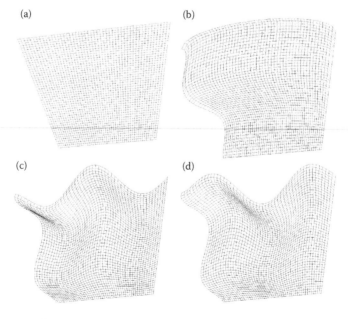

Figure 3.6 Regions meshed by means of transfinite mapping: (a) region bounded by straight edges; (b–d) region with curved boundaries.

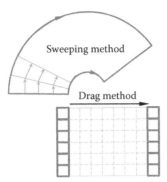

Figure 3.7 **Drag method and sweeping method.**

using the Q4 FE interpolation. Regions bounded by two straight edges and two curves are shown in Figure 3.6b–d. Region (b) is bounded by a sine curve on the left and an elliptical curve on the top, and regions (c) and (d) are each bounded by two sine curves on the left and on the top. While the pattern of subdivision and the topological features are preserved, there is a drastic change of geometry in terms of element size and internal angles of the region by transfinite mapping. Transfinite mapping provides a convenient way in meshing regions bounded by general analytical curves; however, care must be taken to prevent serious mesh distortions, as shown in Figure 3.6c–d, which unfortunately cannot be easily controlled in advance.

3.2.3 Drag method and sweeping method

The drag method and the sweeping method shown in Figure 3.7 can be considered as special cases of the mapping method. These are classical techniques developed at the early stage of FE MG (Park and Washam 1979). With a simple idea and without much computation, the methods were developed for a rapid generation of structured FE meshes over domains of relatively simple geometry. The point in common for meshes generated by these two methods is that the resulting meshes are bounded by four curves (lines). Suppose there are n nodes on the drag section $\{x_i, i = 1,n\}$; a new layer of node $\{x_i^* = x_i + v, i = 1,n\}$ can be generated by adding to each node the incremental vector v. However, for the sweeping method, incremental vector v may be different for each node given by the rotation vector.

3.3 UNSTRUCTURED MESH ON PLANAR DOMAIN

Automatic procedures for the generation of unstructured FE meshes were first developed on 2D domains. Extensive research has been carried out in the past on the formulation of robust and versatile FE MG algorithms. A review and a classification of MG methods were presented by Ho-Le (1988). Unstructured meshes of variable element size and even with anisotropic length requirement are needed for adaptive refinement analysis and to fit complicated geometrical boundaries. However, not all the methods described by Ho-Le are suitable for MG in conjunction with an adaptive refinement process. Adaptive refinement FE analysis first achieved success in 2D problems. Unstructured triangular meshes lend themselves readily to the situation of adaptive analysis, as triangles are less sensitive to shape distortions such that meshes of rapid change of element size can be constructed. Furthermore, many MG schemes have been devised to create high-quality unstructured triangular meshes of variable element size. The following are some popular MG methods that could be used to produce unstructured FE meshes compatible with a given node spacing function on a 2D domain.

i. Delaunay triangulation (DT)
ii. Advancing-front technique (AFT)
iii. MG using contour
iv. Coring technique
v. Quadtree decomposition
vi. Refinement by subdivision

3.3.1 MG using contour

MG using contour is an early attempt for the generation of a triangular mesh of variable size (Lo 1991a), making use of the MG technology available at that time for mesh of uniform size. The contour line of the node spacing function at suitable calculated levels provides the natural division lines of the problem domain into sub-regions, where FEs of different size can be generated using a general-purpose mesh generator. The pictures with some yellowish background shown in Figure 3.8 are scanned copies of the triangular meshes created by this technique in 1989. In Figure 3.8a, the domain is divided into many zones by contour lines in which triangular elements of different size are generated, and in Figure 3.8b, a triangular mesh of rapid-change element size is generated. If a sufficient number of contours are used in the zonal subdivision, meshes of progressive change in element size can be produced after shape optimisation by node shifting.

3.3.2 Coring method

MG is achieved by constructing as many square elements as possible in the interior of the domain, leaving a relatively small region around the boundary that is to be discretised into triangular elements by a standard triangulation procedure (Lo and Lau 1992). As a result,

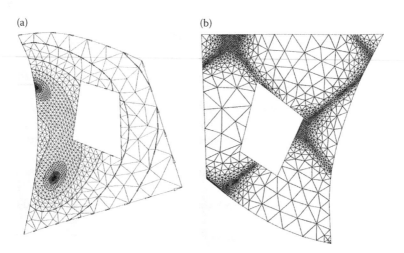

Figure 3.8 Meshes generated by the contour line method: (a) mesh 1; (b) mesh 2.

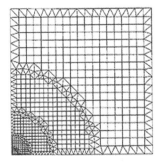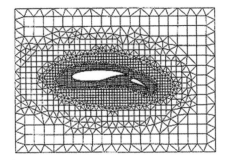

Figure 3.9 Meshes generated by the coring method.

only the quality of the triangular elements near the boundary will be affected by the irregularity of the domain. The main feature of this technique is to extract as much as possible from the interior of the domain to form square elements, as in an excavation process, hence the name *coring method*. The square elements can later be divided into triangular elements to form a pure triangular mesh if necessary. Two examples of the coring method to produce meshes of varying element size are shown in Figure 3.9.

3.3.3 Mesh refinement by subdivision

Generation of adaptive FE meshes by refinement can be achieved by selective subdivision of triangles according to the specified node spacing function. Starting from a coarse mesh, elements can be subdivided progressively until the desired refinement is reached (Wordenweber 1984). The most popular type of division scheme is bisection of the longest edges, for which the maximum and the minimum angles of the resulting triangles are bounded, and the process will terminate in a finite number of steps (Rivara 1997). Many refinement algorithms with various characteristics exist, and a comprehensive account can be found in a paper by Jones and Plassmann (1997). As subdivision of triangles is a local process that does not require too much computation, adaptive meshing by this approach in general is very efficient, although the mesh topology and the general element layout are limited by the previous meshes. With some post-treatment optimisation, the quality of the elements could be drastically improved; nevertheless, the conformity to node spacing function is still restricted and may not be as good as meshes generated from scratch in accordance with the given node spacing specification. A gradation mesh produced by successive refinement is shown in Figure 3.10.

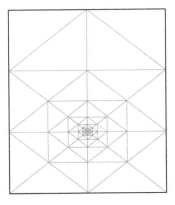

Figure 3.10 Mesh generated by refinement.

3.4 MESHING BY QUADTREE DECOMPOSITION

Quadtree meshing is based on the idea that a partition of space in a progressive manner can produce cells of size compatible with the FE node spacing requirement. The Quadtree technique, which belongs to the class of spatial decomposition methods, was originally employed as a means to approximate the geometry of CAD models (Knuth 1975). The use of Quadtree decomposition for FE MG was developed in the 1980s by Shephard (1988) and his group at Rensselaer. Thereafter, a number of variants of the technique have been developed (for instance, Finnigan et al. 1989; Pressburger and Perucchio 1995; Perucchio et al. 1989). A parallel construction of Quadtree for FE MG was reported by Bern et al. (1999), and a quality MG procedure was developed by Mitchell and Vavasis (2000). Lee and Samet (2000) navigated and labelled triangular meshes by means of a Quadtree construction, and a flow-based grid for finite volume analysis was presented by Edwards (2002). By the Quadtree method, a rectangular grid containing the domain to be discretised is recursively subdivided until a desired resolution is reached. The subdivision can be done in compliance with the complexity of the problem domain and/or with respect to a given element size map for the generation of adaptive meshes. Irregular cells are created where rectangles intersect the object contour, often requiring a substantial amount of calculation for intersections. Four distinct phases can be identified in a typical Quadtree meshing procedure.

1. Boundary specification and discretisation in compliance with the given element size
2. Quadtree spatial partition of the bounding box of the domain
3. Creation of internal points and elements
4. Connection of the interior elements with the boundary segments

3.4.1 Boundary specification

Boundary specification of Quadtree meshing may differ from the other classical mesh generation methods, such as the AFT and DT, in which the boundary of the 2D domain is a collection of directed line segments. The boundary specification for Quadtree meshing may be linked directly to a CAD system in which the boundary is defined implicitly through a geometrical querying process. As how an object is specified in a CAD system is quite open, in order to have a well-defined meshing problem subject to common boundary specifications as the other mesh generation methods, we take for Quadtree meshing the 2D domains also bounded by discrete line segments. As suggested by the method presented in Section 2.4.9, the boundary contour for Quadtree meshing can be divided into segments of size conforming with the specified node spacing function.

3.4.2 Spatial partition of the bounding box

The bounding box of the given 2D domain is a rectangle large enough to contain all the nodal points. An algorithm for the construction of a Quadtree partition for a system of points and the corresponding *pseudo-code* are given in Section 2.7.6. However, this Quadtree partition algorithm cannot be directly applied for two reasons: (i) Interior points are required and are yet to be generated, and (ii) for a smooth grading of the resulting FE mesh, the one-level division rule is adopted, i.e. the level of subdivision between adjacent cells cannot be more than one, as shown in Figure 3.11. A recursive subdivision algorithm is thus proposed in the following sections to discretise a rectangular domain into Quadtree cells (elements) according to the specified element size without violating the one-level restriction.

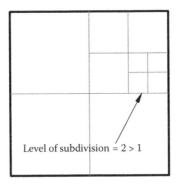

Figure 3.11 **Level of subdivision between neighbouring cells.**

To take the one-level Quadtree decomposition properly as a mesh refinement problem of transition quadrilateral elements, the types of quadrilateral and their node-labelling scheme have to be identified, as shown in Figure 3.12. In the most general situation, there are only six types of quadrilaterals in a transition quadrilateral mesh satisfying the one-level refinement restriction, in which the eight-node element can be divided into four four-node elements once it is formed.

As for the neighbouring relationship of the transition FE mesh, a four-node quadrilateral can have four neighbours along the edges 1–2, 2–3, 3–4 and 4–1, respectively. In case no neighbour exists on an edge, i.e. the edge is only connected to one single quadrilateral, then it is a boundary edge. Quadrilateral elements with more nodes can join with more neighbours; for instance, the seven-node quadrilateral can have as many as seven neighbours, which are numbered sequentially on edges 1–5, 5–2, 2–6, 6–3, 3–7, 7–4 and 4–1. As usual, the number zero can be used to stand for no neighbour (boundary edge), and a positive integer represents a valid neighbour to that edge. In subdividing a transition quadrilateral element, depending on the element type, the number of nodes added may vary, and the neighbouring relationship has to be updated according to the types of quadrilaterals connected to the element.

Let's take a look at how a quadrilateral is subdivided in a mesh and how the adjacency relationship of the relevant elements could be updated. A patch of four four-node quadrilaterals

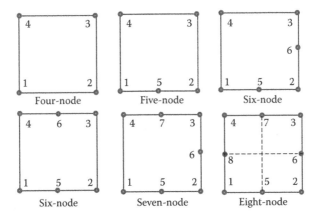

Figure 3.12 **Family of transition quadrilateral elements.**

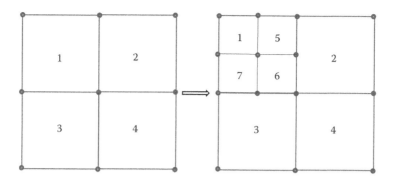

Figure 3.13 Subdividing a quadrilateral element in a patch.

is considered, as shown in Figure 3.13, and suppose element (1) has to be subdivided at the corner. The original neighbours of the elements are given by

Element (1): {3, 2, 0, 0}; Element (2): {4, 0, 0, 1};
Element (3): {0, 4, 1, 0}; Element (4): {0, 0, 2, 3}.

After subdivision, there are seven elements in the patch in which elements (2) and (3) change from four-node elements to five-node elements, and their adjacency relationship becomes

Element (1): {7, 5, 0, 0}; Element (2): {4, 0, 0, 5, 6};
Element (3): {0, 4, 6, 7, 0}; Element (4): {0, 0, 2, 3};
Element (5): {6, 2, 0, 1}; Element (6): {3, 2, 5, 7}; Element (7): {3, 6, 1, 0}.

In a one-level partition, a cell can have no more than two cells adjacent to it on a single side. Before a detailed discussion on this restriction, two concepts are introduced. The first is the concept of neighbouring cells, i.e. two cells sharing a common edge are neighbours. The second concept is the refinement level (RL) of a cell, which indicates the number of subdivisions done so far to the cell. Based on these two concepts, without tedious verification for boundary compatibility between cells, the one-level restriction can be enforced by just checking the RL values of the neighbouring cells.

At the initial stage of one single rectangular cell, or more generally an FE mesh of quadrilateral elements, each cell (element) is assigned an RL value of zero. As the partition progresses, when a cell is indicated to be refined, it is necessary and sufficient to check its neighbouring RL values. If the neighbours' RL values are equal or larger than its own RL value, this cell can be subdivided into smaller cells, and the RL value of the four subdivided cells is increased by one as a result. Otherwise, the cell cannot be subdivided until its neighbours are subdivided first. This simple but effective RL-checking procedure can be easily implemented by a recursive algorithm.

For a rectangular cell, it is subdivided into four cells by adding mid-side nodes and a centre node, as shown in Figure 3.14a. The following are the typical steps in the partition of a rectangular domain with one-level mesh restriction enforced by means of the strategy of RL check. The elements of the initial mesh are each given an RL value of zero. Suppose element (1) has to be refined as indicated by the element size requirement. In order to do so, the RL

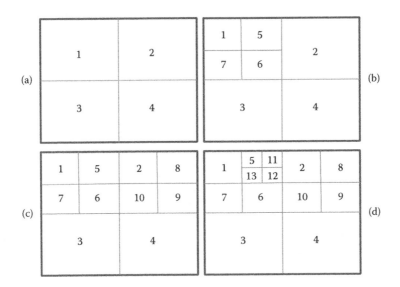

Figure 3.14 Mesh refinement with one-level restriction: (a) initial mesh; (b) first subdivision; (c) second subdivision; (d) third subdivision.

values of its neighbouring elements (2) and (3) are checked to ensure that the one-level mesh restriction is satisfied. Upon dividing element (1), four new elements are generated, which are assigned RL = 1, as shown in Figure 3.14b. Now, element (5) is supposed to be refined in the next step. The RL values of its neighbours, namely, elements (1), (2) and (6), have to be checked. It is found that the RL value of element (2) is smaller than that of element (5). As a result, element (2) has to be subdivided first. Checking the RL values of the neighbours of element (2), the one-level restriction is satisfied for element (2), and it is refined, as shown in Figure 3.14c. The four new elements (2), (8), (9) and (10) have their RL value increased by one, and subdivision of element (5) can now be reactivated, as shown in Figure 3.14d. The recursive refinement algorithm has been applied to the analytical function of the element size distribution as well as practical adaptive refinement analysis, as shown in Figure 3.15.

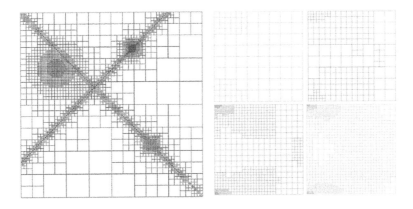

Figure 3.15 Mesh refinements by the recursive subdivision algorithm.

Figure 3.16 Templates to triangulate quadrilaterals.

3.4.3 Creation of internal points and elements

In the earlier classical application of the Quadtree meshing, the refined quadrilateral elements are subdivided into triangular elements to ensure compatibility across the element boundary by means of a template, as shown in Figure 3.16. In the Quadtree mesh with one-level restriction, there are $2^4 = 16$ cases to be considered, which could be reduced to only 6 using various symmetries (Yerry and Shephard 1983). At the turn of the twenty-first century, efficient transition quadrilateral elements were developed (Lo et al. 2005, 2010), and the refined quadrilateral mesh is a valid FE mesh in its own right with full compatibility on the element boundaries. The transition element technology allows the interior part to be kept intact without any further treatment, and what remains to be done is to merge it with the boundary segments in the final step.

3.4.4 Connection of the interior elements with the boundary segments

In the traditional Quadtree meshing, quadrilaterals cut across by boundary segments are divided into triangles, which is achieved in two steps. The first step is to determine all the intersections between the boundary segments and the quadrilaterals in the Quadtree partition. In the second step, a set of templates for the decomposition of a quadrilateral is prepared to subdivide the intersected quadrilaterals into triangles. However, in general, the number of intersections of a quadrilateral cannot be pre-determined, and some boundary segments may also align in the principal directions of the Quadtree grid resulting in very narrow pointed triangles. In order to reduce the number of cases of the intersection to a manageable limit, intersection points have to be snapped onto the $\frac{1}{4}$ positions of the edges of the quadrilaterals. Some of the typical divisions of a quadrilateral into triangles are shown in Figure 3.17, and for a detailed discussion of the templates and the subsequent connections into triangles, readers can refer to the original work of Yerry and Shephard (1983).

There are a number of drawbacks in the triangulation of quadrilaterals by means of templates: (i) The boundary integrity may be jeopardised as intersection points have to be snapped onto quarter points; (ii) determination of intersections is a tedious process prone to numerical errors; (iii) the position and the orientation of the Quadtree grid with respect to the given boundary segments are rather arbitrary; and (iv) narrow pointed triangles

Figure 3.17 Templates for division of quadrilateral at quarter points.

will be generated, and a non-homogeneous division of quadrilaterals may be resulted by joining quarter points. With the development of more versatile MG schemes for planar regions bounded by arbitrary line segments, there exist other strategies to overcome the drawback of the classical Quadtree meshing in the treatment of boundary. One of the possibilities is to remove all the quadrilaterals that are close to the boundary segments. Upon removal of the quadrilaterals, the gap between the remaining quadrilaterals and the boundary segments can be easily filled up by some standard 2D mesh generator such as DT or AFT. This coring scheme has been proposed in a paper by Lo and Lau (1992) in which uniform square elements are taken away as much as possible from the interior of the 2D domain. The remaining part, which is relatively small, is meshed by the AFT. The combined schemes seem to be quite effective in securing good-quality rectangular elements at the interior of the domain leaving only a relatively small portion of the irregular zone for MG, which is never a problem on 2D domains. This idea has been adopted for the Quadtree meshing to avoid low-quality triangles at the boundary, and a detailed discussion will be given in Section 3.8 to elucidate all the steps of such an enhanced Quadtree-meshing scheme.

3.5 DELAUNAY TRIANGULATION

3.5.1 Introduction

At the middle of the nineteenth century, Dirichlet (1850) showed that for a given set of points in two dimensions, it is possible to partition the plane into convex cells based on a proximity criterion. At the turn of the twentieth century, Voronoi (1908) studied the algebraic forms related to this geometrical criterion, which could be extended to a three-dimensional space. The concept of the Voronoi diagram was thus established, i.e. there is a one-to-one correspondence between the cells and the given points of the set in the partition of space into cells, such that any point in a cell is closest to a particular point of the set than any other points in the set. Delaunay in 1934 stated that it is possible to construct a triangulation from the Voronoi diagram by duality, linking up points associated with the cells with those of the neighbouring cells, as shown in Figure 3.18. The triangulation derived from the Voronoi diagram has since been named as DT of the point set.

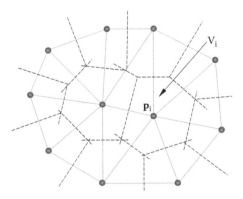

Figure 3.18 Voronoi diagram and Delaunay triangulation.

3.5.1.1 The convex hull of a given point set

Let $\mathbb{P} = \{p_i, i = 1, n\}$ be a set of n distinct node points in the Euclidean 2D plane $\mathbb{R}^2$, and define the set of polygons $V = \{V_i, i = 1, n\}$, where

$$V_i = \left\{ \mathbf{x} \in \mathbb{R}^2 : \left\| \mathbf{x} - \mathbf{p}_i \right\| < \left\| \mathbf{x} - \mathbf{p}_j \right\|, \; \forall j \neq i \right\}$$

and $\|\mathbf{x} - \mathbf{p}\|$ denotes the Euclidean distance between points $\mathbf{x}$ and $\mathbf{p}$.

V_i represents a region on the plane $\mathbb{R}^2$ whose points are nearer to node point $\mathbf{p}_i$ than to any other node points in the given set. Hence, V_i is an open convex polygon (called the Voronoi polygon) whose boundaries are portions of the perpendicular bisectors of the lines joining node $\mathbf{p}_i$ to node $\mathbf{p}_j$ of neighbouring polygons V_i and V_j. The collection of Voronoi polygon V is called the Dirichlet tessellation of the point set $\mathbb{P}$.

In general, the vertex of a Voronoi polygon is shared by two more neighbouring polygons so that connecting the three points associated with such adjacent polygons forms a triangle, as shown in Figure 3.18. The set of triangles so created at each vertex is called the DT. This construction can be shown to be a triangulation of the convex hull for the set of nodal points $\mathbb{P}$. The Dirichlet tessellation is always unique for a given set of points, whereas the DT will be unique as long as the points are in the general positions, or no four points are cyclic. In case there are four points lying on a circle, Delaunay triangles can be formed by cutting across either diagonal of the cyclic quadrilateral, and the DT is only unique up to the division of the cyclic quadrilateral. In the event that there are more than four nodes lying on a circle, the DT is unique up to the division of the cyclic polygon into triangles. If there are more than one spot where points are not at the general positions, the number of DTs depends on the combination of how these cyclic polygons are subdivided into triangles.

3.5.2 Properties of DT

  i. Among all triangulations of a set of points in 2D, the DT maximises the minimum angle and minimises the maximum circumcircle.
 ii. If every triangle in a triangulation is non-obtuse, it is a DT.
iii. Any 2D triangulation can be transformed into a DT by locally flipping of diagonals of adjacent triangles in $O(n^2)$ time (Joe 1993).

In a paper by Rajan (1994), some optimality properties of DT in two or higher dimensions are discussed. It is suggested that the DT is the most compact one in the following sense, which justifies its wide applications in many areas.

  i. The containment sphere is the smallest. It is the smallest sphere that contains the simplex, which may or may not be the circumsphere depending on whether the circumcentre is inside the simplex or not. In a DT, the largest containment sphere among all simplices is minimised.
 ii. The union of the circumspheres of the simplices incident to an interior point is the smallest. That is, the union of circumspheres of simplices incident to an interior point in a DT lies inside the union of the circumspheres of simplices incident to the same point in any other triangulation of the point set.
iii. The weighted sum of the squares of the edge lengths, where the weight is the sum of volumes of the simplices incident on the edge, is the smallest for DT.

A detailed description of the Voronoi diagram, dual of DT, was given by Aurenhammer (1991), and other properties of DT such as the maximum vertex degree have been estimated by Bern et al. (1991).

3.5.3 Time complexity in the construction of DT

Based on a sound geometrical concept and the optimality properties, DT has important applications in many fields, including data visualisation, terrain modelling, FE MG, surface reconstruction and structural networking for arbitrary point sets. The popularity of DT is attributed to its nice geometric properties as a dual of Voronoi tessellation and the speed with which it can be constructed in two or higher dimensions. The existence, uniqueness and other properties of DT have been studied for a long time as formal mathematical topics in computational geometry, and their computation issues and complexity have been interesting problems for computer scientists. In view of its diverse applications, many strategies for its construction have been proposed. In a paper by Su and Drysdale (1997), numerical tests on several DT algorithms, namely, divide-and-conquer, sweep line algorithms, incremental algorithms, a fast incremental construction algorithm, gift-wrapping algorithms and convex hull–based algorithms, were carried out, and their performances were compared.

There are some estimates of the time complexity of DT in 2D and 3D, which could be taken as a reference, but these asymptotic results may not be directly applicable to FE MG as it only makes sense for very large n values, and the algorithm may not be available or could not be easily implemented for practical use; and there are two more important steps in FE MG relative to triangulation, namely, the generation of interior points and boundary recovery.

The complexity of DT is very sensitive to the point distribution and the order of how these points are processed. For instance, for DT in 3D, the estimated time complexity of triangulating n points is $O(n^{4/3})$ for the Bowyer's algorithm (Bowyer 1981) and higher for the algorithms of Watson (1981) and Avis and Bhattacharya (1983). Joe (1989) presented an algorithm that makes use of local transformations to construct a DT of a set of 3D points. The empirical time complexity of the algorithm is $O(n^{4/3})$, and $O(n^2)$ in the worst case. Edelsbrunner et al. (1990) proposed a scheme in the worst case time of $O(n^2)$ for the construction of 3D DT by projecting the given 3D points onto a paraboloid in four dimensions. The convex hull of the 4D points on the paraboloid is constructed, and the 3D DT is then obtained from an appropriate portion of the convex hull. Nevertheless, for practical engineering applications with distance between points not greater than a ratio of, say, $1:10^6$, DTs can be constructed in a virtually linear time complexity, as described in Chapter 8.

3.5.4 FE meshing by DT

By far, the most popular triangular MG schemes are based on the concept of DT (Cavendish et al. 1985; Weatherill 1988; Lo 1989a; Joe 1991a,b,c; Wright and Jack 1994; Lewis et al. 1995; Escobar and Montenegro 1996; Krysl and Ortiz 2001; Nishioka et al. 2001; Quey et al. 2011). The Delaunay criterion, also known as the empty-circle property, states that any node must not be contained within the circumscribing circle of any triangle in the triangulation, as shown in Figure 3.19. Although the Delaunay criterion has been known for many years since the pioneer paper by Delaunay (1934), it was not until the work of Lawson (1977), Bowyer (1981) and Watson (1981) that the criterion was exploited for developing algorithms to form a convex hull of a given set of points. With the rapid development of the FEM in the 1980s, the DT algorithm was further extended to generate valid FE meshes for numerical analysis by Baker (1989a,b), George and Hermeline (1992), Ghosh and Mallett (1994), Weatherill and Hassan (1994), Peterson et al. (1999) and others. For a comprehensive view on the theoretical aspects of DT as well as its applications to FE MG, readers are referred to the book by George and Borouchaki (1998).

Figure 3.19 Delaunay triangulation of 14 points.

In the incremental algorithm of Bowyer and Watson, the points are processed one at a time. In a typical step of point insertion, the triangles whose circumcircle contains the insertion point are identified and deleted. New triangles are constructed in the cavity left behind by the triangles removed. Hence, the efficiency of the triangulation scheme depends on how fast one can identify the triangles to be removed and determine correctly the cavity for insertion and the speed with which the circumcentres, circumradii and adjacency relationship of the new triangles are computed.

The Delaunay criterion itself is not an algorithm for MG. It merely provides a rule to connect a set of existing points in space to form a triangulation. As a result, although the boundary of the domain is well specified, it is necessary to devise a scheme to determine the number and the locations of node points to be inserted within the domain of interest. A typical approach is to first create a triangular mesh large enough to contain the entire domain. The boundary nodes are then inserted and connected according to the Delaunay criterion, and this forms a triangulation of the boundary nodes. More nodes are then inserted progressively into the coarse boundary mesh, refining the triangles as each new node is introduced, until a desirable number of elements are formed at appropriate positions (Borouchaki et al. 1996).

In FE applications, there is a requirement that the triangulation contains the boundary of the domain, so that the boundary integrity can be easily enforced by deleting all the triangles outside the given domain. In most DT processes, before interior nodes are introduced, a tessellation of the nodes on the domain boundary is produced. However, in this process, there is no guarantee that boundary segments will all be present in the triangulation. In many implementations, the approach is to tessellate the boundary nodes using a standard Delaunay point insertion algorithm with no regard to the integrity of the domain boundary. A second step is then employed to force or recover the boundary segments. Of course, by doing so, the triangulation in general is not strictly Delaunay at least locally, hence the term boundary-constrained DT. In 2D, the edge recovery is straightforward simply by swapping diagonals (Weatherill 1990), and in 3D, it is less obvious how boundary integrity could be achieved in general, which will be further elaborated in Chapter 5.

There are a number of ways to create interior points according to the node spacing requirement, which in fact would lead to meshes of different characteristics. Hermeline

(1980) proposed a scheme in which points are inserted at the barycentre of the triangles, and some researchers have suggested to insert points at the circumcentres of triangles (Shenton and Cendes 1985; Holmes and Snyder 1988). Borouchaki and Geroge (1997) advocated the insertion of points along the edges of triangles. Others made use of a set of points at pre-determined positions with the aid of a regular grid, a Quadtree network or some sort of spatial decomposition methods such as the kd-tree. A combined scheme with the AFT was also put forward in which points are inserted at strategic positions as determined in a frontal process, and element connections are modified based on the Delaunay criterion. A quality guaranteed Delaunay refinement of a conforming triangular mesh was presented by Shewchuk (2002) in which the interior angle of the triangles is bounded by some threshold related to the boundary. Based on a hexagonal grid, Süßner and Greiner (2009) produced a DT for a set of 2D points with refinement satisfying a number of conditions.

Borouchaki et al. (1997a) made use of the concept of control space and length measure for the insertion of points according to non-Euclidean metrics to create adaptive meshes of variable element sizes and aspect ratios. The element size map can be explicitly specified as a continuous function over the entire domain or implicitly defined over the domain by means of a background mesh (Borouchaki and George 1996). The idea was later extended with the introduction of a general metric tensor to measure the distance between two points for the generation of anisotropic meshes in which not only element size can vary but also elements are subject to different size requirements along different directions (Borouchaki et al. 1997b).

In the construction of DT, the point insertion algorithm proposed by Bowyer (1981) and Watson (1981) is perhaps the simplest but is the most general in extending to higher dimensions. Moreover, it is also easy to describe as well as to implement in a few clear distinct steps (Borouchaki and Lo 1995). It is also one of the most efficient approaches in terms of serial processing and parallel processing (Lo 2012a,b), such that for evenly distributed randomly generated points, the time complex is basically linear. In the point insertion scheme, the fact that DTs are always constructed for the points already inserted is a nice feature guaranteeing the robustness of the procedure. When the next point is introduced, circumcircles of some triangles may contain the newly inserted points, and hence they become non-Delaunay. According to the lemma of Delaunay, non-Delaunay triangles locally form a patch (connected piece), which could be easily identified by the empty-circle criterion. Upon the removal of the patch of non-Delaunay triangles, and the re-establishment of the connection of the newly inserted points with the boundary edges of the patch (cavity), a DT containing the newly inserted point is thus created. When all the points are processed one by one, a DT of the given point set is constructed. Hence, the point insertion algorithm is very flexible and reliable in dealing with arbitrary point sets or their subsets.

3.5.4.1 Fundamentals and strategy

Before a detailed presentation of the point insertion algorithm, the formal definition and some basic strategies for the construction of DT are discussed. The DT of a set of points on a plane is defined to be a triangulation such that the circumcircle of every triangle in the triangulation contains no point from the set in its interior. Such a triangulation exists for a given set of points, and it is the dual of the Voronoi tessellation. The triangulation is unique if the points are in general position, i.e. no four points are cyclic. A triangle T is said to be Delaunay with respect to a point p if p does not lie inside the circumcircle of T. A triangle T in a triangulation of a set of points is called a Delaunay triangle if T is Delaunay with respect to every point in the set. A triangulation of a set of points is called the DT of the point set if every triangle in the triangulation is a Delaunay triangle, as shown in Figure 3.19. The idea of DT is very general, which can be easily extended to higher dimensions. For instance, the DT in 3D is given by replacing the triangle by a tetrahedron, the circle by a sphere and the 2D plane by a 3D space. The following lemma provides the basis for many algorithms in the construction and verification of DT.

Lemma 3.1: Lemma of Delaunay

Let **T(S)** be a triangulation of the point set **S**. The necessary and sufficient condition that no point of **S** is contained in the circumcircle of any triangle in the triangulation is that any two adjacent triangles in the triangulation are Delaunay with respect to each other's vertices.

The lemma of Delaunay, as depicted in Figure 3.20, is also valid in higher dimensions simply by changing some of the terms in the relevant dimensions. To arrive at an efficient and robust DT scheme, the fundamentals of DT and its implementation must be reviewed. Su and Drysdale (1997) presented a comparison of sequential DT algorithms, including divide-and-conquer, sweep line algorithm, incremental algorithm, fast incremental construction algorithm, gift-wrapping algorithm and convex hull–based algorithm. Based on the numerical tests on SPARCstation 2 and DEC Alpha machines, it was reported that Dwyer's divide-and-conquer algorithm (Dwyer 1987) was the *strongest* overall and was the most resistant to bad data distribution with O(nlogn) as the worst case.

Although the end result of DT of a set of n distinct points is independent of the order following which the points are processed, the speed or the complexity of triangulation is very sensitive to the order of how the points are processed. The divide-and-conquer algorithm is robust as some sorting mechanism is already built into the algorithm, such that the data set is always divided more or less into two equal portions. As for the insertion algorithm, for randomly generated points, the complexity can vary from O(n) to O(n²) depending on the order of point insertion. For the 2D DT of 1 million points on a PC, if points are inserted following the natural order of point generation, triangulation time increases in a quadratic manner with the number of points inserted, as shown in Figure 3.21. However, as also shown in Figure 3.21, if points are sorted into cells, each of which is assigned more or less an equal number of points, a linear time relationship of the same set of points is resulted if cells of points are inserted one by one in a contiguous manner following the natural structural layout of the cells. For a 2D triangulation of 1 million points, O(n²) complexity (149.5 s) represents more than 125 times of an O(n) scheme (1.18 s), and this quasi-linear complexity

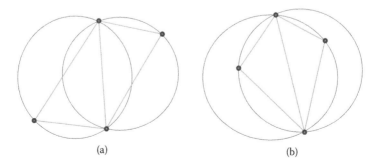

(a)   (b)

Figure 3.20 (a) Delaunay and (b) non-Delaunay triangulations.

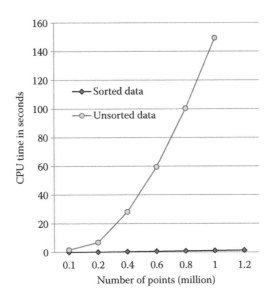

Figure 3.21 Delaunay triangulation for sorted and unsorted data.

can be sustained up to 200 million points in 295 s and beyond to the maximum machine capacity of 16 GB.

This observation is in line with the second interesting remark under Notes and Discussion of the paper by Su and Drysdale (1997), 'A simple enhancement of the naïve incremental algorithm results in an easy to implement algorithm that runs in O(n) expected time for uniformly distributed sites. It is faster than previous incremental variants and competitive with other known algorithms for constructing the DT for all but very bad sites'.

3.5.4.2 *Point insertion algorithm*

Let $\mathbb{S}$ be the set of n distinct points on a 2D plane. A rectangular domain composed of two triangles ($\mathbb{T}_0$) large enough to contain all the points in $\mathbb{S}$ is first constructed. The DT by point insertion algorithm is to insert all the points one by one in $\mathbb{S}$, and each cycle of point insertion can be divided into three steps (Borouchaki and Lo 1995):

i. For a newly inserted point, identify all the triangles whose circumcircles contain the point. The cavity left behind upon the removal of these triangles forms a star-shaped polygon, which will be referred to as the CORE.
ii. Owing to finite precision arithmetic, the boundary edges of the insertion polygon have to be verified with visibility check and corrected before they could be connected with the inserted point to form triangles.
iii. The triangulation of the cavity should be trivial. However, the adjacency relationship among the new triangles and with the old triangles on the boundary edge of the insertion polygon has to be established.

The point-by-point insertion algorithm can be summarised into a point insertion kernel (module). Suppose $\mathbb{T}_k$ is the DT of the first k points in the set $\mathbb{S}$; upon the introduction of the next point **p**, the insertion kernel aims at producing a DT with point **p** included in the triangulation, as shown in Figure 3.22.

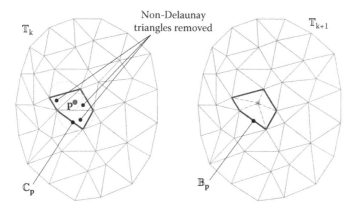

Figure 3.22 Inserting point **p** by the insertion kernel.

$$\mathbb{T}_{k+1} = \mathbb{T}_k - \mathbb{C}_p + \mathbb{B}_p$$

where

$\mathbb{C}_p$ = CORE = cavity of non-Delaunay triangles with respect to **p**
$\mathbb{B}_p$ = ball of triangles = patch of Delaunay triangles connected to **p**
$\mathbb{T}_k$ and $\mathbb{T}_{k+1}$ = the DTs of the first k and k + 1 points of the set $\mathbb{S}$

3.5.4.3 Determination of the CORE

When a new point **p** is inserted in a DT, it is required to find all the triangles whose circum-circle contains the point **p**. A simple method to determine these non-Delaunay triangles is to scan through all the existing triangles in the triangulation for those whose circumcircle contains the point **p**. However, a more efficient approach is to start with the triangle containing the inserted point **p** and to find the others by means of the adjacency relationship. In this way, the boundary of the CORE (insertion polygon) is given by the common edge of two triangles for which one is positive in the circle inclusion test while the other fails.

The triangle containing the insertion point **p** is called the BASE, which is an integral part of the CORE. It seems that finding a triangle whose circumcircle contains **p** is easier than finding the BASE. However, it does not always work in case a wrong decision is made in the circle inclusion test due to numerical error. The difficulty can be elucidated by the following example for the insertion of a new point **p**.

Let **afb** be the triangle that contains **p** in a DT, and **a**, **b**, **c**, **d** and **e** be the five points roughly on the circumference of a circle, as shown in Figure 3.23. In the circle inclusion test, point **p** is found in the circumcircle of triangles **abc** and **cde** but not in that of **ace**. In this case, the CORE is a disconnected piece if one has started building the CORE with a triangle not containing **p**. On the other hand, if one insists that the construction of the CORE has to be started with the containing triangle **afb** (BASE) and builds the CORE by means of adjacency, then triangle **cde** can be excluded from the CORE and a valid CORE composed of triangles **afb** and **abc** can be obtained in one piece.

3.5.4.4 Searching for the BASE

In principle, one would like to start the searching process for the containing triangle of the insertion point **p** from a point as close to point **p** as possible. However, when there is no

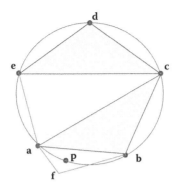

Figure 3.23 Inconsistent triangulation.

better information as to where the point **p** is located with respect to the existing triangles, the search can only be arbitrarily started from the last constructed triangle. Suppose one starts with a point **q** at the centroid of the last constructed triangle. **q** moves across an edge, which separates the moving point **q** and the destination point **p**, as shown in Figure 3.24.

Since the neighbours of each triangle are known, the position of **q** can be easily updated. The process is repeated until point **p** is reached, and the containing triangle is found. Nevertheless, there is a weakness in this scheme as when point **q** moves into a triangle from position (1) as shown in Figure 3.25, it will go around the triangles in a counter-clockwise order and will loop indefinitely. To avoid looping around a close circuit, there is a suggestion that all the triangles visited should be flagged so that point **q** will not go into those triangles

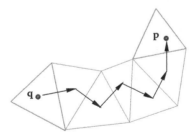

Figure 3.24 Search for the BASE.

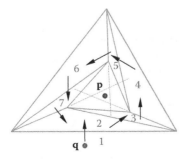

Figure 3.25 Looping around a triangle.

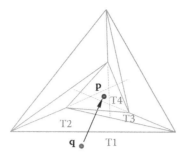

Figure 3.26 Searching along the line of intersection.

again. However, from the same figure, if point **q** starts the journey from position (3), it will go to triangles (4), (5), (6), (7) and (2), and then since (3) is already visited, it is forced to go to (1). Thereafter, point **q** will be separated from point **p** by the visited triangles, which effectively formed a closed ring.

Instead of recording the elements visited, recording the direction (edge) visited may help but still cannot solve the problem, as it is not sure which way to go if the choice is given. Suppose one starts from position (1) and completes a circuit and goes back to triangle (2); all edges have been tried, and again, one does not know what to do and where to go. A solution to overcome the difficulty of looping is by direct crossing from point **q** to point **p** following the line of intersection, as shown in Figure 3.26. The triangle T_1 containing **q** is known. The edge of T_1 that is hit by ray **qp** is determined. Go across the edge along ray **qp** to triangle T_2. Repeat the process until T_4 and point **p** are reached. As it is not often to have a closed circuit, a simple solution both in terms of practical implementation and CPU time is the approach by random walk. Instead of following a systematic rule in approaching **p**, whenever more than one choice is given, invoke the random number generator to decide which way to go. Owing to the random nature of the approach, point **q** will not loop forever in a closed circuit. Implementation experience with work examples of triangulation random samples of more than 1 million points demonstrates that it is efficient and reliable.

Apart from the random walk, another possibility is to change the starting triangle after a certain number of trials. Let n be the average number of visits to locate the BASE in the previous point insertions. If the BASE still cannot be found after, say, 2n visits, change the starting triangle from the last triangle to the second last triangle, and repeat the search from there. As the number of triangles is finite, this procedure of changing the starting triangle will always converge.

3.5.4.5 Steps in locating the BASE

1. Start the search from the last triangle constructed, T.
2. Compute the area co-ordinates of point **p** relative to triangle T.
   The area co-ordinates of **p** with respect to triangle T are

$$L_1 = \frac{\Delta(\mathbf{p}\mathbf{p}_2\mathbf{p}_3)}{\Delta}, \quad L_2 = \frac{\Delta(\mathbf{p}_1\mathbf{p}\mathbf{p}_3)}{\Delta}, \quad L_3 = \frac{\Delta(\mathbf{p}_1\mathbf{p}_2\mathbf{p})}{\Delta}$$

where $\Delta = \Delta(\mathbf{p}_1\mathbf{p}_2\mathbf{p}_3)$ is the area of triangle T with vertices $\mathbf{p}_1$, $\mathbf{p}_2$ and $\mathbf{p}_3$, as shown in Figure 3.27.

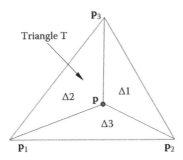

Figure 3.27 Area co-ordinates of **p**.

3. If L_1, L_2 and L_3 are all positive or zero, then the triangle T contains point **p**; otherwise, determine how to update the triangle T. Check how many area co-ordinates are negative.
   a. One negative value, say, L_i, i = 1, 2 or 3. Then update $T \mapsto T_i$, where T_i is the ith neighbour of T opposite to the ith vertex.
   b. Two negative values, say, L_i and L_j, i, j = 1, 2 or 3. Then, randomly select between i and j, and update T in the same way as step 3a.
       In case the scheme of changing the starting triangle is employed, then choose the most negative value between L_i and L_j, and update T in the same way as step 3a.
4. Go back to step 2. For the scheme of changing the starting triangle, check the number of triangles visited to see whether the search has to be started with a new triangle.

3.5.4.6 Circumcentre and circumcircle

Starting from the BASE, the CORE can be easily established by applying the empty-circle test to the neighbouring triangles. Continue applying the test to more adjacent triangles until all non-Delaunay triangles are determined. Following the lemma of Delaunay, a valid insertion CORE consists of non-Delaunay triangles in a connected piece whose boundary is made up of edges shared by two triangles that give a positive and negative response to the circle inclusion test.

The circum-radius of a triangle is given in Appendix A.3. However, the circumcentre is also required in the empty-circle test; the circum-radius and the circumcentre are determined directly using vector algebra from the co-ordinates of a triangle ABC, as shown in Figure 3.28. Let AD be the diameter of the circumcircle passing through point A, and denote vectors AB, AC and AD, respectively, by

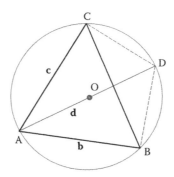

Figure 3.28 Circumcentre and circum-radius of a triangle.

b = AB, **c** = AC and **d** = AD

Since vectors **b** and **c** are linearly independent, there exist real numbers β and γ such that

$$\mathbf{d} = \beta\mathbf{b} + \gamma\mathbf{c}$$

$$AB + BD = AD \Rightarrow BD = AD - AB = \mathbf{d} - \mathbf{b}$$

$$AC + CD = AD \Rightarrow CD = AD - AC = \mathbf{d} - \mathbf{c}$$

Since AB ⊥ BD and AC ⊥ CD, we have

$$AB \cdot BD = 0 \Rightarrow \mathbf{b} \cdot (\mathbf{d} - \mathbf{b}) = 0 \Rightarrow \mathbf{b} \cdot \mathbf{d} = \mathbf{b} \cdot \mathbf{b} \Rightarrow \beta\mathbf{b} \cdot \mathbf{b} + \gamma\mathbf{b} \cdot \mathbf{c} = \mathbf{b} \cdot \mathbf{b}$$

$$AC \cdot CD = 0 \Rightarrow \mathbf{c} \cdot (\mathbf{d} - \mathbf{c}) = 0 \Rightarrow \mathbf{c} \cdot \mathbf{d} = \mathbf{c} \cdot \mathbf{c} \Rightarrow \beta\mathbf{b} \cdot \mathbf{c} + \gamma\mathbf{c} \cdot \mathbf{c} = \mathbf{c} \cdot \mathbf{c}$$

Solving for β and γ, we have

$$\beta = \frac{(\mathbf{c} \cdot \mathbf{c})(\mathbf{b} \cdot \mathbf{b} - \mathbf{b} \cdot \mathbf{c})}{\Delta}, \gamma = \frac{(\mathbf{b} \cdot \mathbf{b})(\mathbf{c} \cdot \mathbf{c} - \mathbf{b} \cdot \mathbf{c})}{\Delta}, \text{ where } \Delta = (\mathbf{b} \cdot \mathbf{b})(\mathbf{c} \cdot \mathbf{c}) - (\mathbf{b} \cdot \mathbf{c})^2$$

The circumcentre O of triangle ABC is given by

$$O = A + \frac{1}{2}\mathbf{d}$$

and the circum-radius r is given by

$$r = \|OA\| = \|OB\| = \|OC\|$$

3.5.4.7 Procedure for the creation of the CORE

The fact that $\mathbb{C}_p$ and $\mathbb{B}_p$ occupy the same space and share a common boundary allows us to construct $\mathbb{B}_p$ rapidly and systematically based on $\mathbb{C}_p$. Given the point inserted **p** and the BASE triangle, the CORE can be established by setting up an initial construction front, which consists of the three edges of the BASE triangle, as shown in Figure 3.29. Each segment on the construction front has to be checked whether it is a boundary edge of the CORE

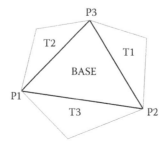

Figure 3.29 Initial boundary of the CORE.

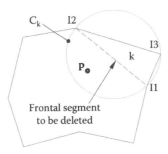

Figure 3.30 Finding the CORE.

or not. If the triangle associated with this edge is zero (no neighbour) or the triangle connected to this edge is Delaunay with respect to the inserted point **p**, then this is a boundary edge of the CORE. Otherwise, the associate non-Delaunay triangle has to be deleted, and the number of frontal segments will be increased by one, replacing the edge under consideration with two edges, which are the other two sides of the associated triangle deleted, as shown in Figure 3.30. This procedure is repeated on each segment of the construction front until there is no more segment left in the set. The following are the detailed steps of this algorithm.

Input: the point inserted, the BASE triangle, the current triangulation
Output: list of boundary segments and the triangles deleted
1. Set up the initial front of the CORE.
    The frontal segments initially are those from the BASE triangle, each of which is yet to be determined whether it is a boundary edge of the CORE. As shown in Figure 3.29, for instance, if triangle T3 is to be retained, P1–P2 will be a boundary segment of the CORE.
    Initial number of boundary segment, m = 0
    Number of frontal segment, n = 3

$P1 = V_{BASE}^1$; $P2 = V_{BASE}^2$; $P3 = V_{BASE}^3$; $T1 = T_{BASE}^1$; $T2 = T_{BASE}^2$; $T3 = T_{BASE}^3$
$F_1^1 = P2$; $F_1^2 = P3$; $F_1^3 = T1$; $F_2^1 = P3$; $F_2^2 = P1$; $F_2^3 = T2$; $F_3^1 = P1$; $F_3^2 = P2$; $F_3^3 = T3$

where P1, P2, P3 and T1, T2, T3 are, respectively, the vertices and the neighbours of the BASE.
    Delete the BASE triangle from the triangulation.
2. Check whether the last frontal segment is a boundary segment or not.
    Triangle associated with the nth frontal segment, $k = F_n^3$
    Vertices of the nth frontal segment are $I1 = F_n^1$ and $I2 = F_n^2$
    If $(k = 0$ or $\mathbf{p} \notin C_k)$ then
    $m = m + 1$; $B_m^1 = I1$; $B_m^2 = I2$; $B_m^3 = k$; $n = n - 1$;
    If $(n > 0)$ go to (2) else return;
    End if
    I3 = vertex of triangle k opposite to edge I1 – I2
    If $(\Delta(\mathbf{p}, I1, I3) < 0$ or $\Delta(\mathbf{p}, I3, I1) < 0)$ then
    $m = m + 1$; $B_m^1 = I1$; $B_m^2 = I2$; $B_m^3 = k$; $n = n - 1$;

If $(n > 0)$ go to (2) else return;
End if
$F_n^2 = I3$; j = neighbour of triangle k opposite to node I2; $F_n^3 = j$
$n = n + 1$; $F_n^1 = I3$; $F_n^2 = I2$; j = neighbour of triangle k opposite to node I1; $F_n^3 = j$
Delete triangle k; go to (2);

Notations used: p = point inserted; C_k = circumcircle of triangle k; $\Delta(I1, I2, I3)$ = area of triangle I1–I2–I3; $V_i^a = a^{th}$ vertex of triangle i, a = 1,2,3; i = 1,N; $T_i^a = a^{th}$ neighbour of triangle i; $\left(F_i^1, F_i^2\right) = i^{th}$ segment of the construction front; F_i^3 = triangle connected to frontal segment i; $\left(B_i^1, B_i^2\right) = i^{th}$ segment of the CORE boundary; B_i^3 = triangle connected to boundary segment i.

Remarks: Apart from the verification of the Delaunay property of the triangle associated with the frontal segment, a visibility check on the other segments of the associated triangle is also carried out to ensure that boundary segments are visible to the insertion point p. One of the major difficulties in the implementation of the DT algorithm by point insertion is to ensure consistency in the circle inclusion test based on finite precision arithmetic. An inconsistent decision in the circle inclusion test may result in a disconnected CORE, as described in Section 3.5.4.3. Imprecise numerical calculation may also introduce triangles with negative area.

DT is not unique unless all the points are in general positions. On a 2D plane, a degeneracy occurs when four points are cyclic giving rise to two different DTs depending on how these four points are connected into triangles, as shown in Figure 3.31. When a fifth point p is introduced, difficulty may arise unless both existing triangles are removed. As shown in Figure 3.31, the algorithm will generate a consistent triangulation if both triangles **abc** and **acd** are regarded as non-Delaunay with respect to point p. A consistent triangulation will also be produced if neither triangle is removed upon the introduction of point p. However, if triangle **acd** is deleted but not triangle **abc**, then the new triangulation will be structurally inconsistent. This situation may arise when four points are cyclic or almost cyclic, because finite precision arithmetic that is used in the Delaunay test may cause one of the two triangles to be accepted into the cavity list.

Using higher precision arithmetic can only postpone the problem to solve more cases, which are near to degeneracy, but not solve it entirely. Baker (1992) proposed a solution in which tolerances depending on the value of the data and the machine precision are applied to all real number calculations. Empty-circle tests based on double precision computations (8-byte variables) can only allow us to obtain a triangulation, which is close to DT.

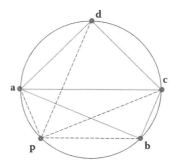

Figure 3.31 Correction of the CORE.

3.5.4.8 Correction of the CORE

Inconsistency of the circle inclusion test is to be supplemented and corrected by the visibility check or the positive volume test. As shown in Figure 3.31, an edge **ba** is visible to point **p** if $\Delta(\mathbf{bap}) > 0$. The boundary of the CORE is a valid one if all the edges on the boundary give a positive response to the visibility test. Suppose we have an inconsistent inclusion test for the cyclic points **a**, **b**, **c**, **d**, such that with respect to point **p**, triangle **acd** is deleted, and triangle **abc** is retained, as shown in Figure 3.31. In this case, the CORE is not a connect piece since it is separated into two parts by triangle **abc**. For the portion of the CORE that contains **p**, there is no problem for the visibility test on the boundary edges. However, as for the boundary edges due to the removal of triangle **acd**, at least one of the edges will form a triangle with a negative area, which is definitely invalid and should be corrected. The boundary edge that forms a triangle of a negative area with point **p** has to be restored. To this end, the triangle connected to this edge has to be reinstated, and the boundary edges are updated taking into account the inclusion of this triangle.

Poorly shaped triangles can also be avoided by the visibility test if it is insisted that their areas have to be greater than a certain threshold value. The condition that every edge on the boundary of the CORE is visible to point **p** also guarantees that the CORE is a single connected piece. The visibility check, albeit simple, is a necessary and sufficient condition to ensure that the CORE is a star-shaped polygon with respect to the insertion point **p**. In fact, other inclusion criteria can be adopted as to which triangles are to be deleted by the insertion point **p** to obtain different triangulations; the closer to the empty-circle criterion, the closer the resulting triangulation to DT. In the extreme case, upon the introduction of point **p**, only one triangle, the BASE triangle that contains point **p**, is deleted. Of course, by this simple rule, one is far away from the DT. Although a valid triangulation of the point set can be obtained rapidly, the triangles so generated are flat and almost degenerated. In summary, the visibility check ensures the validity of the triangulation, and using higher precision arithmetic in the circle inclusion test guarantees that the resulting triangulation is as close to a Delaunay one as possible.

In some rare occasions, an existing point is found at the interior of the CORE as shown in Figure 3.32. As all the triangles connected to the point are removed, this point will also be removed by the newly inserted point. A remedy to this situation is to restore a non-Delaunay triangle connect to the point that is also visible to the inserted point but just barely failed in the Delaunay test. The visibility test that follows will restore more triangles if necessary to ensure the visibility of the CORE and that each triangle will have a positive area.

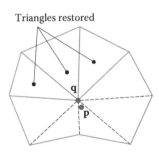

Triangles restored

Figure 3.32 An existing point **q** is deleted by the inserted point **p**.

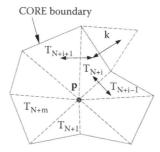

Figure 3.33 Forming triangular elements within the CORE.

3.5.4.9 Construction of triangles in the CORE and
establishment of the adjacency relationship

Given the boundary edges of the CORE, construction of triangular elements within the CORE is a trivial process, as shown in Figure 3.33. Triangular elements are formed by connecting each edge on the boundary in turn and the number of triangles created exactly equal to the number of edges on the CORE boundary. Let $\{(B_i^1, B_i^2), i = 1, m\}$ be the boundary edges of the CORE; the triangles created are given by

$$\{(V_{N+i}^1 = B_i^1, V_{N+i}^2 = B_i^2, V_{N+i}^3 = p), i = 1, m\}$$

N is the number of triangles in the triangulation.

Knowing the associated triangle of each boundary segment of the CORE, it is not difficult to establish the adjacency relationship of the elements inside the CORE and those outside the CORE. The adjacency relationship of triangle T_{N+i} connected to segment i can be established as follows:

$$k = B_i^3, T_{N+i}^1 = N+i+1, T_{N+i}^2 = N+i-1, T_{N+i}^3 = k, T_k^a = N+i$$

where index a corresponds to the ath vertex of triangle k opposite to the boundary edge.

Finally, a correction in the adjacency relationship has to be made between the first and the last triangular elements formed in the CORE as they share a common edge, i.e.

$$T_{N+1}^2 = N+m, T_{N+m}^1 = N+1 \text{ and N is updated to } N = N+m$$

Remarks: Triangles deleted in the CORE are simply flagged but not removed entirely from the triangulation right away. A consolidated list of triangles in the triangulation can be conveniently done at the end of the insertion process when all the points have been inserted.

3.5.5 Details in computer programming

The insertion algorithm can be conveniently programmed in FORTRAN or C++ computer languages. Many algorithms presented in this book have been programmed in Intel FORTRAN VS2010 under operating system Windows 7, which runs on 64 bits for rapid access to the 16-MB RAM on a PC and supports OpenMP parallel directives on a shared memory basis. The DT of a point set by the insertion algorithm can be implemented by proper coding of the insertion kernel, which consists of two parts – the *Base* and the *Core*. The function of the *Base* is to locate the BASE triangles in a triangulation and that of the *Core* is to determine the CORE and form new triangles with the insertion point.

Hence, the function of the insertion kernel is to produce a new triangulation ($\mathbb{T}_k$) of k points from the current triangulation ($\mathbb{T}_{k-1}$) of k – 1 points with the introduction of point P_k, i.e.

$$\mathbb{T}_k = \text{Insertion Kernel}(\mathbb{T}_{k-1}, P_k)$$

$$\text{BASE} = Base(\mathbb{T}_{k-1}, P_k) \text{ and } \mathbb{T}_k = Core(\mathbb{T}_{k-1}, P_k, \text{BASE})$$

Core can again be divided into two parts *Core1* and *Core2*, in which *Core1* is to determine the CORE and *Core2* is to form the patch of triangles with point P_k, i.e.

$$\text{CORE} = Core1(\mathbb{T}_{k-1}, P_k, \text{BASE}) \text{ and } \mathbb{T}_k = Core2(\mathbb{T}_{k-1}, P_k, \text{CORE})$$

In computer programming, a triangulation can be defined by two arrays:

i. $V_i^a = a^{th}$ vertex of triangle i, a = 1,2,3; i = 1,NT.
ii. (X_i, Y_i), i = 1,NN are the (x,y) co-ordinates of the points to be inserted.
   NN is the number of points to be inserted and NT is the number of triangles in a triangulation, which increases from 2 for the initial triangulation of a rectangle of two triangles ($\mathbb{T}_0$) to finally NT when all the points are introduced.
   There are a number of working arrays in the construction of the DT. The adjacency relationship, which is not the fundamental part of a triangulation, can be derived from the triangulation based on the vertices of the triangles, as discussed in Section 2.5.1. However, as it is often referred to in the construction of DT, it is computed simultaneously in the formation of new triangles, and is stored under a separate array.
iii. $T_i^a = a^{th}$ neighbour of triangle i, a = 1,2,3; i = 1,NT.
   While the BASE is an integer, a temporary array is needed to store the CORE in terms of its boundary edges and the associated triangles of the edges.
iv. $(B_i^1, B_i^2) = i^{th}$ edge of the CORE boundary; $B_i^3 =$ associated triangle, i = 1,n.
   n is a variable, which is quite small and independent of the number of points to be inserted, NN.

The circumcentres and the circum-radius squared of the triangles are required in the empty-circle test, and it is recommended to be stored in real number arrays.

$(C1_i, C2_i) =$ circumcentre of triangle i; $R_i =$ circum-radius squared of triangle i, i = 1,NT.

However, as the floating point calculation of PC becomes faster and faster, the bottleneck for the development of large-scale problems is the memory requirement rather than the speed of computation, especially in the memory management for parallel programming introduced in Chapter 7. For randomly generated points and in most practical cases, the complexity of the construction of DT is basically linear, and hence, computation time is not a major problem. In the current computer implementation, the circumcentre and the circum-radius of triangles are not stored, and they are computed whenever necessary. Little difference in computation time is observed between storing and not storing the circumcentre and the circum-radius. Without storing the circumcentre and the circum-radius, 50 million 3D points can be inserted quite comfortably by a parallel insertion based on 16 MB on a PC.

3.5.6 Generation of interior points

The point insertion algorithm and the insertion kernel provide a powerful tool for the construction of the convex hull of DT of a given point set. However, for MG over 2D domains with a well-defined boundary, two more issues are needed to be addressed: (i) the generation of interior points in compliance with the required element-size specification; and (ii) the boundary edge recovery so that all the boundary edges of the given object are present in the triangulation. These two important aspects are to be discussed in Sections 3.5.6.1–3.5.6.5.

Before describing the various methods for the creation of interior points, the specific requirements in the generation of internal points have to be made clear. The geometrical requirement is to create points inside the given domain to produce triangles as equilateral as possible. However, there are other physical requirements and constraints requiring points to be distributed according to a specified nodal spacing, which can be isotropic or anisotropic.

3.5.6.1 Specification of nodal spacing

Nodal spacing is also known as point or element size distribution, which specifies the distance between two points or the size of the resulting FEs anywhere within the given domain to be meshed. The simplest mesh required is the boundary mesh in which an FE mesh is produced based solely on the boundary points, and this is the coarsest (minimum) mesh that can be generated with respect to the given boundary of the domain. In case more triangular elements are needed over empty parts within the domain, a mesh of uniform element size can be generated with the aid of a background grid. For more complicated node distributions with an isotropic or anisotropic variation of element size over the domain arising from, for instance, adaptive refinement analysis, a more general way to specify the element size distribution is required.

3.5.6.2 Control space

The idea of *control space* introduced in Section 2.3.8 is to facilitate MG and the creation of internal points within the problem domain under consideration. The control space provides information whether a given point is inside or outside the problem domain and about the nodal spacing requirement at that point. For this purpose, the control space is a cover of the problem domain so that any point of the problem domain can be found also in the control space. It is natural to use a previous mesh or the minimum mesh as the control space; however, a background grid large enough to cover the problem domain can be used as the control space (Lo and Lee 1994). Based on the work of George and Borouchaki (1998), a control space is defined as follows.

Definition

$(\mathbb{C}, \rho)$ is a control space of a given domain Ω if

   i. Ω is covered by $\mathbb{C}$, i.e. $\Omega \subseteq \mathbb{C}$.
   ii. A function ρ to specify the nodal spacing over the control space $\mathbb{C}$

$$\forall \mathbf{x} \in \mathbb{C}, \rho(\mathbf{x}): \mathbb{C} \rightarrow \mathbb{R}$$

In case of anisotropic element size distribution, the nodal spacing function ρ has to be extended from a scalar quantity to a metric tensor or matrix such that the element size along a particular direction specified by unit vector **u** is given by (Section 4.2.11)

$$\rho = \mathbf{u}^T \mathbf{M} \mathbf{u} = \lambda_1 (\mathbf{u} \cdot \mathbf{u}_1)^2 + \lambda_2 (\mathbf{u} \cdot \mathbf{u}_2)^2$$

where unit vectors $\mathbf{u}_1$ and $\mathbf{u}_2$ are the principal stretch directions (eigenvectors) of metric tensor M, and λ_1 and λ_2 are, respectively, the required element size along these directions. As $\mathbf{u}_1$ and $\mathbf{u}_2$ are orthogonal vectors, it is easy to verify that an element size of λ_i is required if **u** is in the direction of vector $\mathbf{u}_i$. M is a symmetric tensor with positive eigenvalues and orthogonal eigenvectors. It is, in general, a continuous function over the control space $\mathbb{C}$, which can be specified analytically, but it is more often that M is defined only at some nodal points of the control space $\mathbb{C}$ for which intermediate values of M are obtained by means of interpolation.

The nodal spacing function ρ can be specified as a continuous analytical function to create certain node distribution patterns to test various MG algorithms, for example, line distribution, cross distribution, spiral distribution and cluster distribution. In practical FE MG, the nodal spacing function can be specified entirely based on the geometry of the problem domain. When one does not have much idea about the element size of the first mesh, the nodal spacing function can be defined based on the length of the boundary edges of the problem domain, as described in Section 3.5.6.3. The element size can also be estimated based on previous analyses such that smaller elements will be used in an area of high-solution gradient so as to capture the characteristics of the solution with a minimum number of elements, and this procedure is known as FE adaptive refinement analysis presented in Section 8.9.

3.5.6.3 Element size based on domain boundary

From a geometrical point of view, element size can be related to the line segments of the boundary of the domain. One has to devise a way to determine the element size at an interior point of the domain based on the characteristics of the boundary segments. Realising that a small element size has to be used in areas close to boundary segments of short lengths, the distance effect from a boundary segment has to be carefully interpreted in order to obtain a mesh with smooth transition in element size from one end of the domain to the other. For a planar domain Ω with a boundary composed of N_b line segments $\Gamma = \{A_i B_i, i = 1, N_b\}$, the element size or node spacing ρ at any point P(x,y) can be estimated by

$$\rho(x,y) = \frac{\displaystyle\sum_{i=1}^{N_b} \frac{\rho_i}{r_i}}{\displaystyle\sum_{i=1}^{N_b} \frac{1}{r_i}} \tag{3.1}$$

where ρ_i is the node spacing associated with segment i, which can be computed using

$$\rho_i = \|A_i B_i\| \quad i = 1, N_b$$

and r_i is the distance between point P to the midpoint (centre) C_i of segment $A_i B_i$, i.e.

$$r_i = \|P C_i\| \quad i = 1, N_b$$

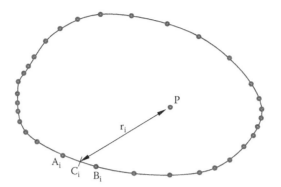

Figure 3.34 Determination of element size at point P.

The effect of a short segment is weighted by the distance away from it such that ρ will be small at the point near the boundary of clustered nodal points, and it will become larger at the point away from short boundary segments, as shown in Figure 3.34. If a background grid is used as the control space $\mathbb{C}$, the element size ρ is evaluated by Equation 3.1 at all the grid points of $\mathbb{C}$, and within the cells of $\mathbb{C}$, ρ is computed by interpolation. On the other hand, if a previous mesh is used as the control space $\mathbb{C}$, element size ρ is evaluated at all the nodal points of the mesh, and within the elements of $\mathbb{C}$, ρ is computed by FE interpolation.

3.5.6.4 Element size based on a previous analysis

The adaptive refinement procedure based on an h-version successive mesh refinement is very popular for its simplicity and efficiency, in which the element size is adjusted for each subsequent analysis, while the polynomial of the interpolation function is kept constant (Lo and Lee 1992). The linear elasticity problem is taken as an example to demonstrate how element size is to be defined based on a previous analysis.

For a linear elasticity problem, the objective of the adaptive refinement is to achieve a solution with relative error norm η smaller than some prescribed value η_0, i.e. $\eta < \eta_0$. As the exact energy norm is not available, at any stage of the adaptive refinement process, exact energy norm $\|\mathbf{u}\|$ is approximated by $\overline{\|\mathbf{u}\|}$ such that

$$\|\mathbf{u}\|^2 \approx \overline{\|\mathbf{u}\|}^2 = \|\hat{\mathbf{u}}\|^2 + \left(\gamma\overline{\|\mathbf{e}\|}\right)^2$$

where $\|\hat{\mathbf{u}}\|$ is the energy norm of the FE solution, and $\overline{\|\mathbf{e}\|}$ is the estimated error in the energy norm obtained from the FE stresses and some enhanced more accurate smoothed stresses. γ is an empirical correction factor suggested by Zienkiewicz and Zhu (1987, 1992), which is set, respectively, to 1.3 and 1.4 for linear triangular element T3 and quadratic triangular element T6. Hence, the relative error norm η is approximated by $\overline{\eta}$ such that

$$\eta \approx \overline{\eta} = \frac{\gamma\overline{\|\mathbf{e}\|}}{\overline{\|\mathbf{u}\|}}$$

To ensure that the error is equally distributed among the elements, a local allowable error e_a is defined as

$$e_a = \eta_0 \left[\frac{\overline{\|u\|}^2}{N} \right]^{1/2}$$

where N is the number of elements in the mesh. It is hoped that

$$\gamma \overline{\|e\|}_i < e_a \text{ for } i = 1,N$$

in which $\gamma \overline{\|e\|}_i$ is the estimated error of element i. Based on the aforementioned quantities, the coefficient defined as

$$\xi_i = \frac{\gamma \overline{\|e\|}_i}{e_a}$$

is a natural indicator for mesh refinement. Let h_i be the current element size of element i; the desirable element size h'_i is given by

$$h'_i = \frac{h_i}{\xi_i^\mu}$$

where μ takes a value of 1/p (p = order of interpolation) if there is no singularity present or where the element is far away from singularity, and μ is set equal to $1/\lambda$ for elements close to a singularity of strength λ (Lee and Lo 1997a,b). If one sets μ equal to 1/p for all the elements, inadequate refinement will occur in areas close to points of singularity, and consequently, more refinement steps will be needed to achieve the target accuracy.

In the above procedure, element size is defined for each element of the mesh, and this information of element size requirement has to be transferred to the grid points of the control space. In the case of a structural background grid, all the grid points within a particular element of size h_i will receive the new element size requirement of h'_i. As for control space based on the last mesh, the element size requirement of h'_i can be directly assigned to all the node points of the element under consideration, and the desirable element size at a node equals the average value of the predicted element size from all the elements connected to it.

3.5.6.5 Creation of interior points

DT provides a general rule to connect points in 2D and in a higher dimensional space. In MG by the concept of DT, usually the boundary points are first inserted to produce an initial triangulation of these points. To complete the FE mesh, more points have to be inserted at the interior of the domain in conformity of the nodal spacing requirement, and the boundary edges have to be recovered so that all boundary edges of the given domain exist in the triangulation. The generation of an interior point will be discussed in Sections 3.5.6.5.1–3.5.6.5.3, and the problem of boundary recovery will be explored in Section 3.5.7. Several methods have been proposed for the creation of interior points for Delaunay meshing and other FE MG schemes, which could be grouped into different categories.

3.5.6.5.1 Refinement method

By means of the refinement method, additional nodes are created with respect to the existing triangular elements. Depending on the strategy and the locations for the creation of additional points, various refinement schemes for 2D and 3D DT have been proposed.

3.5.6.5.1.1 POINT CREATED AT CIRCUMCENTRE

Nodes are created at the circumcentre of those triangles that satisfy certain conditions such as surface area, in-radius, element aspect ratio, etc., to reduce FE solution error based on complementary variation principles (Shenton and Cendes 1985; Holmes and Snyder 1988).

3.5.6.5.1.2 POINT CREATED AT CENTROID

Nodes are created at the centroid of those triangles that are considered to be too large; sometimes, a node can also be positioned at a point that is a weighted sum of the vertices of the triangle (Hermeline 1980) or based on centroidal Voronoi tessellations (Du and Gunzburger 2002; Secchi and Simoni 2003; Du et al. 2010).

3.5.6.5.1.3 POINTS CREATED ON EDGES

Realising that inserting nodes at the interior of a triangle would very often produce triangular elements of poor quality, Borouchaki and George (1997) proposed the creation of points on the edges of the triangles. Points are generated along each edge of the triangles in turn following a geometric progression in compliance with the element size requirement, and those points that are too close to the existing points will be filtered away. Rivara and Inostroza (1997) suggested a refinement scheme of DT based on the longest-side bisection.

Remarks: The refinement methods only suggest where a node can be generated, and a background grid can be employed to provide information about the desirable element size and to control whether a proposed point is acceptable or it is rejected as it is too close to some existing points or simply it is outside the problem domain.

3.5.6.5.2 Use of background grid

The grid points of a background grid are potential positions for point creation in compliance with the given node spacing requirements. Those points can be rapidly and accurately defined without ambiguity for MG by means of the Delaunay insertion kernel or alternatively connected up at the interior of the domain by means of standard templates. Uniform rectangular and triangular grids can be used to generate elements of regular size, and in case of a variable element size requirement, Quadtree or kd-tree recursive spatial partitions can be applied.

3.5.6.5.2.1 REGULAR GRID

A regular grid of appropriate grid spacing can be superimposed with the given domain, as shown in Figure 3.35a. Points at grid positions with a sufficient distance from the boundary edges are inserted one by one to create a mesh of uniform node spacing. The proximity of a point to an edge can be verified by checking the area of the triangle formed between the given point and the boundary edge. As grid points of a rectangular grid are at the corner of a rectangular cell, these cyclic points will cause numerical problems to the Delaunay insertion

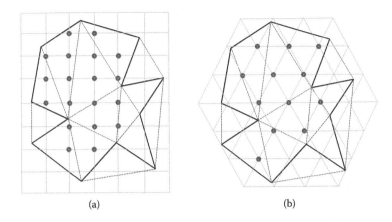

Figure 3.35 Creation of interior points by means of background grid: (a) rectangular grid; (b) triangular grid. Solid line = domain boundary, broken line = triangulation of boundary points.

kernel. Instead of point insertion, points at the interior of the domain can be connected up rapidly by means of standard templates. The remaining points near the domain boundary can be processed by the point insertion kernel. Alternatively, a triangular grid of equilateral triangles can be employed to produce better quality triangles without differentiation between internal nodes or boundary nodes in the insertion process by the Delaunay insertion kernel, as shown in Figure 3.35b (Lo and Liu 2002).

3.5.6.5.2.2 QUADTREE PARTITION

The idea of Quadtree partition of space has been introduced in Section 3.4. A rectangular box that is large enough to contain the given domain is created. A Quadtree partition of the rectangular box is carried out in compliance with the nodal spacing specification, as shown in Figure 3.36a. Corner points and mid-side points of the cells can be triangulated by standard templates or inserted by the Delaunay insertion kernel. As corner points of a cell are cyclic, in the insertion process, the centroids of the cells are also inserted to avoid numerical problems. However, similar to the regular grid, a recursively refined triangular grid analogous to the

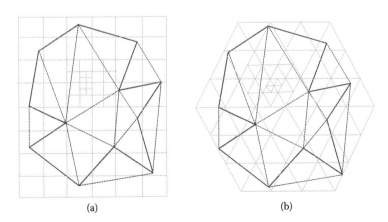

Figure 3.36 Creation of interior points by recursive spatial partition: (a) quadtree decomposition; (b) recursive triangular grid.

Quadtree subdivision can be used to produce triangles with better shape quality without causing numerical difficulties in the Delaunay insertion, as shown in Figure 3.36b.

3.5.6.5.2.3 RANDOM POINT GENERATION

Elements of uniform size will be generated using grid points of a regular grid, and elements of variable size can be created by a recursive subdivision of a rectangular or a triangular grid, as discussed in Section 3.5.6.5.2.2. MG by means of a random point was proposed many years ago by Fukuda and Suhara (1972) and Cavendish (1974) to generate triangles of roughly equal size but of rather arbitrary shapes. However, to generate elements of variable size using a regular grid without recursive spatial partition, the technique of random point generation can also be employed. Points are generated by a random process within a cell, which will be rejected if they are too close to the existing points within the cell or with respect to points of the adjacent cells. Each cell of the grid will be considered in turn, and the point insertion process will pass on to the next cell when an expected number of points have been generated. When all the cells of the grid have been considered, the domain will be filled up with randomly generated points in compliance with the element size requirement in a statistical sense. A grid with recursive spatial partition can also be used in conjunction with the random generation process to produce elements with size in better compliance with the specified node spacing.

3.5.6.5.2.4 'VARIOGRAM' – DISTANCE FROM BOUNDARY

For a planar domain bounded by line segments, points are inserted one by one within the given domain. An inserted point has to be located at a position that is farthest away from all the existing points, including the boundary points and the previously inserted points (Tacher and Parriaux 1996). In the actual implementation, a background grid with grid-point spacing compatible with the smallest element is prepared. The distance of each grid point from the boundary is computed by means of an induction formula based on the previous grid point. A *variogram* is formed by listing the distance of all the grid points in an ascending order of magnitude. Points are accepted as long as the distance to the boundary is greater than some prescribed threshold value.

3.5.6.5.3 *Other techniques*

Apart from the refinement methods and those based on a background grid, many other techniques have been proposed to generate interior points within a given domain. An exhaustive account on all these methods is quite impossible and impractical; apart from the packing of circles, ellipses and spheres (Lo and Wang 2005b,c,d), two more methods that make use of entirely different concepts in point creation are included for a detailed discussion in Sections 3.5.6.5.3.1 and 3.5.6.5.3.2.

3.5.6.5.3.1 CONTOUR LINE

Right from the beginning in the development of FE MG, two major trends were the structured and unstructured meshes. While various mapping methods were available for the generation of structured mesh, the generation of unstructured mesh was still at the infant stage. Apart from the mesh subdivision techniques (Thacker 1980), an early attempt for the automatic generation of unstructured mesh is by means of random point generation (Fukuda and Suhara 1972; Cavendish 1974), which was later modified by Shaw and Pitchen (1978) to generate nodes in a more systematic manner over rectangular cells. When the AFT was proposed in the classical paper by Lo (1985), interior points were generated with the aid of contour lines cutting across the planar regions. This method was also later extended to 3D to generate interior nodes for ADF MG of 3D objects.

Direct deterministic approach and speed are the major advantages of MG by means of contours to produce meshes of uniform element size. Although well-shaped triangles can be produced by using contour, with the emergence of more powerful computers for solutions of ever more challenging physical problems, meshes of non-uniform element size or even anisotropic meshes of directional size specifications are required. Nevertheless, high-quality meshes of uniform element size can be rapidly generated by this simple approach as a first mesh in an adaptive refinement analysis or for further processing.

Let Ω be a planar domain with a boundary composed of N_b line segments $\Gamma = \{A_iB_i, i = 1,N_b\}$. The following are the steps of the contour line method for the generation of interior nodes within Ω.

1. The minimum (y_{min}) and the maximum (y_{max}) of the y-value of the nodal points are determined.
2. Imaginary horizontal lines at different levels are drawn between y_{min} and y_{max} across the domain, as shown in Figure 3.37.
3. The spacing between any imaginary line is exactly equal to the average element size of the region.
4. Determination of the intersection of a horizontal line with the domain boundary is shown in Figure 3.37. The intersections between a horizontal line (y = h) and the domain boundary Γ are determined by considering the line segments of Γ one by one. Consider the intersection of horizontal line y = h with line segment A_iB_i as an example. Write $x_1 = x(A_i)$, $y_1 = y(A_i)$, $x_2 = x(B_i)$ and $y_2 = y(B_i)$. There is intersection if
   i. $(y_1 - h)(y_2 - h) < 0$
   ii. $(y_1 - h)(y_2 - h) = 0$ and $(h > y_1$ or $h > y_2)$

   It is considered as no intersection in other cases. The point of intersection is given by

$$\left(x_1 + \frac{h - y_1}{y_2 - y_1}(x_2 - x_1), h \right)$$

   As the boundary of a region is composed of closed loops, a horizontal line must cut the domain boundary at an even number of points, and the intersection points are arranged in an ascending magnitude of x, i.e. from left to right.

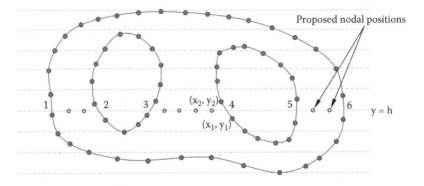

Figure 3.37 Intersection between horizontal lines and the domain boundary.

5. Assume that there are 2n cuts between the horizontal line y = h and the boundary Γ; the cuts are considered two by two, beginning with the first and second cuts. Nodes are generated on this horizontal line segment between the cuts according to the prescribed spacing. This only suggests a series of potential positions where points are to be generated; however, whether a node is finally accepted depends also on whether it is too close to the domain boundary. The distance from a point to a line segment is discussed in Section 2.4.1, and the result could be applied here to check if a point is too close to a boundary edge. In case a point P is too close to a line segment AB such that triangle PAB is likely to form in the mesh, one has to ensure that the quality of triangle PAB is higher than the expected threshold value. The quality measure of triangular and tetrahedral elements will be discussed in Section 6.2.1.3.

6. After the first two cuts, the process is continued with the third and fourth cuts in a similar manner until it terminates with the (2n – 1)th and the 2nth cuts, then one proceeds with the next horizontal line.

Remarks: The method of intersection by using contour lines is simpler and faster than the other methods, as previously generated nodes need not be taken into account in the generation of a new node since its spacing is guaranteed by the generation procedure. A vigorous check on domain boundaries is not necessary, and very often, a uniform distribution of nodes is resulted, which can be directly linked up layer by layer to form well-shaped triangles (Lo et al. 1982). In case the number of cuts recorded on a horizontal line is not an even number, a more robust method can be adopted to ensure that intersection points always come in pairs (Lo 1988a,b). The idea is to assign a value of +1 or –1 to each boundary point depending on whether it is above or below the horizontal line. If a boundary point is on the horizontal line, it doesn't matter whether it is given a positive value or a negative value. The intersection between a boundary edge and a horizontal line is reduced to checking the parities of the two nodal points of the boundary edge. The intersection is recorded if the nodal values are different, i.e. one node is +1, and the other node is –1; otherwise, if nodal values are equal, it is considered as no intersection. No matter how +1 and –1 values are assigned to boundary nodes, by going through segments of a closed loop, the change of parity (+1 and –1 values) is always an even number. However, in practice, a simpler perturbation method can also be adopted, i.e. the level of the horizontal line is slightly adjusted from y = h to y = h +Δh with a typical value of Δh = 1% of the interval between horizontal lines. Instead of cutting the domain with horizontal lines, an interior point can also be generated by cutting the domain with vertical lines. Usually, one can have better results by cutting the domain along the longer dimension, i.e. if $(y_{max} - y_{min}) > (x_{max} - x_{min})$, use horizontal cut lines; on the contrary, if $(y_{max} - y_{min}) < (x_{max} - x_{min})$, use vertical cut lines.

3.5.6.5.3.2 GENERATION OF INTERIOR NODES BY AFT

In terms of quality and speed, the AFT is perhaps one of the most efficient methods to create points in compliance with a given element size specification over 2D or 3D domains. As each point is optimised as far as possible in the ADF meshing, the quality of the elements is verified and guaranteed in the element construction process. The method is also very effective for the generation of isotropic mesh of variable element size (Lo 2013b) as well as meshes with severe anisotropic characteristics (Borouchaki et al. 1997a,b). Moreover, the DT and AFT can be merged harmoniously into one general scheme in which the merits of both methods can be fully exploited, namely, the point generation by the AFT and the point connection following the Delaunay criterion. The details of the ADF–Delaunay method and the Delaunay–ADF method will be described in detail in Section 3.7 after the introduction of DT in this section and AFT in Section 3.6.

3.5.7 Boundary recovery in two dimensions

For MG over a bounded planar domain with a well-defined boundary made up of a series of line segments by means of DT, the procedure can be divided into three distinct steps if boundary integrity is to be respected, namely, (i) insertion of boundary points, (ii) insertion of interior points and (iii) boundary recovery. The problem is known as constrained DT, which is also considered, in general, as a typical MG problem for various engineering applications. Full compliance of the boundary in all aspects is expected for a constrained DT. However, a solution is only proved to exist in 2D where the boundary consists of only edges. Over a 3D space, the boundary of a general polyhedron is made up of edges as well as faces, and it is not difficult to perceive that there is simply no solution for the partition of a twisted pentahedron without adding an interior node. Even if interior nodes are allowed, there is no guarantee with a formal proof that a solution exists for the most general case (Chazelle 1984; Ruppert and Seidel 1992). Hence, solutions in the form of a semi-constrained DT have been proposed (Weatherill 1988; Si 2010) in which the geometry is recovered and the topology is still missing, such that the edges are recovered as a list of line segments and the boundary faces are represented as a concatenation of triangular sub-faces. Nevertheless, theoretically sound heuristic approaches do exist for fully constrained 3D DT for most of the practical cases, which will be discussed in Section 5.3.

Strictly speaking, constrained DT is not a DT in which the Delaunay properties of some triangles are violated, at least locally, in order to satisfy the boundary constraints. Constrained DT is a triangulation derived from the DT of the boundary points and possibly some interior points by properly recovering all the boundary constraints with as little modification as possible. In 2D, such constrained triangulations exist based on the following facts: (1) any bounded region simply or multi-connected can be triangulated (George and Borouchaki 1998), and (2) any triangulation can be converted from one to another by merely swapping the diagonals (Fournier and Montuno 1984). Knowing the existence of solutions for the boundary recovery problems, many methods have been proposed in which boundary edges are recovered one at a time.

Algorithm for recovering an edge. An edge in the DT of the boundary nodes (possibly with some interior nodes) is a line segment connecting any two points in the triangulation, as shown in Figure 3.38. The line segment can be restored in two distinct phases: (i) determination of the polygon associated with the line segment, which is called a *pipe* in the literature

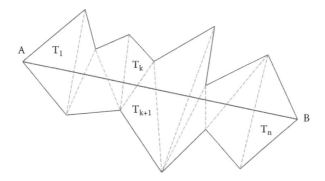

Figure 3.38 The pipe – polygon associated with line AB.

(George et al. 1991), and (ii) modification or reconnection of the triangles of the *pipe* to retrieve the line segment.

3.5.7.1 Determination of the pipe

Given two points A and B in the triangulation, find the *pipe*, which is the polygon of triangular elements intersected by line AB, as shown in Figure 3.38.

    i. Find out triangle T_1, which is connected to point A intersected by line AB.
    ii. From T_k, find triangle T_{k+1}, which is the neighbour of T_k intersected by line AB.
    iii. The *pipe* as a collection of triangles is determined if point B is the opposite node of T_{k+1}; else go to step (ii).

Remarks: For randomly generated points, there is a possibility that a point may lie exactly on the line segment by some numerical interpretation. In this case, the line segment has to be broken up into small sub-segments, each of which is to be recovered in turn. However, in realistic practical situations, no point is allowed to fall within a very small distance from a boundary line segment, and hence, there is always a solution for the boundary recovery problems.

Once the pipe or the associated polygon is determined, line segment AB can be recovered by re-triangulating the pipe. To this end, there are at least two methods that are proved to be pretty effective.

3.5.7.2 Divide-and-conquer

As shown in Figure 3.39, a *pipe* is divided by the line segment AB into two regions: polygon P1 and polygon P2. The two polygons can be dealt with in exactly the same manner by the same procedure. Take for instance the polygon P1 on the left-hand side of line AB. Among all the nodes above line AB, find the node, say C, which is closest to the line AB, as shown in Figure 3.39; in case there are more than one node, any such node can be taken. Form triangle ABC, and polygon P1 is divided effectively into three parts, namely, triangle ABC and two smaller polygons on the left-hand side and the right-hand side of triangle ABC, as shown in Figure 3.39. If C is next to point A or point B, only one polygon will be formed. The same process can be applied repeatedly to the polygons so created until each polygon is reduced to a triangle, and by then, polygon P1 will be triangulated with line AB on its

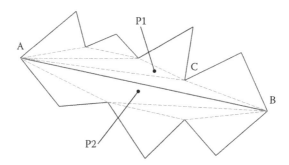

Figure 3.39 Triangulation of the pipe by divide-and-conquer.

boundary. Line AB is retrieved, and the *pipe* is triangulated when the divide-and-conquer process is also applied to polygon P2 on the other side of line AB.

3.5.7.3 Swapping of diagonals

As mentioned in Section 3.5.7.2, one triangulation can be transformed into any triangulation by merely swapping the diagonals between two adjacent triangles. Hence, line AB in the *pipe* can be recovered by swapping the diagonals of adjacent triangles of the *pipe*. However, even if a solution exists, one doesn't know how to arrive at it, i.e. one doesn't know the correct sequence of the diagonal swaps so that line AB can be recovered for the most general case. George and Borouchaki (1998) proposed a swapping process by taking a pair of triangles at random until line AB is recovered. However, a more systematic approach can be formulated, which can be proved to give convergent results (Lo and Liu 2002).

Starting from the first two triangles T_1 and T_2, swap the diagonal within the quadrilateral so formed, as shown in Figure 3.40. Remove the triangle not intersected by line AB, and the number of triangles in the *pipe* is reduced by one. Repeat the same process with the diminished polygon, and if both triangles are intersected by line AB after the diagonal swap or in the case of a concave quadrilateral where diagonal swap is not allowed, pass on to the next pair of triangles. When one comes to the end of the *pipe*, start the entire process over again with the first two triangles. This process has been proved to be convergent, and finally, the *pipe* is reduced into a quadrilateral with line AB on the main diagonal, as shown in Figure 3.41.

Remarks: Comparing between the two methods, the diagonal swap for a recovering line segment is probably simpler as no new object is introduced in the process. The basic idea to implement the diagonal swap algorithm is to maintain a correct sequence of triangles for the shrinking *pipe* when triangles are taken away until the *pipe* is reduced to a quadrilateral. A sequence of contiguous triangles from point A to point B is needed as this would allow adjacent triangles to pair up easily to form quadrilaterals for diagonal swap.

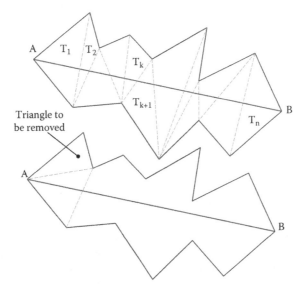

Figure 3.40 Recovering line segment by diagonal swap.

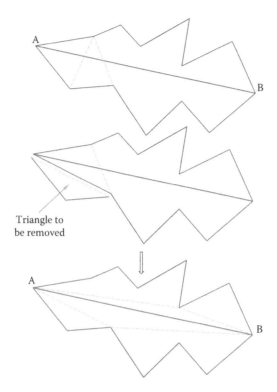

Figure 3.41 The *pipe* is reduced to a quadrilateral.

A DT of 1000 randomly generated points is shown in Figure 3.42, in which two points are picked at random, and the *pipe* of all intersected triangles is determined. The line connecting the two selected points is recovered by the diagonal swap, as shown in Figure 3.42a and b. Line segments recovered by the divide-and-conquer process are shown in Figure 3.42c and d, in which Figure 3.42c shows the intermediate result of one line segment recovered and the *pipe* of another line segment just formed, and the final result for the retrieval of the two line segments is shown in Figure 3.42d. As boundary segments are non-intersecting, using either one of the line recovering methods, the boundary of the planar domain composed of a series of line segments can be easily recovered by taking one edge at a time.

3.5.8 Closure

DT is one of the earliest approaches employed for the generation of unstructured mesh over planar domains and in higher dimensions. With a sound geometrical proximity concept and many optimality properties such as the maximisation of the minimum angle of DT in 2D, robust algorithms can be devised to generate a triangular mesh of well-shaped elements for numerical analysis. The point insertion algorithm of Bowyer and Watson provides a systematic way to introduce a new point in an existing DT to produce a new DT including the inserted point. The point insertion kernel allows points to be processed one after the other to generate DTs on planar domains and in higher dimensions in a flexible and efficient manner.

Triangulation methods only provide a way to link up points to form a triangulation. However, to complete the task of FE MG, two important issues needed to be addressed, namely, the creation of interior points in compliance with the specified node spacing and the

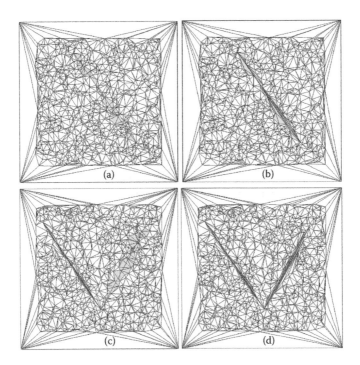

Figure 3.42 Segments recovered by diagonal swap (a, b) and by divide-and-conquer (c, d).

recovery of the domain boundary. Many methods have been proposed for the generation of interior points, which are proved to be effective for Delaunay meshing. As in 2D, any triangulation of a point set can be produced by simple swapping of diagonals of adjacent triangles; the boundary edges of a planar domain can be easily recovered by a sequence of diagonal swaps. Nevertheless, in order to obtain triangular elements of better shape for numerical analysis, the triangles intersected by the boundary edges can be removed, and a local MG process can be applied to fill up the voids, as shown in Section 3.8.

In most practical situations of generating meshes for more or less uniform nodal points, the time complexity of DT is almost linear based on a regular grid insertion algorithm (Lo 2013a). As for highly non-uniform point distributions, it may be a couple of times slower than that of the uniform point distribution of the same number of points. Nevertheless, based on the idea of recursive grid insertion, the CPU time for the generation of highly non-uniform distributed points is not excessive, which exhibits only a mild non-linear growth in the CPU time in practical MG problems for the insertion of more than 100 million points. Details for the DT of non-uniform point distributions will be given in Section 8.2. DT coupled with a generic boundary recovery module can be a powerful MG tool for its robustness and the speed and ease in handling a large quantity of points in two or higher dimensions.

3.6 ADVANCING FRONT APPROACH

3.6.1 Introduction

The AFT, which debuted from the classical paper of Lo (1985), was proposed to generate triangular meshes with pre-placed interior nodes over planar domains of arbitrary boundary characteristics. The method was rapidly developed in the FE communities; it was extended by Zhu et al. (1991a,b), Blacker and Stephenson (1991) and Blacker et al. (1991) to generate unstructured quadrilateral meshes and by Lohner and Parikh (1988), Peraire et al. (1988), Lo (1991b,c), Jin and Tanner (1993), Lohner (1994, 1995, 1996a,b, 1997) and Moller and Hansbo (1995) to generate tetrahedral meshes over 3D domains. The method could also be used to generate FE meshes directly over curved surfaces (Lau and Lo 1996; Lo and Lau 1998; Ito and Nakahashi 2002) or indirectly through a parametric mapping from 2D meshes generated based on an anisotropic metric computed from the surface curvatures (Lo 1988c; Borouchaki et al. 1997a,b; Lee 2003a,b).

It is interesting to note that the AFT can be merged with the well-known DT (George and Seveno 1994) in a harmonious manner to improve efficiency and robustness of the method especially for generating anisotropic meshes over surfaces and volume (Frey et al. 1998; Borouchaki et al. 2000a). There are two possibilities in such a combination: (i) advancing front–Delaunay approach (Frey et al. 1998; Borouchaki et al. 2000a) – in a classical DT procedure, points are created by the ADF strategy; and (ii) frontal Delaunay approach (Lo 1989a) – the ADF procedure is modified such that the Delaunay connection criterion is taken into account to reduce the check for intersections in the phase of element creation. Other applications of the AFT include MG over objects composed of a variety of surfaces derived from surface intersections (Lo 1995; Shostko et al. 1999), to reconstruct surfaces from a seed triangle taken from a collection of triangular facets so as to study its geometry and topology (Lo 1998b), to fill up space with arbitrary-shaped objects (Lohner and Onate 2010), to generate unconstrained hex meshes (Staten et al. 2010a,b) and to build geometric models for granular structures in 2D and 3D to simulate nanostructures (Benabbou et al. 2010).

The AFT is known to be robust and versatile over domains of different dimensions with diverse geometrical and topological characteristics (Owen et al. 1998; Wang et al. 1999), and is able to produce elements of various shapes such as triangles, quadrilaterals, tetrahedra and hexahedra close to the well-shaped ideal geometry in compliance with the specified node spacing function (Owen 1999; Owen et al. 1999; Owen and Saigal 2000). However, the main difficulty with the AFT is its efficiency in dealing with a large number of elements. Whenever a new element is formed with a node on the generation front or with an inserted interior node, one has to ensure that the segments created do not cut into the generation front. As the generation front during the meshing process is quite unpredictable and is allowed to evolve in whatever ways necessary for the generation of the most optimal elements, to ensure no intersection, a rigorous search has to be performed over all the line segments on the generation front, which is obviously quite a tedious time-consuming process. Nevertheless, the searching process could be made localised, and some parallel procedures have been proposed to speed up the method in which the domain to be meshed is subdivided spatially into the cells by means of the Octree technique (Rassineux 1998; deCougny and Shephard 1999a,b).

Apart from a regular grid, the Quadtree partition of space is often used in MG over 2D domains. However, the main purpose is to provide a static background grid for a given object, which can be refined locally to capture the boundary and features to any desirable level of details. Based on a set of standard templates for inclined edges, quadrilateral elements can be generated to fit boundaries (Liang et al. 2009; Ebeida et al. 2010). Adaptive refinement by means of progressive Quadtree subdivision can be applied to problems governed by the Poisson equation over a regular planar domain (Tabarraei and Sukumar 2005). Regular grids can also be employed to store nodes and other geometrical quantities to facilitate searching for the intersection of triangular facets (Ito and Nakahashi 2002). The intersected triangles are removed leaving gaps that could be filled up with new triangles by means of the AFT (Chand 2005). Anisotropic quadrilateral elements can be generated by an indirect approach based on a triangulated surface (Merhof et al. 2007). Discrete fracture modelling by unstructured triangular mesh was generated by the Delaunay–AFT with data points being stored in a uniform grid (Sahimi et al. 2010). An Octree partition for parallel meshing rather than to facilitate local searching has also been proposed (Lohner 2001). The Octree subdivision has to be updated as the generation front progresses, and the sizes of the boxes and the smallest element have to be carefully controlled to ensure that the elements do not protrude into other zones. Octree as a background grid can also be used to discern boundaries of different material types, and tetrahedral elements are generated at the boundary between materials (Zhang et al. 2010). From this brief survey of recent publications on the use of background grids, it is clear that most of the applications are made to provide support for some geometrical issues, and very often the grid will turn into a mesh or part of the final mesh after some local modifications. The Delaunay–AFT has been developed to generate high-quality tetrahedral meshes (Frey et al. 1998) and anisotropic meshes specified by a general metric for computational fluid dynamics (Borouchaki et al. 1997b). In the Delaunay–ADF meshing, the DT, which has been generated earlier, in fact, serves as an unstructured background grid (control space) to generate interior points by the AFT. A grid is often employed to reduce the CPU time for searching and matching of geometrical quantities; however, it is usually used in a static manner under the condition that the position of every item is known *a priori*.

After a brief introduction of the basic idea of AFT, the procedure of ADF meshing will be presented in 10 steps. In Section 3.6.2, the generation of adaptive mesh governed by a specified nodal spacing by the AFT will be discussed. The implementation details for each step of the AFT will also be given so as to explain when and why a search on the generation front would be needed. In order to speed up the check for intersections between the proposed element and the generation front, which is perhaps the weakest part in the ADF approach, the idea of using a dynamic grid is presented in Section 3.6.3. The setting of the background grid is to facilitate a search on the generation front for possible intersections in which details will be given on how the size of an individual cell in the partition can be determined, and formulas and procedure as to how segments are stored and deleted from intersecting cells will also be given. The dynamic grid has been designed to deal with an unknown number of quantities assigned to a cell, and those items could be added or deleted from time to time in a continuous manner throughout the MG process.

Test examples are given in Section 3.6.4 in which two series of meshes are generated with full statistics on their characteristics depicted in Tables 3.1 and 3.2 to show the characteristics of meshes generated by the AFT. Two examples on adaptive refinement meshing are also provided to assess the performance of the dynamic grid in practical applications. Finally, closing remarks and discussions are presented in which a brief note on how a straightforward extension of the idea to three dimensions is also included.

3.6.2 Adaptive meshing by the AFT

The MG problem is completely defined by the boundary segments Γ, which is derived from the discretised boundary $\mathbf{B}$ of domain $\mathbf{\Omega}$, along with the node spacing function ρ, which governs the size of the elements to be generated. By virtue of the counter-clockwise order assigned to the nodes on the exterior boundary and clockwise order to the nodes on the internal boundaries, the domain to be meshed always situates to the left of a boundary segment. Following this convention, the list of segment in Γ need not follow any sequential or particular order as long as the segments are oriented correctly as exterior and interior boundary segments. This flexibility allows boundary segments to be entered and prepared independently in the collection of domain boundary edges for MG.

Table 3.1 Statistics for meshes of the first node spacing function

Mesh (ρ_{min}) δ	Grid	N_n	N_e	CPU Time(s)	α	N_b	N_c	N_s	Efficiency
1A	–	15412	30772	2.23	0.8931	3589	–	8.023	1
(0.1)	30 × 20	15525	31009	0.55	0.8959	3454	66.29	7.971	4.02
0.866	60 × 40	15218	30384	0.52	0.8952	3342	37.75	7.862	4.25
1B	–	30514	60976	7.51	0.8971	6287	–	8.440	1
(0.05)	30 × 20	31303	62554	2.16	0.8958	7406	122.7	8.597	3.48
0.869	60 × 40	31306	62560	2.12	0.8956	7406	66.71	8.597	3.54
1C	–	75740	151428	47.25	0.8925	16148	–	9.250	1
(0.02)	30 × 20	76395	152749	11.2	0.8950	16051	109.8	9.249	4.22
0.864	60 × 40	76067	152082	10.93	0.8931	16004	110.3	9.283	4.32
1D	–	149559	299066	196.23	0.8930	33810	–	9.748	1
(0.01)	30 × 20	150923	301794	44.01	0.8930	32914	55.59	9.788	4.46
0.863	60 × 40	150935	301818	42.73	0.8930	32917	55.59	9.787	4.59
1E	–	303577	607102	750.1	0.8934	63426	–	10.48	1
(0.005)	30 × 20	298292	596532	184.8	0.8937	67466	28.12	10.30	4.06
0.862	60 × 40	298292	596532	174.2	0.8936	67468	28.12	10.30	4.31

Note: N_n, number of nodes in the mesh; N_e, number of elements in the mesh; N_b, average number of boundary segments on the generation front; N_c, average number of neighbouring segments from the background grid; N_s, average number of segments for intersection tests; CPU Time, computer time in seconds for MG and graphics display; α, mean element shape factor calculated by Equation 3.2; δ, mean spacing conformity coefficient of line segments by Equation 3.3; Efficiency, CPU time for meshing without grid/CPU time for meshing with grid.

Table 3.2 Statistics for meshes of the second node spacing function and other meshes

Mesh (ρ_{min}) δ	Grid	N_n	N_e	CPU Time(s)	α	N_b	N_c	N_s	Efficiency
2A	–	22592	45132	4.39	0.9062	4912	–	7.254	1
(0.1)	30 × 20	22527	45002	0.99	0.9046	4980	71.75	7.277	4.43
0.871	60 × 40	22491	44937	0.95	0.9039	4873	38.81	7.164	4.62
2B	–	46256	92460	17.49	0.9031	9992	–	7.825	1
(0.05)	30 × 20	46338	92624	3.52	0.9039	10125	131.74	7.758	4.97
0.872	60 × 40	46305	92565	3.29	0.9044	9901	67.68	7.659	5.32
2C	–	117496	234940	111.5	0.9013	24893	–	8.351	1
(0.02)	30 × 20	117160	234268	19.37	0.9011	24723	71.62	8.315	5.76
0.870	60 × 40	117400	234755	18.7	0.9019	24987	71.47	8.279	5.96
2D	–	236052	472052	458.9	0.9009	50544	–	8.842	1
(0.01)	30 × 20	235684	471316	76.9	0.9010	50258	35.60	8.878	5.97
0.870	60 × 40	235698	471351	69.6	0.9014	50131	35.59	8.836	6.59
2E	–	472442	944832	1945	0.9015	101777	–	9.361	1
(0.005)	30 × 20	471906	943760	374.2	0.9013	101245	17.78	9.346	5.20
0.870	60 × 40	471909	943766	340.4	0.9013	101251	17.78	9.346	5.71
3	–	66379	132428	29.25	0.9205	11570	–	6.091	1
(0.05)	30 × 20	66381	132432	7.86	0.9206	11564	126.7	6.091	3.72
0.878	60 × 40	66377	132424	7.75	0.9206	11562	67.55	6.091	3.77
4	–	155783	311304	151.7	0.9202	25322	–	6.442	1
(0.03)	30 × 20	155761	311260	28.9	0.9200	25301	53.90	6.444	5.25
0.879	60 × 40	155772	311282	28.3	0.9200	25309	53.90	6.444	5.36

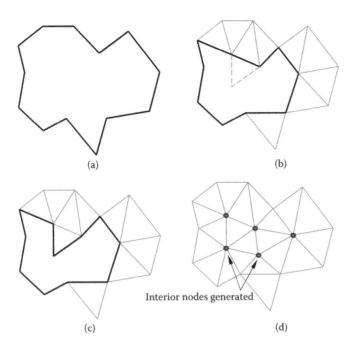

(a)   (b)

(c)   Interior nodes generated   (d)

Figure 3.43 Mesh generation by the advancing-front method: (a) initial front $\Gamma = \mathbf{B}$; (b) a new element is proposed on Γ; (c) Γ is updated to include the new element; (d) $\Gamma = 0$, mesh generation finished.

A very important concept is that at the beginning of MG, the generation front Γ is exactly equal to the list of boundary segments. While the domain boundary $\mathbf{B}$ remains always the same, the generation front Γ changes continuously throughout the MG process and has to be updated from time to time whenever a new element is formed, as shown in Figure 3.43a and b. MG terminates when there is no more line segment left in Γ, or Γ is reduced to an empty set. The essential steps for the generation of adaptive meshes over a 2D domain in compliance with a specified node spacing function are given in the following.

Step 1: *Input the boundary segments of domain Ω to be meshed*
The domain Ω is a bounded region defined by a set of line segments $\mathbf{B} = \{\mathcal{L}_i, i = 1, N_B\}$ in which each line segment is connected to two nodal points, i.e. $\mathcal{L}_i = (P_i, Q_i)$. In order to deal with multi-connect regions, the internal and external boundary loops have to be distinguished. Following the usual mathematical notations, the exterior boundary will be entered in a counter-clockwise sense, whereas interior boundaries will be entered in the clockwise sense, as shown in Figure 3.44, so that the part to be meshed lies always on the left-hand side of a segment.

Step 2: *Specify the node spacing function ρ over Ω, $\rho: \Omega \to \mathbf{R}^+$*
The nodal spacing ρ can be specified either explicitly by some global function covering the entire domain Ω, or by adaptive meshing, in which it is given implicitly by interpolation over the nodal values of a previous mesh. For the mesh of non-uniform element size, ρ varies from point to point within Ω, and for any given point $\mathbf{x}$ in Ω, $\rho(\mathbf{x})$ will return a positive value as the specified element size at point $\mathbf{x}$.

Step 3: *Boundary discretisation*
A set of new nodes have to be generated along segments in $\mathbf{B}$, such that node distribution on the generation front complies with the given node spacing function ρ before MG.

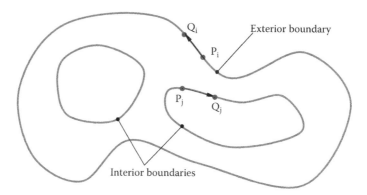

Figure 3.44 Boundary of a domain to be meshed.

$$B = \{\mathcal{L}_i, \; i = 1, N_B\} \xrightarrow{\text{New nodes based on } \rho} \Gamma = \{L_i, i = 1, N_b\}$$

The domain boundary **B** is first decomposed into straight lines composed of one or more segments, each of which is bounded by two consecutive sharp corners (turns) found on boundary **B**. The task of boundary discretisation is to generate intermediate nodes along the straight lines retrieved from **B** in consistency with the given node spacing function. A simple method assuming a linear node spacing variation between nodes described in Section 2.4.9 is adopted for generating intermediate nodes on the boundary segments. The result is a list of refined boundary segments $\Gamma = \{L_i, i = 1, N_b\}$ whose length is in compliance with the required nodal spacing, which will form the initial front for MG by the AFT.

Step 4: *Selection of base segment*

Select a line segment L from the generation front $\Gamma = \{L_i, i = 1, N_b\}$, and this segment will be called the base segment as it forms the basis for the creation of a new triangular element. Usually, the last segment on the generation front could be used as the base segment, i.e. $L = L_{N_b}$. However, for a mesh with rapid change in element size, the stability of the meshing process as well as the quality of the elements can be enhanced if the shortest segment in Γ is chosen as the base segment. Once line segment L is taken, it is removed from the generation Γ.

Step 5: *Identify an optimal node on the generation front*

Search on the generation front Γ a node C, which forms the *best* triangular element with L without crossing the generation front. In adaptive meshing, the ideal triangular element has to be optimal in terms of nodal spacing conformity and element geometrical

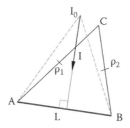

Figure 3.45 Forming triangle with base segment AB.

shape factor. Consider triangle ABC as shown in Figure 3.45. The geometrical shape factor α of triangle ABC (Lo 1985) is defined as

$$\alpha(ABC) = \frac{\text{signed area}}{\text{sum of edges squared}} = 2\sqrt{3}\frac{AB \times AC}{\|AB\|^2 + \|BC\|^2 + \|CA\|^2} \qquad (3.2)$$

in which $2\sqrt{3}$ is a normalising factor so that an equilateral triangle will have a maximum α value equal to 1. Node spacing conformity δ of a line segment is measured by the deviation of the edge relative to the prescribed length as imposed by the node spacing function, i.e.

$$\delta = \min\left(\frac{\ell}{\rho}, \frac{\rho}{\ell}\right) \qquad (3.3)$$

where ℓ is the length of the edge, and ρ is the node spacing function evaluated at the mid-point of the line segment. From Equation 3.3, it is seen that $0 < \delta \leq 1$. As for triangle ABC created on the base segment AB, there are two newly created edges, namely, AC and CB. The coefficients of node spacing conformity of edges AC and CB are, respectively, given by

$$\delta_1 = \min\left(\frac{\|AC\|}{\rho_1}, \frac{\rho_1}{\|AC\|}\right) \text{ and } \delta_2 = \min\left(\frac{\|CB\|}{\rho_2}, \frac{\rho_2}{\|CB\|}\right)$$

in which ρ_1 and ρ_2 are values of node spacing function evaluated at the mid-points of AC and CB, respectively, as shown in Figure 3.45. Hence, the node spacing conformity of triangle ABC can now be defined as

$$\delta(ABC) = \delta_1\delta_2$$

An overall optimality coefficient λ combining the effects of shape factor α and the node spacing conformity δ can be written as

$$\lambda = \alpha\delta = \alpha\delta_1\delta_2$$

λ will be used for judging the overall quality of a triangle in the adaptive meshing process.

Step 6: *Create an interior node at strategic location*

Create an interior node I, which has the largest λ value without cutting into the generation front. In general, the position of node I has to be determined by iteration starting from some ideal point. As shown in Figure 3.45, locate I on the normal bisector of segment AB such that λ is maximised. Verify if node I intersects with the generation front; I is accepted if there is no intersection; otherwise, adjust I downwards towards line segment AB until there is no more intersection. To ensure good-quality elements to be generated in the subsequent MG process, the new node I created should not be too close to the generation front, and this could be easily verified in the checking process for intersections. As to what threshold should be adopted, a simple rule is that the potential element likely to be formed with the nearest edge should not be inferior to the element proposed.

Step 7: *Formation of new element*

Compare the λ values of node C and node I. The node that gives a higher λ value is selected to form a new element with the base segment AB.

Step 8: *Boundary update*

 i. Boundary node C is selected:

   If CA belongs to Γ, delete CA from Γ; otherwise, add AC to Γ.

   If BC belongs to Γ, delete BC from Γ; otherwise, add CB to Γ.

 ii. Interior node I is selected:

   Two new segments are formed; add new segments AC and CB to Γ.

Step 9: *If Γ ≠ 0, i.e. N_b > 0, go to step 4.*

Step 10: *Meshing by AFT completed*

Carry out the analysis with the mesh generated. Revise the node spacing function ρ based on the results of analysis, and go to step 1 for the generation of a new mesh.

The only item that has not been fully covered in the MG by the AFT is the determination of the intersection between two line segments. There are many simple methods for checking the intersection between two line segments. For instance, in Section 2.4.4, detailed formulas for the determination of intersections by four different methods are given, and the efficiency and merits of these methods have also been compared.

3.6.3 Use of background grid

In MG by the AFT, the check for intersections and the update of the boundary upon the creation of a new element are the most time-consuming steps (Kwok et al. 1995). As shown in examples of adaptive meshes listed in Tables 3.1 and 3.2, in a typical triangular mesh of 152,749 triangular elements, the average number of line segments on the generation front is 16,051. The intersection check on these line segments for the generation of the best valid triangle involves operations of an order $16,051^2 \approx 2.5 \times 10^8$, which is a rather large number to be efficiently handled by a PC. To reduce the amount of computation in geometrical calculations, a background grid to localise the searching process could be the most effective, as shown in Figure 3.46, in which line segments associated with a cell are only a small portion of the total population. For an even distribution of line segments, the extent for geometrical check can be much reduced to within a small zone adjacent to the edge under consideration.

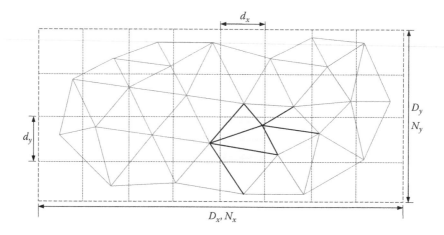

Figure 3.46 Line segments intersected by a cell are shown as thicker lines.

3.6.3.1 Construction of the background grid

In 2D, a background grid of $N = N_x \times N_y$ cells (boxes, zones, bins) subdivides regularly the domain of interest into a number of regions, such that for any given point, the cell that contains it can be determined rapidly without much calculation. As shown in Figure 3.46, the dimensions of a typical cell are given by

$$d_x = \frac{D_x}{N_x}, \quad d_y = \frac{D_y}{N_y}$$

where
D_x and D_y are the lengths of the domain along the x- and y-directions, respectively.
N_x and N_y are the numbers of division along the x- and y-directions, respectively.
$N = N_x \times N_y$ is the number of cells in the background grid.

However, in the actual applications, in particular, for adaptive meshing, elements are not evenly distributed, which means that an even distribution of line segments in the cells is not an efficient scheme. A variable allocation scheme of line segments into cells is preferred in which the number of line segments contained in each cell varies from cell to cell dependent on the node spacing function ρ over the given domain.

3.6.3.2 Setting the size of each cell in the grid

Let **L** be an integer array of size N_m assigned for recording line segments in the cells, and ρ be the node spacing function over domain Ω to be meshed such that

$$\rho: \Omega \to \mathbf{R}^+$$

The problem is how to determine the size (number of segments to be stored) in each cell k, $k = 1 \sim N$, according to the given node spacing function ρ in the most optimal manner. The size of cell k, n_k, can be estimated based on a weight inversely proportional to the square of the spacing function, i.e.

$$n_k \propto \frac{1}{\rho^2} = \frac{c}{\rho^2} \qquad \text{for some constant } c$$

ρ^2 has to be taken simply because ρ only prescribes the linear distance between nodes, whereas the number of line segments in a cell should be related to the area of the cell. Having adopted the weighting relationship, the following can be written:

$$\sum_{k=1}^{N} n_k = N_m \Rightarrow c \sum_{k=1}^{N} \frac{1}{\rho_k^2} = N_m \quad \text{or} \quad c = \frac{N_m}{\sum_{k=1}^{N} \frac{1}{\rho_k^2}} \tag{3.4}$$

where ρ_k is the node spacing function computed at the centre of cell k, and a more accurate estimation based on more Gaussian points should be employed if the cells of a large area are used. Hence,

$$n_k = \frac{c}{\rho_k^2}$$

By construction,

$$\sum_{k=1}^{N} n_k = \sum_{k=1}^{N} \frac{c}{\rho_k^2} = \frac{N_m}{\sum_{k=1}^{N} \frac{1}{\rho_k^2}} \sum_{k=1}^{N} \frac{1}{\rho_k^2} = N_m$$

However, usually it is necessary to allow for a minimum number of line segments, say $n_0 = 10$, in a cell, and Equation 3.4 has to be modified as follows:

$$c = \frac{N_m^*}{\sum_{k=1}^{N} \frac{1}{\rho_k^2}} \text{ and } n_k = \frac{c}{\rho_k^2} + n_0 \text{ with } N_m^* = N_m - N n_0 \tag{3.5}$$

where $N = N_x \times N_y$ is the number of cells in the background grid.

To complete the grid system with cells of variable size, an auxiliary integer array P of size $N + 1$ is also needed as a pointer to mark off the number of line segments in each cell, i.e. the set of line segments contained in cell k, k = 1 ~ N, is given by

$$\{L_i, i = P_k + 1, P_{k+1}\} \text{ with } P_{k+1} = P_k + n_k$$

Set $P_1 = 0$. It is easy to see that $P_{k+1} = \sum n_k$ and $P_{N+1} = N_m$, the length of the integer array L holding all the boundary segments recorded in the cells, $\{L_i, i = 1, N_m\}$. In case ρ is not known for the first mesh, the line segments could be evenly distributed to all the cells, as a relatively coarse mesh is used for the first preliminary analysis. Hence, the number of segments allowed for cell k in the initial mesh $n_k^0 = \frac{N_m}{N}$.

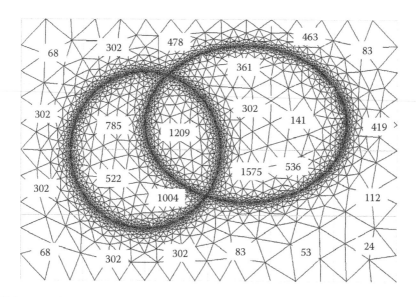

Figure 3.47 Rectangular domain partitioned into cells, $N_e = 10,054$.

Consider a 150×100 rectangular domain partitioned into $6 \times 4 = 24$ cells, as shown in Figure 3.47. The node spacing function is characterised with elements clustered along the perimeters of a circle and an ellipse, which could be expressed mathematically as follows:

$$r = \sqrt{(x-50)^2 + (y-50)^2}$$

$$\rho_1 = 0.5 |r - 30| + \rho_{min}, \quad \rho_2 = 20 \left| \sqrt{\left(\frac{x-90}{40}\right)^2 + \left(\frac{y-60}{30}\right)^2} - 1 \right| + \rho_{min} \qquad (3.6)$$

$$\rho = \min(\rho_1, \rho_2, \rho_{max}), \quad \rho_{max} = 20$$

Suppose $N_m = 10{,}000$, which is roughly equal to the number of elements ($N_e = 10{,}054$) to be generated. Based on the node spacing ρ given in Equation 3.6 with $\rho_{min} = 0.3$, n_k, $k = 1{,}24$ could be easily determined by Equation 3.5. For $n_0 = 10$, the number of segments allowed for the cells are calculated using four Gaussian points, and the values are listed in Figure 3.47.

3.6.3.3 Marking and unmarking cells intersected by a line segment

As described in Section 2.7.3, there are many ways in finding out the cells intersected by a line segment L for which the following three methods are commonly considered.

   i. Method 1: Search by neighbourhood
   ii. Method 2: Bounding the cells plus a distance check
   iii. Method 3: By elimination

In the actual implementation, method 2 is adopted as the distance calculation is also needed in the determination of the set of boundary segments potentially having an intersection with the base segment L under consideration. Unlike many background-grid constructions, the objects contained in each cell are constant throughout for the inquiring need. In MG by means of the AFT, the boundary segments have to be updated whenever a new element is formed. As a consequence, the line segments contained in each cell are not a static process, and a dynamic marking and unmarking of cells have to be done whenever there is a change of the segments on the generation front.

Whenever new segments are formed with the creation of a new node for the formation of triangle ABC with base segment AB, segments AC and CB are recorded, respectively, to the cells being cut across by the segments, as shown in Figure 3.48a. On the other hand, as

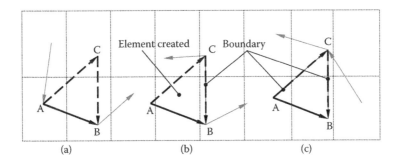

(a)   (b)   (c)

Figure 3.48 Updated boundary segment upon the creation of triangle ABC: (a) interior point C; (b, c) point C on the front.

shown in Figure 3.48b, when a boundary segment is absorbed as one of the edges of a triangular element, the new segment AC has to be marked on the cells being intersected, and at the same time, segment CB has to be deleted by removing it from all the cells containing it in a reverse unmarking process. Similarly, two line segments are to be deleted when boundary segments are used as the sides of a newly formed triangular element with the base segment; as shown in Figure 3.48c, line segments AC and CB have to be deleted from the boundary by removing them from the cells containing them.

As the generation front changes continuously throughout the MG process, the marking and unmarking of boundary segments have to be done accurately in a fully consistent manner. Otherwise, the slightest error in the determination of intersecting cells would very rapidly lead to serious boundary inconsistency, which would only be rectified by examining all the line segments on the current generation front. Fortunately, the unmarking of a line segment is exactly the inverse process of recording a line segment in all the cells being intersected. Hence, also for the rigorous requirement of absolute self-consistency, an identical procedure ought to be applied in marking and unmarking cells being intersected by a particular line segment.

3.6.3.4 Marking cells intersected by a line segment L

Method 2 – bounding the cells plus a distance check will be employed to determine all the cells cut across by line segment L. The essential steps are summarised as follows.

Let (x_1, y_1) and (x_2, y_2) be the co-ordinates of the end points of line segment L; calculate

$$x_{min} = \min(x_1, x_2), \quad x_{max} = \max(x_1, x_2)$$
$$y_{min} = \min(y_1, y_2), \quad y_{max} = \max(y_1, y_2)$$

from which the range of the cells being intersected is determined, i.e.

$$I_1 = \text{Int}(x_{min}/d_x) + 1, \quad I_2 = \text{Int}(x_{max}/d_x) + 1$$
$$J_1 = \text{Int}(y_{min}/d_y) + 1, \quad J_2 = \text{Int}(y_{max}/d_y) + 1$$

Consider a cell $k = (j - 1)N_x + i$ within the range $i = I_1 \rightarrow I_2, j = J_1 \rightarrow J_2$.

Let r be the radius of the circle circumscribing cell k; cell k will be included if $d \le r$, where d is the projected distance from the centre of the circle to line segment L, as shown in Figure 3.49. All cells adjacent to line segment L will be collected to form a list S, such that

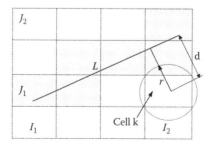

Figure 3.49 Cells intersected by line L.

$$k \in S \quad \text{iff} \quad k = (j-1)N_x + i \quad \text{for } i = I_1 \rightarrow I_2, j = J_1 \rightarrow J_2 \quad \text{and} \quad d \leq r$$

The set of cells intersected by L (cells shaded in Figure 3.49) is denoted by $\mathbf{C} = \{C_k, k \in S\}$. For each cell C_k, check the line segments contained in it, $\{L_i, i = P_k + 1, P_{k+1}\}$, one by one starting from L_{P_k+1}. If $L_i = 0$ for some $i \in \{P_k + 1, P_{k+1}\}$, set $L_i = L$.

3.6.3.5 Unmarking cells intersected by a line segment L

In the unmarking phase, set $\mathbf{C}$ is obtained by collecting all the cells that are intersected by line segment L based on exactly the same procedure for marking cells. For each cell $C_k \in \mathbf{C}$, check the line segments contained in it, $\{L_i, i = P_k + 1, P_{k+1}\}$, one by one starting from L_{P_k+1}. If $L_i = L$ for some $i \in \{P_k + 1, P_{k+1}\}$, then assign $L_i = L_{i^*}$ and $L_{i^*} = 0$, where i^* is the largest value in the list $\{P_k + 1, P_{k+1}\}$ for which $L_{i^*} \neq 0$.

3.6.3.6 Search for nearby line segments with the help of the background grid

In MG on 2D by the AFT, a typical scenario for the construction of a new element is to consider a base segment on the generation front. This base segment has to form a triangle with a node on the generation front or with a node strategically created at the interior of the unmeshed region. Whichever the case, the line segments close to the base segment need to be identified to ensure that the new element does not cut into the generation front. With the help of the background grid, the searching process for the neighbouring frontal segments could be made localised, which tremendously speeds up the entire process. Following the procedure of marking and unmarking of cells for a given line segment, the cells intersected by the base segment L could be determined. Let $\mathbf{C} = \{C_k, k \in S\}$ be the cells intersected by the base segment L; the set of boundary segments adjacent to L is given by

$$\Lambda = \bigcup_{k \in S} \{L_i, i = P_k + 1, P_{k+1}\}$$

Obviously, Λ is much smaller than the total number of segments on the generation front. The typical value of the size of Λ is denoted by N_c in Tables 3.1 and 3.2, whereas the number of boundary segments is denoted by N_b. For the largest mesh 2E listed in Table 3.2, the average value of N_b for MG is 101,245 and that for N_c is only 17.78, showing a substantial reduction in searching time especially for dense complicated meshes.

Further screening by simple geometrical checks could be applied to Λ to reduce it to a set Υ of potential line segments for rigorous yet more time-consuming line intersection tests to ensure that the generation front is not cut across by the new element to be created. As shown in Tables 3.1 and 3.2, the size of Υ denoted by N_s is just a tiny fraction of N_c, indicating that rigorous intersection tests are performed only on those boundary segments closest to the base segment with a fair chance of intersection.

3.6.3.7 Updating boundary segments

One characteristic of the AFT is the boundary update whenever a new element is generated. Without a background grid, a thorough search on the generation front has to be carried out to determine which line segment has been used as an edge of the new element, which has to be identified and removed from the generation front. As the node on the boundary used for the creation of the new element always comes from the set Υ of potential line segments for intersection, for the boundary update, it is sufficient to search over the set Υ in lieu of the entire set of line segments on the generation front. The set Υ, which has been obtained earlier in the creation of the new element, is a much reduced set compared to all the line segments on the generation front (N_s vs. N_b). The gain in computation time for the boundary update as a by-product in boosting intersection checks is also very significant.

3.6.4 Test examples

The effectiveness of the dynamic grid was tested systematically using two series of adaptive meshes progressively refined with an increasing number of elements and nodes. The MG was coded on a standard PC, Intel(R) Core™2 i7, CPU870 at 2.93 GHz, running on Windows 7 with Intel FORTRAN VS2010 on XP mode. The domain employed is a 150×100 rectangle whose boundary is decomposed uniformly into line segments of length 10 units. The first node spacing function has been given by Equation 3.6, which specifies a higher-element density around a circle and an ellipse, as shown in Figure 3.50.

As for the second node spacing function, it requires elements to be concentrated on the perimeter of five intersecting circles, as shown in Figure 3.51. The mathematical expressions for this node spacing function are given by

$$\rho_i = \frac{1}{2}\left|r_i - R\right| + \rho_{\min}, \quad r_i = \sqrt{(x - x_i)^2 + (y - y_i)^2},$$

$$x_i = x_0 + R_x \cos\left(\frac{\pi}{2} + i\alpha\right), \quad y_i = y_0 + R_y \sin\left(\frac{\pi}{2} + i\alpha\right), \quad i = 1,5$$

$$\alpha = \tfrac{2}{5}\pi, \quad R = 18, \quad R_x = 30, \quad R_y = 24, \quad x_0 = 75, \quad y_0 = 48$$

$$\rho = \min_{i=1,5}(\rho_i, \rho_{\max}), \quad \rho_{\max} = 20$$

For the series of five meshes corresponding to the first node spacing function, $\rho_{\max} = 20$ and the minimum node spacing $\rho_{\min}$ varies from 0.1 to 0.005, which represents a variation

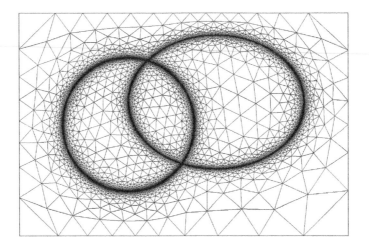

Figure 3.50 FE mesh for the first nodal spacing function, N_n = 75,740, N_e = 151,428.

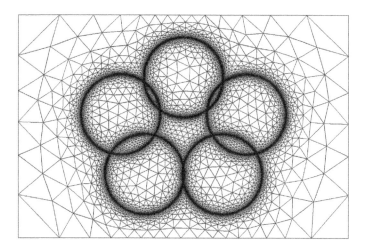

Figure 3.51 FE mesh for the second nodal spacing function, N_n = 117,496, N_e = 234,940.

from 200 to 4000 in the linear size ratio between the smallest elements and the largest elements. The number of elements, N_e, increases from the coarsest mesh of 30,772 to the finest mesh of 607,102 for which the CPU time for the generation of nodal points, the construction of the triangular elements and the graphics display has also increased from 2.23 to 750 s.

Two partitions of background grid 30 × 20 and 60 × 40 have been tested, and it is found that, in general, the CPU time could be reduced by more than four times depending on the complexity of the mesh, as shown in the column of efficiency in Table 3.1. As expected, the major gain in the CPU time is attributed to searching on the boundary segments for intersections and locating a candidate node for the formation of a new element. Referring to Table 3.1, by means of the background grid, N_c is usually a small fraction of N_b, which accounts for the drastic reduction in the searching time. However, the difference between the two grids 30 × 20 and 60 × 40 is pretty small, showing that efficiency is not sensitive to grid resolution as long as it is sufficiently fine and reasonable.

For the two grids, the segments retrieved in the intersecting cells are more or less the same, indicating that potential line segments could be accurately determined by a relatively coarse grid, even though the number of cells intersected by the base line segment might not be the same for the two different grid settings. Meshes generated with and without a background grid are not identical as the sequence of line segments, and hence, the sequence of element construction might be altered in the presence of a background grid or using grids of different resolutions. The number of elements and nodes and the resulting mean element shape factor α are, however, more or less the same for meshes generated with and without a background grid, as shown in Table 3.1, indicating that mesh characteristics are not affected by the use of a background grid.

Mean spacing conformity coefficients of line segments for the meshes are also given in Tables 3.1 and 3.2, which almost did not change whether a background grid has been used or not in MG. This implies that the node spacing conformity only depends on how elements are created but not on how the validity of the elements is assured in the formation process. For the example meshes that have been generated in this study, the mean value of the node spacing conformity of the line segments in the meshes is approximately 12% short of the ideal situation, whereas the mean element shape factor for the mesh is fairly high at about 0.9. For the given domain and the nodal spacing functions tested, there is no analytical theoretical value for optimal node spacing conformity, and the values quoted could be taken only for reference; a direct comparison with other meshes might not be appropriate.

Although the pattern of nodal density distribution is more complicated than that of the first node spacing function, trends and observations for the second node spacing function are very much similar to those that have been found for the first series of meshes. The number of elements, N_e, varies from the coarsest mesh of 45,132 to the finest mesh of 944,832, for which the CPU time has also increased from 4.39 to 1945 s, as shown in Table 3.2. Two background grids 30 × 20 and 60 × 40 have also been tested with which the MG time could be reduced by four to six times depending on the mesh characteristics. Again, the reduction in the CPU time is mainly ascribed to the reduced search on boundary segments in the element construction. As shown in Table 3.2, N_c is only a tiny fraction of N_b, which accounts for the substantial reduction in the CPU time for the MG. For the second series of meshes based on a different node spacing distribution, the difference between the two grids 30 × 20 and 60 × 40 is again very minor, and both grids are equally apt for the purpose of improving efficiency in ADF MG.

High-quality gradation meshes could be generated for rather complicated node spacing functions, as shown in Figures 3.52 and 3.53. The results for MG with and without the use

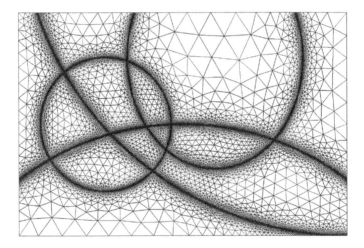

Figure 3.52 Mesh 3: N_n = 66,379, N_e = 132,428.

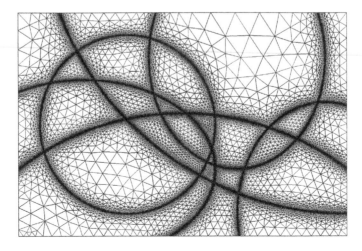

Figure 3.53 Mesh 4: N_n = 155,783, N_e = 311,304.

of a background grid are listed in Table 3.2. The gains in CPU time are 3.7 times and 5.3 times, respectively, for these two meshes, showing that the efficiency of the background grid reduces for more complicated node spacing functions whose computation may take a substantial portion of the overall MG time.

Upon further analysis on the use of the two background grids, it is found that identical sets of potential segments susceptible to intersections adjacent to the base segment could be accurately retrieved by both grids, verifying the reliability of the procedure in marking and unmarking of cells intersected by line segments. Similar to the first series of meshes, the use of background grid has no bearing on the characteristics of the resulting meshes, such that the number of elements N_e, the number of nodes N_n and the mean shape factor α are almost exactly the same.

As a practical example of adaptive refinement FE analysis, the electromagnetic wave distribution over a planar domain is considered. A uniform plane wave is scattered by a circular cylinder and propagates freely towards infinity, as shown in Figure 3.54. The analytical solution expressed in cylindrical co-ordinates to this problem was given by Balanis (1989).

$$E_z(\rho,\phi) = E_0 \sum_{-\infty}^{+\infty} j^{-n} \left[J_n(k_0\rho) - \frac{J_n(k_0 r)}{H_n^{(2)}(k_0 r)} H_n^{(2)}(k_0\rho) \right] e^{jn\phi}$$

Linear triangular elements T3 were used in the adaptive refinement analysis, and the required accuracy was set equal to 1%. Four analyses in three steps of refinements were required to bring the energy error norm from 22.7% to about 1%, as shown in Table 3.3. The refinement adaptive triangular meshes generated by the proposed dynamic grid technique are shown in Figure 3.55. It is seen that there is little difference between the exact error norm and the estimate error norm ($\bar{\eta}/\eta$), showing that the meshes are of very high quality, and the required nodal spacing over the domain is well respected.

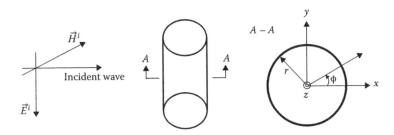

Figure 3.54 Wave direction and the cylindrical domain and its cross section.

Table 3.3 Adaptive refinement analysis of the electromagnetic wave problem

Mesh	DOF	NN	NE	‖e‖*	η(%)	‖e‖	η(%)	$\bar{\eta}/\eta$
1	308	324	576	0.089047	24.10	0.084717	22.69	1.06
2	3695	3755	7338	0.013265	3.43	0.011950	3.09	1.11
3	7069	7135	14,092	0.009856	2.54	0.009569	2.47	1.03
4	13,872	13,985	27,350	0.004561	1.17	0.004472	1.15	1.02

Note: DOF = number of degrees of freedom; NN = number of nodes; NE = number of elements; ‖e‖* = estimated energy norm; ‖e‖ = exact energy norm; $\bar{\eta}$ = estimated % error; η = exact % error.

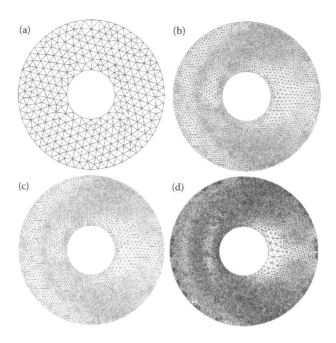

Figure 3.55 Adaptive refinement meshes of hollow cylinder: (a) mesh 0, NN = 324, NE = 576; (b) mesh 1, NN = 3755, NE = 7338; (c) mesh 2, NN = 7135, NE = 14092; (d) mesh 3, NN = 13985, NE = 27350.

In the second example of adaptive refinement analysis, two co-planar but non-overlapping wedges are considered. As shown in Figure 3.56, the geometry is composed of two cylindrical wedges, which are prismatic along the z-direction. The problem was analysed in a domain of radius R = 1.5λ using quadratic triangular elements T6. For adaptive analysis, the target accuracy of the FE solution is set at 2%. The relative error of the initial mesh

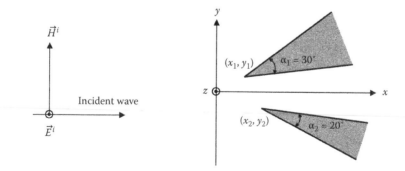

Figure 3.56 Co-planar wedges.

Table 3.4 Adaptive refinement analysis of co-planar wedges

Mesh	DOF	NN	NE	$\eta(\%)$	$\eta(\%)$	$\bar{\eta}/\eta$
1	1203	1275	594	13.24	16.50	0.80
2	4487	4621	2228	6.93	8.81	0.79
3	13,153	13,376	6530	2.03	2.33	0.87
4	17,107	17,456	8563	1.65	1.72	0.96

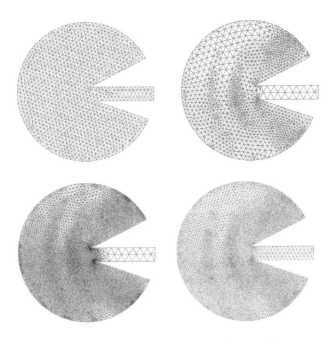

Figure 3.57 Adaptive refinement finite element meshes for co-planar wedges.

was about 16.5%, and the meshes achieved the required accuracy in three refinements, as shown in Table 3.4. The meshes generated by the AFT in the adaptive procedure are given in Figure 3.57.

3.6.5 Closure

A dynamic process has been developed such that the number of objects recorded in each cell of the background grid is a variable depending on the actual need as required by the node spacing function. The cells have to be marked and unmarked whenever there is a change in the generation front upon the formation of a new element; thus, the line segments contained in each cell change continuously throughout the entire process of MG. This scheme drastically reduces the memory requirement for the implementation of a background grid of a higher resolution of cell partitions. From meshes of various characteristics, it is found that an integer array of size equal to the number of elements to be generated is all that is required for the background grid construction, which is quite worthwhile for the drastic reduction in CPU time for MG by more than five times, which could have been even higher if the nodal point generation time were excluded in the mesh-construction time.

The dynamic background grid scheme could be easily added on to any existing MG programs based on AFT to boost efficiency, as the main stream flow of the MG procedure would not be affected at all, except that the search for entities could now be localised and be made much more efficient than before with the aid of a dynamic grid. Obviously, the idea could be further extended to the case of MG in 3D and over curved surfaces following exactly what is done in 2D. The only modification is that the domain of interest has to be partitioned into 3D cells whose number is given by $N = N_x \times N_y \times N_z$, where N_x, N_y and N_z are, respectively, the subdivisions along the three principal directions.

3.7 MESHING BY A COMBINED SCHEME OF DT AND ADF APPROACH

3.7.1 Introduction

The AFT and the DT are, so far, the most popular approaches developed for the generation of unstructured meshes on a planar domain, on curved surfaces and over volumes for constrained meshing. However, it is interesting to note that the two approaches are not contradicting to each other and can be put together quite naturally in harmony to form an overall more efficient scheme, in which the speed and the robustness of the DT and the quality in the generation of interior nodes of the AFT can all be retained in the combined scheme.

There are two ways in merging these two methods, which are to be referred to as the ADF–Delaunay scheme and the Delaunay–ADF scheme. In the ADF–Delaunay scheme, the set of interior points is already known before MG. The segments on the generation front are classified into Delaunay and non-Delaunay segments depending on whether the circumcircle of the supporting triangle of the segment contains any points in its interior. Given a base segment on the generation front, a point on the generation front or at the interior of the domain is selected to form a triangle based on the Delaunay criterion. As segments on the generation front are classified into Delaunay and non-Delaunay segments, intersection check is reduced only to those non-Delaunay segments, which are relatively few on the generation front.

For the Delaunay–ADF scheme, interior points are not known in advance, which are to be generated one by one as new elements are created at the generation front. At a typical stage including the initial stage, the unmeshed part of the domain is a constrained DT of the generation front. A point is created following the same procedure of the AFT, which is inserted into the constrained DT by means of the Delaunay-insertion kernel. The Delaunay triangle formed with the base segment and the inserted point is taken away from the unmeshed region, and the generation front is updated accordingly. In case no interior points can be generated for a given frontal segment, the supporting Delaunay triangle associated with this segment is deleted from the unmeshed part.

3.7.2 Advancing-front–Delaunay scheme

3.7.2.1 DT of non-convex planar domains

Let $\partial\Omega$ be the boundary of planar domain Ω and Λ be the set of interior nodes. A constrained DT for the domain Ω and interior nodes Λ is a collection of triangles $\{T_k\}$ satisfying the following properties:

i. Each T_k is formed by three nodes from $\partial\Omega \cup \Lambda$.
ii. Each T_k lies completely inside the domain Ω.
iii. The circumcircle associated with each T_k contains, in its interior, no other node point, which forms a valid triangle with any edge of T_k satisfying (i) and (ii).
iv. Ω is totally covered by $\{T_k\}$, and no two triangles of $\{T_k\}$ overlap.

3.7.2.2 Delaunay and non-Delaunay triangles

i. Delaunay triangles are those triangles whose associated circumcircles contain no node points on their circumferences or in their interiors.
ii. Triangle T_k will be said to be *semi-Delaunay* if there are node(s) lying on the circumference of the circumcircle of T_k.
iii. Triangle T_k will be said to be *non-Delaunay* if there are node(s) lying inside the circumcircle associated with T_k.

In the sequel, semi-Delaunay triangles will be considered as a special case of non-Delaunay triangles.

3.7.2.3 Delaunay and non-Delaunay segments

Line segments on the generation front are classified into Delaunay and non-Delaunay segments according to the Delaunay property of the triangles associated with them. At the beginning of triangulation, all the line segments on the domain boundary are assumed to be non-Delaunay since they are not connected to any triangle. Whenever a triangular element is formed, the Delaunay property of the triangle is examined to see if the circumcircle of this triangle contains any node(s) in its interior or on its circumference. The edges of a Delaunay triangle are Delaunay segments, and the edges of a non-Delaunay triangle are non-Delaunay segments.

The algorithm presented in the following can generate triangular meshes over general planar domains that are simply connected or multi-connected. The boundary of the domain is represented by a disjoint union of simple closed loops of straight line segments. For simply connected regions, there is only one closed loop, whereas for multi-connected regions, there can be as many interior loops as the number of openings inside the domain. The nodes on the exterior boundary are entered in a counter-clockwise order, while the nodes on the interior boundaries are entered in a clockwise order. The construction and the verification of the domain boundary are usually done more or less automatically by means of a geometric modelling interactive module. Nodes on the boundary need not be made following any particular numerical order; this flexibility allows us later to generate triangular meshes from one sub-region to another without bothering to identify the common boundaries between sub-regions. The algorithm first generates additional interior points according to the specified node spacing requirement. The algorithm then connects the boundary nodes and the interior nodes in such a way that no elements overlap, and the entire region is covered. The resulting mesh is a constrained DT, as defined in Section 3.7.2.1.

3.7.2.4 Triangulation process

Suppose that a complete nodal system has been generated according to the specified node spacing function, the next step is to connect these interior nodes with the boundary node to form a valid FE mesh. Similar to the classical AFT, at the beginning of triangulation, the generation front is exactly equal to the domain boundary. While the given domain boundary remains always the same, the generation front evolves continuously throughout the triangulation process and has to be updated whenever a new element is formed.

Let Γ_1 be the set of non-Delaunay line segments and Γ_2 be the set of Delaunay line segments on the generation front Γ. Since Γ_1 and Γ_2 is a partition of Γ, we always have $\Gamma_1 \cup \Gamma_2 = \Gamma$ and $\Gamma_1 \cap \Gamma_2 = \emptyset$. At the beginning of meshing, $\Gamma_1 = \Gamma = \partial\Omega$ and $\Gamma_2 = \emptyset$. Let Σ be the set of nodes on the generation front Γ and Λ be the set of interior nodes within the unmeshed region bounded by Γ. The triangulation is initiated by taking the last segment $AB \in \Gamma_1$. A node $C \in \Sigma \cup \Lambda$ has to be found such that triangle ABC lies within the domain Ω and its associated circumcircle is the smallest, as shown in Figure 3.58.

A node $C_i \in \sum \cup \Lambda$ is said to be a candidate node if it satisfies

   i. $C_iA \times C_iB > 0$
   ii. $C_iA \cap \Gamma_1 = \{\emptyset, A, \{C_i, A\}, C_iA\}$ and $C_iB \cap \Gamma_1 = \{\emptyset, B, \{C_i, B\}, C_iB\}$

Let $\mathbb{C} = \{C_i\} \subset \Sigma \cup \Lambda$ be the set of candidate nodes. The node to be selected is a node $C_m \in \mathbb{C}$ such that $C_m \in C_{ABC_i}, \forall C_i \in \mathbb{C}$, where C_{ABC_i} is the circumcircle of triangle ABC_i, which

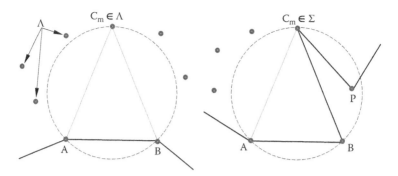

Figure 3.58 Forming Delaunay triangle with base segment AB. Thick line = boundary segment.

means that C_m is inside or on the circumference of circumcircle C_{ABC_i}. Condition (i) ensures that the proposed triangle lies on the left-hand side of the line segment AB. As no intersection check is needed for Delaunay segments Γ_2 (property of DT), condition (ii) ensures that the proposed triangle does not cut across the generation front Γ. As shown in Figure 3.58, C_m forms the largest angle with base segment AB; however, point P is not selected even though it is inside the circumcircle of triangle ABC_m as line segment AP has intersection with the generation front, i.e. condition (ii) is violated.

When there are no more segments left in Γ_1, the construction process continues with segments of Γ_2 until the generation front is reduced to zero, i.e. $\Gamma = \emptyset$. Calculations for intersections are kept to a minimum by first exhausting all the segments in Γ_1. Most of the intersection check is done at the beginning of the triangulation process since the boundary segments are assumed to be non-Delaunay. As more and more Delaunay triangles are constructed, the number of non-Delaunay segments decreases rapidly. From time to time, semi-Delaunay triangles are encountered; however, in general, their formation is not so numerous except for the special case of regular rectangular grids.

3.7.2.5 Updating Γ_1 and Γ_2

The generation front $\Gamma = \Gamma_1 \cup \Gamma_2$ has to be updated whenever a triangular element is formed. Let C_{ABC} be the circumcircle of triangle ABC; Γ can be updated by

    i. Removing line segment AB from the generation front.
    ii. If triangle ABC is non-Delaunay, i.e. $P \in C_{ABC}$ for some $P \in \Sigma \cup \Lambda$

    Set $\Gamma_1^* = \Gamma_1 + \{AC, CB\}$; otherwise, set $\Gamma_2^* = \Gamma_2 + \{AC, CB\}$.

3.7.2.6 Existence and Delaunay property of the triangulation

The existence and the Delaunay property of the triangulation will be briefly discussed in the following paragraphs, and for a formal definition and a more detailed discussion on 2D and 3D constrained DT, readers can consult the text by George and Borouchaki (1998). With regard to the validity of the mesh construction, it can be shown that for each line segment AB on the generation front, a node C can always be found without ambiguity such that ABC is the best triangle in the sense of DT.

Proposition 3.1: For each line segment $AB \in \Gamma$, there exists node $C_m \in \mathbb{C}$ such that $C_m \in C_{ABC_i} \ \forall C_i \in \mathbb{C}$, where C_{ABC} is the circumcircle associated with triangle ABC.

Proof: Define an ordering relation for the set $\mathbb{C}$,

$$C_i, C_j \in \mathbb{C}, \quad C_i \geq C_j \Leftrightarrow C_i \in C_{ABCj} \Leftrightarrow \angle AC_iB \geq \angle AC_jB$$

i. Exclusivity

Given $C_i, C_j \in \mathbb{C}$, we have either $\angle AC_iB \geq \angle AC_jB$ or $\angle AC_jB \geq \angle AC_iB$, hence, any two points of $\mathbb{C}$ can be compared.

ii. Transitivity

$$C_i, C_j, C_k \in \mathbb{C}, C_i \geq C_j \text{ and } C_j \geq C_k \Rightarrow \angle AC_iB \geq \angle AC_jB \text{ and } \angle AC_jB \geq \angle AC_kB$$
$$\Rightarrow \angle AC_iB \geq AC_kB \Rightarrow C_i \geq C_k$$

iii. Non-uniqueness

$$C_i, C_j \in \mathbb{C}, \quad C_i \geq C_j \text{ and } C_j \geq C_i \nRightarrow C_i = C_j$$

With (i) and (ii), since $\mathbb{C}$ is a non-empty finite set, there exists $C_m \in \mathbb{C}$ such that $\angle AC_mB$ is a maximum, or $C_m \geq C_i \ \forall C_i \in \mathbb{C}$.

Remarks: For the set of candidate nodes $\mathbb{C}$, nodes having intersection with the generation front are excluded from the set. However, this will not affect the setting up of an ordering relationship for the points in the set with respect to the base line segment AB, such that an optimal point C_m can always be identified in $\mathbb{C}$ for which $\angle AC_mB$ is the largest.

3.7.2.7 Delaunay property of triangulation

By the lemma of Delaunay, the triangulation will be Delaunay if for any diagonal AC shared by triangle ABC and ACD, D lies outside or on the circumference of the circumcircle of triangle ABC and vice versa. Consider quadrilateral ABCD composed of two triangles ABC and ACD of the mesh, as shown in Figure 3.59. If D′ lies inside the circumcircle of triangle ABC, $\angle AD'B$ is greater than $\angle ACB$, and point D′ would have been chosen rather than point C in the construction of triangle ABC. Similarly, if B lies at the interior of the circumcircle of triangle ACD, $\angle ABD$ is greater than $\angle ACD$, and in the construction of triangle ACD, node B would have been selected. In both cases, the Delaunay property of the triangulation is established by selecting the point making the largest angle possible with the base segment for each triangular element constructed throughout the MG process.

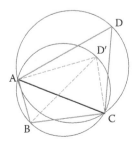

Figure 3.59 Delaunay property of triangulation.

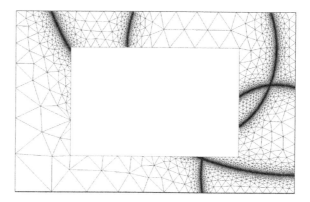

Figure 3.60 Triangular mesh generated by the ADF–Delaunay scheme.

A triangular mesh of a multi-connect planar domain is generated by the ADF–Delaunay scheme, as shown in Figure 3.60. There are 253 line segments on the interior and the exterior boundary loops, and the final mesh consists of 9793 nodal points and 19,333 triangles. The α mesh quality after two cycles of smart Laplace smoothing improves to 0.924 from the original value of 0.89 before optimisation. It is seen that all triangles in the mesh are Delaunay triangles except a few of those constrained by the boundary of the domain. It is noted that the system of interior nodes is generated from some mathematical functions by a separate program, and the ADF–Delaunay scheme has been applied to connect the nodes on the boundary with those at the interior of the domain.

3.7.3 Delaunay–advancing-front scheme

Another possibility to merge the DT and the AFT is to insert interior points proposed by the AFT using the Delaunay insertion kernel. A constrained DT is constructed using nodes on the boundary of the domain. Analogous to the classical ADF approach, the initial generation front consists of all the boundary segments of the planar domain. A line segment is taken from the generation front as the base segment. Based on the specified element size, an interior node is generated by the AFT, which is then inserted to the constrained DT of frontal nodes by means of the Delaunay insertion kernel. In case the proposed node is too close to the generation front, no interior node will be generated. Whichever the case, the triangle connected to the base segment is deleted from the DT, and the generation front is updated accordingly resulting in another constrained DT bounded by the generation front without any nodes in its interior. Obviously, this procedure can be repeated until there is no more line segment on the generation front, and the region bounded by the generation front reduces to nothing. The final mesh is given by all the triangles removed one by one from the system for each frontal base segment considered.

The procedure of the Delaunay–ADF scheme is elucidated by an example of triangular mesh of non-uniform element size. An FE mesh of variable element size is going to be generated over a rectangular domain. Following the specified nodal spacing, the boundary of the rectangular domain is decomposed into line segments of different lengths. Using nodes on the domain boundary, a DT is constructed, as shown in Figure 3.61.

An intermediate stage of MG is depicted in Figure 3.62 where the boundary segments are represented by arrows, which divided the rectangular domain into the meshed region and the unmeshed region. The meshed region consists of those triangles connected to the base segments removed one by one from the unmeshed region, and the unmeshed region is

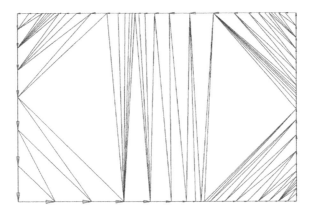

Figure 3.61 Delaunay triangulation of the boundary nodes.

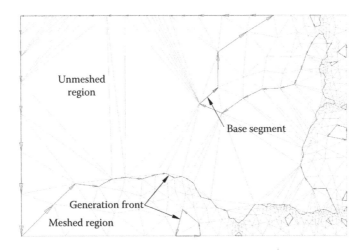

Figure 3.62 An intermediate stage of mesh generation.

a constrained DT of the generation front. At a typical stage, a line segment is selected from the generation front as the base segment for the construction of a triangular element.

An interior node is generated, which forms the best triangle with the base segment in terms of element shape quality and node spacing requirement, as shown in Figure 3.63. This node is inserted to the unmeshed region of a constrained DT by means of the standard Delaunay insertion kernel. The non-Delaunay triangles of the unmeshed region with respect to the inserted node are removed from the triangulation, and the cavity left behind is filled up by triangles formed by the interior node and the boundary of the cavity, as shown in Figure 3.64. The triangle connected to the base segment is transferred from the unmeshed region to the meshed region. In case no interior node is needed for the base line segment under consideration, the triangle connected to the base segment is simply removed from the unmeshed region to the meshed region, and the generation front is updated accordingly in the same way as the insertion of an interior node. Line segments on the generation front are considered in turn until there is no more line segment left on the generation front, and by then, the unmeshed region will shrink to nothing, and the meshed region is a triangular mesh of the given domain, as shown in Figure 3.65. In this example, a very simple node generation scheme has been employed for the generation of the interior nodes to show the basic

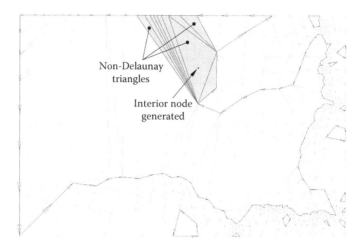

Figure 3.63 Delaunay insertion of an insertion node.

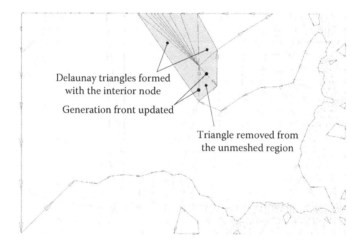

Figure 3.64 Delaunay triangles formed with the inserted node.

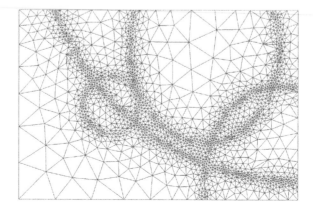

Figure 3.65 Triangular mesh generated by the Delaunay–ADF scheme.

principle and the robustness of the scheme, and hence, the quality of the triangular mesh is not the most optimal.

From this example, it can be seen that the DT and the AFT can be easily merged together to form an overall scheme in which the merits of the two individual schemes can be retained, namely, the boundary treatment and the generation of interior nodes by the AFT and the robustness and efficiency of the DT by the point insertion kernel. In fact, the two schemes can be put together in a fairly natural manner such that the given domain is partitioned into two regions: the meshed part and the unmeshed part. At the beginning of MG, the unmeshed part is a constrained DT of the entire domain bounded by the initial generation front of all the boundary segments. Based on a selected line segment, an interior node is generated and inserted to the unmeshed region, and the meshed region grows in size by taking one triangle at a time from the unmeshed region until there are no more triangles remaining in the unmeshed region. As nodes are created by the AFT and the triangles are formed following the Delaunay criterion, the resulting triangular mesh consists of well-shaped Delaunay triangles conformable with the specified nodal spacing. This is an elegant MG scheme for the construction of unstructured meshes, and the idea can be extended to 3D as long as we have an efficient scheme for the generation of the constrained DT of the boundary nodes. Here a constrained DT is a triangulation of the boundary nodes (possibly with some additional Steiner interior nodes to facilitate the boundary recovery), which contains all the triangular boundary facets of the given object. It is remarked that the introduction of interior points is merely to improve the quality of the mesh, and in the worst case that not even a single point is inserted, the initial constrained DT will be the final mesh.

3.8 ENHANCED QUADTREE MESHING

With the introduction of transition quadrilateral elements, the Quadtree subdivision can be applied to the domain interior to produce quadrilateral elements in compliance with the given node spacing specification. Alternatively, in case transition elements are not available, the quadrilateral elements of a Quadtree partition can also be divided rapidly into triangular elements. Instead of using templates for the treatment of the boundary, an enhanced Quadtree-meshing scheme by means of the coring technique in conjunction with the AFT is presented. A general multi-connected planar domain and its bounding box are shown in Figure 3.66, and it is required to mesh the domain into quadrilaterals and triangles with the element size distribution similar to that shown in Figure 3.15.

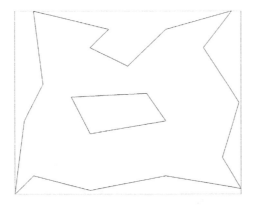

Figure 3.66 Planar domain and its bounding box.

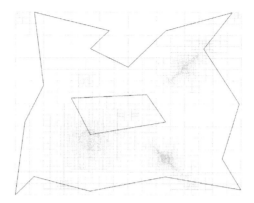

Figure 3.67 Planar domain and its bounding box.

3.8.1 Quadtree partition of the bounding box

Similar to the classical approach, given the planar domain and the node spacing specification, the first step of the enhanced scheme is to subdivide the rectangular bounding box into quadrilaterals by means of the classical Quadtree decomposition. This can be achieved using the recursive subdivision method described in Section 3.4, and the result of the Quadtree partition of the bounding box is shown in Figure 3.67. It is seen that element concentrations are found at the interior of the domain and around the interior boundary of the planar domain.

3.8.2 Removal of quadrilaterals near domain boundary

The next step is to remove those quadrilaterals that are outside the bounded region or too close to the domain boundary. Quadrilaterals that are too close to the boundary can be removed simply by checking the distance to the boundary segments. Suppose that the domain boundary consists of N_b line segments; then quadrilateral Q will be removed if

$$\min_{i=1,Nb} (\text{distance between Q and segment i}) < \text{size of Q}$$

where distance is the distance between the centre of Q and the ith line segment under consideration, as discussed in Section 2.4.1, and the size of Q can be taken as the longest edge of Q. Upon the removal of the quadrilaterals near the boundary by the distance check, the remaining quadrilaterals are well away from the boundary line segments.

However, those quadrilaterals that are outside the domain boundary still have to be removed. The Inside_or_Outside check provided in Section 2.4.11 can be applied here to determine which quadrilaterals are to be retained. A more consistent treatment is to find out the status of one quadrilateral and determine the patch of quadrilaterals by means of the neighbouring relationship. Whichever method is used, the Quadtree partition within the planar domain upon the removal of outside and nearby quadrilaterals is shown in Figure 3.68. It is observed that the quadrilaterals are all removed over regions of large element size, showing that quadrilaterals cannot be easily fitted to an irregular boundary, and it is also an inherent difficulty of Quadtree decomposition for a systematic and strategic positioning and orientation of the grid with respect to the given domain.

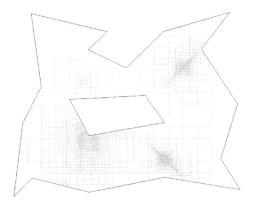

Figure 3.68 Quadtree partition within the planar domain.

3.8.3 Boundary recovery for triangulation

The boundary of the triangulation problem consists of two parts – the given boundary line segments and the boundary of the Quadtree quadrilaterals. The part to be meshed is the void between these two boundaries. The original given boundary segments may be too long or too short to be consistent with the given element size specification. Hence, each boundary line segment of the planar domain has to be verified and subdivided into more segments to conform with the node spacing requirements. The division of a line segment into smaller segments according to a given density function has been discussed in Section 2.4.9, which can now be applied to the boundary segments in turn to produce a consistent boundary for MG. As the one-level rule has been enforced in the generation of the transition quadrilateral mesh, the node spacing requirement is not strictly conforming with the quadrilateral decomposition. In order to have a better fit with the quadrilateral elements, in dividing a boundary edge, apart from the node spacing requirement, one can also take reference of the size of the adjacent quadrilaterals. As for the boundary of the internal Quadtree partition, the size of the segments is already in line with the element size specification; what has to be done is to retrieve the boundary of the patch of quadrilaterals. This can be easily done based on the neighbouring relationship. Following the neighbouring relationship of the transition quadrilateral elements, an edge of a quadrilateral is a boundary segment if there is no neighbour on this edge, or the neighbouring quadrilateral has been removed earlier in Section 3.8.2. The orientation of the boundary line segments is crucial for the ADF MG, which determines the nature of a boundary loop, i.e. counter-clockwise represents exterior boundary and clockwise represents interior boundary, as shown in Figure 3.69.

3.8.4 Advancing-front MG

The AFT is very efficient in handing irregular planar domains. The boundary segments with correct orientations collected in the last step are fed into a 2D ADF mesh generator to produce the triangular mesh to fill the gap between the Quadtree partition and the original domain boundary. While the segment orientation is important for a correct triangulation, the order of the segments does not matter even though they are usually prepared following a natural sequential order. The result of ADF meshing of the given bounding segments is shown in Figure 3.70. It is seen that some internal nodes have been generated over empty areas to produce triangles of better quality. The boundary mesh along with the Quadtree

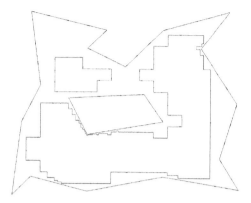

Figure 3.69 Retrieved boundary for mesh generation.

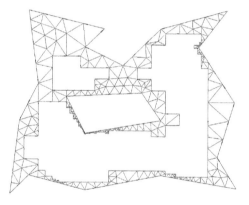

Figure 3.70 Mesh generation along domain boundary.

meshing at the interior part are put together to obtain the final mesh by the enhanced Quadtree meshing scheme, as shown in Figure 3.71.

Making use of a combination of existing techniques, the meshes produced by the enhanced Quadtree meshing scheme seem to be quite promising in terms of element shape quality and size distribution. It can be made fairly automatic, and all the steps starting from the

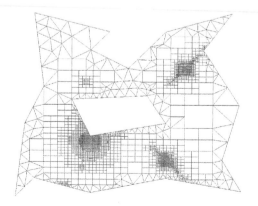

Figure 3.71 Final mesh by the enhanced Quadtree meshing.

input of the boundary segments to the final mixed mesh, as shown in Figure 3.71, have all been incorporated into a single FORTRAN program of about 1200 lines excluding graphics commands.

3.9 QUADRILATERAL MESH

In the early development of FE MG, quadrilateral meshes were generated by the mapping method and the drag or extrusion method for which a brief account has been given in Section 3.2. However, these methods are only applicable to relatively simple regular domains, and only structured meshes are produced in general, which have difficulties in complying with the requirement of rapid change in element size for adaptive FE analysis. As a result, adaptive meshes of quadrilateral elements are mainly unstructured, and they are more flexible in coping with the node spacing specification and fitting domain boundaries.

Unstructured quadrilateral meshing algorithms can be grouped into two categories – direct and indirect methods. By means of the direct method, quadrilateral elements are generated directly without first going through the process of triangular meshing. As for the indirect approach, the physical domain is first meshed into triangles, which are then converted fully or partially into quadrilaterals by means of various algorithms.

3.9.1 Direct method

The methods of directly generating quad meshes can be classified into two categories. The first category consists of those that rely on some form of decomposition of the domain into simpler regions possibly convex in shape, which allow a straightforward natural division into quads. The second category consists of those that make use of a moving front for the direct placement of nodes and elements.

The Quadtree technique based on a spatial decomposition, described in Section 3.4, employed by Baehmann et al. (1987) to generate quad meshes belongs to the first category. Following an initial decomposition of the 2D space into a Quadtree subdivision, taking into account of the local features, quads are fitted into the Quadtree cells, and the nodes are adjusted wherever necessary to conform to the domain boundary. Talbert and Parkinson (1991) introduced another decomposition technique in which the domain is recursively subdivided into simpler polygonal shapes to fit boundaries defined by Bezier curves. With the aid of standard templates, quad elements are generated on the resulting polygons. Chae and Jeong (1997) and Dietrich (1997) proposed an improvement over Talbert's algorithm. Generation of quad meshes by means of medial axis decomposition was introduced by Tam and Armstrong (1991). The medial axis can be taken as the path generated from the centre of a maximal circle as it is rolled through the area (Section 5.8.9). Having decomposed the area into simpler regions, quad elements are generated in each region to cover the original domain. Joe (1995a) described a decomposition algorithm to divide a given 2D domain into convex polygons. Using boundary recovery techniques formerly developed for DT, a boundary-constrained quad mesh was constructed within each convex region. Quad elements are generated by means of recursive domain decompositions into quads (Sarrate and Huerta 2000). The given planar domain is first partitioned into convex polygons by Voronoi tessellation, and the convex polygons are each divided into quads (Weyer et al. 2002).

Zhu et al. (1991a,b) were among the pioneers to propose a quadrilateral meshing algorithm by means of the AFT. Starting with an initial placement of nodes offsetting from a boundary edge, elements are formed by projecting edges towards the interior of the domain. Two triangles can be created using standard AFT, which are then combined to form one single quadrilateral. The paving method introduced by Blacker and Stephenson (1991) offers a way to form complete rows of quadrilaterals starting from the boundary and working inwards. Methods of projection of nodes, handling of special geometric situations and intersection of opposing fronts are discussed. White and Kinney (1997) proposed improvements to the paving method, suggesting individual placement of quadrilaterals rather than row by row.

3.9.2 Indirect method

A simple way to convert a triangular mesh into a quadrilateral mesh is to divide each triangle in the mesh into three quadrilaterals by inserting a node at the centroid. This method guarantees an all-quadrilateral mesh, but a large number of irregular nodes are introduced into the mesh, resulting in poor element quality. Moreover, the boundary of the domain cannot be kept intact as the mid-side nodes are added to each boundary segment. An alternative way is to combine pairs of adjacent triangles to form quadrilaterals, and the resulting mesh is a mixture of triangles and quadrilaterals. The method of combining triangles can be improved if some care is taken in the order in which the triangles are merged. In an effort to maximise the number of quadrilaterals, Lo (1989b) defined a shape measure for quadrilaterals in an algorithm that suggested several heuristic procedures for the sequence in which triangles are combined. The result is a quad-dominated mesh containing a small number of triangles.

Lee and Lo (1994) later proposed a frontal merging scheme that enhanced Lo's strategy by including local splitting of triangles. Furthermore, an ADF approach is employed to divide the triangular mesh into meshed and unmeshed zones for systematic merging of triangles. An initial set of merging front is set up, which consists of all the boundary segments. Triangles are systematically combined at the front, advancing towards the interior of the triangular mesh. At an intermediate stage, the front is a collection of contour lines separating the quadrilaterals already formed and the triangles yet to be combined. Based on this scheme, it is possible to convert a given triangular mesh into an all-quadrilateral mesh, provided the number of segments on the boundary is even.

As the number and the position of the boundary nodes are not altered during the merging process, the integrity of the domain boundary can be guaranteed. Since the quadrilateral mesh is derived from a background triangular mesh, quadrilateral meshes with a nodal space compatible with the specified node spacing function can be expected, making it very attractive for adaptive meshing. Since all operations are local, indirect methods enjoy the merit of being very fast. Global intersection checks, which are required by many direct methods, are not necessary. The drawback of indirect method is that there are typically many irregular nodes left in the mesh, and there is no guarantee that the quadrilateral elements are in alignment with the domain boundary, a desirable feature for some applications and easy visualisation. Irregular nodes can be reduced, to a large extent, by geometric and topological clean-up operations, hence the improvement in the mesh quality.

A modification proposed by Owen et al. (1999) known as quad-morphing (Q-Morph) also utilises an AFT to convert triangles into quadrilaterals. This is basically a remeshing process in which quadrilaterals are formed at the construction front, while triangular elements are removed and modified (Lee et al. 2003). In this approach, local edge swaps are performed, and additional nodes are introduced to ensure boundary alignment and mesh orthogonality. Figure 3.72 shows two quadrilateral meshes, one by Q-Morph and the other by Lee and Lo's algorithm developed in 1994. It appears that meshes by Q-Morph have a better alignment with the domain boundary; however, the size variation across the domain is also less significant compared to meshes by direct merging of triangular elements.

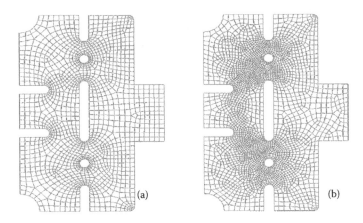

Figure 3.72 Comparing (a) Q-Morph and (b) Lee and Lo's algorithm.

By the indirect method, quadrilateral meshes are generated from triangular meshes, and thus, it can be very effective for irregular or multi-connect domains. With the advancement in the techniques for element shape optimisation, fairly good quality quadrilateral elements can be produced even for a rapid change of element size. Hence, apart from regular domains where mapping methods can be applied to generate high-quality quadrilateral elements, for irregular domain and adaptive FE analysis, indirect methods are commonly adopted to generate the required quadrilateral meshes. In Sections 3.93 to 3.96, general considerations to produce quadrilaterals from triangular meshes will be discussed, and the detailed procedures to convert a triangular mesh into an all-quad mesh will also be presented.

3.9.3 Quadrilateral-dominated mesh

An algorithm is presented in this section to generate quad elements from a triangular mesh by selectively removing the diagonals between adjacent triangles. The shape quality of a triangular element is measured by the α-quality coefficient. As for quad elements, a distortion coefficient β is introduced, with which the quality of the quad can be compared to those of the two triangles arising from a cut along either diagonal. A parameter γ can be specified by the user to give higher preference either to triangles or to quads. As a result, a careful selection of γ would lead to an optimised mixed mesh consisting of both triangular and quad elements.

The generation of quadrilateral elements is based on the simple fact that a quadrilateral will be produced whenever a diagonal between two triangles is removed. The quality of the quadrilateral and hence the quality of the resulting FE mesh will depend highly upon the way in which the diagonals are removed. In order to facilitate the mesh updating work for each diagonal removal, prior to the removal of diagonals, some computations on the topology of a triangular element mesh have to be done. These include (1) determination of edges in the triangulation (Section 2.5.4), (2) determination of the neighbouring triangles (Section 2.5.1) and (3) for each triangle, determination of the three edges associated with it.

In order to determine the exact sequence in which diagonals are to be removed, one should have some global measure of each edge of the mesh on the quality of the quadrilateral that it will generate and how the neighbouring edges are affected upon its removal. A β value will be attached to each edge, which determines the quality of the quadrilateral that will be generated as a result of its removal. A new β^* value for each edge can then be computed by subtracting from the β value the β values of its neighbouring edges. Thus, the β^* value is an overall measure on the value of the diagonal and the influence of its removal on neighbouring edges. The diagonal that bears the highest β^* value will be the first to be removed. After the removal of a diagonal and the generation of a quadrilateral, the mesh as well as the β^* values of the neighbouring edges have to be updated. Removal of diagonals continues as long as there is a diagonal whose β value is greater than the prescribed γ value. For a given triangulation, more quadrilateral elements will be produced for a small γ value and vice versa; and in the extreme case that $\gamma = 1$, no diagonal will be removed. The generation process stops when no more edges can be removed, or the β values of all edges fall below γ. The resulting mesh will be a mixed mesh of triangular and quadrilateral elements.

Definition

Recall the dimensionless α-quality of a triangle,

$$\alpha(ABC) = \frac{\text{signed area}}{\text{sum of edge squared}} = 2\sqrt{3}\,\frac{CA \times CB}{CA^2 + AB^2 + BC^2}$$

3.9.3.1 Distortion coefficient β of a quadrilateral

Four triangles can be obtained by cutting the quadrilateral ABCD along the diagonals AC and BD, as shown in Figure 3.73. Let α_1, α_2, α_3 and α_4 be the α values of the triangles arranging in a descending order of magnitude, i.e.

$$\{\alpha_1, \alpha_2, \alpha_3, \alpha_4\} = \{\alpha(ABC),\ \alpha(ACD),\ \alpha(ABD),\ \alpha(BCD)\}\ \text{and}\ \alpha_1 \geq \alpha_2 \geq \alpha_3 \geq \alpha_4$$

Definition

The distortion coefficient β of quadrilateral ABCD is given by

$$\beta(ABCD) = \frac{\alpha_3\alpha_4}{\alpha_1\alpha_2}$$

β would lie between 0 and 1 for a convex (valid) quadrilateral. For rectangles, β attains a maximum value of 1, whereas for quadrilaterals degenerated into triangles, β approaches 0. The higher the β value, the better the shape of the quadrilateral relative to a rectangle. Hence, quadrilaterals can be preferably generated by removing diagonals having large β values. The β values of some typical quadrilateral are shown in Figure 3.74.

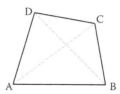

Figure 3.73 Quadrilateral ABCD.

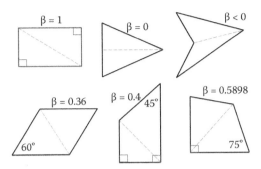

Figure 3.74 β values of typical quadrilaterals.

3.9.3.2 Merging of triangles to form quadrilaterals

Triangular elements can be converted to quadrilateral elements by removing the diagonal between a pair of adjacent triangles. The resulting mesh consists of both quadrilaterals and triangles, and their proportion is governed by a user-specified parameter γ. The diagonal removal and the forming of quadrilateral elements can be carried out by the following steps.

1. *Retrieval of unique lines (edges) from the triangular mesh*
   The list of edges and the two triangles connecting to each edge are needed to set up the priority sequence in the removal of diagonals between adjacent triangles. An efficient method to retrieve edges from a triangular mesh has been discussed in Section 2.5.4, and with little care, the triangles attached to each edge can also be identified.
2. *Removal of diagonals*
   Before the removal of edges, a check is made on the removability and the β value of each edge. In general, one can only remove those lines shared by two triangles, and these removable edges will be referred to as diagonals. Hence, at the beginning of the merging process, all boundary segments have to be flagged so as not to be removed. To determine which diagonal is to be removed, the β* value of each diagonal is computed, which is equal to the β value of the diagonal minus the β values of the neighbouring diagonals. The diagonal having the greatest β* value will be removed, and a quadrilateral element will be generated. If no more edges can be removed and the β values of the diagonals are less than the prescribed γ value, the merging process terminates.
3. *Updating the mesh*
   Whenever a quadrilateral is generated, the mesh has to be updated following steps i to iii:
   i.   Suppose that line AC is to be removed as shown in Figure 3.75. The two triangles ABC and ACD are identified and deleted from the mesh, whereas a quadrilateral element ABCD is added to the mesh.
   ii. The edges associated with the quadrilateral, i.e. the four edges of the quadrilateral AB, BC, CD and DA, have to be flagged so that these edges will not be removed as the merging process continues.
   iii. The β* values of the neighbouring edges are updated; at most eight lines will be affected, as shown in Figure 3.75. Return to step 2 for the creation of more quads.

The DT of 1000 randomly generated points is shown in Figure 3.76a. The merging algorithm is applied to the triangular mesh with γ = 0.9, and the resulting mesh is shown in Figure 3.76b. It is seen that with a high γ value of 0.9, very few quadrilaterals closed to rectangles are

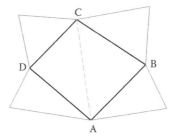

Figure 3.75 A quadrilateral is created by removing diagonal AC between triangles ABC and ACD.

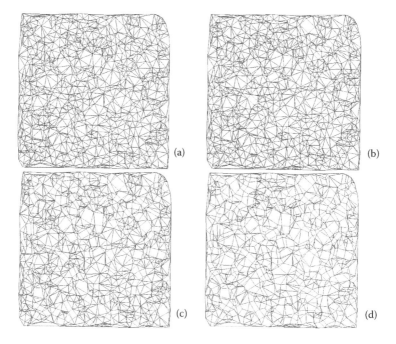

Figure 3.76 Merging triangles into quadrilaterals: (a) Delaunay triangulation of 1000 points; (b) mesh with
$\gamma = 0.9$; (c) mesh with $\gamma = 0.6$; (d) mesh with $\gamma = 0.3$.

retrieved from the triangulation by the algorithm. When γ is set to a lower value of 0.6, many more quadrilaterals are created, as shown in Figure 3.76c. With a further reduction of γ to 0.3, most of the triangles are converted to quadrilaterals, as shown in Figure 3.76d.

Remarks: A new criterion, the distortion coefficient β, is introduced for merging triangles with which the suitability of dividing a quadrilateral into two triangles can be quantified. This also provides a systematic way to determine the priority sequence for the removal of diagonals between triangles. A γ parameter, which is specified by the user to exercise higher preference to either triangles or quadrilaterals, and a careful choice of γ would lead to high-quality FE mesh optimised in both element types. The merging algorithm can be considered as a post-treatment of a triangular mesh, which offers an efficient way for the generation of a mixed mesh of triangular and quadrilateral elements. The method is simple but effective if a quadrilateral-dominated mesh is required. Apart from the information on the edges, which is well prepared in advance prior to mesh merging, only a couple of β^* values need to be updated for each diagonal removal. These new β^* values are compared with the current β^*_{max} for the selection of the next diagonal to be removed. Triangulation of a simply or multi-connected domain on a plane and over surfaces can now be done fairly easily; the algorithm offers an efficient way to produce unstructured mixed mesh of triangular and quadrilateral elements. In Section 3.9.4, a systematic merging procedure will be presented to convert an unstructured triangular mesh completely into an all-quadrilateral mesh.

3.9.4 All-quad unstructured mesh

If the grading and the size of the quadrilateral elements can be carefully controlled, an increase in efficiency over a pure triangular mesh can be achieved (Lo and Lee 1992; Lee and Lo 1995). However, relative to triangular meshing, algorithms for unstructured quadrilateral meshing were developed at a much later stage. Apart from the method of region sub-division (Joun and Lee 1997), an effective way in generating gradation quadrilateral meshes based on the indirect method to convert a triangular mesh completely into quadrilateral elements by means of a systematic merging and splitting process was presented by Lee and Lo (1994) and Masa et al. (1999). The element size density and grading of the triangular mesh will be preserved in the merging process. Hence, taking a graded triangular mesh as the initial mesh, an equally well-graded quadrilateral mesh will be resulted. This conversion scheme is applicable to arbitrary unstructured triangular meshes on a plane and over curved surfaces, provided that the triangular meshes to be processed are bounded by an even number of boundary segments. A complete conversion into quadrilaterals is possible but is not guaranteed unless the merging of triangles is conducted carefully in a controlled manner.

For multi-connected regions, it is further imposed that the number of boundary segments of each closed boundary loop must be even. This condition allows us to devise a scheme in which cut lines (Johnston et al. 1991; Johnston and Sullivan 1992) can be avoided in the mesh merging. The initial triangular mesh can be generated by any standard triangulation schemes such as DT or AFT, as described in Sections 3.5 and 3.6. As the number of boundary edges is even, the triangular mesh must consist of an even number of triangles.

The basic idea for the generation of quadrilaterals lies in the fact that a triangular mesh consisting of an even number of triangles can always be converted into an all-quad mesh by carefully controlled merging and triangle splitting operations even for the extreme case of a concave quadrilateral of two triangles, as shown in Figure 3.77. The AFT is employed to ensure that unconverted regions of triangles are always bounded by an even number of

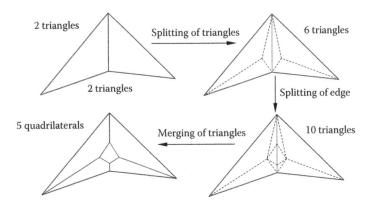

Figure 3.77 Forming of quadrilaterals by splitting and merging of triangles.

segments. The merging front will shrink whenever two triangles are combined to form a quadrilateral, and the entire process terminates when the merging front is reduced to zero, and all the triangles have been processed. The conversion process can be divided into three distinct phases, which are described in Sections 3.9.4.1 to 3.9.4.4.

3.9.4.1 Initialisation of the merging front

Before the merging process starts, an initial merging front has to be set up, which is the collection of all the boundary segments, as shown in Figure 3.78. The orientation of the merging front follows that of the boundary segments, such that the unmerged triangles are lying on the left-hand side of the merging front. The merging front will be updated during the conversion process by adding and deleting line segments to and from the merging front.

Sub-fronts may also be formed from time to time, and the merging process is completed when no more triangles are left, and all fronts are reduced to nothing.

3.9.4.2 Merging of triangles

A quadrilateral is formed when two triangles on the merging front are combined, and the details of such a procedure are given as follows.

i. A segment AB on the merging front is selected as the base segment for the construction of a quadrilateral, as shown in Figure 3.79a.
ii. Let triangle ABC be the triangle connected to segment AB; two triangular elements ACL and CBR, which can possibly merge with triangle ABC on the two sides, are identified.
iii. If node C does not divide the merging front into two sub-fronts, either triangle ACL or CBR can be considered to combine with triangle ABC to form a quadrilateral. Suppose that triangle ACL is taken; if node L does not divide the merging front into sub-fronts

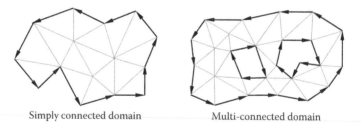

Simply connected domain   Multi-connected domain

Figure 3.78 The domain boundary is taken as the initial merging front.

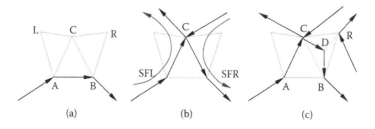

(a)   (b)   (c)

Figure 3.79 Merging of triangles into quadrilaterals: (a) node C does not divide the merging front; (b) node C divides the merging front; (c) node R divides the merging front.

or if both these two sub-fronts consist of an even number of segments, quadrilateral ABCL can be formed. Otherwise, triangle CBR will be considered. In case R divides the merging front into sub-fronts having an odd number of segments, go to step v.

iv. Triangle ABC divides the merging front into two sub-fronts SFL on the left and SFR on the right, as shown in Figure 3.79b. As there are always an even number of segments on a merging front, excluding segment AB, one sub-front should have an even number of segments and the other sub-front an odd number of segments. To maintain an even number of segments on each boundary loop, the triangle selected to merge with triangle ABC is the one lying on the sub-front having an odd number of segments, as shown in Figure 3.79b.

v. The triangle CBR is split into three triangles by inserting a point at its centroid D, as shown in Figure 3.79c. Diagonal BC between triangles ABC and CBD is removed to create quadrilateral ABDC, and the merging front is split into two sub-fronts of even number of segments.

3.9.4.3 Updating merging front

Whenever a quadrilateral is constructed, the merging front has to be updated in a way similar to the update in the AFT, as described in Section 3.6.2. The segments that have been used to form the sides of the new quadrilateral element will be deleted from the merging front, and new edges created in the merging process will be added to the merging front, as shown in Figure 3.79.

3.9.4.4 Complete conversion to quadrilateral mesh

Steps i to v are repeated to convert triangles into quadrilaterals until the number of segments in the merging front is reduced to zero. By this time, all the triangular elements are converted to quadrilaterals. It is remarked that by selecting a triangle for merging with the base triangle as described in steps iii to v, all the merging fronts and sub-fronts will, at any time, remain as closed loops of an even number of line segments. The flowchart for the merging of a triangular mesh into a quadrilateral mesh is shown in Figure 3.80.

3.9.5 Mesh quality enhancement

Some of the quadrilaterals produced by the mesh merging algorithm may be distorted, and elements with internal angles greater than or equal to 180° may be generated. Hence, mesh quality enhancements are required to bring the mesh quality up to an acceptable level. Mesh quality improvement can be achieved by geometrical node shifting or by topological modification on the ways elements are connected (Canann et al. 1998). Mesh enhancement by node shifting and topological modification will be discussed in detail in Chapter 6, and in the following, some simple topological operations pertinent to quadrilateral meshes will be discussed.

In most cases, the information in judging whether a swap would improve the overall quality of the mesh is the number of elements connected to a node. This information has to be prepared beforehand, and an algorithm has already been presented in Section 2.5.3 to find out the elements connected to each node of the mesh. The number N, which is the number of elements surrounding a given node, indicates the degree of uniformity of the quadrilateral mesh. When N equals to four for most of the interior nodes, the mesh is highly uniform. However, if N is much greater or less than four, severe distortion will result. Therefore, it is

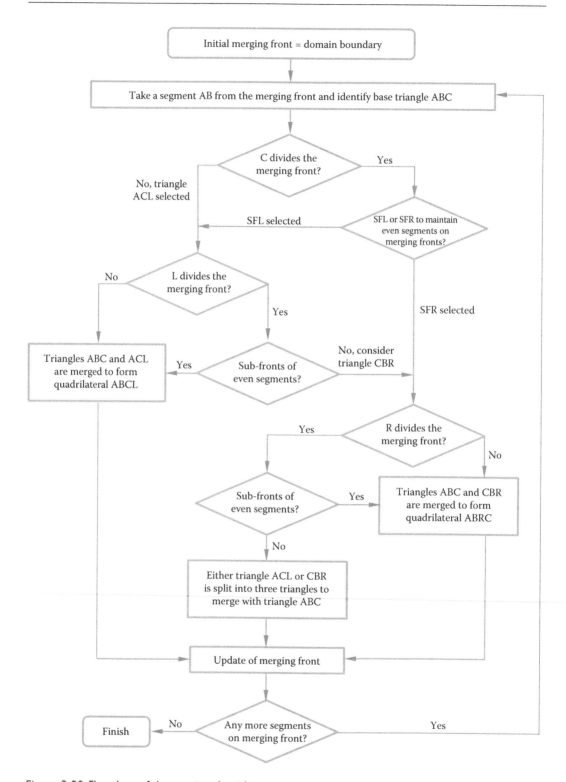

Figure 3.80 Flowchart of the merging algorithm.

expected that for a good-quality quadrilateral mesh, the number N for all the interior nodes should fall within the range

$$3 \leq N \leq 6 \tag{3.7}$$

The objective of mesh modification is to rearrange the topological structure of the mesh so that all the interior nodes will have N satisfying Equation 3.7. Node repositioning in the next phase of mesh quality improvement can be most effectively applied to the interior nodes with connectivity number N governed by Equation 3.7.

3.9.5.1 Elimination of node

All interior nodes of the mesh will be examined in turn. An interior node A will be eliminated if the connectivity number N is equal to two. The two elements connected to node A will be combined to a single element, as shown in Figure 3.81.

3.9.5.2 Elimination of element

All quadrilaterals in the mesh are inspected in turn. Quadrilateral Q will be eliminated if the connectivity number N of a pair of opposite interior nodes A and B is equal to three for each of the nodes. One of the nodes will also be eliminated as shown in Figure 3.82.

3.9.5.3 Swapping of diagonals

The aim of diagonal swap is to even out the connectivity numbers of the interior nodes. Let's consider the diagonal swapping of the hexagon formed by quadrilaterals ABCD and ADEF sharing a common edge AD, as shown in Figure 3.83. Compute the sum of the connectivity numbers of each pair of opposite nodes of hexagon ABCDEF.

$$N_1 = N_A + N_D, \qquad N_2 = N_B + N_E, \qquad N_3 = N_C + N_F$$

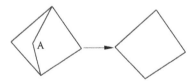

Figure 3.81 Node A is eliminated.

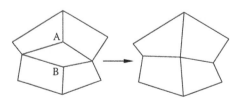

Figure 3.82 Elimination of element.

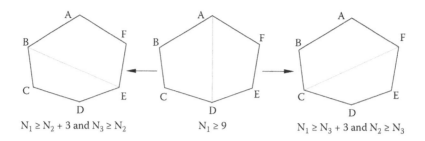

$$N_1 \geq N_2 + 3 \text{ and } N_3 \geq N_2 \qquad N_1 \geq 9 \qquad N_1 \geq N_3 + 3 \text{ and } N_2 \geq N_3$$

Figure 3.83 Swapping of diagonal.

If $N_1 \geq 9$, diagonal AD will be swapped with BE if

$$N_1 \geq N_2 + 3 \text{ and } N_3 \geq N_2 \qquad (3.8)$$

or with CF if

$$N_1 \geq N_3 + 3 \text{ and } N_2 \geq N_3 \qquad (3.9)$$

There will be no diagonal swap if neither Equation 3.8 nor 3.9 is satisfied.

3.9.5.4 Elimination of segment

All the line segments connected to two interior nodes are examined in turn. Line segment AB will be eliminated if the connectivity numbers of both nodes A and B are equal to three, i.e. $N_A = 3$ and $N_B = 3$. The segment elimination is completed by the elimination of quadrilateral ABCD and AFGB, as shown in Figure 3.84. A new diagonal DG will be formed if

$$N_C + N_F \geq N_D + N_G$$

Otherwise, diagonal CF will be formed.

3.9.6 Examples of quadrilateral meshes

The conversion of the triangular mesh of 13,171 nodes and 26,094 triangles into a quadrilateral mesh of 15,463 nodes and 15,342 elements is shown in Figure 3.85. The increase in the number of nodes is probably due to the additional nodes introduced in the splitting of

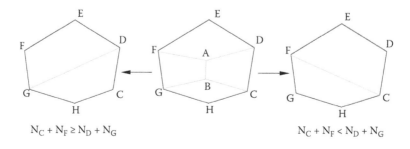

$$N_C + N_F \geq N_D + N_G \qquad\qquad N_C + N_F < N_D + N_G$$

Figure 3.84 Elimination of segment.

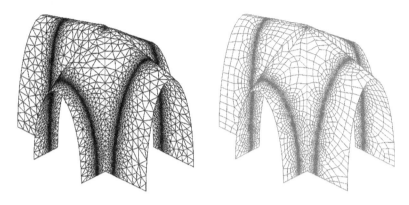

Figure 3.85 Triangular mesh converted to quadrilateral mesh.

Figure 3.86 Splitting boundary triangles to avoid large obtuse angles.

triangular elements. With post-processing mesh optimisation, the quality of the resulting quadrilateral mesh is quite promising with a β coefficient at 0.706 (Section 3.9.3.1).

Remarks: In order to avoid the formation of quadrilaterals with large obtuse angles, adjacent boundary triangles with a combined angle greater than some threshold, say $3\pi/2$, are split, as shown in Figure 3.86. This process of splitting boundary triangles is repeated until no more combined angle from two adjacent boundary triangles is greater than the threshold value of $3\pi/2$.

As shown in Figure 3.79, in case both triangles ACL and CBR are feasible options to merge with base triangle ABC without violating the parity requirement (even number of segments on a closed loop) of the merging fronts, the triangle that results in a quadrilateral of better quality will be selected. During the process of merging triangles, the merging fronts will be combined or further split into sub-fronts, and it is important to ensure that an even number of segments on each frontal closed loop is always maintained.

Chapter 4

Mesh generation over curved surfaces

A surface is curved and non-Euclidean, and not every point is visible, but all these are negated by the fact that locally it has a topological structure that resembles that of a plane.

4.1 INTRODUCTION

There is a growing demand for robust and efficient discretisation algorithms for general curved surfaces into finite elements of variable sizes and shapes. The boundary of a physical object is generally made up of patches of curved surfaces, which can be typically represented by NURBS surfaces generated in a commercial CAD environment. The surface discretisation itself is a surface model of the object for rapid visualisation, and it can also be used for the FE analysis, or it serves as an input for FE mesh generation within a volume. While many techniques developed in planar mesh generation are still applicable to surface meshing, there are difficulties and different geometric characteristics over curved surfaces that require modification, extension and special considerations.

There are, in general, two ways to represent a surface for the purpose of visualisation and computation. One way is to use analytical surface patches via functions such as Coon's patches, B-splines or NURBS surfaces (Kobbelt et al. 1997; Frey and George 2000). Another way is to use discrete data structure such as quadrilateral facets and triangular facets (Lohner 1996a). For engineering applications, surfaces defined by discrete data are widely adopted, for instance, in the presentation of complex molecule (Akkiraju and Edelsbrunner 1996; Laug and Borouchaki 2000, 2001), engineering design (Shostko et al. 1999), visualisation (Bruyns and Senger 2001; Dillard et al. 2007) and computational models for analysis by the FE method (Lo 1988c, 1995). According to the type of surface representation, many surface mesh generation schemes are available, and they can be broadly classified into either parametric mapping approaches or direct surface mesh generation in 3D space.

4.1.1 Parametric meshing for curved surfaces

By this approach, the surface to be discretised is represented by a bivariate analytical function such that any point on the 3D surface is mapped to a 2D parametric space (Canann et al. 1997; Cuillière 1998; Lee and Hobbs 1998; Shimada and Gossard 1998; Borouchaki et al. 2000a; Sherwin and Peiro 2002; Lee 2003a,b; Cherouat et al. 2010). The mesh generation process is carried out entirely on the parametric space by a 2D general-purpose mesh generator. The final surface mesh is obtained by proper transformation of the mesh generated on the parametric space back to the 3D space, as shown in Figure 4.1. This method gives reasonably good meshes for simple surfaces that are sufficiently smooth with respect to a planar

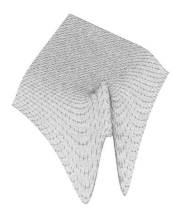

Figure 4.1 **A structured surface mesh by parametric mapping.**

domain. For the more complex curved surfaces, the resulting meshes are usually of poor quality, owing to the element distortions induced by the mapping (Burouchaki et al. 2005).

An improved version of the mapping method is to first decompose the complex surfaces into a union of simple surfaces, and then meshes are generated on each of the sub-surfaces (Lo 1988c, 1995; Owen 1998). Such a modification does give much improved results; however, for general free-form surfaces with a large variation of curvatures, the problem of decomposing the surface into simpler smooth sub-surfaces may be just as hard as the mesh generation of the original surface, and special care has to be taken on the boundaries to ensure compatibility among patches. Hence, in order to obtain an FE mesh satisfying the user specifications, it is necessary to control the shape and size of the elements in the parametric space (Remacle et al 2010). To generate an FE mesh over curved surfaces in compliance with nodal spacing specifications, we have to modify the mesh generation algorithm so that stretched or anisotropic elements generated on the 2D parametric domain will map to well-shaped elements on the 3D curved surface.

A method commonly employed in practice is to take advantage of the surface derivatives easily computed from NURBS surfaces. Borouchaki and Laug (2004) and Laug (2010) proposed the use of a general anisotropic metric derived from the surface curvature. The metric represented by a 2×2 matrix can be used to transform vectors and distances in the parametric space onto the surface. For anisotropic meshing based on DT, the empty circle property effectively becomes an empty-ellipse criterion, and the meshing of analytical surfaces is achieved through a 2D mesh generation process governed by an anisotropic metric. Without any additional effort, the option to incorporate the element size and the stretching properties of the surface mesh can also be included within the metric. Thus, with a slight modification in the metric tensor, anisotropic meshes of variable element size can be easily constructed over curved surfaces through a mapping process.

An approach to parametric Delaunay surface meshing has been presented by Borouchaki et al. (2000a) and Valette et al. (2008). ADF surface mesh generation algorithms, which employ a metric derived from the first fundamental form of the surface, were presented by Cuillere (1998), Lee and Hobbs (1998), Lee (1999), George et al. (2002) and Aubry et al. (2011). Unstructured surface meshes generated by the parametric mapping are shown in Figure 4.2. Much more complicated curved surfaces can be represented as a union of surface patches, each of which can be discretised by the mapping technique (Inoue et al. 2001; Laug and Borouchaki 2004). As mentioned before, it is interesting to note that unstructured anisotropic meshes over curved surfaces can be easily achieved by properly modifying the

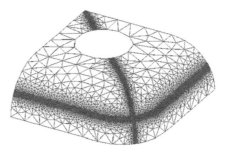

Figure 4.2 Unstructured mesh by parametric mapping.

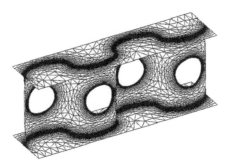

Figure 4.3 Anisotropic mesh by parametric mapping.

underlying metric used in the parametric space (Borouchaki and Frey 1998; Lee and Lee 2002). Figure 4.3 shows some anisotropic meshes on curved surfaces generated by a parametric mapping. Apart from those versatile mesh generation schemes on 2D domains, given a parametric mapping, surface meshes can also be generated by packing ellipses according to the specified metric (Lo and Wang 2005; Wang et al. 2007a). What we have discussed thus far are mesh generation techniques for unstructured triangular meshes of variable size and shape over curved surfaces.

4.1.2 Direct mesh generation on surfaces

Direct 3D surface mesh generation forms elements directly on the surface without the need for a parametric representation of the underlying geometry. In the case where a parametric representation is not available or where the surface parameterisation is poor, direct 3D surface mesh generation can be applied. Lau and Lo (1996) and Lo and Lau (1998) presented an ADF scheme for the generation of triangular meshes on analytical curved surfaces in space. In this method, surface normals and tangents are computed to determine the direction of the generation front. A number of surface projections are required to bring the new nodes generated back on the surface. Intersection checks are required to make sure that triangles on the surface do not overlap. Direct mesh generation on surfaces will also be useful in situations of remeshing an existing surface triangulation of discrete elements for which surface parameterisation is not readily available or for the generation of hybrid meshes with surface and 3D solid elements (Lee et al. 2010). If an adaptive refinement mesh is required for the current FE model, defined in terms of element connections and nodal co-ordinates, direct 3D mesh construction is the most suitable. Figure 4.4 shows some triangular meshes generated by a direct 3D generation method.

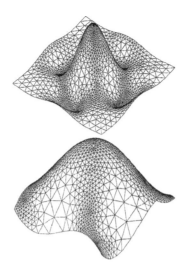

Figure 4.4 Surface meshes by 3D direct construction.

Direct 3D surface mesh generation of quadrilateral elements is also possible. A direct 3D implementation of the *paving* algorithm over curved surfaces was presented by Cass et al. (1996). Figure 4.5 shows a quadrilateral mesh of two intersecting cylinders by direct construction of quadrilateral elements. Heuristic *sticky space* is introduced to detect intersection and overlapping quadrilaterals. Nevertheless, it is considerably much more flexible to generate triangular meshes over curved surfaces, and by the systematic merging of triangles, unstructured quadrilateral meshes can be conveniently generated by a two-step process with much better control on the size and shape of the quadrilateral elements.

Quadrilateral-dominated meshes for curved surfaces can also be obtained by combining pairs of adjacent triangles. Gradation quadrilateral meshes over curved surfaces can be obtained by conversion of surface triangulation with particular care taken to maintain correct surface curvature. Similar to planar meshes discussed in Section 3.9, all-quadrilateral meshes on curved surfaces can be generated by systematic merging of triangular meshes. Figure 4.6 shows examples of quadrilateral meshes from the conversion of triangulated surfaces. The Q-morph scheme proposed by Owen et al. (1999) can also be used to generate quadrilateral meshes on triangulated curved surfaces. Q-morph can be considered as a remeshing process, which converts a background triangulation into a quadrilateral mesh. This method is applicable to 2D domains as well as over curved surfaces to generate gradation unstructured quadrilateral meshes, as shown in Figure 4.7.

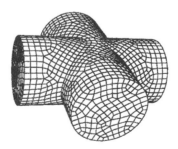

Figure 4.5 A quadrilateral mesh by 3D direct construction.

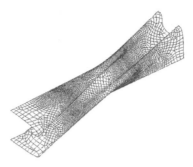

Figure 4.6 Quadrilateral mesh by merging of triangles.

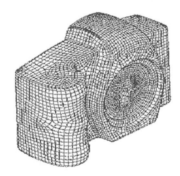

Figure 4.7 Quadrilateral mesh by Q-Morph scheme.

4.1.3 Surface meshing by means of intersection

Apart from the two approaches discussed, there is a third approach for the construction of complex surface models of triangular elements. The first approach is the mapping approach, which generates a mesh on the parametric domain followed by a mapping process of the resulting mesh onto the surface. The second approach is to form triangles directly on the surface by the well-known mesh generation techniques, such as the Octree method, the AFT, the paving method and so forth. The third approach is to construct models by Boolean operations such as intersection and union between groups of surfaces composed of triangular facets. A great variety of meshes can be easily created by selectively putting together various surface parts derived from surface intersections. This approach has been explored and applied in the field of FE MG (Lo 1995; Coelho et al. 2000; Cebral et al. 2001; Lo and Wang 2003, 2004, 2005a), molecular modelling (Laug and Borouchaki 2000, 2001, 2002, 2003b), engineering design (Bonet and Peraire 1991; Aftosmis et al. 1998), etc.

4.2 PARAMETRIC MAPPING METHOD

4.2.1 Introduction

4.2.1.1 The mapping φ from planar domain Ω to the surface S

Let ϕ be the mapping from a two-dimensional parametric domain $\Omega \subset \mathbb{R}^2$ onto a curved surface $S \subset \mathbb{R}^3$ in the three-dimensional space such that $\phi: \Omega \to S$.

2D planar domain Ω can be readily meshed by the methods, as discussed in Chapter 3, and when points on Ω are mapped by ϕ onto S in 3D space, an FE mesh is created on the surface S. However, in general ϕ is not isometric (length-preserving), and the edges and hence the shape and the area of the triangular and quadrilateral elements would change by the mapping ϕ. As a result, the quality of well-shaped triangles and quadrilaterals in Ω may deteriorate upon the transformation from Ω to S. The geometrical approximation may not be adequate either, as straight line segments are used in places of large curvatures on the surface.

4.2.1.2 Gap between a triangular facet and the curved surface

Let S be the curved surface of parameterisation $\phi(u,v)$ with parametric variables (u,v) on planar domain Ω, and T be a triangle in Ω with vertices mapped on the surface S. The gap between the triangle and the surface S, denoted by δ, is given by Filip et al. (1986)

$$\delta = \sup_{(u,v)\in T} \left\| \phi(u,v) - T(u,v) \right\|$$

where $T(u,v)$ is a point on triangle T, and $\phi(u,v)$ is the corresponding point on surface S.

$$\delta \leq \frac{2}{9} h^2 (I_1 + 2I_2 + I_3)$$

in which h is the longest edge of triangle T, and I_1, I_2 and I_3 are given by

$$I_1 = \sup_{(u,v)\in T} \left\| \phi''_{uu}(u,v) \right\|, \quad I_2 = \sup_{(u,v)\in T} \left\| \phi''_{uv}(u,v) \right\|, \quad I_3 = \sup_{(u,v)\in T} \left\| \phi''_{vv}(u,v) \right\|$$

Hence, the gap between the surface and a triangular facet is governed by the longest edge of the triangle and the second derivatives of the surface.

4.2.1.3 Metric for curved surface geometry

The idea to be explored in detail is the design of a mesh generated on the parametric planar domain such that the resulting mesh on surface S upon the transformation by ϕ is a close geometrical approximation to the surface. This characteristic can be achieved by controlling the mesh construction on the parametric space ensuring that the gap between an edge and the surface is within any specified tolerance.

4.2.2 Fundamental forms and the related metric

ϕ is a mapping from Ω to S and is of class C^2. For curved surfaces with ridges, the C^2 requirement can be relaxed for a finite number of points, where the surface characteristics can be estimated or interpolated from the neighbouring points (Lee and Lee 2003; Clemencon et al. 2006). Two fundamental forms are defined at every point $p \in S$ allowing the length of a curve on S and the curvature along the curve to be determined.

4.2.2.1 Tangent and normal vectors

$p = \phi(u,v)$. The tangent plane T_p at p is spanned by the basis (co-ordinate) vectors.

$$\mathbf{u} = \frac{\partial \mathbf{p}}{\partial u} = \frac{\partial \phi}{\partial u} = \phi_u \quad \text{and} \quad \mathbf{v} = \frac{\partial \mathbf{p}}{\partial v} = \frac{\partial \phi}{\partial v} = \phi_v$$

The unit normal to the surface S at $\mathbf{p}$ is given by

$$\mathbf{n} = \frac{\mathbf{u} \times \mathbf{v}}{\|\mathbf{u} \times \mathbf{v}\|}$$

Hence, $(\mathbf{u}, \mathbf{v}, \mathbf{n})$ represents a local base at $\mathbf{p}$, as shown in Figure 4.8.

As $\mathbf{u}$ and $\mathbf{v}$ are basis vectors of T_p, every vector V on T_p can be written as

$$V = \alpha \mathbf{u} + \beta \mathbf{v} \text{ with } \alpha, \beta \in \mathbb{R}$$

and the first fundamental form Φ_1 is given by

$$\Phi_1 = V \cdot V = E\alpha^2 + 2F\alpha\beta + G\beta^2$$

where

$$E = \mathbf{u} \cdot \mathbf{u}, F = \mathbf{u} \cdot \mathbf{v}, G = \mathbf{v} \cdot \mathbf{v}$$

Let $\Lambda = \begin{bmatrix} \alpha \\ \beta \end{bmatrix}$ and $M = \begin{bmatrix} E & F \\ F & G \end{bmatrix}$ then $V \cdot V = \Lambda^T M \Lambda$.

Matrix M, which measures the length of vector V, is symmetric and positive-definite called the tangent plane metric at $\mathbf{p}$. By means of metric M, the length of a line segment created on the parametric space can be evaluated. As shown in Figure 4.8, a curve Γ on S is defined by

$$\gamma: (u(t), v(t)) \rightarrow S \quad t \in [a,b]$$

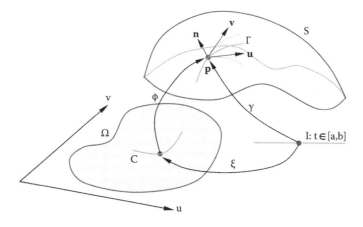

Figure 4.8 Curved surface S produced by parametric mapping ϕ.

The length of curve Γ is given by

$$L(\Gamma) = \int_a^b \left\| \gamma'(t) \right\| dt = \int_a^b \sqrt{\gamma'(t) \cdot \gamma'(t)} \, dt$$

From the interval $I = [a,b]$, a curve C can be created by the mapping $\xi: I \rightarrow \Omega$, such that

$$\gamma = \phi \circ \xi \Rightarrow \gamma' \cdot \gamma' = \gamma'^T \gamma' = (\phi' \circ \xi')^T (\phi' \circ \xi')^T = \xi'^T \phi'^T \phi' \xi'$$

However,

$$\phi' = \begin{bmatrix} \dfrac{\partial \phi}{\partial u} \\[2mm] \dfrac{\partial \phi}{\partial v} \end{bmatrix} = \begin{bmatrix} \dfrac{\partial p}{\partial u} \\[2mm] \dfrac{\partial p}{\partial v} \end{bmatrix} = \begin{bmatrix} u \\ v \end{bmatrix} \Rightarrow \phi'^T \phi' = \begin{bmatrix} u & v \end{bmatrix} \begin{bmatrix} u \\ v \end{bmatrix} = \begin{bmatrix} u \cdot u & u \cdot v \\ u \cdot v & v \cdot v \end{bmatrix} = M$$

$$\gamma'^T \gamma' = \xi'^T M \xi' \text{ or } L(\Gamma) = \int_a^b \sqrt{\xi'^T M \xi'} \, dt$$

In case C is a line segment from point A to point B, we have

$$\xi(t) = A + tAB, \quad t \in [0,1], \quad \xi' = AB \text{ and } L(\Gamma) = \int_0^1 \sqrt{AB^T M(A + tAB) AB} \, dt$$

Geometric interpretation of M: Let A be a point on Ω and $M(\phi(A))$ be the metric at point A. For arbitrary real value ε, the locus of point B on Ω such that

$$AB^T M(\phi(A)) AB = \varepsilon^2$$

is an ellipse. In other words, an ellipse on Ω will map to a circle on T_p of S and vice versa. In general, an ellipse on Ω will map to an ellipse of different size and orientation on the tangent plane of S, and by controlling the shape and size of the elements on Ω, meshes of various characteristics on S can be created.

4.2.3 Principal curvatures

Principal curvatures (the maximum and minimum curvatures at a point) are of fundamental importance to the curved surface under consideration, which can also be used to define natural characteristic lines on the surface. Let $V = \alpha u + \beta v$ be a vector on the tangent plane T_p of S at p and n be the unit normal at p. Set

$$L = n \cdot \frac{\partial^2 \phi}{\partial u^2}, \quad M = n \cdot \frac{\partial^2 \phi}{\partial u \partial v}, \quad N = n \cdot \frac{\partial^2 \phi}{\partial v^2}$$

The second fundamental form, which is the projection of n onto the second derivatives of ϕ, is given by

$$\Phi_2 = L\alpha^2 + 2M\alpha\beta + N\beta^2$$

The curvature κ of S at $\mathbf{p}$ associated with tangent vector $V = \alpha\mathbf{u} + \beta\mathbf{v}$ is given by (Lelong-Ferrand and Arnaudiès 1977)

$$\kappa = \frac{\Phi_2}{\Phi_1}$$

Relative to the basis vectors $\mathbf{u}$ and $\mathbf{v}$, the principal curvature at $\mathbf{p}$ is given by

$$\frac{\partial\kappa}{\partial\alpha} = 0 \quad \text{and} \quad \frac{\partial\kappa}{\partial\beta} = 0$$

or

$$(FL - EM)\alpha^2 + (GL - EN)\alpha\beta + (GM - FN)\beta^2 = 0$$

Let $\lambda = \dfrac{\alpha}{\beta}$. We have

$$a\lambda^2 + b\lambda + c = 0$$

in which

$$a = FL - EM, \, b = GL - EN, \, c = GM - FN$$

In case all the coefficients a, b and c of the quadratic equation are all zero, Φ_2 is proportional to Φ_1 along any direction at $\mathbf{p}$, and the curvature is constant in any direction (locally spherical). In such a case, the curvature is given by

$$\kappa = \frac{L}{E}$$

In general, the quadratic equation gives two distinct roots λ_1 and λ_2 such that the orthogonality relationship holds, i.e. $V_1 \cdot V_2 = 0$, where V_1 and V_2 are tangent vectors in the directions λ_1 and λ_2. Define unit vectors in the directions of V_1 and V_2 such that

$$\hat{\mathbf{v}}_1 = \frac{V_1}{\|V_1\|} \quad \text{and} \quad \hat{\mathbf{v}}_2 = \frac{V_2}{\|V_2\|}$$

The triple $(\hat{\mathbf{v}}_1, \hat{\mathbf{v}}_2, \mathbf{n})$ forms an orthonormal basis at $\mathbf{p}$. Let κ_1 and κ_2 be the principal curvatures and ρ_1 and ρ_2 be the corresponding principal radii of curvatures. Then

$$\rho_1 = \frac{1}{\kappa_1} \quad \text{and} \quad \rho_2 = \frac{1}{\kappa_2}$$

The surface turns more rapidly at points of large curvature or small radius of curvature where finite elements of reduced size have to be used to approximate the surface geometry.

4.2.3.1 Gaussian curvature and mean curvature

The Gaussian curvature K at a point on a surface is the product of the principal curvatures, and its value depends only on the geometrical shape of the surface but not on the way it is embedded in space. On the other hand, the mean curvature H is the arithmetic mean of the two principal curvatures. Like the principal curvatures, the Gaussian curvature and the mean curvature are intrinsic properties of a curved surface at a point, and the relationships between principal curvatures, Gaussian curvature and mean curvature are given by

$$K = \kappa_1 \kappa_2 \quad \text{and} \quad H = \frac{\kappa_1 + \kappa_2}{2}$$

Based on these relationships, the principal curvatures can also be determined as follows:

$$K = \frac{LN - M^2}{EG - F^2} \quad \text{and} \quad H = \frac{GL - 2FM + EN}{2(EG - F^2)} \quad \kappa_1, \kappa_2 = H \pm \sqrt{H^2 - K}$$

Classification of surface by means of Gaussian curvature K (Smith and Farouki 2001):

K > 0: elliptic, K < 0: hyperbolic, K = 0: parabolic, $\quad \kappa_1 = \kappa_2$: umbilic point

Most surfaces consist of regions of positive Gaussian curvature of elliptical points and regions of negative Gaussian curvature of hyperbolic points separated by curves of points with zero Gaussian curvature of parabolic lines.

4.2.4 Metric and principal curvatures

Let V be a vector on the tangent plane T_p making an angle θ with $\hat{v}_1$. V can be written as $V = \|V\|(\hat{v}_1 \cos\theta + \hat{v}_2 \sin\theta)$. The curvature along vector V with angle θ from $\hat{v}_1$ is given by

$$\kappa = \kappa_1 \cos^2\theta + \kappa_2 \sin^2\theta$$

This result also suggests that the surface curvature is a second-order tensor following the usual rule of tensor transformation with respect to angle θ. In terms of tensor notation, the surface curvature tensor $\boldsymbol{\kappa}$ is written as

$$\boldsymbol{\kappa} = \kappa_1 \hat{v}_1 \hat{v}_1 + \kappa_2 \hat{v}_2 \hat{v}_2$$

Curvature along vector V is given by

$$(\boldsymbol{\kappa} \cdot \hat{v}) \cdot \hat{v} = \kappa_1 \hat{v}_1 (\hat{v}_1 \cdot \hat{v}) \cdot \hat{v} + \kappa_2 \hat{v}_2 (\hat{v}_2 \cdot \hat{v}) \cdot \hat{v} = \kappa_1 (\hat{v}_1 \cdot \hat{v})^2 + \kappa_2 (\hat{v}_2 \cdot \hat{v})^2$$
$$= \kappa_1 \cos^2\theta + \kappa_2 \sin^2\theta$$

where $\hat{v} = \dfrac{V}{\|V\|}$ is the unit vector along direction vector V.

Let $\mathbf{p}$ be a point on S, $\hat{\mathbf{v}}_1$ and $\hat{\mathbf{v}}_2$ be the two principal directions and κ_1 and κ_2 be the respective principal curvatures. If $\mathbf{q}$ is a point on the ellipse on the tangent plane T_p with axes κ_1 and κ_2, we have $\|\mathbf{pq}\| \geq \kappa$, as shown in Figure 4.9a.

$$\|\mathbf{pq}\|^2 = \kappa_1^2 \cos^2\theta + \kappa_2^2 \sin^2\theta \quad \text{and} \quad \kappa^2 = (\kappa_1 \cos^2\theta + \kappa_2 \sin^2\theta)^2$$

$$\|\mathbf{pq}\|^2 - \kappa^2 = \kappa_1^2 \cos^2\theta(1 - \cos^2\theta) + \kappa_2^2 \sin^2\theta(1 - \sin^2\theta) - 2\kappa_1\kappa_2 \cos^2\theta \sin^2\theta$$
$$= \kappa_1^2 \cos^2\theta \sin^2\theta + \kappa_2^2 \sin^2\theta \cos^2\theta - 2\kappa_1\kappa_2 \cos^2\theta \sin^2\theta = (\kappa_1 - \kappa_2)^2 \sin^2\theta \cos^2\theta \geq 0$$

This result shows that the curvature at a point $\mathbf{p}$ in any direction is bounded by the ellipse centred at $\mathbf{p}$ with κ_1 and κ_2 as axes. Similarly, as shown in Figure 4.9b, for ellipse drawn with axes in terms of radius of curvature $\rho = 1/\kappa$, we have $\|\mathbf{pq}\| \leq \rho$.

As the locus of a metric is an ellipse, a metric M can be defined on the tangent plane T_p at $\mathbf{p}$ such that the length in any direction from $\mathbf{p}$ will not exceed the radius of curvature in that direction. Such a metric is given by

$$M = \begin{bmatrix} \hat{\mathbf{v}}_1 & \hat{\mathbf{v}}_2 \end{bmatrix} \begin{bmatrix} \kappa_1^2 & 0 \\ 0 & \kappa_2^2 \end{bmatrix} \begin{bmatrix} \hat{\mathbf{v}}_1 \\ \hat{\mathbf{v}}_2 \end{bmatrix} = \begin{bmatrix} \hat{\mathbf{v}}_1 & \hat{\mathbf{v}}_2 \end{bmatrix} \begin{bmatrix} \dfrac{1}{\rho_1^2} & 0 \\ 0 & \dfrac{1}{\rho_2^2} \end{bmatrix} \begin{bmatrix} \hat{\mathbf{v}}_1 \\ \hat{\mathbf{v}}_2 \end{bmatrix}$$

Let AB be a straight line segment on the tangent plane T_p. The length of edge AB with respect to $M(\mathbf{p})$ is given by

$$L = \sqrt{AB^T M AB} = \|AB\| \sqrt{\mathbf{v}^T M \mathbf{v}} = \|AB\| \sqrt{\begin{bmatrix} \hat{\mathbf{v}}_1 \cdot \mathbf{v} & \hat{\mathbf{v}}_2 \cdot \mathbf{v} \end{bmatrix} \begin{bmatrix} \kappa_1^2 & 0 \\ 0 & \kappa_2^2 \end{bmatrix} \begin{bmatrix} \mathbf{v} \cdot \hat{\mathbf{v}}_1 \\ \mathbf{v} \cdot \hat{\mathbf{v}}_2 \end{bmatrix}}$$

$$= \|AB\| \sqrt{\begin{bmatrix} \cos\theta & \sin\theta \end{bmatrix} \begin{bmatrix} \kappa_1^2 & 0 \\ 0 & \kappa_2^2 \end{bmatrix} \begin{bmatrix} \cos\theta \\ \sin\theta \end{bmatrix}} = \|AB\| \sqrt{\kappa_1^2 \cos^2\theta + \kappa_2^2 \sin^2\theta} \geq \kappa \|AB\|$$

$$L \geq \kappa \|AB\| \Rightarrow L \geq \frac{\|AB\|}{\rho} \quad \text{where } \mathbf{v} = \frac{AB}{\|AB\|}$$

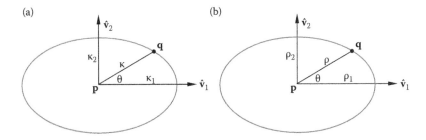

Figure 4.9 Curvature at a point is bounded by an ellipse: (a) ellipse of surface curvature; (b) ellipse of radius of curvature.

If edge AB is of unit length with respect to M, i.e. L = 1, we have $\rho \geq \|AB\|$. The metric M allows us to control the edges of a mesh on S such that the unit length in any direction is always smaller than the radius of curvature at the point under consideration.

4.2.5 Geometrical control

Two common measures can be applied to control the geometrical closeness to a curved surface (Miranda et al. 2009). The first criterion is to specify the gap between an edge and the curved surface, as shown in Figure 4.10. Let h be the length of the edge and ρ be the radius of curvature in the direction of the edge. We have

$$d^2 = \rho^2 - \left(\frac{h}{2}\right)^2 \Rightarrow d = \sqrt{\rho^2 - \frac{h^2}{4}}$$

$$\delta = \rho - d = \rho - \sqrt{\rho^2 - \frac{h^2}{4}} = \rho\left(1 - \sqrt{1 - \frac{\alpha^2}{4}}\right) \quad \text{or} \quad \frac{\delta}{\rho} = 1 - \sqrt{1 - \frac{\alpha^2}{4}} \leq \varepsilon \quad \alpha = h/\rho$$

$$1 - \varepsilon \leq \sqrt{1 - \frac{\alpha^2}{4}} \Rightarrow \frac{\alpha^2}{4} \leq 2\varepsilon - \varepsilon^2 \Rightarrow \alpha \leq 2\sqrt{\varepsilon(2-\varepsilon)} \Rightarrow h \leq 2\sqrt{\varepsilon(2-\varepsilon)}\rho$$

Hence, in order that the gap δ between a line segment and the surface be bounded by the ratio $\varepsilon = \delta/\rho$, h = $\alpha\rho$ could not exceed the value given by

$$\alpha = 2\sqrt{\varepsilon(2-\varepsilon)}$$

For example, if ε is set at 1%, α = 0.282, i.e. h cannot exceed 0.282ρ; otherwise, the gap between the line segment and the curved surface will be more than 1% of ρ.

Alternatively, the ratio of the length of a chord to that of the corresponding curve can be specified, as shown in Figure 4.11. Let h be the length of an edge (chord) and s be that of the shortest curve on the surface connected to the same end points of the edge. As the curve is always longer than the chord, we can control the length of an edge by specifying a value ε such that

$$\frac{s-h}{s} \leq \varepsilon \quad \text{or} \quad h \geq (1-\varepsilon)s$$

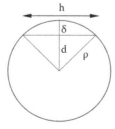

Figure 4.10 Gap between an edge and a curved surface.

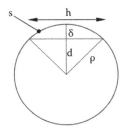

Figure 4.11 **Compare lengths of chord and curve.**

In the limiting case,

$$\varepsilon = \frac{s-h}{s} = \frac{2\rho\theta - 2\rho\sin\theta}{2\rho\theta} = 1 - \frac{\sin\theta}{\theta} = 1 - \left(1 - \frac{\theta^2}{3!} + \frac{\theta^4}{5!} - \cdots\right)$$

Taking one term in the expansion, we have

$$\varepsilon = \frac{\theta^2}{3!} \Rightarrow \theta^2 = 6\varepsilon \quad \text{or} \quad \theta = \sqrt{6\varepsilon}$$

$$h = (1-\varepsilon)s = 2(1-\varepsilon)\rho\theta = 2(1-\varepsilon)\sqrt{6\varepsilon}\rho = \beta_1\rho$$

Taking two terms in the expansion, we have

$$\varepsilon = \frac{\theta^2}{3!} - \frac{\theta^4}{5!} \Rightarrow 120\varepsilon = 20\theta^2 - \theta^4$$

Let $\mu = \theta^2$. We have

$$\mu^2 - 20\mu + 120\varepsilon = 0$$

Solving for μ and taking the smaller root, we get

$$\mu = 10 - \sqrt{100 - 120\varepsilon} \Rightarrow \theta = \sqrt{\mu} = \sqrt{10 - \sqrt{100 - 120\varepsilon}}$$

$$h = (1-\varepsilon)s = 2(1-\varepsilon)\rho\theta = 2(1-\varepsilon)\sqrt{10 - \sqrt{100 - 120\varepsilon}}\,\rho = \beta_2\rho$$

For instance, set $\varepsilon = 1\%$; we have $\beta_1 = 0.485$ and $\beta_2 = 0.48573$, i.e. h cannot exceed 0.485ρ.
  Whichever criterion we would like to apply for a close approximation of a linear FE mesh of triangular facets to an analytical curved surface, we can simply modify the metric tensor by replacing the principal curvatures with the reduced principal curvatures according to the chosen criterion, such that

$$\bar{M} = \begin{bmatrix} \hat{v}_1 & \hat{v}_2 \end{bmatrix} \begin{bmatrix} \bar{\kappa}_1^2 & 0 \\ 0 & \bar{\kappa}_2^2 \end{bmatrix} \begin{bmatrix} \hat{v}_1 \\ \hat{v}_2 \end{bmatrix} \quad \text{where } \bar{\kappa}_1 = \frac{\kappa_1}{\lambda} \text{ and } \bar{\kappa}_2 = \frac{\kappa_2}{\lambda}$$

in which the value of λ depends on the criterion adopted:

    i. The gap between the edge and the surface is specified, i.e. $\dfrac{\delta}{\rho} \leq \varepsilon$, $\lambda = \alpha = 2\sqrt{\varepsilon(2-\varepsilon)}$.

    ii. The ratio of the lengths of the edge and the curve is specified, i.e. $h \geq (1-\varepsilon)s$.

$$\lambda = \beta_1 = 2(1-\varepsilon)\sqrt{6\varepsilon} \quad \text{or} \quad \lambda = \beta_2 = 2(1-\varepsilon)\sqrt{10-\sqrt{100-120\varepsilon}}$$

In terms of the principal radii of curvatures, the metric tensor can be modified as

$$\overline{M} = \begin{bmatrix} \hat{v}_1 & \hat{v}_2 \end{bmatrix} \begin{bmatrix} \dfrac{1}{\overline{\rho}_1^2} & 0 \\[2mm] 0 & \dfrac{1}{\overline{\rho}_2^2} \end{bmatrix} \begin{bmatrix} \hat{v}_1 \\ \hat{v}_2 \end{bmatrix} \quad \text{where } \overline{\rho}_1 = \lambda\rho_1 \text{ and } \overline{\rho}_2 = \lambda\rho_2$$

In the practical implementation, once a criterion is adopted and the value of ε is specified, the value of λ can be evaluated in advance before mesh generation. By a similar argument, the modified metric $\overline{M}$ allows us to control the edges of a mesh on surface S such that edges of unit length in any direction measured by the modified metric $\overline{M}$ are smaller than the modified radius of curvature $\overline{\rho}$ at the point under consideration.

4.2.6 Metric on parametric planar domain

$\overline{M}$ is the metric defined on the tangent plane T_p at point $\mathbf{p}$ on surface S. Hence, $\overline{M}$ is the metric for length measure of vectors on T_p, and the length element measured along direction vector $V = d\phi$ is given by

length measure of $d\phi$ on $T_p = d\phi^T \overline{M} d\phi$

$$d\phi = \begin{bmatrix} \dfrac{\partial\phi}{\partial u} & \dfrac{\partial\phi}{\partial v} \end{bmatrix} \begin{bmatrix} du \\ dv \end{bmatrix} = \begin{bmatrix} \phi_u & \phi_v \end{bmatrix} \begin{bmatrix} du \\ dv \end{bmatrix} = \begin{bmatrix} \mathbf{u} & \mathbf{v} \end{bmatrix} \begin{bmatrix} du \\ dv \end{bmatrix}$$

length measure of $\begin{bmatrix} du \\ dv \end{bmatrix}$ on $\Omega = \begin{bmatrix} du & dv \end{bmatrix} \begin{bmatrix} \mathbf{u} \\ \mathbf{v} \end{bmatrix} \overline{M} \begin{bmatrix} \mathbf{u} & \mathbf{v} \end{bmatrix} \begin{bmatrix} du \\ dv \end{bmatrix} = \begin{bmatrix} du & dv \end{bmatrix} M_\Omega \begin{bmatrix} du \\ dv \end{bmatrix}$

where

$$M_\Omega = \begin{bmatrix} \mathbf{u} \\ \mathbf{v} \end{bmatrix} \overline{M} \begin{bmatrix} \mathbf{u} & \mathbf{v} \end{bmatrix} = \begin{bmatrix} \mathbf{u} \\ \mathbf{v} \end{bmatrix} \begin{bmatrix} \hat{v}_1 & \hat{v}_2 \end{bmatrix} \begin{bmatrix} \overline{\kappa}_1^2 & 0 \\ 0 & \overline{\kappa}_2^2 \end{bmatrix} \begin{bmatrix} \hat{v}_1 \\ \hat{v}_2 \end{bmatrix} \begin{bmatrix} \mathbf{u} & \mathbf{v} \end{bmatrix}$$

$$M_\Omega = \begin{bmatrix} \mathbf{u}\cdot\hat{v}_1 & \mathbf{u}\cdot\hat{v}_2 \\ \mathbf{v}\cdot\hat{v}_1 & \mathbf{v}\cdot\hat{v}_2 \end{bmatrix} \begin{bmatrix} \overline{\kappa}_1^2 & 0 \\ 0 & \overline{\kappa}_2^2 \end{bmatrix} \begin{bmatrix} \mathbf{u}\cdot\hat{v}_1 & \mathbf{v}\cdot\hat{v}_1 \\ \mathbf{u}\cdot\hat{v}_2 & \mathbf{v}\cdot\hat{v}_2 \end{bmatrix}$$

$$= \begin{bmatrix} \overline{\kappa}_1^2(\mathbf{u}\cdot\hat{v}_1)^2 + \overline{\kappa}_2^2(\mathbf{u}\cdot\hat{v}_2)^2 & \overline{\kappa}_1^2(\mathbf{u}\cdot\hat{v}_1)(\mathbf{v}\cdot\hat{v}_1) + \overline{\kappa}_2^2(\mathbf{u}\cdot\hat{v}_2)(\mathbf{v}\cdot\hat{v}_2) \\ \overline{\kappa}_1^2(\mathbf{u}\cdot\hat{v}_1)(\mathbf{v}\cdot\hat{v}_1) + \overline{\kappa}_2^2(\mathbf{u}\cdot\hat{v}_2)(\mathbf{v}\cdot\hat{v}_2) & \overline{\kappa}_1^2(\mathbf{v}\cdot\hat{v}_1)^2 + \overline{\kappa}_2^2(\mathbf{v}\cdot\hat{v}_2)^2 \end{bmatrix}$$

Hence, the metric tensor on Ω can be represented by a symmetrical positive-definite matrix with coefficients that are functions of the reduced principal curvatures and the projections of $\mathbf{u}$ and $\mathbf{v}$ onto the principal curvature directions $\hat{\mathbf{v}}_1$ and $\hat{\mathbf{v}}_2$.

Metric tensor $\bar{M}$ on tangent plane T_p can be extended to 3D space by a simple modification:

$$M_{3D} = \begin{bmatrix} \hat{\mathbf{v}}_1 & \hat{\mathbf{v}}_2 & \mathbf{n} \end{bmatrix} \begin{bmatrix} \bar{\kappa}_1^2 & 0 & 0 \\ 0 & \bar{\kappa}_2^2 & 0 \\ 0 & 0 & c \end{bmatrix} \begin{bmatrix} \hat{\mathbf{v}}_1 \\ \hat{\mathbf{v}}_2 \\ \mathbf{n} \end{bmatrix} \quad \text{where c is an arbitrary constant}$$

In terms of M_{3D}, the new M_Ω is given by

$$M_{\Omega 3D} = \begin{bmatrix} \mathbf{u} \\ \mathbf{v} \end{bmatrix} M_{3D} \begin{bmatrix} \mathbf{u} & \mathbf{v} \end{bmatrix} = \begin{bmatrix} \mathbf{u} \\ \mathbf{v} \end{bmatrix} \begin{bmatrix} \hat{\mathbf{v}}_1 & \hat{\mathbf{v}}_2 & \mathbf{n} \end{bmatrix} \begin{bmatrix} \bar{\kappa}_1^2 & 0 & 0 \\ 0 & \bar{\kappa}_2^2 & 0 \\ 0 & 0 & c \end{bmatrix} \begin{bmatrix} \hat{\mathbf{v}}_1 \\ \hat{\mathbf{v}}_2 \\ \mathbf{n} \end{bmatrix} \begin{bmatrix} \mathbf{u} & \mathbf{v} \end{bmatrix}$$

$$M_{\Omega 3D} = \begin{bmatrix} \mathbf{u} \cdot \hat{\mathbf{v}}_1 & \mathbf{u} \cdot \hat{\mathbf{v}}_2 & \mathbf{u} \cdot \mathbf{n} \\ \mathbf{v} \cdot \hat{\mathbf{v}}_1 & \mathbf{v} \cdot \hat{\mathbf{v}}_2 & \mathbf{v} \cdot \mathbf{n} \end{bmatrix} \begin{bmatrix} \bar{\kappa}_1^2 & 0 & 0 \\ 0 & \bar{\kappa}_2^2 & 0 \\ 0 & 0 & c \end{bmatrix} \begin{bmatrix} \mathbf{u} \cdot \hat{\mathbf{v}}_1 & \mathbf{v} \cdot \hat{\mathbf{v}}_1 \\ \mathbf{u} \cdot \hat{\mathbf{v}}_2 & \mathbf{v} \cdot \hat{\mathbf{v}}_2 \\ \mathbf{u} \cdot \mathbf{n} & \mathbf{v} \cdot \mathbf{n} \end{bmatrix}$$

$$= \begin{bmatrix} \bar{\kappa}_1^2(\mathbf{u} \cdot \hat{\mathbf{v}}_1)^2 + \bar{\kappa}_2^2(\mathbf{u} \cdot \hat{\mathbf{v}}_2)^2 + c(\mathbf{u} \cdot \mathbf{n})^2 & m_{12} \\ m_{21} & \bar{\kappa}_1^2(\mathbf{v} \cdot \hat{\mathbf{v}}_1)^2 + \bar{\kappa}_2^2(\mathbf{v} \cdot \hat{\mathbf{v}}_2)^2 + c(\mathbf{v} \cdot \mathbf{n})^2 \end{bmatrix}$$

with $m_{21} = m_{12} = \bar{\kappa}_1^2(\mathbf{u} \cdot \hat{\mathbf{v}}_1)(\mathbf{v} \cdot \hat{\mathbf{v}}_1) + \bar{\kappa}_2^2(\mathbf{u} \cdot \hat{\mathbf{v}}_2)(\mathbf{v} \cdot \hat{\mathbf{v}}_2) + c(\mathbf{u} \cdot \mathbf{n})(\mathbf{v} \cdot \mathbf{n})$.

For vector V on T_p, $V \cdot \mathbf{n} = 0$, and there is no difference for vectors on T_p measured by either M_Ω or $M_{\Omega 3D}$. However, if A and B are two distinct points on a planar domain Ω, and $\phi(A)$ and $\phi(B)$ are two points on surface S such that edge $V = \phi(A)\phi(B)$ is not in S and does not lie on the tangent planes $T_{\phi(A)}$ or $T_{\phi(B)}$, then $V \cdot \mathbf{n} \neq 0$ and in fact $V \cdot \mathbf{n}$ accounts for the contribution of the component of V out of the tangent plane T_p towards the length measure. The parameter c is used to adjust the significance of the out-of-plane component in length measure.

4.2.7 Metric tensor and Green–Cauchy deformation tensor

The metric tensor M can be interpreted in terms of the Green–Cauchy deformation tensor C (Marsden and Hughes 1983) in which the length of infinitesimal fibre dx upon deformation is given by $ds^2 = dx \cdot C \cdot dx$, in which ds is the deformed length of dx. Based on the length measure by metric tensor M for vector dx, we have $dr^2 = dx \cdot M \cdot dx$.

From this, we can see that M is identical to C mathematically, although they have different physical interpretations; dr represents the length of dx as measured with respect to metric M, and ds represents the length of dx after deformation. Many results established in the theory of large deformation can be applied to the geometrical quantities governed by the metric M.

Metric tensor M can be expressed in terms of the principal base vectors as follows.

$$M = \begin{bmatrix} \hat{e}_1 & \hat{e}_2 \end{bmatrix} \begin{bmatrix} \lambda_1^2 & 0 \\ 0 & \lambda_2^2 \end{bmatrix} \begin{bmatrix} \hat{e}_1 \\ \hat{e}_2 \end{bmatrix} = R^T U U R = (UR)^T U R = F^T F$$

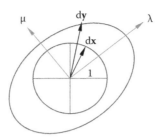

Figure 4.12 A unit circle is mapped to an ellipse by the metric tensor M.

in which

$$F = UR, \quad \text{stretch matrix } U = \begin{bmatrix} \lambda_1 & 0 \\ 0 & \lambda_2 \end{bmatrix}, \quad \text{rotation matrix } R = \begin{bmatrix} \hat{e}_1 \\ \hat{e}_2 \end{bmatrix}$$

and $\hat{e}_1, \hat{e}_2$ are unit eigenvectors of M.

$$dr^2 = dx \cdot M \cdot dx = dx \cdot F^T \cdot F \cdot dx = (F \cdot dx) \cdot (F \cdot dx) = dy \cdot dy \quad \text{with } dy = F \cdot dx$$

As shown in Figure 4.12, a unit circle is mapped onto an ellipse of stretches λ and μ along the principal directions along with a rotation, and dy represents the transformed vector of dx by metric M. Along the principal directions, there is no change in the direction for vector dx but just a change in length by a scalar factor λ_i; however, in any other directions, there are rotation as well as scaling for vector dx. The transformation of vector dx into dy = F·dx = U·(R·dx) can be interpreted geometrically as a rotation R to the direction of dx followed by a stretching to the length of dy by U. Alternatively, dx can also be transformed to dy by first a stretch followed by a rotation.

4.2.7.1 Change in length by metric M

Define fibre extension α by the ratio of the deformed length to the original length, i.e.

$$\alpha = \frac{dr}{\|dx\|} = \left[\frac{dx \cdot M \cdot dx}{\|dx\| \|dx\|} \right]^{\frac{1}{2}} = [u \cdot M \cdot u]^{\frac{1}{2}} \quad \text{where } u = \frac{dx}{\|dx\|}$$

We can see that the change in length depends on M and the direction of the fibre u.

4.2.7.2 Change in area by metric M

Let M be the metric tensor at point P. The area of the triangle spanned by vectors dx_1 and dx_2 measured with respect to metric M is given by

$$\text{area } A_M = (F \cdot dx_1) \times (F \cdot dx_2) \cdot n = \det(F)(dx_1 \times dx_2 \cdot n) = (\lambda \mu) A_E$$

$$\det(F) = \det(UR) = \det(U)\det(R) = \lambda \mu, \quad A_E = dx_1 \cdot dx_2 \cdot n = \text{area in Euclidean metric}$$

Hence, $\lambda\mu$ governs the size of the metric, and it is also the magnification factor for the transformed area by the metric M. This result has also been established by Borouchaki et al. (1997a,b) by directly applying the metric measure to the edges of a triangle.

The change in area and shape of a triangle by the metric M is shown in Figure 4.13.

Area of unit circle $= \pi$, area of ellipse $= \pi\lambda\mu$

$$\text{Area of triangle ABC} = \frac{1}{2}(2\times 1) = 1, \quad \text{are of deformed triangle} = \frac{1}{2}(2\lambda\times\mu) = \lambda\mu$$

Now suppose triangle ABC is stretched by metric tensor M in the direction 45° with respect to the base of the triangle. We have

$$\hat{v}_1 = \frac{1}{\sqrt{2}}(1,1),\ \hat{v}_2 = \frac{1}{\sqrt{2}}(-1,1),\ \text{rotation matrix R} = \frac{1}{\sqrt{2}}\begin{bmatrix}1 & 1\\-1 & 1\end{bmatrix}\text{and U} = \begin{bmatrix}\lambda & 0\\0 & \mu\end{bmatrix}$$

Point A $= (-1,0)$ and point A′ is given by

$$A' = F\cdot A = U\cdot R\cdot A = U\cdot\frac{1}{\sqrt{2}}\begin{bmatrix}1 & 1\\-1 & 1\end{bmatrix}\begin{bmatrix}-1\\0\end{bmatrix} = \frac{1}{\sqrt{2}}\begin{bmatrix}\lambda & 0\\0 & \mu\end{bmatrix}\begin{bmatrix}-1\\1\end{bmatrix} = \frac{1}{\sqrt{2}}\begin{bmatrix}-\lambda\\\mu\end{bmatrix}$$

Point B $= (1,0)$ and point B′ is given by

$$B' = F\cdot B = U\cdot R\cdot B = U\cdot\frac{1}{\sqrt{2}}\begin{bmatrix}1 & 1\\-1 & 1\end{bmatrix}\begin{bmatrix}1\\0\end{bmatrix} = \frac{1}{\sqrt{2}}\begin{bmatrix}\lambda & 0\\0 & \mu\end{bmatrix}\begin{bmatrix}1\\-1\end{bmatrix} = \frac{1}{\sqrt{2}}\begin{bmatrix}\lambda\\-\mu\end{bmatrix}$$

Point C $= (0,1)$ and point C′ is given by

$$C' = F\cdot C = U\cdot R\cdot C = U\cdot\frac{1}{\sqrt{2}}\begin{bmatrix}1 & 1\\-1 & 1\end{bmatrix}\begin{bmatrix}0\\1\end{bmatrix} = \frac{1}{\sqrt{2}}\begin{bmatrix}\lambda & 0\\0 & \mu\end{bmatrix}\begin{bmatrix}1\\1\end{bmatrix} = \frac{1}{\sqrt{2}}\begin{bmatrix}\lambda\\-\mu\end{bmatrix}$$

The area of triangle A′B′C′ is given by

$$\frac{1}{2}A'B'\times A'C' = \frac{1}{2}\frac{1}{\sqrt{2}}\begin{bmatrix}2\lambda\\-2\mu\end{bmatrix}\times\frac{1}{\sqrt{2}}\begin{bmatrix}2\lambda\\0\end{bmatrix} = \lambda\mu$$

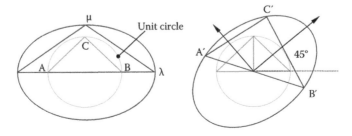

Figure 4.13 **Shape of the triangle may change, but the area of the deformed triangle remains the same transformed by the same metric in different directions.**

This demonstrates that the change in area depends only on the stretching part of metric tensor M but not on the rotational part of metric tensor M. The ratio of the area between the transformed object and the original object of any shape and size is always equal to $\det(F) = \det(U) = \lambda\mu$. Indeed, by the property of rotational (orthogonal) matrix R, we have

$$(R \cdot u) \cdot (R \cdot v) = u \cdot v \text{ and } (R \cdot u) \times (R \cdot v) = u \times v \quad \forall u, v \in \mathbb{R}^2$$

In other words, the length and the area are preserved under rotation.

4.2.8 Interpolation of metric

For complicated surfaces, the evaluation of metric is quite computationally intensive, and very often the metric is only evaluated and available at some strategic nodal points of a mesh. For sufficient fine meshes, at the interior of an FE, the metric can be estimated by interpolation (Laug and Borouchaki 2003a).

4.2.8.1 Metric interpolation over a line segment

The metric tensors are given at the end points of a line segment; our task is to estimate the metric tensor at any point on the line segment, as shown in Figure 4.14.

Metric tensor M, which is symmetric and positive definite, admits a polar decomposition into two fundamental parts, pure stretch U and rotation R. Let M_1 and M_2 be the metrics at the end points of line segment AB; metric M at point $P = (1 - t)A + tB, t \in [0,1]$ can be defined by the following interpolation scheme. By polar decomposition of metric M_1, we have

$$M_1 = F_1^T F_1, \quad F_1 = U_1 R_1, \quad U_1 = \begin{bmatrix} \lambda_1 & 0 \\ 0 & \mu_1 \end{bmatrix}, \quad R_1 = \begin{bmatrix} \cos(\theta_1) & \sin(\theta_1) \\ -\sin(\theta_1) & \cos(\theta_1) \end{bmatrix}$$

where λ_1 and μ_1 are the principal stretches, and $\theta_1 \in \left[0, \dfrac{\pi}{2}\right]$ characterises the principal direction associated with stretch λ_1, and similarly we have for metric tensor M_2

$$M_2 = F_2^T F_2, \quad F_2 = U_2 R_2, \quad U_2 = \begin{bmatrix} \lambda_2 & 0 \\ 0 & \mu_2 \end{bmatrix}, \quad R_2 = \begin{bmatrix} \cos(\theta_2) & \sin(\theta_2) \\ -\sin(\theta_2) & \cos(\theta_2) \end{bmatrix}$$

The rotation part for metric M at point P can be defined by a linear interpolation such that

$$\theta = s\theta_1 + t\theta_2 \quad \text{where } s = 1 - t$$

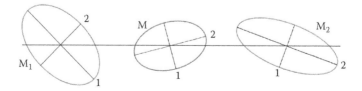

Figure 4.14 Interpolation of metric on a line segment.

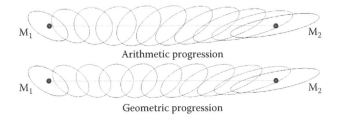

Arithmetic progression

Geometric progression

Figure 4.15 Intersection of metrics along a straight line.

However, for the interpolation of the stretch part, we have two choices.

 i. Arithmetic progression: $\lambda = s\lambda_1 + t\lambda_2$ and $\mu = s\mu_1 + t\mu_2$
 ii. Geometric progression: $\lambda = \lambda_1^s\lambda_2^t$ and $\mu = \mu_1^s\mu_2^t$

Hence, metric M at point P is given by

$$M = F^T F, \quad F = UR, \quad U = \begin{bmatrix} \lambda & 0 \\ 0 & \mu \end{bmatrix}, \quad R = \begin{bmatrix} \cos(\theta) & \sin(\theta) \\ -\sin(\theta) & \cos(\theta) \end{bmatrix}$$

Metric interpolations of M_1 and M_2 along a straight line for arithmetic progression and geometric progression are plotted in Figure 4.15. It can be seen that the difference between the two progressions is quite small, and the metrics interpolated by the arithmetic progression appear to be only slightly larger than those interpolated by the geometric progression.

4.2.8.2 Metric interpolation within a triangular element

By means of polar decomposition of a metric tensor, the metric interpolation within a triangle can be easily defined by means of area co-ordinates of a triangular element. Let M_1, M_2 and M_3 be, respectively, the metric at the vertices of triangle ABC. By polar decomposition, metric M_i (i = 1,2,3) at the nodal points of the triangle can be expressed as

$$M_i = F_i^T F_i, \quad F_i = U_i R_i, \quad U_i = \begin{bmatrix} \lambda_i & 0 \\ 0 & \mu_i \end{bmatrix}, \quad R_i = \begin{bmatrix} \cos(\theta_i) & \sin(\theta_i) \\ -\sin(\theta_i) & \cos(\theta_i) \end{bmatrix} \quad \theta_i \in \left[0, \frac{\pi}{2}\right)$$

where λ_i and μ_i are the principal stretches, and θ_i characterises the principal directions of metric tensor M_i. Let L_1, L_2 and L_3 be the area co-ordinates of point P within triangle ABC, as shown in Figure 4.16. Then the metric at point P is defined as

$$L_1 = \frac{\Delta_1}{\Delta}, L_2 = \frac{\Delta_2}{\Delta}, L_3 = \frac{\Delta_3}{\Delta}, \quad \Delta = \text{area of triangle ABC}$$

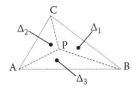

Figure 4.16 Interpolation of metric within a triangle.

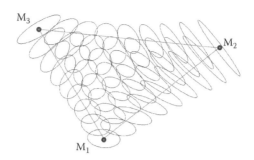

Figure 4.17 Interpolation of metrics over a triangle.

The rotation part for metric M at point P can be defined using area co-ordinates such that

$$\theta = L_1\theta_1 + L_2\theta_2 + L_3\theta_3$$

Again, for the interpolation of the stretch part, we have two choices:

   i. Arithmetic progression: $\lambda = L_1\lambda_1 + L_2\lambda_2 + L_3\lambda_3$ and $\mu = L_1\mu_1 + L_2\mu_2 + L_3\mu_3$
   ii. Geometric progression: $\lambda = \lambda_1^{L_1}\lambda_2^{L_2}\lambda_3^{L_3}$ and $\mu = \mu_1^{L_1}\mu_2^{L_2}\mu_3^{L_3}$

Knowing θ, λ and μ, the metric M at point P is given by

$$M = F^T F, \quad F = UR, \quad U = \begin{bmatrix} \lambda & 0 \\ 0 & \mu \end{bmatrix}, \quad R = \begin{bmatrix} \cos(\theta) & \sin(\theta) \\ -\sin(\theta) & \cos(\theta) \end{bmatrix}$$

The interpolation of metrics M_1, M_2 and M_3 at the vertices of a triangle by means of area co-ordinates is shown in Figure 4.17. Since interpolated metrics are computed by the convex sum formulas, fairly smooth metrics are produced at the interior of the triangle, which are bounded by the metrics at the vertices.

4.2.9 Lengths controlled by multiple metrics

In adaptive refinement analysis over curved surfaces, the sizes of the finite elements are controlled not only by the geometry of the curved surfaces but also by the physical require-ments, e.g. solution errors have to be within a specified amount. Over regions of high stress or solution gradients, it is necessary to control the size of the finite elements as specified by the error indicators. Based on the solution errors, a physical metric tensor $M_{\Omega P}$ can be defined to govern the size and shape of the finite elements on Ω.

In case geometrical metric M_Ω and $M_{\Omega P}$ tensors are having the same set of eigenvectors, or they are defined based on the same set of basis vectors, then the intersection of M_Ω and $M_{\Omega P}$ is simply given by the smaller eigenvalue in each principal direction (eigenvector). In practice, at the expense of a bit more computation, the length limit, d, of an edge along the direction of unit vector V governed by a number of metrics M_1, M_2, ..., M_n is given by

$$d = \min_{i=1,n} d_i \quad \text{where } d_i = V^T M_i V$$

If all metrics M_i are defined following the idea of unit length mesh control, then for any edge of the mesh, we have to ensure that $d_i = V^T M_i V \le 1$. As shown in Figure 4.18, the locus

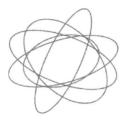

Figure 4.18 **Region of intersection of metrics is not an ellipse.**

of d or the boundary of the intersected region of several metrics is not an ellipse; thus, the direct minimisation of metrics does not, in general, give rise to another metric. To define a metric that is a subset of all the given metrics, we have to find the largest ellipse within the regions of intersection. Here we will explore a less general procedure by examining the intersection of two metrics at a time.

Our problem is as follows: given two metrics M_1 and M_2, find the metric M that is bounded by both M_1 and M_2. For metrics M_1 and M_2 without common eigenvectors ($\theta_1 \neq \theta_2$), introduce auxiliary matrix $A = M_1^{-1}M_2$ (A exists as M_1 and M_2 are positive definite). Let e_1 and e_2 be the eigenvectors of A, which may not be orthogonal vectors as A is, in general, not symmetric. Metrics M_1 and M_2 are simultaneously diagonalised by the eigenvectors of A such that $e_1^T M_1 e_2 = 0$ and $e_1^T M_2 e_2 = 0$

Let

$$P = \begin{bmatrix} e_1 & e_2 \end{bmatrix}, \quad Q = P^T = \begin{bmatrix} e_1^T \\ e_2^T \end{bmatrix}, \quad QM_1P = \begin{bmatrix} a_1 & 0 \\ 0 & b_1 \end{bmatrix} \quad \text{or} \quad M_1 = Q^{-1}\begin{bmatrix} a_1 & 0 \\ 0 & b_1 \end{bmatrix}P^{-1}$$

Similarly,

$$QM_2P = \begin{bmatrix} a_2 & 0 \\ 0 & b_2 \end{bmatrix} \quad \text{or} \quad M_2 = Q^{-1}\begin{bmatrix} a_2 & 0 \\ 0 & b_2 \end{bmatrix}P^{-1}$$

The intersection of metrics M_1 and M_2 is another metric given by

$$M = M_1 \cap M_2 = Q^{-1}\begin{bmatrix} a & 0 \\ 0 & b \end{bmatrix}P^{-1} \quad \text{where } a = \min(a_1, a_2) \text{ and } b = \min(b_1, b_2)$$

Let $\mathbf{v} = v_1 e_1 + v_2 e_2$ be a vector expressed in the basis (e_1, e_2),

$$\mathbf{v}^T M_1 \mathbf{v} = (v_1 e_1 + v_2 e_2)^T M_1(v_1 e_1 + v_2 e_2)$$
$$= v_1^2 e_1^T M_1 e_1 + v_2^2 e_2^T M_1 e_2 = a_1 v_1^2 + b_1 v_2^2$$

Similarly, $\mathbf{v}^T M \mathbf{v} = a v_1^2 + b v_2^2$ and $\mathbf{v}^T M_2 \mathbf{v} = a_2 v_1^2 + b_2 v_2^2$.

Hence, $\mathbf{v}^T M \mathbf{v} \leq \mathbf{v}^T M_1 \mathbf{v}$ and $\mathbf{v}^T M \mathbf{v} \leq \mathbf{v}^T M_2 \mathbf{v}$. Thus, M is bounded by M_1 and M_2.

The intersection of two metrics M_1 and M_2 by simultaneous diagonalisation is shown in Figure 4.19, in which $M_1 = M_1(\theta_1 = 1, \lambda_1 = 2, \mu_1 = 3)$, $M_2 = M_2(\theta_2 = 2 - \pi/2, \lambda_2 = 5, \mu_2 = 1)$ and the metric of intersection $M_I = M_I(\theta_I = 0.3645, \lambda_I = 2.2022, \mu_I = 0.9903)$.

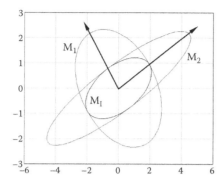

Figure 4.19 Intersection of two metrics.

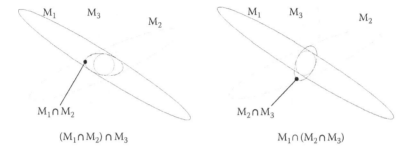

Figure 4.20 Intersection of two metrics is not associative.

In case there are several metrics governing the size of an edge, metric intersection can be applied recursively two at a time to reduce to a single metric, which is bounded by all the given metrics, i.e. $M = M_1 \cap M_2 \cap \ldots \ldots M_n$. It is obvious that metric intersection is communicative; however, it is not associative, i.e. $(M_1 \cap M_2) \cap M_3 \neq M_1 \cap (M_2 \cap M_3)$, as shown in Figure 4.20. Metric intersection tends to produce a metric bounded by the given metrics, which is, in general, smaller than the largest metric that could be fitted in the region of intersection. Although we can try to find the largest metric of intersection by altering the sequence of intersection, in practice, we simply follow the natural sequence in the intersection process to avoid excessive computation. For the intersection of metric tensors in general, especially in three dimensions, readers can refer to the optimisation procedure proposed in the paper by Alauzet (2010).

4.2.10 Element shape measure with respect to anisotropic metric

There is no universally accepted criterion to assess the shape of a triangle in a parametric anisotropic metric domain. Borouchaki et al. (2000a) extended the α-quality of triangle on a planar domain proposed by Lo (1985) to judge the shape of a triangle over a non-Euclidean domain. The α-quality of a triangle over a planar domain is the ratio of the signed area over the sum of edges squared, and mathematically for triangle ABC, it is given by

$$\alpha = \frac{\text{signed area}}{\text{sum of edges squared}} = 2\sqrt{3}\frac{AB \times AC}{\|AB\|^2 + \|BC\|^2 + \|CA\|^2}$$

where vector cross product AB × AC represents twice the signed area, and $2\sqrt{3}$ is a normalising factor such that an equilateral triangle on a planar domain will have a maximum value of 1.

Borouchaki et al. (2000a) tried to generalise the α-quality by evaluating the signed area of the triangle and the length of the edges with respect to the given metric field, and the α-quality for triangle ABC under metric M is given by

$$\alpha = 2\sqrt{3}\,\frac{\|AB\times AC\|_M}{\|AB\|_M^2 + \|BC\|_M^2 + \|CA\|_M^2}$$

Since M varies over the triangular element, Borouchaki et al. (2000a) further proposed that the α-quality of triangle ABC be sampled only at the vertices A, B and C such that

$$\alpha_A = 2\sqrt{3}\,\frac{\|AB\times AC\|_{M(A)}}{\|AB\|_{M(A)}^2 + \|BC\|_{M(A)}^2 + \|CA\|_{M(A)}^2}$$

where α_A is the α-quality, and M(A) is the metric at point A. Similarly, α-qualities of the triangle are evaluated at points B and C, and a minimum norm for the α-quality of triangle ABC for metric M is given by $\alpha = \min(\alpha_A, \alpha_B, \alpha_C)$.

However, there is an obvious weakness in this definition. Consider triangles ABC and ABD, as shown in Figure 4.21, in which the ellipse represents the locus of unit length from the centre O for constant metric field M. With reference to Section 4.2.7, triangles ABC and ABD are of the same area $\lambda\mu$ for metric M with principal stretches λ and μ, and as for edges $\|AC\|_M = \|BD\|_M = 2$, $\|AD\|_M = \|BC\|_M$; hence, they are of the same α-quality. Nevertheless, triangle ABD is preferred over triangle ABC in terms of shape quality as it possesses larger internal angles.

Shape quality is something that can easily be appreciated and judged in the Euclidean space, and a quality measure of triangles with respect to a general metric may not be that appropriate and be well-defined with a solid mathematical basis. Therefore, instead of defining a shape measure for triangles generated over a parametric domain with anisotropic metric, the shape of a triangle is assessed by the actual transformed geometry in the 3D Euclidean space, as shown in Figure 4.22. Hence, the shape measure of a triangle on a parametric domain is given by

$$\alpha = \frac{2\sqrt{3}\phi(A)\phi(B)\times\phi(A)\phi(C)}{\left\|\phi(A)\phi(B)\right\|^2 + \left\|\phi(B)\phi(C)\right\|^2 + \left\|\phi(C)\phi(A)\right\|^2} = \frac{2\sqrt{3}A'B'\times A'C'}{\left\|A'B'\right\|^2 + \left\|B'C'\right\|^2 + \left\|C'A'\right\|^2}$$

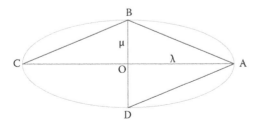

Figure 4.21 α-Quality of triangles in metric field.

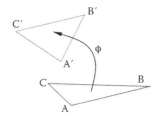

Figure 4.22 ΔABC is mapped by φ to ΔA′B′C′ on the surface.

where apostrophe stands for the mapping of the point from parametric domain Ω to surface S by the mapping φ such that A′ = φ(A), B′ = φ(B) and C′ = φ(C).

4.2.11 Metric tensor field of parametric curved surfaces and its characteristics

By means of the metric tensor equations for parametric surfaces summarised in Section 4.2.6, the metric field for a given curved surface can be easily evaluated at a parametric point (u,v). Typical curved surfaces along with their metric field plots are presented, and their characteristic features are also discussed in this section. The first curved surface is a simple parabolic surface with parametric equations given by

$$x = u; \ y = v; \ z = u^2 + v^2; \quad u, v \in [-5,5]$$

The plot of the surface and the unit metric field are shown in Figure 4.23. The unit metric is the contour of the unit circle with respect to the metric, and it is the inverse of the given metric, i.e. small ellipses indicate regions of large curvature or small radius of curvature. Even though the surface metric is to be verified for the target mesh, in the phase of mesh generation, it is the unit metric that we will be using for mesh generation. In other words, based on the metric field of a parametric surface, we have to construct a mesh whose edges are of unit length with respect to the given metric. The unit metric field for this example is symmetric along any radial direction from the centre as $u^2 + v^2$ is the equation of a circle. The second example is a curved surface of double peak with parametric equations:

$$x = u, \quad y = v, \quad z = (u^2 + 3v^2)e^{1-u^2-v^2} \quad u, v \in [-5,5]$$

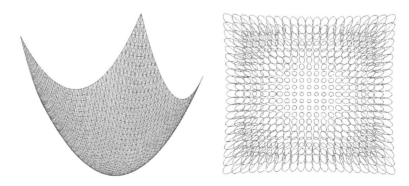

Figure 4.23 Parabolic curved surface and its unit metric field.

Even though the two peaks are not identical, it can be seen that the unit metric plot is more or less symmetric about these two peaks, as shown in Figure 4.24. Owing to the large curvature at the peaks and the trough, ellipses of very small size are found at these positions; that is to say that triangles of relatively small size have to be used at these points in order to have a close approximation of the surface geometry. The third example is a wavy surface with a sharp peak at the middle, whose parametric equations are given by

$$x = u; \quad y = v; \quad z = \frac{\sin(3u)\sin(3v)}{uv}; \quad u, v \in [-1, 1]$$

There are sharp changes of curvature from the peak towards the rest of the domain. As shown in Figure 4.25, a small ellipse of unit metric is found at the peak, and elongated ellipses are found on the four sides of the peak showing a large difference in the principal curvatures along the side of the peak.

The fourth example is a wavy surface characterised with a series of peaks of more or less equal height, and the surface parameterisation is given by

$$x = u^3 + 10u; \quad y = v^3 + 10v; \quad z = \sin(2u)\cos(2v); \quad u, v \in [-5, 5]$$

Owing to the wavy nature of the surface, the surface curvatures vary substantially from peak to trough and across ridges, as shown in Figure 4.26. Small ellipses are found at peak and trough positions, and elongated ellipses are found along ridges, as shown in Figure 4.27. As this wavy surface will be taken as an example for anisotropic mesh generation of parametric surfaces in Section 4.2.12, contour lines around points are shown in Figure 4.28,

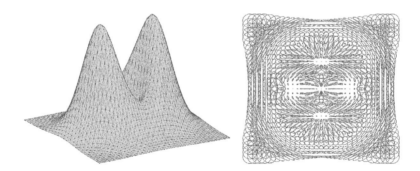

Figure 4.24 **Curved surface of double peak and its unit metric plot.**

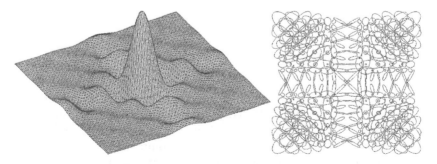

Figure 4.25 **Ripple surface of single peak and its unit metric field.**

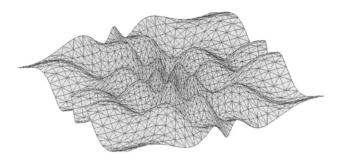

Figure 4.26 Wavy surface of many peaks.

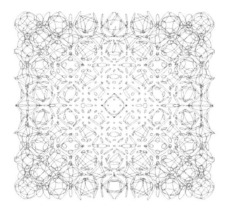

Figure 4.27 Unit metric plot of the wavy surface.

Figure 4.28 Contour lines around points.

which are constructed by locating points of equal metric distance along lines in various directions. The shape and the pitch of the contour lines around points at different positions over a general metric field are quite different. It is only of oval shape for contour close to the point, and contours away from the point can take quite different patterns, which may be even non-convex in shape over regions of rapidly changing metric. Indeed, there is not

Figure 4.29 Intersection of contours.

much significance for contour far away from the point passing through peaks and troughs. Intersection of contour lines is shown in Figure 4.29, which allows us to locate points of equal metric distance from two points.

4.2.12 Generation of anisotropic mesh by the Delaunay–ADF method

The Delaunay–ADF method possesses the merits from both the DT for its efficiency and robustness and the ADF method for its mesh quality in placing interior nodes. With modifications in the Delaunay criterion for the given metric specification, the Delaunay–ADF method can be readily extended to the generation of triangular meshes compatible to the given metric field (Borouchaki et al. 1997a,b). Such a mesh generation procedure will be presented in detail, in which the steps are elucidated by the example of the wavy surface in Section 4.2.11. As shown in Figure 4.26, the metric tensor field for the wavy surface at any point can be computed using the metric tensor equations given in Section 4.2.6 with scaling factor $\lambda = 0.02$ corresponding to a mesh close to a gap-to-curvature ratio of 0.01% on the curved surface. The task of mesh generation for a parametric curved surface is to create a triangular mesh whose edges are of unit length with respect to the corresponding metric field M.

4.2.12.1 Steps for anisotropic meshing

1. *Data for mesh generation.* Analogous to advancing-front meshing, the boundary of the domain for mesh generation consists of closed loops. The closed loop of the exterior boundary is entered in a counter-clockwise direction, whereas the closed loops of interior boundaries are entered in a clockwise sense, as shown in Figure 4.30.
2. *Subdivision of boundary segments.* In order to generate a mesh of compatible size, the boundary segments have to be divided into sub-segments of size compatible to the given metric, as shown in Figure 4.31. The generation of intermediate nodes along a straight line segment can be done following the procedure as described in Section 2.4.9. Alternatively, nodes can also be directly generated along a curve using a unit length function q,

Figure 4.30 Domain boundary.

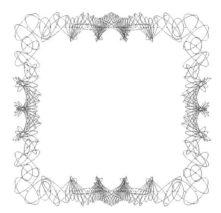

Figure 4.31 Dividing boundary segments.

$\mathbf{q} = \mathbf{q}(\mathbf{p}, M, \mathbf{u})$ such that $\|\mathbf{pq}\|_M = 1$

Given metric M, starting at point $\mathbf{p}$ along direction $\mathbf{u}$, unit length function $\mathbf{q}(\ )$ will locate point $\mathbf{q}$ such that the metric distance between points $\mathbf{p}$ and $\mathbf{q}$ is equal to unity. This unit-length function $\mathbf{q}$ is one of the most important operations in anisotropic meshing as unit-length measure from a point is often required throughout the mesh generation process. From a corner point of the domain, apply function $\mathbf{q}$ repeatedly to generate nodes of unit spacing along a boundary edge, as shown in Figure 4.31, such that

$\mathbf{p}_{k+1} = \mathbf{q}(\mathbf{p}_k, M, \mathbf{u})$ such that $\|\mathbf{p}_k\mathbf{p}_{k+1}\|_M = 1,$ $k = 1, n$

where $\mathbf{p}_1$ is the starting point, and $\mathbf{p}_{n+1}$ is the ending point. The unit distance between two points A and B measured with respect to metric field M is given by

$$\|AB\|_M = \int_0^1 \sqrt{AB^T M(A + tAB)AB}\ dt = 1$$

Given point A and unit vector **u**, point B can be determined by iteration:

i.   Get unit length s at A in the direction **u**, $(s\mathbf{u})^T M_A (s\mathbf{u}) = 1 \Rightarrow s = \dfrac{1}{\sqrt{\mathbf{u}^T M_A \mathbf{u}}}$
ii. Compute point B = A + s**u**
iii. Evaluate the distance between points A and B, $r = \|AB\|_M$
iv. Update $s \mapsto 2s/(1 + r)$
v.   If $\|1 - r\| > \epsilon$, go to step ii

In general, there is a residue between point P_{n+1} and the ending point of the line segment E. However, this residue cannot be distributed to the points by direct relocation similar to the case of isotropic metric as the distance measure is position-dependent for general metric field. Instead, we have to compute the discrepancy for each point given by

$$\delta = \frac{\|P_{n+1}E\|_M}{n}$$

The points have to be regenerated along the line segment following iteration steps i to v with adjusted distance $1 + \delta$ or $1 - \delta$ if P_{n+1} goes beyond E, and the last point P_{n+1} should coincide with ending point E.

3. *Triangulation of the boundary points.* The initial triangulation of the boundary points can be constructed by means of the classical DT using Euclidean metric, as shown in Figure 4.32.
4. *Insertion of interior points.* Generation of interior points follows the procedure of the Delaunay–ADF method, as elaborated in Section 3.7.3, in such a way that the underlying metric is respected. Given a frontal (boundary) line segment AB, which is of more or less unit length, a point C is created, which is of unit length from points A and B. For anisotropic metric field, the distance measure is non-linear, and point C, in general, cannot be determined analytically; hence, an iterative procedure is adopted in locating point C.

The initial position of point C can be estimated by means of a simple Euclidean norm such that

$$\|AC\| = \|AB\| \quad \text{and} \quad \|BC\| = \|AB\|$$

A more sophisticated procedure for locating the initial position of point C is by means of the intersection of the unit metric at points A and B. Ellipses of the unit

Figure 4.32 Triangulation of boundary points.

metric are drawn centred, respectively, at points A and B, as shown in Figure 4.33. For the determination of intersection points between two ellipses, an order-four quartic equation has to be solved as there can be as many as four intersection points between two ellipses. Instead, we can obtain an approximate position by the intersection of the bounding rectangles of the ellipses, as shown in Figure 4.33.

Starting from the initial position of point C, an iteration scheme can be devised to locate the point that is of unit distance from points A and B; such a point is given by the intersection of unit contours around points A and B, as shown in Figure 4.33. As shown in Figure 4.34, an improved position of point C from an initial point is given by Borouchaki et al. (1997a,b)

$$AC_1 = \frac{AC}{\|AC\|_M} \quad \text{and} \quad BC_2 = \frac{BC}{\|BC\|_M}$$

$$C_1 = A + AC_1 = A + \frac{AC}{\|AC\|_M}, \quad C_2 = B + BC_2 = B + \frac{BC}{\|BC\|_M} \quad \text{and} \quad C' = \frac{1}{2}(C_1 + C_2)$$

Iteration on point C is repeated until $\|AC\|_M \approx 1$ and $\|BC\|_M \approx 1$.

5. *Creation of insertion cavity.* For Euclidean metric, the insertion cavity is created based on the empty circumcircle criterion, i.e. a triangle is removed if its circumcircle contains the inserted point. However, for a non-Euclidean metric field, the point that is of equal distance to the three vertices of a triangle is not given by the circumcircle. In fact,

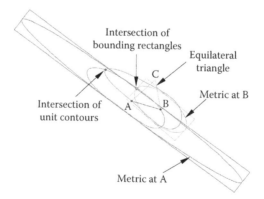

Figure 4.33 Initial position of point C.

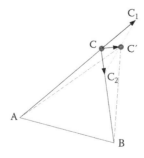

Figure 4.34 Improve point C by iteration.

there is no proof that such a point exists in a general metric field; and in practice, this point exists for a mildly varying metric field in the vicinity of a unit-length triangle. Take for instance the wavy surface metric as an example, as shown in Figure 4.35. The circumdisk of a triangle resembles an ellipse in a region of mildly varying metric, and the circumdisk of a triangle can take an arbitrary shape in a region of rapidly varying metric, which, in general, may even be non-convex. Nevertheless, there is no need to trace out circumdisks of triangles covering a large area of rapidly changing metric (peaks and troughs), and the circumdisk of unit-length triangles resembles an ellipse whose centre can be determined by iteration.

The initial position of centre O of the circumdisk of triangle ABC can be estimated using the Euclidean metric, i.e. the initial position of centre O is given by the circum-centre of triangle ABC or simply the centroid of the triangle, which is to be improved by iteration, as shown in Figure 4.36. Suppose O is the current position of the centre of the circumdisk, which is to be relocated to O′ such that $\|O'A\|_M \approx \|O'B\|_M \approx \|O'C\|_M$.

Let $d = \dfrac{d_A + d_B + d_C}{3}$ where $d_A = \|OA\|_M$, $d_B = \|OB\|_M$, $d_C = \|OC\|_M$.

Set $AO_A = \dfrac{d}{d_A}AO$, $\quad BO_B = \dfrac{d}{d_B}BO$, $\quad CO_C = \dfrac{d}{d_C}CO$

$$O_A = A + AO_A = A + \frac{d}{d_A}AO$$

$$O_B = B + BO_B = B + \frac{d}{d_B}BO$$

$$O_C = C + CO_C = C + \frac{d}{d_C}CO$$

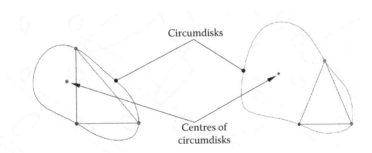

Figure 4.35 Circumdisks of triangles for non-Euclidean metric field.

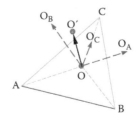

Figure 4.36 Point O improved by iteration.

We have $O' = \dfrac{1}{3}(O_A + O_B + O_C)$.

Alternatively, the centre of the circumdisk can also be determined by the Newton–Raphson iteration (Appendix A.6). Define $F = d_A - d_B$ and $G = d_A - d_C$. Then compute numerically the derivatives of F and G with respect to x and y.

$$F_x = \frac{\partial F}{\partial x} \approx \frac{F(x+h) - F(x)}{h}, \quad F_y = \frac{\partial F}{\partial y} \approx \frac{F(y+h) - F(y)}{h},$$

$$G_x = \frac{\partial G}{\partial x} \approx \frac{G(x+h) - G(x)}{h}, \quad G_y = \frac{\partial G}{\partial y} \approx \frac{G(y+h) - G(y)}{h},$$

where (x,y) are the co-ordinates of O, and h can be taken as 0.0001 relative to the size of the triangle. The new position of centre O is given by

$$O' = \begin{bmatrix} x \\ y \end{bmatrix} + \begin{bmatrix} \Delta x \\ \Delta y \end{bmatrix} = \begin{bmatrix} x \\ y \end{bmatrix} - \begin{bmatrix} F_x & F_y \\ G_x & G_y \end{bmatrix}^{-1} \begin{bmatrix} F \\ G \end{bmatrix}$$

Iteration cycles are repeated as necessary until $F \approx 0$ and $G \approx 0$. Starting from the base triangle, which contains the inserted point P, triangles are deleted following the adjacency relationship. Let O be the centre of the circumdisk of triangle $P_1P_2P_3$; triangle $P_1P_2P_3$ will be deleted if $\|OP\|_M \leq \max \|OP_i\|_M$ i = 1, 2, 3.

6. *Updating the generation front.* The generation front can be updated in exactly the same manner as the classical ADF approach before the next point is introduced; refer to Section 3.6.2 for details. The mesh generation is completed when there are no more segments on the generation front; otherwise, go to step 4.

4.2.12.2 The completed mesh

Figure 4.37 shows the construction of the first 1000 triangular elements by the Delaunay–ADF method for the anisotropic meshing according to the wavy surface metric. The pattern and sequence of element formation are similar to those of the ADF meshing of isotropic meshes except that elongated triangles are generated over some parts of the domain as required. Figure 4.38 shows another intermediate stage of anisotropic mesh near completion in which 5000 triangles have been generated. Smaller and elongated elements are created at the centre of the domain where there are many peaks clustered closely together. The α-quality of triangles on parametric curved surfaces has been discussed and defined in Section 4.2.10, whereas the conformity coefficient of a unit metric mesh is given by

$$\delta = \min\left(\|AB\|_M, \frac{1}{\|AB\|_M} \right), \quad \text{length of edge AB,}$$

$$\|AB\|_M = \int_0^1 \sqrt{AB^TM(A+tAB)AB}\, dt$$

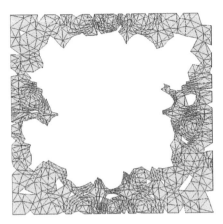

Figure 4.37 Mesh of 1000 triangles.

Figure 4.38 Mesh of 5000 triangles.

By the definition of the conformity coefficient, it is always smaller than or equal to 1, and sometimes it appears to be much smaller than 1 as short edges are often used to improve the quality of the mesh. The completed anisotropic mesh shown in Figure 4.39 consists of 4556 nodes and 8934 triangles. As for the quality of the mesh: the geometric mean α-quality = 0.518, the minimum α-quality = 0.0117, the mean conformity coefficient = 0.852 and the minimum conformity coefficient = 0.0434.

The parametric curved surface produced by the mapping of the anisotropic unit metric triangular mesh is shown in Figure 4.40. It is seen that relatively well-shaped triangles are resulted in mapping the triangles of the parametric domain onto the curved surface, which could be further improved by an optimisation process described in Section 4.2.13. Even though many iterations are used in various stages including the subdivision of boundary segments, the introduction of interior points and the creation of the insertion cavity, anisotropic meshing is basically linear as searching is not necessary in DT, and the number of iterations for each individual process is more or less a constant.

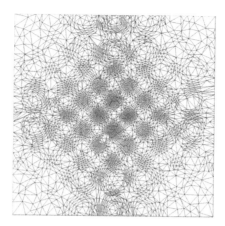

Figure 4.39 Completed mesh of 8934 triangles.

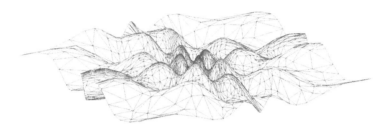

Figure 4.40 Parametric curved surface of 8934 triangles.

4.2.13 Optimisation of anisotropic meshes

Unlike the optimisation of isotropic meshes of which the shape of the element is the only concern, for anisotropic meshes, both the shape of the elements and how close the edges are in compliance with the unit metric have to be considered. Hence, the quality of a triangle consists of two parts: the α-value of the triangle and the δ-coefficient of all its edges, i.e.

quality of metric triangle ABC, $\psi = \alpha\delta_1\delta_2\delta_3$

where the α-value of metric triangle ABC on a parametric domain is defined in Section 4.2.10, and δ_1, δ_2 and δ_3 are the conformity coefficients of the edges discussed in Section 4.2.12. In general, the optimisation of an FE mesh can be achieved by geometrical and topological means. For the optimisation of unit metric meshes, one geometrical operation and one topological operation are proposed, which are performed in an alternative manner to achieve the best results.

4.2.13.1 Node smoothing

The quality of the mesh is to be improved by shifting each node to a more strategic position by means of the smart Laplacian method and its variation. For each interior node I (nodes generated by insertion in the triangulation process), determine the polygon of all the n surrounding triangles, as shown in Figure 4.41. Compute the average and minimum qualities of the triangles in this polygon:

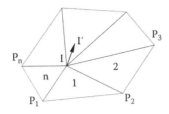

Figure 4.41 Shifting of interior node I to I'.

$$\overline{\psi} = (\psi_1 \psi_2 \cdots \psi_n)^{1/n} \quad \text{and} \quad \psi_{min} = \min(\psi_1, \psi_2, \cdots \psi_n)$$

1. *Laplacian smoothing.* The node is shifted to the centre of the polygon if there is improvement both for $\overline{\psi}$ and ψ_{min}; otherwise, the shifting is ignored for this interior node.
2. *Gradient-guided shifting.* Compute $\nabla \overline{\psi} = \left(\dfrac{\partial \overline{\psi}}{\partial x}, \dfrac{\partial \overline{\psi}}{\partial y} \right)$ by numerical differentiation; the node is shifted along $\nabla \overline{\psi}$ a small distance (about 5% of the size of the polygon) if there is an improvement both for $\overline{\psi}$ and ψ_{min}; otherwise, the movement of the node is reduced. This node-shifting process can be repeated a number of times until an improvement can no longer be made for this node.

Either by Laplacian smoothing or gradient-guided shifting, each interior node is processed in turn until all the interior nodes are treated. Usually, two cycles of smoothing can be carried out before moving on to the next phase of diagonal swapping.

4.2.13.2 Diagonal swapping

Diagonal swapping is a topological operation in which the diagonal of a quadrilateral formed by two adjacent triangles is swapped to the other position to improve the overall quality of the two triangles. The unique lines or edges of the mesh are extracted, as described in Section 2.5.4, and each interior edge shared by two triangles is examined in turn. For each edge, compute the overall quality of the triangles sharing it to see if any improvement can be made by a diagonal swap. Diagonal BD is swapped to AC if $\psi(ABD)\psi(BCD) < \psi(ACD)$ $\psi(ABC)$, and the minimum ψ-value is not compromised in the swapping process, as shown in Figure 4.42. Although the best results are expected by working with edges in the sequence

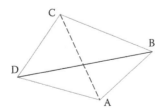

Figure 4.42 **Diagonal swap.**

of the larger gains in ψ-values, a sorting process will be involved, and it is up to the user to decide whether a marginal gain is worth the additional cost for sorting.

4.2.13.3 Optimisation of the wavy surface

Even though two or three cycles are usually needed in practice, five cycles of gradient-guided nodal smoothing and diagonal swapping without sorting have been performed on the raw mesh generated by the extended Delaunay–ADF method to see how the mesh progresses with geometrical and topological optimisations. The resulting mesh after five cycles of optimisation is shown in Figure 4.43, and the corresponding surface mesh is shown in Figure 4.44. Compared to the mesh of Figure 4.40, the shapes of the triangles have been much improved by optimisation. The statistics of the five cycles of nodal smoothing and diagonal swapping are given in Table 4.1. There is a sharp increase in the α-values of the triangles both in the average and minimum norms. While the minimum conforming coefficient is substantially improved from 0.0434 to 0.156, there is a slight reduction in the average value. As expected, the largest gain in the α and δ values is achieved in the first two or three cycles of optimisation. While there is plenty of room for improving the shape of the triangles, the conformity coefficient δ is more rigid, and there is only an improvement in the minimum norm. However, this is quite reasonable as during the insertion process, nodes are created solely based on the unit length metric without considering the shape of the triangular elements.

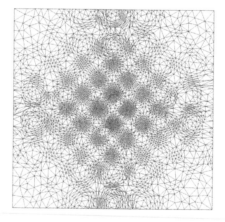

Figure 4.43 Mesh improved by five optimisation cycles.

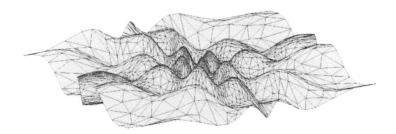

Figure 4.44 Parametric curved surface after optimisation.

Table 4.1 α and δ values improved by node smoothing and diagonal swap

Cycle	Operation	Mean α	Minimum α	Mean δ	Minimum δ
0		0.518	0.0117	0.852	0.0434
1	Node shifting	0.578	0.0285	0.840	0.122
	Diagonal swap	0.659	0.0509	0.818	0.134
2	Node shifting	0.692	0.0519	0.817	0.135
	Diagonal swap	0.712	0.0796	0.812	0.158
3	Node shifting	0.724	0.0825	0.811	0.151
	Diagonal swap	0.730	0.0825	0.810	0.151
4	Node shifting	0.733	0.0897	0.810	0.157
	Diagonal swap	0.734	0.0897	0.809	0.157
5	Node shifting	0.736	0.104	0.809	0.157
	Diagonal swap	0.737	0.105	0.809	0.157

4.3 MESH GENERATION BY PACKING ELLIPSES

As anisotropic meshes are governed by the unit metric of locally elliptical contour, packing ellipses, of size and shape compatible with the metric requirement, tightly together offers a practical means of generating metric meshes (Lo and Wang 2005b,c,d; Wang et al. 2007a). The effectiveness of ellipse packing is attributed to its flexibility in locating points in compliance with the underlying metric, which could be linked up fairly easily to form triangular elements of required shape and size. Packing of objects of various shapes as a means of mesh generation has been employed for many years over 2D and 3D space bounded or unbounded (Bossen and Heckbert 1996; Yamada et al. 1999; Shimada et al. 2000). By tightly packing objects of simple geometry, nodal points can be located in compliance with the required element size specification, which can be connected to form an FE mesh by means of the DT or the AFT.

In the following, an ellipse-packing algorithm will be introduced for the generation of anisotropic triangular meshes on a parametric domain. An algorithm for the generation of anisotropic mesh of variable element size over an unbounded 2D domain by using the ADF ellipse-packing technique is presented. Unlike the conventional frontal method, the procedure does not start from the object boundary but from a convenient point within an open domain. The sequence of packing ellipses is determined by the shortest distance from the fictitious centre in such a way that the generation front is more or less a circular loop with occasional minor concave parts due to element size variation. Whenever an ellipse is added to the generation front, triangular elements are directly constructed by properly connecting frontal segments with the centre of the new ellipse. Ellipses are packed closely and in contact with the existing ellipses by an iterative procedure in accordance with the unit metric length measure.

The anisotropic meshes generated by ellipse packing can also be used through a mapping process to produce parametric surface meshes of various characteristics. The size and the orientation of the ellipses in the pack are controlled by the unit metric tensor as derived from the principal surface curvatures, as described in Section 4.2.6. In contrast to other mesh generation schemes, the domain boundary is not considered in the packing process; this reduces a lot of geometric checks for intersections with frontal segments for each ellipse inserted. The boundary of the curved surface, if available, can be easily incorporated into the unbounded planar anisotropic mesh by intersection and node shifting, or the boundary lines of the curved surfaces are to be defined through an intersection process with other curved surfaces, as shown in the example surface meshes of Section 4.3.3.

4.3.1 Ellipse-packing algorithm

The idea of anisotropic mesh generation is to connect the centres of ellipses of variable size and orientation packed together by the ADF approach. Unlike the conventional frontal technique, the procedure does not start from the boundary of the object but starts from a convenient point within an open domain. The initial front consists of three tangent ellipses, which are to be expanded towards the exterior and has to be updated whenever a new ellipse is introduced. Triangular elements are subsequently formed when the centres of the ellipse are connected systematically along the generation front.

4.3.1.1 Data structure

The structure of the generation front is very simple, which is a closed loop of line segments. As a result, in the packing process, only those ellipses on the generation front need to be considered. In the implementation, the generation front is represented by an ordered list of ellipses on the boundary of the pack. For each ellipse, the centre, the length of the two principal axes, the direction of the major principal axis and the distance from the origin are stored. The process of front updating is just to apply the operations of inserting (deleting) a point to (from) the generation front. It is interesting to note that the ellipses along the generation front are tightly packed and in contact with each other.

4.3.1.2 Three criteria for ellipse packing

  i. *Densest – Ellipses are to be packed as close to one another as possible.* Ideally all ellipses are packed closely together so that the gaps between them are minimised. However, it is sometimes difficult to pack them together as the size and the orientation of a given ellipse may not always fit its neighbours. Hence, not only the position of the inserted ellipse but also the size and the orientation of the ellipse have to be adjusted to fit the local site in order to achieve a dense packing.
 ii. *No overlapping – There is no or little overlapping between any two ellipses.* It is best that any new ellipse added on the generation front does not overlap with the existing ellipses. The overlapping of ellipses leads to perhaps not only bad mesh quality but also faulty connections such as mesh intersection, overlapping and holes.
iii. *Nearest – A new ellipse is always generated at the location closest to the origin.* The front nearest to the origin is often the most concave part. Packing new ellipses at the concave part can fill gaps, which reduces the chance of forming large holes or voids. By packing ellipses nearest to the origin, the shape of the generation front is basically convex like a circle with minor concave parts.

4.3.1.3 Unit metric

The unit metric field M for a curved surface has been discussed in detail in Section 4.2.6, and the unit length requirement for arbitrary vector $\mathbf{v}$ in any direction is given by $\mathbf{v}^T M \mathbf{v} \leq 1$.

The length of vector $\mathbf{v}$ on the 2D plane is given by the Euclidean distance, i.e. $h = \|\mathbf{v}\|$. Alternatively, if metric M is represented by (λ, μ, θ) as described in Section 4.2.8, then length of unit vector $\mathbf{v}$ making an angle α with the x-axis is given by (Figure 4.45)

$$h^2 = \lambda^2 \cos^2(\alpha - \theta) + \mu^2 \sin^2(\alpha - \theta)$$

Given the unit length metric, the above equation allows us to compute the required distance of a proposed point in any direction.

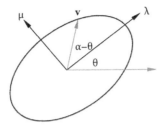

Figure 4.45 Unit length metric.

4.3.1.4 Initial pack and coefficient β

To simplify, the first three ellipses, denoted by C_1, C_2 and C_3, are placed tangent to each other around the origin $O(0,0)$, as shown in Figure 4.46. The size and the orientation of the three ellipses are determined according to the specified metric field. Triangle $C_1C_2C_3$ is the first element of the mesh. The initial front $\{C_1, C_2, C_3\}$ including the information of the centre, the length of the axes, the orientation of the ellipse and the distance to the origin is stored as an ordered list $\{C_1 - C_2 - C_3 - C_1\}$.

The parameter described below indicates how close two ellipses are.

$$\beta_{C_1C_2} = \min\left(\lambda, \frac{1}{\lambda}\right) \quad \text{with } \lambda = \frac{h_1 + h_2}{d}$$

where h_1 and h_2 are the Euclidean lengths in the direction C_1C_2 for ellipses C_1 and C_2, respectively, as shown in Figure 4.47, and d is the distance between the two centres of ellipses C_1 and C_2. The larger the value of $\beta_{C_1C_2}$ $(0 \leq \beta_{C_1C_2} \leq 1)$ is, the closer the two ellipses are between each other. In case $\beta_{C_1C_2} = 1$ i.e. $d = h_1 + h_2$, the two ellipses are in contact.

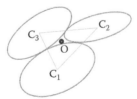

Figure 4.46 Initial front of three ellipses.

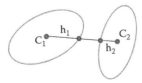

Figure 4.47 β Coefficient between two ellipses.

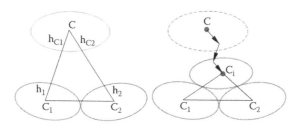

Figure 4.48 Determine C_i by iteration.

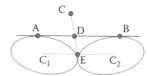

Figure 4.49 Locating initial point C.

4.3.1.5 Fitting an ellipse to the existing pack

Let C_1 and C_2 be the two neighbouring ellipses on the generation front. C is an arbitrary point near C_1C_2. An ellipse starting from point C can be packed to C_1 and C_2 following these steps:

   i. Compute the metric at point C.
   ii. Calculate the β-value $(\beta = \beta_{C_1C}\beta_{C_2C})$ at point C.
   iii. If $\beta > \eta$ $(\eta \approx 0.9)$, the ellipse at C can be fitted to C_1C_2, end; otherwise
   iv. Calculate β_x at point $(C + \Delta x)$, β_y at point $(C + \Delta y)$, $\Delta\beta_x(\Delta\beta_x = \beta_x - \beta)$ and $\Delta\beta_y(\Delta\beta_y = \beta_y - \beta)$ for arbitrary small Δx and Δy.

   v. Move C to C′ along the direction $\left(\dfrac{\Delta\beta_x}{\Delta x}, \dfrac{\Delta\beta_y}{\Delta y} \right)$ for a distance $(\lambda + \mu)\delta\%$, where λ and

   μ are the lengths of the principal axes of the ellipse at C, and δ takes progressively descending values of 10, 8, 6, 4, 2 and 1 in the iterative process to satisfy (iii).

According to the above procedure, the ellipse moves towards C_1 and C_2 gradually, as shown in Figure 4.48, and takes up position C_i as the final position for the ith inserted point when the closeness criterion (iii) is satisfied. The ellipse C_i is in contact with ellipses C_1 and C_2 to within the specified tolerance η. As the position of the initial point C would significantly affect the rate of convergence in fitting C to C_1 and C_2, it is important to have an accurate estimate of the initial position of point C. The initial position of point C can be estimated as follows. First, a line is drawn tangent to ellipse C_1 and C_2 touching the ellipses at points A and B, respectively, as shown in Figure 4.49. The initial position of C is given by extending line segment ED such that ED = DC where D is the midpoint of AB, and E is the mid-point of C_1 and C_2.

4.3.1.6 Checking intersection and mesh generation

Let C_m be the nearest ellipse to the origin on the front and C_{m+1} be the ellipse on the left of C_m. Following the procedure discussed in Section 4.3.1.5, C_i is the ellipse to be inserted into the pack at the site $C_m C_{m+1}$. The coefficient $f_{ij} = d_{ij} - (h_i + h_j)$ indicates whether C_i intersects with the generation front, where d_{ij} is the distance between the centres of C_i and C_j, h_i and h_j are the Euclidean lengths of the two ellipses C_i and C_j, respectively, in the direction $C_i C_j$, and C_j is an ellipse on the front with $j \neq m$ and $j \neq m + 1$. If ellipses C_i and C_j overlap, f_{ij} is less than zero; and if ellipses C_i and C_j are far apart, f_{ij} is greater than zero. Hence, the coefficient f_{ij} measures the relative positions of C_i and C_j.

Usually, $f_{ij} > 0$, i.e. C_i is not overlapping any ellipse C_j on the front, a new element $C_m C_i C_{m+1}$ can be directly generated, and the generation front has to be updated to include C_i such that $\{... C_m - C_{m+1} ...\}$ becomes $\{... C_m - C_i - C_{m+1} ...\}$, as shown in Figure 4.50.

However, when a large ellipse is required at a relatively small insertion site, C_i will possibly penetrate into some neighbouring ellipses, as shown in Figure 4.51a. To remedy this situation, we have to push C_i away from the line segment $C_m C_{m+1}$ until C_i is tangent with the two outermost ellipses without overlapping with other ellipses on the front, as shown in Figure 4.51b. The following are the detailed steps of the pushing-out operation.

    i. Calculate C_i touching C_m and C_{m+1} according to the procedure in Section 4.3.1.5.
    ii. A search on the neighbouring ellipses of $C_m C_{m+1}$ is conducted to find out if any ellipse near $C_m C_{m+1}$ intersects with C_i.
    iii. If C_i does not intersect the generation front, element $C_m C_{m+1} C_i$ is directly generated and packing ellipse C_i to ellipses C_m and C_{m+1} terminates.
    iv. Suppose C_i penetrates into C_j, such that $f_{ij} < 0$; the process of pushing out C_i is invoked.

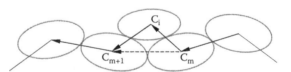

Figure 4.50 Updating generation front with the introduction of C_i.

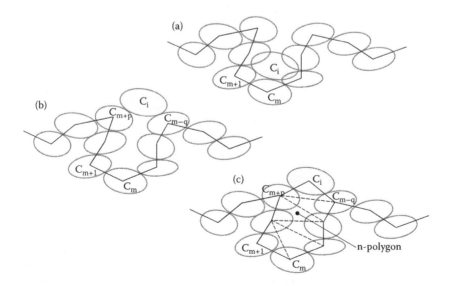

Figure 4.51 Push-out operation for serious overlapping: (a) insertion of large ellipse C_i at a concave site; (b) push C_i from the segment $C_m C_{m+1}$; (c) triangulation of n-polygon.

v. If ellipse C_j intersecting with C_i is on the left-hand side of C_mC_{m+1}, $j = m + p$, with $p \geq 2$. On the other hand, if ellipse C_j is on the right-hand side, $j = m - q$, with $q \geq 1$.

vi. Updating the supporting line segment C_mC_{m+1} for the construction of C_i, such that C_i is now generated from $\{C_{m+p}, C_{m-q}\}$. It is noted that either p or q will be updated within each cycle of the pushing-out operation. Go to step (i).

After the pushing-out operation, an n-polygon $\{C_m - C_{m+1} \dots C_{m+p} - C_i - C_{m-q} \dots \}$ can be defined as shown in Figure 4.51c. The problem now under consideration is the triangulation of a simple n-polygon, which can be easily done by dividing the narrow-shaped polygon into triangles of the best possible quality. As for the generation front, it could be updated by including two new segments $C_{m-q}C_i$ and C_iC_{m+p}, and, at the same time, segments from C_{m-q} to C_m, C_mC_{m+1} and from C_{m+1} to C_{m+p} have to be deleted from the generation front.

The consequence of the push-out operation is that triangles of slightly larger size are constructed at the concave sites; however, this is not a frequent operation, and the concave parts are, in general, pretty shallow that the inserted ellipse is only pushed up by one or two segments as we always search for the segment nearest to the origin to avoid serious concave parts being formed at the generation front.

There are two ways to control the termination of the packing of ellipses by the ADF approach. The first way is to check the number of ellipses packed, and the second way is based on the area of the mesh generated. If the number of ellipse is more than a given number or if the shortest distance to the origin is larger than a specified value, the procedure terminates. At this stage, the final generation front can be taken as the boundary of the triangular mesh generated.

4.3.2 Efficiency and complexity

There are three main operations in the ellipse-packing process. The first operation is to determine the ellipse insertion site nearest to the origin. The second operation is to determine the location, size and orientation of the ellipse to be packed at the insertion site. The third operation is to check whether there are any intersections between the inserted ellipse and the existing ellipses on the generation front.

As the distance of the ellipses from the origin are stored when they are created, the searching for the nearest ellipse is only a simple comparison. The computation cost grows with the number of ellipses on the generation front. However, the number of ellipses on the front increases rather slowly as ellipses are deleted and added to the front. From the examples in Section 4.3.3, we can see that the number of ellipses on the front is quite small compared to the number of ellipses in the pack, and the ellipses on the front are always closely packed together to form a circular ring structure. Upon further investigation into the searching process, we found that the searching time could be much reduced by restricting the distance check to a few ellipses near the current insertion site. In the present implementation, only five ellipses to the left and five ellipses to the right of the current insertion site are checked to locate the nearest point from the origin for the next insertion. The meshes generated are virtually the same compared to the scheme in which all the ellipses on the front are tested for the absolute minimum distance point.

The time of iteration to determine the position, the size and the orientation of the inserting ellipse is controlled by two major factors. The first factor is the initial position of the ellipse where iteration starts, and the second is the precision adopted for the convergence. A scheme has been proposed in Section 4.3.1.5 to estimate the initial position of the ellipse to be inserted. Following this simple procedure, the number of iterations has been greatly reduced, and the average number of iterations for various metric fields tested in the examples

Table 4.2 CPU time and statistics of example meshes

	1. 'X' curve	2. 'α' curve Origin outside α	Origin inside α	3. Two peaks	4. Klein bottle	5. Wavy surface
N	4000	3091	3117	7171	23991	29362
M	7865	6106	6159	14185	46624	58117
Max1	2.386	2.533	2.583	2.591	2.660	2.628
Avg1	1.163	1.157	1.151	1.120	1.120	1.121
Max2	4.882	3.237	3.196	3.057	4.003	2.750
Avg2	1.133	1.121	1.121	1.107	1.060	2.750
Max3	6.372	2.985	2.955	2.399	4.189	6.564
Avg3	1.611	1.499	1.485	1.099	1.046	1.135
Max4	100.0	49.91	49.95	10.24	6.653	6.723
Avg4	25.22	8.738	8.453	3.155	4.817	3.193
Iteration	9	7	7	4	4	4
CPU(s)	1.495	1.126	1.137	3.953	13.52	15.98

Note:

$$Max1 = \max_{k=1,N_L}\left(\frac{d_{ij}}{\sqrt{4h_ih_j}}, \frac{\sqrt{4h_ih_j}}{d_{ij}}\right); \quad Avg1 = \frac{1}{N_L}\sum_{k=1}^{N_L}\max\left(\frac{d_{ij}}{\sqrt{4h_ih_j}}, \frac{\sqrt{4h_ih_j}}{d_{ij}}\right)$$

$$Max2 = \max_{k=1,N_L}\left(\frac{\lambda_i}{\lambda_j}, \frac{\lambda_j}{\lambda_i}\right); \quad Avg2 = \frac{1}{N_L}\sum_{k=1}^{N_L}\max\left(\frac{\lambda_i}{\lambda_j}, \frac{\lambda_j}{\lambda_i}\right)$$

$$Max3 = \max_{k=1,N_L}\left(\frac{\mu_i}{\mu_j}, \frac{\mu_j}{\mu_i}\right); \quad Avg3 = \frac{1}{N_L}\sum_{k=1}^{N_L}\max\left(\frac{\mu_i}{\mu_j}, \frac{\mu_j}{\mu_i}\right)$$

$$Max4 = \max_{k=1,N}\left(\frac{\lambda_k}{\mu_k}, \frac{\mu_k}{\lambda_k}\right); \quad Avg4 = \frac{1}{N}\sum_{k=1}^{N}\max\left(\frac{\lambda_k}{\mu_k}, \frac{\mu_k}{\lambda_k}\right)$$

where

i and j are ellipses connected to edge k.
N = number of nodes, N_L = Number of edges.
d_{ij} = the distance of neighbouring nodes C_i and C_j.
h_i, h_j = Euclidean lengths in the direction of the neighbouring ellipses i and j.
λ_i, μ_i = lengths of principal axes of ellipse i.
Iteration = average number of iterations for ellipse packing.

is given in Table 4.2. A tighter control on the condition of touching between ellipses would require more iterations; however, a tolerance of 0.9 or above adopted in the examples is usually sufficient to produce a pack of ellipses virtually touching one another.

Overlapping between the proposed ellipse and the existing ellipses on the front has to be checked. If all the ellipses on the front are verified, the cost could be very high. However, because of the convex shape of the generation front, the number of frontal ellipses involved in the overlapping check could be reduced to a fixed number n, resulting in a linear time complexity in the checking process. In the examples presented in Section 4.3.3, n has been set to 10, i.e. checking of overlapping is conducted for 10 ellipses on the left and 10 ellipses on the right of the insertion site. Indeed, in 90% of the cases, there is no overlapping. The justification for the use of a small number n = 10 is that the front is more or less convex and is of the shape of a circular ring with occasional minor concave parts. Hence, the newly inserted ellipse does not overlap with ellipses far away from the insertion site. The result, however, is a great reduction of mesh generation time, and a linear relationship between the

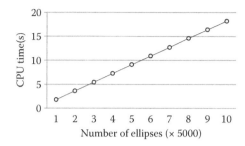

Figure 4.52 **CPU time vs number of ellipses.**

CPU time and the number of elements generated can be observed, as shown in Figure 4.52. The CPU time quoted is just for reference as these examples were done on a relatively slow machine around 2001 with a speed of 150 MHz and 128-MB RAM.

4.3.3 Examples of surface meshing by ellipse packing

Five examples of anisotropic mesh generation of variable element size over a 2D unbounded domain by ellipse packing are presented in this section. In two of the examples, the element size is controlled by a distance function, and in the other three examples, the metric field for element size control is derived from the surface curvatures, which are mapped to the 3D space to produce the required surface meshes.

Table 4.2 shows the statistics of the example meshes. M and N are, respectively, the number of nodes and the number of elements in the mesh. *Max1* and *Avg1* are, respectively, the maximum and the average of the ratio of the required sizes (h_i, h_j) at two points C_i and C_j to the actual length (d_{ij}) of the edge C_iC_j. *Max2* and *Avg2* are, respectively, the maximum and the average of the ratio of the major principal stretch between neighbouring elements. *Max3* and *Avg3* are, respectively, the maximum and the average of the ratio of the minor principal stretch between neighbouring elements. *Max4* and *Avg4* are, respectively, the maximum and the average of the ratio of the major principal axis to the minor principal axis of the packed ellipses. It is noted that the meshes presented in this section are raw meshes without optimisation by node shifting and diagonal swap.

Example 1 shows an anisotropic mesh whose element size is controlled by the distance to two crossing lines, and the principal directions of the ellipses are set along the direction of these two lines. The maximum ratio of the major principal axis to the minor principal axis is as large as 100 (*Max4*), and the average of this ratio is about 25 (*Avg4*). The difference in the size between neighbouring elements can be as large as 6.37 (*Max3*), and the average change between neighbouring elements is 1.61 (*Avg3*). Under the extreme situation, the error in the required sizes between neighbouring points to the actual length is about 16% (*Avg1*). The average number of iteration is 9. Figure 4.53a and b shows the ellipse packing and the associated triangular mesh of the crossing lines. Figure 4.53c and d shows the magnified views at the central part of Figure 4.53a and b.

Figure 4.54a and b shows the ellipse packing and the corresponding mesh of a curve in the shape of Greek alphabet α, in which the ratio between the major principal axis to the minor principal axis is about 50 (*Max4*). The average change in size between the neighbouring ellipse is 1.49 (*Avg3*). The discrepancy in the required sizes between the neighbouring points to the actual length is about 15% (*Avg1*). One remarkable fact worth mentioning is that by this frontal mesh generation procedure, the origin or the centre of radiation is not obvious at all, and indeed this point, in general, cannot be discerned in the mesh by visual inspection.

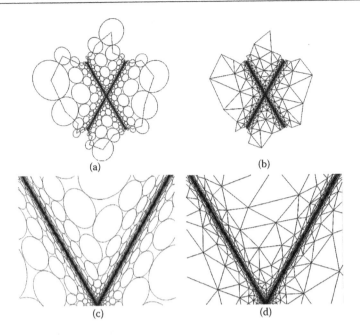

Figure 4.53 Ellipse packing and meshing generation around a cross line pattern: (a) ellipses packed along intersecting lines; (b) anisotropic triangular mesh; (c) a magnified view of (a) at the centre; (d) a magnified view of (b) at the centre.

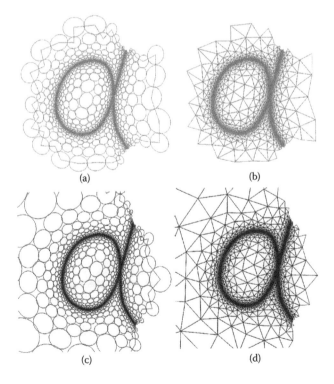

Figure 4.54 Ellipse packing and mesh generation for pattern governed by Greek letter α: (a) ellipses packed around α, origin inside α; (b) anisotropic triangular mesh of (a); (c) ellipses packed around α, origin outside α; (d) anisotropic triangular mesh of (c).

This important characteristic ensures that the choice of origin can be arbitrary and would not affect the resulting mesh, which is governed only by the underlying metric requirement. The meshes shown in Figure 4.54b and d is generated with the origin located at two different positions, one inside the α-curve and the other outside the α-curve.

The next examples are about the application of anisotropic meshing to curved surfaces. In general, curved surfaces can be represented by a bivariate mapping of the form

$$\mathbf{S} = \mathbf{S}\,(x, y, z) = \mathbf{S}(u, v) \text{ such that } x = x(u, v),\ y = (u, v),\ z = z(u, v)$$

The size and the orientation of the ellipses in the pack are controlled by the unit metric field of curved surface **S**, which can be computed following the procedure described in Section 4.2.6. The parametric surface shown in Figure 4.55a and b with two distinct peaks is defined by

$$x = u, \quad y = v, \quad z = (u^2 + 3v^2)e^{1-u^2-v^2}$$

In this example, the discrepancy in the required sizes at neighbouring points to the actual length is about 12% (*Avg1*). The average number of iteration is relatively few, which is equal to 4. Figure 4.55c shows the surface mesh mapped from the anisotropic mesh of Figure 4.55b. In 3D space, the boundary of the surface is defined by a surface intersection process, as shown in Figure 4.55d.

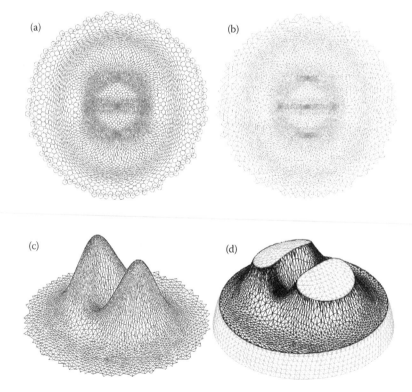

Figure 4.55 Ellipse packing and mesh generation for curved surface 'Two Peaks': (a) ellipses packed on parametric domain; (b) anisotropic mesh on parametric domain; (c) curved surface mapped from mesh (b); (d) boundary created by surface intersection.

Figure 4.56a and b shows the ellipse packing and the corresponding anisotropic triangular mesh of the Klein bottle. The Klein bottle is defined by the following parametric surface (Borouchaki et al. 2000a)

$$x = \begin{cases} 6\cos u(1+\sin u) + r\cos u\cos v & \text{if } 0 \le u \le \pi \\ 6\cos u(1+\sin u) - r\cos v & \text{if } \pi \le u \le 2\pi \end{cases}$$

$$y = \begin{cases} 16\sin u + r\sin u\cos v & \text{if } 0 \le u \le \pi \\ 16\sin u & \text{if } \pi \le u \le 2\pi \end{cases}$$

$$z = r\sin v, \quad r = 4 - 2\cos u, \quad \text{over the square } 0 \le u, v \le 2\pi$$

For this example, the error in the required sizes at neighbouring points to the actual length is also 12% ($Avg1$). The average number of iterations is also equal to 4. Figure 4.56c shows the surface mesh obtained through the mapping of the anisotropic mesh of Figure 4.56b.

Figure 4.57a and b shows the ellipse packing and the corresponding anisotropic triangular mesh of a wavy curved surface. The parametric surface is defined by Borouchaki et al. (1999)

$$x = u^3 + 10u, \quad y = v^3 + 10v, \quad z = 100\sin u\cos v$$

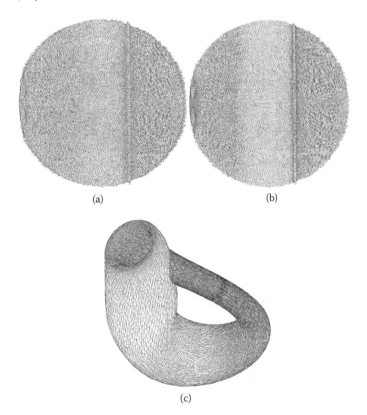

(a)   (b)

(c)

Figure 4.56 Ellipse packing and mesh generation for curved surface Klein bottle: (a) ellipses packed on parametric domain; (b) anisotropic mesh on parametric domain; (c) surface mesh of Klein bottle.

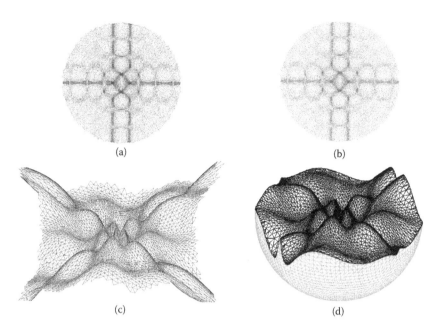

(a)

(b)

(c)

(d)

Figure 4.57 **Ellipse packing and mesh generation for the wavy surface: (a) packing ellipses; (b) triangulation; (c) curved surface mapped from mesh (b); (d) boundary created by surface intersection.**

In the last example, the discrepancy in the required sizes at neighbouring points to the actual length is about 12% (*Avg1*). The average number of iteration is equal to 4. Figure 4.57c shows the meshed curved surface produced by mapping the anisotropic mesh of Figure 4.57b. In 3D space, the boundary of the surface can be conveniently defined through a surface intersection process, as shown in Figure 4.57d.

Remarks: The ellipse packing has turned the mesh generation problem into a problem of fitting objects of simple geometrical shape. The size and the orientation of the ellipses are governed by the local metric, which provides a visual inspection of the point distribution with respect to the specified metric field (unit metric contour). Owing to the simple geometry of an ellipse, determination of intersection has been reduced to a distance check by a simple function f_{ij} between adjacent ellipses C_i and C_j on the generation front. Moreover, relative to the usual intersection check, the distance function used is much less stringent for which there is a large tolerance for the fitting process resulting in a robust and efficient alternative other than a direct generation of anisotropic meshes by proper well-established mesh generation schemes.

4.4 DIRECT MESH GENERATION ON SURFACE

Direct three-dimensional surface mesh generation forms elements directly on the curved surface without the need for a parametric representation of the underlying geometry. In the case where a parametric representation is not available or where the surface parameterisation is poor, direct 3D surface mesh generation can be applied. Lau and Lo (1996) presented an ADF scheme for the generation of triangular meshes on curved surface in 3D space. By this method, surface normal and tangents are computed to determine the direction of the generation front. A number of surface projections are required to bring a proposed node point back on to the surface. Intersection checks are also needed to make sure that triangles on the surface do not overlap. Direct surface mesh generation will also be useful in situation of remeshing an existing surface triangulation of discrete elements for which surface param- eterisation is not readily available. In case an adaptive refinement mesh is required for the current FE model, defined in terms of element connections and nodal co-ordinates, direct 3D meshing is the most suitable.

Direct 3D surface mesh generation of quadrilateral elements is also possible. A direct 3D implementation of the paving technique over curved surfaces was presented by Cass et al. (1996). Heuristic *sticky space* is defined to detect intersection and overlapping quadrilater- als. However, similar to the merging of triangles over planar domains, more flexibility and controls can be retained by merging triangles on the curved surface to form quadrilater- als, resulting, in general, higher-quality quadrilateral meshes (Lau et al. 1997). For such a scheme, we can basically follow the procedure for the merging of triangles, as described in Section 3.9.4, except that we have to take also the surface orientations (normals) of the adjacent triangles into consideration in the merging process.

4.4.1 Initial generation front

Finite elements are generated directly on the curved surface one by one based on the ADF approach. Similar to the ADF method over planar domains, the boundary of the surface is taken as the initial front, which is a union of disjoint closed loops of straight line segments approximating the curved boundary of the domain. To start meshing from the boundary, a systematic way for the identification of the interior of the surface to be discretised is required. One simple method that has been used for many years is to enter the exterior loop in a counter-clockwise order and interior loops in a clockwise order. As a result, the interior part of the surface is always lying on the left-hand side of the generation front, as shown in Figure 4.58. However, for general curved surfaces, one more condition is needed before we can fix the orientation of the segments so that there is no ambiguity in the surface orienta- tion. The following convention based on the direction of the surface normal is adopted for the orientation of the boundary curves.

> The orientation of the boundary curves is defined such that the cross product of the surface outward normal and boundary curve tangent vector always points towards the interior of the surface.

It is assumed that the normal is well defined over the entire curved surface, so that sur- faces like the Mobius strip and the Klein bottle are excluded. If the surface normal changes gradually but vanishes at a finite number of points, which usually happens for surfaces with

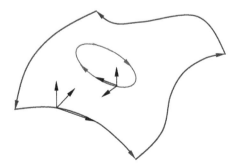

Figure 4.58 Orientation of boundary curves.

poor parameterisation and/or at singular points, it is still possible to define the orientation of the surface by considering a close neighbourhood around a singular point (Section 2.4.6).

4.4.2 Forming triangular elements on a surface

In the mesh generation process, a line segment AB is taken randomly from the generation front Γ. A node C_Γ on the generation front is selected, which forms the *best* triangle with the base segment AB. The possibility of forming a better triangle with the introduction of an interior point C_I is then explored. If the quality of triangle ABC_I is superior to that of triangle ABC_Γ, judging from a number of factors including the surface curvature, the element shape and the size requirement, triangle ABC_I will be formed with interior node C_I to replace triangle ABC_Γ. Otherwise, the best triangle ABC_Γ formed with a node C_Γ on the generation front will be taken as the element for base segment AB.

Unlike the case of the two-dimensional domain where discretisation error only occurs at the domain boundary, discretisation error will also occur at the interior of the curved surface due to the surface curvature. In order to control the geometrical discrepancy induced by the surface curvature, the angle between the triangle connected to segment AB, say AOB, and the new triangle ABC ($C = C_\Gamma$ or C_I) is checked to ensure that it is smaller than the allowable tolerance ϕ_ε before it is accepted. This can be done fairly easily by computing the cosine of the angle between the normals to the triangles AOB and ABC, as shown in Figure 4.59.

$$\frac{\mathbf{n}_1 \cdot \mathbf{n}_2}{\|\mathbf{n}_1\|\|\mathbf{n}_2\|} > \cos(\phi_\varepsilon) \quad \text{where } \mathbf{n}_1 = AO \times AB \quad \text{and} \quad \mathbf{n}_2 = AB \times AC$$

By setting a small ϕ_ε, there is a limit to the size of the largest triangle that can be formed with a base segment, thus resulting in a better approximation of the curved surface. For the generation of elements with relatively small sizes over a curved surface with a large radius of curvature, such a test might not be necessary. However, for graded meshes on general curved surfaces, this element-to-element curvature control is essential.

To facilitate the search for an interior node, a reference frame is constructed at the midpoint M of base segment AB, as shown in Figure 4.60. The orthonormal base vectors $\mathbf{n}$, $\mathbf{e}_1$ and $\mathbf{e}_2$ are defined as follows:

1. $\mathbf{e}_1$ is the unit vector along AB, $\mathbf{e}_1 = \dfrac{AB}{\|AB\|}$.
2. $\mathbf{n}$ is the surface normal at M perpendicular to $\mathbf{e}_1$.
3. $\mathbf{e}_2$ is the unit normal vector to the plane spanned by $\mathbf{n}$ and $\mathbf{e}_1$, i.e. $\mathbf{e}_2 = \mathbf{n} \times \mathbf{e}_1$.

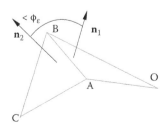

Figure 4.59 Angle constraint between elements.

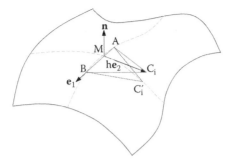

Figure 4.60 Local reference frame and estimated position of a new node.

The unit vector **n** is a vector normal to the surface and perpendicular to base segment AB. In the actual implementation, the average of the surface normals at A and B is sufficient for the purpose of mesh generation. The interior of the region is bounded by the current generation front where new elements are to be generated in the direction e_2.

4.4.2.1 Find the best node on the generation front

An extension of the method discussed in Section 3.6.2 for planar domain is used to find the best node on the generation front. Let Λ be the set of node points on the generation front Γ. A node $C_\Gamma \in \Lambda$ is said to be a candidate node of AB if the following conditions are satisfied.

  i. The triangle ABC_Γ lies on the interior bounded by the generation front.
  ii. The triangle ABC_Γ does not cut into the generation front.

The first condition can be ensured by comparing the normal **n** and normal n_Γ of triangle ABC_Γ, which is defined as

$$n_\Gamma = \frac{AB \times AC_\Gamma}{\left\| AB \times AC_\Gamma \right\|}$$

If the angle between **n** and n_Γ is less than 90°, triangle ABC_Γ is considered to be lying at the interior of the unmeshed region. Mathematically, this can be expressed as the dot product of the vectors **n** and n_Γ.

$$n \cdot n_\Gamma = n \cdot \frac{AB \times AC_\Gamma}{\left\| AB \times AC_\Gamma \right\|} > 0 \quad \text{or simply} \quad n \cdot (AB \times AC_\Gamma) > 0$$

The second condition involves the intersection check over a curved surface, which is slightly more complicated than a similar process on a planar domain. The standard intersection test of line segments using the spatial nodal points is not sufficient, as the line segments forming the generation front are not necessarily lying on the curved surface (in fact, only the end points of the line segments are on the surface). Two line segments may not intersect in space even if their corresponding surface curves are crossing each other.

Let $\widetilde{AC_\Gamma}$ and $\widetilde{BC_\Gamma}$ be the surface curves corresponding to line segments AC_Γ and BC_Γ, respectively. The second condition can be expressed as

$$\widetilde{AC_\Gamma} \cap \Gamma \in \left\{\widetilde{AC_\Gamma}, \{A, C_\Gamma\}\right\} \quad \text{and} \quad \widetilde{BC_\Gamma} \cap \Gamma \in \left\{\widetilde{BC_\Gamma}, \{B, C_\Gamma\}\right\}$$

For analytical surfaces, we can assume that the surface is given by the parametric equation $S(u,v)$, and a surface curve $\tilde{C}(t)$ on surface $S(u,v)$ can be written as

$$\tilde{C}(t) = S(u(t), v(t))$$

Thus, two surface curves $\tilde{C}$ and $\tilde{C}' = S(u'(s), v'(s))$ intersect with each other if

$$u(t) = u'(s) \quad \text{and} \quad v(t) = v'(s) \quad \text{for some s and t within the range of the curves}$$

Depending on how the surface is defined, intersection check can be done by a projection of the line segments on a local tangent plane. For sufficiently smooth surfaces, check for frontal intersections can be conveniently done between line segments on the parametric domain.

4.4.2.2 Locate interior node

Apart from taking a node from the generation front, very often, a new node created at the surface interior can form a well-shaped triangular element with the base segment. Given the nodal spacing specification at each point on surface S, the position of the interior point forming the best element with the base line segment can be determined. Let h be the required element height for the base segment AB. As shown in Figure 4.61, from the mid-point of AB, M, an approximate spatial position C_I for the new interior node is given by

$$C_I = M + h e_2$$

The spatial point C_I has to be relocated back to the curved surface S by means of nearest point (normal) projection. Furthermore, the distance from surface point C_I' on the curved surface to the generation front Γ should be at least half the required element height, i.e.

$$\text{Distance}\left(C_I', E_k\right) > \frac{h}{2} \quad \text{Edge } E_k \in \Gamma$$

This ensures that poor-shaped triangles will not be formed with the frontal segments in the subsequent mesh generation. In case any one of the conditions cannot be satisfied by the candidate node, the required element height has to be reduced progressively until the quality

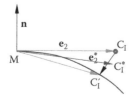

Figure 4.61 Closest point projection.

of triangle ABC_I' is inferior to the best triangle ABC_Γ that could be formed with a node C_Γ on the generation front.

4.4.2.3 Space-to-surface projection

The purpose of space-to-surface projection is to map a point C_I in the 3D space onto the surface such that the size and the shape of triangle ABC_I' are preserved after projection. Two algorithms, one using the closest point projection for simple surfaces and the other using a more general scheme based on surface derivatives, will be described.

4.4.2.3.1 Closest-point projection

For simple surfaces such as cylinder and sphere whose closest-point projection is available in explicit form or surfaces whose closest-point projection can be computed readily, this mapping is a convenient choice for the space-to-surface projection. Although the size and the shape of the resulting triangle need not be the same as those of the original one, the closest-point projection does give a reasonably close approximation of the required triangle.

Depending on how far the spatial point C_I is from the surface, the quality of the projected triangle may differ from the required one by an unacceptable amount if relatively large elements are generated at places of small radius of curvature. A remedy for this is to construct a better spatial point C_I^* closer to the surface. This can be achieved by replacing vector e_2 with a vector e_2^*, which better approximates the direction of the new node relative to the surface. Let C_I' be the projection of C_I; the updated direction vector e_2^* is given by

$$e_2^* = \frac{e_2 + e'}{\|e_2 + e'\|} \quad \text{where unit vector } e' = \frac{v}{\|v\|} \quad \text{with } v = MC_I' - \left(MC_I' \cdot e_1\right)e_1$$

As shown in Figure 4.61, a spatial point closer to the surface is given by

$$C_I^* = M + he_2^*$$

v is the projection of vector MC_I' onto the plane spanned by vectors n and e_2 normal to e_1, and the updated direction e_2^* is the average of the original direction e_2 and the projected vector v. Since the new direction is on the plane of vectors n and e_2, the size and the shape of triangle ABC_I^* are preserved. Geometrically, it is equivalent to rotating the direction of the vector towards the surface. The process can be repeated in an iterative manner until the projected point on the surface is of a distance close to the specified element size.

4.4.2.3.2 Projection for general surface

Let $p = p(u,v)$ be a parametric representation of curved surface S, where p is the position vector of the points on S. Assuming that S is regular, i.e. for all the points on S,

$$\frac{\partial p}{\partial u} \times \frac{\partial p}{\partial v} = u \times v \neq 0$$

where u and v are, respectively, basis vectors associated with co-ordinates u and v.

Given a point $p(u,v)$ on S, a close neighbour, say, q on S at $(\Delta u, \Delta v)$ from p, can be expressed as a Taylor's expansion about p:

$$\bar{q} = p + (\Delta u)u + (\Delta v)v + \cdots$$

Hence, given a point p on S and the associated tangent vectors u and v, a first-order approximation of a point $q = q(u + \Delta u, v + \Delta v)$ can be computed. However, the point $\bar{q}$, in general, will not lie on the surface S.

Now consider a point r on S whose spatial co-ordinates are known but not its parametric co-ordinates. If a point $p = p(u,v)$ on S sufficiently close to r is known, then a first-order approximation of the increment of the parametric co-ordinates from p to r $(\Delta u, \Delta v)$ can be determined by solving the equation

$$\Delta p = r - p = (\Delta u)u + (\Delta v)v$$

However, a solution may not exist as there are more equations than the number of unknowns (three equations with two unknowns). In fact, it is not possible to find Δu and Δv because Δp is not lying on the tangent plane at point p. A solution for this is to use the projection of Δp onto the tangent plane spanned by vectors u and v, i.e.

$$\Delta p' = \text{Projection of } \Delta p = (\Delta u)u + (\Delta v)v \tag{4.1}$$

Better approximation can be obtained by iteration with updated co-ordinates of point p, i.e.

$$p \mapsto p(u + \Delta u, v + \Delta v)$$

Repeatedly solving Equation 4.1 with updated point p until Δp is normal to the tangent plane at p, or $\Delta p'$ is smaller than a specified tolerance ε, i.e. $\|\Delta p'\| < \varepsilon$.

For mesh generation based on space-to-surface projection, we have to solve a slightly different problem: determine the point C on S, or more strictly speaking, find its parametric co-ordinates (u,v) such that the size and the shape of the triangle formed with the base segment is preserved. This leads to the condition that the projected point C on S should lie also on the plane normal to the segment AB at M and at a distance h from M.

The initial guess C_o on S can be obtained from the closest point projection (Equation 4.1) of a spatial point r on the tangent plane at M:

$$r = M + he_2$$

An improvement of C_i based on a first-order approximation is given by

$$C_{i+1} = p(u_{i+1}, v_{i+1}) = p(u_i + \Delta u, v_i + \Delta v) \approx C_i + (\Delta u)u + (\Delta v)v$$

Δu and Δv are to be solved based on the following conditions:

    i. $e_1 \cdot MC_{i+1} = 0$
    ii. $\|MC_{i+1}\| = h$

The first condition ensures that C_{i+1} is on the plane normal to segment AB, and the second condition ensures that C_{i+1} is of a distance h from M. Iteration is needed as there is a discrepancy between $C_{i+1} = (u_i + \Delta u, v_i + \Delta v)$ and $C_i + (\Delta u)\mathbf{u} + (\Delta v)\mathbf{v}$, which is not on the surface. Given an initial guess C_o on surface S, a series of approximations $\{C_1, C_2, ..., C_i, ...\}$ can be constructed such that it converges to a point C on S satisfying conditions (i) and (ii) provided that the portion of the surface around C is sufficiently smooth. The iteration process for surface point C is summarised as follows.

For i = 0, 1, 2, ...
Compute tangent vectors $\mathbf{u}$ and $\mathbf{v}$ at C_i;
Solve for Δu and Δv from $e_1 \cdot M\overline{C}_{i+1} = 0$ and $\left\| M\overline{C}_{i+1} \right\| = h, \overline{C}_{i+1} = C_i + (\Delta u)\mathbf{u} + (\Delta v)\mathbf{v}$;
Compute the updated co-ordinates of C_{i+1}, $u_{i+1} = u_i + \Delta u$, $v_{i+1} = v_i + \Delta v$;
If C_{i+1} is sufficiently close in satisfying conditions (i) and (ii), break;
Next i

Upon the generation of a triangle at the base segment with a frontal node or an interior node, the generation front has to be updated to include the newly generated element following exactly the same procedure of the planar domain. The element formation process on the curved surface is repeated by taking another base segment from the updated generation front until there is no more segment left as the generation front shrinks to zero.

4.4.3 Examples of direct construction

Several examples of triangular meshes over surfaces of different geometrical characteristics are shown in Figures 4.62 to 4.67. The element density distribution is a specified piecewise continuous function, so that the grading effect of the resulting meshes can be easily checked. Some statistics about the sample meshes are listed in Table 4.3.

In the first two meshes shown, respectively, in Figures 4.62 and 4.63, the exact formula of the closest-point projection is available for the space-to-surface mapping. Due to the simplicity of these two surfaces, the least amount of CPU time was required for mesh generation. A triangulation of the ruled surface is shown in Figure 4.64. More CPU time was required in this case because of the rapid change of curvature. The proposed general projection algorithm was used in this example and the other examples. The triangulation of a free-form surface modelled by Bezier patches is shown in Figures 4.65 and 4.66. Lastly, a mesh of a multi-connected curved surface is shown in Figure 4.67. More CPU time required for the last four examples is not due to the use of the more general projection algorithm but the more complex geometry of the surface and their derivatives. In fact, a few iterations,

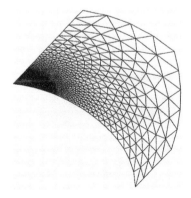

Figure 4.62 **Part of a cylindrical surface.**

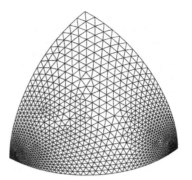

Figure 4.63 Part of a spherical surface.

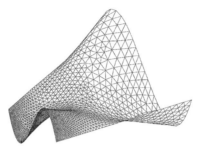

Figure 4.64 Ruled surface.

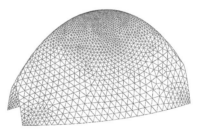

Figure 4.65 Spline surface 1.

Figure 4.66 Spline surface 2.

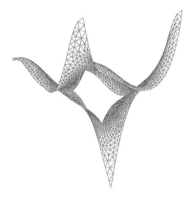

Figure 4.67 Multi-connected surface.

Table 4.3 Statistics of the sample meshes

Figure	NN	NE	Time	$\bar{\alpha}$	α_{min}	$\alpha_{0.9}$	$\bar{\delta}$
5	1827	3533	13.4	0.983	0.778	99.0%	0.883
6	1508	2849	10.9	0.994	0.791	99.9%	0.943
7	2416	4600	137.3	0.983	0.745	99.2%	0.887
8	2965	5826	75.2	0.986	0.804	99.6%	0.894
9	793	1412	26.8	0.983	0.643	99.0%	0.909
10	4942	9592	163.5	0.978	0.630	98.0%	0.883

Note: NN = number of nodes; NE = number of elements; Time = CPU time of IBM 350 Powerstation in seconds; $\bar{\alpha}$ = geometrical mean α-quality; α_{min} = minimum α-quality; $\alpha_{0.9}$ = percentage of elements with α-quality > 0.9; $\bar{\delta}$ = geometrical mean δ-quality.

usually about one to three, were sufficient to produce a projected triangle having the same size and shape as the expected one.

To assess the quality of the graded meshes, both the element shape and the size distribution are measured. For measuring the shape of the triangular elements, the α-quality described in Section 3.7.2 is employed:

$$\alpha(ABC) = \frac{\text{signed area}}{\text{sum of edges squared}} = \frac{2\sqrt{3}AB \times AC}{\|AB\|^2 + \|BC\|^2 + \|CA\|^2}$$

Several measures showing the distribution of the α-values of the triangular elements are listed in Table 4.3. It can be seen that the elements in all meshes are of very high quality. The poor elements are usually found on the boundary due to the geometrical constraint of the domain.

The δ-coefficient defined as

$$\delta = \min\left(\frac{\text{actual element size}}{\text{required element size}}, \frac{\text{required element size}}{\text{actual element size}}\right)$$

is used to measure the compliance in element size. A value of $\delta = 1$ represents that the actual element size is of perfect match with the required value. For meshes of continuously varying element density, there is no guarantee that $\delta = 1$ for all elements and all edges, and thus, any value above 90% should be considered satisfactory. Here, the required element size is taken as the average of the element size density at the three vertices of the triangle. From Table 4.1, the element size distribution of the meshes is in good agreement with the specified one with a mean δ value greater than 88%. Such a difference is probably due to the rather strict element-to-element curvature control, which prevents some relatively large elements of the required size from being formed at places of rapid change of curvature.

Remarks: Direct mesh generation on curved surfaces offers additional flexibility in the size and shape control of the elements and in the definition or modelling of the curved surface. The amount of discretisation error induced by the surface curvature can be easily controlled by limiting the maximum allowable element to an element-turning angle. As the mesh generation is carried out in the Euclidean space, the shape of triangular elements can be accurately specified and measured. For simple analytical surfaces, maximum efficiency can be attained by using the explicit formula for the closest-point projection. For arbitrary curved surfaces, the general projection algorithm provides a practical means to produce the required projected triangle.

4.5 MESH GENERATION BY SURFACE INTERSECTION

4.5.1 Introduction

The primary objective is to merge two meshed surfaces into a single FE mesh in a fully automatic and robust manner. The types of surfaces to be dealt with are surfaces already discretised into triangular facets or elements. This definition of surface intersection is broad enough to cover the intersection of artificial surfaces that can be discretised into triangular facets and FE meshes made up of surfaces of triangular and quadrilateral elements.

The intersection lines between surfaces are represented in terms of line segments, which are to be determined by considering a pair of triangular facets at a time. Intersection line segments are determined one by one in sequence to form structural loops and chains. An intersection line will be formed progressively by connecting line segments end to end. Two situations can be identified: (i) the line segment terminates on the boundary of the intersection surfaces to form an open chain, and (ii) it goes back to the starting point to form a closed loop, as shown in Figure 4.68.

An accurate and efficient process for the determination of intersection chains and loops is of utmost importance to the robustness of the intersection of discretised surfaces. A neighbour-tracing technique proposed in the work of Lo and Wang (2003, 2004, 2005a) is perhaps the most effective for this purpose, which greatly enhances the reliability and the

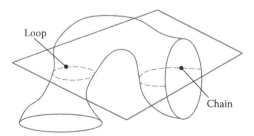

Figure 4.68 Intersection between surfaces.

efficiency in the determination of surface intersections as the neighbours for each triangle are known, and connections of segments are done by means of continuity. The following are the steps for the construction of intersection chains/loops by the neighbour-tracing technique.

1. Find out and record the neighbours of each triangle on the two surfaces.
2. Reduce the population of candidate triangles by means of a background grid filter.
3. Identify the first intersection segment from one of the partition cells.
4. From the seed intersection segment and by means of continuity, the entire chain/loop is determined by tracing neighbouring triangles one after the other.

4.5.1.1 The determination of neighbours

The neighbours of each triangle are useful topological information of the triangulated surface. The searching time can be much reduced by making use of the continuity as provided by the neighbourhood relationship. For a given surface composed of triangular facets, apart from those on the boundary, each triangle on the surface connects to three distinct adjacent triangles. If two triangles share a common edge, then they are neighbours to each other. A linear algorithm in determining the neighbours of a triangular mesh is given in Section 2.5.1. The notation adopted for naming the surfaces are as follows:

First surface –
Triangles on the surface = $\{S_k, k = 1, N_S\}$;
Vertices of the triangles = $\left\{ S_k^i \quad i = 1, 2, 3 \quad k = 1, N_S \right\}$
Neighbours of the triangles = $\left\{ M_k^i \quad i = 1, 2, 3 \quad k = 1, N_S \right\}$
Second surface –
Triangles on the surface = $\{T_k, k = 1, N_T\}$;
Vertices of the triangles = $\left\{ T_k^i \quad i = 1, 2, 3 \quad k = 1, N_T \right\}$
Neighbours of the triangles = $\left\{ N_k^i \quad i = 1, 2, 3 \quad k = 1, N_T \right\}$

4.5.2 Background grid

A direct computation of the intersection between two surfaces by considering all the triangles on one surface against all the triangles on the other surface is not an economical process. As intersections only occur in a few parts of the surfaces, it is necessary to filter out triangles that do not possibly have any intersection. A background grid is introduced for this purpose in which the search for the intersection of a triangular facet can be localised to within a small volume as given by the size of a typical cell of the background grid. The time complexity of the process is linear for triangular facets of homogeneous size. A more sophisticated spatial partition into cells of variable sizes can be employed for triangular facets of a large difference in element size, such as the Octree partition described in Section 2.7.7 and the kd-tree partition described in Section 2.7.8.

Following the natural sequence of the cells of the background grid, the triangles having intersections with the cells are recorded. For a given cell, there may be intersection if there is at least one triangular facet from each surface. There will not be any intersection if a triangular facet from either surface is missing. This condition will be used as a preliminary check for possible intersections. Figure 4.69 shows triangular facets of a typical cell in which

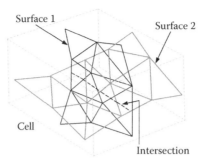

Figure 4.69 Intersection of triangles in a typical cell.

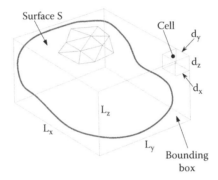

Figure 4.70 Background grid for triangulated surfaces.

thinner lines represent facets of the first surface, and thicker lines represent facets from the second surface.

For completeness, the regular background grid introduced in Section 2.7 is also presented here following the adopted notations. The dimensions of a typical cell is given by

$$d_x = \frac{L_x}{N_x}, \quad d_y = \frac{L_y}{N_y}, \quad d_z = \frac{L_z}{N_z}$$

where L_x, L_y and L_z represent, respectively, the lengths on the three sides of the bounding box, and N_x, N_y and N_z represent, respectively, the number of divisions along the three principal directions, as shown in Figure 4.70. The number of cells $N = N_x \times N_y \times N_z$ depends on the amount of computer memory available. The size of a cell can be set equal to the average length of the edges on the surface. However, the dimensions of the cells can be also adjusted so that the use of available memory can be optimised.

4.5.2.1 Determination of cells intersected by a triangular facet

A typical triangular facet will intersect more than one cell in the background grid. Considering each triangular facet in turn, the cells cut by the triangular facet under consideration are recorded. At the end of the process, the list of triangular facets intersected with each cell can be determined. The cells cut across by a triangular facet can be found by the following procedure.

a. Calculate the bounding box B of the triangular facet T with vertices v_1, v_2, v_3.

$$x_{min} = \min(a_1, a_2, a_3), \quad x_{max} = \max(a_1, a_2, a_3)$$
$$y_{min} = \min(b_1, b_2, b_3), \quad y_{max} = \max(b_1, b_2, b_3)$$
$$z_{min} = \min(c_1, c_2, c_3), \quad z_{max} = \max(c_1, c_2, c_3)$$

where $v_1 = (a_1, b_1, c_1)$, $v_2 = (a_2, b_2, c_2)$, $v_3 = (a_3, b_3, c_3)$.

b. Calculate the cells intersected by triangular facet T.

Let $C = [X_{min}, X_{max}] \times [Y_{min}, Y_{max}] \times [Z_{min}, Z_{max}]$ be the bounding box of the smaller surface. If B is outside C, no cell in C will intersect with the triangular facet; otherwise, record the intersecting cells with address (i, j, k) using the following formulas:

$$i = \left[\text{Int}\left(\frac{x_{min} - X_{min}}{d_x} \right) + 1 \right] \to \left[\text{Int}\left(\frac{x_{max} - X_{min}}{d_x} \right) + 1 \right]$$

$$j = \left[\text{Int}\left(\frac{y_{min} - Y_{min}}{d_y} \right) + 1 \right] \to \left[\text{Int}\left(\frac{y_{max} - Y_{min}}{d_y} \right) + 1 \right]$$

$$k = \left[\text{Int}\left(\frac{z_{min} - Z_{min}}{d_z} \right) + 1 \right] \to \left[\text{Int}\left(\frac{z_{max} - Z_{min}}{d_z} \right) + 1 \right]$$

4.5.3 Find all the candidate triangles

i. Define a background grid $\mathbb{G}$ on the smaller surface, named as surface $\mathbb{S}$.

ii. For all the triangular facets in $\mathbb{S}$: determine the cells intersected by each triangular facet $S_j \in \mathbb{S}$; hence, record the intersecting facets S_j in each cell $C_i \in \mathbb{G}$.

iii. For all the triangular facets in the larger surface $\mathbb{T}$: determine the cells intersected by each triangular facet $T_k \in \mathbb{T}$; hence, record the intersecting facets T_k in each cell $C_i \in \mathbb{G}$.

iv. For all the cells C_i in $\mathbb{G}$ from 1 to N: each cell C_i will be examined in turn. Cell C_i will be ignored if either S_j from surface $\mathbb{S}$ or T_k from surface $\mathbb{T}$ is absent. Triangles that may have a possibility of intersection are recorded as candidate elements for each surface.

Two spherical surfaces are shown in Figure 4.71a, on each of which are 672 triangular facets. Figure 4.71b shows the candidate triangles determined by using a background grid.

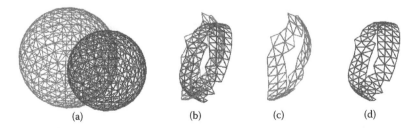

(a) (b) (c) (d)

Figure 4.71 Candidate triangles of intersecting surfaces: (a) spherical surfaces; (b) candidate triangles; (c) loop of triangles of the first surface; (d) loop of triangles of the second surface.

As shown in Figure 4.71c and d, the number of the candidate triangles for the first surface is 88 and that for the second surface is 157. Obviously, the candidate triangles are just a small subset of the original surfaces.

4.5.3.1 Calculating the intersection between a pair of triangular facets

The two surfaces $\mathbb{S} = \{S_k^i, i = 1, 2, 3; k = 1, N_S\}$ and $\mathbb{T} = \{T_k^i, i = 1, 2, 3; k = 1, N_T\}$ are the two given sets of triangular facets. The intersection between the surfaces of $\mathbb{S}$ and $\mathbb{T}$ can be determined by considering a pair of triangles in turn, taken from the two surfaces. The intersection between typical triangular facets ABC from surface S and DEF from surface T can be determined as follows.

In order to find out where triangle DEF cuts triangle ABC, we have to consider the intersection between triangle ABC and edges DE, EF and FD of triangle DEF, as shown in Figure 4.72a. Line segment DE will have intersection with triangle ABC if

    i. Points D and E are on the opposite sides of the plane containing triangle ABC.
    ii. Intersection point P lies inside triangle ABC.

Condition (i) can be verified by checking if

$$(AD \cdot \mathbf{n})(AE \cdot \mathbf{n}) \le 0$$

where normal vector $\mathbf{n}$ of triangle ABC is given by $\mathbf{n} = AB \times AC$.

If condition (i) is satisfied, compute the point of intersection P.

$$P = tE + (1 - t)D \quad \text{where } t = \frac{d}{d - e} \quad \text{with } d = AD \cdot \mathbf{n} \text{ and } e = AE \cdot \mathbf{n}$$

Finally, P is inside triangle ABC if

$$(AB \times AP) \cdot \mathbf{n} \ge 0, \quad (BC \times BP) \cdot \mathbf{n} \ge 0, \quad (CA \times CP) \cdot \mathbf{n} \ge 0$$

Similarly, the second point where triangle ABC cuts triangle DEF can be determined by considering the intersection between triangle DEF and edges AB, BC and CA of triangle ABC. Let Q be the second intersection point where triangle ABC cuts triangle DEF; then the line segment PQ will be the intersection between the two triangles, as shown in Figure 4.72. An alternative method to calculate the intersection line segment is to use signed vol-

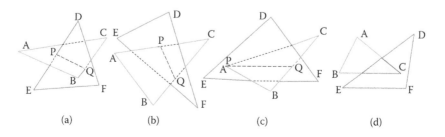

Figure 4.72 Types of intersection between two triangular facets: (a) type 1; (b) type 2; (c) type 3; (d) type 4.

ume (Aftosmis et al. 1998; Shostko et al. 1999). For the intersection of a pair of triangular facets, four general cases can be identified as follows:

1. The intersection point P is inside triangle ABC, as shown in Figure 4.72a.
2. The intersection point P is on an edge of triangle ABC, as shown in Figure 4.72b.
3. The intersection point P is on a vertex of triangle ABC, as shown in Figure 4.72c.
4. The intersection is more than one point and is a planar region, as shown in Figure 4.72d.

Intersections are first classified into one of these four types, which will then be treated accordingly. This classification enables a consistent treatment for the same intersection on the two surfaces.

4.5.4 Tracing neighbours of intersecting triangles

As explained before, the intersection between surfaces is best represented by structural elements of chains and loops rather than individual unconnected line segments. Along each intersection line segment, the intersecting triangles on each surface are neighbours to one another. Making use of this neighbouring relationship, an intersection line can be constructed by tracing neighbouring triangles one after the other.

In the neighbour-tracing process, the type of intersection will indicate to which neighbour the intersection line will extend. The following are the details of how neighbouring triangles are traced for the four types of intersections: (1) inside the triangle, (2) on an edge, (3) at a vertex and (4) planar zone of intersection.

1. If the intersection point is inside a triangle, there is no need to trace its neighbour. As shown in Figure 4.73a for intersection type 1, point P is inside triangle S_1. In this case, there is no need to trace for the neighbour of S_1. Instead, P is on the common edge of T_1 and T_2; hence, by neighbour tracing, the intersection line continues into T_2.
2. As shown in Figure 4.73b, the intersection point P is on an edge of S_1. The triangle having this common edge with S_1 is S_2, and the intersection line grows from S_1 to neighbouring triangle S_2.
3. If the intersection point is on a vertex of triangle S_1, all the triangles connected to this node have to be examined. Consider the intersection between triangles S_1 and T_1, as shown in Figure 4.73c. The intersection point P is on a vertex of S_1; triangles S_2, S_3, S_4 and S_5 having P as a common vertex will all be examined for intersection with triangle T_1.

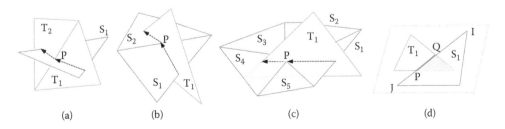

(a)   (b)   (c)   (d)

Figure 4.73 Construction of intersection line by tracing of neighbours: (a) type 1; (b) type 2; (c) type 3; (d) type 4.

4. If the two intersecting triangles are on a common plane, there is no need to compute the intersection. In other words, intersections will only be calculated if the two triangles concerned are not on a common plane, in which case they can be again classified into types 1, 2 and 3, as described above. However, if the boundary edge is involved, the intersection has to be calculated along the boundary part, as shown in Figure 4.73d. Two triangles are on a common plane, IJ is a boundary edge, and the intersection line segment PQ between triangles S_1 and T_1 has to be recorded. Figure 4.74 shows the intersection between a planar surface with an internal opening (Figure 4.74a) and a cylindrical surface with a closed bottom plate (Figure 4.74b). The outer loop is the result of the intersection between the vertical cylindrical surface and the flat surface. The inner loop is the intersection between the interior boundary of the flat surface and the bottom plate of the cylinder, as shown in Figure 4.74c and d.

Intersection lines can be constructed by linking up individual line segments to form open chains or closed loops. In the tracing process, if the intersection line goes back to the starting point, a closed loop will be formed; otherwise, a chain can be defined with the end points on the boundary of the surface. Figure 4.75a shows the loop of intersection between two spheres of Figure 4.71a. There are 114 line segments in the intersection loop, as shown in Figure 4.75b. Figure 4.75c and d shows the intersection loop and the neighbouring triangles on the two surfaces. Figure 4.76 shows the intersection of two open surfaces. The intersection line in the form

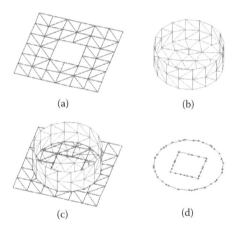

(a)   (b)

(c)   (d)

Figure 4.74 Intersection of planar surfaces: (a) planar surface with an internal opening; (b) cylindrical surface with a bottom plate; (c) cylindrical surface on top of the planar surface; (d) resulting intersection loops.

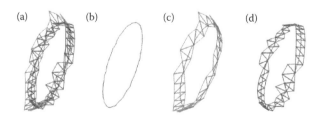

(a)   (b)   (c)   (d)

Figure 4.75 Intersection loop and neighbouring triangles: (a) triangles collected by neighbour tracing; (b) loop of intersection segments; (c) ring of intersecting triangles of the first surface; (d) ring of intersecting triangles of the second surface.

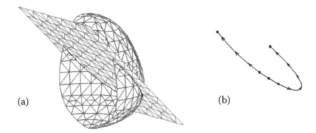

Figure 4.76 Intersection of open surfaces: (a) surfaces; (b) intersection chain.

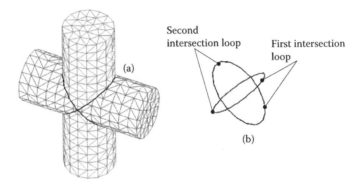

Figure 4.77 Intersection of cylindrical surfaces: (a) surfaces; (b) two intersection loops.

of an open chain terminates on the boundary of the hemisphere. An intersection chain can be defined in the neighbour-tracing process when no neighbour can be found on a boundary edge.

When two surfaces intersect, the intersection may consist of more than one loop/chain, as shown in Figure 4.77. In this case, the intersection loops/chains can be traced out one by one. To improve efficiency, the pairs of candidate triangles already considered for intersections are flagged. In the construction of a new chain/loop, what is needed is the first (seed) intersection segment, and the line of intersection can be easily determined by neighbour tracing. When all pairs of intersection candidate triangles are flagged indicating there are no more intersections, the entire process terminates.

4.5.5 Time complexity and memory management

Let's examine the time complexity for all the process involved in the surface intersection algorithm.

1. *Establishing the neighbouring relationship of the elements.* The neighbouring relationship can be determined by scanning through all the elements twice (Section 2.5.1); hence, this is a process of linear time complexity.
2. *Defining the background grid and determining the elements in a cell.* This can be achieved by calculating the cells intersected by each element. As the cells are contiguous and sequential, this is also a process of linear time complexity.
3. *Intersection between elements.* This is a more complicated process depending on the element distribution and the number and characteristics of the intersections. Only a rough estimate of time complexity for surfaces of more or less homogeneous element size is given.

Let N_1 and N_2 be, respectively, the number of elements in the two surfaces. For even element distribution in a grid of N cells, the average number of elements in a cell is given by $k_1 = N_1/N$ for surface 1 and $k_2 = N_2/N$ for surface 2. The number of combinations for the intersection between elements within a cell is $k = k_1 k_2 = N_1 N_2/N^2$.

If we have to look into all the cells for intersections, the total number of operations is

$$N_{Total} = \frac{N_1 N_2}{N^2} N = \frac{N_1 N_2}{N}$$

Assuming $N_1 = N_2 = M$ and N is of the same order of M, i.e. $N = \lambda M$, then

$$N_{Total} = \frac{M^2}{N} = \frac{M}{\lambda}$$

Thus, N_{Total} is of the same order as M, that is, if the elements are evenly distributed or made into such a distribution by means of advanced spatial partition schemes, the search for intersection is of linear time complexity. However, other measures could be and have been taken to speed up the element intersection process in general and in the worst scenario.

   i. The min/max comparison will be applied between elements before rigorous intersection tests are performed.
   ii. Intersection chain/loop is recovered by neighbour tracing, and all intersected elements are flagged and deleted from the list of candidate elements, which reduces in number as more intersection lines are retrieved. In fact, each intersection line (chain or loop) can be recovered by just locating one intersection point as the seed starting point.

For the examples of different characteristics, the time taken for searching for neighbours, construction of a background grid and surface intersection are more or less the same, indicating a linear time complexity of all these processes. Moreover, the algorithm can deal with generalised surfaces consisting of several disjoint patches of triangles, and the number of surface components in a surface group will have little influence on the computational cost, as this information is never used, and individual surface patches need not be extracted or identified in the intersection process.

The main additional memory requirement is for the construction of the background grid. An efficient scheme for recording the list of elements for each cell is to make use of a pointer, as described in Section 2.5.3. Suppose the background grid consists of N cells; then a single vector **M** is needed to store the information of elements in each cell. For i = 1,N, elements in cell i = {**M**(j), j = P_i + 1, P_{i+1}}. Auxiliary pointer P is of size N + 1, and the size of **M** equals the total number of intersections between elements and cells. For instance, if each element intersects with three cells on average, then the size of **M** equals three times the number of elements.

4.5.6 Mesh generation along intersection lines

The intersection lines have to be incorporated into each of two triangulated surfaces. Based on Lo (1995), by keeping the features of the intersection lines, the nodes on the intersection lines are repositioned according to the local element size. Elements cut by the intersection lines are removed, and the void as a result of element removal is meshed by proper connection of nodes between the intersection line and the surface boundary (Hartmann 1998). While this method is simple, due care is needed to maintain the original geometry of the surface

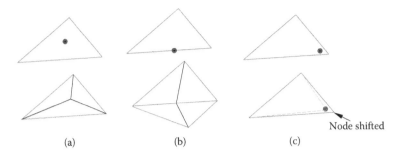

Figure 4.78 Insertion of intersection nodes: (a) interior node; (b) edge node; (c) corner node.

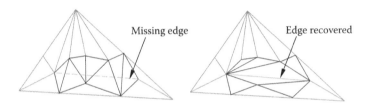

Figure 4.79 Segment recovered by edge swaps.

during mesh generation. Alternatively, a robust method is available ensuring that elements generated are always on the triangulated surface. The steps are summarised as follows.

i. *Incorporating the intersection line to each surface.* This process is easy and fast as the element associated with each intersection node is recorded in the neighbour-tracing process. An intersection node can be classified into three cases according to its position relative to the element, as shown in Figure 4.78. (a) The node is at the interior of the element – the element is divided into three triangles; (b) the node is on an edge of the element – the edge will be split, and two new elements are generated; and (c) the node is close to a vertex of the element – the vertex is shifted to the intersection node, and no element is created. Intersection line segments are incorporated into the surface by inserting intersection nodes one by one following the sequence along the intersection line (loop/ chain). In case some intersection line segments are missing after the node insertion process, they can be easily recovered by swapping of edges, as shown in Figure 4.79.

ii. *Mesh optimisation along the intersection line.* Excessive nodes are removed or repositioned along the intersection lines. Elements near the intersection lines are improved by node repositioning and a swap of diagonals. In fact, all the techniques employed to improve a planar triangular mesh can be applied. However, in the optimisation, due care has to be exercised to maintain surface features and surface curvatures, and this can be achieved by simply verifying the geometry of the patch of elements involved in the process.

4.5.7 Work examples

Six examples of various surface characteristics are presented in this section to illustrate how intersections are determined by tracing neighbours of intersecting triangles (TNOIT) on a relatively slow PC machine with a CPU speed of 150 MHz and 128-MB RAM. In the first example, three spheres consisting, respectively, of 192, 672 and 3584 triangular facets are

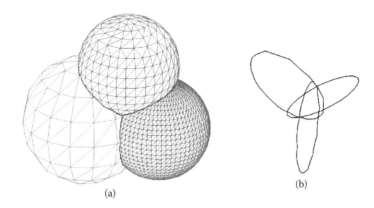

(a)

(b)

Figure 4.80 Intersection of surfaces discretised into elements of different sizes: (a) three surfaces of 4448 triangles; (b) intersection loops of 422 segments.

considered as shown in Figure 4.80a. Different sizes of the triangles on the spheres were adopted to test the algorithm. The intersection consists of three loops in 422 line segments, as shown in Figure 4.80b.

In the second example, 100 spheres are randomly placed in a box one by one, as shown in Figure 4.81. Each sphere consists of 192 triangular facets. The intersection of two spheres is first considered, and the resulting surface consists of 384 triangles. Then the third sphere is introduced, which interacts with the previously merged spheres, and this process is repeated until the last sphere is placed in the box. At the final stage, the updated surface of 19,008 triangles intersects with the last sphere to produce a surface composed of 19,200 triangles. For the intersection of these 100 spheres, there are 13,539 intersection line segments from which 277 loops can be traced out. Figure 4.82 shows the statistics of the CPU time for the major operations in the intersection process. Referring to Figure 4.82, the CPU time for calculating neighbours, for the determination of candidate triangles by background grid and for computing intersection loops by TNOIT and the overall CPU time are shown by four separate graphs. From the graphs, the CPU time increases with the number of spheres

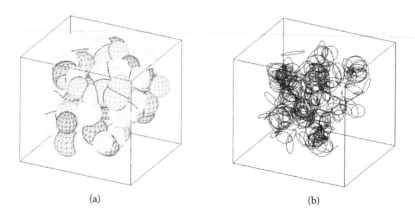

(a)

(b)

Figure 4.81 100 randomly placed spherical surfaces and their intersections: (a) 100 triangulated spherical surfaces; (b) 277 intersection loops of 13,549 segments.

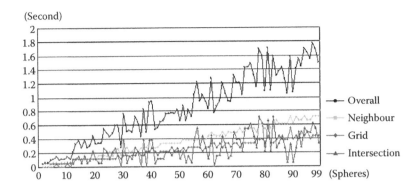

Figure 4.82 CPU time plots for the intersection of 100 spheres.

added, and a quasi-linear relationship can be observed. The local fluctuation is probable due to the random nature of how the spheres are placed in the box.

One thousand spheres are studied in the third example as shown in Figure 4.83a. These 1000 spheres were divided into 10 groups of 100 spheres each. A group of 100 spheres as a single surface was put in the box at a time until all 10 groups of spheres were processed. A magnified view of the intersection between the spheres is shown in Figure 4.83b. The intersection of these spheres consists of 138,807 intersection line segments, which form 1965 loops. The CPU times for the various stages of the intersection process are shown in Figure 4.84.

Figure 4.85a shows the finite element mesh of a plant modelled by 49,235 triangular elements and 24,936 nodes. The model is constructed by placing tree branches arbitrarily in space. The algorithm determines all the 4378 intersection segments and combines all individual surfaces into one single model. Similarly, a blood vessel system can also be built by the same technique, as shown in Figure 4.85b. Individual blood vessels in the form of curved cylindrical tubes with a variable diameter as determined in an x-ray or CT scan process are placed in space. The algorithm automatically determined all the intersection segments and

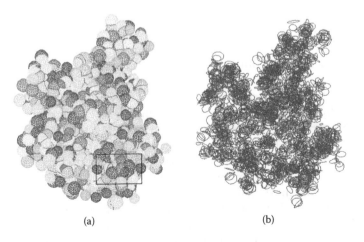

(a)   (b)

Figure 4.83 1000 randomly placed spherical surfaces and their intersections: (a) 1000 triangulated spherical surfaces; (b) 1965 intersection loops of 138,807 segments.

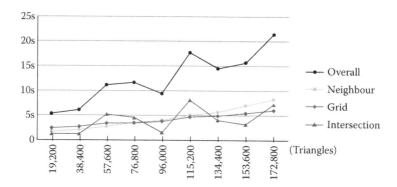

Figure 4.84 CPU time plots for the intersection of 100 spheres.

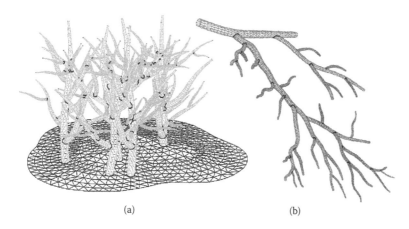

(a)   (b)

Figure 4.85 Biological systems modelled by surface intersection: (a) plants modelled by 49,235 triangles; (b) blood vessel system created by surface intersection.

recorded them as intersection loops, which joined separate components together to form the entire blood vessel system.

The fifth example is the intersection of a space shuttle and an analytical curved surface. The space shuttle was arbitrarily immersed into the curved surface to produce very irregular lines of intersection around the body of the space shuttle, as shown in Figure 4.86a. There are 6419 nodes and 12,843 elements on the surface of the space shuttle, and the curved surface is represented by 13,999 nodes and 27,548 elements. The surface interaction produces 1573 nodes from intersection segments, and 9948 elements are generated in the surface-merging process. Figure 4.86b shows a magnified view of the intersection, and it can be seen that sharp elongated elements have been used on the analytical curved surface to test the algorithm. The total CPU time taken for this example is about 1 s.

The last example is the most challenging. Two triangulated surfaces constructed from a surface scan were downloaded from the website http://lodbook.com/models/, as shown in Figure 4.87a. There are 1,309,332 elements in the two triangulated surfaces in which there are many pointed triangles connecting points from scanned images to model the finger bone geometry. As shown in Figure 4.87b, 54,966 intersection segments in 16 loops are

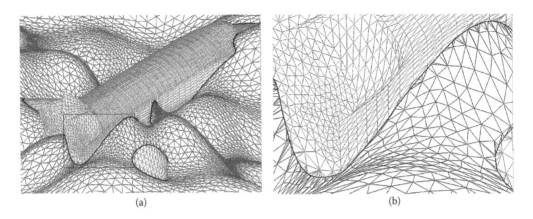

(a)   (b)

Figure 4.86 Intersection of space shuttle and wavy curved surface: (a) overall intersection view; (b) magnified view.

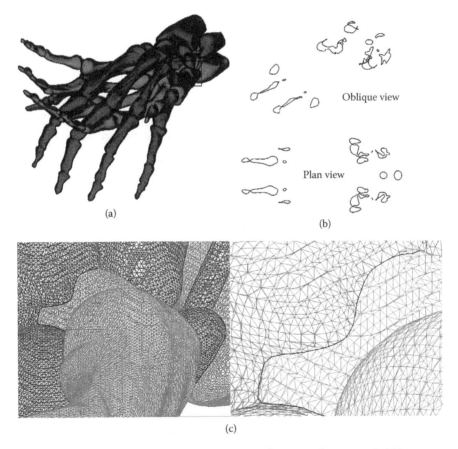

(a)   (b)

(c)

Elements = 1,309,332, Nodes = 654,646, Intersection loops = 16, Segments = 54,966
Search for neighbours: 7.43 s, Background grid: 5.09 s, Intersections by TNOIT: 6.61 s
Local mesh generation: 1.70 s, Overall CPU time = 20.83 s

Figure 4.87 Intersection of complicated surface models of a large number of elements: (a) hands (Stereo Lithography Archive at Clemson University); (b) 16 intersection loops; (c) magnified views.

recovered, and the total CPU time required is 20.83 s. Figure 4.87c shows the magnified views of the hands. It can be seen that intersection lines have already been incorporated into the surface by a local mesh modification procedure. After mesh optimisation, elements of size and shape compatible with those on the surface are generated along the lines of the intersection.

4.5.8 Intersection of surfaces of quadrilateral elements

Surfaces discretised into quadrilateral and/or triangular elements can also be merged by the TNOIT algorithm (Lo and Wang 2003, 2004). All steps for intersection remain the same except that in the neighbour-tracing process, quadrilaterals have to be first divided into two triangles. Quadrilaterals are divided into triangles according to the following:

1. The surface curvature has to be maintained by dividing quadrilaterals such that the dihedral angle between adjacent triangles is as close to 180° as possible.
2. In case surface curvature does not indicate any preferential subdivision into triangles, the quadrilateral is divided into triangles along the shorter diagonal.

An aeroplane and a space shuttle consisting of, respectively, 2891 and 7087 triangular and quadrilateral elements are intersected, as shown in Figure 4.88; 674 intersection line segments can be traced out in the neighbour-tracing process.

4.5.8.1 Closure

An algorithm based on TNOIT for the determination of intersection lines between surfaces of triangular and quadrilateral elements has been presented in detail. Along with the

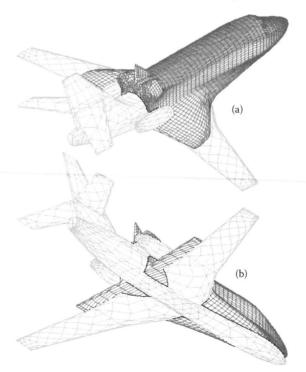

Figure 4.88 Intersection between aeroplane and space shuttle: (a) view from top; (b) view from bottom.

insertion of intersection nodes and local mesh optimisation, meshed surfaces can be directly merged into a single surface mesh in a completely automatic and robust manner. The reliability of the intersection process is greatly enhanced as intersections are classified into four general types based on which a clear direction as to how the intersection line should progress following the element adjacency relationship can be determined. While TNOIT ensures a consistent treatment of intersection on the two intersecting surfaces, the determination of intersections is also tremendously speeded up, as the search for intersection is kept to a minimum using the adjacency information.

Based on TNOIT, the intersection between surfaces is represented as structural loops and chains following a natural order rather than individual intersection segments. This not only can save a lot of searching work for each intersection point, but more importantly, it also enhances the overall reliability of the process as undetermined degenerated and branching cases can all be detected and dealt with in a consistent manner. In order to speed up the process of detecting intersections, the use of background grid is introduced. In this way, the search for intersection is localised to within cells, each of which contains, in general, only a few triangles from the surfaces. The determination of triangular facet cutting partition cells is also of linear time complexity; hence, little overhead is added in the intersection procedure.

4.6 QUADRILATERAL SURFACE MESH

Similar to 2D mesh generation, generation of quadrilateral curved surfaces can be classified into direct and indirect approaches. Direct approaches such as mapping, co-ordinate transformation, drag and sweeping methods, etc., are the common techniques employed for the generation of structured meshes rapidly over smooth regular surfaces. As for meshes subject to element size variation for FE adaptive refinement analysis, for other reasons such as to match the surface curvature or to capture the geometric characteristics of the surface, unstructured meshes have to be generated. In view of the development of efficient surface triangulation algorithms as described in Sections 4.2–4.4, the indirect method seems to be the most effective approach in producing high-quality quadrilateral meshes of variable element size. Provided that the surface is bounded by an even number of edges, the schematic merging of triangles into quadrilaterals proposed by Lee and Lo (1994) and presented in Section 3.9.4 is relatively simple and effective in producing high-quality meshes on 2D and over surfaces. Two examples of quadrilateral mesh by schematic merging of triangles are shown in Figures 4.89 and 4.90. A frontal merging by the *Blossom* algorithm in the graph

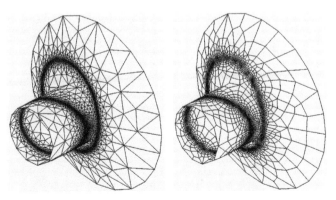

Figure 4.89 All-quadrilateral mesh of a funnel by merging of triangles.

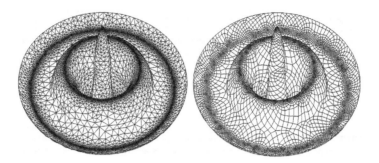

Figure 4.90 All-quadrilateral mesh of a hat by merging of triangles.

theory of triangular meshes generated by DT with the L^∞ metric was proposed by Remacle et al. (2013) to produce more regular quadrilaterals. A direct generation of quadrilateral and hexahedral meshes by applying the Octree decomposition with standard templates to volumetric data was also presented by Zhang and Bajaj (2006). The quality of the quad surface mesh can be further improved by optimisation techniques that will be discussed in Chapter 6 (Wang and Yu 2009).

Chapter 5

Mesh generation in three dimensions

The main challenge of FE mesh generation in 3D comes from the boundary constraints.

5.1 INTRODUCTION

From Chapters 3 and 4, it appears that automatic mesh generation (MG) has reached such a mature stage that efficient algorithms are available to generate high-quality meshes in compliance with the metric requirement on arbitrary 2D domains and over general curved surfaces in a robust manner. However, when we look at the MG for 3D solid objects, we immediately notice that the problem becomes much more complex, and many skills that work pretty well in 2D simply cannot be extended to a higher dimension. In 2D, the boundary-constrained MG is a deterministic process so that solutions are always guaranteed. On the other hand, in 3D, constrained MG algorithms are more iterative in nature. The fundamental difference between meshing a 2D domain and a 3D domain is that a 2D boundary can always be meshed without the need for additional nodes. However, there are geometries in 3D that cannot be discretised without introducing interior points, and a twisted pentahedron is a well-known example. Since there is no systematic way to decide where points should be introduced, analytical solutions are not available, leading to the development of iterative algorithms of heuristic nature for specific applications.

The problem becomes even more complicated when meshes of variable element size are required for isotropic and anisotropic metric specifications. Nevertheless, after years of dedicated research, many practical algorithms for meshing 3D solid objects are quite reliable such that the integrity of the domain boundary can be preserved, unless extremely poor boundary conditions are encountered, characterised by the presence of many elongated facets with a large aspect ratio and sharp dihedral angles between adjacent faces. The two popular techniques, namely, the Delaunay triangulation (DT) (George 1997; Johnson and Tezduyar 1997) and the AFT (Lo 1991b; Lohner 1996c; Rassineux 1997), again play an important role in 3D FE MG and will be discussed in detail in this chapter. Other techniques such as the Octree method, the adaptive refinement, the medial surface method, the plastering method, the whisker weaving method and the H-morph algorithm for the generation of hex meshes will be briefly described at the end of this chapter.

Over a 3D domain, it is necessary to employ different strategies for the MG of solid objects and for the fluid mechanics problems (Baker 1997; Mavriplis 1997; Morgan and Peraire 1998; Johnson and Tezduyar 1999). As stress and strain are likely to concentrate on the solid boundaries and solid objects are often meshed by components, boundary surface conformity in terms of both geometry and topology is strictly required leading to a (fully) constrained boundary MG problem. On the other hand, for fluid mechanics problems, emphasis is more on the speed of

MG and the quality of the elements rather than on the boundary conditions, which leads to a semi-constrained boundary MG problem in which only geometrical conformity is required or simply an open boundary without any constraints. Various methods (Viceconti et al. 1998) will be presented in this chapter and in Chapter 8 targeted for these two situations with appropriate measures to address the specific needs for these two problem types.

5.2 DELAUNAY TRIANGULATION (3D)

5.2.1 Introduction

The fundamentals of the DT have been discussed in Section 3.5. As the DT of 3D points is one of the most useful techniques in the triangulation of general 3D domains subject to various boundary requirements, the generation of tetrahedral elements by DT has been proved to be one of the fastest and reliable means in producing FE meshes of different characteristics. Since the insertion algorithm is known to be robust, efficient and versatile, a detailed implementation of the insertion algorithm of 3D points will be described in Section 5.2.2. Most of the steps are just simple extensions from the 2D situation, except in the determination of the adjacency relationship of the new tetrahedra formed inside the CORE with respect to those outside the CORE. A rotation scheme has to be employed to establish the adjacency relationship of the tetrahedral elements without much searching and matching. For clarity and easy reference, some of the concepts presented in Chapter 2 are repeated here in the context for 3D applications.

5.2.2 The insertion algorithm

For the construction of DT in three or higher dimensions, point insertion algorithm is the most popular, and many interesting methods have been proposed (Bern et al. 1994; Borouchaki and Lo 1995; Borouchaki and George 1996; Cortis and Friesner 1997; Boissonnat et al. 1998; Attali and Boissonnat 2004; Lo and Wang 2005d; Devillers and Teillaud 2011). For a set of 3D points, the initial triangulation is a cuboid consisting of five or six Delaunay tetrahedra large enough to contain all the given points, as shown in Figure 5.1. The DT is achieved by inserting points one by one into the initial triangulation. Each cycle of point insertion can be divided into three steps.

　i. For a newly inserted point, identify all the tetrahedra whose circumsphere contains the point in its interior. The cavity left behind upon removal of these tetrahedra forms a star-shaped insertion polyhedron (CORE).

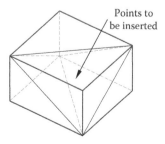

Figure 5.1 Initial triangulation of five tetrahedra.

ii. Owing to the finite precision arithmetic, the triangulation facets on the boundary of the cavity have to be verified with the visibility check and corrected before they are connected to the inserted point to form tetrahedra.

iii. The triangulation of the insertion polyhedron should be trivial. However, the adjacency relationship of the tetrahedra has to be established, which will be frequently referred to throughout the triangulation process.

5.2.2.1 Determination of the CORE

When a new point P is inserted in a DT, we have to find the insertion CORE or the polyhedron of all the tetrahedra whose circumsphere contains the point P. A simple method to determine these non-Delaunay tetrahedra is to scan through all the existing tetrahedra in the triangulation for those whose circumsphere contains the point P. However, a more efficient approach is to start with the tetrahedron that contains the inserted point P and find the others by means of the adjacency relationship. By the lemma of Delaunay, the boundary of the CORE is given by the common faces of a pair of tetrahedra for which one is positive and the other is negative to the sphere inclusion test.

The tetrahedron that contains the insertion point P is referred to as the BASE, which is an integral part of the CORE. It seems that finding a tetrahedron whose circumsphere contains P is easier than finding the BASE. Nevertheless, it does not always work in case a wrong decision is made in the sphere inclusion test due to numerical errors. The difficulty has been explained in terms of a 2D example, as discussed in Section 3.5.4.

5.2.2.2 Search for the BASE

In principle, we would like to start searching for the BASE from a tetrahedron as close to the newly inserted point P as possible. When there is no better clue as to where the point P is with respect to the existing tetrahedra in the triangulation, we can only arbitrarily start the search from the last constructed tetrahedron. This is the general situation of Delaunay point insertion that point P is independent of the current triangulation; however, in case insertion (interior) points are created in relation to the tetrahedra in the triangulation as in some local refinement scheme, there is no need to search for the BASE.

5.2.2.2.1 Steps in locating the BASE

1. Start the search from the last tetrahedron constructed, T.
2. Compute the volume co-ordinates of point P relative to tetrahedron $T = T(P_1P_2P_3P_4)$. The volume co-ordinates of P with respect to tetrahedron T are given by

$$L_1 = \frac{V(PP_2P_3P_4)}{V}, \quad L_2 = \frac{V(P_1PP_3P_4)}{V},$$

$$L_3 = \frac{V(P_1P_2PP_4)}{V}, \quad L_4 = \frac{V(P_1P_2P_3P)}{V},$$

where $V = V(P_1P_2P_3P_4)$ is the volume of tetrahedron T, as shown in Figure 5.2.
3. If L_1, L_2, L_3 and L_4 are all positive or zero, then tetrahedron T contains point P; otherwise, determine how to update tetrahedron T moving towards point P. Check how many volume co-ordinates are negative.

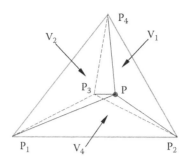

Figure 5.2 **Volume co-ordinates of P.**

    a. One negative value, say L_i, i = 1, 2, 3 or 4. Then update T ↦ T_i, where T_i is the ith neighbour of T opposite to the ith vertex.

    b. More than one negative value, say L_i, L_j and L_k, i, j, k = 1, 2, 3 or 4. Then randomly select among i, j and k, and update T in the same way as step 3(a). In case the scheme of changing the starting tetrahedron is employed, then choose the most negative value among L_i, L_j and L_k, and update T in the same way as step 3(a).

  4. Go back to step 2. For the scheme of changing the starting tetrahedron, as described in Section 3.5.4.4, check the number of tetrahedra visited to see whether we have to start the search with a new tetrahedron.

5.2.2.3 Determination of the CORE

Starting from the BASE, which contains point P, the CORE can be easily established by applying the sphere inclusion test to the neighbouring tetrahedra. Continue applying the test to more adjacent tetrahedra until all non-Delaunay tetrahedra with respect to point P are determined. The boundary of the CORE is made up of triangular facets, which are the faces between two tetrahedra that give positive and negative response to the sphere inclusion test.

5.2.2.3.1 Origin of inconsistency

Once the boundary of the CORE is established by the sphere inclusion test, the CORE consists of all the non-Delaunay tetrahedra with respect to the insertion point P. One of the major difficulties in the implementation of the DT by point insertion is to ensure consistency in the sphere inclusion test based on finite precision computations. An inconsistent decision in the sphere inclusion test may result in a disconnected CORE, as explained in Section 3.5.4. Imprecise numerical calculation may also introduce tetrahedral with zero or negative volumes. Another type of degeneracy in three dimensions, which is due to the nature of DT rather than the result of numerical error, can occur when four points are co-planar and cyclic. It is possible for these four points to form a tetrahedron of zero volume with non-zero edges and faces, which is known as a *sliver* (Tournois et al. 2009).

Using higher-precision arithmetic can only postpone the problem to solve more cases that are near degeneracy, but not to solve it completely. Baker (1992) proposed a solution in which tolerances depending on the values of the data and the machine precision are applied to all real number calculations. A more elegant solution is to use adaptive precision floating point arithmetic (Shewchuk 1997; Devillers and Preparata 1999), which in theory would not introduce any numerical error for a wrong decision in case of nearly co-spherical points (i.e. five or more points on the circumsphere of a tetrahedron). However, exact integer arithmetic is not the panacea to all these problems, and the situation of exactly co-spherical points and natural degeneracy of *sliver* in DT are still not resolved with robust geometric predicates (Goliaz and Dutton 1997). Hardware and compilers may also prevent the algorithm from functioning properly, and even though a right decision is arrived at for the sphere inclusion test, the connection is not unique; moreover, we do not have a systematic way in joining up any number of points on the surface of a sphere to produce a consistent triangulation, and *sliver* elements of zero or nearly zero volume will still be formed. Nevertheless, the visibility or the positive volume test appears to be simple and reliable in meshing domains for a wide range of point distributions.

5.2.2.3.2 Correction of the CORE

Inconsistency of the sphere inclusion test is to be supplemented and corrected by the visibility check or the positive volume test. As shown in Figure 5.3, a triangular face ABC is visible to point P if $\mathbf{a} \cdot \mathbf{n} > 0$, where $\mathbf{a}$ is a vector joining a point of triangle ABC to point P, and $\mathbf{n}$ is a vector normal to the triangle. The volume of tetrahedron ABCP is given by

$$V(ABCP) = AP \cdot (AB \times AC) = \mathbf{a} \cdot \mathbf{n}$$

Poorly shaped tetrahedra can also be avoided at the same time if we insist that their volume ought to be greater than a certain threshold value. The condition that every triangular facet on the boundary of the CORE is visible to P guarantees that the CORE is a single connected piece. The visibility check, although simple, is a necessary and sufficient test to ensure that the CORE is a star-shaped polyhedron with respect to the insertion point P. In fact, we can adopt other inclusion rules as to which tetrahedra are to be removed by the insertion point P to obtain different triangulations. In the extreme case, upon the introduction of point P, only one tetrahedron is deleted: the tetrahedron that contains P or the BASE. Four new tetrahedra are created in the cavity by joining P to the four faces of the deleted tetrahedron. Of course, by this simple rule, we are far away from the DT. Though we can obtain a fast and valid triangulation of the given set of points, the tetrahedra so generated are flat and almost degenerated. In summary, the visibility test ensures the validity of the triangulation, and using higher-precision computations in the sphere inclusion test guarantees that the resulting triangulation is as close to the required DT as possible.

5.2.2.4 Triangulation of the CORE

As all boundary facets of the CORE are visible to point P, the triangulation of the CORE can be easily constructed by connecting P to each triangular facet on the boundary. This

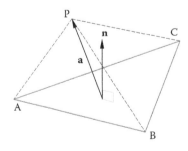

Figure 5.3 **Visibility test.**

local triangulation of the CORE together with the existing tetrahedra outside the CORE forms a new triangulation of all the inserted points including the $(k + 1)^{th}$ point P, i.e.

$$\mathbb{T}_{k+1} = \mathbb{T}_k - \mathbb{C}_p + \mathbb{B}_p$$

where

$\mathbb{C}_p \subset \mathbb{T}_k$ = CORE = cavity of non-Delaunay tetrahedra with respect to P
$\mathbb{C}_p = \{T \in \mathbb{T}_k : \|OP\|^2 \le r^2, O$ and r are the circumcentre and circumradius of T$\}$
$\mathbb{B}_p$ = ball of tetrahedra = patch of tetrahedra connected to P
$\mathbb{B}_p$ = set of tetrahedra formed by joining P to each triangle on the CORE boundary
$\mathbb{T}_k$ and $\mathbb{T}_{k+1}$ = the DTs of the first k and $k + 1$ points inserted

The enclosing cube of five or six tetrahedra serves as the initial triangulation $\mathbb{T}_0$, and the DT $\mathbb{T}_n$ will be constructed when all the n points are inserted sequentially one after the other by the point insertion kernel.

5.2.2.5 Adjacency relationship

As the element adjacency is frequently referred to in various steps of the DT, a more delicate issue is to determine the element adjacency relationship for the new tetrahedra in the CORE and update those that are attached to the CORE. There is not much a problem in establishing the adjacency relationship between the existing tetrahedra attached to the boundary facets of the CORE and the new tetrahedra that fill up the interior of the CORE. However, to determine the adjacency relationship between the new tetrahedra inside the CORE is less straightforward.

In the triangulation over a 2D domain, the insertion CORE is a polygon, and there is no difficulty in establishing the adjacency relationship between triangles by creating elements following the boundary contour of the polygon. In three or higher dimensions, the situation is quite different. The boundary of the CORE is a surface, and there is no obvious order for the boundary faces following which the adjacency relationship of the tetrahedra so generated could be established in a more or less natural manner without much calculation.

From the star-shaped connection of point P with boundary faces of the CORE, each common face between two tetrahedra within the CORE can be associated with an edge on the boundary, as shown in Figure 5.4. Tetrahedra ABCP and BADP are neighbours if they share common edge AB on the boundary of the CORE. Hence, the adjacency relationship for each tetrahedron can be established by identifying the three edges on the boundary of the CORE and determining the three neighbours through a matching process of common

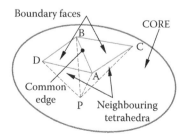

Figure 5.4 **Tetrahedra ABCP and BADP are neighbours.**

edges. Nevertheless, in a typical point insertion, the number of tetrahedra in the CORE is in the order of a hundred, and searching and matching can be time-consuming compared to the other steps of the triangulation.

The adjacency relationship of the new tetrahedra inside the CORE is closely related to the adjacency relationship of the triangular facets on the boundary of the CORE. In fact, there is a one-to-one correspondence between the two topological structures, i.e. the three neighbours of a new tetrahedron can be identified with the three neighbouring triangles of the triangular facet to which the tetrahedron is attached. The fourth neighbour is, of course, the one opposite to point P, joining to the same boundary facet from outside the CORE. As a result, search for the neighbours of the new tetrahedra can be avoided if the neighbours of a triangular facet on the boundary of the CORE can be directly determined.

The answer to this question is quite positive if we take a closer look at the existing triangulation outside the CORE. The neighbouring triangles of a triangular facet on the CORE boundary can be determined by rotating about the common edge between the triangles through the tetrahedra connected to that edge, as shown in Figure 5.5. Based on the adjacency relationship, the tetrahedra connected to an edge can be determined by a simple rotation about the edge, as described in Section 2.5.9; the identification of neighbouring triangles by this method is much faster, and no searching and matching is required.

An even better solution exists if we can make use of the tetrahedra to be deleted inside the CORE. Using the tetrahedra inside rather than outside the CORE offers at least two advantages: (i) the path of rotation is, in general, shorter as the angle of turn is usually smaller; and (ii) the interior of the CORE is a continuous piece, and there is no void inside, whereas there may be void outside the CORE for certain applications such as constrained DT insertion. In the actual implementation, the tetrahedra are constructed directly on the boundary surface of the CORE using a technique similar to the ADF approach. The construction process can be initiated by taking any triangular facet ABC on the boundary of the CORE. Tetrahedron ABCP is formed by joining point P to triangular facet ABC. The construction front for this initial tetrahedron construction consists of three edges AB, BC and CA. To advance across edge AB, the neighbouring triangle connected to AB has to be determined. Neighbouring triangle BAD can be determined by rotating about edge AB through the tetrahedra connected to this edge inside the CORE. The adjacency relationship between tetrahedra ABCP and BADP is established, and the construction front is updated to four segments, as shown in Figure 5.6.

The construction process can be repeated until all the triangular facets on the boundary of the CORE are processed, and the construction front will be reduced to zero. It is noted that the construction front can be closed in a fairly natural manner. An edge will be deleted

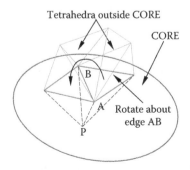

Figure 5.5 Matching neighbouring triangles of edge AB by rotation through tetrahedra outside CORE.

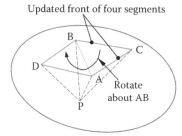

Figure 5.6 Evolution of construction front.

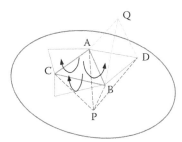

Figure 5.7 Tetrahedra BADP and ABDQ are neighbours.

from the front if a tetrahedron has already been generated in that direction. For instance, edge AB will be deleted if element BADP has already been constructed. To check whether face BAD has been used or not, simply refer to the element connected to the face BAD, tetrahedron ABDQ on the opposite side of the CORE, as shown in Figure 5.7. Element BADP is the neighbour of element ABDQ sharing common face BAD. If the element number of this neighbour of ABDQ is in the list of newly formed elements, then face BAD has already been processed, and element BADP has already been constructed; face BAD has to be ignored, and no element should be created.

An alternative scheme slightly easier to code for establishing the adjacency relationship of new tetrahedra within the CORE is to identify the neighbours of each tetrahedron in turn. Taking any triangular facet on the boundary of the CORE, construct a new tetrahedron by joining point P to the triangular facet. In fact, the number of new tetrahedra is equal to the number of triangular facets on the boundary, and that's the basis for a one-to-one correspondence between boundary facets and new tetrahedra, which can also be labelled following the same sequence of the boundary facets. Identify the neighbours of the new tetrahedra one after the other by rotating about the three edges of the associated triangular facet through the old tetrahedra within the CORE. The adjacency relationship of the new tetrahedra of the CORE will be established when each boundary facet is processed in turn. As adjacency is a reciprocal relationship, a little care in recording mutual neighbours can save some repeated computations in neighbour identification.

5.2.2.6 Heredity of geometrical quantities

Many geometrical quantities of a new tetrahedron such as the circumcentre, the circumradius and the normal to the faces can be easily calculated based on the same quantities of the

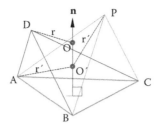

Figure 5.8 **Transfer of geometrical quantities.**

old tetrahedron. It is reminded that a new tetrahedron is constructed by joining the insertion point P with a triangular facet on the boundary of the CORE. However, the triangular facet to which P is connected is a face of a deleted tetrahedron. It is from this old tetrahedron that the relevant geometrical quantities are derived for the new tetrahedron.

Let (O', r') be the circumcentre and circumradius of tetrahedron ABCP and (O, r) be those of an old tetrahedron ABCD, as shown in Figure 5.8. As circumcentres O and O' are lying on the normal of the common face ABC of tetrahedra ABCD and ABCP, we have

$$O' = O + \lambda \mathbf{n} \text{ where } \lambda \in \mathbb{R}, \quad \mathbf{n} = AB \times AC \text{ is a normal vector to face ABC}$$

$$\|O'A\| = \|O'P\| \Rightarrow MO' \perp AP \text{ where M is the midpoint of AP}$$

$$MO' \perp AP \Rightarrow AP \cdot (MO + \lambda \mathbf{n}) = 0 \Rightarrow \lambda = \frac{AP \cdot OM}{AP \cdot \mathbf{n}}$$

$$AP \cdot OM = (OP - OA) \cdot \frac{OA + OP}{2} = \frac{\|OP\|^2 - \|OA\|^2}{2} = \frac{\|OP\|^2 - r^2}{2}$$

The value $\|OP\|^2 - r^2$ is already calculated in the sphere inclusion test, and $AP \cdot \mathbf{n}$ is the volume of tetrahedron ABCP, which is also available from the visibility test. λ and hence the circumcentre O' of the new tetrahedron ABCP can be readily computed without forming and solving a linear 3×3 system. Knowing O', r' is given by the Euclidean distance between O' and P. Usually, the square of distance (r^2) is computed and stored as almost all calculations are based on the square of the distance rather than the distance itself.

Let $\mathbf{n}$, $\mathbf{n}_1$ and $\mathbf{n}_2$ be the normal to the faces ABC, BCD and BCP, respectively, as shown in Figure 5.9. Since $\mathbf{n}$, $\mathbf{n}_1$ and $\mathbf{n}_2$ are normal to the line segment BC, they are lying on the

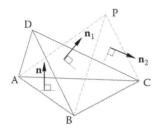

Figure 5.9 **Transfer of normal to tetrahedron ABCP.**

plane normal to BC. Hence, $\mathbf{n}$, $\mathbf{n}_1$ and $\mathbf{n}_2$ are not linearly independent vectors, and a linear relationship of the vectors exists, which can be written as $\mathbf{n}_2 = \mathbf{n}_1 + \mu\mathbf{n}$.

$$CP \cdot \mathbf{n}_2 = 0 \Rightarrow CP \cdot (\mathbf{n}_1 + \mu\mathbf{n}) = 0 \Rightarrow CP \cdot \mathbf{n}_1 + \mu CP \cdot \mathbf{n} = 0 \Rightarrow \mu = -\frac{\text{vol}(BCDP)}{\text{vol}(ABCP)}$$

The volume function for each face is available, and vol(ABCP) is already calculated in the visibility test; hence, μ and $\mathbf{n}_2$ can be readily computed from $\mathbf{n}$ and $\mathbf{n}_1$. The normals to the other faces of tetrahedron ABCP can be determined in a similar manner. While the transfer of geometrical quantities from the existing tetrahedra to the new tetrahedra can save some computations, numerical errors do accumulate in the process, and as a result, from time to time, the circumcentre and the circumradius of a new tetrahedron have to be calculated directly from its vertices.

5.2.2.7 Memory management

For each tetrahedron, four vertices, four neighbours, the circumcentre, the circumradius and the normal to the face are stored. The tetrahedra to be deleted are flagged and are not removed right away as they are still needed in the neighbour identification process, and those newly created in the CORE are appended at the end of the list of tetrahedra. For higher efficiency in computation and a better use of computer memory, flagged tetrahedra are deleted, and the list of tetrahedra is consolidated only when there is no more room to accommodate any new tetrahedra, as described in Section 2.5.10. As the computer is becoming faster and faster, to make the best use of the memory to generate the largest possible meshes, we can consider to store only the essential data, namely, the vertices and the neighbours of the tetrahedra and the nodal co-ordinates, but not the circumcentres or the circumradii of the tetrahedra. Computer programing of a 3D DT insertion algorithm can be coded following exactly the same logical sequence of the 2D DT program presented in Section 3.5.5. Such a computer program in Fortran is given in Appendix A13.

5.2.3 Examples

Two simple examples of different point distributions are presented in this section. In the DT of a given set of points, an enclosing cube large enough to contain all the points is first constructed. The points are then processed one after the other by the insertion algorithm until all the points in the set are dealt with. Theoretically, the resulting mesh is the convex hull of the point sets. However, as the enclosing cube is just large enough to contain all the points and the corner points of the cube are not placed at infinity, the tetrahedral meshes excluding the vertices of the enclosing cube are not strictly convex, as shown in Figures 5.10 and 5.11. The first example consists of 2000 points randomly generated within the cube, which are then inserted one by one to generate a tetrahedral mesh of 12,069 tetrahedral elements, as shown in Figure 5.10. There are also 2000 points in the second example; however, half of the points in this example are concentrated along the diagonals of the cube, as shown in Figure 5.11. For an equal number of points, there are only 11,534 tetrahedral elements in the second mesh, which is considerably fewer in number compared to the first example of a uniform point distribution. As for the CPU time of DT, over 1 million tetrahedral elements per second can be generated on today's PCs by an insertion algorithm with partition grids coded in Visual Fortran 2010 using Intel® Core™ i7-3770 CPU at 3.40 GHz with 32 GB and 64-bit operating system. Over 5 million tetrahedral elements per second up to half a billion

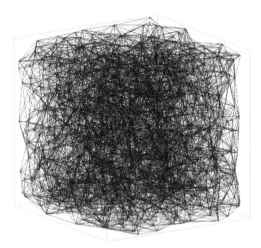

Figure 5.10 Delaunay triangulation of 2000 randomly generated points.

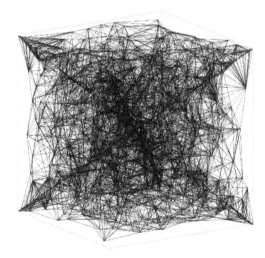

Figure 5.11 Delaunay triangulation of points concentrated along diagonals.

tetrahedra can be generated by a parallel triangulation algorithm on a PC; please refer to Chapter 7 for details.

5.3 BOUNDARY RECOVERY FOR 3D DT

5.3.1 Introduction

Taking the triangular surface mesh as input, *boundary recovery* requests all triangles on the input surface mesh be preserved without introducing Steiner points on the boundary. The integrity of the boundary (both geometry and topology) is crucial in meshing by components, parallel MG, partial re-meshing, multi-interface problems, etc. With the development of the FEM and its applications, how to generate a tetrahedral mesh with an integral boundary is increasingly important (Joe 1992; Wright and Jack 1994; George et al. 2003; Du and Wang 2004; Si and Gaertner 2010). However, while it is rapid and robust, DT of spatial points, in general, will only define a convex hull of the given points. Hence, for irregular non-convex objects or objects with interior boundaries, the DT has to be properly modified so as to recover those missing boundary edges or faces. As the missing edges and faces are not present in the original DT, they are not Delaunay and they violate the empty sphere criterion with respect to the system of spatial points. As a result, strictly speaking, the modified triangulation is no longer a DT, and this problem is also known as constrained DT.

In 2D, boundary edge recovery is quite easy as an edge joining any two points in the mesh can always be recovered simply by the swap of diagonals, as elucidated in Section 3.5.7. In 3D, the situation is quite different. First, not all the triangulations of the point set can be reproduced by a series of valid element transformations of edge/face swaps (Joe 1993, 1995b), and second, there exists a twisted pentahedron for which a conforming triangulation of the boundary points does not exist without introducing additional points, the so-called Steiner points; in other words, a twisted pentahedron cannot be triangulated without introducing an interior point. However, there is no formal rigorous proof that a fully conformable FE mesh exists for any general boundary surface, even though Steiner points are allowed at the interior of the domain. Arguably, if tetrahedra of zero volumes (*sliver*) are permitted, there is always a solution; indeed, *slivers* exist anyhow in a valid DT. In practice, from the research over the last 20 years or so, we are now able to produce a topologically and geometrically valid conforming FE mesh from the DT for boundary surfaces of rather complex industrial objects or biological models with some flat elements of a very small volume at some critical sites. These flat elements can possibly be opened up to attain a finite volume provided that relatively poorly shaped tetrahedral elements are still acceptable for difficult boundary conditions.

In general, the boundary recovery approaches developed so far can be grouped into two categories: (1) local mesh reconnection and (2) introduction of Steiner points. In the first approach, no Steiner point will exist in the final tetrahedron mesh. However, there is no guarantee for the success of such methods due to the existence of a twisted pentahedron in Schönhardt configuration. In the second approach, Steiner points are introduced to assist boundary recovery. Although they could recover the geometry and the topology of the missing quantities, how to open up the flat tetrahedral elements so created systematically to ensure topological integrity is still an open issue. Nevertheless, effective heuristic schemes were developed for practical difficult 3D domains, as elaborated in Section 5.3.3.

5.3.2 Boundary recovery by local mesh reconnection

As no Steiner points are inserted in the final mesh, approaches by local mesh reconnection attract the interest of many researchers. Borouchaki et al. (2000b) split the constrained boundary by inserting Steiner points on edges and facets, and then suppressed the inserted points by locally remeshing the tetrahedral elements linked to them. Nevertheless, local remeshing could not guarantee a valid boundary-recovered topological structure even though it exists, and very often, Steiner points remain on some of the boundary faces. Liu et al. (2007) employed an exhaustive method named *small polyhedron reconnection* to achieve the boundary recovery. For a *small* polyhedron with no more than 20 triangular facets, all possible topological structures are evaluated to detect if the missing quantities could be recovered in one of the configurations. Ghadyani et al. (2010) carved out a hole in the vicinity of missing boundary facets, and the hole along with the missing boundary facets formed a polyhedron, which is to be meshed by a method known as the LAST RESORT.

5.3.3 Boundary recovery by introducing Steiner points

5.3.3.1 Introduction

With the help of additional points, missing edges and facets could be recovered. An edge is represented as a series of broken line segments separated by Steiner points, and a face is represented by a concatenation of sub-triangles supported on Steiner points. Such meshes are known as conforming DT (Karamete et al. 2000; Shewchuk 2008), and if topological integrity of boundary edges and facets are not required, the boundary recovery process can stop at this point. However, if a *constrained* DT including topological integrity is required, all Steiner points have to be removed or re-positioned towards the interior of the domain. George et al. (2003) improved their previous work (George et al. 1991) to propose a method to mesh an arbitrary polyhedron. By this method, Steiner points are first inserted on the constrained boundary edges and faces, which are suppressed one by one later in a subsequent process. A triangulation of the missing edges/facets is constructed by connecting those Steiner points introduced at the surface, and flat elements of zero volume are created to recover the related topological structure. Du and Wang (2004) inserted Steiner points on boundaries through a heuristic approach to reduce the number of Steiner points. They used an edge-swap procedure on the missing boundary facets to remove some of the Steiner points. However, there is a drawback in the method proposed by George et al. (2003) and Du and Wang (2004): a number of *locked* Steiner points are generated, which could not be easily removed. Chen et al. (2011) combined the work of Du and Wang and Liu et al. (2007) by employing the *small polyhedron reconnection* to reduce the number of Steiner points in the final mesh. However, their method still faces the problem of *locked* Steiner points. Guan et al. (2006) proposed a technique named *dressing wound*. Compared to the method proposed by George et al. or Du and Wang, it introduced more Steiner points; however, this method provided a new perspective for the boundary recovery problem in the way how Steiner points could be removed. Si and Gaertner (2011) introduced Steiner points to edges and faces so as to preserve them in the Delaunay point insertion process, and boundary faces will exist as a concatenation of small triangular facets. These Steiner points are to be removed as far as possible at a later stage to ensure topological integrity with the given boundary surface.

To sum up, there are still rooms for improvement in boundary recovery for meshing arbitrary polyhedrons. Ideas for the boundary recovery of the 3D DT will be discussed to address some of the difficulties. The method described in Section 5.3.3.2 focuses on a systematic removal of Steiner points. As a departure from the previous works, in the process of removing Steiner points, the sequence and the locations in the removal of these points are optimised to reduce the number of *locked* Steiner points as much as possible. Moreover, a linear programming optimisation is adopted to determine the feasible space in relocating Steiner points. Compared with Laplacian smoothing–based methods, it guarantees finding feasible positions for Steiner points should they exist.

5.3.3.2 Insertion algorithm and boundary recovery

For a polyhedron bounded by a set of triangular facets, the 3D DT of the boundary points will generate the convex hull of the points, but there is no guarantee that the boundary facets of a concave or multi-connected polyhedron will be present in the DT. For the boundary integrity of a general polyhedron, additional work is needed to be done after point insertion. Following the partial recovery of some edges and facets by local element swaps 2–3, 3–2 and 4–4 (Lo 1997), Steiner points can be inserted to recover the missing boundary edges and facets, which have to be removed or repositioned to ensure the topological integrity of the boundary surface. The following algorithm presents the main ideas of a constrained boundary recovery procedure.

Algorithm: 3D-constrained DT

Step 1: Take the given polyhedron as input. Set the original triangular facets on the boundary as constraints.

Step 2: Perform Delaunay insertion for points on the boundary of the polyhedron.

Step 3: Carry out local element swaps to recover boundary edges.

Step 4: Carry out local element swaps to recover boundary facets.

Step 5: Insert Steiner points to missing boundary edges/faces.

Step 6: Remove Steiner points on boundary edges.

Step 7: Remove Steiner points on boundary facets.

Step 8: Delete tetrahedra outside the model to obtain the final tetrahedral mesh.

5.3.3.2.1 Step 1: Restrained edges and faces

This step is trivial; the triangles on the boundary surface are properly recorded as restricted boundary to be retrieved in the recovery process, and the boundary edges of a triangulated surface can easily be extracted, as described in Section 2.5.8.

5.3.3.2.2 Step 2: Initial DT

An initial DT (the convex hull) of the boundary points is created by the 3D Delaunay insertion kernel presented in Section 5.2.2.

5.3.3.2.3 Step 3: Recovering missing edges by element swaps

If any boundary edge is missing, it is not an edge of a tetrahedron in the DT, and such an edge will intersect with at least one face in the triangulation. Usually, longer edges will cut through more tetrahedra and are, in general, more difficult to recover. The missing line segment along with the associated intersected tetrahedra is called a *pipe* in the paper (George et al. 1991), though the assembly looks more like a *brochette*. Our task is to recover edge PQ by 2–3 element swaps, and the basic idea is to swap a pair of tetrahedra to three tetrahedra about the common face intersected by the missing edge so as to reduce intersections. Consider adjacent tetrahedra ABCD and CBAE in a *pipe* intersected by the missing edge PQ on three faces ABE, ABC and ABD, as shown in Figure 5.12. By a 2–3 element swap,

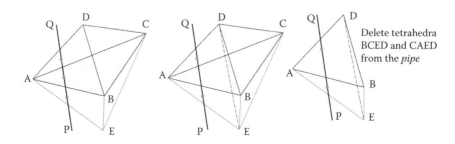

Figure 5.12 Swap tetrahedra ABCD + CBAE to ABED + BCED + CAED.

tetrahedra ABCD and CBAE are converted to three tetrahedra ABED, BCED and CAED. As face ABC is also removed in the element swap process, the number of intersection with PQ is reduced by one. Two of the three tetrahedra, namely, BCED and CAED, are not intersected by PQ, and they can be removed from the *pipe*. Effectively, by a proper 2–3 element swap, both the number of intersections and the number of tetrahedra in the *pipe* are reduced by one. In principle, by means of a sequence of element 2–3 swaps on adjacent tetrahedra, a *pipe* with n intersections and n + 1 tetrahedra can be recovered in n swaps.

Let PQ be a missing boundary edge cutting through eight tetrahedra and seven faces, and the *pipe* associated with edge PQ is shown in Figure 5.13; (1) CBAP on CBA; (2) ABCD on CBA and BCD; (3) BCDE on BCD and BDE; (4) BEDF on BDE and EDF; (5) DEGF on EDF and FEG; (6) FEGI on FEG and FGI; (7) FGHI on FGI and GHI; and (8) GHIQ on GHI. Starting from the first two tetrahedra on the side of point P, CBAP and ABCD are converted to tetrahedra CAPD, ABPD and BCPD, as shown in Figure 5.14a. Tetrahedra CAPD and ABPD, which are not intersected by PQ, are deleted from the *pipe*, as shown in Figure 5.14b.

The number of tetrahedra and the number of intersections are reduced by one as face ABC is also removed in the element swap process. More element swaps on adjacent tetrahedra are done to remove the remaining six intersected faces BCD, BDE, DEF, FEG, FGI and GHI, as shown in Figure 5.14c–h.

We would like to ask the following question: Can missing edges always be recovered by merely element swaps? The answer is no. While element swap is an effective tool in recovering many missing edges, there are cases that cannot be reduced by element swaps. Element swap will lead to invalid elements of zero or negative volumes if the two adjacent tetrahedra are not convex, as shown in Figure 5.15, such that diagonal DE does not intersect with common face ABC. George et al. (1991) proposed a scheme to remedy this situation. As shown in Figure 5.16, ADCE is the adjacent tetrahedron of the non-convex assembly of tetrahedra CBAP and ABCD, and I, J and K are, respectively, the intersection points on faces ABC, ADC and ADE.

CPF is the extension of face CPB onto face ADC with CF cutting JD at R, so that any point between J and R will be visible to P. Let's introduce a new point S between J and R to divide tetrahedron ABCD into three tetrahedra ADBS, DCBS and CABS, and tetrahedron ADCE into AEDS, DECS and CEAS, as shown in Figure 5.17. Intersection point I can now be eliminated by performing 2–3 swap to elements CBAP and ABCS. The number of intersection points will be reduced if J and K are in the same sub-tetrahedron without cutting any new faces created in the subdivision of tetrahedron ADCE by point S. In case intersection points J and K are in two sub-tetrahedra, PQ must cut one of the newly created faces to generate an extra intersection point, say, L on face CSE, as shown in Figure 5.18.

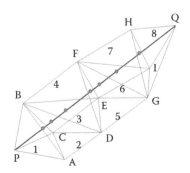

Figure 5.13 **A missing edge cutting through eight tetrahedral elements.**

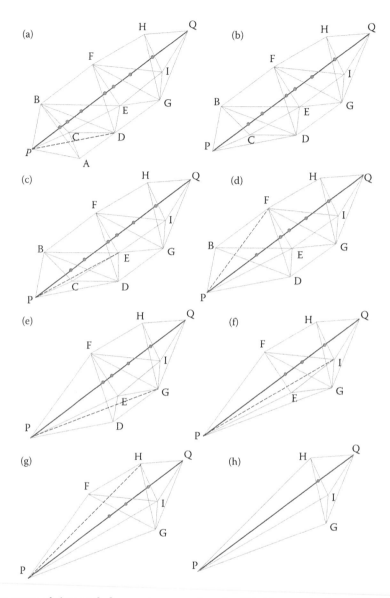

Figure 5.14 Sequence of element 2–3 swaps in recovering edge PQ: (a) removing face ABC; (b) delete tetrahedra CAPD and ABPD; (c) removing face BCD; (d) removing face BDE; (e) removing face DEF;
(f) removing face FEG; (g) removing face FGI; (h) removing face GHI.

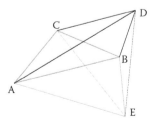

Figure 5.15 Tetrahedra ABCD and CBAE are not convex.

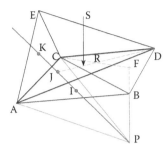

Figure 5.16 Tetrahedra ABCD and CBAE are not convex.

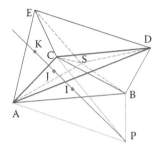

Figure 5.17 Subdividing tetrahedra ABCD and ADCE.

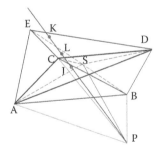

Figure 5.18 Points J and K are in different tetrahedra.

In case a new intersection point is created, there is no actual reduction in the number of intersection points (tetrahedra) in the *pipe*, which will be referred to as irreducible. George and Borouchaki (1998) further argued that S can be so located such that the swap of tetrahedra CBAP and ABCS is still valid yet allows more space to eliminate L by repeating exactly the same procedure. However, in considering all the tetrahedra centred at a point (D), not only more Steiner points will be created but also many subdivisions of tetrahedra may be required before the process converges to some valid numerical solutions, which inevitably would have generated many elongated tetrahedra. An alternative, which may be simpler, is to just keep those irreducible intersection points and insert them as Steiner points to the mesh.

5.3.3.2.4 Step 4: Recovering missing faces by element swaps

Following the edge recovery by element swaps, if the three edges of a boundary face are present, the face, in general, will also exist. However, there are situations where the face is intersected by other edges, even though its three boundary edges have already been restored, as shown in Figure 5.19. Edges cutting through face ABC can be detected by considering the three *shells* associated with the edges of the triangle, namely, *shells* generated by rotating about edges AB, BC and CA, respectively. Penetrating edges can be identified if any edge of these *shells* cut across the face. These intersecting edges, to a certain extent, can be removed by proper element swaps. In case there is only one penetrating edge, the *shell* associated with this edge consists of exactly three elements with a proper connection to the three vertices of the face, as shown in Figure 5.20. The penetrating edge can be easily removed by a 3–2 element swap, which is unconditional and applicable to any *shell* with exactly three tetrahedra. Suppose that there are more than one penetrating edges, and they have to be considered one by one. Let PQ be one of the penetrating edges and $\{R_1, R_2,..., R_n\}$ be the nodes of the associated polyhedron opposite to PQ, as shown in Figure 5.21. The ring of n nodes $\{R_1, R_2,..., R_n\}$ can be triangulated into n – 2 triangles, and tetrahedra can be formed by joining points P and Q to the triangles of the triangulated ring. If in any one of the valid triangulations, there is no internal edge passing through the missing face ABC and each triangle of the ring is visible to both P and Q, then a swap of a *shell* of n tetrahedra into 2(n – 2) tetrahedra will remove the penetrating edge PQ. Each penetrating edge and its associated *shell* can be considered, in turn, for its possible removal by the same procedure. It is noted that George and

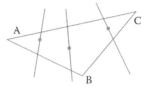

Figure 5.19 **Face ABC intersected by edges.**

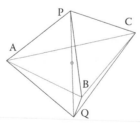

Figure 5.20 **Face ABC intersected by one edge.**

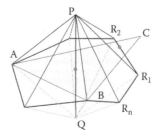

Figure 5.21 **Penetrating edge PQ and its associated shell.**

Borouchaki (1998) proposed a more complicated scheme to remove a penetrating edge in which three Steiner points are introduced at various positions on the cutting edge; however, while this offers a plausible solution in removing a penetrating edge, it cannot be considered as an optimal process as far as the number of Steiner points is concerned.

5.3.3.2.5 Step 5: Insert Steiner points to edges and faces

For those intersection points on boundary edges being cut across by a face or on boundary faces being penetrated through by an edge, which cannot be simply removed by element swaps, Steiner points are introduced on the face to restore the geometry of the missing quantities. For a missing boundary edge, Steiner points are inserted at the position of all the irreducible intersection points breaking each of the cutting faces into three triangles, as shown in Figure 5.22. Each missing boundary edge is recovered as a series of line segments separated by Steiner points. As for missing boundary faces, Steiner points are introduced on the face concerned, and the face is triangulated with the Steiner points as internal nodes such that the boundary face is represented as a concatenation of smaller triangles, as shown in Figure 5.23. Nevertheless, a more elegant way to introduce Steiner points is to insert them by means of the Delaunay point insertion kernel.

5.3.3.2.6 Step 6: Removing Steiner points on boundary edges

Steiner points inserted on boundary edges for their geometrical recovery (geometrically present) have to be completely removed (relocated) for topological integrity with the original boundary surface. This task, however, is relatively easier than originally thought as we only have to deal with one side of the boundary surface, i.e. the side within the object; as for the outside part, we don't have to care much as elements outside the boundary surface will be discarded anyway to reveal the final meshed object. There should be at least two Steiner points on an edge for recovery as one intersection point is always reducible. Hence, suppose there are three Steiner points on a boundary edge, which can be lifted off the surface one by one, as shown in Figure 5.24a. Upon the removal of the Steiner points, edge AB and

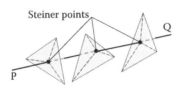

Figure 5.22 Inserting Steiner points to an edge.

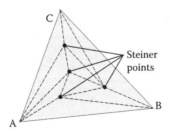

Figure 5.23 Inserting Steiner points on a triangular facet.

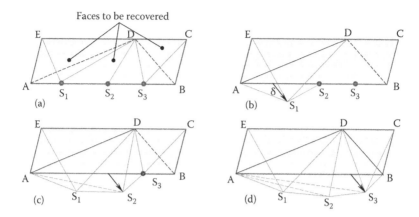

Figure 5.24 Sequence in removing Steiner points on a boundary edge: (a) Steiner points on edge AB; (b) first point lifted; (c) second point lifted; (d) all points lifted.

boundary facets ADE, ABD and BCD will be recovered. Only the upper part of the boundary surface is shown; the lower part of the boundary surface on the other side of edge AB is not shown for clarity. The other part of the boundary surface may be of the same or a different connection pattern, but the same idea and procedure can always be applied. As shown in Figure 5.24b, a small displacement δ normal to the boundary surface is applied to the first Steiner point S_1 to create two new tetrahedra $ADES_1$ and AS_2DS_1, and similar connections to produce new tetrahedra can be done on the other side of the edge according to the pattern of the boundary triangles. The amount of displacement δ depends on the size of all the tetrahedra connected to S_1, and a non-zero value always exists as tetrahedra in a triangulation should have a finite volume. In case of *slivers* (zero volume degenerated tetrahedra) present in a DT, they can be eliminated in advance by some mesh optimisation techniques described in Chapter 6. Similar to S_1, Steiner point S_2 is then lifted off the surface to generate tetrahedron AS_3DS_2, as shown in Figure 5.24c. Finally, Steiner point S_3 is also lifted to create tetrahedra $ABDS_3$ and $BCDS_3$, as shown in Figure 5.24d.

Alternatively, we can lift S_1 and S_3 at the same time and then lift S_2 lastly to generate tetrahedra $ADES_1$, $ABDS_1$, $BCDS_3$, BDS_1S_3 and $DS_1S_3S_2$, as shown in Figure 5.25a and b. Let

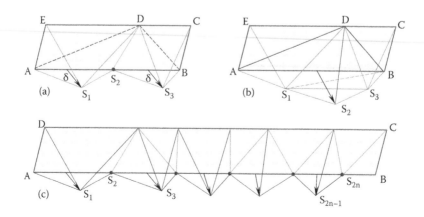

Figure 5.25 Lifting Steiner points on a boundary edge simultaneously: (a) two points lifted simultaneously; (b) lifting the remaining point; (c) points lifted group by group.

there be 2n Steiner points on an edge following a natural sequence from one end to another $\{S_1, S_2, ..., S_{2n}\}$; in general, more than one Steiner point can be lifted simultaneously if Steiner points on an edge are divided into two groups according to their even and odd sequence on the list, i.e. group G1 = $\{S_1, S_3, ..., S_{2n-1}\}$ and group G2 = $\{S_2, S_4, ..., S_{2n}\}$. All the Steiner points on the edge can then be processed in two steps: in the first step, points in group G1 can be lifted off the surface simultaneously, as shown in Figure 5.25c, and in the second step, points in G2 are also lifted. Tetrahedral elements are created when points are moved towards the interior of the domain by a proper connection of each point lifted with its associated triangular facets on the boundary surface. More tetrahedra are generated when points in G2 are also lifted leaving the required boundary edge AB at the middle of a boundary surface patch, which has to be triangulated in compliance with the prescribed surface boundary, as shown in Figure 5.26. Similar procedure can also be applied if the number of Steiner points is odd; in this case, just divide the points into two groups such that G1 = $\{S_1, S_3, ..., S_{2n+1}\}$ and group G2 = $\{S_2, S_4, ..., S_{2n}\}$. Lifting points group by group can much improve the quality of the resulting tetrahedra, as in a sequential node lifting process, each node has to be moved by an amount less than the preceding nodes, as shown in Figure 5.24d, producing flat tetrahedra of rather poor quality towards the end of the process.

As the optimal amount of node lifting δ cannot be determined in general, a better approach to recover a boundary edge is to first create a valid topological structure before moving the nodes towards the interior of the domain. Take for instance the recovery of edge AB in Figure 5.25b; flat tetrahedra $ADES_1$, $ABDS_1$, $BCDS_3$, BDS_1S_3 and $DS_1S_3S_2$ of zero volume are first created to form a supporting topological structure of a valid tetrahedral mesh, as shown in Figure 5.27a. The Steiner points are allowed to move freely in the feasible space to take strategic positions so as to optimise the shape of the tetrahedra, as shown in Figure 5.27b. This Steiner point relocation process is similar to the problem of mesh optimisation by shifting of nodal points, and sometimes, neighbouring nodes and tetrahedra can also be included to form an optimisation patch of a larger scale for even better results.

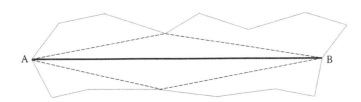

Figure 5.26 **Patch of boundary surface recovered containing edge AB.**

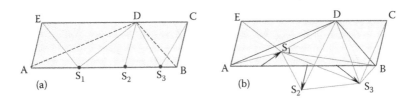

Figure 5.27 **Steiner points are so located to produce the *best* tetrahedra: (a) flat tetrahedra created; (b) flat tetrahedra are opened up.**

5.3.3.2.7 Step 7: Removing Steiner points on boundary faces

Similar to the removal of Steiner points from a boundary edge, the elimination of Steiner points from a boundary face is achieved by relocating all the Steiner points towards the interior of the object. Again, we only have to focus on the tetrahedral elements on one side of the boundary face. Let's consider face ABC whose geometry has been recovered by the introduction of four Steiner points, as shown in Figure 5.28a. Following the natural sequence of the Steiner points, S_1 is raised by an amount δ normal to the face to create tetrahedron $ABCS_1$, as shown in Figure 5.28b. By lifting S_1, S_2 and S_4 are also raised on the triangular face BCS_1, and S_3 is raised on face CAS_1. When S_2 is lifted, tetrahedra BCS_1S_2 and $CS_3S_1S_2$ are created, and S_4 is further raised on the face CS_3S_2, as shown in Figure 5.28c. Finally, S_3 and S_4 are lifted at the same time to create tetrahedra CAS_1S_3 and $CS_3S_2S_4$, as shown in Figure 5.28d.

There is a possibility to improve the quality of the tetrahedral elements (Liu et al. 2014). The Steiner points are to be lifted following the order based on a neighbouring coefficient, which is equal to the number of adjacent Steiner points. Take the above face recovery with four Steiner points as an example; the neighbouring coefficients of the Steiner points are, respectively, $N(S_1) = 2$, $N(S_2) = 3$, $N(S_3) = 3$ and $N(S_4) = 2$. Accordingly, Steiner points S_1 and S_4 are lifted up first to create tetrahedra $AS_2S_3S_1$, ABS_2S_1 and $CS_3S_2S_4$, as shown in Figure 5.29a. The Steiner points S_3 and S_2 remaining on the face ABC can now be lifted to produce tetrahedra CAS_2S_3 and $CABS_2$ in turn, as shown in Figure 5.29b. Similar to the edge recovery, the lifting of Steiner points off a face can be converted into a mesh optimisation problem

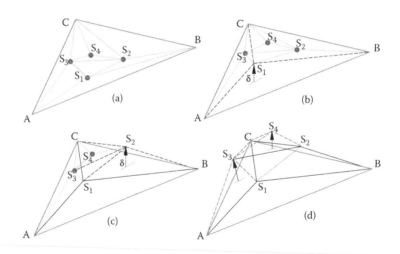

Figure 5.28 Lifting Steiner points off a boundary face: (a) Steiner points on triangle ABC; (b) point S_1 lifted; (c) point S_2 lifted; (d) points S_3 and S_4 are lifted.

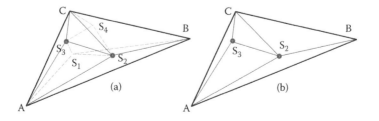

Figure 5.29 Steiner points prioritised with neighbouring coefficient: (a) points S_1 and S_4 are lifted; (b) points S_2 and S_3 are lifted.

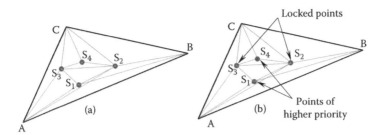

Figure 5.30 Mesh topology based on the priority of the Steiner points: (a) Steiner points S_1, S_2, S_3 and S_4; (b) S_1 and S_4 are lifted first.

by first creating a supporting topologically structure of flat tetrahedral elements, which can be opened up to form a valid FE mesh by relocating the Steiner points within the feasible space. To this end, before the relocation of Steiner points, tetrahedra $AS_2S_3S_1$, ABS_2S_1, $CS_3S_2S_4$, CAS_2S_3 and $CABS_2$ are formed, as shown in Figure 5.30a. To open up these five tetrahedra by shifting of nodes, it is noticed that only points S_1 and S_4 are allowed to move first; otherwise, if one tries to move S_2 or S_3 before S_1 and S_4, tetrahedra $AS_2S_3S_1$ and $CS_3S_2S_4$ will be inverted to have a negative volume as shown in Figure 5.30b, which may not be convergent or permitted in most mesh optimisation procedures. To relate the priority sequence of the Steiner points with the neighbouring coefficient is to minimise the so-called *locked* points, which can be defined as follows:

Locked point – A point will be said to be *locked* if a movement in any direction will render one or more tetrahedral element(s) connected to it to be negative (inverted).

Mesh optimisation by an iterative procedure can be made easier and is more likely to yield promising results if *locked* Steiner points are minimised. The construction of a valid mesh topology of flat tetrahedra based on the neighbouring coefficients provides a simple way in minimising the *locked* Steiner points; and for a facet with four Steiner points, there is an 83.3% chance of producing three *locked* points based on a random or natural sequence in lifting Steiner points instead of only two following a priority order. In a natural-order node lifting of four Steiner points, as shown in Figure 5.28, to create a set of five tetrahedra $ABCS_1$, BCS_1S_2, $CS_3S_1S_2$, CAS_1S_3 and $CS_3S_2S_4$, three Steiner points S_1, S_2 and S_3 are *locked*.

5.3.3.2.7.1 REMOVAL OF FLAT TETRAHEDRAL ELEMENTS BY RELOCATION OF STEINER POINTS

To establish a valid topological structure to eliminate Steiner points, flat elements of zero volume are created. These elements have to be opened up by a combination of face swap and node reposition. Reposition method is an essential part to open up flat elements. Among existing reposition approaches (Freitag and Ollivier-Gooch 1997; Alliez et al. 2005; Klingner and Shewchuk 2008), a new attempt to relocate nodes based on linear programming is presented here in which the optimal position of a Steiner point could be determined such that all tetrahedra connected to it are ensured to be positive. The method is illustrated with the reposition of nodal points in two dimensions.

As shown in Figure 5.31a, P is a point in a triangular mesh with adjacent elements bounded by polygon $V_1V_2V_3V_4V_5$. Let $\mathbf{n}_1$ be a normal vector to edge V_1V_2 pointing towards the interior of polygon $V_1V_2V_3V_4V_5$. If $V_1P \cdot \mathbf{n}_1 > 0$, triangle V_1V_2P is positive; hence, P is a valid point if it satisfies the following conditions,

$$V_iP \cdot \mathbf{n}_i > 0 \quad (i = 1,2,\ldots m)$$

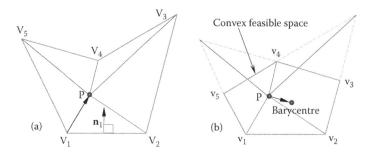

Figure 5.31 Relocation of point P within a concave polygon: (a) non-convex polygon; (b) convex feasible space.

where m is the number of bounding edges, n_i is the unit normal vector to edge E_i, and V_i is the first node of edge E_i. The constraints can be represented by the following inequalities:

$$(P \cdot n_i - s_i)\delta_i = c_i\,\delta_i \quad (i = 1,2,\dots m)$$

with

$$c_i = V_i \cdot n_i, \quad \delta_i = \begin{cases} 1 & c_i \geq 0 \\ -1 & c_i < 0 \end{cases} \quad (i = 1,2,\dots m)$$

The inequalities can be rewritten in the form of a linear programming problem, which can be readily solved by the simplex method. We thus have the following minimisation problem:

$$\text{minimise } f = \sum_{i=1}^{m} S_i \quad \text{s.t. } AX = b, \ \alpha, \beta \geq 0, \ s_i \geq 0 \tag{5.1}$$

$$A = \begin{bmatrix} \delta_1 n_{1x} & \delta_1 n_{1y} & -\delta_1 & & 0 \\ \vdots & \vdots & & \ddots & \\ \delta_m n_{mx} & \delta_m n_{my} & 0 & & -\delta_m \end{bmatrix}$$

$$X = [\alpha \quad \beta \quad s_1 \quad \cdots \quad s_m]^T, \ b = [c_1\delta_1 \quad c_2\delta_2 \quad \cdots \quad \cdots \quad c_m\delta_m]^T$$

As we aim at obtaining the feasible region, the objective function f could be an arbitrary linear polynomial of the slack variables. A feasible region exists if f > 0, and it degenerates into a line or a point if f = 0. The feasible region $v_1v_2v_3v_4v_5$ of formulation 5.1 for the concave pentagon of Figure 5.31a is shown in Figure 5.31b. As the feasible region is always convex, the barycentre of $v_1v_2v_3v_4v_5$ can be the new position for the relocation of P.

In 3D, all tetrahedra connected to a point form a polyhedron. As shown in Figure 5.32a, node P is shared by eight tetrahedra formed with the faces of a twisted Schönhardt polyhedron. Tetrahedron ABCP is a flat element as P is on the face ABC. The problem of removing flat element ABCP is equivalent to finding a feasible position for P in the polyhedron ABCDEF, as shown in Figure 5.32b. The node reposition problem can be transformed into

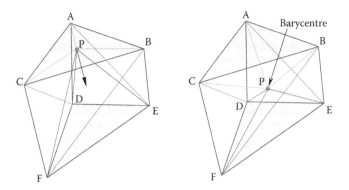

Figure 5.32 **Relocation point P to the barycentre of the feasible region.**

an optimisation problem similar to the 2D case. The constraints to locate point P are given by

$$V_iP \cdot n_i > 0 \quad (i = 1,2,...m) \tag{5.2}$$

where m is the number of bounding faces, n_i is the unit normal vector to face F_i and V_i is one of the vertices of face F_i. Inequality 5.2 can be rewritten in a standard form for linear programming, as explained in Section 5.3.3.1. In case the feasible region is not empty, the objective function f is greater than 0, and P will be relocated to the barycentre of the feasible region. In fact, the feasible space is given by the convex region bounded by the extended boundary facets (hyper-planes) of the given polyhedron.

The drawback of Laplacian-based methods is that a valid position for a Steiner point cannot be determined inside a non-convex polygon, whereas the proposed reposition method by means of linear programming could always give the optimal solution. As shown in Figure 5.33a, there are six tetrahedra, PCBA, PECA, PBDA, PDEA, CBEP and EBDP, where CBEP and EBDP are flat elements. To relocate point P, Laplacian-based methods move P downwards outside the polyhedron, as shown in Figure 5.33b.

However, the linear programming reposition method finds a valid position that is well within the polyhedron, as shown in Figure 5.33c. In the actual implementation, the linear programming reposition method and the smart Laplacian method are combined to relocate Steiner points. Specifically, the Laplacian method is first applied to determine a new position for a Steiner point, and if this fails, the linear programming reposition method is invoked.

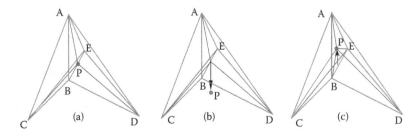

Figure 5.33 **Relocation of point P within a concave polyhedron: (a) point P at edge BE; (b) point P relocated by Laplacian method; (c) point P relocated by linear programming.**

5.3.3.2.8 Step 8: The final mesh within the boundary surface

The final mesh of the object consists of those tetrahedral elements within the boundary surface. Indeed, this should be a trivial process, as a usual practice in MG by DT or Delaunay–ADF approach is that elements inside and outside the domain or elements belonging and not belonging to the object are always distinguished by giving different zonal labels. In case labelling has not been done, elements within the boundary surface can be conveniently determined by the adjacency relationship, which is available in any DT. Just start with any seed element, which will grow in size by attaching adjacent tetrahedral elements on all the non-boundary faces (see *Algorithm RBR3D* in Section 2.5.11). This procedure will partition all the tetrahedral elements in the mesh into two groups, one inside and the other outside the object separated exactly by the domain boundary. If we have started with an element inside the object, we will automatically get the FE mesh of the object; on the other hand, if we have chosen an element outside the boundary, then we shall get all the elements outside the object by connecting neighbouring elements. Always use topological criteria but not geometrical criteria in separating elements to avoid numerical error and enhance the overall robustness of the MG scheme.

5.3.4 Worked examples and industrial applications

A number of worked examples and industrial applications are presented in this section to show the technical details in recovering boundary edges and faces. The results in attaching priority to Steiner points according to the number of neighbouring Steiner points are studied with an aeroplane bounded by 141,470 poorly shaped triangles with a lot of sharp angles. Complicated industrial objects and biological models are also included to show the characteristics and the statistics of the boundary recovery procedure in realistic applications.

5.3.4.1 Worked examples

Four numerical examples are presented to illustrate some key features of the boundary recovery procedure. The first example is a mechanical support model shown in Figure 5.34a. One hundred fifty-two boundary facets were missing in the initial DT. To recover these missing boundary facets, 50 Steiner points were inserted including 43 points on the boundary edges. The largest number of Steiner points inserted on a single edge was 5, as shown in Figure 5.34b and c. The missing edge is drawn in blue colour, and the five Steiner points are marked with different colours. To remove these Steiner points, only two iterations were required by the method of dividing the Steiner points into groups. As shown in Figure 5.34d, in the first iteration, three *unlocked* points are first relocated; then the other two *locked* points are repositioned in the second iteration. On the other hand, a random procedure would need five iterations to remove all these five Steiner points as four out of five Steiner points were *locked*. As shown in Figure 5.34e, only one Steiner point could be relocated at a time.

The second example is a screw model shown in Figure 5.35a. In this model, 433 Steiner points were introduced including 40 points on the boundary faces in which the largest number of Steiner points inserted on a single face was four. One face marked in blue with four Steiner points is shown in Figure 5.35b and c, where Figure 5.35c is the mesh of Figure 5.35b in a different view. To remove these Steiner points, Figure 5.35d depicts the process following the suppression order optimisation (i.e. higher priority to points with a small number of Steiner point neighbours). As expected, the random order required more iterations to suppress Steiner points on the missing boundary face. As shown in Figure 5.35e, although it is not the worst case (four iterations), it still needed one more iteration compared to the prioritised sequence.

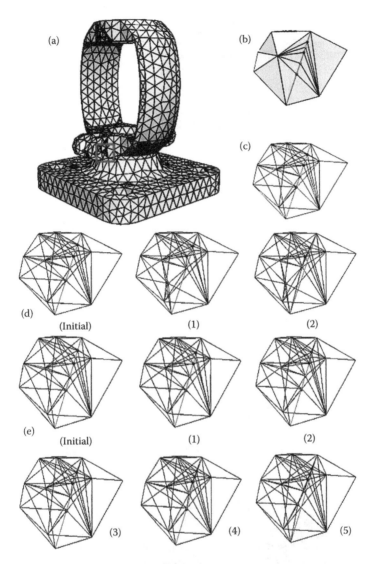

Figure 5.34 Boundary edge recovery: (a) the support model; (b and c) missing edge with five Steiner points; (d) Steiner points removed simultaneously in a group; (e) Steiner points removed by a random order.

The third example is an impeller model with 172 Steiner points inserted, as shown in Figure 5.36a. Point relocation is an essential step in removing flat elements. Magnified views of a critical recovery zone in the form of a concave polyhedron are shown in Figure 5.36b and c. The node reposition aims at relocating the Steiner point shown in red in Figure 5.36b and c so as to remove the flat elements connected to it. The feasible region of the Steiner point is a polyhedron that is marked green in Figure 5.36d. The Steiner point is repositioned at the barycentre of the region, which is marked red in Figure 5.36e. Consequently, all flat elements connected to the Steiner point are opened up, as shown in Figure 5.36f.

The well-known falcon model, as shown in Figure 5.37a, is quoted here to discuss the main features of the boundary recovery procedure. There are 141,470 triangles on the boundary, including 19,738 (14%) sharp triangles with an angle less than 5°. The result of MG is shown in Figure 5.37b. To compare with available tetrahedral mesh generators,

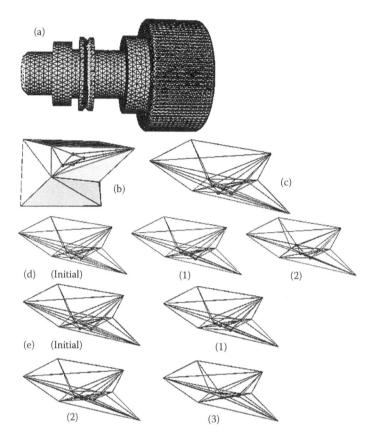

Figure 5.35 Boundary edge recovery: (a) the screw model; (b and c) missing edge with four Steiner points, (d) Steiner points relocated with a prioritised order; (e) Steiner points lifted by a random order.

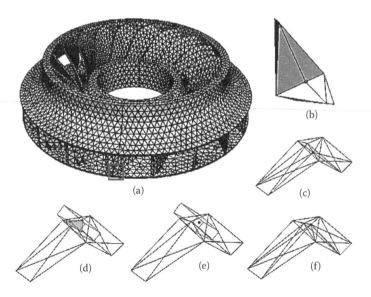

Figure 5.36 Boundary edge recovery: (a) the impeller model; (b and c) concave polyhedron; (d) feasible region and its bounding box; (e) new position for the Steiner point; (f) Steiner point relocated.

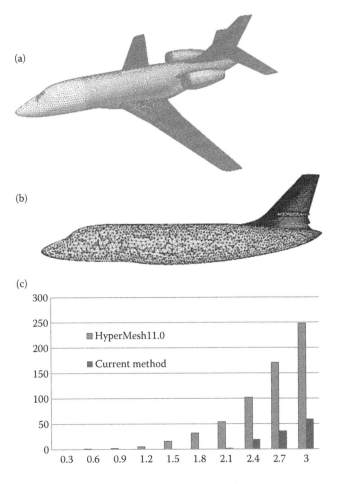

Figure 5.37 (a) The falcon model, (b) tetrahedral mesh after boundary recovery and (c) sharp angles in the falcon mesh.

ANSYS13.0, HyperMesh10.0 and HyperMesh11.0 were selected to mesh the falcon model. If Steiner points were permitted to stay on the boundary, all mesh generators succeeded in producing a valid tetrahedral mesh; otherwise, only HyperMesh11.0 was able to produce a valid mesh, and ANSYS13.0 and HyperMesh10.0 crashed in the process of boundary recovery. Comparing the tetrahedral mesh generated by the boundary recovery procedure described in Section 5.3.3 with the one by HyperMesh11.0, the smallest dihedral angle in the mesh by the current approach is 2.09°, whereas in the mesh by HyperMesh11.0, it is 0.513°. Considering the distribution of dihedral angles from 0° to 3° with an increment interval of 0.3°, there are much fewer sharp angles in the mesh generated by the enhanced scheme with prioritised Steiner points and node shifting by linear programming compared to the mesh of HyperMesh11.0, as shown in Table 5.1 and Figure 5.37c.

5.3.4.2 Industrial applications

As shown in Figures 5.38–5.41, four industrial models are included to demonstrate the capability and the characteristics of the enhanced boundary recovery procedure in dealing with complex practical applications. The general description and the statistics in the boundary

Table 5.1 Sharp angles in falcon

Angle (degree)	HyperMesh11.0	Enhanced method
0.3	0	0
0.6	1	0
0.9	2	0
1.2	5	0
1.5	16	0
1.8	32	0
2.1	54	2
2.4	102	19
2.7	171	36
3.0	248	59

recovery process of the model are listed in Table 5.2. For the quoted examples, in each case, there are about 1000 missing edges and a couple of thousands of missing faces; however, the Steiner points needed to assist the edge and face recovery are relatively few, showing that most of the missing quantities have already been recovered by some topological operations involving the swapping of element edges and faces. The seahorse model is the largest example with 173,160 boundary faces meshed in 833,309 tetrahedral elements. It is not the most difficult example as the shape qualities of the boundary triangles are not that bad, even though 11 Steiner points are needed in the recovery of one boundary face. The tyre and engine models can be considered as more challenging as many Steiner points are required in recovering boundary edges and faces. The minimum γ-quality of the tetrahedral elements

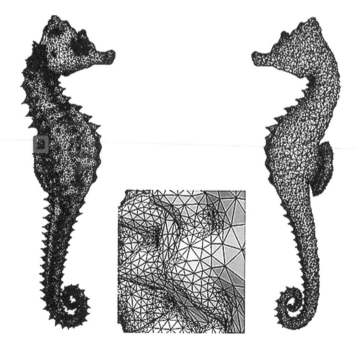

Figure 5.38 Seahorse of 173,160 boundary triangles meshed in 833,309 tetrahedra.

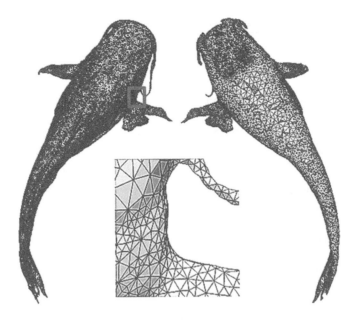

Figure 5.39 Fish of 93,054 boundary triangles meshed in 517,766 tetrahedra.

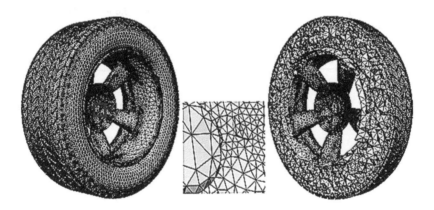

Figure 5.40 Tyre of 99,198 boundary triangles meshed in 463,147 tetrahedra.

Figure 5.41 Engine of 25,502 boundary triangles meshed in 51,766 tetrahedra.

Table 5.2 Statistics of the boundary recovery process for industrial applications

Model	Seahorse	Fish	Tyre	Engine
Figure	5.38	5.39	5.40	5.41
Model description				
Number of boundary faces	173,160	93,054	99,198	25,502
Number of nodal points	86,576	46,519	49,581	12,703
Edge aspect ratio (worst/average)	0.192/0.697	0.270/0.721	0.012/0.733	0.022/0.580
Radius aspect ratio (worst/average)	0.069/0.857	0.176/0.878	2.08e–4/0.892	1.39e–4/0.712
Minimum angle	8.24	15.18	0.28	0.36
Statistics of the boundary recovery process				
Number of missing edges	2161	897	1963	2611
Number of missing faces	4226	1798	3835	4168
Number of Steiner points on edges	26	11	40	306
Max. Steiner points on one edge	5	4	4	4
Number of Steiner points on faces	62	45	1975	875
Max. Steiner points on one face	11	8	5	5
Characteristics of output tetrahedral mesh				
Number of tetrahedral elements	833,309	517,766	463,147	51,766
Number of nodal points	184,611	111,161	98,599	14,670
γ-quality (worst/average)	0.064/0.776	0.064/0.809	3.85e–5/0.738	3.4e–4/0.477
Edge aspect ratio (worst/average)	0.139/0.629	0.118/0.657	0.008/0.620	0.002/0.445
Radius aspect ratio (worst/average)	0.038/0.805	0.037/0.836	2.54e–4/0.770	2.04e–4/0.508
Dihedral angle (minimum/average)	10.03/47.33	5.42/49.23	0.27/44.71	0.24/32.43

is quite low, probably due to the poor boundary facets, and more Steiner points are used in the boundary recovery process.

5.4 BOUNDARY PROTECTION IN DT

5.4.1 Introduction

As we can see in Section 5.3, boundary recovery in a triangulation is still a difficult problem in three dimensions. Are there any alternatives in which the boundary can be preserved in the triangulation process? One idea is that if we can verify that a boundary face or edge is Delaunay, then it will exist in the DT. Hence, if we can ensure that all the boundary faces are Delaunay, the boundary surface will be present in the triangulation. It sounds simple, but in reality, this is not easier than a proper boundary recovery process. The reason is that Delaunay empty sphere test is a global criterion, and a thorough search has to be done before we can be sure about the Delaunay property of a given element. Hence, it may not be easy to determine the Delaunay property of each boundary element, except by a direct construction of the DT of the boundary nodes by point insertion, which can be done in a linear time for practical point distributions. However, the major difficulty is that even though we have detected non-Delaunay edges and faces, there is just no simple way to restore them without losing other edges and faces already present. Du and Wang (2006) and Si (2008, 2010) presented several theorems related to conforming and constrained DT and discussed how Steiner points are added to edges and faces for their recovery.

5.4.2 2D conforming DT

Take, for instance, a 2D example of 86 boundary edges. A simple way to determine which are the Delaunay edges is to check whether there are points falling in the minimum bounding circle of the edges, as shown in Figure 5.42. This is, however, a sufficient condition but not a necessary one, i.e. the edge is Delaunay if the bounding circle contains no other points, but it may still be Delaunay if there are points in the bounding circles. As shown in Figure 5.43, only 22 edges are missing in the DT of the boundary points out of 59 potentially non-Delaunay edges whose bounding circle contains one or more points.

The missing edges are not Delaunay edges with respect to the system of boundary points, and there is no way to convert them into Delaunay edges by introducing points anywhere in the domain, as additional points can only turn Delaunay triangles into non-Delaunay ones, but not the other way around. Bear in mind that the constrained DTs constructed by a boundary recovery procedure are not strictly DTs; in particular, those edges or faces retrieved are non-Delaunay elements. However, missing edges in a triangulation can be recovered in the form of separate line segments by adding points on those missing edges. There are at least two simple ways in achieving this: (i) introduce points at the middle of the missing edges, and (ii) introduce nodes at all the intersection points. More sophisticated

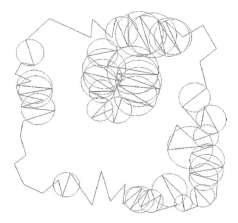

Figure 5.42 Simple Delaunay check by bounding circles.

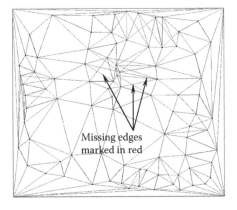

Figure 5.43 Missing edges in a Delaunay triangulation.

point insertion procedures can be found in the works of Yang et al. (2005), Si (2009, 2010) and Si and Gaertner (2011).

5.4.2.1 Insert Steiner points at the mid-points of missing edges

As shown in Figure 5.44, nodes are introduced at the mid-points of all the 22 missing edges, and a new DT is constructed. Not all the boundary edges are recovered in the new DT, and quite surprisingly, there are 23 missing edges in the new DT, one more than we had in the first triangulation. There are still missing boundary edges, because some originally Delaunay boundary edges are rendered non-Delaunay by the introduction of Steiner points, as shown in Figure 5.45. The same step was repeated five more times, and the process did converge to produce DTs with 15, 10, 6, 1 and 0 missing edges. After six iterations, the final DT containing all the boundary edges is shown in Figure 5.46. A triangular mesh of the domain can now be retrieved by collecting all the triangles within the bounded region, as shown in Figure 5.47. There are 163 nodes and 328 triangles in the entire mesh from which 163 triangles are retrieved to form the final mesh of the object. Steiner points have to be removed from the boundary edges if a fully constrained DT is required; however, the primary purpose of introducing the boundary protection scheme is to show some possibilities to generate conforming DTs rapidly in a robust manner.

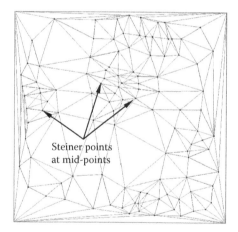

Figure 5.44 Second Delaunay triangulation with 23 missing edges.

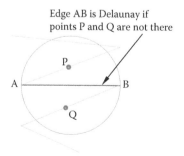

Figure 5.45 Delaunay edges turned non-Delaunay by Steiner points.

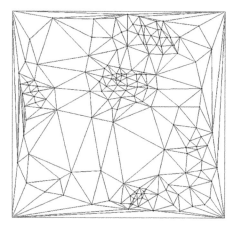

Figure 5.46 Final Delaunay triangulation after six iterations.

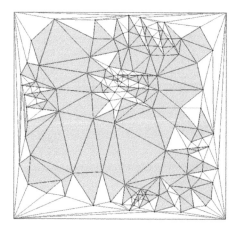

Figure 5.47 Triangles within the bounded region are marked yellow.

5.4.2.2 Insert Steiner points at the intersections of missing edges

Alternatively, Steiner points are introduced at the intersections of the missing edges for their recovery. Accordingly, points are added to the intersections of the 22 missing edges, as shown in Figure 5.48. Again, not all the boundary edges are recovered in the new DT, and there are still 20 missing edges in the second DT. The same step was repeated five more times, and the process converged to produce DTs with 9, 4, 1, 1 and 0 missing edges; the final DT containing all the boundary edges is generated in six iterations. A mesh of the domain was retrieved by collecting all the triangles within the bounded region, as shown in Figure 5.49. There are 172 nodes and 338 triangles in the entire mesh from which 168 triangles are collected to form the final mesh of the object.

Intuitively, we expected much better outcome for inserting Steiner points at the intersections of the missing edges. However, the results turned out to be a bit disappointing: first, additional calculations are required in the determination of intersection points; second, triangles of poor shape are created in the final mesh; and third, the process does not converge at a faster rate. Based on this example and the experience of the other researchers (George and Borouchaki 1998), mid-point insertion is preferred, in general, over intersection point

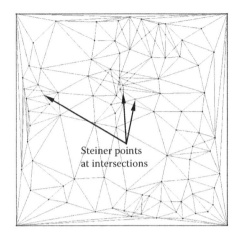

Figure 5.48 Inserting Steiner points at intersections of missing edges.

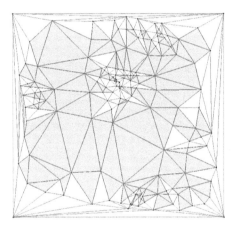

Figure 5.49 Final mesh and bounded region recovered.

insertion for the generation of conforming DTs. The main drawback of the boundary protection scheme is that an excessive number of Steiner points may be created for complex boundary surfaces as the final conforming mesh can only be generated after a couple of iterations. However, this approach can be useful for boundary surfaces, for which only geometrical integrity is of primary concern but not topological integrity, and Steiner points are allowed to stay on the boundary surface. On completion of the boundary recovery, the triangular mesh within the boundary region can be retrieved by collecting triangles inside the boundary contour, and such an element collection scheme for a bounded contour or surface is given by the following algorithm.

5.4.3 Algorithm RBR: Retrieving bounded region

The following is a generic algorithm in collecting elements within a bounded region Ω.

Input: Mesh $\mathbf{M} = \{T_i, i = 1, N_T\}$, boundary $\mathbf{B} = \{E_i, i = 1, N_B\}$;

Output: Zone label on $\mathbf{M}$, $\mathbf{Z} = \{Z_i, i = 1, N_T, Z_i = 1 \text{ if } T_i \in \Omega, Z_i = 0 \text{ if } T_i \notin \Omega\}$;

where $\mathbf{M}$ is the mesh containing the bounded region $\boldsymbol{\Omega}$, and $\mathbf{B}$ contains the boundary edges of $\boldsymbol{\Omega}$.

1. Pick any point within the bounded region $\boldsymbol{\Omega}$.
2. Identify the triangle T_k containing the point (BASE) and include triangle T_k to the bounded region, i.e. set $Z_k = 1$.
3. For each edge of triangle T_k, if it is not a boundary edge ($\notin \mathbf{B}$), add it to the list of frontal segments $\mathbf{F} = \{F_i, i = 1,n\}$, n = 1, 2 or 3, in setting up the initial front.
4. Take the last frontal segment F_n and identify the two triangles connected to F_n. One triangle T_i connecting F_n from inside of the construction front $\mathbf{F}$ should already been known; and the other triangle T_j opposite to T_i can be found by adjacency, as shown in Figure 5.50. Delete F_n from $\mathbf{F}$ by simply updating n, $n \mapsto n - 1$.
5. Include triangle T_j to $\boldsymbol{\Omega}$ by setting $Z_j = 1$.
6. For each edge E of triangle T_j not connected to F_n, if it is not a boundary edge ($E \notin \mathbf{B}$), update it with the construction front $\mathbf{F}$, i.e. $\mathbf{F} = \mathbf{F} \cup \{E\}$ if $E \notin \mathbf{F}$, $\mathbf{F} = \mathbf{F} - \{E\}$ if $E \in \mathbf{F}$.
7. If $\mathbf{F}$ is not empty, go to step 4.

If an exterior point close to the border of the square is picked, the elements between the boundary of the domain and the edges of the square are collected, as shown in Figure 5.51.

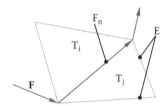

Figure 5.50 Triangle T_j absorbed by the generation front F.

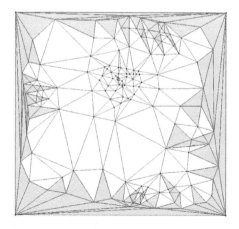

Figure 5.51 Elements outside the domain are collected by picking an exterior point.

5.4.4 3D conforming DT

The idea of boundary protection can also be extended to three or higher dimensions, in which nodes are inserted to the missing geometrical quantities until they appear in the DT of the points. In 3D, objects are bounded by closed surfaces of triangular facets. Unlike the boundary recovery of 2D domains, whose boundaries are composed of only line segments and points can be inserted on the boundary edges without much issue of convergence, in 3D, the boundary of the domain consists of edges and triangular faces (Rand and Walkington 2009).

Similar to the boundary recovery of 3D objects presented in Section 5.3, boundary edges are first recovered, followed by the boundary triangular faces. If we follow the procedure of boundary recovery in 2D by simply inserting nodes to the missing edges until they are recovered in the DT, this process may fail by not converging as more new missing boundary edges are created at the same time. As shown in Figure 5.52, nodes C and D are inserted in order to recover edge AB; however, elongated boundary line segments are formed as a result which may not be the edges also found in the new DT with additional points C and D. It is very likely that the process may not converge as more new missing boundary edges are formed than the number of edges being recovered.

5.4.4.1 Recovery of boundary edges

In the recovery process, we have to control the number of missing edges being formed as boundary edges are recovered. The bisection of the longest edges described in Section 8.4.2 ensures that relatively long edges are not generated in the division process. Moreover, the bisection process will terminate in a finite step, and the minimum quality of the boundary triangles is guaranteed. As shown in Figure 5.53, edge AB is divided only if it is the longest edge shared by the triangles. In case AB is not the longest edge, other edges connected to the triangles have to be divided first. As AB is not the longest edge of triangles ACB and ABH, before AB is divided, points P and Q are introduced to divide edges BD and BG, which are the longest edges shared, respectively, by triangles BCD and BDE and triangles BFG and BGH, as shown in Figure 5.53a. As shown in Figure 5.53b, point R is inserted to divide BC,

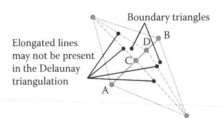

Figure 5.52 Edge recovery by inserting nodes on the missing edges.

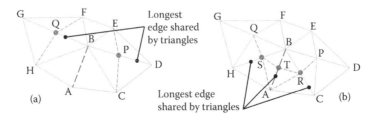

Figure 5.53 Edge recovery by bisection of the longest edges: (a) AB is not the longest edge; (b) neighbouring edges are first subdivided.

which is the longest edge shared by triangles ACB and BCP; and point S is inserted to divide BH, which is now the longest edge shared by triangles ABH and BQH. After dividing edges BD, BG, BC and BH, AB is now the longest edge shared by triangles ARB and ABS, and point T can now be inserted to bisect AB. A detailed algorithm of the bisection of the longest edges over a planar domain and triangulated curved surfaces is given in Section 8.4.2.2.

5.4.4.2 Recovery of boundary faces

Similar to the fully constrained boundary recovery, as described in Section 5.3, boundary faces will usually be present if all the boundary edges exist in the DT. However, there are cases in which a few faces are still missing, even though all the edges are present. In this circumstance, instead of dealing with the faces directly, we simply apply the bisection of the longest edge procedure to the missing triangles until the triangles are recovered. This uniform treatment not only is efficient but also can save a lot of programming efforts by applying the same procedure to both missing edges and missing faces. Suppose face BAG is missing in the DT, and in order to recover face BAG, a point has to be inserted on the longest edge of triangle BAG. Let AB be the longest edge shared by triangles BAG and ABC. While AB is the longest edge for triangle BAG, AB is not the longest edge of triangle ABC. Before AB is divided, nodes have to be introduced to edges FD, AD and AC in turn, as shown in Figure 5.54. As AB is the longest edge of triangle ABR, edge AB can now be divided, and BAG will possibly be recovered as two triangles BSG and SAG.

5.4.4.2.1 Algorithm for the semi-constrained boundary recovery

By repeatedly applying the bisection of the longest edges and DT of the resulting point set to recover missing edges and missing faces, each boundary edge is recovered as one single line segment or a list of broken segments, and a face may be intact as one single triangle or may be represented as a concatenation of smaller triangles, as shown in Figure 5.55. Obviously, at the end of the recovery process, many Steiner points have been introduced on the boundary; unlike the fully constrained boundary compatibility requirement where additional points are not allowed to stay on the boundary surface, these extra points are allowed to stay on the

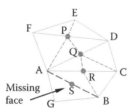

Figure 5.54 **Recovery of the missing face.**

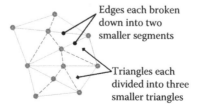

Figure 5.55 **Edges and faces are recovered as smaller segments and triangles.**

boundary surface. Hence, only the geometry of the boundary surface is recovered, but not the original topology of the boundary mesh, and usually, the recovered boundary surface consists of more triangular facets connected in a different mesh pattern. Relative to a fully constrained boundary recovery, boundary surface recovered only in geometry without identical element connections is termed as a semi-constrained or conforming boundary mesh.

Algorithm: Semi-constrained boundary recovery

1. Input boundary surface triangular mesh $S = \{T_i, i = 1, N_s\}$.
2. Create a DT using nodes on the boundary surface S.
3. Collect missing edges $ME = \{E_i, i = 1, N_E\}$.
4. If $N_E > 0$, then (i) divide missing edges by bisection of longest edges; (ii) update the boundary surface S taking into the account the division of edges; (iii) go to step 2 with additional nodes introduced to missing edges and other edges in the bisection process.
5. Collect missing faces $MF = \{F_i, i = 1, N_F\}$.
6. If $N_F > 0$, then (i) divide missing faces by bisection of longest edges; (ii) update the boundary surface S taking into the account the division of edges; (iii) go to step 2 with additional nodes introduced to missing faces and other edges in the bisection process.
7. Semi-constrained boundary surface tetrahedral mesh completed.

Remarks: Missing edges can be determined by first finding out all the tetrahedral elements connected to each node (Section 2.5.3). To check whether edge AB exists in the DT, simply verify if B is in the tetrahedra connected to node A. Similarly, to check whether face ABC is in the DT, all we have to do is to verify if nodes B and C are in one of the tetrahedra connected to node A. For each division of line segment by the introduction of a mid-side node, as shown in Figure 5.56, the boundary surface mesh S has to be updated as follows: delete triangles ABC and ACD from surface S, and add triangles PAB, PBC, PCD and PDA to surface S. It is noted that the number of triangles in S will increase by two for each edge division by introducing an additional point P. That's why the number of points and triangles in S may increase quite rapidly in the boundary recovery process.

In some fluid mechanics and field problems, the adjacent space has to be meshed as well. In this case, the MG can be taken as a constrained Delaunay insertion problem by introducing nodes to the ambient space according to the node space function taking into account the constraints of the boundary surface, as discussed in Section 5.6. Alternatively, the boundary surface S can also be inserted directly to an existing mesh M, which is typically a DT prepared based on the specified element-size distribution. Let there be N_M points in mesh M and N_B points on boundary surface S; a Delaunay mesh can be created for this set of points $N = N_M + N_B$ sequentially. The process can be repeated with the updated set of points $N^* = N_M + N_B^*$ in which N_B^* is the number of points in the modified (subdivided) boundary surface S^* for edge and face recovery. If computer memory is not an issue, mesh M needs to be generated only once, which can be stored and retrieved as the starting point for each Delaunay insertion of

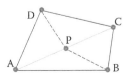

Figure 5.56 Introducing point P and surface mesh updating.

the subdivided boundary surface S*. Effectively, the semi-constrained boundary recovery procedure provides a systematic robust approach in introducing geometrical constraints in arbitrary structural forms of triangular facets to an existing mesh or DT. Algorithm RBR3D can be applied to retrieve all the tetrahedral elements within a bounded surface.

5.4.5 Practical examples

Three example meshes are presented in this section with objects downloaded from the public domain. The ambient space is meshed by Delaunay insertion of points, which are distributed in a way likely to produce tetrahedra as close to equilateral as possible. The points filling up the space are generated layer by layer, and a typical assembly of three layers of points with a uniform spacing of unity marked with different colours is shown in Figure 5.57. The layered structure of points is built up in a zigzag manner along the vertical direction such that corner nodes are on the same vertical of the mid-points of the edges of the adjacent layers below and above.

The first worked example is a string whose boundary surface consists of 2613 nodes, 5226 triangles and 7839 edges, as shown in Figure 5.58a. Four hundred fifty-six points are generated in the adjacent space of the string, for which 156 points are filtered away as they are too close to the string and 300 points are retained to form the initial DT, as shown in

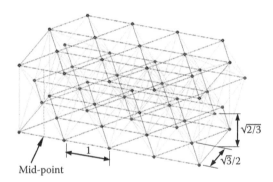

Figure 5.57 **Spatial points generated layer by layer.**

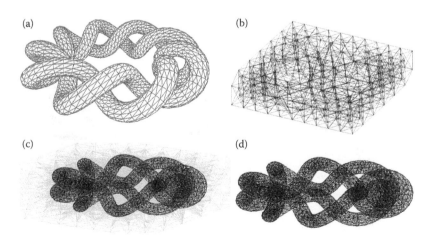

Figure 5.58 **Example 1, boundary recovery of a string by surface refinement: (a) string model; (b) background mesh; (c) final Delaunay mesh; (d) tetrahedral mesh of the string.**

Figure 5.58b. The boundary surface of the string is then inserted into the tetrahedral mesh. The first DT of the boundary surface and the spatial points consists of 2913 points and 23,503 tetrahedral elements, in which 1074 edges are found missing. Nodes are introduced to the missing edges and other related edges following the bisection of the longest edges, and as a result, there are 3987 nodes and 32,804 tetrahedra in the second DT. The number of missing edges is reduced to 462, for which exactly 462 nodes are needed for edge recovery. In the fifth and the last DT of 4525 points and 35,535 tetrahedra, there is no more missing edge, as shown in Figure 5.58c. Coincidentally, there is also no missing face in the DT, and the final recovered boundary surface after refinement consists of 8450 triangles for which it is discretised into 20,078 tetrahedral elements, as shown in Figure 5.58d.

The second example is a hand model defined by a closed surface of 6222 nodes and 12,440 triangular facets. There are 967 points in the initial background Delaunay mesh, to which the boundary surface of the hand model is inserted to produce a Delaunay mesh of 7189 nodes and 43,205 tetrahedra. Only 17 boundary edges out of 18,660 are missing in the first triangulation. The boundary recovery takes altogether four iterations or DTs to arrive at a final tetrahedral mesh, in which all the 12,554 boundary faces are present, as shown in Figure 5.59. The third example is a simpler model of a screw, which is composed of

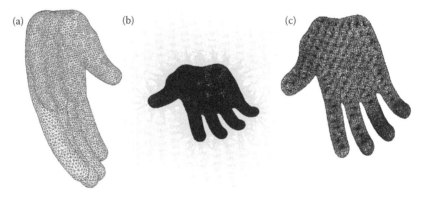

Figure 5.59 Example 2, boundary recovery of a hand model: (a) hand model; (b) final Delaunay mesh; (c) tetrahedral mesh.

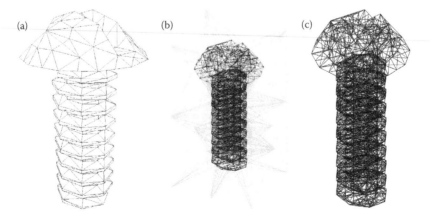

Figure 5.60 Example 3, boundary recovery of a screw model: (a) screw model; (b) final Delaunay mesh; (c) tetrahedral mesh.

413 nodes and 822 triangles. There are merely 11 points in the initial background Delaunay mesh, to which the boundary surface of the screw model is inserted to produce a Delaunay mesh of 424 nodes and 2807 tetrahedra. Seventy-two boundary edges out of 1273 are missing in the first triangulation. In spite of fewer triangles on the initial boundary surface, the recovery process requires 11 triangulations before all 1950 boundary faces are recovered in the final Delaunay mesh of 988 nodes and 5896 tetrahedral elements, as shown in Figure 5.60. The convergence characteristics and the statistics of the important parameters of the three examples are listed in Table 5.3.

Table 5.3 Convergence statistics of examples 1, 2 and 3

Object	Example 1 String	Example 2 Hand	Example 3 Screw
Number of nodes	2613	6222	413
Number of triangles	5226	12,440	822
Number of edges	7839	18,660	1233
Number of spatial points	300	967	11
DT1: *Number of points*	2913	7189	424
Number of tetrahedra	23,503	43,205	2807
Number of missing edges	1074	17	72
DT2: *Number of points*	3987	7206	601
Number of tetrahedra	32,804	43,306	3824
Number of missing edges	462	2	150
DT3: *Number of points*	4449	7208	763
Number of tetrahedra	35,072	43,315	4631
Number of missing edges	61	1	30
DT4: *Number of points*	4510	7209	798
Number of tetrahedra	35,434	43,320	4842
Number of missing edges	15	0	44
Number of missing faces	0	0	16
Number of triangles	8420	12,554	1570
DT5: *Number of points*	4525		864
Number of tetrahedra	35,535		5192
Number of missing edges	0		30
Number of missing faces	0		12
Number of triangles	8450		1702
DT = *Delaunay triangulation*			
DT10: *Number of points*			987
Number of tetrahedra			5889
Number of missing edges			1
DT11: *Number of points*			988
Number of tetrahedra			5896
Number of missing edges			0
Number of missing faces			0
Number of triangles			1950

Note: Number of missing faces = missing faces not connected to missing edges; Number of triangles = number of triangles on the refined boundary surface.

Remarks: It is found that the initial background DT mesh has little impact on the outcome of the boundary recovery. Indeed, the number of points introduced to the surface and the number of iterations by DT for boundary recovery are more or less the same even if the background mesh is only a cube of five tetrahedral elements just like the standard triangulation of the nodal points on the boundary surface. However, the number of points needed for boundary recovery depends very much on the shape or the aspect ratio of the triangular facets on the boundary surface. By the nature of surface refinement based on the bisection of the longest edge, the process tends to produce well-shaped triangles. In order to turn a poorly shaped triangle into a well-shaped triangle, a lot of nodes and iterations are required as the adjacent triangles are inevitably affected and are accordingly refined as well. With reference to Example 2, for a surface of 6222 nodes, only 20 additional points are required for boundary recovery, whereas for Example 3, 564 points are needed for the recovery of a supposedly simpler surface of only 413 nodes. The boundary recovery by surface refinement is therefore very attractive for the recovery of boundary surfaces or any geometrical constraints composed of triangular elements of more or less uniform size. Nevertheless, a more intelligent scheme ought to be devised for the recovery of boundary surfaces consisting of poorly shaped triangles. Another observation is that the scheme is relatively simple and robust as it relies only on the repeated applications of the DT and an edge refinement scheme, and an efficient DT procedure or DT by parallel processing would have a great impact on the success of the method.

5.5 GENERATION OF TETRAHEDRAL MESH BY ADF APPROACH

5.5.1 Introduction

The volume to be discretised is defined by closed surfaces composed of triangular facets. The boundary specifications imposed on the volume to be meshed are much more stringent than in many triangulation schemes in which object boundaries are piecewise smooth analytical surfaces or loosely defined by scattered spatial points. Consequently, two important characteristics can be observed from this boundary definition:

i. The volume is uniquely defined without ambiguity.
ii. Whenever necessary, complex objects can be meshed sub-region by sub-region, and the connectivity among individual pieces is guaranteed provided that correct boundary surfaces of the sub-regions are collected for MG at the sub-region level.

Realising that DT in 3D does not necessarily produce the type of tetrahedral element mesh required by the FE analysis, to ensure that tetrahedral elements generated are well proportioned, the ratio of volume to the sum of squares of edges put into a dimensionless form is adopted to judge the quality of a tetrahedral element. It can be shown that this ratio, when normalised, attains a value of 1 for regular equilateral tetrahedral elements and 0 for degenerated elements. The quality of the mesh can thus be ensured if the shape of each tetrahedral element is carefully controlled in the MG process.

The procedures of 3D triangulation by the ADF approach are very similar to those applied to the triangulation of planar domains and curved surfaces. Nevertheless, owing to the more complicated nature of a 3D problem, many steps of the MG process have to be revised, modified and extended to suit new situations. The algorithm described in this section can discretise volumes that are simply connected, multi-connected or even composed of several disjoint pieces. Based on the general principles of ADF, no special treatment is required for internal openings as long as correct orientations are applied when interior boundary surfaces are included as part of the domain boundary. The MG time depends on the number of elements to be generated and the number of nodes present in the system. Irregular domains with complex geometry and difficult boundary conditions may also take more computation time.

The boundary of the solid object to be meshed is represented by a collection of closed surfaces discretised into triangular elements of size in compliance with the required nodal spacing. For simple domains, the volume is bounded by one single closed surface, whereas for the more complicated domains, there can be as many internal surfaces as the number of openings inside the object. Multi-connected bodies in the form of a torus or rubber tube can be handled in exactly the same manner as a simply connected region. The triangular facets on the boundary surface are oriented in such a way that their normal vectors, as defined by the right-hand grip rule, are always pointing inwards towards the object. Node and element numbering on the boundary surfaces need not follow any particular order; and this flexibility allows us later to generate a tetrahedral mesh from one sub-region to another without bothering to identify the common boundaries between sub-regions. Similar to 2D MG, a node on the generation front is selected to form the *best* tetrahedron with the base triangular facet, and this boundary tetrahedron is to be compared with the *best* tetrahedron that can be created with an interior node; the overall best tetrahedron will be the element to be constructed with the given base triangular facet. The generation front is updated to include the new element, and the process is repeated until no more triangular facets are left on the generation front.

5.5.1.1 γ-quality of tetrahedral element

The quality of a tetrahedral element ABCD, as shown in Figure 5.61, is measured by the coefficient γ (Liu and Joe 1994b), which is the ratio between the volume of the tetrahedron and the sum of squares of edges put in a dimensionless form:

$$\gamma = \frac{72\sqrt{3} \text{ Volume of tetrahedron}}{(\text{sum of squares of edges})^{3/2}} = \frac{12\sqrt{3}(AB \times AC) \cdot AD}{(AB^2 + BC^2 + CA^2 + AD^2 + BD^2 + CD^2)^{3/2}}$$

$12\sqrt{3}$ is a normalising factor so that equilateral tetrahedral elements will have a maximum value of 1.

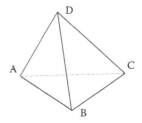

Figure 5.61 **Tetrahedron ABCD.**

5.5.2 ADF meshing procedures

5.5.2.1 The generation front

Let Ω be the given domain (object) and $B = \partial\Omega$ be the discretised boundary surface of Ω. By virtue of the orientation of the triangular facet on B, the domain to be triangulated has an interior volume always situated in the direction of the normal vectors of the boundary triangular facets, as shown in Figure 5.62. At the beginning of the triangulation, the generation front Γ is exactly equal to the boundary of the object, i.e. $\Gamma = B$. While the domain boundary remains the same, the generation front Γ evolves continuously throughout the MG process and has to be updated whenever a new tetrahedral element is created.

5.5.2.2 Generation of interior node

For ADF on 2D domains, interior nodes are created simultaneously as triangular elements are formed. For tetrahedral meshes of more or less uniform element size, in an earlier attempt, a convenient and efficient way for ADF meshing over 3D domains is to generate a system of uniformly spaced interior nodes within the problem domain before the construction of tetrahedral elements (Lo 1991c; Bajaj et al. 1999). Interior nodes within the given domain Ω are generated layer by layer. Planar cross sections can be defined when a series of parallel planes cut across the 3D object to be discretised. Interior nodes are generated on these cut sections, as shown in Figure 5.63 (refer to Section 3.5.6 for details). The following are the procedures to obtain parallel cut sections of a bounded volume, as shown in Figure 5.64.

i. The z_{min} and z_{max} of the domain are determined.
ii. Imaginary horizontal planes at various levels between z_{min} and z_{max} are allowed to cut across the 3D domain.

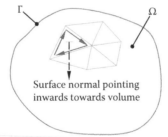

Figure 5.62 Orientation of boundary triangular facets.

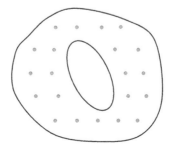

Figure 5.63 Generation of interior points on a cut section.

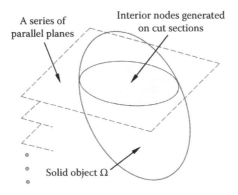

Figure 5.64 **Volume cut across by a series of parallel planes.**

   iii. The space between any two imaginary planes is equal to the average size of the triangular elements on the boundary surface.

   iv. The intersection between each horizontal plane and the solid object is a plane section bounded by closed loops.

   v. Interior nodes are generated on each cut section using the 2D node generation scheme described in Section 3.5.6.5.

The cut section corresponding to a horizontal plane at level h can be found by considering the intersections between the plane and the triangular facets on the boundary surface. Let $B = \{\Delta_i, i = 1, N_B\}$ be the set of triangular elements making up the boundary of solid volume Ω. The contour at a given level h can be determined by computing the intersection of horizontal plane z = h with all the triangular facets in B. Suppose J_1, J_2 and J_3 are the three vertices of triangle Δ_i, and h_1, h_2 and h_3, are, respectively, the heights at nodes J_1, J_2 and J_3. The intersection between the horizontal plane and triangle Δ_i can be determined by considering the intersection of the three edges of the triangle with the horizontal plane, as shown in Figure 5.65. Considering edge J_1J_2, there is intersection if

$$(h_1 - h)(h_2 - h) < 0 \text{ or } (h_1 - h)(h_2 - h) = 0 \text{ and } (h_1 < h \text{ or } h_2 < h)$$

The point of intersection P = (u, v) is given by

$$u = (1 - t)x_1 + tx_2 \text{ and } v = (1 - t)y_1 + ty_2$$

where (x_1, y_1) and (x_2, y_2) are, respectively, the x- and y-co-ordinates of nodes J_1 and J_2, and the parameter t can be calculated from $t = (h - h_1)/(h - h_2)$.

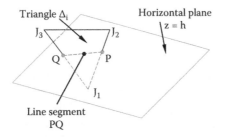

Figure 5.65 **Intersection between triangle Δ_i and horizontal plane z = h.**

Similarly, the intersections between the plane and the other two edges J_2J_3 and J_3J_1 are determined. In case of intersection, the horizontal plane should intersect with exactly two of the three edges of triangle Δ_i. Let P and Q be the two points of intersection; then line segment PQ will be the intersection of the horizontal plane and triangle Δ_i. Having considered all the triangular elements in B in turn for intersection with the horizontal plane z = h, the contour line at level h is given by the set of all these individual line segments, $S = \{P_jQ_j, j = 1, N_S\}$.

The order of these line segments of intersection, which appears to be random, in fact related directly to the numbering of triangular elements Δ_i in B. However, as the order of the line segments composing the boundary of the cut section is not important in the generation of interior nodes, retrieval of structural forms in terms of closed loops for the segments in S by proper renumbering is not necessary. Points generated on a cut section using the 2D node generation scheme represent only potential positions at which nodes can be generated. However, whether a node is actually generated depends also on how close it is to the domain boundary B.

5.5.2.3 Construction of tetrahedral elements

The 3D meshing process is initiated by selecting a triangular facet on the generation front. The choice is rather arbitrary; however, there are reports suggesting taking the smallest facet on the front for additional stability (Peraire et al. 1988). Let Σ be the nodal points on the generation front Γ, Λ be the set of interior nodes remaining inside the generation front and $J_1J_2J_3 \in \Gamma$ be the selected base triangular facet on the generation front for the construction of a new tetrahedral element. To create a tetrahedron at the base triangle, we have to find a point C from Σ and Λ such that tetrahedron $J_1J_2J_3C$ lies completely within the generation front, and its γ value is optimised.

1. *Selection of a node on the generation front.* A node $C \in \Sigma \cup \Lambda$ is said to be a candidate node if it satisfies

   i. $J_1J_2 \times J_1J_2 \cdot J_1C > 0$

      and

   ii. $J_iC \cap \Gamma \in \{\{J_i, C\}, J_iC\}, i = 1,2,3$

      Condition (i) ensures that the proposed tetrahedron possesses a positive volume, and condition (ii) makes sure that the proposed tetrahedron does not cut across the generation front. Let $\mathbb{C} = \{C_i\} \subset \Sigma \cup \Lambda$ be the set of candidate nodes. The node to be selected is therefore a node $C_m \in \mathbb{C}$ such that the γ value of tetrahedron $J_1J_2J_3C_m$ is maximised, i.e.

   $$\gamma(J_1J_2J_3C_m) \geq \gamma(J_1J_2J_3C_i) \quad \forall C_i \in \mathbb{C}$$

2. *Shape optimisation of tetrahedral elements.* The choice of node $C_m \in \mathbb{C}$ by maximising the γ value of tetrahedron $J_1J_2J_3C_m$ is simple enough, but it may not be sufficient to guarantee the *best* triangulation for domains of very irregular boundaries. Under such circumstances, a more sophisticated procedure has to be adopted. The idea is that in the selection of node C_i, instead of considering merely the γ quality of tetrahedron $J_1J_2J_3C_i$, we ought to consider also the quality of the future tetrahedral elements generated on the triangular facets $J_1J_2C_i$, $J_2J_3C_i$ and $J_3J_1C_i$, as shown in Figure 5.66.

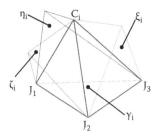

Figure 5.66 Tetrahedron $J_1J_2J_3C_i$.

Let γ_i be the γ quality of tetrahedron $J_1J_2J_3C_i$, i.e. $\gamma_i = \gamma(J_1J_2J_3C_i)$, and ξ_i, η_i and ζ_i be, respectively, the γ quality of the best tetrahedral element (without cutting into the generation front) that can be constructed on facets $J_2J_3C_i$, $J_3J_1C_i$ and $J_1J_2C_i$ with a frontal node or an interior node. Should triangular facet $J_2J_3C_i$ happen to be on the generation front, i.e. $J_2J_3C_i \in \Gamma$, set $\xi_i = 1$; if $J_3J_1C_i \in \Gamma$, set $\eta_i = 1$; and if $J_1J_2C_i \in \Gamma$, set $\zeta_i = 1$. We can now define

$$\lambda_i = \gamma_i\xi_i\eta_i\zeta_i$$

which measures the overall quality of the tetrahedron formed on the triangular facet $J_1J_2J_3$ with node C_i and the potential tetrahedral elements built on the triangular facets as a consequence of C_i being selected. Hence, in the shape optimisation process, with the quality of the tetrahedral elements possibly formed in the subsequent generation taken into account, the node to be selected is the node $C_m \in \mathbb{C}$ such that $\lambda_m = \gamma_m\xi_m\eta_m\zeta_m$ is maximised.

3. *Updating generation front.* The generation front has to be updated whenever a tetrahedral element $J_1J_2J_3J_4$ ($J_4 = C_m$) is formed. The four faces of the tetrahedral element have to be considered in turn, and Γ can be updated by
   i.   Removing triangular facet $J_1J_2J_3$ from the generation front
   ii. Removing triangular facet $J_1J_2J_4$ if $J_4J_2J_1 \in \Gamma$; otherwise adding $J_1J_2J_4$ to Γ
   iii. Removing triangular facet $J_2J_3J_4$ if $J_4J_3J_2 \in \Gamma$; otherwise adding $J_2J_3J_4$ to Γ
   iv. Removing triangular facet $J_3J_1J_4$ if $J_4J_1J_3 \in \Gamma$; otherwise adding $J_3J_1J_4$ to Γ
4. *Updating the set of interior nodes* Λ. As a flexibility in MG to guarantee quality, not each of the interior nodes generated in Section 5.5.2.2 needs to be used to form tetrahedral elements. Accordingly, the set of interior nodes Λ has to be updated upon the formation of a new tetrahedral element by eliminating nodes from Λ that are found inside or within a close distance to tetrahedron $J_1J_2J_3J_4$. A point $P \in \Lambda$ will be deleted from Λ if

$$L_i \geq 0, \quad i = 1, 2, 3, 4 \text{ where barycentre co-ordinates } L_1 = \frac{\text{volume } (PJ_2J_3J_4)}{\text{volume } (J_1J_2J_3J_4)}, \text{ etc.}$$

Analogous to ADF in 2D, steps 1–4 are repeated until all the triangular facets on the generation front Γ are exhausted, i.e. the generation front is reduced to zero with no more triangular facets, and the MG is completed.

5.5.2.4 No tetrahedron found on triangle $J_1J_2J_3$

As there is no guarantee that tetrahedral meshes exist for any volume bounded by discretised surface with full conformity in geometry and topology for all the boundary triangles, ADF meshing over 3D volumes may encounter, from time to time, difficulty in convergence. A typical scenario is that no valid tetrahedral element can be formed with a frontal node or an interior node on the selected triangular facet. In this case, a simple remedy is to generate additional interior nodes normal to the triangular facet. Let $\mathbf{n}$ be the unit normal at the centroid M of triangular facet $J_1J_2J_3$ and h be the height of the ideal tetrahedron that could be formed with the base triangle, as shown in Figure 5.67. The initial position of interior node I is located at

$$I = M + h\mathbf{n}$$

The following are the steps to judge whether node I can be accepted:

1. Verify if tetrahedron $J_1J_2J_3I$ penetrates into the generation front Γ; if yes, lower the height h by 10% and go back to step 1 to check again; else go to the next step.
2. Compute the λ value of tetrahedron $J_1J_2J_3I$; if $\lambda_I = \gamma_I\xi_I\eta_I\zeta_I$ is greater than some given threshold hold value, point I will be accepted; else further lower the height h to see if the situation improves.

In case no such interior node I exists, the last resort is to remove some tetrahedral elements in the vicinity of the troubled facet and reconstruct the local site with additional interior nodes generated normal to the frontal facets. While there is no guarantee for success in every case, fairly complicated practical 3D domains can be meshed by this strategy. The way in placing interior nodes with respect to a base triangle is crucial in producing high-quality tetrahedral meshes in the ADF scheme. Many strategies have been proposed in the positioning of interior nodes to enhance stability and efficiency for ADF meshing (Lohner and Parikh 1988; George and Seveno 1994; Borouchaki et al. 1996; Lohner 1997; Lohner and Onate 1998; Frey et al. 1998; Zuo et al. 2005).

5.5.2.5 Check for intersections

In the 3D ADF meshing, one important step is to ensure that the tetrahedral element formed with the base triangle $J_1J_2J_3$ is valid without penetrating into the generation front. To this end, the intersection between tetrahedron $J_1J_2J_3P$, $P \in \Sigma \cup \Lambda$, and the generation front has to be verified. The intersection check between a tetrahedron and the generation front can be reduced to a series of intersection checks between a line segment and a triangular facet,

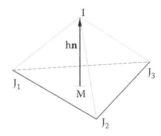

Figure 5.67 Forming a tetrahedron with an interior node I.

as described in Section 2.4.7. More explicitly, the intersection between tetrahedron $J_1J_2J_3P$ and the generation front Γ can be determined by checking the intersections of line segments $\{J_1P, J_2P, J_3P\}$ with $\Gamma = \{\Delta_i, i = 1, N_\Gamma\}$ and line segments $\{L_j, j = 1, N_L\}$ in Γ with $\{\Delta J_1J_2P, \Delta J_2J_3P, \Delta J_3J_1P\}$, where Δ_i and L_j are, respectively, the triangular facets and the line segments (edges) on the generation front Γ.

5.5.3 Efficiency consideration and mesh quality

The most time-consuming step in ADF meshing is to check whether the proposed tetrahedral elements have penetrated into the generation front. As all the triangles on the generation front have to be examined, the computation may become excessive if there are many nodes in the system. Supposing there are N nodes in the mesh, since the generation front is one dimension less, a rough estimate of the number of triangles on the generation front is $N^{2/3}$, and the processing time is of order $N^{5/3}$. As a result, the computation time will increase rapidly with the number of nodes in the system unless the intersection checks can be made localised. Similar to 2D ADF meshing presented in Section 3.6, a dynamic grid scheme can be applied to facilitate 3D ADF meshing for fairly large systems within a reasonable meshing time. As the shape and size of each tetrahedral element are well controlled in the MG, the quality of the resulting mesh is *guaranteed*. Usually, tetrahedral elements of good quality tend to form at the boundary as MG is operated from the surface of the volume towards its interior, which is quite an important factor for some finite element applications where boundary stresses or gradient values are of primary concern.

5.5.4 ADF meshing of 3D objects

Four volumes of machine parts bounded by triangulated surfaces meshed by the ADF method are presented in this section, as shown in Figures 5.68–5.71. The statistics of these example meshes are listed in Table 5.4. ADF meshing was carried out by reading in the co-ordinates of the nodes on the boundary surface and the node numbers of the boundary triangles. Interior nodes were first generated before ADF meshing, and for some meshes, remedial actions were needed with additional interior nodes generated as indicated in Table 5.4. Objects 1 and 4 characterised with many internal features and elongated boundary

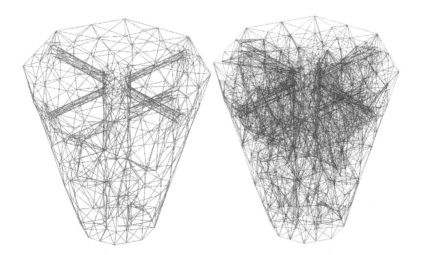

Figure 5.68 **Object with many internal features.**

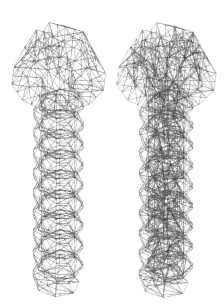

Figure 5.69 **Screw with many turns.**

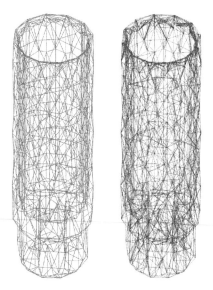

Figure 5.70 **Container with thin wall.**

triangles turned out to be slightly more difficult to mesh with additional interior nodes generated as required. The minimum γ-quality of 0.00038 for object 4 is pretty small; however, the shape quality of the boundary triangles of object 4 is also very poor with $\alpha_{min} = 0.0056$. Referring to Section 2.4.10, as the γ-quality of a tetrahedron cannot exceed the α-value of its faces, this is quite acceptable.

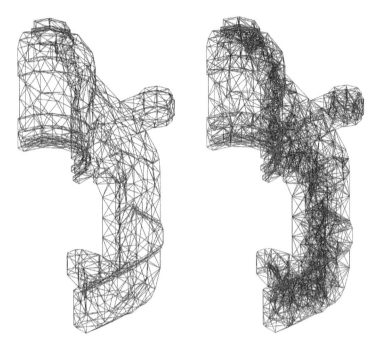

Figure 5.71 Machine part with many tiny features.

Table 5.4 Statistics of objects meshed by the ADF method

Object	1	2	3	4
NN	548	413	556	815
NB	1128	822	1112	1626
α_{min}	0.0358	0.147	0.0445	0.00562
$\bar{\alpha}$	0.499	0.57	0.63	0.553
NN*	852	499	608	1075
NE	3443	1643	1965	3836
γ_{min}	0.00111	0.0234	0.00127	0.00038
$\bar{\gamma}$	0.275	0.363	0.243	0.313
N	24	0	6	25

Note: NN = number of nodes on the domain boundary; NB = number of triangles on the domain boundary; α_{min} = minimum α-quality of boundary triangles; $\bar{\alpha}$ = geometrical mean α-quality of boundary triangles; NN* = number of nodes in the mesh; NE = number of tetrahedral elements in the mesh; γ_{min} = minimum γ-quality of tetrahedral elements; $\bar{\gamma}$ = geometrical mean γ-quality of tetrahedral elements; and N = number of frontal facets required remedial action.

5.6 DELAUNAY–ADF MESHING

DT is known to be robust and fast for the triangulation of a large system of spatial points, and the AFT is able to keep the boundary intact and generate well-shaped elements in compliance with the specified nodal space requirements. As shown in Chapters 3 and 4 on planar domain and for surface meshing, DT and AFT can be merged into a robust scheme to generate high-quality isotropic and anisotropic meshes meeting the required size and shape specifications. In 3D, Delaunay–ADF meshing is even more popular as ADF has convergence issue over difficult boundary conditions, and there is no simple strategy in placing interior nodes for DT to produce meshes of various characteristics.

5.6.1 Delaunay–ADF mesh procedure

The essence of AFT is not where MG is started, whether it is from the boundary or radiating from an interior point, but the partition of the problem domain into a meshed zone and an unmeshed zone clearly delineated by the generation front, which is the common moving boundary between the zones. While the meshed and unmeshed regions can take any flexible shape and form and each of which may consist of several disconnected pieces, the frontal process allows us to concentrate on element generation at the front, which is one dimension less than the problem domain, and to pay no more attention to the meshed zones in which mesh has already been generated.

The meshing problem by the Delaunay–ADF procedure can be stated as follows:

Given an object Ω and its boundary $\mathbf{B}$, which is a closed surface of triangular facets $\{B_i, i = 1, N_B\}$ and nodal spacing function $\rho(\mathbf{x})$, $\mathbf{x} \in \Omega$, generate tetrahedral mesh $\mathbf{T}$, $\{T_i, i = 1, N_T\}$, such that $\partial \mathbf{T} = \mathbf{B}$ and the edges of the mesh are as close to the specified lengths as possible.

5.6.1.1 Initial generation front

The initial generation front is given by the boundary of the domain $\mathbf{B}$. However, for element consistency, similar to the ADF meshing on a planar domain, the boundary $\mathbf{B}$ has to be re-meshed into triangles in conformity with the given nodal spacing function ρ. While surface re-meshing will generate meshes of the best quality, surface mesh refinement is a simple efficient procedure to produce triangular elements of the required size without violating the original geometry as defined by the boundary triangular facets $\{B_i, i = 1, N_B\}$. Following the surface refinement algorithm presented in Section 8.4, the refined boundary surface $\{\Delta_i, i = 1, N_\Delta\}$ or the initial generation front of a ball-shaped object is shown in Figure 5.72.

5.6.1.2 Boundary triangulation

A DT is created by the insertion algorithm for nodes on the refined boundary surface. Boundary facets $\{\Delta_i, i = 1, N_\Delta\}$ are restored by the boundary recovery technique or kept intact by the

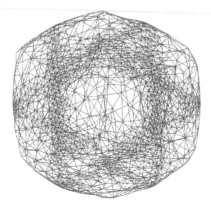

Figure 5.72 Initial generation front.

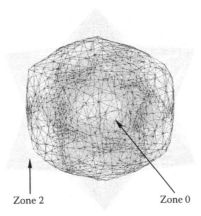

Zone 2   Zone 0

Figure 5.73 **Mesh of boundary nodes.**

boundary protection scheme as described in Sections 5.3 and 5.4. The DT of the cube consists of two zones separated by the boundary surface as shown in Figure 5.73; zone 0 consisting of those tetrahedra within the boundary surface is a region to be meshed by the Delaunay–ADF procedure, and zone 2 consists of the auxiliary tetrahedra outside the boundary surface. The tetrahedral elements in zone 0 will be given a label 0, and those of zone 2 will be given a label 2 for easy identification. Indeed, the boundary can be implicitly defined once we have properly labelled the tetrahedral elements of the two zones. It is important to keep all the tetrahedral elements within the cube as the search for the BASE tetrahedron of an insertion point merely by the tetrahedra in zone 0 is rather difficult and time consuming, as zone 0 is non-convex and even disconnected, whereas the union of zone 0 and zone 2, i.e. the entire cube, is always convex.

5.6.1.3 Zonal division and MG front

Similar to the classical frontal process, tetrahedral elements are created with frontal base triangles one by one until zone 0 is filled up with newly created tetrahedral elements. It is noted that the object has already been discretised into a valid tetrahedral mesh after boundary surface recovery, and any element creation with the frontal triangles by means of point insertion is just a mesh modification to improve the quality of the mesh to fulfil the mesh characteristic requirements. From the last step, where is zone 1? The answer is that zone 0 will be split into two zones, zone 0 and zone 1, in the process of Delaunay–ADF meshing. At the beginning of Delaunay–ADF mesh, all tetrahedral elements within the boundary surface belong to zone 0, and whenever a new tetrahedral element is formed with a boundary triangular facet, it will be assigned to zone 1. Hence, the current generation front is given by the boundary between elements of zone 0 and elements in either zone 1 or zone 2. While the number of elements in zone 2 is fixed, in the progress of Delaunay–ADF meshing, the number of elements in zone 1 will keep on increasing, whereas elements in zone 0 will decrease. At the end of Delaunay–ADF meshing, all the tetrahedral elements in zone 0 will be converted to elements in zone 1.

5.6.1.4 Generation of tetrahedral elements on a frontal triangle

Let's talk about the basic mechanics for the creation of a tetrahedral element with a base triangle on the generation front by means of Delaunay insertion. The strategic position of such an insertion point and the shape optimisation of the elements will be discussed later in Section 5.6.1.7. Let Δ be the base triangle and T be the associated tetrahedron in zone 0

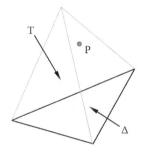

Figure 5.74 Front triangle and its associated tetrahedron.

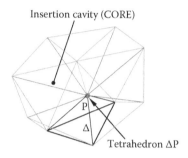

Figure 5.75 Insertion cavity associated with point P.

as shown in Figure 5.74 (such a tetrahedron T exists and is unique). A point P is inserted at the vicinity of triangular facet Δ, which can be at the centroid of tetrahedron T or on the normal to the triangle Δ. The BASE tetrahedron is determined using tetrahedra in all the zones 0, 1 and 2. Although T is likely to be the BASE tetrahedron in most cases, other tetrahedron in zone 0 can also be the BASE depending on where point P has been inserted. The basic conditions that point P could be accepted are that (i) the BASE tetrahedron is in zone 0 and (ii) tetrahedron T is non-Delaunay with respect to point P, i.e. ΔP is a tetrahedron to be constructed in the insertion process, as shown in Figure 5.75. Starting from the BASE tetrahedron, all the non-Delaunay tetrahedra in zone 0 are identified, which form an insertion cavity (CORE) for the insertion point P. The tetrahedral elements in the CORE are deleted from zone 0, and tetrahedral elements formed with the boundary triangles of the CORE are added to zone 0 except tetrahedron ΔP, which is assigned to zone 1 as it is a new tetrahedral element generated at the base facet Δ. As only non-Delaunay tetrahedra in zone 0 are taken to form the CORE, it is a form of restricted or constrained DT.

5.6.1.5 Updating the generation front

The generation front is implicitly and consistently defined by the moving boundary between the elements in zone 0 and the elements in other zones 1 and 2. However, for easy monitoring of the progress of MG and the strategic placement of insertion points, a list of frontal triangular facets, $\Gamma = \{\Delta_i, i = 1, N_\Delta\}$, is explicitly maintained. The generation front Γ has to be updated whenever a new tetrahedral element is created with a frontal triangle in Γ. Let A, B and C be the vertices of triangle Δ; updating the generation front can be easily done, as shown in Figure 5.76.

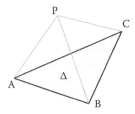

Figure 5.76 **Updating generation front.**

    i. Delete Δ = ABC from Γ.
   ii. Delete PAB from Γ if PAB ∈ Γ; otherwise, add PAB to Γ.
  iii. Delete PBC from Γ if PBC ∈ Γ; otherwise, add PBC to Γ.
  iv. Delete PCA from Γ if PCA ∈ Γ; otherwise, add PCA to Γ.

5.6.1.6 Termination of the meshing process

The Delaunay–ADF meshing is completed if there is no more triangular facet left in the generation front Γ or there is no more tetrahedron in zone 0; else, go to Section 5.6.1.4 to generate more tetrahedral elements.

5.6.1.7 Strategy in placing interior nodes

It is noted that Delaunay–ADF meshing is as robust as DT, as a valid FE mesh of the object is always maintained throughout the MG process. The introduction of interior nodes is just to improve the quality of the mesh and to satisfy the element size and shape specifications. In the introduction of a point P to a frontal triangular facet Δ, if the quality of tetrahedron ΔP is inferior to the tetrahedron T associated with Δ, simply take T as the newly created element for Δ and update the generation front. In doing so, it is guaranteed that the quality of the FE mesh of the object is ever improving. The quality of the mesh is closely related to how nodes are created for a given frontal triangular facet, and it is the very reason why a frontal guiding process is involved for the Delaunay point insertion. Before we could have a detailed discussion on the various possibilities for the introduction of interior points, we have to clearly define the quality of a tetrahedral element with respect to the given nodal spacing function ρ. A combined measure on the shape quality of the tetrahedron and the conformity to the required length for its edges will be considered as the overall quality of the tetrahedron. Thus, the λ-quality of tetrahedron PABC is given by

$$\lambda = \gamma \delta_a \delta_b \delta_c$$

where shape quality

$$\gamma = \frac{72\sqrt{3} \text{ Volume of tetrahedron}}{(\text{sum of squares of edges})^{3/2}}$$

and conformity coefficients are given by

$$\delta_a = \min\left(\frac{\|PA\|}{\rho_{AP}}, \frac{\rho_{AP}}{\|PA\|}\right) \quad \rho_{AP} = \text{desirable length between A and P}$$

$$\delta_b = \min\left(\frac{\|PB\|}{\rho_{BP}}, \frac{\rho_{BP}}{\|PB\|}\right) \quad \rho_{BP} = \text{desirable length between B and P}$$

$$\delta_c = \min\left(\frac{\|PC\|}{\rho_{CP}}, \frac{\rho_{CP}}{\|PC\|}\right) \quad \rho_{CP} = \text{desirable length between C and P}$$

and the desirable length between two points P and Q can be estimated simply by

$$\rho_{PQ} = \sqrt{\rho(P)\rho(Q)}$$

The conformity of other edges AB, BC and CA is not included as those are existing edges on the generation front. As all the parameters are between 0 and 1, the overall quality measure λ will also be within 0 and 1.

5.6.1.7.1 Evaluation of an insertion point

The quality of an insertion point P can be evaluated by considering the tetrahedra deleted by the point against those created by the point in the Delaunay insertion process. Let $\{T_i, i = 1, N_T\}$ be the non-Delaunay tetrahedra in the CORE and $\{E_i, i = 1, N_E\}$ be the new elements formed with the boundary triangles of the CORE. Then the quality of point P, μ_P, is given by

$$\mu_P = \frac{\min_{i=1,NE}\{\lambda(E_i)\}}{\min_{i=1,NT}\{\lambda(T_i)\}}$$

Other convenient norms such as the geometrical mean value can also be used. μ_P will be greater than 1 if the minimum quality of the newly created tetrahedra is larger than that of the old tetrahedra in the CORE. Greater μ_P represents better improvement by the introduction of point P, and hence, we can select an optimal insertion point by comparing the μ-values of the points inserted. This is a very rigorous scheme for the identification of a strategic insertion point; yet, it is rather expensive if many points have to be evaluated to obtain high-quality tetrahedral elements.

5.6.1.7.2 Some suggested locations to insert point P

As far as the element shape quality is concerned, the best tetrahedral element that could be formed with a triangle ABC is to pick a point along the normal passing through the centroid of the triangle, O, as shown in Figure 5.77. From Section 2.4.10, the optimal point is found at a height, h, given by

$$h^2 = \frac{2}{9}(AB^2 + BC^2 + CA^2)$$

Given a node spacing requirement ρ, the overall quality of a tetrahedron depends also on the conformity coefficients; thus, two more locations at $\pm20\%h$ relative to the ideal

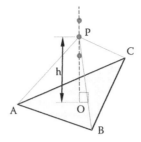

Figure 5.77 Locating insertion point P.

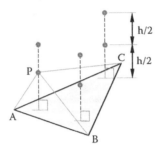

Figure 5.78 Integration points of triangle ABC.

point P are also suggested to allow for a progressive change in the element size. In case of a rapid change in the element size and shape, more strategic locations have to be checked. As shown in Figure 5.78, points at the normal to the three integration points can sometimes produce promising results. The area co-ordinates of the integration points are, respectively, $(3/5,1/5,1/5)$, $(1/5,3/5,1/5)$ and $(1/5,1/5,3/5)$.

5.6.2 Example

The Delaunay–ADF meshing of the object shown in Figure 5.72 is taken as an example. Sections 5.6.1.4 and 5.6.1.5 are applied repeatedly to zone 0 tetrahedral elements within the boundary surface as shown in Figure 5.73. By always taking the last frontal triangle for MG, an intermediate stage after generating 2500 elements is shown in Figure 5.79. It

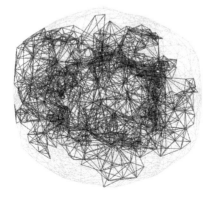

Figure 5.79 Intermediate stage of 2500 tetrahedral elements.

can be seen that the shape and the structure of the generation front are quite arbitrary and unpredictable consisting of triangles of various sizes and shapes in several disconnected pieces. If elements are generated by taking the smallest triangle on the generation front, the generation front looks more regular, though it is still disconnected as shown in Figure 5.80. Usually, more elements of slightly better quality will be generated using the smaller triangles on the generation front. The final mesh of 23,085 elements is shown in Figure 5.81, and a

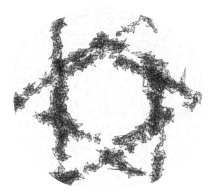

Figure 5.80 Mesh generation at frontal facets of the smallest size.

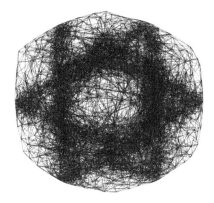

Figure 5.81 Tetrahedral mesh of 23,085 elements by Delaunay–ADF meshing.

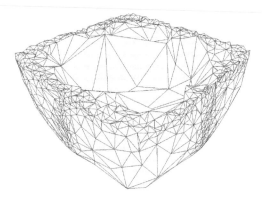

Figure 5.82 Cut section of the mesh by a horizontal plane.

cross section of the mesh along a horizontal plane is shown in Figure 5.82. The geometrical mean γ-shape quality of the mesh is 0.45, even though there are many poorly shaped triangles on the boundary. The mean conformity coefficient is approximately 80% for a node spacing function with small elements at various locations on the boundary whose sizes are to be increased based on a function of distance to the power 4.

5.7 GENERATION OF TETRAHEDRAL MESH BY SPHERE PACKING

5.7.1 Introduction

FE MG has evolved rapidly over the last two decades, and meshing techniques seem to have reached a high level of maturity; moreover, different methods with their variations have been developed for the generation of tetrahedral meshes. The Octree technique, DT and AFT and their combination are admittedly thought as classical methods widely employed by the meshing community today. The Quad/Octree method, which belongs to a class of spatial decomposition, was proposed more than three decades ago. By this method, a square grid containing the object to be discretised is recursively subdivided until a desired resolution is reached. The computational geometrical properties of the DT have been investigated for many years. With the rapid development of the FEM, the DT algorithm was further extended to generate valid FE meshes for practical engineering problems. AFT, once developed on the planar domains, was later modified for the generation of quadrilateral elements and extended into 3D, surface meshing and other applications such as filling volume and space with arbitrary polyhedral solids (Lohner and Onate 2004, 2010).

The circle packing technique as a tool for the placement of nodes within a bounded domain and possibly for FE MG coupled with existing node connection schemes such as DT has aroused much attention (Bern et al. 1995; Eppstein 1997; Bern and Eppstein 2000; Li et al. 2000; Shimada et al. 2000; Chung and Kim 2003; Kim et al. 2003; Liu 2003; Sowa and Koch 2004; Jerier et al. 2010). The circle packing method fills a bounded domain with a set of circles whose centres provide valid nodal point locations for a quality Delaunay mesh. For instance, Bern et al. (1995) and Eppstein (1997) triangulated an n-vertex polygonal region (with holes) by packing circles of different radii in it so that no element has an angle large than π/2. However, it is not sure whether pointed triangles can still be formed in the packing process. Through an iteration process or with the aid of differential equations for the determination of the centre positions of the packing circles, the circle packing technique can be used to generate nodal points for the construction of adaptive meshes and anisotropic meshes (Shimada et al. 2000). Li et al. (2000) made use of the ADF approach to construct quality circle packing; with the first circle found on the boundary, the packing grows gradually towards the interior of the domain. Feng et al. (2003) extended the idea of ADF to construct packs of circles of different sizes with consideration for application to discrete element methods. However, no discussion has been given as to how FE mesh is generated from the packed circles, even though DT could possibly be applied for this purpose.

Sphere packing has also been studied by mathematicians emphasising on various theoretical aspects. Muses (1997a,b) studied the packing of equal spheres at higher dimensions. Cockayne and Mihalkovic (1999), Radin (2004) and Sowa and Koch (2004) studied the symmetry and stability of various sphere packings. Sutou and Dai (2002) and Li and Ng (2003a,b) described algorithms to deal with the problem of packing unequal spheres. Donev et al. (2004) attempted the problem of fitting spheres in a bounded space using linear programming. However, all these results could not be directly applied to the FE MG as the sizes of the spheres are governed by the local geometry, which may not be compatible with the specified nodal spacing function.

A scheme for the generation of FE meshes of variable element size over a 2D unbounded domain was proposed later by Lo and Wang (2003). The idea of MG is to connect centres of packing circles efficiently and robustly by the AFT. Nevertheless, unlike the conventional frontal method, the procedure does not start from the boundary of the object but from a convenient point (origin) in an open domain containing the object. The sequence of construction of the packing circles is determined by the shortest distance from the origin in such a way that the generation front is more or less circular in shape with minor local concave parts due to element size variation. As soon as a circle is added to the generation front, finite elements are directly generated by proper connection of the frontal segments to the centre of the circle just packed. The size of the triangular elements is controlled by packing circles of various sizes as specified by the nodal spacing function such that, as much as possible, circles are tangent to each other to minimise the gap between them. The MG can be terminated by specifying a maximum distance from the origin or by limiting the number of circles packed in the cluster. In case the MG of a physical object is required, the object boundary can be introduced in the unbounded mesh by any well-established procedure as described in Section 3.5.7 (Lo and Liu 2002).

Generation of large-scale tetrahedral meshes of variable element size fully conformable with the boundary surface is still a tedious and time-consuming task. However, there are problems such as fluid flow mechanics in which the focus is on the mesh size variation and shape quality at the interior of the domain rather than the far field boundary constraints. Even in case certain boundary conditions need to be introduced, this could be achieved by some boundary recovery techniques described in Sections 5.3 and 5.4 or by removing those tetrahedral elements in the vicinity of the domain boundary and refilling the cavity by ADF meshing presented in Sections 5.5 and 5.6. The isolation of MG at the interior of the domain and the treatment of boundary conformity offer a number of advantages: (i) well-shaped elements of required size can be formed freely, (ii) no need to check for boundary conformity for each element constructed and (iii) by making full use of the space provided without any constraints, MG is much faster and simpler.

An algorithm for the generation of tetrahedral mesh of specified size over an unbounded 3D domain is presented in this section. Starting from an interior point (defined as the origin) within the problem domain and guided by the ADF concept, spheres of size compatible with the prescribed element size are packed tightly together one by one to form a cluster of spheres of variable sizes. The compactness of the cluster is achieved by packing spheres at a site closest to the origin in a dense manner, tangent but not overlapping with as many adjacent spheres as possible. In view of these criteria, a rotational mechanism between spheres is innovated, which allows the newly inserted sphere to follow the path by rotation between existing spheres until the lowest point is reached. The centres of the packed spheres provide ideal locations for Delaunay point insertion to form a triangulation of tetrahedral elements of size compatible to the specified value.

Over a 2D unbounded domain, the circles on the generation front are tangent to one another and form a water-tight closed loop, which divides the unbounded domain into two parts, namely, the meshed region and the unmeshed region. However, in 3D, the spheres on the front do not form a water-tight surface, even though all the spheres are of the same size. That is to say, given a cluster of spheres, no matter how tightly packed they are, it is always possible to find a continuous path from a distant point at the far field to the centre of the cluster. In other words, the unbounded domain is connected and is not divided into two regions by the spheres at the generation front. This situation was given some thought; it came to the conclusion that there is no need to define a generation front in a way how the

ADF method operates in order to pack spheres. In fact, what is needed is a point among all the tetrahedral elements on the surface that is relatively close to the origin. Tight packing of spheres is achieved by a descending mechanism following the path between existing spheres by rotation until the lowest point is reached. By its very design, the deepest descending mechanism by rotation about axes between spheres naturally fulfils the criteria of being the nearest to the origin, the densest and tangent without overlapping in sphere packing.

5.7.2 Sphere packing and MG algorithm

The idea of MG is to connect centres of tightly packed spheres of variable size by Delaunay point insertion scheme. Unlike the conventional ADF approach, the procedure does not start from the object boundary but at a convenient point within the 3D open space. The initial pack consists of four spheres tangent to each other as shown in Figure 5.83, which will expand towards the exterior and has to be updated whenever a new sphere is added. Tetrahedral elements are subsequently generated when the centres of the spheres are connected to form tetrahedra. The data structure and the rules of sphere packing are to be described in Sections 5.7.2.1 and 5.7.2.2.

5.7.2.1 Data structure

The data structure requirements for MG are very similar to those required by the DT. For a typical Delaunay point insertion scheme, the four vertices and the four neighbours of each tetrahedral element have to be stored. In addition, for packing of spheres, the centre, the radius and the distance from the origin of each inserted sphere have to be stored as well. As the nodal points of the tetrahedral mesh are the centres of the packed spheres, a neighbouring relationship of the spheres can be defined based on the adjacency of the tetrahedral elements.

5.7.2.2 Criteria for sphere packing

1. *Nearest.* New spheres have to be generated at locations as close to the origin as possible. Packing spheres as near to the origin as possible ensures compactness and reduces the chance of forming holes and voids. By packing spheres close to the origin, the shape of the cluster of spheres is basically convex and spherical with minor concave parts.
2. *Densest.* Spheres are to be packed as close to one another as possible. It is best that spheres are packed tightly together so that the gaps between them are minimised, and preferably be surrounded by four or five tangential spheres. However, it is sometimes difficult to pack them densely if we have to fit a sphere of specified size into a given

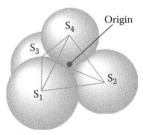

Figure 5.83 Initial pack of four spheres.

site. Hence, a dynamic scheme has to be developed in which the radii of the packing spheres can be slightly adjusted for a tight packing. Of course, the algorithm can also pack spheres with a stricter size requirement that adjustment to radius is not allowed.

3. *Tangent and no overlapping.* There is no overlapping between any two spheres. A newly added sphere has to be tangent to the existing spheres on the front without overlapping. Overlapping spheres lead to poor mesh quality and violation of the specified element size.

5.7.2.3 The controlled space

Similar to the construction of DT for a set of random points, the space of interest is bounded by a cube of five tetrahedra, which is large enough to contain all the points. For the purpose of MG over an unbounded volume, we can devise a cube based on the estimated diameter of the target cluster of spheres. In case we find the cube to be too small during the MG process, it really doesn't matter, as the size of the containing cube can be adjusted by moving its vertices away from the origin. Moreover, a dynamic cube would also enhance the numerical stability of the Delaunay point insertion.

5.7.2.4 Generation of the initial pack

The first four spheres S_1, S_2, S_3 and S_4 are placed tangent to each other such that the centroid of the tetrahedron formed by joining the centres of the spheres coincides with the origin $O(0,0,0)$ as shown in Figure 5.83. The radii of the spheres are determined from the size control function. Tetrahedral elements are formed by the Delaunay insertion algorithm, so that the centres of the spheres define a tetrahedron containing the origin and other tetrahedral elements formed by joining the spheres and the vertices of the enclosing cube.

5.7.2.5 Packing spheres

5.7.2.5.1 Locating a point for possible insertion of a new sphere

As mention in Section 5.7.1, comparing with packing circles in 2D, there is no analogous definition of a frontal surface for a cluster of spheres as there are always gaps between spheres. It was discovered later that, indeed, such a front surface is not required in packing spheres, though a rough idea about the nodes on the front would be helpful in facilitating the packing process. In fact, operating without a frontal structure can reduce the memory requirement and the computational time in the management and the updating of the front. However, we still have to make use of the frontal concept in order to locate possible insertion sites for sphere packing. The front of a cluster of spheres can be defined as the set of triangular facets connected to the *high points*, which are the eight vertices of the bounding cube. This definition gives us all the necessary information about the front including its faces, edges and nodal points. However, unlike a typical ADF process, there is no need to follow the frontal surface structure and update it each time a new element is constructed. As tetrahedral elements are constructed and recorded in the DT in the course of sphere packing, the frontal surface is readily available with little additional memory and computations.

5.7.2.5.2 Descending between spheres by rotation

The problem we have is to find a location in a cluster of spheres to accommodate a new sphere of specified size that best verifies the criteria of being the closest, the densest, and tangent without overlapping to existing spheres. Imagine that the surface of the cluster is bent and spread onto a flat bed, which is same as the process of developing the earth surface into maps; the problem is similar to tossing a ball into a pool of balls of various sizes and then finding out where the tossed ball will come to land. The agent that set up the motion is the gravitational force, which pulls the ball to the lowest possible position, and the geometry permitting this to happen is the path between existing spheres about which the tossed ball rotates around and descends on the way to the lowest point. Following the same principle for the packing of spheres, the newly inserted sphere will move down the track by rotation starting from an edge on the front relatively close to the origin. The inserted sphere will acquire a lower position each time it rotates between two spheres, and it may change axis along its way until further descent is not possible where rotation about any one of the three axes that block the sphere will raise it to a higher position.

5.7.2.5.3 Rotate between spheres

As shown in Figure 5.84, rotation between two spheres S_1 and S_2 with, respectively, radii r_1 and r_2 is possible provided that the new sphere S with radius r satisfies a simple condition:

$$r_1 + r_2 + 2r > d_{12} \text{ where } d_{12} \text{ is the distance between spheres } S_1 \text{ and } S_2 \qquad (5.3)$$

Descent by rotation starts from the nearest edge from the front. Let S_1 be the frontal point closest to the origin. Examine all the frontal nodes connected to S_1 and select S_2, which is closest to the origin. Then $S_1 S_2$ can serve as the initial rotation axis for the new sphere S. In case condition 5.3 cannot be satisfied, pick another sphere connected to S_1 to form the axis of rotation.

5.7.2.5.4 Rotation about an axis

Since we would like to pack spheres of different size specified by the node spacing function, the radius of the newly inserted sphere has to be computed according to its current physical position in space. As the sphere moves around in the rotation process, its radius has to be updated from one axis to the next in order to be consistent with the size requirement. The radius of the inserted sphere can be estimated by evaluating the desirable size at the midpoint of S_1 and S_2. The radius of S has to be updated as it moves away from the mid-point to a new position. A couple of iterations may be necessary to set up the initial position with a radius compatible with the node spacing function and tangent to spheres S_1 and S_2 as shown in Figure 5.84. As the radius r of the proposed sphere S tangent to S_1 and S_2 is often needed, the details for its determination are given as follows. Spheres S_1, S_2 and S are lying on the same plane, and hence, two simple conditions have to be satisfied.

$$\|SS_1\| = r + r_1 \text{ and } \|SS_2\| = r + r_2 \qquad (5.4)$$

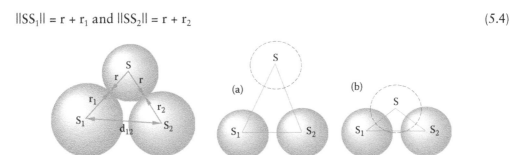

Figure 5.84 Proposed sphere S touching existing spheres S_1 and S_2: (a) S is away from S_1 and S_2; (b) S is close to S_1 and S_2.

Let (x,y) be the centre of the sphere S. As $r = r(x,y)$ is a function of co-ordinates x and y on the plane SS_1S_2, (x,y) or position of S and radius r can be determined based on the two conditions in Equation 5.4. As long as $r = r(x,y)$ is continuous and sufficiently smooth, solution for Equation 5.4 exists. Consider the following extreme cases: (i) that S is far away from S_1 and S_2, as shown in Figure 5.84a, and (ii) that S is close to S_1 and S_2, as shown in Figure 5.84b. If r is a continuous function of position, i.e. $r = r(S(x,y))$, then there exists S that is tangent to S_1 and S_2 on a given plane satisfying conditions 5.4.

As shown in Figure 5.85, an orthonormal basis can be set up on the plane OS_1S_2 where O is the origin, such that **w** is the unit vector along S_1S_2, **u** is the unit vector on the OS_1S_2 plane perpendicular to **w** and **v** is the unit vector normal to the OS_1S_2 plane. Before rotation, sphere S is placed at the highest point with its centre being on the OS_1S_2 plane in such a way that it is farthest away from the origin O. For a rotation governed by an angle θ about the **w** axis as shown in Figure 5.86, sphere S assumes a new position S' as given by

$$AS' = d\cos\theta\ \mathbf{u} + d\sin\theta\ \mathbf{v} \text{ where } d = \|AS\| \text{ and A is the projection of S onto } \mathbf{w} \text{ axis}$$

The sphere S is rotated about axis S_1S_2 until it is in touch with another sphere S_3. This situation is not difficult to detect as the list of tetrahedra connected to S_1S_2 is readily available from the adjacency relationship of the DT. Suppose S_3 is the sphere on the list of tetrahedra connected to S_1S_2 making the smallest angle α measured from the initial position of S, i.e. α is the angle between planes OS_1S_2 and $S_1S_2S_3$. On the plane $S_1S_2S_3$, recalculate sphere S, which is tangent to spheres S_1 and S_2 with the centre lying on $S_1S_2S_3$ and the radius given by the nodal spacing function. This is similar to setting up the initial position for S on OS_1S_2, except that it is now done on the $S_1S_2S_3$ plane. As shown in Figure 5.87, the condition for intersection with sphere S_3 is given by

$$\|SS_3\| < r + r_3 \text{ where r and } r_3 \text{ are, respectively, the radii of S and } S_3.$$

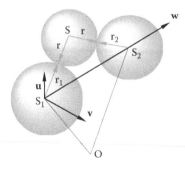

Figure 5.85 Initial position of the inserted plane.

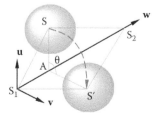

Figure 5.86 Rotate about an axis.

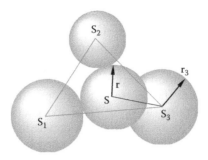

Figure 5.87 **Updating sphere S on plane S₁S₂S₃.**

The angle θ that S has to turn through just touching sphere S3 on the surface has to be determined by an iterative process. Such a θ exists as there is no intersection at the initial position where θ = 0, and there is intersection when S is rotated through an angle α. In fact, θ lies between 0 and α. A simple iterative procedure can be devised as follows.

$$e = \|SS_3\| - (r + r_3) \tag{5.5}$$

When θ = 0, e > 0, and when θ = α, e < 0; hence, θ can be computed by linear interpolation. With a new value for θ, say θ′, the position of S and its radius r are updated on the new θ′ plane, and a new error e′ can be obtained from Equation 5.5. Further iteration by means of the secant method using this new pair of values (θ′, e′) can be conducted until error e is within the tolerance of some specified value. In the present implementation, a tolerance of 5% has been adopted. However, tighter control, say, 1%, can be applied if a stricter compliance of the specified size must be observed.

More rotations about new axes S_1S_3 or S_2S_3 are performed until no further descent is possible as shown in Figure 5.88. From the worked examples shown in Section 5.7.4, a sphere can rotate as many as nine times before finding its parking place. However, on average, a sphere only takes two rotations to find the insertion site. In the descending process, colliding with non-neighbouring spheres is possible and could happen. However, checking for this to happen would not be done during the process of descending by rotation for efficiency consideration. Instead, in the phase of Delaunay point insertion, all connections will be verified to make sure that edges are of size compatible to the node spacing function. Should there be any discrepancy detected, the descent by rotation has to be done again with additional care to avoid intersection.

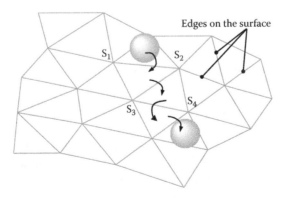

Figure 5.88 **Descent by rotation between planes.**

5.7.2.6 MG by Delaunay point insertion

Once the location for point insertion is determined, the MG part is quite straightforward as 3D DT by point insertion is now a fairly mature technique in terms of robustness and efficiency. For the sake of completeness, the essential steps of the point insertion algorithm with special emphasis on the situation of sphere packing will be briefly described.

For a new insertion point P at the centre of the proposed sphere, identify all the tetrahedra whose circumsphere contains the point. The cavity left behind upon the removal of these non-Delaunay tetrahedra forms the insertion polyhedron, which is referred to as the CORE in Section 5.2.2.1. The CORE can be determined by a reliable and efficient procedure starting with the tetrahedron (BASE), which contains the inserted point, and the other tetrahedra can be found by means of the adjacency relationship. In a way, the boundary of the CORE is given by the common face of two tetrahedra for which one is positive in the sphere inclusion test while the other fails. During rotation, the position of S can be easily monitored, and hence, the tetrahedral element containing the centre of sphere S, i.e. the BASE for point insertion in DT, can be readily available without further searching.

Owing to the finite precision arithmetic, the triangular facets on the boundary of the CORE have to be verified and corrected before they are connected to the insertion point to form tetrahedral elements. Inconsistency of the sphere inclusion test is to be supplemented and corrected by the visibility check or the positive volume test as discussed in Section 5.2.2.3. Poorly shaped tetrahedra or slivers can also be avoided at the same time if we insist that the volume of an element has to be greater than a certain threshold value. The newly generated tetrahedral elements inside the CORE along with the existing tetrahedral elements outside the CORE form a new DT containing the inserted point P. The adjacency relationship of the new tetrahedra inside the CORE is established, and that of the existing tetrahedra attached to the boundary of the CORE has to be updated.

5.7.2.7 Termination of the meshing process

Similar to the packing of circles or ellipses, there are two ways to control the termination of the MG process over an unbounded domain. One way is to set a number for the packing spheres, and the other is based on the size of the cluster or the diameter of the pack. If the number of spheres is more than a given number and/or the shortest distance to the origin is larger than a specified value, the process terminates. At this stage, the final frontal surface can serve as the mesh boundary. However, if the MG of a physical object is required, the object boundary can be introduced by well-established boundary recovery procedures as described in Section 5.3.

5.7.3 Efficiency and time complexity

There are two main computational costs for packing spheres. The first one is to determine the position of the insertion sphere, and the second one is to construct the DT containing the new sphere by point insertion at its centre. As the distance of the spheres from the origin is stored when they are created, searching for the nearest frontal sphere is only a simple comparison. The computational cost grows as the number of spheres on the front increases. However, the number of spheres on the front increases very slowly as spheres are covered when new spheres are added to the front. From the examples in Section 5.7.4, we can see that the number of spheres on the front is quite small compared to the number of spheres in the pack. The cost can be reduced by employing an advanced searching scheme, such as a grid-based procedure. As the descent by rotation is very robust, we can arrive at an insertion

Table 5.5 CPU time of sphere packing

Nodes	Elements	CPU time(s)
10,000	59,725	62.438
20,000	119,801	146.922
30,000	178,997	237.470
40,000	239,038	328.891
50,000	299,286	426.984

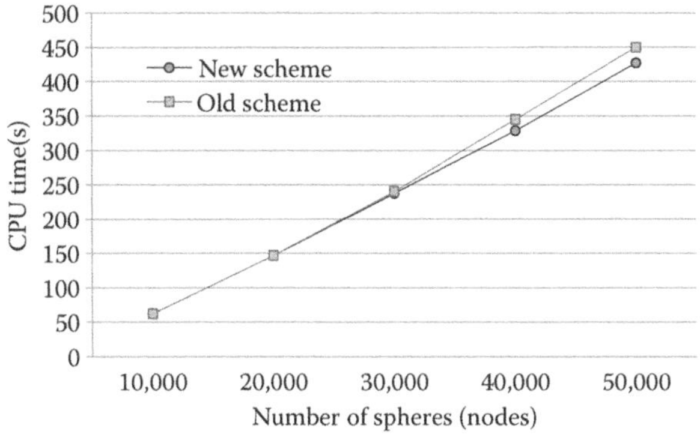

Figure 5.89 Graph of CPU time vs packing spheres.

site starting from a suitable close point on the front without a need for intensive searching for the absolute minimum.

The Delaunay point insertion without searching for the BASE is a process of linear time complexity. In the earlier implementation, we searched for the nearest sphere on the front, and a quasi-linear time complexity was observed for the worked examples. Combining the packing sphere and Delaunay point insertion, we would expect a linear time complexity for MG if the insertion site can be located based on descent by rotation without searching for the nearest sphere on the front. Table 5.5 shows the CPU times of a series of examples of sphere packing and MG. The meshes were generated on an old PC more than 10 years ago with CPU speed 1.7 GHz and 512-MB RAM running on VC++ 6.0 program. From Figure 5.89, the CPU time increased slowly with the number of spheres packed, and a quasi-linear relationship can be observed. It is remarked that computer codes for many parts in the early development had not been optimised, and the purpose of the plot is to show the CPU time trend. A new version had been worked out later, in which only a few adjacent tetrahedra on the front from the current insertion site were checked for the next insertion point. Compared with the initial scheme, for which all the surface tetrahedra are examined, the new scheme is a much improved version. As all the processes are now localised in the new packing scheme, a linear time complexity relationship is observed as shown in Figure 5.89.

5.7.4 Examples of sphere packing

Four examples of sphere packing and MG over a 3D unbounded domain have been worked out for discussion. The statistics of the examples are listed in Table 5.6. N and M are, respectively, the number of spheres (nodes) and the number of elements in the mesh. Max1 and Avg1 are, respectively, the maximum and the average number of rotations to locate the insertion site in the sphere packing process. Max2 and Avg2 are, respectively, the maximum and the

Table 5.6 CPU time, shape coefficients and other statistics of the examples

Examples	Random	Planes	Surfaces	Curve 1	Curve 2
N	30,000	35,000	30,000	25,000	25,000
M	178,451	208,075	175,691	150,046	150,877
Max1	9	8	6	6	7
Avg1	2	2	2	2	2
Max2	10.94	3.58	3.17	5.53	6.08
Avg2	2.72	1.24	1.22	1.27	1.28
Max3	8.99	1.47	1.31	1.60	1.78
Avg3	0.324	0.115	0.118	0.119	0.120
Max4	–	2.09	2.12	2.24	2.24
Avg4	–	1.04	1.04	1.06	1.06
Max5	–	2.48	2.25	2.61	2.88
Avg5	–	1.12	1.12	1.13	1.14
Min6	0.0034	0.00196	0.00793	0.00240	0.00671
Avg6	0.642	0.786	0.79	0.770	0.765
CPU time(s)	171.72	227.35	197.08	219.66	220.98

Note:

Max1: Maximum number of rotations

Avg1: Average number of rotations

Max2: $\max(f_i)$, $f_i = \max_j \left(f_{ij}, \dfrac{1}{f_{ij}} \right)$ $f_{ij} = \dfrac{r_i}{r_j} (i, j = 1, 2, \ldots, N)$

Avg2: $\dfrac{1}{N} \sum_{i=1}^{N} f_i$

Max3: $\max(g_i)$, $g_i = \max_j \dfrac{d_{ij} - (r_i + r_j)}{r_i + r_j} (i, j = 1, 2, \ldots, N)$

Avg3: $\dfrac{1}{N} \sum_{i=1}^{N} g_i$

Max4: $\max \left(\dfrac{r_i}{\rho_i}, \dfrac{\rho_i}{r_i} \right)$, $(i = 1, 2, \ldots, N)$

Avg4: $\dfrac{1}{N} \sum_{i=1}^{N} \left[\max \left(\dfrac{r_i}{\rho_i}, \dfrac{\rho_i}{r_i} \right) \right]$

Max5: $\max(h_i)$, $h_i = \max_j \dfrac{d_{ij}}{\sqrt{4 \rho_i \rho_j}} (i, j = 1, 2, \ldots, N)$

Avg5: $\dfrac{1}{N} \sum_{i=1}^{N} h_i$

Min6: $\min(\gamma_k)$, $\gamma_k = 72\sqrt{3} \dfrac{\text{volume}}{(\text{sum of squares of edges})^{3/2}} (k = 1, 2, \ldots, M)$

Avg6: $\left(\prod_{k=1}^{M} \gamma_k \right)^{1/M}$

where

    N is the number of spheres (nodes);
    M is the number of tetrahedral elements;
    i and j are neighbouring spheres;
    d_{ij} is the distance between spheres i and j;
    ρ_i is the specified radius of sphere i;
    r_i is the actual radius of sphere i; and
    r_j is the radius of the neighbouring spheres of i.

average of the ratio of radii between adjacent spheres. Max3 and Avg3 are, respectively, the maximum and the average gap space between adjacent spheres. With an average gap space of 12%, it can be seen that the spheres are tightly packed together. Max4 and Avg4 are, respectively, the maximum and the average ratio of the actual size of the spheres to the required size. For the three examples of specified node spacing, the mean deviation is approximately 5%. A tighter tolerance could have been applied in the sphere packing process to further reduce this deviation. Max5 and Avg5 are, respectively, the maximum and the average ratio of the edge length of the tetrahedral elements to the specified value. For the three examples where a node spacing function is specified, the average deviation is approximately 13%.

Finally, Min6 and Avg6 are, respectively, the minimum and the average γ-quality of the tetrahedral elements, as defined in Section 5.5.1. The Min6 values are quite low for all the examples as they are calculated from the raw mesh directly from sphere packing without mesh quality optimisation to remove sliver elements. In the DT process, the threshold for accepting a face on the CORE boundary to construct a tetrahedral element was set at $\gamma = 0.001$. This could probably be increased slightly to 0.01 to remove potential slivers. The mesh quality can be further improved by means of any existing optimisation techniques as described in Chapter 6. Nevertheless, the average γ-quality of the tetrahedral mesh is already quite high at about 0.78 even without post-generation enhancement. A more detailed breakdown of element γ-quality of the examples is shown in Table 5.7 and Figure 5.90. It can be seen that most tetrahedral elements are having a γ-value greater than 0.7.

Table 5.7 γ-quality of tetrahedral meshes of the examples

γ	Random	Planes	Surfaces	Curve 1	Curve 2
0.1	1181	479	348	371	378
0.2	3923	1360	1154	1043	1143
0.3	7774	2467	2019	2004	2100
0.4	12,099	3927	3123	3080	3423
0.5	17,208	6146	5078	5121	5474
0.6	23,717	10,090	8143	8850	9635
0.7	31,538	19,306	15,614	17,158	18,047
0.8	36,610	42,635	35,382	33,741	33,596
0.9	31,849	72,418	61,844	47,256	46,433
1.0	12,552	49,247	42,986	31,422	30,648

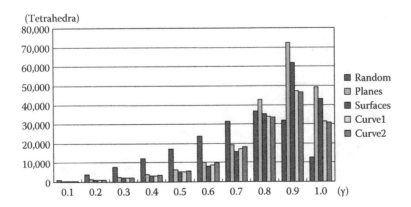

Figure 5.90 Histogram of γ-quality of the tetrahedral mesh.

Example 1 is about a rather difficult case of generating a finite element mesh of random element size with a range of 1 to 11, that is, the difference in size between neighbouring spheres can be as large as 11. A size ratio larger than 11 could have been tested as well. However, in three dimensions, generation of spheres of random sizes is not a well-defined problem, because small spheres can penetrate through the cluster of existing spheres between gaps to any possible locations. For the packing of circles over a 2D unbounded domain, circles of random size in a range of 1 to 10,000 have been packed (Lo and Wang 2003). The packing of the first 30,000 spheres of random size is shown in Figure 5.91a, and the corresponding mesh of 178,451 tetrahedral elements is shown in Figure 5.91b. In this example of packing spheres of random size, the average size ratio between adjacent spheres is 2.7, and the gap spacing is approximately 32%. Nevertheless, the average element γ-quality is quite acceptable at $\gamma = 0.64$. Figure 5.91c shows a cross section of the spheres packed, and a layered structure is revealed when spheres at different distance from the origin are displayed with different colours. A cut open section of the tetrahedral mesh is shown in Figure 5.91d.

Spheres over three intersecting planes are packed densely in Example 2. As shown in Figure 5.92a, characteristic lines of small sphere concentration are well recognised on the surface of the cluster, and the spheres grow into larger size when they are away from the intersecting planes. A transverse cross section of the corresponding tetrahedral mesh is depicted in Figure 5.92b. It can be seen that elements of smaller size are generated close to the intersecting planes. In spite of a strong variation in element size, a high average shape quality of the mesh with $\gamma = 0.79$ was attained without mesh optimisation.

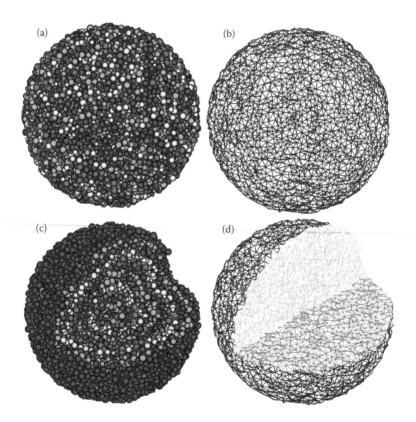

Figure 5.91 (a) Packing of spheres of random size, (b) tetrahedral elements of random size, (c) a cut section of the packed spheres and (d) open cut of the tetrahedral mesh.

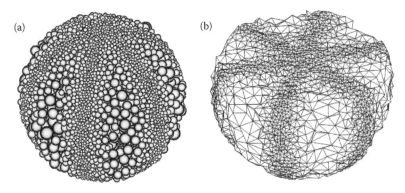

Figure 5.92 (a) Packing of spheres on intersecting planes and (b) section of tetrahedral mesh.

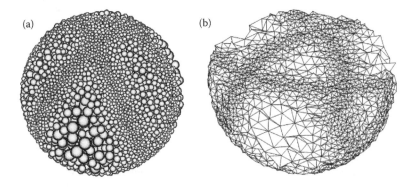

Figure 5.93 (a) Packing of spheres on intersecting surfaces and (b) section of tetrahedral mesh.

Example 3 is to pack spheres densely over three intersecting curved surfaces in space. This example is similar to the last example except that the geometry required for dense packing is different. As shown in Figure 5.93a, characteristic lines of small spheres can be identified on the surface of the cluster. A transverse cross section of the corresponding tetrahedral mesh reveals that small elements are generated close to the curved surface, as shown in Figure 5.93b. The average γ-quality of this example is also $\gamma = 0.79$.

The last example is to pack spheres along an arbitrary curve in space. As shown in Figure 5.94a, two lines of spheres of very small size can be distinguished on the surface of the cluster. The associated tetrahedral mesh is shown in Figure 5.94b. In spite of a large change in element size moving away from the space curve, a fairly high mean γ-quality of 0.77 can be achieved. A cut section by removing one quarter of the elements to reveal the internal structure is depicted in Figure 5.94c. It can be seen that there are regions of element concentration along the path of the spatial curve. If the elements within a short distance from the curve are retained, elements of consistent small size are generated all along the curve as shown in Figure 5.94e and f. Similar to the 2D case, it appears that the origin for sphere packing can be set at any convenient point in space without much impact to the final results. Figure 5.94c shows the mesh of packing spheres densely over the same curve in space from an origin placed at the centre of the mesh (curve 1), whereas the mesh shown in Figure 5.94d is generated from an origin outside the mesh (curve 2).

Remarks: The number of tetrahedral elements or the size of the cluster that could be generated depends solely on the capacity of the computer and the time allowed for MG. The

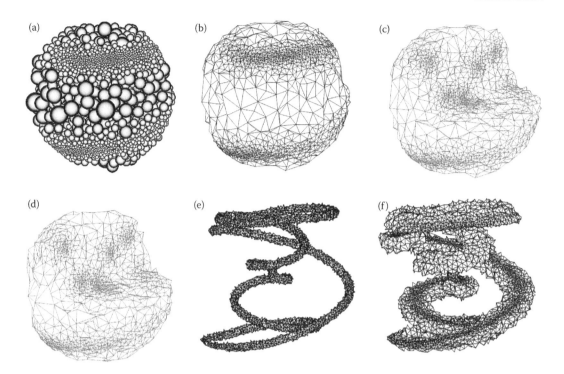

Figure 5.94 (a) Packing of spheres along a spatial curve, (b) tetrahedral mesh of (a), (c) tetrahedral mesh: origin for sphere packing at the centre of the curve, (d) tetrahedral mesh: origin for sphere packing away from the centre of the curve, (e) small elements are generated along the curve as required and (f) element size variation over a cut section along the curve.

sphere packing algorithm is of linear time complexity if there is no need to search for the insertion site and the BASE tetrahedron for Delaunay point insertion. While the BASE tetrahedron is known once we can identify the insertion site, the search for the insertion site (point nearest to the origin) could be speeded up by limiting the search to a close neighbourhood of the last insertion site. What we have to do is to search from the current insertion site within an area of a fixed number of frontal triangles, which is readily available from the adjacency relationship of the DT. From the examples tested, a search over a radius of 5 to 10 layers of triangles, there are little differences in the final resulting mesh compared to a thorough search for the nearest point over the frontal surface. In the event that an anisotropic mesh is needed, the problem of fitting spheres of varying size has to be changed to the problem of fitting ellipsoids of different sizes along the three principal axes in arbitrary orientations. This is a much more difficult problem, and the descent by rotation has to be rigorously revised and modified substantially before it could possibly be applied in this case.

5.8 GENERATION OF HEXAHEDRAL MESH

5.8.1 Introduction

Many physical phenomena and objects are 3D in nature, and a rigorous analysis and meaningful simulation could only be done in terms of a solid mesh of tetrahedral and/

or hex elements. Along with the FE technology, there are significant advances in the generation of tetrahedral and hex meshes (Cook 1974; Benzley et al. 1995; Schneiders and Bunten 1995; Melander et al. 1997; Pavlakos et al. 1997; Tautges et al. 1997). Hex elements are superior to tetrahedral elements in the following aspects: (i) high-quality structured regular hex elements can be generated rapidly by means of mapping or transformation; (ii) hex elements could be generated following the domain boundary in layers and in alignment with important geometrical features for easy visualisation and inspection, whereas the validity of tetrahedral meshes can only be verified by a computer; and (iii) low-order high-performance hybrid-stress hex elements are available (Sze 1992; Sze and Ghali 1993a,b; Lee and Lo 1997a,b; Sze and Lo 1999; Sze et al. 2002; Ramos and Simoes 2006), which are considered to be more computationally efficient than the tetrahedral element counterparts.

Hex meshes of various characteristics can be generated by different approaches as proposed by researchers working in diverse scientific and engineering fields, and a comprehensive account of the current status and difficulties of hex meshing can be found in the works of Shepherd and Johnson (2008) and Staten et al. (2010a,b). Tautges (2001b), on the other hand, not only reviewed the hex meshing strategies but also laid down some evaluation criteria for hex meshing. Automatic unstructured MG algorithms usually refer to the generation of tetrahedral meshes, as mapping techniques based on regular grid will, in general, give rise to structured meshes. While most of the literature and software on unstructured meshes are about triangulation methods, there have been continuous research efforts on structured and unstructured hex MG (Li and Cheng 1998; Mitchell 1998; Muller-Hannemann 1998; Kraft 1999; Shepherd et al. 1999; Staten et al. 1999; Tautges 1999; Trease and Barrett 1999; Wada et al. 1999; White and Tautges 1999; Dhondt 2001; Lu et al. 2001; Baker 2005; Zhao et al. 2008; Ruiz-Girones et al. 2009, 2012; Ran et al. 2012). Advanced meshing algorithms are available for the generation of isotropic and anisotropic tetrahedral meshes over complex 3D industrial and biomechanical objects; a versatile mesh generator capable of producing high-quality unstructured hex elements is yet to be developed. Compared to tetrahedral meshing, these meshing techniques are still at the development stage; further improvement and refinement are expected, and verifications are required before any conclusion can be drawn on the merits and drawbacks of the various schemes put forward.

The situation is compounded by the fact that answers to many fundamental questions are still outstanding. For instance, what are the generally acceptable boundary conditions for a hex meshing problem? Given a closed surface meshed in quads, what can you say about the possibility of having an all-hex mesh? What would hex meshes of variable element sizes as specified by a node spacing function look like? Anyway, hex meshing will remain as one of the most interesting topics in the world of MG, and answers to some of these questions along with better meshing algorithms will emerge as time goes by. Unlike triangular meshes and tetrahedral meshes, extension from a 2D quad mesh to a 3D hex mesh is by no means straightforward, and very often, a completely novel approach has to be adopted. Owing to the methods employed and since the boundary condition requirements are quite different, it is necessary to categorise hex meshing into two broad types: (I) conforming mesh bounded by smooth surfaces and (II) constrained mesh bounded by surfaces discretised into quads. Most hex meshing problems and the methods developed today only address problem type I. Similar to the quad MG, there are both direct and indirect methods for the construction of structured and unstructured hex meshes. Very often, several methods can be applied in a combined mode or one after the other to generate hex meshes of various characteristics for objects of difficult geometry and topology. In summary, hex meshing is still a challenging task, and no single method appears to dominate over all situations.

5.8.2 Direct methods

Direct methods refer to those techniques that are applied directly to the problem domain to generate hex or hex-dominated meshes without the need of a background grid. Most of the direct methods are quite straightforward, which are usually applied to objects with relatively simple boundary and topology to generate structured meshes. Mapping (Li and Cheng 2000), transformation (Su et al. 2005), decomposition (Liu and Rajit 1997), subdivision (Bajaj et al. 2002; Stadler and Holzapfel 2004; Marechal 2009), drag method (Park and Washam 1979), extrusion (Chalasani and Thompson 2004) and sweeping (Lai et al. 2000; White and Tautges 2000; Roca and Sarrate 2010b) are among the classical techniques in generating well-shaped structured hex meshes rapidly over simple regular domains. Rypl (2010) extended the sweeping process to construct hex meshes over volumes bounded by arbitrary curved surfaces. As the topological structure of the mesh will not be affected by a change of geometry, these methods can operate based solely on the mapping of the nodal points from a reference domain to a physical domain.

The grid-based approach proposed by Schneiders (1996) and Camacho et al. (1997) involves the fitting of a 3D grid of hex elements into the interior of a volume. Decomposition method coupled with the basic techniques has been developed to subdivide complicated objects into simpler blocks, possibly convex in geometry, so that they could be easily meshed into hex elements (Joe 1994; White et al. 2004). Object decomposition can also be based on Voronoi graph (Sheffer et al. 1999) or medial axis/surface construction (Price and Armstrong 1997; Xia and Tucker 2009), which could in turn be defined with the aid of a DT (Benabbou et al. 2009). The whisker weaving method (Murdoch and Benzley 1995) is based on the concept of a spatial twisting continuum (STC), which provides hints as where hex element could be formed. Unconstrained plastering hex MG by means of ADF approach on 3D decomposition primitives was proposed by Staten et al. (2010a). Hex meshes of variable element size to fit a generally curved boundary surface were generated by means of the Octree technique with a set of refinement templates (Ito et al. 2009). Staten et al. (2010b) proposed an algorithm to match and modify non-conforming quad interfaces so as to link up individual hex meshes. Hex meshes were generated from MR imaging data onto a multi-block grid to fit required curved surfaces (Ji et al. 2011). With the introduction of transition elements, quad and hex meshes of variable element sizes were generated by recursive element subdivision for adaptive refinement analysis (Wu et al. 2009, 2010; Lo et al. 2010).

5.8.3 Indirect methods

Indirect methods are those techniques applied to the problem domain that is already discretised into a valid FE mesh to generate structured, unstructured or hex-dominated meshes. The background mesh can be the initial mesh, from which hex elements are generated through modification and enhancement of the existing elements, and it can also serve as the control space with or without a possible natural partition into simpler regions over which hex elements are directly generated by some well-controlled localised procedures. A simple way to convert a tetrahedral mesh into a hex mesh is to divide each tetrahedron into four hex elements. However, the elements generated by this method are of very poor quality and not suitable for FE analysis. The boundary integrity is also violated as nodes are introduced at the centre of each triangular face. Another possibility of converting tetrahedra into hex elements is to combine several tetrahedra to form a hex. As at least five tetrahedra are needed to form a hexahedron, a method that combines tetrahedra to form hexahedra would therefore need to look for a combination of five or more tetrahedra to form a single hex. However, unless nodes are placed with some regularity, it is rather difficult to retrieve hex from a general tetrahedral mesh. This idea to date has not yet been proved to be a viable option for hex meshing.

Since most methods for all-hex meshing seem to be less versatile, some researchers have advocated the generation of a mixed mesh of hex and tetrahedral elements along with any other element types such as pyramids and wedges to serve as liaison elements as necessary. One approach introduced by Owen et al. (1997) is to manually subdivide the volume into regions that could be meshed into hex through a mapping process or meshed into tetrahedral elements by a standard triangulation procedure. Pyramid elements can be used to link up hex and tetrahedral elements at the interfaces. Tuchinsky and Clark (1997) have presented an algorithm that combines plastering and 3D triangulation. Applying the plastering method, hex elements are generated as far as possible into the volume similar to the coring method in 2D described in Sections 3.3.2 and 3.8. The remaining voids within the volume are meshed into tetrahedral elements by a constrained triangulation procedure.

Other sophisticated procedures for hex-dominated meshing include the use of graph theory to represent possible polyhedron decompositions into tetrahedra, based on which Meshkat and Talmor (2000) proposed a method to generate a mixed mesh of hex, pentahedra and tetrahedra from an underlying tetrahedral mesh. On the other hand, Min (1997) presented a hex-dominated meshing technique by making offsets from the boundary towards the interior of the domain to form layers of hex. After shells of hex elements are peeled off as far as possible, the shrunken volume is then triangulated. Employing a grid to extract the basic geometrical features of the existing mesh, Fernandes and Martins (2007) presented an all-hex remeshing procedure for the FE analysis in metal forming. Hex-dominated meshes were generated over irregular geometries by means of packing rectangular solid cells (Yamakawa and Shimada 2003). The H-morph approach proposed by Owen (1999, 2001) and Owen and Saigal (2000) is an indirect method operating on a triangulated volume in such a way that hex elements are created while tetrahedra are removed and modified. Finally, high-quality hex-dominated meshes can also be conveniently generated in a systematic manner by merging hex meshes first converted into tetrahedral elements (Lo 2012c).

5.8.4 Subdivision, mapping and transformation

Regular hex element mesh can be easily generated from a cube by dividing each edge along three principal directions into a number of segments as shown in Figure 5.95a. In the first

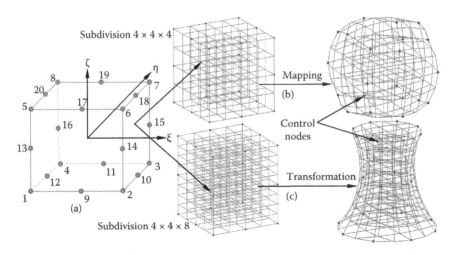

Figure 5.95 Subdividing a cube and mapping it into different shapes: (a) H20 hexahedral element; (b) sphere; (c) cooling tower.

subdivision, the cube is divided into $4 \times 4 \times 4 = 64$ hex elements, which are mapped onto the surface of a sphere by the FE interpolation of the H20 isoparametric element as shown in Figure 5.95b. In the second subdivision, the cube is divided into $4 \times 4 \times 8 = 128$ hex elements, which are mapped onto a shape similar to a cooling tower by the FE interpolation as shown in Figure 5.95c. Transformation usually refers to a change in geometry due to deformation from one shape to another, and mapping is a change of co-ordinates based on some functional relationship from one set of co-ordinates to another. However, in MG, effectively, mapping and transformation are more or less performing the same task of a continuous change from one geometrical shape to another.

Many functions or CAD-based systems can be used to define new shapes (Cuillière et al. 2009), such as the transfinite mapping as described in Section 3.2.2. The simplest method, however, is based on the FE interpolation for the transformation of a cube into arbitrary curved objects. The 20-node hex element H20 is shown in Figure 5.95a, whose interpolation functions in reference element co-ordinates (ξ, η, ζ) are given as follows:

$$H_1 = (1-\xi)(1-\eta)(1-\zeta)(-\xi-\eta-\zeta-2)/8, \quad H_2 = (1+\xi)(1-\eta)(1-\zeta)(+\xi-\eta-\zeta-2)/8,$$
$$H_3 = (1+\xi)(1+\eta)(1-\zeta)(+\xi+\eta-\zeta-2)/8, \quad H_4 = (1-\xi)(1+\eta)(1-\zeta)(-\xi+\eta-\zeta-2)/8,$$
$$H_5 = (1-\xi)(1-\eta)(1+\zeta)(-\xi-\eta+\zeta-2)/8, \quad H_6 = (1+\xi)(1-\eta)(1+\zeta)(+\xi-\eta+\zeta-2)/8,$$
$$H_7 = (1+\xi)(1+\eta)(1+\zeta)(+\xi+\eta+\zeta-2)/8, \quad H_8 = (1-\xi)(1+\eta)(1+\zeta)(-\xi+\eta+\zeta-2)/8,$$
$$H_9 = (1+\xi^2)(1-\eta)(1-\zeta)/4, \quad H_{11} = (1-\xi^2)(1+\eta)(1-\zeta)/4,$$
$$H_{17} = (1-\xi^2)(1-\eta)(1+\zeta)/4, \quad H_{19} = (1-\xi^2)(1+\eta)(1+\zeta)/4,$$
$$H_{10} = (1-\eta^2)(1-\zeta)(1+\xi)/4, \quad H_{12} = (1-\eta^2)(1-\zeta)(1-\xi)/4,$$
$$H_{18} = (1-\eta^2)(1+\zeta)(1+\xi)/4, \quad H_{20} = (1-\eta^2)(1+\zeta)(1-\xi)/4,$$
$$H_{13} = (1-\zeta^2)(1-\xi)(1-\eta)/4, \quad H_{14} = (1-\zeta^2)(1+\xi)(1-\eta)/4,$$
$$H_{15} = (1-\zeta^2)(1+\xi)(1+\eta)/4, \quad H_{16} = (1-\zeta^2)(1-\xi)(1+\eta)/4.$$

The co-ordinate transformation of a standard $2 \times 2 \times 2$ cube to a physical object is given by

$$(\xi, \eta, \zeta) \mapsto (x, y, z) : \mathbf{x} = \sum_{i=1}^{20} H_i(\xi, \eta, \zeta)\mathbf{x}_i$$

where $\mathbf{x}_i = (x_i, y_i, z_i)$ are the control points on the physical object.

5.8.5 Block decomposition

Non-convex objects or objects not able to be mapped directly onto a standard cube can be first decomposed into convex hex blocks, which are then meshed individually by the mapping method as shown in Figure 5.96. More sophisticated decomposition strategies and procedures were also proposed based on the geometrical characteristics of the object. As an attempt for the automatic decomposition of industrial objects, geometrical identification algorithms are devised such as edge and face detections to define special features of an object, which can be detached from the object as a separate block (Ruiz-Girones and Sarrate 2010; Shivanna et al. 2010; Tam and Armstrong 1993). Based on an underlying triangulation, discretised dual surfaces are constructed and intersected to define zones for block decomposition (Roca and Sarrate 2010a).

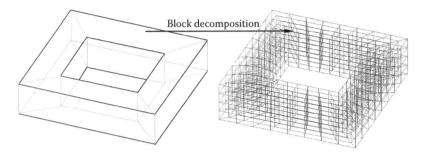

Figure 5.96 Meshing by block decomposition and mapping.

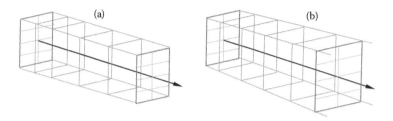

Figure 5.97 Section of quadrilaterals dragged along an axis: (a) uniform sections; (b) variable sections.

5.8.6 Drag method and extrusion

By the drag method, planar quad mesh is dragged along a straight line normal to or sometimes making an oblique angle with the planar mesh to produce hex elements as shown in Figure 5.97. In Figure 5.97a, elements of different size can be generated by dragging the section with a different pitch, and on the other hand, elements of different cross sections can be produced by dragging the section along non-parallel edges as shown in Figure 5.97b. Extrusion shares the same idea with the drag method (Vassberg 1999), except that it is more general in the way it operates; instead of following a straight line drag path, the section is dragged along a curve, which can be defined analytically or by a list of control points as shown in Figure 5.98.

5.8.7 Meshing by revolution

Sweeping or meshing by revolution is to generate a 3D mesh by rotating a quad mesh about a rotation axis as shown in Figure 5.99. What we have to do to generate a hex mesh by rotating a planar quad mesh is to produce a new set of nodes based on those of the quad

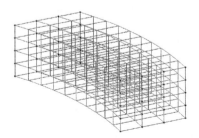

Figure 5.98 Extruding a face of 12 quadrilaterals along a curve.

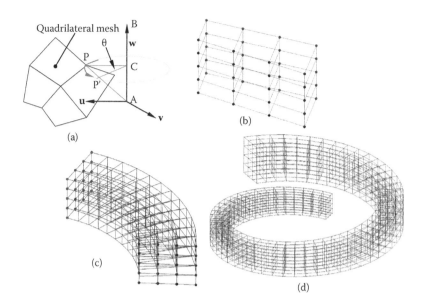

Figure 5.99 Generation of hexahedral mesh by rotation: (a) rotate about axis AB; (b) hexahedral elements generated by linking up two layers of nodes; (c) 12 layers of hexahedra are generated by rotation; (d) sweeping sections of increasing size.

mesh. This can be easily achieved by rotating each node on the mesh one at a time. Before we apply rotation to the nodes of the quad mesh, we have to set up an orthonormal basis $(\mathbf{u}, \mathbf{v}, \mathbf{w})$ using the given rotational axis AB. Take for instance the rotation of point P of the quad mesh about axis AB by an angle θ is given by the following as shown in Figure 5.99a:

$$\mathbf{w} = \frac{AB}{\|AB\|}, \mathbf{u} = \text{unit vector normal to } \mathbf{w} \text{ on the plane of the quad mesh,}$$

$$\mathbf{v} = \mathbf{w} \times \mathbf{u}$$

$$CP = AP - AC = AP - s\mathbf{w}, \text{ where } s = AP \cdot \mathbf{w}$$

$$CP' = r\cos(\theta)\mathbf{u} + r\sin(\theta)\mathbf{v}, \text{ where } r = \|CP\|$$

$$P' = A + AC + CP' = A + r\cos(\theta)\mathbf{u} + r\sin(\theta)\mathbf{v} + s\mathbf{w}$$

For each quad, a hex element can be created by properly connecting the nodes of the quad to the corresponding nodes produced by rotating the quad as shown in Figure 5.99b, and 12 hex elements are generated from 12 quads in this example. More hex elements can be generated by rotating the same quad mesh a number of times to produce several layers of nodes, which are connected properly between layers to form hex elements as shown in Figure 5.99c. If in each rotation, the distance along the rotational axis is adjusted to avoid overlapping on completing a revolution, a spiral of many hex elements can be generated, as shown in Figure 5.99d. The sweeping process can be generalised to sweep a cross section of quad elements or between two different planar/non-planar cross sections along curved sweep paths defined analytically or by a list of control points (Roca and Sarrate 2010b; Rypl 2010). Another extension of the method is to define objects by sweeping cross sections from multiple sources at different position and orientation (Scott et al. 2006).

5.8.8 Grid-based or voxel-based method

The grid-based or voxel-based approach presented by Schneiders (1997), Lee and Yang (2000), Kim and Swan (2003a,b), Kaminsky et al. (2005), Teran et al. (2005) and Owen and Shepherd (2009) involves the fitting of a 3D grid of hex elements into the interior of a volume, as shown in Figure 5.100. More hex or tetrahedral elements have to be added to fill the gaps between the regular grid and the boundary surface of the solid. Very much limited by the geometry of the solid, poor elements of irregular shapes are almost inevitable in the boundary fitting process. As shown in Figure 5.100a and b, the hex elements retained within the same ellipsoid are quite different if the major principal axis of the ellipsoid is pointing at a different direction. Hex elements are, in general, not in good alignment with the domain boundary, and the resulting mesh is rather sensitive to the orientation and the positioning of the interior grid, which can, however, be improved by the *marching cube* and other smoothing techniques (Samani et al. 2001; Boyd and Muller 2006; Labelle and Shewchuk 2006; Zhang et al. 2006, 2007; Young et al. 2008). Owing to the use of a regular grid, the element size at the interior of the volume will be approximately the same. Frey et al. (1994), Weiler and Schneiders (1996), Greaves and Borthwick (1999), Schneiders (1997) and Zhang et al. (2005) have made modifications that allow for significant change in the element size based on an Octree decomposition, which can also be combined with DT (Schroeder and Shephard 1990; Jung and Lee 1993) or with AFT to mesh regions with cracks (Neto et al. 2001). While the grid-based or voxel-based method is less convenient for domains with prescribed discretised boundary, it is pretty effective to mesh biomedical volumes bounded by smooth surfaces or scattered data points from MRI images.

5.8.9 Medial surface method

The *medial axis* of an object is the set of all points having more than one closest point on the object's boundary (Gursoy and Patrikalakis 1992; Sherbrooke et al. 1996a,b). In 2D, the medial axis of a planar object is given by the locus of the largest circle rolling along the boundary of the object, as shown in Figure 5.101. As a direct extension of the medial axis method for quad meshing, the domain is subdivided by a set of medial surfaces, which can be thought of as surfaces generated from the mid-point of a maximal sphere allowed to roll through the volume. The medial surface can also be constructed with the aid of a DT of the boundary points of the solid object (Sheehy et al. 1996; Turkiyyah et al. 1997). The medial

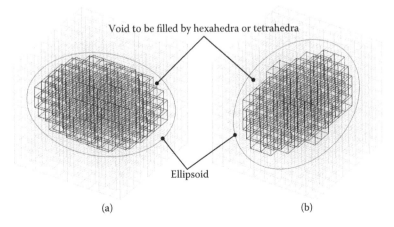

Figure 5.100 Fitting hexahedral element within a bounding surface: (a) major principal axis normal to a face; (b) major principal axis along a diagonal.

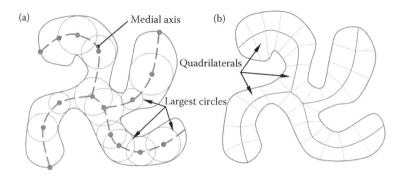

Figure 5.101 Mesh generation by means of medial axis construction: (a) medial axis of an object; (b) object divided using medial axis as a guide.

surface method (Price and Armstrong 1995, 1997) involves an initial decomposition of the volume into sub-regions.

The decomposition of the volume by medial surfaces will define regions to be meshed by means of a mapping procedure. A set of templates for the expected topology of the regions formed by the medial surfaces are employed to fill the volume with hex mesh. Linear and integer programming is used to ensure that element divisions match from one region to another (Li et al. 1995). This method, while proving useful for some geometry, has been less than reliable for meshing general 3D objects with irregular non-smooth boundaries. Robustness issues in generating medial surfaces, as well as in providing, for all cases, regions that are readily defined for simple hexahedral meshing, appear to be rather difficult issues.

5.8.10 Plastering method

The paving method introduced by Blacker and Stephenson (1991) and Blacker et al. (1991) offers a simple way to form complete rows of quads, starting from the boundary and working inwards, as shown in Figure 5.102a (Thompson and Soni 1999). Front closure remains a major issue when the offset contour intersects itself, and very often, poor quads have to be used to fill up the interior just for building up a quad mesh of correct topological structure, as shown in Figure 5.102b. White and Kinney (1997) proposed an improvement to the paving procedure, suggesting individual placement of quads rather than a complete row. Plastering is an attempt to extend the paving method in 2D to generate hex meshes in 3D (Canann 1991; Johnston and Sullivan 1993; Cass et al. 1996; Meyers et al. 1998). Within a

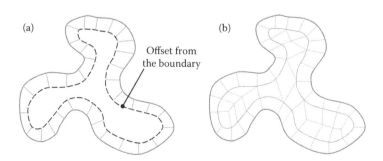

Figure 5.102 Quadrilateral mesh generated by the paving method: (a) layer of quadrilaterals generated along the boundary; (b) irregular quadrilaterals are found at the interior of the domain.

bounded volume, Vyas and Shimada (2009) employed a metric tensor field to generate hex meshes. By this method, elements are formed from the boundary quad facets, and the construction advances towards the volume interior. A set of heuristic procedures are devised to determine the order in which elements should be created. Similar to the AFT for tetrahedral meshing, the generation front for plastering is composed of quads. Hex elements are defined by projecting quads on the generation front towards the interior of the volume. Of a more complicated nature compared to the classical AFT, intersecting faces have to be checked and carefully controlled, and the problems of when and how to connect existing nodes and to seam faces have to be resolved. As the front advances, complex irregular internal voids may occur, which in some cases cannot be filled with hex elements. Existing elements already placed have to be removed or modified from time to time to cater for the formation of new elements. The plastering algorithm has not yet been proved to be reliable and versatile enough for general applications.

5.8.11 Whisker weaving method

The whisker weaving method (Murdoch and Benzley 1995; Murdoch et al. 1997), which is based on the concept of the spatial twisting continuum (STC), is an attempt to mesh objects with a prescribed boundary of quad facets (hex problem type II). Tautges et al. (1996) described the STC as a dual of the hex mesh represented by intersecting surfaces, which bisect hex elements in each direction. The principle behind the whisker weaving method is to first construct the STC from the boundary quad faces. By means of the STC, hex elements can be defined within the volume using the STC as a guide. The intersection of the twisting planes with the volume will form a closed loop on the surface, which can be deduced from the boundary quads. The objective of the whisker weaving algorithm is to determine where the intersection of the twisting planes will occur within the volume, as shown in Figure 5.103. This is entirely based on the topological consideration, and there are no actual geometrical calculations involved (Folwell and Mitchell 1999; Calvo and Idelsohn 2000; Kawamura et al. 2008). Once a valid topological representation of the twisting plane model has been established, hex elements are formed inside the volume at places where three twist planes meet. Nodes can then be created within the volume afterwards to complete the MG. Indeed, it is a method based on the STC loops on the boundary surface to deduce the topological structure of the corresponding hex mesh. Given a hex mesh, it is easy to find its boundary surface of quads; however, the inverse problem is less obvious, and it may not be always possible to find out the internal structure of a mesh based merely on its boundary surface. The whisker weaving algorithm faces problems of unresolved topological situations and the formation of degenerated hex elements of zero volume. However, the method could help in establishing the topological structure of a non-conforming hex mesh (Shepherd 2009).

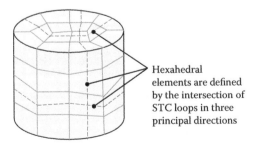

Hexahedral elements are defined by the intersection of STC loops in three principal directions

Figure 5.103 **Whisker weaving method.**

5.8.12 H-morph approach

The H-morph approach proposed by Leland et al. (1998), Owen (1999) and Owen and Saigal (1999) is an indirect method operating on a triangulated volume in such a way that hex are created while the tetrahedra are removed and modified. Tetrahedra are transformed and combined starting from the boundary surface of quads derived from the triangulation and working inwards towards the interior of the volume. H-morph, which can be considered as a 3D version of the Q-morph method (Owen et al. 1998, 1999), is essentially a re-meshing process generating hex elements by techniques common to the plastering method. However, unlike the Q-morph procedure, which produces all-quad meshes, H-morph does not always guarantee an all-hex mesh for arbitrary geometry. It will give meshes ranging from hex-dominated to all-hex meshes, depending on the complexity of the domain and the quality of the hex elements required. Starting from a tetrahedral mesh of the volume, the algorithm systematically transforms tetrahedra into hex. At any instant of the transformation, a valid mixed mesh of hex and tetrahedral elements is maintained. At some stage when no reasonably shaped hex could be formed within the volume, the procedure stops resulting in a hex-dominated mesh. Employing the same concept of ADF, the H-morph algorithm operates more or less the same way as the plastering method. Similar to plastering, H-morph works on an initial front consisting of quads on the surface and systematically projects hex elements towards the interior in an attempt to fill the volume completely with hex. Operating within the meshed space, the H-morph method alleviates to a large extent the error-prone process of checking intersections and front closure inherent in plastering. Both stability and flexibility are much enhanced in the H-morph procedure as a valid FE mesh is maintained throughout the generation process. A hex-dominated mesh of an object generated by H-morph adapted from Owen and Saigal (2000) and Owen (2001) is shown in Figure 5.104 in which hex elements are aligned well along the domain boundary.

5.8.13 Generation of transition elements

In adaptive refinement analysis, hexahedral elements of variable size are required so that small elements can be placed at regions of relatively large discretisation error. However, elements will inevitably be severely distorted changing from one size to another, and quadrilateral and hexahedral elements are rather sensitive to shape distortion; moreover, a large numerical error will be introduced due to a deviation from the regular rectangular shape. In order to preserve the regular shape of the quadrilateral and the hexahedral elements while a change in element size is allowed, the idea of transition elements was introduced, and the corresponding efficient transition finite elements were developed for the adaptive refinement

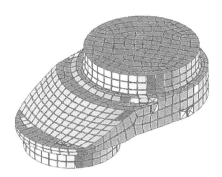

Figure 5.104 Hex-dominant mesh by H-Morph.

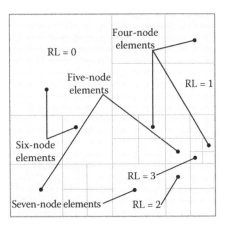

Figure 5.105 **Transition quadrilateral elements.**

analysis (Lo et al. 2010). In 2D, quadrilaterals of four to seven nodes have to be used so that large elements can be connected to smaller elements without a change in shape, as shown in Figure 5.105. The one-level restriction (1-LR) rule between adjacent elements will be enforced to ensure a gradual change in element size, and the types of transition quadrilaterals will be limited to only four, namely, four-, five-, six- and seven-node quadrilaterals.

5.8.14 Generation of transition quadrilateral mesh

In the 1-LR mesh for a 2D adaptive refinement, each edge of a 2D transition element can interface with at most two adjacent elements, as shown in Figure 5.105. Neighbouring elements refer to elements that share at least two common nodes. Hence, elements sharing a common edge or part of a common edge are neighbouring elements. Furthermore, the refinement level (RL) of an element is the number of refinements done to attain the current configuration or size of the element. Thus, in the initial mesh in which no subdivision has been performed, RL = 0 for all the elements. As the refinement process goes on, when an element is indicated to be refined, it is necessary to check the RLs of its neighbouring elements. If their values are equal to or greater than the element's own RL, the element can be subdivided into smaller elements, and their RLs are increased by one. Otherwise, the element cannot be subdivided until its neighbours are subdivided first. This RL check can be done locally and is easily implemented in code by a recursive sub-routine.

A regular element, which refers to a four-node quadrilateral element, can only be divided into four smaller regular elements by adding four mid-side nodes and a centroidal node, as shown in Figure 5.105. Elements of different RLs are connected by transition elements, which are quadrilateral elements of five, six and seven nodes with a different number of mid-side nodes, as shown in Figure 5.105. The following is an example of mesh refinement with transition quadrilaterals following the 1-LR refinement rule. Mesh i in Figure 5.106 shows the initial mesh of only four-node quadrilaterals. All their RLs are set to zeros. Now we try to subdivide element 1. Before doing so, the RLs of its neighbouring elements, i.e. elements 2, 3 and 4, are checked to ensure that the 1-LR mesh refinement rule is satisfied. By dividing element 1, four new quadrilateral elements are generated, and their RLs are equal to one, as shown in mesh ii of Figure 5.106. Elements 2 and 4 become transition elements of five nodes, and yet their RL remains to be zero. In the next stage of refinement, element 5 is supposed to be refined. The RLs of its neighbouring elements, i.e. elements 1, 2 and 6, are

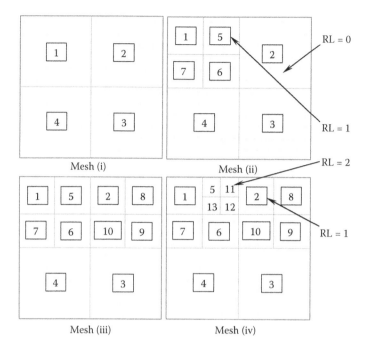

Figure 5.106 Mesh refinement following the 1-LR rule.

checked. It is found that the RL (=0) of element 2 is lower than that (RL = 1) of element 5. As a result, subdivision of element 5 has to be suspended, and element 2 ought to be divided first. After checking the RLs of the neighbours of element 2, the restriction criterion is satisfied for element 2, and it is divided as shown in mesh iii of Figure 5.106. RLs of the four smaller quadrilateral elements 2, 8, 9 and 10 are set to 1, which is the result of increasing the RL of the previous element 2 in mesh ii. The subdivision procedure for element 5 can now be reactivated to generate elements 5, 11, 12 and 13 with RL = 2, as shown in mesh iv of Figure 5.106. In summary, before dividing an element, we have to make sure that the 1-LR mesh refinement criterion is satisfied by checking the RL values of its neighbouring elements, which can be done locally without much effort.

Although the 1-LR refinement scheme based merely on a check on the refined levels of neighbouring elements is simple enough, it is quite efficient even for the most general node

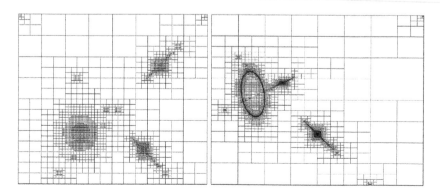

Figure 5.107 Refinement of quadrilateral meshes based on different node-spacing functions.

spacing functions in the finite element adaptive analyses. Two quadrilateral meshes refined based on rather different element size specifications are shown in Figure 5.107. There are many levels of refinement between the largest elements and the smallest elements, which are linked up by transition elements of various node patterns.

5.8.15 Generation of transition hexahedral mesh

The refinement strategy for 2D meshes can be extended into 3D hexahedral meshes in exactly the same way that transition elements are used between elements of different RLs. In 3D, the regular element refers to an eight-node hexahedral element, and the 1-LR mesh allows a transition element to have at most two elements adjacent to anyone of its edges and four elements adjacent to one face, as shown in Figure 5.108. The concept of RL in 2D is also applicable to the refinement of hexahedral meshes, and the check for 1-LR rule follows a similar procedure. Before an element is divided, we have to make sure that the 1-LR mesh refinement criterion is satisfied by checking the RL values of all its neighbouring elements,

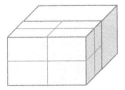

Figure 5.108 Dividing a hexahedral element.

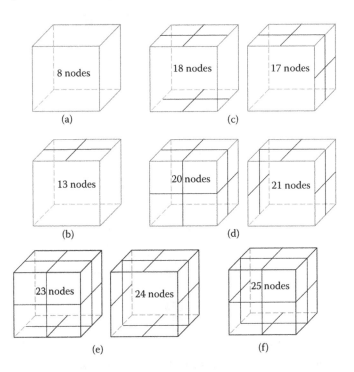

Figure 5.109 Node patterns for various face refinements: (a) regular element without refinement; (b) refinement on one face; (c) refinement on two faces; (d) refinement on three faces; (e) refinement on four faces; (f) refinement on five faces.

which cannot be lower than that of the element to be refined. The possible nodal configurations in 3D transition hexahedral elements are much more complex in 3D than those of 2D transition elements. Besides mid-edge nodes, there can also be mid-face nodes in 3D transition elements. All possible node patterns of eight types of transition hexahedral elements for refinements on one, two, three, four or five faces are shown in Figure 5.109.

In an adaptive refinement analysis of a 3D cantilever beam discretised into regular hexahedral elements, after three cycles of adaptive refinements to achieve an error in energy norm of less than 10%, the end section of the final refined transition hexahedral mesh consisting of 15,812 elements is shown in Figure 5.110. Figure 5.111 shows a joint connecting structural beams and columns of a building, which is refined to a mesh of 25,060 transition hexahedral elements to achieve an error of less than 8% in energy norm in four cycles of adaptive refinement. Readers are also referred to the classical papers on FE MG by Octree decompositions by Yerry and Shephard (1984) and Shephard and Georges (1991, 1992) for possible comparison and enhancement.

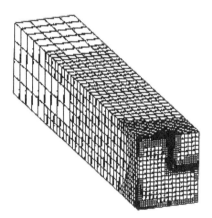

Figure 5.110 Transition hexahedral mesh of a beam.

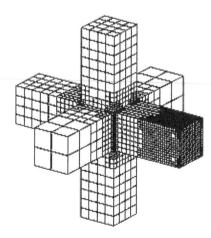

Figure 5.111 Transition hexahedral mesh of a structural joint of a building.

Chapter 6

Mesh optimisation

Nothing is immaculate, and there is always room to improve a finite element mesh.

6.1 INTRODUCTION

Various methods have been presented in Chapters 3–5 to generate unstructured meshes on a planar domain, over curved surfaces and within volumes. By the ADF approach, each node has been created to form the *best* element with an edge/face, and hence, the positions of the nodes are only optimised locally with respect to a few elements. For the Octree/Quadtree method, triangles or tetrahedra are inserted to fill the gaps between the domain boundary and the grid created by Quadtree/Octree decomposition, and very often, grid cells have also to be subdivided to produce elements of the required size. In the merging of triangles to generate pure quadrilateral unstructured meshes, nodes that support well-shaped triangular elements may not be well positioned for the formation of regular quadrilaterals close to a square shape. As for the DT, the empty circumcircle criterion optimises the minimum angle of a triangular mesh; however, the nodes may not have been placed at the most strategic locations to produce as many equilateral triangles as possible. In 3D, the empty sphere criterion has no direct relationship with the shapes of the elements and may produce degenerate tetrahedra known as *sliver*. Similar to the 2D case, the position of the spatial points can also be relocated so as to improve the shape quality of the mesh.

In view of the difficulty in placing nodes at the most strategic positions and the subsequent connection of nodes to form unstructured meshes in the mesh generation (MG) phrase by some of the most popular methods, the shape of the elements and hence the quality of the mesh have to be further improved after MG, and this process is known as *mesh optimisation*. There are many techniques to improve the shape quality of the finite elements (FEs), and as an FE mesh is defined by the nodal connections of the elements and the co-ordinates of the nodes, optimisation methods can be broadly grouped into two categories – (i) geometrical methods: those that involve a change of element shapes by means of shifting of nodal points; and (ii) topological methods: those that involve a change in the element connections to the nodal points.

Before we can work on improving a mesh by means of geometrical and/or topological operations, we have to ask ourselves what is meant by a *good* mesh that is aesthetically pleasing and also apt for numerical computations. We all agree that regular elements such as equilateral triangles, equilateral tetrahedra, squares and cubes are the desirable shapes for an FE mesh. Accordingly, shape measures have been introduced to measure the deviation of an element from its regular counterpart, such that a valid shape measure μ will attain a maximum value for elements in their regular form and will give zero values for degenerate elements with no volume. Many valid shape measures have been proposed, which include the mean ratio η, radius ratio ρ, minimum solid angle σ and volume and edges ratio γ. These shape measures serve as the objective function for the optimisation of a mesh by means of geometrical and topological operations. To name but just the more important ones, we have Laplace smoothing and its variants, local and global nodal optimisations and the recently proposed geometric element transformation method (GETMe) for simplicial and non-simplicial elements as popular geometrical optimisation techniques. As for topological optimisation operations, swap of diagonals in two dimensions, face and edge swaps in 3D and elimination of short edges and small elements are the common strategies employed.

6.2 SHAPE MEASURE AND QUALITY COEFFICIENT

In the optimisation of FE meshes by a node-smoothing scheme, nodal points are shifted to different positions so as to improve the quality of the shapes of the elements. Shape measure of FEs provides a mathematical basis and an objective measure on the overall quality of an FE mesh. In the pioneer papers on the relationship between tetrahedron shape measures by Liu and Joe (1994a,b), the essential properties of a valid shape measure are as follows: it is invariant under translation, rotation and uniform scaling; it attains a maximum value only for the regular tetrahedron; and it approaches zero for a degenerated tetrahedron. However, Dompierre et al. (2005) rephrased these ideas and presented a formal definition for shape measures of simplices as follows:

> A simplex shape measure is a continuous function that evaluates the shapes of a simplex. It must be invariant under translation, rotation, reflection and valid uniform scaling. It must be maximum for the regular simplex and it must be minimum for all degenerated simplices. For ease of comparison, it should be scaled to the interval [0,1], and 1 for the regular simplex and 0 for all degenerate simplices.

In the definition by Dompierre et al., compared to a valid shape measure by Liu and Joe, that simplices only attain the maximum value in the regular form is not emphasised. However, regular-shaped elements are what we are looking for in MG, and a shape measure at its maximum value still cannot guarantee that elements in their regular form may not be appropriate for mesh optimisation. Furthermore, for consistency and unified treatment, one additional property should also be considered, i.e. shape measures ought to be generic across dimensions. This property allows us to relate and compare the quality of the simplices on the boundary to that of the simplices at the interior of the domain, which is one dimension higher.

As far as mathematical rigor is concerned for the analysis of shape measures, the papers on algebraic quality metrics have to be cited (Field 2000; Knupp 2001; Branets and Garanzha 2002). The mathematical relationship of the associated transformation (distortion) matrix of a simplex on element quality measure was studied in which the properties of shape measures and possible algebraic metrics defined based on the associated transformation matrix were discussed with mathematical formalism and rigorous algebraic treatment. As the Jacobian matrix for the transformation of a simplex to its regular form contains all the information for skew, length ratio, shape, distortion, volume change and orientation, the notion of algebraic mesh quality metric was introduced in which the shape of a simplex is to be computed by scalar functions of the determinant, trace and norms of the associated transformation matrix. Shape measures were defined for triangular, tetrahedral, quadrilateral and hexahedral elements in terms of the Jacobian matrix, and contours of some of the shape measures

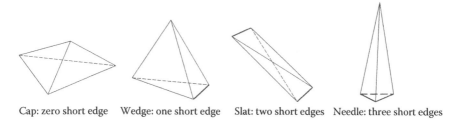

Cap: zero short edge   Wedge: one short edge   Slat: two short edges   Needle: three short edges

Figure 6.1 **Degenerate tetrahedra with zero, one, two and three short edges.**

were plotted for easy visualisation (Knupp 2003a). A shape measure for tetrahedral elements based on the condition number of the transformation matrix was also proposed, which can, in turn, be used to define a shape measure for hexahedral elements (Knupp 2003b).

Degenerate tetrahedral elements. A tetrahedron is said to be degenerated when it shrinks to zero volume, where all its vertices are either on the same plan, on the same line or all in one point. Some researchers would only take degenerate tetrahedral elements as tetrahedra with zero volume but non-zero faces or edges; hence, the big crunch of a tetrahedron shrinking to a point is not considered as a proper case of degeneracy (Knupp 2001). Liu and Joe (1994a) listed a number of practical degenerate cases by considering the number of short edges in a tetrahedron, as shown in Figure 6.1. By a progressive reduction of the length of the short edges, the behaviour of some common shape measures was monitored and compared. A systematic classification of ten degenerate cases for tetrahedral elements degenerating on a plane and on a line can be found in Dompierre et al. (2005) in which the number of short edges, how faces are collapsed and the limiting circumradius for degenerate tetrahedra were studied.

6.2.1 Common simplex shape measures

Simplex shape measures aroused much interest of Liu and Joe (1994a,b) when they worked on establishing a bound for the quality of tetrahedral elements subject to repeated subdivisions. The solid angle in three dimensions, with natural geometrical relationship for object visualisation, measures how large an object appears to an observer looking it from a point. The solid angle and the ratio between the in-radius and the circumradius of a tetrahedron are among the earliest shape measures used for simplices. The mean ratio defined by the geometric mean and the arithmetic mean of the eigenvalues of the transformation matrix turns out to have also a strong geometrical interpretation related to the volume and the edges of a simplex. With the introduction of the distortion matrix between a simplex and its regular counterpart, other shape measures, following more on algebraic derivations rather than geometric considerations, can be defined in terms of some scalar functions of the associated transformation matrix.

6.2.1.1 Minimum solid angle θ

The solid angle θ_i at vertex $\mathbf{x}_i$ of tetrahedron $T(\mathbf{x}_1, \mathbf{x}_2, \mathbf{x}_3, \mathbf{x}_4)$ is given by the surface area formed by projecting each point on the face opposite to $\mathbf{x}_i$ to the unit sphere centred at $\mathbf{x}_i$, as shown in Figure 6.2. Since the area of a unit sphere is 4π, we have

$$0 \leq \theta_i \leq 2\pi$$

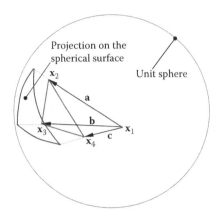

Figure 6.2 **Solid angle measured by projection onto a unit sphere.**

The minimum solid angle θ is defined as

$$\theta = \min(\theta_1, \theta_2, \theta_3, \theta_4)$$

A formula to compute the solid angles adapted from Liu and Joe (1994a) is given by

$$\sin\left(\frac{\theta_1}{2}\right) = \frac{\mathbf{a} \times \mathbf{b} \cdot \mathbf{c}}{\sqrt{2(ab + \mathbf{a} \cdot \mathbf{b})(bc + \mathbf{b} \cdot \mathbf{c})(ca + \mathbf{c} \cdot \mathbf{a})}}$$

where $\mathbf{a}$, $\mathbf{b}$ and $\mathbf{c}$ are vectors from vertex $\mathbf{x}_1$ to other vertices, and $a = \|\mathbf{a}\|$, $b = \|\mathbf{b}\|$ and $c = \|\mathbf{c}\|$. Similarly, the formulas for calculating θ_2, θ_3 and θ_4 are given by vectors emanating from the respective vertices. On the other hand, Van Oosterom and Strackee (1983) proposed a formula to compute solid angles in terms of $\tan(\theta/2)$:

$$\tan\left(\frac{\theta_1}{2}\right) = \frac{\mathbf{a} \times \mathbf{b} \cdot \mathbf{c}}{abc + (\mathbf{a} \cdot \mathbf{b})c + (\mathbf{b} \cdot \mathbf{c})a + (\mathbf{c} \cdot \mathbf{a})b}$$

It can be shown that

$$\theta \leq 3\operatorname{acos}\left(\frac{1}{3}\right) - \pi$$

and the equality holds only for regular tetrahedra.

As a shape measure, it is more convenient to use the sine of the angle rather than the angle itself. Hence, a shape measure σ based on the minimum solid angle normalised to 1 for equilateral tetrahedron is defined as

$$\sigma = \frac{9}{\sqrt{6}} \sin\left(\frac{\theta}{2}\right) = \frac{9\mathbf{a} \times \mathbf{b} \cdot \mathbf{c}}{\sqrt{12(ab + \mathbf{a} \cdot \mathbf{b})(bc + \mathbf{b} \cdot \mathbf{c})(ca + \mathbf{c} \cdot \mathbf{a})}} \leq 1$$

The counterpart in two dimensions is the minimum angle of the triangle, which is between 0 and $\pi/3$. In two dimensions, Delaunay triangulation maximises the minimum angle of the triangles, whereas in three dimensions, the solid angle of the triangulation may not be maximised as degenerated and almost degenerated tetrahedral elements known as *slivers* with zero or very small solid angles can be found in a Delaunay triangulation. Hence, Delaunay triangulations in three or higher dimensions may not be optimal in terms of shape measures.

6.2.1.2 Radius ratio ρ

The radius ratio ρ of a tetrahedron T is defined as $\rho = 3r/R$, where r and R are, respectively, the in-radius and the circumradius of T, as shown in Figure 6.3. In terms of the geometry of tetrahedron T, the in-radius r and the circumradius R are given by

$$r = \frac{3v}{S_1 + S_2 + S_3 + S_4}$$

and

$$R = \frac{\sqrt{(a+b+c)(a+b-c)(a+c-b)(b+c-a)}}{12v}$$

where v is the volume of tetrahedron T, S_i are the areas of the four faces of T, and a, b, c are the products of the lengths of opposite edges of T. In two dimensions, the radius ratio $\rho = 2r/R$, where r and R of a triangle are given in Appendix A.3.

$$r = \frac{A}{S} \text{ and } R = \frac{S_1 S_2 S_3}{4A}$$

where A is the area of the triangle, S_1, S_2, S_3 are the sides of the triangle, and half-perimeter $S = (S_1 + S_2 + S_3)/2$.

6.2.1.3 Mean ratio η

There are a number of ways in defining the mean ratio, and as suggested by the name, it is natural to define η by the ratio of the geometric mean to the arithmetic mean of the transformation matrix. Liu and Joe (1994b) and Knupp (2001) developed expressions and relationships for the mean ratio based on the linear algebra matrix approach. However, an equivalent approach following the concept of large deformation will be presented here.

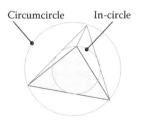

Circumcircle   In-circle

Figure 6.3 **Radius ratio of a tetrahedron.**

While there is not much difference in higher dimensions, over two and three dimensions, it is more intuitive, and many established results in large deformation theory can be applied right away.

Consider the linear transformation of a unit cube to a parallelepiped as shown in Figure 6.4. The transformation matrix from a unit cube to the parallelepiped can be represented by the deformation gradient tensor F_a such that $a_i = F_a \cdot e_i$, $i = 1, 2, 3$. The 3×3 transformation matrix F_a, which carries vector e_i to a_i, can be easily constructed as follows.

$$F_a = [a_1 \quad a_2 \quad a_3] = [x_2 - x_1 \quad x_3 - x_1 \quad x_4 - x_1]$$

In computing the vectors a_1, a_2 and a_3 from the vertices of tetrahedron T, apart from x_1, other vertex can also be chosen as the origin as long as they are so oriented to give a positive volume following the right-hand system (Knupp 2001). As e_1, e_2 and e_3 are vectors along the sides of a unit cube, $[e_1 \quad e_2 \quad e_3]$ forms the identity matrix (tensor) such that

$$F_a \cdot [e_1 \quad e_2 \quad e_3] = [a_1 \quad a_2 \quad a_3] \cdot \begin{bmatrix} 1 & 0 & 0 \\ 0 & 1 & 0 \\ 0 & 0 & 1 \end{bmatrix} = [I \cdot a_1 \quad I \cdot a_2 \quad I \cdot a_3] = [a_1 \quad a_2 \quad a_3]$$

Similarly, if r_1, r_2 and r_3 are the vectors along the sides of a regular tetrahedron, the transformation matrix F_r, which maps the right-angled tetrahedron spanned by vectors e_1, e_2 and e_3 onto the regular equilateral tetrahedron spanned by vectors r_1, r_2 and r_3, is given by

$$F_r = [r_1 \quad r_2 \quad r_3] = \begin{bmatrix} 1 & 1/2 & 1/2 \\ 0 & \sqrt{3}/2 & \sqrt{3}/6 \\ 0 & 0 & \sqrt{2/3} \end{bmatrix}$$

Hence, the transformation matrix from a regular tetrahedron to tetrahedron T is given by

$$F \cdot F_r = F_a \Rightarrow F = F_a \cdot F_r^{-1}$$

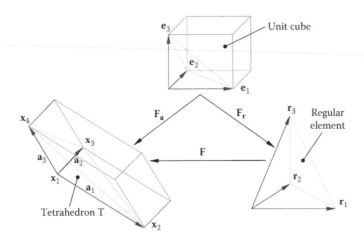

Figure 6.4 **Transformation from Tetrahedron T to the regular element.**

Based on the polar decomposition in the theory of large deformation, deformation tensor **F** can be uniquely decomposed into a rotation part **R** and a pure deformation part **U** such that

$$\mathbf{F} = \mathbf{R} \cdot \mathbf{U} \quad \mathbf{R}^{\mathrm{T}} \cdot \mathbf{R} = \mathbf{R} \cdot \mathbf{R}^{\mathrm{T}} = \mathbf{I} \text{ (Identity tensor) and } \mathbf{U} \text{ is symmetric}$$

The deformation **F** between tetrahedron T and the regular element can be measured by the Green–Cauchy deformation tensor **C** given by

$$\mathbf{C} = \mathbf{F}^{\mathrm{T}} \cdot \mathbf{F} = (\mathbf{R} \cdot \mathbf{U})^{\mathrm{T}} \cdot (\mathbf{R} \cdot \mathbf{U}) = \mathbf{U}^{\mathrm{T}} \cdot \mathbf{R}^{\mathrm{T}} \cdot \mathbf{R} \cdot \mathbf{U} = \mathbf{U}^2$$

By this definition, **C** is symmetric and positive definite for non-degenerate tetrahedron T. Let λ_1, λ_2 and λ_3 be the eigenvalues of **C**; deformation from the regular element to tetrahedron T can be measured by the mean ratio η defined as the ratio between the geometric mean and the arithmetic mean of the eigenvalues of **C**.

$$\eta = \frac{\sqrt[3]{\lambda_1 \lambda_2 \lambda_3}}{\frac{1}{3}(\lambda_1 + \lambda_2 + \lambda_3)}$$

η attains the minimum value of 0 for degenerated tetrahedra, for which one or more λ values equal to 0, and it reaches the maximum value of 1 when $\lambda_1 = \lambda_2 = \lambda_3 = \lambda$; in such a case, we have

$$\mathbf{C} = \begin{bmatrix} \lambda_1 & & \\ & \lambda_2 & \\ & & \lambda_3 \end{bmatrix} = \begin{bmatrix} \lambda & & \\ & \lambda & \\ & & \lambda \end{bmatrix} = \lambda \mathbf{I}$$

For $\eta = 1$, tetrahedron T and the regular element are similar and only differ by a scaling factor $\lambda^{1/2}$. On the other hand, if not all the λ values are equal, η will be less than 1, i.e. $\eta = 1$ only for regular tetrahedral elements. According to Knupp (2001), among shape measures σ, ρ and η, only η is an algebraic mesh quality metric. As a result, mean ratio η can also be expressed in terms of the determinant and the Frobenius norm of transformation matrix **F** as follows:

$$\eta = \frac{\sqrt[3]{\lambda_1 \lambda_2 \lambda_3}}{\frac{1}{3}(\lambda_1 + \lambda_2 + \lambda_3)} = \frac{3\sqrt[3]{\det(\mathbf{C})}}{\mathrm{tr}(\mathbf{C})} = \frac{3(\det(\mathbf{F}))^{2/3}}{\mathrm{tr}(\mathbf{F}^{\mathrm{T}} \cdot \mathbf{F})} = \frac{3(\det(\mathbf{F}))^{2/3}}{\mathbf{F} : \mathbf{F}} = \frac{3(\det(\mathbf{F}))^{2/3}}{\|\mathbf{F}\|^2}$$

$\det(\mathbf{C}) = \det(\mathbf{F}^{\mathrm{T}} \cdot \mathbf{F}) = (\det(\mathbf{F}))^2$; $\mathbf{F} : \mathbf{F} = F_{ij}F_{ij}$ (sum over i, j); $\|\cdot\|$ = Forbenius norm

In terms of shape measure regularity, η is preferred over σ and ρ in mesh optimisation as η is more smooth, differentiable everywhere and symmetric about its peak (Dompierre et al. 2005). Alternatively, a more geometrically based quality coefficient γ related to the mean ratio η was proposed by Lo (1991c),

$$\gamma(\mathrm{T}) = \frac{72\sqrt{3} \times \text{signed volume of T}}{(\text{sum of edges}^2)^{3/2}}$$

The equivalence between γ and η was proved by Liu and Joe (1994b) such that $\gamma = \eta^{3/2}$. γ can be directly computed entirely from the geometry of T, and strictly speaking, by the definition of Dompierre et al., it is not a shape measure as it can take positive and negative values. That's why it is called the quality coefficient rather than the shape measure of tetrahedron T. As FE analysis does not allow for inverted elements, reflected tetrahedral elements with a negative volume cannot be accepted in the FE mesh, even though they are well balanced in shape. Extending the shape measure to the negative regime allows the inverted element to be detected by the quality coefficient; for instance, in the construction of Delaunay triangulation, invalid elements can be avoided by the visibility (positive volume) test; hence, γ-quality can ensure the validity of a triangulation as well as the shape quality of the tetrahedral elements. In the Quality Laplace (QL) smoothing to be presented in Section 6.3.1.1, γ-quality can also prevent invalid inverted elements from being formed in the iteration process. For MG based on large deformation in geometry such as the Mecanno method (Section 8.4) in transforming a cube to an object of the same topology through continuous deformations, inverted elements are created in the intermediate stages, which have to be untangled based on some versatile quality-measure coefficients. In the boundary recovery process, almost degenerated elements have to be used to first create a valid topology to cater for difficult boundary surfaces of complicated industrial and biomechanical objects. These poorly shaped elements can be eliminated in a subsequent optimisation process. However, the signed volume is crucial in this situation for elements close to degeneracy, and γ-quality is more convenient in providing additional information to ensure that a valid mesh is always maintained. Finally, the γ-quality coefficient is also generic across dimensions, and the two-dimensional counterpart α-quality coefficient for triangle ABC is defined as (Lo 1985)

$$\alpha(ABC) = \frac{4\sqrt{3} \text{ signed area}}{AB^2 + BC^2 + CA^2} = \frac{2\sqrt{3}(AB \times AC)}{AB^2 + BC^2 + CA^2}$$

In Section 2.4.10, it has been shown that the best tetrahedron that can be built on an α-quality triangular face is γ = α, and this result provides an upper bound on the quality of the tetrahedral elements for a given boundary surface. Given the sound geometrical concept, fast evaluation for large meshes and versatility in MG and mesh optimisation, α- and γ-quality coefficients are systematically used as quality measures for triangular and tetrahedral elements in the MG and mesh optimisation algorithms here presented.

6.2.1.4 Shape measures based on condition number κ

Let λ_1, λ_2 and λ_3 be the eigenvalues of deformation tensor **C** associated with tetrahedron T, and without loss of generality, arrange the eigenvalues in an ascending order of magnitude such that $\lambda_1 \leq \lambda_2 \leq \lambda_3$; then the condition number κ given by the ratio of the smallest eigenvalue to the largest eigenvalue is a shape measure (Formaggia and Perotto 2000; Knupp 2000a,b)

$\kappa = \lambda_1/\lambda_3$

Let **F** be the linear transformation of a regular tetrahedron to tetrahedron T, as shown in Figure 6.4; then the condition number of the transformation matrix between the equilateral tetrahedron and tetrahedron T is given by

$\kappa(\mathbf{F}) = \|\mathbf{F}\|\|\mathbf{F}^{-1}\|$ $\|\cdot\|$ = Frobenius norm

Freitag and Knupp (2002) proposed and proved that $\tau = 3/\kappa(\mathbf{F})$ is a valid shape measure.

Figure 6.5 Dihedral angles of a needle-shaped tetrahedron.

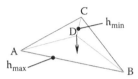

Figure 6.6 Volume of a tetrahedron ABCD approaches zero when D comes close to triangular face ABC.

6.2.1.5 Minimum dihedral angle is not a valid shape measure

The dihedral angle ϕ at an edge of a tetrahedron between two triangular faces is given by

$$\phi = \pi - \text{acos}(\mathbf{n}_1 \cdot \mathbf{n}_2)$$

where $\mathbf{n}_1$ and $\mathbf{n}_2$ are the unit normals to the faces sharing the same edge under consideration.

The minimum dihedral angle is not a valid shape measure as it cannot detect the degenerate case of a needle-shaped tetrahedron, as shown in Figure 6.5. The dihedral angles of the tetrahedron remain more or less the same as the triangular face opposite to the pointed node is getting smaller and smaller. Refer to Dompierre et al. (2005) for more degenerate cases, which cannot be detected by measuring the dihedral angles of a tetrahedron.

6.2.1.6 Edge ratio is not a valid shape measure

The edge ratio of a tetrahedron is defined as the ratio between the shortest edge and the longest edge of the tetrahedron, i.e.

$$\text{edge ratio} = \frac{h_{min}}{h_{max}}$$

The edge ratio is not a valid shape measure as it does not vanish for all the degenerate cases, as shown in Figure 6.6, for instance, the flattening of a cap element. The edge ratio does not change much for the flattened tetrahedron ABCD when D reaches face ABC. *Sliver* tetrahedra in Delaunay triangulation are another example, which cannot be detected by the edge ratio.

6.2.2 Relationship between shape measures

Liu and Joe (1994a) systematically compared shape measures σ, ρ and η and established the relationship among these shape measures as follows:

$$\eta^3 \leq \rho \leq \eta^{3/4}, \quad \rho^{4/3} \leq \eta \leq \rho^{1/3}$$

$$0.2296\eta^{3/2} \leq \sigma \leq 1.1398\eta^{3/4}, \quad 0.8399\sigma^{4/3} \leq \eta \leq 2.6667\sigma^{2/3}$$

$$0.2651\rho^2 \leq \sigma \leq \rho^{1/2}, \quad \sigma^2 \leq \rho \leq 1.942\sigma^{1/2}$$

More relationships on shape measures based on the condition number were given by (Dompierre et al. 2005):

$$\kappa^{1/2} \leq \tau \leq 3\kappa^{1/2}, \quad \kappa \leq \eta \leq 3\kappa^{1/3}, \quad \sqrt{2/3}\eta^{3/2} \leq \tau \leq 3\eta^{1/2}, \quad \tau = 3/\|\mathbf{F}\|\|\mathbf{F}^{-1}\|$$

Liu and Joe (1994a) further proposed an equivalent relationship for shape measures. Let λ and μ be two different shape measures scaled in the interval [0,1]; λ and μ are equivalent if there exist positive constants a, b, p and q such that

$$a\mu^p \leq \lambda \leq b\mu^q \tag{6.1}$$

λ is related to $\mu(\lambda \sim \mu)$ if Equation 6.1 holds for λ and μ. This is an equivalence relation since it is reflexive, symmetric and transitive, i.e. (i) $\lambda \sim \lambda$, (ii) $\lambda \sim \mu \Rightarrow \mu \sim \lambda$ and (iii) $\lambda \sim \mu$ and $\mu \sim \nu \Rightarrow \lambda \sim \nu$. By this definition of equivalence for shape measures, all valid shape measures are indeed equivalent in the sense of Equation 6.1. However, equivalence of shape measures does not result in the same ordering of simplices based on two different shape measures. Equivalence simply implies that if the shape of a degenerate element approaches zero measured by one shape measure, so will the others. On the other hand, if one shape measure gives a value of 1 for a perfect regular element, so do the other measures. As how shape measures approach these two extremes, the rate and the manner vary from one measure to another, and not much conclusion can be drawn from meshes evaluated by different shape measures.

6.2.3 Extension to Riemann space

There were attempts to extend the shape measures defined in the Euclidean space to Riemann manifold (George and Borouchaki 1998; Formaggia and Perotto 2000; Frey and George 2000; Dompierre et al. 2002, 2005; Sirois et al. 2005). In general, the transformation of a simplex in the Euclidean space to the Riemann space is non-linear, and the treatment on shape measures based on the linear algebraic approach cannot be directly applied. Instead, attentions were paid to the geometrical quantities such as the length, surface area and volume of the simplex in the Riemann space, which can all be evaluated through the metric tensor. Shape measures can then be defined based on these geometrical quantities evaluated in the Riemann space. For example, the γ-quality of tetrahedron T measured in the Riemann space is given by

$$\gamma_R(T) = \frac{72\sqrt{3} \times \text{signed volume of T in Riemann space}}{(\text{sum of edges measured in Reimann space}^2)^{3/2}}$$

The length of a parametric curve $\mathbf{x} = \mathbf{x}(t)$ in the Riemann space is given by

$$\text{length of curve} = \int_0^1 \sqrt{(\mathbf{F} \cdot d\mathbf{x})^T (\mathbf{F} \cdot d\mathbf{x})} = \int_0^1 \sqrt{\mathbf{x}'(t)^T \mathbf{M}(t)\mathbf{x}'(t)} \, dt \qquad \text{where } \mathbf{M} = \mathbf{F}^T \cdot \mathbf{F}$$

$$\text{surface area} = \int_A \|(\mathbf{F} \cdot d\mathbf{x}) \times (\mathbf{F} \cdot d\mathbf{y})\| = \int_A \det(\mathbf{F})\|\mathbf{F}^{-T} \cdot d\mathbf{a}\| \quad \text{with } d\mathbf{a} = d\mathbf{x} \times d\mathbf{y}$$

where deformation gradient (transformation) tensor $\mathbf{F}$ maps a vector in the Euclidean space to the corresponding vector in the Riemann space, vector $\mathbf{da}$ is a surface element in the Euclidean space and A is the area domain of the element in the Euclidean space. The volume of a simplex K, V, in the Riemann space is given by

$$V = \int_K (\mathbf{F} \cdot d\mathbf{x}_1) \times (\mathbf{F} \cdot d\mathbf{x}_2) \cdot (\mathbf{F} \cdot d\mathbf{x}_3) = \int_K \det(\mathbf{F})(d\mathbf{x}_1 \times d\mathbf{x}_2 \cdot d\mathbf{x}_3) = \int_K \det(\mathbf{F}) dv$$

where dv is the volume element spanned by vectors $d\mathbf{x}_1$, $d\mathbf{x}_2$ and $d\mathbf{x}_3$. Alternatively, the shape measure on the Riemann space can be approximately evaluated by taking sample points over the simplex; in a more rigorous manner, we can take the Gaussian points, and for the sake of convenience, the vertex points are taken for the evaluation (Alauzet et al. 2003).

A direct extension to Riemann metric may not be consistent with the original idea of shape measure in the Euclidean space as the element with the highest shape measure may not be a regular element in the sense that the angles may not be equal and optimal; the faces may not have the same surface area either, even though all the edges are of the same length. Regular simplex with equal edges, equal faces and equal angles well defined in the Euclidean space simply does not exist in the Riemann space, and the shape measure may not attain the higher value of 1 for simplex with edges of equal lengths; neither can we prove that there is only a single peak for shape measures evaluated with the Riemann metric. For anisotropic meshes, the lengths of the edges or distance measure are of primary concern rather than the shape of the elements as the definition of regular simplex is no longer applicable in the Riemann space. In other words, anisotropic meshes are generated such that all the edges ought to be conformable with the specified length requirement measured by the given metric. As the simplex is actually generated in the Euclidean space in which regular element and shape measure are defined, another approach is to adopt a criterion λ combining shape measure η in Euclidean metric and an edge conformity coefficient δ to assess the quality of a tetrahedron in the Riemann space. The conformity coefficient δ_i of the ith edge is given by

$$\delta_i = \min\left(\frac{r_i}{\rho_i}, \frac{\rho_i}{r_i}\right) \qquad r_i = \text{length of edge i in Riemann metric}, \quad \rho_i = \text{specified length}$$

Combined measure λ for tetrahedral elements can now be defined as

$$\lambda = \eta\delta, \quad \text{where } \eta = \text{mean ratio measured with Eucliden metric}, \delta = \delta_1\delta_2\delta_3\delta_4\delta_5\delta_6$$

η is included in the combined measure to ensure that the simplex will not be degenerate in the Riemann space, as a degenerate element in the Euclidean space will map into a degenerate element in the Riemann space and vice versa. Shape measure in the Euclidean space is still meaningful as well-shaped tetrahedra may only be slightly distorted under the transformation of a smooth metric. Combined measure λ is relatively less costly to compute compared to the shape measure evaluated in Riemann metric.

6.2.4 Shape measure for polyhedron

In three dimensions, apart from tetrahedral elements, structured and unstructured solid FE meshes of hexahedral, pentahedral and pyramid elements are quite common. As these non-simplicial elements are generated by some automatic processes and are very often mixed up with tetrahedral elements and other element types, it is necessary to assess the quality of these elements so as to have an overall evaluation of the FE mesh used in the analysis. Trilinear eight-node hexahedral elements are the most popular, and it is crucial that we could develop a generic scheme to measure its shape properties.

For the four-node tetrahedral element, the deformation is homogeneous and the Jacobian (transformation) matrix is a constant, which can be evaluated at any point within the element. As for distorted hexahedral elements, the deformation is not homogeneous, and the Jacobian matrix depends on the point where it is evaluated. This was the major difficulty defying early attempts in defining logical and consistent shape measures for hexahedral elements. For simplicity, shape measured can be defined based on the Jacobian matrix evaluated at some interior points of the element such as the centroid or the Gaussian points. However, this computation is more expensive, and worse still, the Jacobian at the centroid may not be sensitive enough to detect degenerate cases on the boundary, as shown in Figure 6.7. By extending the shape measure for simplices, Knupp (2003a,b) proposed a genius shape measure for quadrilaterals and hexahedra, which turns out to be a generic scheme applicable to general polyhedron. For each vertex of a hexahedral element H, there is an associated tetrahedron formed by the vertex and the three neighbouring vertices (neighbouring vertices are on the same edge of the hexahedron). As shown in Figure 6.8, the associated tetrahedron for vertex 1 is tetrahedron $T_1 = T(1,2,4,5)$, and similarly, the associated tetrahedron for vertex 8 is tetrahedron $T_8 = T(8,7,5,4)$. As the reference tetrahedron for a cube is given by the identity matrix, the transformation F_1 from the reference tetrahedron to tetrahedron T_1 is given by

$$F_1 = [a_1 \quad a_2 \quad a_3] = [x_2 - x_1 \quad x_4 - x_1 \quad x_5 - x_1]$$

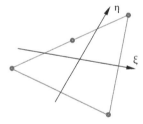

Figure 6.7 Degeneracy on the boundary cannot be detected by Jacobian at an interior point.

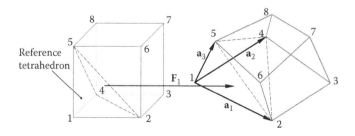

Figure 6.8 Associated tetrahedron at vertices.

A shape measure of the deviation of tetrahedron T_1 from the reference tetrahedron is given by the mean ratio η_1 of transformation $\mathbf{F}_1$, i.e.

$$\eta_1 = \frac{3(\det(\mathbf{F}_1))^{2/3}}{\|\mathbf{F}_1\|^2}$$

Shape measure other than the mean ratio could also have been used as long as it is a valid measure of the deviation of $\mathbf{F}_1$ from the identity matrix. Similarly, η_k is the η value of tetrahedron T_k at vertex k, and the arithmetic mean η_a, geometric mean η_g and minimum value η_m of the mean ratios η_k for hexahedral element H can now be defined as

$$\eta_a(H) = \frac{1}{8}\sum_{k=1,8}\eta_k, \quad \eta_g(H) = \left(\prod_{k=1,8}\eta_k\right)^{1/8}, \quad \eta_m(H) = \min_{k=1,8}\eta_k$$

As a polyhedron is completely defined by its boundary, the shape measure of hexahedron H is based on the argument that if all the corners of a hexahedron are identical to those of a cube (i.e. $\eta_k = 1$), it must be a regular hexahedron similar to a cube but only possibly differs by a scaling factor. Element inversion or degeneracy at one or more corner points can be detected by the η values computed for each vertex of the polyhedron. η_a and η_g can also be extended into the negative range to detect inverted polyhedral elements by simply setting $\eta_a = \eta_g = \eta_m$ if $\eta_m \le 0$. In two dimensions, a shape measure for a quadrilateral element Q can be defined in an analogous manner such that

$$\eta(Q) = \frac{1}{4}\sum_{k=1,4}\eta_k, \quad \text{with } \eta_k = \frac{2\det(\mathbf{F}_k)}{\|\mathbf{F}_k\|^2}$$

where $\mathbf{F}_k$ is the transformation matrix of the associated triangle at vertex k.

Following the same principle, shape measures can also be defined for a general polyhedron. What we have to do is to define the reference tetrahedron for the vertex (corner) under consideration. For instance, there is only one type of reference tetrahedron for a cube as all corners of a cube are identical, each of which is a right-angled tetrahedron whose matrix representation is the identity matrix. Vartziotis and Wipper (2012) gave the matrix representation $\mathbf{F}_r$ of reference tetrahedron for regular tetrahedron, hexahedron, pyramid and pentahedron, as shown in Figure 6.9.

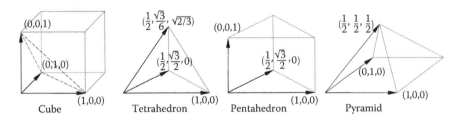

Figure 6.9 Reference tetrahedron of regular polyhedral elements.

Hexahedron: $F_r = \begin{bmatrix} 1 & 0 & 0 \\ 0 & 1 & 0 \\ 0 & 0 & 1 \end{bmatrix}$ = Identity Matrix

Tetrahedron: $F_r = \begin{bmatrix} 1 & 1/2 & 1/2 \\ 0 & \sqrt{3}/2 & \sqrt{3}/6 \\ 0 & 0 & \sqrt{2/3} \end{bmatrix}$, $F_r^{-1} = \begin{bmatrix} 1 & -\sqrt{3}/3 & -\sqrt{6}/6 \\ 0 & 2\sqrt{3}/3 & -\sqrt{6}/6 \\ 0 & 0 & \sqrt{6}/2 \end{bmatrix}$

Pyramid: $F_r = \begin{bmatrix} 1 & 0 & 1/2 \\ 0 & 1 & 1/2 \\ 0 & 0 & 1/\sqrt{2} \end{bmatrix}$, $F_r^{-1} = \begin{bmatrix} 1 & 0 & -\sqrt{2}/2 \\ 0 & 1 & -\sqrt{2}/2 \\ 0 & 0 & \sqrt{2} \end{bmatrix}$

Pentahedron: $F_r = \begin{bmatrix} 1 & 1/2 & 0 \\ 0 & \sqrt{3}/2 & 0 \\ 0 & 0 & 1 \end{bmatrix}$, $F_r^{-1} = \begin{bmatrix} 1 & -\sqrt{3}/3 & 0 \\ 0 & 2\sqrt{3}/3 & 0 \\ 0 & 0 & 1 \end{bmatrix}$

The transformation F_k from the reference tetrahedron to the associated tetrahedron T_k at vertex k of a polyhedron is given by

$$F_k = T_k \cdot F_r^{-1}, \qquad \text{where } T_k \text{ is the matrix representation of } T_k$$

The mean ratio of F_k, $\eta_k = \dfrac{3(\det(F_k))^{2/3}}{\|F_k\|^2}$; γ – quality, $\gamma_k = \dfrac{3\sqrt{3}\det(F_k)}{\|F_k\|^3}$

Similar to the hexahedron, a shape measure for a polyhedron can now be defined based on some average or minimum values of the mean ratios (γ-qualities) of the associated tetrahedra at all the vertices. This shape measure scheme for a polyhedron induced by the shape measures of a tetrahedron is generic as an associated tetrahedron can always be defined without ambiguity for a given vertex of the polyhedron. The induced relationship is intuitive with sound geometrical interpretations, and it also provides a link between the quality of tetrahedral and polyhedral elements for the optimisation of meshes of mixed element types.

6.3 OPTIMISATION BY SHIFTING OF NODES

The shape quality of an FE mesh can be improved by shifting the interior points within the domain boundary, and usually, boundary points have to be kept intact for compatibility with other meshes or for imposing the required boundary conditions for the FE analysis. Without a proper idea of shape measure, an early attempt to improve a triangular mesh was to shift each interior node in turn to the centroid of the polygon formed by all the nodes connected to it through an edge of the mesh, and this iteration process is known as Laplace smoothing (Herrmann 1976). Inspired by the α-quality measure introduced in the MG by the ADF approach, Lo (1985, 1992) proposed a smart Laplace smoothing scheme in which node shifting is carried out only when there is an improvement in the overall quality of the triangular elements connected to the node. A quality Laplace (QL) scheme has also been developed later in which the node shifting between the original position and the centroid of the surrounding polygon only indicates the direction along which a node has to be moved. As to the distance ought to be moved, it is controlled and guided by a shape measure on all the elements affected (Lo 1997). A comparison of node-smoothing algorithms of triangular meshes was given by Erten et al. (2009) and Sastry and Shontz (2009).

A distortion coefficient β derived from the α-quality of the triangles formed by cutting across the diagonal was proposed to assess the shape quality of quad elements (Lo 1989b; Lo and Lee 1992). Before the development of valid shape measure for quad elements, apart from Laplace smoothing, heuristic procedures were proposed to shift nodal points so as to increase the internal angles of quad elements (Lee and Lo 1994; Lau et al. 1997; Zhou and Shimada 2000; Lee and Lee 2002, 2003; Xu and Newman 2006). Based on a shape measure, a cost function can be defined for an FE mesh, which can be maximised as a local/ global optimisation problem to improve the overall quality of the mesh (Parthasarathy and Kodiyalam 1991; Amezua et al. 1995; Buscaglia and Dari 1997; Freitag and Plassmann 2000; Branets and Garanzha 2002; Garimella et al. 2004; Sirois et al. 2005; Knupp 2006; Sastry and Shontz 2009; Branets and Carey 2010; Qian et al. 2010; Knupp 2012). Sarrate and Huerta (2001) put forward a rezoning scheme for the optimisation of graded quad meshes. Combined schemes based on topological restructuring and geometrical node shifting on FE meshes were also presented (Freitag and Ollivier-Gooch 1996, 1997; Lee and Hobbs 1999; Pain et al. 2001). High-quality meshes are obtained when a few cycles of topological and geometrical smoothing operations are applied most effectively in an alternative manner (Lo 1997). Theoretical developments on the local optimisation (LO) problem based on shape measures for triangular and tetrahedral elements were presented by Aiffa and Flaherty (2003).

Laplace smoothing and global optimisation had been the dominating techniques in mesh optimisation based on node shifting in the early days of MG. These are algebraic computations without a sound geometrical interpretation for individual elements, especially for non-simplices such as quad and hex elements. The lack of a valid shape measure for non-simplicial elements was one of the reasons, but the major difficulty was that we did not have any systematic procedure to improve the quality of non-simplicial elements. A breakthrough was made when Vartziotis et al. (2008) proposed the GETMe method, which is purely a geometric process to move the nodes of a triangle so as to improve its quality. The method was extended to optimise tetrahedral meshes (Vartziotis et al. 2009) in which the mean ratio was employed as the shape measure. GETMe turned out to be a generic scheme, which is also applicable to non-simplicial elements such as quads and hex (Vartziotis and Wipper 2009). By means of the associated octahedron constructed using the mid-points of the six faces of a hexahedron, node-shifting vectors can be defined by the normals to the faces of the octahedron (Vartziotis and Wipper 2011). The same idea of auxiliary polyhedron built on the centroids of the faces of a given polyhedron can be applied to improve the mean ratio quality of general polyhedral elements, such as tetrahedron, hexahedron, pentahedron and pyramid (Vartziotis and Wipper 2012). As a practical application to FE analysis, GETMe was applied to tetrahedral and hexahedral meshes to solve the Poisson equation, with the resulting error norms indicating that GETMe meshes are not only superior in mean ratio shape measure but also apt in adaptive refinement analysis (Vartziotis et al. 2013).

6.3.1 Optimisation of triangular meshes

Three optimisation schemes for triangular meshes by node shifting, namely, QL smoothing, local quality optimisation and GETMeT3, will be presented in this section. Examples of various characteristics are given, and their performance will be assessed in terms of the shape quality of the meshes and the CPU time taken in the optimisation process.

6.3.1.1 QL smoothing

Given a triangular mesh of N triangle, $T = \{\Delta_i, i = 1,N\}$, for a given node x, the patch of triangles surrounding node x, as shown in Figure 6.10, is given by

$$P(x) = \{\Delta_k \in T; x \in \Delta_k\}, \quad x \in \Delta_k \text{ means that } x \text{ is one of the vertices of triangle } \Delta_k$$

The *centroid* of polygon P is given by

$$c = \frac{\sum_{\Delta_k \in P} \left(x_1^k + x_2^k + x_3^k - x \right)}{2n}$$

where n is the number of triangles in P. Here, *centroid* does not mean the centre of gravity of polygon P, but the average of the co-ordinates of the nodes on the boundary of P. By the classical Laplace smoothing, node x is shifted to c, and this node-smoothing process can be applied sequentially to each node in turn until all the nodes in the mesh are treated. A number of cycles of Laplace smoothing can be applied to a triangular mesh until no further improvement can be made. However, node shifting to the centroid of the surrounding polygon does not guarantee that all triangles are valid elements, especially for concave polygon P; some of the triangles may become inverted by the nodal displacement. In order to prevent inverted elements from being formed in the Laplace smoothing process, the α-qualities of the triangles in P are computed, which are to be compared to those after shifting node x to centroid c. The displacement of x to c is only executed when there is an improvement in the overall quality of the triangles in P. As a quality measure of the triangles in P, the geometric mean α-quality and the minimum α-value of the triangles can be used.

$$\bar{\alpha} = \left(\prod_{\Delta_k \in P} \alpha_k \right)^{1/n}; \quad \alpha_{min} = \min_{\Delta_k \in P} \alpha_k$$

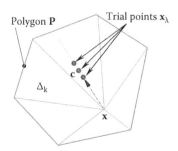

Figure 6.10 Quality Laplace smoothing.

Node shifting will be carried out only if $\bar{\alpha}^* > \bar{\alpha}$ and α^*_{min} is greater than α_{min}, where α^* is the α-quality after node shifting. Element inversion can be prevented as some triangles will then have negative α^* values after node shifting. This Laplace smoothing scheme, which involves the checking of the shape quality of the elements before and after node shifting, is known as *smart Laplace smoothing*. This idea can be further refined to check the shape quality of the triangles in **P** at more than one point. Points on the line joining **x** and **c** can be conveniently specified by parameter λ.

$$\mathbf{x}_\lambda = (1 - \lambda)\mathbf{x} + \lambda\mathbf{c} \quad \text{such that } \mathbf{x}_0 = \mathbf{x} \text{ and } \mathbf{x}_1 = \mathbf{c}$$

Let α^*_λ be the geometric mean α-quality of the triangles in **P** corresponding to a shift of node **x** to $\mathbf{x}_\lambda$; then by maximising α^*_λ, the best location of the shift for **x** can be determined. To limit the points to be evaluated to within a reasonable number, only $\lambda = 0.9, 1.0$ and 1.1 are tested. The enhanced Laplace smoothing scheme in which the node shift is based on the quality assessment at some strategic points is known as *QL smoothing*.

6.3.1.2 LO of triangular mesh

Based on shape measures such as mean ratio η or α-quality coefficient on triangular elements, the quality of a triangular mesh $\mathbf{T} = \{\Delta_i, i = 1,N\}$ can be evaluated. Let α_i be the α-quality of triangle Δ_i; then a measure of the quality of triangular mesh **T** is given by

$$\bar{\alpha} = \left(\prod_{\Delta_i \in T} \alpha_i\right)^{1/N} \quad ; \quad \alpha_{min} = \min_{\Delta_i \in T} \alpha_i$$

The geometric mean $\bar{\alpha}$ can be used as the cost function of the triangular mesh to be optimised by shifting of interior nodes subject to the constraint that nodal shifting should not reduce the α-quality of the triangles in **T** lower than α_{min}. The condition that **T** being optimal with respect to α-quality measure is that for each point **x** surrounded by polygon **P**,

$$\frac{\partial \alpha_P}{\partial x} = 0 \text{ and } \frac{\partial \alpha_P}{\partial y} = 0, \quad \text{where } \alpha_P = \left(\prod_{\Delta_k \in P} \alpha_k\right)^{1/n} \quad n = \text{number of triangles in P}$$

As a global optimisation on α is very costly, this local condition for being optimal in α measure can be used to improve the quality of a triangular mesh. Analytical expressions for the derivatives of α_P in closed form are not available and, in general, rather tedious. Hence, numerical differentiation is applied to obtain approximate solutions for $\frac{\partial \alpha_P}{\partial x} = 0$ and $\frac{\partial \alpha_P}{\partial y} = 0$.

Let $\mathbf{x}^* = \mathbf{x} + (\Delta x, \Delta y)$ be the new position of node **x**; then the α_P value at $\mathbf{x}^*$ is given by

$$\alpha_P(\mathbf{x}^*) \approx \alpha_P(\mathbf{x}) + \frac{\partial \alpha_P}{\partial x}\Delta x + \frac{\partial \alpha_P}{\partial y}\Delta y$$

α_P can always be improved by setting $(\Delta x, \Delta y) = \lambda \nabla \alpha_P = \lambda\left(\frac{\partial \alpha_P}{\partial x}, \frac{\partial \alpha_P}{\partial y}\right)$ for some positive λ:

$$\alpha_P(\mathbf{x}^*) = \alpha_P(\mathbf{x}) + \lambda\left[\left(\frac{\partial \alpha_P}{\partial x}\right)^2 + \left(\frac{\partial \alpha_P}{\partial y}\right)^2\right]$$

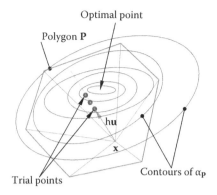

Figure 6.11 Local optimisation scheme.

α_P always increases unless the derivatives both vanish. By numerical differentiation, we have

$$\frac{\partial \alpha_P}{\partial x} \approx \frac{\alpha_P(x+h) - \alpha_P(x)}{h} \quad \text{and} \quad \frac{\partial \alpha_P}{\partial y} \approx \frac{\alpha_P(y+h) - \alpha_P(y)}{h}$$

where h can be set equal to the nodal shift, which is approximately 10% of the local element size.

$$(\Delta x, \Delta y) = \lambda \left(\frac{\partial \alpha_P}{\partial x}, \frac{\partial \alpha_P}{\partial y} \right) \Rightarrow \lambda^2 \left[\left(\frac{\partial \alpha_P}{\partial x} \right)^2 + \left(\frac{\partial \alpha_P}{\partial y} \right)^2 \right] = (\Delta x)^2 + (\Delta y)^2 = h^2$$

$$\Rightarrow \lambda = \frac{h}{\sqrt{\left(\frac{\partial \alpha_P}{\partial x} \right)^2 + \left(\frac{\partial \alpha_P}{\partial y} \right)^2}}; \quad (\Delta x, \Delta y) = \lambda \nabla \alpha_P = h\mathbf{u} \quad \text{where unit vector } \mathbf{u} = \frac{\nabla \alpha_P}{\| \nabla \alpha_P \|}$$

As numerical differentiation has been used, λ has to be varied in order to optimise $\alpha_P(x^*)$. Typically, we can adjust λ in 0.1λ intervals between [a, b], with a = 0.8λ and b = 1.2λ, as shown in Figure 6.11. Unless $\Delta\alpha_P = (\alpha_P(x+h) - \alpha_P(x))^2 + (\alpha_P(y+h) - \alpha_P(x))^2$ is greater than some threshold $\varepsilon = 10^{-6}$, node shifting will not take place to enhance numerical stability; and this node is deemed to be optimised since α_P can hardly be increased any more. For small $\Delta\alpha_P < 10^{-4}$ the shift $h\mathbf{u}$ can also be reduced from 10% to say 1%. Obviously, the interval [a, b] and the increment of 0.1λ are rather arbitrary without any theoretical basis, and readers can formulate their own schemes in the light of more experience. Like the QL smoothing, each interior node in the mesh will be processed in turn following the natural order of the nodes until all the nodes are processed. A number of iteration cycles can be performed until no more improvement could be made.

6.3.1.3 GETMe (2D)

The GETMe is a geometry-based smoothing method in which nodes are shifted entirely based on the geometry of the polygon (polyhedron) in such a way that in each cycle of node shifting, the polygon (polyhedron) will become more and more regular. A generic scheme has been proposed for node shifting of a general polygon (polyhedron) in which the directions for node shifting are based on an auxiliary polygon (polyhedron) formed by connecting the centroids of the faces of the original polygon (polyhedron) (Vartziotis and Wipper 2012). However, for the T3 triangular elements and Q4 quadrilateral elements, an earlier version by Vartziotis and Wipper (2009) without reference to an auxiliary polygon will be introduced in this section.

Let $\{X_1, X_2, X_3\}$ be the vertices of triangle T. The transformation of triangle T will be done in two steps. In the first step, triangle T will be changed to triangle $T' = \{Y_1, Y_2, Y_3\}$ by the following vertex transformation, as shown in Figure 6.12a:

$$Y_1 = X_3 + \frac{L_{31}(\mathbf{u}_3 + \tan\theta\mathbf{v}_3)}{2}, \quad Y_2 = X_1 + \frac{L_{12}(\mathbf{u}_1 + \tan\theta\mathbf{v}_1)}{2}, \quad Y_3 = X_2 + \frac{L_{23}(\mathbf{u}_2 + \tan\theta\mathbf{v}_2)}{2}$$

$$Y_{k+1} = X_k + \frac{L_{k(k+1)}(\mathbf{u}_k + \tan\theta\mathbf{v}_k)}{2} \tag{6.2}$$

where $\mathbf{u}_k$ is a unit vector along X_kX_{k+1}, $\mathbf{v}_k$ is a unit vector normal to X_kX_{k+1}, $L_{k(k+1)}$ is the length between vertices X_k and X_{k+1} and $k + 1$ follows the modulus arithmetic such that $k + 1 \equiv \text{mod}(k,3) + 1$. Comparing T' with T, we can see that the shape of the triangle has been much improved by the transformation. Vartziotis and Wipper (2009) showed that if we keep applying transformation 6.2 to triangle T, it will converge to an equilateral triangle, i.e.

$T \mapsto T^n \approx$ Equilateral triangle for large n by transformation 6.2

It is most optimal to set θ equal to $\pi/3$, for which the convergence is pretty fast, and usually, a few cycles will bring triangle T close to equilateral. However, as you can see, there is a rotation movement between T and T', which is not a problem for a single element, but it may not be convenient to apply the process to all the elements in a triangular mesh. Hence, it is necessary to minimise the torsional effect in the transformation. To this end, another similar transformation can be carried out but somehow in a reverse order. In the second step of reverse transformation, T' will be changed to triangle $T'' = \{Z_1, Z_2, Z_3\}$ by the following vertex transformation, as shown in Figure 6.12b:

$$Z_1 = Y_1 + \frac{L_{12}(\mathbf{u}_1 + \tan\theta\mathbf{v}_1)}{2}, \quad Z_2 = Y_2 + \frac{L_{23}(\mathbf{u}_2 + \tan\theta\mathbf{v}_2)}{2}, \quad Z_3 = Y_3 + \frac{L_{31}(\mathbf{u}_3 + \tan\theta\mathbf{v}_3)}{2}$$

$$Z_k = Y_k + \frac{L_{k(k+1)}(\mathbf{u}_k + \tan\theta\mathbf{v}_k)}{2} \tag{6.3}$$

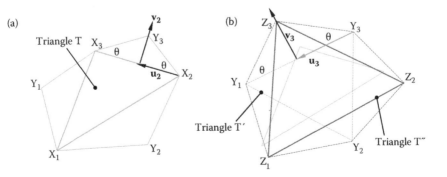

Figure 6.12 **Transforming triangle T to T″ in two steps: (a) step 1; (b) step 2.**

The transformation from $\{Y_1, Y_2, Y_3\}$ to $\{Z_1, Z_2, Z_3\}$ is similar to that from $\{X_1, X_2, X_3\}$ to $\{Y_1, Y_2, Y_3\}$, except that there is a difference in node labels so that no rotation of the triangle for the transformation from $\{X_1, X_2, X_3\}$ to $\{Z_1, Z_2, Z_3\}$ will be induced. Hence, in the GETMe node smoothing of triangular meshes, the two steps will always be combined into one single operation. The triangles have to be transformed without a change in area; however, in the transformation, the nodes are displaced normal to the edges, and the area of the triangle will be increased. As the centroid of the triangle will not move in the transformation, the area can be restored by scaling the transformed vertices with a factor ρ given by

$$\text{scaling factor,} \quad \rho = \sqrt{\frac{\text{Area}(T)}{\text{Area}(T'')}}$$

and the scaled triangle T^* whose vertices after scaling are adjusted to

$$X_k^* = C + \rho(Z_k - C), \quad k = 1, 2, 3; \quad \text{Centroid } C = (X_1 + X_2 + X_3)/3 \qquad (6.4)$$

Let's take a look at the transformation of triangle $T = \{(0,0), (10,0), (1,2)\}$ with area = 10 units and α-quality = 0.3646, as shown in Figure 6.13. After the first step, T is transformed to T' with α increased to 0.7915, and the second step further brings α to 0.9434 for T'' whose area has been increased to 29.1 times that of the original triangle T. The area of triangle T'' is brought back to 10 units by adjusting its vertices with the scaling factor ρ without altering the α-quality, as shown in Figure 6.13.

Given a triangle T, the GETMe transformation is formally defined as

Transformation GETMe: $T \mapsto T^*$,   with T^* being computed using formula 6.4

Similar to QL smoothing and LO scheme, GETMe transformation works on each interior node in turn. Given a triangular mesh of N triangles, $\mathbf{T} = \{\Delta_i, i = 1, N\}$, for a given node $\mathbf{x}$, the patch of triangles surrounding node $\mathbf{x}$ is given by

$P(\mathbf{x}) = \{\Delta_k \in \mathbf{T}; \mathbf{x} \in \Delta_k\}$, $\mathbf{x} \in \Delta_k$ means that $\mathbf{x}$ is one of the vertices of triangle Δ_k

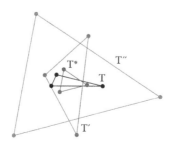

Figure 6.13 GETMe transformation.

Apply GETMe transformation to each triangle Δ_k in the polygon $\mathbf{P}$, and let $\mathbf{x}_k$ be the new position of node $\mathbf{x}$ of Δ_k after transformation. As $\mathbf{x}_k$ are not the same for all the triangles in $\mathbf{P}$, the final position of node $\mathbf{x}$ is given by a weighted sum of $\mathbf{x}_k$ such that

New position of node $\mathbf{x}$, $\quad \bar{\mathbf{x}} = \dfrac{\sum_{\Delta_k \in P} w_k \mathbf{x}_k}{\sum_{\Delta_k \in P} w_k}$

in which the weight for each triangle Δ_k can be computed in a number of ways. Vartziotis and Wipper (2009) proposed the following weight for Δ_k:

$$w_k = \eta_o + (1 - \eta_k)$$

where $\eta_o \in [0,1]$ prevents division by zero and η_k is a shape measure of Δ_k.

However, promising results can also be obtained by a weight given by the ratio of shape qualities after and before transformation of triangle Δ_k:

$$w_k = \eta_k^* / \eta_k$$

This weighting formula suggests that higher weights are attached to triangles with greater improvement in the transformation. Other weights of interest include $w_k = \eta_k^* - \eta_k$. In case invalid elements are formed in repositioning $\mathbf{x}$ to $\bar{\mathbf{x}}$, a relaxation process was proposed by Vartziotis and Wipper (2009) such that the new position of node $\mathbf{x}$ is given by

$$\mathbf{x}_{new} = \xi\mathbf{x} + (1 - \xi)\bar{\mathbf{x}}, \quad \text{Relaxation factor } \xi \in [0,1]$$

The GETMe transformation can be applied sequentially to each node in turn to complete one cycle of nodal smoothing. However, with an additional vector (V) to store the new positions of the nodes, the smoothing process can be carried out in parallel. Under this circumstance, in lieu of considering interior nodes and their surrounding polygons in turn, attention is focused on the triangles in which GETMe transformation is applied to all the triangles in the mesh. For a given triangle, new positions of its three vertices multiplied by an appropriate weight are added to vector V, and the weights of the respective nodes are summed up in another vector W. When all the triangles are processed, the new positions of the nodes are given by

New position of nodes: $\quad \bar{\mathbf{x}}_i = \dfrac{V_i}{W_i} \quad i = 1, n$ where n is the number of interior nodes

In the parallel transformation, new node positions are computed at the same time, and GETMe transformation is applied to each element only once to establish vectors V and W for the new positions of the nodes. On the other hand, in the sequential version, GETMe transformation has to be applied repeatedly on the triangles as they belong to many nodal patches.

6.3.1.4 Examples: Node smoothing for triangular meshes

Delaunay triangulations of randomly generated points are taken as test meshes for the three node-smoothing schemes. Uniformly distributed meshes are obtained by extracting the central part of Delaunay triangulations of randomly generated points, and these meshes are easily reproduced to assess the performance of a node-smoothing scheme. A relatively coarse mesh of random point distribution before and after optimisation is shown in Figure 6.14. Ten such meshes of more triangles are prepared, and the node-smoothing schemes, namely, QL, LO and GETMeT3 parallel smoothing, are applied to these meshes.

For each optimisation scheme, 10 cycles of smoothing are conducted to each mesh, and the results are presented in Table 6.1 in which NE = number of triangles, α_{min} = minimum α-quality, α_{mean} = mean α-quality and Aver = average values of the ten test runs. It can be seen that the performance of all three node-smoothing schemes is very promising and similar in which LO gives the best α_{min} value, while GETMe gives the best α_{mean} value. As for the average CPU time for ten cycles of node smoothing, QL takes 0.305 s, LO takes 2.06 s and GETMe takes 0.702 s on a PC i7 CPU 870 at 2.93 GHz running on XP mode with

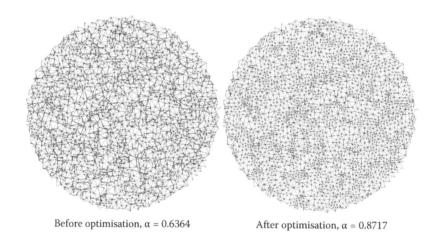

Before optimisation, α = 0.6364   After optimisation, α = 0.8717

Figure 6.14 Central part extracted from Delaunay triangulation of random points.

Table 6.1 Results of node smoothing by QL, LO and GETMe

Test	NE	Initial mesh α_{min}	Initial mesh α_{mean}	QL α_{min}	QL α_{mean}	LO α_{min}	LO α_{mean}	GETMe α_{min}	GETMe α_{mean}
1	58,712	0.002934	0.6364	0.1033	0.8696	0.154	0.8535	0.154	0.8717
2	58,782	0.001558	0.6344	0.06006	0.87	0.06006	0.8529	0.06006	0.8712
3	59,111	0.001251	0.6358	0.09901	0.8693	0.1278	0.8526	0.1278	0.871
4	58,400	0.001392	0.6373	0.08131	0.8697	0.08845	0.8523	0.08845	0.8711
5	58,780	0.00297	0.6369	0.03019	0.8715	0.0189	0.854	0.103	0.8721
6	58,736	0.00176	0.6359	0.07378	0.8695	0.1922	0.8528	0.1294	0.8715
7	59,049	0.000494	0.6351	0.04016	0.8681	0.1721	0.8516	0.03943	0.8699
8	58,768	0.002593	0.6355	0.02925	0.8688	0.02925	0.8514	0.02925	0.8704
9	59,268	0.002943	0.6354	0.04467	0.8698	0.05312	0.8521	0.05312	0.8706
10	59,015	0.00157	0.636	0.05398	0.8701	0.1896	0.8517	0.1393	0.8711
Aver	58,862	0.001947	0.6359	0.06157	0.8696	0.1086	0.8525	0.09238	0.8711

Visual Fortran QuickWin graphics supports. The CPU time can only be taken for reference in relative performance, as it varies significantly on many factors including the speed and the available memory of the machine, the operating system, the programming language, the coding skills and style, the algorithm and the parameters selected, etc.

In the second test, the three node-smoothing schemes are applied to meshes of non-uniform point distribution in which points are generated clustered on curves of 0.1% width relative to the diameter of the mesh, as shown in Figure 6.15. Due to the rapid change in size for triangles near the curves, α_{min} and α_{mean} are much lower than those of meshes of uniform distribution of random points, as shown in Table 6.2. Again, the meshes are optimised by the three node-smoothing schemes with ten cycles of iteration for each scheme, and the results are listed in Table 6.2. A similar trend is observed for meshes of non-uniform point distribution such that LO gives the best α_{min} value and GETMe gives the best α_{mean} value.

As for the characteristics of convergence, the α_{min} value, the α_{mean} value and the number of nodes moved (NM) are recorded for each cycle of node smoothing for a typical run on the non-uniform mesh, as shown in Table 6.3. QL is the most efficient if only a few cycles of iteration are allowed in the smoothing process, as the NM decreases rapidly after the

Figure 6.15 Delaunay triangulation of non-uniformly distributed points.

Table 6.2 Non-uniform mesh optimised by QL, LO and GETMe

Test	NE	Initial mesh α_{min}	Initial mesh α_{mean}	QL α_{min}	QL α_{mean}	LO α_{min}	LO α_{mean}	GETMe α_{min}	GETMe α_{mean}
1	59,006	0.00268	0.5821	0.02212	0.7866	0.08805	0.7731	0.0578	0.7949
2	58,102	0.002471	0.5822	0.006538	0.7832	0.1283	0.7701	0.08445	0.7927
3	58,396	0.001233	0.5813	0.02093	0.7827	0.09708	0.7696	0.04395	0.7925
4	58,399	0.002802	0.5807	0.01947	0.783	0.02628	0.7703	0.06385	0.7921
5	58,489	0.001125	0.5822	0.001614	0.7848	0.000938	0.77	0.01607	0.7939
6	58,302	0.001212	0.5816	0.02844	0.7866	0.02337	0.7738	0.06416	0.7962
7	58,313	0.000468	0.581	0.02025	0.7815	0.107	0.7709	0.03237	0.7924
8	58,594	0.001515	0.5819	0.01487	0.7827	0.1712	0.7701	0.03724	0.7922
9	59,058	0.000967	0.5795	0.01984	0.7853	0.03421	0.772	0.01984	0.7948
10	58,212	0.001441	0.5808	0.02632	0.7869	0.1474	0.7738	0.05748	0.7957
Aver	58,487	0.001591	0.5813	0.01804	0.7843	0.08238	0.7714	0.04772	0.7937

Table 6.3 Convergence characteristics of QL, LO and GETMe

Cycle	QL			LO			GETMe		
	α_{min}	α_{mean}	NM	α_{min}	α_{mean}	NM	α_{min}	α_{mean}	NM
0	0.000468	0.581		0.000468	0.581		0.000468	0.581	
1	0.01877	0.7321	37,654	0.02629	0.6769	59,343	0.01909	0.6838	51,539
2	0.02025	0.769	27,158	0.02513	0.723	59,246	0.01908	0.7345	50,287
3	0.02025	0.7781	16,117	0.02678	0.7456	59,086	0.03237	0.7591	48,200
4	0.02025	0.7806	7837	0.06244	0.757	58,935	0.03237	0.7722	46,067
5	0.02025	0.7812	3124	0.08931	0.7631	58,820	0.03237	0.7797	44,022
6	0.02025	0.7814	1045	0.1015	0.7666	58,683	0.03237	0.7845	42,017
7	0.02025	0.7815	317	0.107	0.7687	58,730	0.03237	0.7876	40,317
8	0.02025	0.7815	98	0.107	0.7698	58,731	0.03237	0.7898	38,670
9	0.02025	0.7815	38	0.107	0.7706	58,765	0.03237	0.7913	37,315
10	0.02025	0.7815	12	0.107	0.7709	58,846	0.03237	0.7924	36,150

first three cycles. As for the other two smoothing schemes, there is still consistent improvement in each cycle of iteration, even though the rate is diminishing to a very little progress after, say, ten cycles. From the convergence characteristics of the three smoothing methods, a combined scheme might be the most effective. As a numerical experiment, a combined scheme (4 + 2 + 4) consisting of four cycles of QL, followed by two rounds of LO and further optimised with four cycles of GETMe is tested. The average results on a number of meshes are given by $\alpha_{min} = 0.08448$, $\alpha_{mean} = 0.8033$ and the CPU time taken = 0.766 s. Relative to the three smoothing methods, the combined scheme gives the best performance on both α_{min} and α_{mean}, and in terms of CPU time, it is similar to GETMe.

The performance of the three schemes is plotted in Figure 6.16. We can see that the QL smoothing reaches the converged value basically in the first four iterations; as for the other two schemes, improvement could still be made in the tenth cycle, though it is already pretty small. QL smoothing is the most efficient in the first five cycles of iteration, whereas GETMe is the most consistent, and all three schemes can be used, in general, to improve triangular meshes of various characteristics.

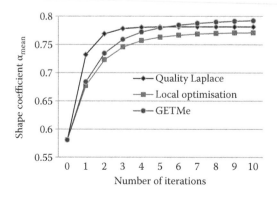

Figure 6.16 Convergence characteristics.

6.3.2 Optimisation of quadrilateral and mixed meshes

The three optimisation schemes for triangular meshes by node shifting, namely, QL smoothing, local quality optimisation and GETMeT3, are generic and can be applied to quadrilateral or mixed meshes of triangles and quadrilaterals. Nothing needs to be changed for the QL smoothing and the LO schemes except a proper shape measure for quadrilateral elements. As for the GETMe, only a slight modification is needed for the transformation of quadrilaterals.

6.3.2.1 Shape quality of a mixed mesh of triangles and quadrilaterals

The shape measure η for a quadrilateral element Q can be defined as follows.

$$\eta(Q) = \left(\prod_{k=1,4} \eta_k \right)^{1/4} \text{ with } \eta_k = \frac{2\det(F_k)}{\|F_k\|^2}$$

where F_k is the transformation matrix at the kth vertex. Let a_k be the vector along edge $X_k X_{k+1}$ and b_k be the vector along edge $X_k X_{k-1}$, as shown in Figure 6.17; then F_k is given by

$$F_k = [a_k \quad b_k]$$

$\det(F_k)$ = area of the parallelogram spanned by vectors a_k and b_k, and $\|F_k\|^2 = \|a_k\|^2 + \|b_k\|^2$. Without changing the magnitude of vectors a_k and b_k, $\det(F_k)$ attains the largest value when a_k and b_k are perpendicular to each other forming a rectangle. Moreover, η_k will have the largest value of 1 when vectors a_k and b_k are of the same magnitude forming the sides of a square. Hence, quadrilateral Q will be regular (a square) if η_k at the four vertices are all equal to 1.

Given a mixed mesh of N triangular and quadrilateral elements, $M = \{E_i, i = 1,N\}$, for a given node x, the patch of elements surrounding node x, as shown in Figure 6.18, is given by

$$P(x) = \{E_k \in M; x \in E_k\} \quad x \in E_k \text{ means that } x \text{ is a vertex of element } E_k$$

To measure the quality of the elements in P, the geometric mean μ-quality and the minimum μ-value of the elements can be used:

$$\bar{\mu} = \left(\prod_{E_k \in P} \mu_k \right)^{1/n}; \quad \mu_{min} = \min_{E_k \in P} \mu_k$$

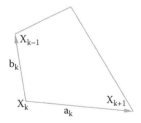

Figure 6.17 Transformation matrix at vertex K.

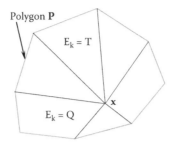

Figure 6.18 Surrounding polygon for a mixed mesh of quadrilaterals and triangles.

where

$$\mu_k = \begin{cases} \alpha_k & \text{if } E_k \text{ is a triangle} \\ \eta_k & \text{if } E_k \text{ is a quadrilateral} \end{cases}$$

6.3.2.2 GETMe transformation for quadrilaterals

Let $\{X_1, X_2, X_3, X_4\}$ be the vertices of quadrilateral Q. The transformation of quadrilateral Q will be done in two steps. In the first step, quadrilateral Q will be changed to quadrilateral $Q' = \{Y_1, Y_2, Y_3, Y_4\}$ by the following vertex transformation, as shown in Figure 6.19a:

$$Y_{k+1} = X_k + \frac{L_{k(k+1)}(\mathbf{u}_k + \tan\theta\mathbf{v}_k)}{2} \tag{6.5}$$

where $\mathbf{u}_k$ is a unit vector along $X_k X_{k+1}$, $\mathbf{v}_k$ is a unit vector normal to $X_k X_{k+1}$, $L_{k(k+1)}$ is the length between vertices X_k and X_{k+1} and $k + 1$ follows the modulus arithmetic such that $k + 1 \equiv \text{mod}(k,4) + 1$. Comparing Q' with Q, we can see that the shape of the quadrilateral has been much improved by the transformation. Vartziotis and Wipper (2009) showed that if we keep applying transformation 6.5 to quadrilateral Q, it will converge to a square, i.e.

$$Q \mapsto Q^n \approx \text{square for large n by transformation 6.5}$$

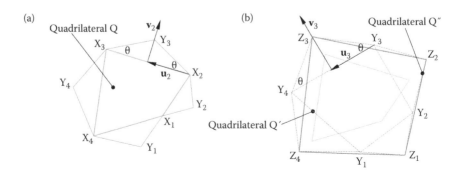

Figure 6.19 Transforming quadrilateral Q to Q'' in two steps: (a) step 1; (b) step 2.

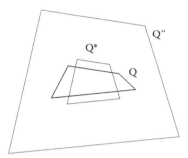

Figure 6.20 **GETMe transformation of quadrilateral element Q to Q*.**

It is most optimal to set θ equal to $\pi/4$ for which the convergence is pretty fast, and usually, a few cycles will bring quadrilateral Q close to a square. However, as you can see, there is a rotation between Q and Q'. Hence, it is necessary to minimise the torsional effect in the transformation. To this end, another similar transformation can be carried out in a reverse order. In the second step of reverse transformation, Q' will be changed to quadrilateral $Q'' = \{Z_1, Z_2, Z_3, Z_4\}$ by the following vertex transformation, as shown in Figure 6.19b:

$$Z_k = Y_k + \frac{L_{k(k+1)}(u_k + \tan\theta v_k)}{2} \tag{6.6}$$

To preserve the area of quadrilateral Q, Q'' has to be scaled down to Q* as follows:

$$X_k^* = C + \rho(Z_k - C), \quad k = 1,2,3,4; \quad \text{Centroid } C = (X_1 + X_2 + X_3 + X_4)/4 \tag{6.7}$$

where scaling factor ρ = square root of the ratio of areas. Given a quadrilateral Q, the GETMe transformation is formally defined as

Transformation GETMe: $Q \mapsto Q^*$,   with Q* being computed using formula 6.7

Let's take a look at the transformation of quadrilateral Q = {(0,0), (10,1), (8,3), (2,4)} with an area of 24 units and η-quality = 0.5662, as shown in Figure 6.20. After GETMe transformation, Q is transformed to Q'' with η being increased to 0.9229 and the area increased to 256.9 units. Quadrilateral Q'' is scaled to Q* to bring its area back to 24 units by adjusting its vertices with scaling factor $\rho = 0.3057$ without altering the η-quality, as shown in Figure 6.20. Five more GETMe transformations with proper scaling are applied to Q* to further increase the η values to 0.9633, 0.9827, 0.9919, 0.9962 and 0.9982.

6.3.2.3 Examples: Node smoothing for mixed meshes

A coarse mesh of only 80 triangles and 12 quadrilaterals for easy visual inspection, as shown in Figure 6.21, is employed as the first test of the QL smoothing, the GETMe and a combined scheme in which QL smoothing and GETMe are applied once in each cycle of iteration. The results of applying ten cycles of QL, ten cycles of GETMe and five cycles of the combined scheme are presented in Table 6.4. It is seen that for this simple example, the QL performs slightly better than GETMe, and the combined scheme records the best performance in terms of both μ_{min} and μ_{mean} values. Similar to the case of triangular meshes, QL

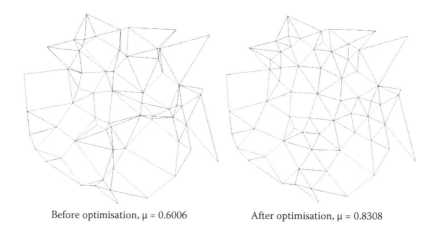

Before optimisation, μ = 0.6006   After optimisation, μ = 0.8308

Figure 6.21 Optimisation of quadrilateral and triangular elements.

Table 6.4 Smoothing a mixed mesh by QL, GETMe and a combined scheme

Cycle	QL			GETMe			Combined scheme		
	μ_{min}	μ_{mean}	NM	μ_{min}	μ_{mean}	NM	μ_{min}	μ_{mean}	NM
0	0.10014	0.6006		0.10014	0.6006		0.10014	0.6006	
1	0.39294	0.7985	32	0.19559	0.7133	32			32
2	0.43302	0.8232	25	0.27505	0.7610	37	0.43302	0.81118	31
3	0.43302	0.8239	15	0.39369	0.7851	35			26
4	0.43302	0.8236	8	0.40485	0.7985	34	0.43302	0.82921	27
5	0.43302	0.8236	3	0.41122	0.8067	35			17
6	0.43302	0.8236	0	0.41122	0.8109	29	0.43302	0.83077	15
7	0.43302	0.8236	0	0.41817	0.8135	29			12
8	0.43302	0.8236	0	0.41817	0.8154	28	0.43302	0.83077	13
9	0.43302	0.8236	0	0.41817	0.8165	25			9
10	0.43302	0.8236	0	0.41817	0.8177	26	0.43302	0.83077	12

achieves the converged μ_{min} and μ_{mean} values virtually in the first four iterations. Hence, if QL is used for mesh optimisation, five cycles will, in general, be sufficient.

There are 3450 triangles and 1233 quadrilaterals in the second mixed mesh of randomly generated points, as shown in Figure 6.22, to test the performance of QL smoothing, the GETMe and a combined scheme for mixed meshes of more elements. The results of μ_{min} and μ_{mean} values for triangles and quadrilaterals of applying five cycles of QL, five cycles of GETMe and the combined scheme consisting of five cycles of QL followed by five cycles of GETMe are presented in Table 6.5, in which T_i and Q_i are, respectively, the quality of triangles and quadrilaterals after the ith smoothing cycle. QL achieves higher values in μ_{mean} both for triangles and quadrilaterals; on the other hand, GETMe achieves better results in μ_{min} values. Extremely high-quality mesh is produced by the combined scheme with the best results in μ_{min} and μ_{mean} values for both triangles and quadrilaterals. Hence, for efficiency consideration, we can just apply a few cycles of QL to a mixed mesh, and in case

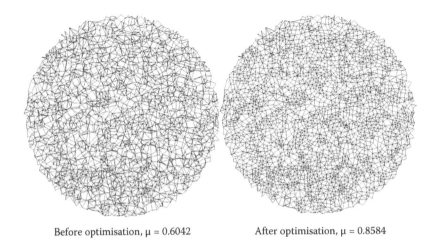

Before optimisation, μ = 0.6042   After optimisation, μ = 0.8584

Figure 6.22 **Optimisation of a mixed mesh of more elements.**

Table 6.5 **Second mesh optimised by QL, GETMe and a combined scheme**

	QL			GETMe			Combined scheme		
Cycle	μ_{min}	μ_{mean}	NM	μ_{min}	μ_{mean}	NM	Cycle	μ_{min}	μ_{mean}
T0	0.02884	0.6042		0.02884	0.6042		Quality	0.05640	0.8732
Q0	0.1523	0.7083		0.1523	0.7083		Laplace	0.2676	0.8535
T1	0.05640	0.8372		0.1028	0.7611		GETMe1	0.1934	0.8784
Q1	0.2644	0.8080	2195	0.2686	0.7880	2460		0.3833	0.8562
T2	0.05640	0.8659		0.1507	0.8189		GETMe2	0.2928	0.8796
Q2	0.2674	0.8438	1730	0.3421	0.8200	2375		0.3964	0.8578
T3	0.05640	0.8716		0.1507	0.8438		GETMe3	0.2928	0.8799
Q3	0.2674	0.8516	1197	0.3421	0.8344	2220		0.3964	0.8580
T4	0.05640	0.8729		0.1507	0.8555		GETMe4	0.2928	0.8802
Q4	0.2676	0.8534	747	0.3421	0.8413	2011		0.3964	0.8582
T5	0.05640	0.8732		0.1507	0.8620		GETMe5	0.2928	0.8804
Q5	0.2676	0.8535	233	0.3421	0.8449	1858		0.3964	0.8584

higher mesh quality is required, a few more cycles of GETMe can then be applied to further enhance the mesh quality.

The ability of QL and GETMe in untangling inverted quadrilateral elements is tested using a quadrilateral mesh of 100 elements. A rectangular domain of length 100 and height 80 is divided into a 10 × 10 grid, such that there are 100 rectangular elements represented by the cells of the grid. A perturbation is then applied to all the interior nodes randomly along the horizontal and vertical directions with a magnitude as large as 30 units, as shown in Figure 6.23a. The results of a typical test of applying ten cycles of QL smoothing and ten cycles of GETMe smoothing are shown in Figure 6.23b–g. It is found that the inverted mesh is always untangled by the QL in two cycles, and an η-quality of 0.9676 is achieved in ten cycles of smoothing. On the other hand, the inverted mesh is untangled by GETMe only in three iteration cycles, and an η-quality of 0.9425 is achieved in ten cycles of smoothing.

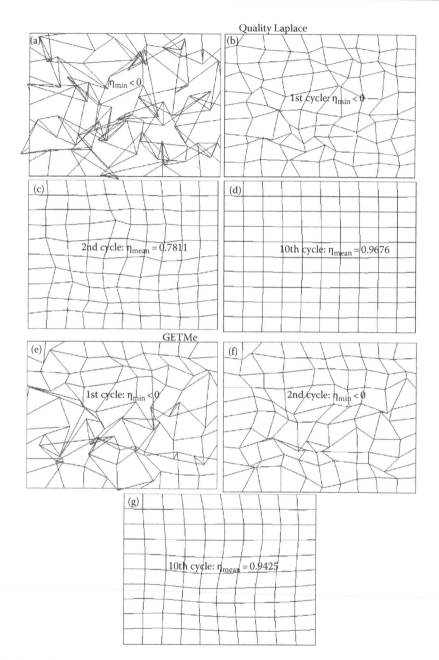

Figure 6.23 Untangling inverted elements by quality Laplace and GETMe: (a) inverted mesh; (b) one cycle of QL; (c) two cycles of QL; (d) ten cycles of QL; (e) one cycle of GETMe; (f) two cycles of GETMe; (g) ten cycles of GETMe.

6.3.3 Node smoothing for 3D meshes

The three optimisation schemes for 2D meshes by node shifting, namely, QL, LO and GETMe smoothing schemes, are generic across dimensions and can be applied to 3D meshes without much modification. Nothing needs to be changed for the QL smoothing and the LO schemes except a proper shape measure for the FEs of the 3D mesh. As for the GETMe, a general transformation operation involving an auxiliary polyhedron defined by connecting the centroids of the faces of the 3D solid elements will be introduced.

6.3.3.1 QL smoothing (3D)

Given a 3D mesh of N polyhedral elements, $M = \{E_i, i = 1,N\}$, for a given node $\mathbf{x}$, the patch of polyhedral elements surrounding node $\mathbf{x}$, as shown in Figure 6.24, is given by

$$P(\mathbf{x}) = \{E_k \in M; \mathbf{x} \in E_k\}, \quad \mathbf{x} \in E_k \text{ means that } \mathbf{x} \text{ is one of the vertices of element } E_k$$

The *centroid* of polygon $\mathbf{P}$ is given by

$$c = \frac{1}{m-n}\left[\left(\sum_{E_k \in P}\sum_{\mathbf{x}_i^k \in E_k} \mathbf{x}_i^k\right) - n\mathbf{x}\right]$$

where $m = \sum_{E_k \in P}$ (number of nodes in E_k) is the sum of the number of nodes in each element of P, and n is the number of elements in P. This is an approximation formula for the average of the nodes on the surface of polyhedron P. However, this discrepancy may not be a draw-back, as c will automatically lean towards nodes shared by many elements, hence enhancing the quality of most of the elements in P.

By the QL smoothing for polyhedral meshes, the shape qualities of the elements in **P** at more than one point are evaluated. Points on the line joining $\mathbf{x}$ and $\mathbf{c}$ can be conveniently specified by parameter λ:

$$\mathbf{x}_\lambda = (1 - \lambda)\mathbf{x} + \lambda\mathbf{c} \quad \text{such that } \mathbf{x}_0 = \mathbf{x} \text{ and } \mathbf{x}_1 = \mathbf{c}$$

Let η_λ^* be the geometric mean η-quality of the elements in **P** corresponding to a shift of node $\mathbf{x}$ to $\mathbf{x}_\lambda$; then by maximising η_λ^*, the best location of the shift for $\mathbf{x}$ can be determined. To limit the points to be evaluated to within a reasonable number, only $\lambda = 0.9, 1.0$ and 1.1 are tested. In the light of more experience and at the expense of more computations, sophisticated schemes can be formulated by evaluating more points around centroid **c**.

6.3.3.2 LO of polyhedral mesh

The geometric mean $\bar{\eta}$ can be used as the cost function of a polyhedral mesh **M** to be optimised by shifting of interior nodes subject to the constraint that nodal shifting should

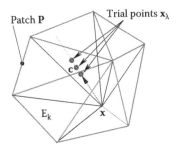

Figure 6.24 **A patch of tetrahedral elements around node x.**

not reduce the η-quality of the elements in $\mathbf{M}$ lower than η_{min}. The condition that $\mathbf{M}$ being optimal with respect to η-quality measure is that for each point $\mathbf{x}$ surrounded by patch $\mathbf{P}$,

$$\frac{\partial \eta_P}{\partial x} = 0, \frac{\partial \eta_P}{\partial y} = 0 \text{ and } \frac{\partial \eta_P}{\partial z} = 0 \text{ where } \eta_P = \left(\prod_{E_k \in P} \eta_k \right)^{1/n} \quad n = \text{number of elements in } \mathbf{P}$$

As a global optimisation on η is very costly, this local condition for being optimal in η measure can be used to improve the quality of a polyhedral mesh. Let $\mathbf{x}^* = \mathbf{x} + (\Delta x, \Delta y, \Delta z)$ be the new position of node $\mathbf{x}$; then the η_P value at $\mathbf{x}^*$ is given by

$$\eta_P(\mathbf{x}^*) \approx \eta_P(\mathbf{x}) + \frac{\partial \eta_P}{\partial x} \Delta x + \frac{\partial \eta_P}{\partial y} \Delta y + \frac{\partial \eta_P}{\partial z} \Delta z$$

Setting $(\Delta x, \Delta y, \Delta z) = \lambda \nabla \eta_P = \lambda \left(\frac{\partial \eta_P}{\partial x}, \frac{\partial \eta_P}{\partial y}, \frac{\partial \eta_P}{\partial z} \right)$ for some positive λ, we have

$$\eta_P(\mathbf{x}^*) = \eta_P(\mathbf{x}) + \lambda \left[\left(\frac{\partial \eta_P}{\partial x} \right)^2 + \left(\frac{\partial \eta_P}{\partial y} \right)^2 + \left(\frac{\partial \eta_P}{\partial z} \right)^2 \right]$$

η_P always increases unless all the derivatives vanish. By numerical differentiation, we have

$$\frac{\partial \eta_P}{\partial x} \approx \frac{\eta_P(x+h) - \eta_P(x)}{h}, \quad \frac{\partial \eta_P}{\partial y} \approx \frac{\eta_P(y+h) - \eta_P(y)}{h}, \quad \frac{\partial \eta_P}{\partial z} \approx \frac{\eta_P(z+h) - \eta_P(z)}{h}$$

where h can be set equal to the nodal shift, which is about 10% of the local element size.

$$(\Delta x, \Delta y, \Delta z) = \lambda \left(\frac{\partial \eta_P}{\partial x}, \frac{\partial \eta_P}{\partial y}, \frac{\partial \eta_P}{\partial z} \right) \Rightarrow \lambda^2 \left[\left(\frac{\partial \eta_P}{\partial x} \right)^2 + \left(\frac{\partial \eta_P}{\partial y} \right)^2 + \left(\frac{\partial \eta_P}{\partial z} \right)^2 \right]$$

$$= (\Delta x)^2 + (\Delta y)^2 + (\Delta z)^2 = h^2 \quad \text{or} \quad \lambda = \frac{h}{\sqrt{\left(\frac{\partial \eta_P}{\partial x} \right)^2 + \left(\frac{\partial \eta_P}{\partial y} \right)^2 + \left(\frac{\partial \eta_P}{\partial z} \right)^2}}$$

$$(\Delta x, \Delta y, \Delta z) = \lambda \nabla \eta_P = h\mathbf{u} \quad \text{where unit vector } \mathbf{u} = \frac{\nabla \eta_P}{\|\nabla \eta_P\|}$$

As numerical differentiation has been used, λ has to be varied in order to optimise $\eta_P(\mathbf{x}^*)$. Typically, we can adjust λ in 0.1λ intervals between $[a, b]$, with $a = 0.8\lambda$ and $b = 1.2\lambda$, as shown in Figure 6.25. Node shifting will only take place unless

$$\Delta \eta_P = (\eta_P(x+h) - \eta_P(x))^2 + (\eta_P(y+h) - \eta_P(y))^2 + (\eta_P(z+h) - \eta_P(z))^2 > \varepsilon \approx 10^{-6}$$

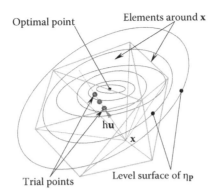

Figure 6.25 Shifting **x** along the gradient of η_P.

Otherwise, this node is deemed to be optimised since η_P can hardly be increased any more. For small $\Delta \eta_P < 10^{-4}$, the shift hu can also be reduced from 10% to say 1%. Obviously, the interval [a, b] and the increment of 0.1λ are rather arbitrary without any theoretical basis, which can be further refined in the light of more experience. Like the 2D case, each interior node in the mesh will be processed in turn until all the nodes are processed. A number of iteration cycles can be performed until η-quality can no longer be improved for mesh **M**.

6.3.3.3 GETMe (3D)

The GETMe is a nodal optimisation method in which nodes are shifted entirely based on the geometry of the polyhedron in such a way that for each cycle of shifting of nodes, the polyhedron will become more and more regular. A generic scheme has been proposed for node shifting of a general polyhedron in which the directions for node shifting are based on an auxiliary polyhedron formed by connecting the centroids of the faces of the original polyhedron (Vartziotis and Wipper 2012). However, for a tetrahedron, a simpler procedure is available in the transformation of a vertex in which the normal of the opposite face represents the direction of shift (Vartziotis et al. 2009).

6.3.3.3.1 Tetrahedral element

Let $\{\mathbf{x}_1, \mathbf{x}_2, \mathbf{x}_3, \mathbf{x}_4\}$ be the vertices of tetrahedron T. The normal vectors to the faces are given by

$$\mathbf{n}_1 = \mathbf{x}_2\mathbf{x}_4 \times \mathbf{x}_2\mathbf{x}_3, \quad \mathbf{n}_2 = \mathbf{x}_3\mathbf{x}_4 \times \mathbf{x}_3\mathbf{x}_1, \quad \mathbf{n}_3 = \mathbf{x}_4\mathbf{x}_2 \times \mathbf{x}_4\mathbf{x}_1, \quad \mathbf{n}_4 = \mathbf{x}_1\mathbf{x}_2 \times \mathbf{x}_1\mathbf{x}_3$$

As shown in Figure 6.26, tetrahedron T is transformed to $T' = \{\mathbf{y}_1, \mathbf{y}_2, \mathbf{y}_3, \mathbf{y}_4\}$, which is given by

$$T' = \begin{bmatrix} \mathbf{y}_1 \\ \mathbf{y}_2 \\ \mathbf{y}_3 \\ \mathbf{y}_4 \end{bmatrix} = \begin{bmatrix} \mathbf{x}_1 \\ \mathbf{x}_2 \\ \mathbf{x}_3 \\ \mathbf{x}_4 \end{bmatrix} + \lambda \begin{bmatrix} \mathbf{n}_1/\sqrt{\|\mathbf{n}_1\|} \\ \mathbf{n}_2/\sqrt{\|\mathbf{n}_2\|} \\ \mathbf{n}_3/\sqrt{\|\mathbf{n}_3\|} \\ \mathbf{n}_4/\sqrt{\|\mathbf{n}_4\|} \end{bmatrix}$$

where parameter $\lambda \approx 0.8$ will only affect the convergence rate of the GETMe transformation.

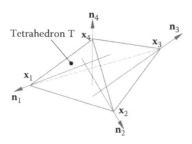

Figure 6.26 GETMe transformation following the normal vectors.

The centroid will not be preserved in the transformation as scaled normal vectors $n_i/\sqrt{\|n_i\|}$ rather than just the normal vectors are used in the transformation; on the other hand, the transformation is scale invariant, i.e. for $\sigma > 0$, $(\sigma T)' = \sigma T'$. A numerical test was conducted by Vartziotis et al. (2009) to show that the GETMe transformation will converge to a regular tetrahedron, and it is the most efficient to take λ roughly equal to 0.78. As the tetrahedron will expand in the transformation process following the directions of the normal vectors, the transformed element has to be scaled down so that its volume remains unchanged.

$$x_i^* = c + \rho(y_i - c), \quad i = 1,2,3,4; \quad \text{centroid } c = (x_1 + x_2 + x_3 + x_4)/4$$

where scaling factor $\rho = \sqrt[3]{\text{volume}(T)/\text{volume}(T')}$.

For the smoothing of an interior node connected to many elements, a weighted average can be used. Given mesh M, the patch of elements P connected to interior node x is given by

$$P(x) = \{E_k \in M; x \in E_k\}, \quad x \in E_k \text{ means that } x \text{ is one of the vertices of element } E_k$$

The new position for interior node x is a weighted average given by

$$\text{New position of node } x, \quad \bar{x} = \frac{\sum_{E_k \in P} w_k x_k}{\sum_{E_k \in P} w_k}$$

in which x_k is the new position of node x for element E_k after transformation, and the weight for E_k can be computed using

$$w_k = \eta_k^* / \eta_k$$

where η_k and η_k^* are, respectively, the η values of E_k before and after transformation.

In case that invalid elements are formed in repositioning x to $\bar{x}$, a relaxation process was proposed by Vartziotis et al. (2009) such that the new position of node x is given by

$$x_{new} = \xi x + (1 - \xi)\bar{x}, \quad \text{Relaxation factor } \xi \in [0,1]$$

The GETMe transformation can be applied sequentially to each node in turn to complete one cycle of nodal smoothing or in parallel for the new positions of all the interior nodes simultaneously. No change will be needed for the GETMe smoothing of other polyhedral element types except in the way the transformation is to be carried out. A generic scheme for GETMe transformation of a general polyhedron making use of an auxiliary polyhedron will be presented in Sections 6.3.3.3.2–6.3.3.3.5.

6.3.3.3.2 Hexahedral element

Let $\{x_i, i = 1,8\}$ be the vertices of a hexahedral element H labelled in a usual manner, as shown in Figure 6.27a. An auxiliary octahedron $\{y_i, i = 1,6\}$ can be defined by connecting the centroids of the six faces of hexahedron H, as shown in Figure 6.27b, such that

$$\text{octahedron}\begin{bmatrix} y_1 \\ y_2 \\ y_3 \\ y_4 \\ y_5 \\ y_6 \end{bmatrix} = \frac{1}{4}\begin{bmatrix} x_1 + x_2 + x_3 + x_4 \\ x_1 + x_5 + x_6 + x_2 \\ x_2 + x_6 + x_7 + x_3 \\ x_3 + x_7 + x_8 + x_4 \\ x_1 + x_4 + x_8 + x_5 \\ x_5 + x_8 + x_7 + x_6 \end{bmatrix}$$

Vectors normal to the faces and the centroid at each face of the octahedron are given by

$$\text{normal vectors:}\begin{bmatrix} n_1 \\ n_2 \\ n_3 \\ n_4 \\ n_5 \\ n_6 \\ n_7 \\ n_8 \end{bmatrix} = \begin{bmatrix} y_1y_2 \times y_1y_5 \\ y_1y_3 \times y_1y_2 \\ y_1y_4 \times y_1y_3 \\ y_1y_5 \times y_1y_4 \\ y_6y_5 \times y_6y_2 \\ y_6y_2 \times y_6y_3 \\ y_6y_3 \times y_6y_4 \\ y_6y_4 \times y_6y_5 \end{bmatrix}, \quad \text{centroids:}\begin{bmatrix} c_1 \\ c_2 \\ c_3 \\ c_4 \\ c_5 \\ c_6 \\ c_7 \\ c_8 \end{bmatrix} = \frac{1}{3}\begin{bmatrix} y_1 + y_2 + y_5 \\ y_1 + y_3 + y_2 \\ y_1 + y_4 + y_3 \\ y_1 + y_5 + y_4 \\ y_6 + y_5 + y_2 \\ y_6 + y_2 + y_3 \\ y_6 + y_3 + y_4 \\ y_6 + y_4 + y_5 \end{bmatrix}$$

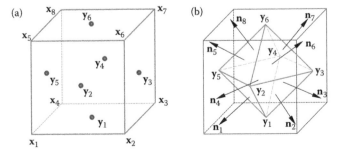

Figure 6.27 **A hexahedral element and its auxiliary octahedron: (a) centroids on faces; (b) auxiliary octahedron.**

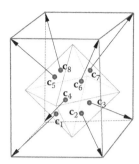

Figure 6.28 Transformation of H by the auxiliary octahedron.

As shown in Figure 6.28, hexahedron H is transformed to $H' = \{x_i', i = 1,8\}$, whose vertices are given by

$$H' = \left[x_i' \right] = \left[c_i + \lambda n_i / \sqrt{\|n_i\|} \right] \quad i = 1,8; \quad \lambda \in \left[\frac{1}{2}, 1 \right]$$

The properties of the GETMe transformation, i.e. invariant with respect to translation, rotation and scaling, and preserves the centroid of the initial hexahedron, are discussed in Vartziotis and Wipper (2011).

6.3.3.3.3 Pentahedral (wedge) element

Let $\{x_i, i = 1,6\}$ be the vertices of a pentahedral element P labelled in a usual manner, as shown in Figure 6.29a. An auxiliary dual polyhedron $\{y_i, i = 1,5\}$ can be defined by connecting the centroids of the five faces of pentahedron P, as shown in Figure 6.29b, such that

$$\text{auxiliary polyhedron} \begin{bmatrix} y_1 \\ y_2 \\ y_3 \\ y_4 \\ y_5 \end{bmatrix} = \begin{bmatrix} (x_1 + x_2 + x_3)/3 \\ (x_1 + x_2 + x_5 + x_4)/4 \\ (x_2 + x_3 + x_6 + x_5)/4 \\ (x_3 + x_1 + x_4 + x_6)/4 \\ (x_4 + x_5 + x_6)/3 \end{bmatrix}$$

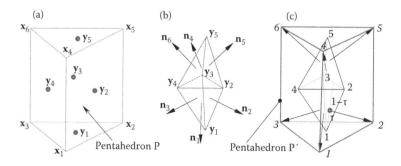

Figure 6.29 GETMe transformation of a pentahedral element: (a) centroids of faces; (b) dual polyhedron; (c) special base points.

The normal and the base point on each face of the dual polyhedron are given by

$$\text{normals:}\begin{bmatrix} \mathbf{n}_1 \\ \mathbf{n}_2 \\ \mathbf{n}_3 \\ \mathbf{n}_4 \\ \mathbf{n}_5 \\ \mathbf{n}_6 \end{bmatrix} = \begin{bmatrix} \mathbf{y}_1\mathbf{y}_2 \times \mathbf{y}_1\mathbf{y}_4 \\ \mathbf{y}_1\mathbf{y}_3 \times \mathbf{y}_1\mathbf{y}_2 \\ \mathbf{y}_1\mathbf{y}_4 \times \mathbf{y}_1\mathbf{y}_3 \\ \mathbf{y}_5\mathbf{y}_4 \times \mathbf{y}_5\mathbf{y}_2 \\ \mathbf{y}_5\mathbf{y}_2 \times \mathbf{y}_5\mathbf{y}_3 \\ \mathbf{y}_5\mathbf{y}_3 \times \mathbf{y}_5\mathbf{y}_4 \end{bmatrix}, \quad \text{base points:}\begin{bmatrix} \mathbf{b}_1 \\ \mathbf{b}_2 \\ \mathbf{b}_3 \\ \mathbf{b}_4 \\ \mathbf{b}_5 \\ \mathbf{b}_6 \end{bmatrix} = \begin{bmatrix} (1-\tau)\mathbf{y}_1 + \tau((\mathbf{y}_4 + \mathbf{y}_2)/2 \\ (1-\tau)\mathbf{y}_1 + \tau(\mathbf{y}_2 + \mathbf{y}_3)/2 \\ (1-\tau)\mathbf{y}_1 + \tau(\mathbf{y}_3 + \mathbf{y}_4)/2 \\ (1-\tau)\mathbf{y}_5 + \tau(\mathbf{y}_4 + \mathbf{y}_2)/2 \\ (1-\tau)\mathbf{y}_5 + \tau(\mathbf{y}_2 + \mathbf{y}_3)/2 \\ (1-\tau)\mathbf{y}_5 + \tau(\mathbf{y}_3 + \mathbf{y}_4)/2 \end{bmatrix}$$

The centroids on the faces of the dual polyhedron cannot be used as base points for the face normals in the transformation as this will not satisfy the basic requirement that regular elements should be reproduced in the transformation. Special base points located between the triangular face apex and the mid-point of the opposite edge could be used for which the parameter τ is given by Vartziotis and Wipper (2012)

$$\tau = \frac{4}{5}\left(1 - \sqrt[4]{\frac{4}{39}}\lambda\right)$$

As shown in Figure 6.29c, pentahedron P is transformed to $P' = \{\mathbf{x}'_i, i = 1,6\}$ such that

$$P' = [\mathbf{x}'_i] = \left[\mathbf{b}_i + \lambda \mathbf{n}_i / \sqrt{\|\mathbf{n}_i\|}\right], \quad i = 1,6; \quad \lambda \in \left[\frac{1}{2}, 1\right]$$

6.3.3.3.4 Pyramid element

Let $\{\mathbf{x}_i, i = 1,5\}$ be the vertices of a pyramid element Y with nodes properly labelled, as shown in Figure 6.30a. An auxiliary dual pyramid $\{\mathbf{y}_i, i = 1,5\}$ can be defined by connecting the centroids of the five faces of pyramid Y (Vartziotis and Wipper 2012), as shown in Figure 6.30b, such that

$$\text{auxiliary polyhedron}\begin{bmatrix} \mathbf{y}_1 \\ \mathbf{y}_2 \\ \mathbf{y}_3 \\ \mathbf{y}_4 \\ \mathbf{y}_5 \end{bmatrix} = \begin{bmatrix} (\mathbf{x}_1 + \mathbf{x}_2 + \mathbf{x}_3 + \mathbf{x}_4)/4 \\ (\mathbf{x}_1 + \mathbf{x}_2 + \mathbf{x}_5)/3 \\ (\mathbf{x}_2 + \mathbf{x}_3 + \mathbf{x}_5)/3 \\ (\mathbf{x}_3 + \mathbf{x}_4 + \mathbf{x}_5)/3 \\ (\mathbf{x}_4 + \mathbf{x}_1 + \mathbf{x}_5)/3 \end{bmatrix}$$

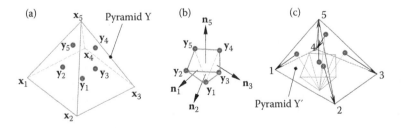

Figure 6.30 GETMe transformation of a pyramid element: (a) centroids of faces; (b) dual polyhedron; (c) base points.

The normal and the base point on each face of the dual polyhedron are given by

$$
\text{normals:}
\begin{bmatrix}
\mathbf{n}_1 \\
\mathbf{n}_2 \\
\mathbf{n}_3 \\
\mathbf{n}_4 \\
\mathbf{n}_5
\end{bmatrix}
=
\begin{bmatrix}
\mathbf{y}_1\mathbf{y}_2 \times \mathbf{y}_1\mathbf{y}_5 \\
\mathbf{y}_1\mathbf{y}_3 \times \mathbf{y}_1\mathbf{y}_2 \\
\mathbf{y}_1\mathbf{y}_4 \times \mathbf{y}_1\mathbf{y}_3 \\
\mathbf{y}_1\mathbf{y}_5 \times \mathbf{y}_1\mathbf{y}_4 \\
\dfrac{1}{2}\mathbf{y}_2\mathbf{y}_4 \times \mathbf{y}_3\mathbf{y}_5
\end{bmatrix},
\quad
\text{base points:}
\begin{bmatrix}
\mathbf{b}_1 \\
\mathbf{b}_2 \\
\mathbf{b}_3 \\
\mathbf{b}_4 \\
\mathbf{b}_5
\end{bmatrix}
=
\begin{bmatrix}
(1-\tau)\mathbf{y}_1 + \tau(\mathbf{y}_5 + \mathbf{y}_2)/2 \\
(1-\tau)\mathbf{y}_1 + \tau(\mathbf{y}_2 + \mathbf{y}_3)/2 \\
(1-\tau)\mathbf{y}_1 + \tau(\mathbf{y}_3 + \mathbf{y}_4)/2 \\
(1-\tau)\mathbf{y}_1 + \tau(\mathbf{y}_4 + \mathbf{y}_5)/2 \\
(\mathbf{y}_2 + \mathbf{y}_3 + \mathbf{y}_4 + \mathbf{y}_5)/4
\end{bmatrix}
$$

Like the case of a pentahedron, a pyramid is not completely symmetric, and the base points to support the face normal in the transformation located between the triangular face apex and the mid-point of the opposite edge have to be used, for which the parameter $\tau = \lambda + \dfrac{1}{2}$.

As shown in Figure 6.30c, pyramid Y is transformed to $Y' = \{\mathbf{x}_i', i = 1,5\}$ such that

$$
Y' = \left[\mathbf{x}_i'\right] = \left[\mathbf{b}_i + \lambda\mathbf{n}_i / \sqrt{\|\mathbf{n}_i\|}\right], \quad i = 1,5; \quad \lambda \in \left[\frac{1}{2}, 1\right]
$$

As polyhedral element E will expand in the GETMe transformation, the element E′ after transformation has to be scaled down so that its volume remains unchanged.

$$
\mathbf{x}_i^* = \mathbf{c} + \rho\left(\mathbf{x}_i' - \mathbf{c}\right), \quad i = 1, n(E); \quad \text{centroid } \mathbf{c} = \frac{1}{n(E)} \sum_{\mathbf{x}_i \in E} \mathbf{x}_i
$$

where scaling factor $\rho = \sqrt[3]{\text{volume}(E)/\text{volume}(E')}$ and n(E) = number of vertices in E.

6.3.3.3.5 GETMe transformations on polyhedra

Poorly shaped tetrahedral, hexahedral, pentahedral and pyramid elements are subjected to a number of GETMe transformations, as shown in Figure 6.31. The co-ordinates of the polyhedral elements before transformations are given in Table 6.6. It is seen that the shape quality of the elements can be rapidly improved by even one single GETMe transformation, and among all the elements tested, the rate of convergence of the pyramid element is relatively slow in approaching the regular form.

6.3.3.4 Examples: Node smoothing for tetrahedral meshes

Delaunay triangulations of randomly generated points are taken as test meshes for the three node-smoothing schemes. Meshes are obtained by extracting the *convex hull* of DTs, and these meshes are easily reproduced to assess the performance of any node-smoothing scheme that is newly developed. A relatively coarse triangulation of 1000 random points before and after optimisation is depicted in Figure 6.32. Ten more meshes of 100,000 points are generated, and the node-smoothing schemes, namely, QL, LO and GETMeT4 parallel smoothing, are applied to these meshes. For each optimisation scheme, ten cycles of smoothing are

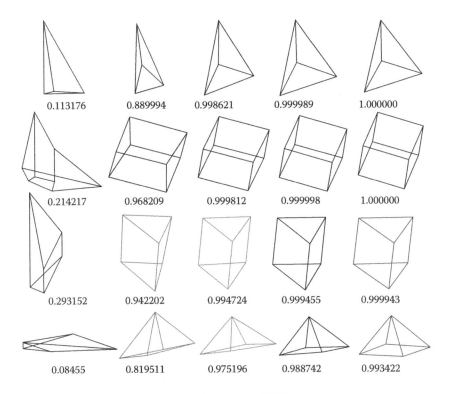

Figure 6.31 Polyhedral elements and their γ-qualities under GETMe transformations.

Table 6.6 Co-ordinates of polyhedral elements before GETMe transformations

Node	Tetrahedron	Hexahedron	Pentahedron	Pyramid
1	(0, 0, 0)	(0, 1, 0)	(0, 0, 0)	(0, 0, 0)
2	(20, 0, 0)	(3, 0, 0)	(1, 0, 0)	(5, 0, 0)
3	(5, 2, 0)	(2, 2, 0)	$(1, \sqrt{3}, 1)$	(15, 5, 0)
4	(0, 0, 30)	(0, 2, 0)	(0, 0, 4)	(5, 10, 0)
5		(0, 0, 2)	(1, 0, 1)	(0, 0, 1)
6		(8, 0, 2)	$(1, \sqrt{3}, 2)$	
7		(2, 2, 4)		
8		(0, 2, 8)		
λ	4/5	1/2	1/2	1/2

carried out to each mesh, and the results are presented in Table 6.7, in which NE = number of tetrahedra, γ_{min} = minimum γ-quality, γ_{mean} = mean γ-quality and Aver = average values of the ten test runs. It can be seen that the performance of all three node-smoothing schemes is quite different in which LO gives the best results in terms of both γ_{min} and γ_{mean} values, followed by GETMe and then QL. As for the average CPU time for ten cycles of node smoothing, QL took 6.14 s, LO took 42.69 s and GETMe took 21.94 s on a PC i7 CPU 870 at 2.93 GHz running on XP mode Compaq Visual Fortran with QuickWin graphics supports.

There are three major parameters in the GETMe scheme, namely, λ to control the rate of convergence of transformation, weight factor w for elements connected to the node and

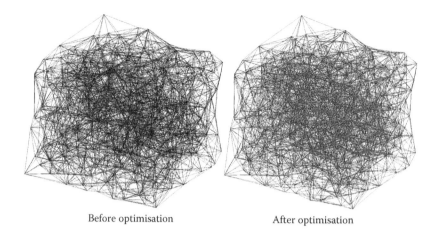

Before optimisation　　　　After optimisation

Figure 6.32 Convex hull extracted from Delaunay triangulation of random points.

Table 6.7 Results of node smoothing by QL, LO and GETMe

Test	NE	Initial mesh		QL		LO		GETMe	
		γ_{min}	γ_{mean}	γ_{min}	γ_{mean}	γ_{min}	γ_{mean}	γ_{min}	γ_{mean}
1	658,915	0.000483	0.43471	0.001116	0.5000	0.003957	0.58923	0.001618	0.53563
2	658,781	0.000398	0.43545	0.001394	0.5007	0.001901	0.59009	0.000741	0.53642
3	658,835	0.000493	0.43534	0.00126	0.5002	0.00126	0.58998	0.001260	0.53604
4	658,933	0.000869	0.43558	0.003242	0.5013	0.002128	0.59028	0.003242	0.53689
5	659,150	0.000337	0.43505	0.000674	0.50086	0.001336	0.58998	0.001336	0.53626
6	658,461	0.000912	0.43618	0.000912	0.50196	0.004119	0.59054	0.001836	0.53655
7	659,084	0.000767	0.43596	0.001214	0.50101	0.001214	0.58996	0.001214	0.53601
8	659,094	0.000278	0.43517	0.001286	0.50055	0.000889	0.59003	0.002422	0.53606
9	658,398	0.000959	0.43558	0.001486	0.50138	0.003593	0.59052	0.002574	0.53691
10	658,999	0.000538	0.43558	0.000538	0.5006	0.001453	0.58992	0.001453	0.5366
Aver	658,865	0.000603	0.43546	0.001312	0.500856	0.002185	0.59005	0.001770	0.53634

relaxation factor ξ to avoid inverted elements. GETMe is not very sensitive to λ, which was set to 0.8 in the test examples, and since interior nodes are surrounded by elements on all sides, the effect of λ somehow cancels out summing over the elements around a node. Vartziotis et al. (2009) proposed to set the weight w = 1 – η, where η is a quality measure of tetrahedra. However, there are a few other schemes worthwhile for a test, namely, w = $(1 - \gamma)^2$, γ'/γ and $(\gamma'/\gamma)^2$, where γ' is the γ value after transformation. It was found that γ'/γ and $(\gamma'/\gamma)^2$ produced slightly better results on the average for meshes of randomly generated points, and hence, w was set equal to γ'/γ in the tests. As for the relaxation factor ξ, at the expense of more CPU time, the overall quality of a patch is evaluated for the interior node being shifted to several locations around the target position computed by the weighted average, and the point that gives the highest element qualities will be chosen as the new position for the interior node. By this local mini optimisation process, there is no need to guess for the relation factor, and element inversion can be detected at the same time. As there are so many ways in fixing these parameters for GETMe, which may not have been optimised in the tests presented here, the results can only be taken as a reference.

In the second test, the three node-smoothing schemes are applied to meshes of non-uniform point distribution, in which points are generated clustered on a diagonal of 0.1% width relative to the diameter of the mesh, as shown in Figure 6.33. Due to the rapid change in size for tetrahedral elements near the diagonal, γ_{min} and γ_{mean} are much lower than those of the meshes of uniform distribution of random points, as shown in Table 6.8. Again the meshes are optimised by the three node-smoothing schemes with ten cycles of iteration for each scheme, and the results are listed in Table 6.8. Similar trend is observed for meshes of non-uniform point distribution such that the LO scheme gives the best γ_{min} and γ_{mean} values.

As for the characteristics of convergence, the γ_{min} value, the γ_{mean} value and the number of nodes moved NM are recorded for each cycle of node smoothing for a typical run on the non-uniform mesh of 100,000 points, as shown in Tables 6.9 and 6.10. QL is the most efficient if only a few cycles of iterations are allowed in the smoothing process, as NM decreases rapidly after the first three cycles. As for the other two smoothing schemes, there is still consistent improvement in each cycle of iteration, for which the LO scheme shows better performance at a slightly more CPU time. From the convergence characteristics of the three smoothing

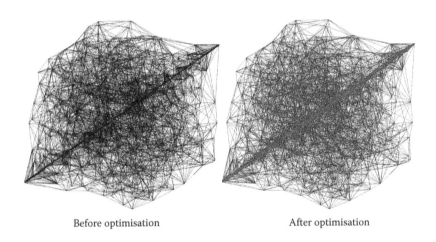

Before optimisation   After optimisation

Figure 6.33 Delaunay triangulation of 2000 non-uniformly distributed points.

Table 6.8 Non-uniform mesh optimised by QL, LO and GETMe

Test	NE	Initial mesh		QL		LO		GETMe	
		γ_{min}	γ_{mean}	γ_{min}	γ_{mean}	γ_{min}	γ_{mean}	γ_{min}	γ_{mean}
1	653,138	0.000352	0.39181	0.000352	0.45075	0.000465	0.55464	0.000755	0.4942
2	653,110	0.000055	0.39153	0.000055	0.44967	0.000248	0.55379	0.000055	0.49349
3	653,888	0.000615	0.39139	0.000615	0.44979	0.000963	0.55349	0.000615	0.49258
4	653,077	0.00009	0.39147	0.00009	0.45082	0.00068	0.55468	0.000417	0.49401
5	653,545	0.000199	0.3918	0.000424	0.44953	0.00062	0.5544	0.000508	0.4929
6	653,223	0.000276	0.39174	0.000315	0.45051	0.00104	0.55375	0.000315	0.4927
7	653,232	0.000281	0.39222	0.000289	0.45081	0.000689	0.55492	0.000549	0.49373
8	653,130	0.000084	0.39234	0.000084	0.45107	0.000455	0.55504	0.000226	0.49456
9	653,488	0.000243	0.39243	0.000243	0.44988	0.000215	0.55396	0.000268	0.49296
10	653,265	0.00008	0.39269	0.000161	0.45129	0.000161	0.55475	0.000161	0.49344
Aver	653,310	0.000228	0.39194	0.000263	0.45041	0.000554	0.55434	0.000387	0.49346

Table 6.9 Convergence characteristics of QL, LO and GETMe

Cycle	QL			LO			GETMe		
	γ_{min}	γ_{mean}	NM	γ_{min}	γ_{mean}	NM	γ_{min}	γ_{mean}	NM
0	0.000015	0.39299		0.000015	0.39299		0.000015	0.39299	
1	0.000268	0.4391	30,018	0.000551	0.47608	96,392	0.000268	0.45755	65,062
2	0.000268	0.44857	16,849	0.001008	0.51875	96,174	0.000268	0.47924	53,434
3	0.000268	0.4505	7491	0.000418	0.53743	95,854	0.000268	0.4873	45,048
4	0.000268	0.45087	2417	0.001154	0.5456	95,648	0.000268	0.49082	38,971
5	0.000268	0.45094	617	0.001154	0.54961	95,778	0.000268	0.49258	34,284
6	0.000268	0.45096	135	0.001154	0.55188	95,950	0.000268	0.49356	30,482
7	0.000268	0.45096	26	0.001154	0.5533	96,092	0.000268	0.49415	27,592
8	0.000268	0.45096	6	0.001154	0.55424	96,141	0.000268	0.49451	25,018
9	0.000268	0.45096	1	0.001154	0.55488	96,188	0.000268	0.49473	23,177
10	0.000268	0.45096	1	0.001154	0.55532	96,194	0.000268	0.49487	21,456

Table 6.10 Combined scheme

γ_{min}	γ_{mean}	NM
0.000015	0.39299	
0.000268	0.4391	30,018
0.000268	0.44857	16,849
0.000268	0.4505	7491
0.000268	0.45087	2417
0.000268	0.48072	55,836
0.000268	0.48962	45,156
0.000268	0.49292	38,396
0.000856	0.53103	96,227
0.00052	0.54503	95,994
0.001154	0.55055	95,844

methods, a combined scheme might be the most effective. As a numerical experiment, a combined scheme (4 + 3 + 3) consisting of four cycles of QL, followed by three rounds of GETMe and then further optimised with three cycles of LO is tested, as shown in Figure 6.34. The average results on a number of meshes are given by $\gamma_{min} = 0.001154$ and $\gamma_{mean} = 0.55055$. Relative to the three smoothing methods, the combined scheme gives the best performance on both γ_{min} and γ_{mean}, and in terms of the CPU time, it is similar to that of GETMe.

6.3.3.5 Examples: Node smoothing for hexahedral meshes

Examples of various size and characteristics using the three generic smoothing schemes, namely, QL, LO and GETMe, will be presented in this section. The quality of a hexahedral mesh depends on its elements, which are measured in turn based on the associated tetrahedra at the vertices. The η value presented in Section 6.2.4 can be used in general; however, in order to extend the quality measure also to inverted elements, the γ-quality of tetrahedral elements will be used.

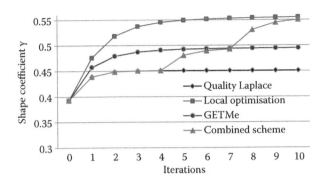

Figure 6.34 Convergence characteristics.

Let $\mathbf{a}_k$, $\mathbf{b}_k$ and $\mathbf{c}_k$ be the vectors emanating from vertex k to the neighbouring vertices along the edges of the hexahedron, as shown in Figure 6.35. For instance, the transformation matrix $\mathbf{F}_1$ of tetrahedron T_1 at vertex $\mathbf{x}_1$ is given by

$$\mathbf{F}_1 = [\mathbf{a}_1 \quad \mathbf{b}_1 \quad \mathbf{c}_1] = [\mathbf{x}_2 - \mathbf{x}_1 \quad \mathbf{x}_4 - \mathbf{x}_1 \quad \mathbf{x}_5 - \mathbf{x}_1]$$

The γ-quality of tetrahedron T_k is defined by means of $\mathbf{F}_k$ (k = 1,8) as follows.

$$\gamma_k = \frac{3\sqrt{3}\det(\mathbf{F}_k)}{\|\mathbf{F}_k\|^3} \qquad \|\cdot\| = \text{Frobenius norm}$$

γ_k is equal to 1 when vectors $\mathbf{a}_k$, $\mathbf{b}_k$ and $\mathbf{c}_k$ span the edges of a cube (orthogonal vectors of equal length), and it will be negative if the volume of tetrahedron T_k at vertex k, $\det(\mathbf{F}_k)$, is negative; hence, inverted hexahedral elements can also be detected by checking the γ-qualities of their vertices. Accordingly, the γ-quality of a hexahedron H can be defined as

$$\gamma(H) = \left(\prod_{k=1,8} \gamma_k \right)^{1/8} \text{ and } \gamma(H) = \gamma_{\min} \text{ if } \gamma_{\min} = \min_k \gamma_k < 0$$

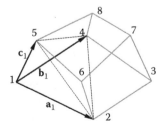

Figure 6.35 Associated tetrahedron at each vertex of a tetrahedron.

The γ-quality of a mesh $\mathbf{M}$ of n hexahedral elements can then be computed by

$$\gamma(\mathbf{M}) = \left(\prod_{i=1,n} \gamma_i \right)^{1/n} \text{ and } \gamma_{min} = \min_{i=1,n} \gamma_i$$

In case there are some inverted elements in the mesh $\mathbf{M}$ for which $\gamma_i < 0$, they will be skipped in the calculation of the geometric mean, such that $\gamma(\mathbf{M})$ is always greater than zero, and γ_{min} will be greater than zero only if there are no inverted or degenerate elements. For the smoothing of an interior node, we always have to evaluate the quality of the element connected to the node before and after node shifting. Given mesh $\mathbf{M}$, the patch of elements $\mathbf{P}$ connected to interior node $\mathbf{x}$ is given by

$$\mathbf{P(x)} = \{H_i \in \mathbf{M}; \mathbf{x} \in H_i\}, \quad \mathbf{x} \in H_i \text{ means that } \mathbf{x} \text{ is one of the vertices of hexahedron } H_i$$

The quality of a patch of elements $\mathbf{P(x)}$ is given by

$$\gamma(\mathbf{P}) = \left(\prod_{H_i \in \mathbf{P}} \gamma_i \right)^{1/n} \text{ and } \gamma_{min} = \min_{i=1,n} \gamma_i \quad \text{where n = number of hexahedra in } \mathbf{P}$$

Again only those hexahedral elements in $\mathbf{P}$ with positive γ values are included in the computation of the geometric mean $\gamma(\mathbf{P})$.

i. QL. For each interior point $\mathbf{x}$ in hexahedral mesh $\mathbf{M}$, determine the surrounding polyhedron $\mathbf{P}$ of all the hexahedra connected to $\mathbf{x}$. Compute the *centroid* (average of the nodal co-ordinates) of polyhedron $\mathbf{P}$, and $\mathbf{x}$ will be shifted to centroid $\mathbf{c}$ if

$$\gamma(\mathbf{P(x)}) < \gamma(\mathbf{P(c)})$$

The smoothing process can be accelerated by making the shift a little bit larger, say, by 5%.

ii. LO. Given an interior point $\mathbf{x}$ and its associated patch $\mathbf{P(x)}$, evaluate $\nabla\gamma_\mathbf{P}$ by numerical differentiation; the target point $\mathbf{x}'$ is given by

$$\mathbf{x}' = \mathbf{x} + h\mathbf{u}, \mathbf{u} = \frac{\nabla\gamma_\mathbf{P}}{\|\nabla\gamma_\mathbf{P}\|}$$

where h depending on the diameter of patch $\mathbf{P}$ is the amount of shift, which is also used as a reference distance in numerical differentiation. In LO, more points in the direction h$\mathbf{u}$ at the vicinity of $\mathbf{x}'$ will be evaluated to determine the best location for the shift, as shown in Figure 6.36. Furthermore, unlike QL and GETMe, the shift h, which depends only on the diameter of patch $\mathbf{P}$, will not automatically be reduced when the interior point $\mathbf{x}$ is near its optimal position. Hence, it is necessary to dynamically vary h with the number of iteration cycles as only minor adjustment is expected after cycles of smoothing. In the light of the numerical examples done so far, promising results have been obtained without extra effort by just reducing h by 10% for each cycle of iteration made.

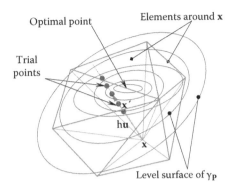

Figure 6.36 Shifting **x** along the gradient of γ_P.

iii. GETMe. GETMe smoothing will be applied according to the procedure detailed in Section 6.3.3.3. As we have introduced γ-quality to hexahedral elements, patches and meshes, meshes of inverted elements can also be handled by GETMe using a weight given by $w = 1 - \gamma$, such that more weights are attached to open up negative elements. In order to ensure that the quality of the mesh will be improved for each nodal movement and to avoid the formation of inverted elements, the γ-quality of the patch before and after the shifting of nodes are calculated, and the shift will only be materialised should there be an increase in the γ-quality. As it is quite expensive to evaluate the γ-quality of a patch of hexahedral elements, it is a trade-off between cost and quality, and in the present version of GETMe, three strategic points are evaluated.

A hexahedral mesh of series (1) is generated by the sweeping method such that the cross section increases with the angle of rotation. Elements of different size are created by subdividing hexahedral elements at random locations, as shown in Figure 6.37a, and there are 4820 nodes and 2400 hexahedral elements in the resulting mesh. Element irregularity is introduced by a perturbation of all the interior nodes with random movement of nodes for an amount as large as the diameter of the elements. Inverted elements are often found in the mesh with negative γ values (Escobar et al. 2003). For each optimisation scheme, ten cycles of smoothing are carried out to each of the ten test meshes, and the results are presented in Table 6.11 in which

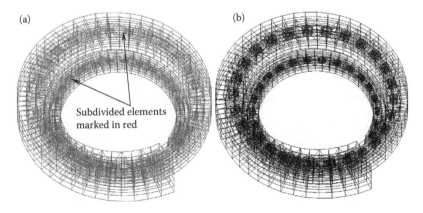

Figure 6.37 Hexahedral mesh with random subdivisions around interior nodes: (a) before optimisation; (b) after optimisation.

Table 6.11 Mesh (1) smoothed by QL, LO and GETMe

		Initial mesh		QL		LO		GETMe	
Test	IE	γ_{min}	γ_{mean}	γ_{min}	γ_{mean}	γ_{min}	γ_{mean}	γ_{min}	γ_{mean}
1	4	−0.18985	0.52584	0.31651	0.57433	0.33033	0.64927	0.31376	0.64000
2	2	−0.00525	0.52736	0.27840	0.57421	0.32428	0.64939	0.25270	0.63953
3	5	−0.18309	0.5258	0.26625	0.57380	0.34587	0.64956	0.26462	0.63982
4	3	−0.06860	0.52825	0.26272	0.57314	0.33957	0.64970	0.25697	0.63914
5	6	−0.36447	0.52516	0.29070	0.57296	0.36381	0.64996	0.29586	0.64001
6	1	−0.00888	0.52741	0.27978	0.57325	0.34848	0.64957	0.27558	0.63982
7	6	−0.06837	0.52633	0.27338	0.57327	0.33670	0.64978	0.26305	0.63903
8	2	−0.02993	0.52433	0.28888	0.57215	0.35749	0.64988	0.28573	0.63878
9	5	−0.13186	0.52795	0.23240	0.57275	0.31992	0.64950	0.24920	0.63868
10	7	−0.12670	0.52232	0.15337	0.57136	0.34179	0.64928	0.24865	0.63902
Aver		−0.11770	0.52608	0.26424	0.57312	0.34082	0.64959	0.27061	0.63938

IE = inverted elements, γ_{min} = minimum γ-quality, γ_{mean} = mean γ-quality and Aver = average values of the ten test runs. It can be seen that the performance of all three node-smoothing schemes is quite different, in which LO gives the best results in terms of both γ_{min} and γ_{mean} values, followed by GETMe and then QL. The resulting mesh after ten iteration cycles of node smoothing is shown in Figure 6.37b.

In the second test, the three node-smoothing schemes are applied to mesh (2) generated by mapping a unit cube of uniform subdivisions onto a sphere using H20 isoparametric interpolation, as shown in Figure 6.38, in which there are 1331 nodes and 1000 elements. Similar meshes of much larger size of 39,304 nodes and 35,937 hexahedral elements are generated, and the interior nodes are perturbed by a random displacement as large as twice the element size, such that many inverted elements are created as a consequence. Again, the meshes are optimised by the three node-smoothing schemes with ten cycles of iteration for each scheme, and the results are listed in Table 6.12. For this set of meshes of more elements and nodes with severe irregularity of a large number of inverted elements, the performance of LO and of GETMe are excellent and are really close with each other, followed by that of QL, which is lacking behind only in γ_{min} measure. Due to the fluctuation of the CPU time taken, by dropping the two largest and the two smallest values, the average is computed

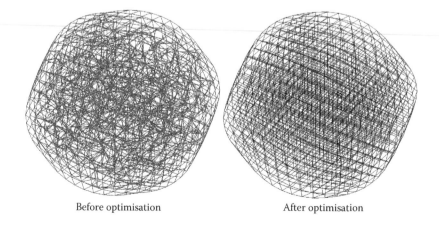

Before optimisation   After optimisation

Figure 6.38 Hexahedral mesh generated by finite element interpolation onto a sphere.

Table 6.12 Node smoothing of mesh (2) by QL, LO and GETMe

Test	IE	Initial mesh γ_{min}	Initial mesh γ_{mean}	QL γ_{min}	QL γ_{mean}	LO γ_{min}	LO γ_{mean}	GETMe γ_{min}	GETMe γ_{mean}
1	237	−0.5871	0.77758	0.24490	0.91143	0.27313	0.90712	0.26337	0.91176
2	215	−0.6035	0.77683	0.24442	0.91131	0.25067	0.90698	0.26362	0.91168
3	236	−0.4979	0.77706	0.24425	0.91132	0.27353	0.90709	0.26287	0.91171
4	246	−0.6320	0.77744	0.24426	0.91139	0.27298	0.90713	0.26341	0.91175
5	227	−0.5748	0.77654	0.24443	0.91141	0.26644	0.90710	0.26278	0.91176
6	249	−0.6049	0.77696	0.24422	0.91139	0.26958	0.90718	0.26341	0.91175
7	270	−0.5259	0.77813	0.24435	0.91143	0.26566	0.90714	0.26289	0.91179
8	209	−0.5044	0.77654	0.24441	0.91141	0.27192	0.90711	0.26330	0.91178
9	264	−0.8062	0.77783	0.24411	0.91139	0.27250	0.90709	0.26280	0.91174
10	217	−0.6054	0.77715	0.24443	0.91139	0.27180	0.90711	0.26391	0.91177
Aver	237	−0.5942	0.77721	0.24438	0.91139	0.26882	0.90711	0.26324	0.91175

using the middle six values. With regards the average CPU time for ten cycles of iteration, QL took 2.69 s, LO took 31.06 s and GETMe took 13.09 s.

As for the characteristics of convergence, the γ_{min} value, the γ_{mean} value and the number of nodes moved NM are recorded for each cycle of node smoothing for a typical run, as shown in Tables 6.13 and 6.14. QL is the most efficient if only a few cycles of iteration are allowed in the smoothing process. As for the other two smoothing schemes, there is still consistent improvement in each cycle of iteration, for which the LO scheme shows better performance in terms of $\gamma_{min} + \gamma_{mean}$ at the expense of more CPU time. From the convergence characteristics of the three smoothing methods, a combined scheme might be more effective. As a numerical experiment, a combined scheme (2 + 4 + 4) consisting of two cycles of QL, followed by four rounds of GETMe and then further optimised with four cycles of LO is tested, as shown in Figure 6.39. Relative to the three smoothing methods, the combined scheme gives the best performance in the combined measure of $\gamma_{min} + \gamma_{mean}$, and in terms of CPU time, it is similar to that of GETMe.

We sum up the three generic node-smoothing schemes, each of which possesses its own merits and characteristics. QL is the least expensive, but it is only effective in the first few

Table 6.13 Convergence characteristics of QL, LO and GETMe

Cycle	QL γ_{min}	QL γ_{mean}	QL NM	LO γ_{min}	LO γ_{mean}	LO NM	GETMe γ_{min}	GETMe γ_{mean}	GETMe NM
0	−0.6452	0.77692	IE = 227	−0.6452	0.77692	IE = 227	−0.6452	0.77692	IE = 227
1	−0.0177	0.89197	28,792	−0.1222	0.87222	31,994	−0.1211	0.89270	31,558
2	0.23141	0.90353	25,998	0.20618	0.89429	25,089	0.21625	0.90267	29,647
3	0.23676	0.90588	29,627	0.23734	0.89966	22,258	0.23308	0.90537	30,955
4	0.23909	0.90712	31,036	0.25722	0.90168	19,304	0.24507	0.90687	31,926
5	0.24072	0.90807	31,449	0.25753	0.90284	17,531	0.25086	0.90798	32,341
6	0.24176	0.90889	31,623	0.26195	0.90373	18,558	0.25442	0.90891	32,541
7	0.24255	0.90960	31,678	0.26195	0.90455	21,572	0.25709	0.90973	32,590
8	0.24320	0.91024	31,730	0.27033	0.90543	26,143	0.25936	0.91047	32,651
9	0.24372	0.91083	31,699	0.27033	0.90643	30,570	0.26132	0.91115	32,679
10	0.24414	0.91138	31,681	0.27410	0.90718	27,210	0.26304	0.91178	32,693

Table 6.14 Combined scheme

γ_{min}	γ_{mean}	NM
−0.6452	0.77692	IE = 227
−0.0177	0.89197	28,792
0.23141	0.90353	25,998
0.24071	0.90607	31,642
0.24679	0.90736	32,254
0.25046	0.90837	32,492
0.25372	0.90925	32,615
0.26495	0.90952	4793
0.27170	0.90987	10,749
0.27603	0.91033	19,389
0.27807	0.91072	18,649

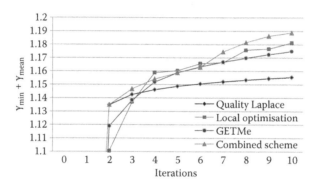

Figure 6.39 Convergence characteristics.

cycles of iteration. LO is relatively more expensive; however, promising results are ensured, in general, as the mesh quality is set as the objective function for optimisation. GETMe is the most interesting node-smoothing scheme as it is purely geometric in nature in transforming elements into regular forms without a direct link to shape quality measure. It is robust in removing inverted and poorly shaped elements rapidly and is consistent in diverse applications to produce quality meshes.

6.4 OPTIMISATION BY TOPOLOGICAL OPERATIONS

Local topological operations such as edge/face swaps, elimination of nodes, elimination of elements and elimination of edges are effective means to improve the shape quality of FE meshes. The operations can be carried out based on an appropriate shape measure and/or according to some topological considerations so as to create as much as possible a balanced structure in the connection among nodes, edges, faces and elements. We would like to mention edge refinement as well, but it is also a powerful tool in the generation of FE meshes in compliance with a specified node-spacing function, and accordingly, it is not covered here but will be presented in detail in Section 8.3.

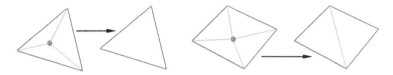

Figure 6.40 Elimination of nodes surrounded by three or four triangles.

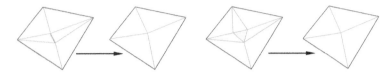

Figure 6.41 Elimination of a short edge and a small triangle.

Figure 6.42 Swap of diagonal between two triangles.

6.4.1 Triangular meshes

For efficient operations, some topological structures such as element adjacency relationship and the patch of elements surrounding each node have to be first retrieved from the mesh, which have to be updated whenever there is a structural change in the mesh. In a triangular mesh, nodes surrounded by three triangles are eliminated, as shown in Figure 6.40. For nodes surrounded by four triangles, we have to consider the shape quality of the elements and the length conformity of all the edges before and after the transformation. Short edges smaller than a certain threshold are identified, which are eliminated by shrinking the two related triangles to line segments, as shown in Figure 6.41. Similarly, small triangles are detected by checking their areas relative to those of their neighbours, and a triangle with too small an area is eliminated by shrinking it to a point, as shown in Figure 6.41. Diagonal swap within a quadrilateral formed by a pair of adjacent triangles is a common and effective measure in creating meshes of different characteristics and in enforcing boundary compatibility requirements. In shape optimisation, a diagonal swap can be carried out if the overall quality of the resulting triangles is superior to that before transformation, as shown in Figure 6.42.

6.4.2 Quadrilateral meshes

When nodes surrounded by three quadrilaterals are detected in a mesh, as shown in Figure 6.43, we have to consider the shape quality of the elements and the length conformity of all the edges before and after the transformation. Short edges much smaller than their

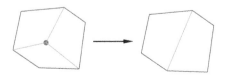

Figure 6.43 Elimination of nodes surrounded by three quadrilaterals.

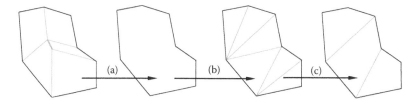

Figure 6.44 Elimination of a short edge: (a) delete segment; (b) triangulation; (c) merging.

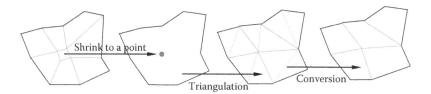

Figure 6.45 Elimination of a small quadrilateral.

Figure 6.46 Swap of diagonal between two quadrilaterals.

neighbours are identified, which are eliminated (a) by deleting all the edges connected to the short edge; (b) the resulting empty polygon is triangulated; and (c) triangles are merged to form quadrilaterals, as shown in Figure 6.44. This process can also be applied to eliminate a small quadrilateral by first shrinking it to a point, followed by the triangulation of the associated polygon and the conversion of triangles into quadrilaterals, as shown in Figure 6.45. Diagonal swap between two quadrilaterals is also an effective means in improving the quality of a mesh. Within the hexagon formed by two adjacent quadrilaterals, there are altogether three ways in dividing it into two quadrilaterals. Apart from the original configuration, an edge swap for the other two alternatives can be carried out if the overall quality of the resulting quadrilaterals is superior to that before transformation, as shown in Figure 6.46.

6.4.3 Tetrahedral meshes

In a tetrahedral mesh, a node on the average is surrounded by some 27 tetrahedra, and a node can be deleted if it is surrounded by only a few tetrahedra, as shown in Figure 6.47. In

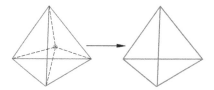

Figure 6.47 Elimination of nodes surrounded by four tetrahedral elements.

the node elimination process, we have to consider the shape quality of the elements and the length conformity of all the edges before and after the transformation. Edges shorter than a certain threshold are identified. For each short edge, the surrounding polyhedron, which consists of all the tetrahedra connected to it, is determined. The short edge can be collapsed if there is a point on the edge that is visible to all the faces of the surrounding polyhedron, as shown in Figure 6.48. A small element can also be eliminated if there is a point within the element that is visible to all the faces of the bounding polyhedron, as shown in Figure 6.49.

There are many possibilities of face/edge swap or local remeshing (Misztal et al. 2009) for the optimisation of tetrahedral meshes. The most well-known operations are the transformations T_{23} and T_{32}, which swap two tetrahedra for three tetrahedra and vice versa within the same space. Let $P_1P_2P_3J$ and $P_3P_2P_1I$ be two tetrahedra sharing common face $P_1P_2P_3$ with nodes I and J on opposite sides of face $P_1P_2P_3$. If segment IJ intersects the interior of triangle $P_1P_2P_3$, the convex hull of the two tetrahedra can be transformed by T_{23} into another configuration consisting of three tetrahedra IJP_1P_2, IJP_2P_3 and IJP_3P_1, as shown in Figure 6.50. T_{32} is the inverse transformation of T_{23}. When a line segment IJ in a mesh is shared by exactly three tetrahedra IJP_1P_2, IJP_2P_3 and IJP_3P_1, the ring of these three tetrahedra can be transformed into a second configuration consisting of two tetrahedra $P_1P_2P_3J$ and $P_3P_2P_1I$.

Consider a polyhedron formed by putting two pyramids back to back to each other, as shown in Figure 6.51a. Owing to the symmetry in topology along the three axes, there are three ways in dividing the octahedron into four tetrahedra by introducing a diagonal joining any pair of opposite nodes, namely, P_1P_2, Q_1Q_2 or R_1R_2, as shown in Figure 6.51b–d. The transformation T_{44} converts these configurations from one to another. Transformation T_{56}

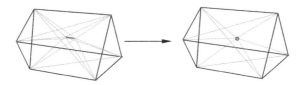

Figure 6.48 **A short edge shrinks to a point.**

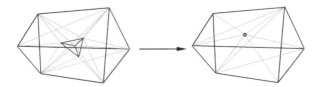

Figure 6.49 **A small element shrinks to a point.**

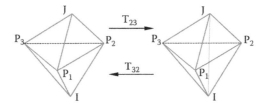

Figure 6.50 Transformation of tetrahedral elements T_{23} and T_{32}.

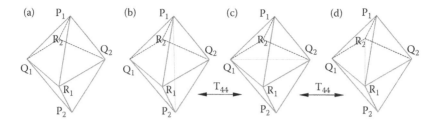

Figure 6.51   Three configurations in transformations T_{44}: (a) octahedron; (b) joining P_1P_2; (c) joining Q_1Q_2; (d) joining R_1R_2.

involves the replacement of a ring of five tetrahedra around line segment IJ by six tetrahedra, three on each side of the polygon $P_1P_2P_3P_4P_5$ made, respectively, with nodes I and J, as shown in Figure 6.52a. There are five ways in dividing the polygon $P_1P_2P_3P_4P_5$ into three triangles by selecting each vertex and its two opposite vertices in turn, as shown in Figure 6.52b. The desired configuration is the one in which the sum of the diagonals is minimised.

There are other transformations of higher order involving more tetrahedral elements, but only transformations T_{23}, T_{32}, T_{44} and T_{56} will be considered in the mesh optimisation, which are relatively simple and effective in removing poorly shaped elements and rapidly improve the quality of the mesh. The triangular faces in the tetrahedral mesh are examined one by one; T_{23} transformation is performed if the mean shape quality of the two original elements is less than that of the three new elements. As for the other transformations T_{32}, T_{44} and T_{56}, they are carried out on a line-by-line basis. The edges (line segments) of the mesh are examined in turn, and the ring of tetrahedral elements around the edge is identified. The edge will be skipped if there are more than five tetrahedra attached to it; otherwise, the quality of the elements is evaluated, and appropriate transformations T_{32}, T_{44} or T_{56} will be performed according to the number of tetrahedra around the edge.

The preferred configurations for local transformations are those whose mean geometric shape qualities are maximised. Although the resulting mesh would certainly be different, the optimisation by local transformations can be done with respect to any valid shape measure μ ($\mu = \gamma$, ρ or θ). A minimum shape quality, μ_{min}, can also be imposed such that no element whose μ value is less than μ_{min} would be generated by local transformations to assure the minimum quality of the elements. A single pass consists of scanning through all the edges in the mesh, and the edges affected by a local transformation will be kept in a separate list, which has to be updated throughout the transformation process. As the shape quality of a mesh cannot be improved indefinitely by shifting from one configuration to another, at a certain stage, no transformation could be done to further improve the quality of the mesh.

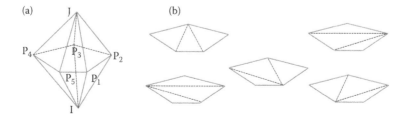

Figure 6.52   Transformations T_{56} and five ways of forming six tetrahedra: (a) ring of five tetrahedra; (b) five ways of dividing polygon $P_1P_2P_3P_4P_5$.

Hence, the process will converge for meshes of finite size if we impose a minimum increase in mesh quality for each cycle of local transformations. The optimisation procedure can actually be speeded up if transformations are only carried out when there is an increase of at least 1% in the mean quality of the elements involved.

6.4.4 Examples of optimisation by face/edge swap

Delaunay triangulations of 10,000 randomly generated points are constructed, which are optimised by topological transformations T_{23}, T_{32}, T_{44} and T_{56}, as shown in Figure 6.53. As T_{32} and T_{44} are the more frequently used operations, two cycles of each swapping scheme will be applied to a number of tetrahedral meshes, and the results are shown in Tables 6.15 and 6.16. As seen from these tables, on the average of 7842 scans for swap

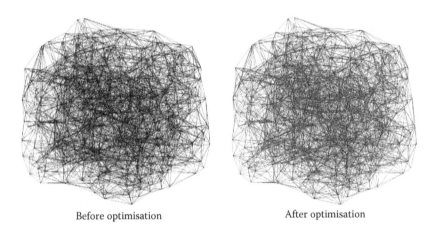

Before optimisation   After optimisation

Figure 6.53 Optimisation of Delaunay triangulation of 1000 points.

Table 6.15 Optimisation of Delaunay triangulation by swap T_{32}

Mesh	Scan	T_{32}	γ_{omin}	γ_{omean}	γ_{min}	γ_{mean}
1	7795	7254	0.000168	0.42538	0.0005664	0.49239
2	7799	7278	0.000147	0.42552	0.0002921	0.49307
3	7977	7365	0.000276	0.42064	0.0006907	0.48786
4	7820	7249	0.000232	0.42694	0.0004755	0.49456
5	7820	7309	0.000148	0.42667	0.0006892	0.49373
Aver	7842	7291	0.000194	0.42503	0.0005428	0.49232

Table 6.16 Optimisation of Delaunay triangulation by swap T_{44}

Test	Scan	T_{44}	γ_{omin}	γ_{omean}	γ_{min}	γ_{mean}
1	40,137	1902	0.0000580	0.42644	0.0000580	0.43027
2	39,242	1813	0.0003305	0.42238	0.0003305	0.42605
3	39,587	1826	0.0001507	0.42742	0.0001507	0.43103
4	39,479	1826	0.0001166	0.42106	0.0001166	0.42491
5	39,766	1837	0.0001956	0.42669	0.0001956	0.43032
Aver	39,642	1841	0.0001703	0.42482	0.0001703	0.42852

T_{32}, 7291 were actually carried out, which brought the initial minimum γ-value (γ_{omin}) of 0.000194 to 0.0005428 and the initial mean γ-value (γ_{omean}) of 0.42503 to 0.49232. As for the T_{44} transformation, out of the average of 39,642 scans for swap T44, only 1841 were actually performed, which improved the initial mean γ-value (γ_{omean}) by a very small margin from 0.42482 to 0.42852, and there was hardly any change for the initial minimum γ-value (γ_{omin}) of 0.0001703.

In the second test, tetrahedral meshes generated by Delaunay insertion of 10,000 randomly generated points are each optimised by two cycles of all the local topological transformations developed, T_{23}, T_{32}, T_{44} and T_{56}, and the results are listed in Table 6.17. We can see from Table 6.17 that, on average, T_{23} was only carried out once out of 279,129 scans, and similarly, only 33 T_{56} were performed for 75,669 scans; this indicates that if we are not aiming for the minute improvement of the mesh quality, a great deal on CPU time can be saved by omitting the swaps T_{23} and T_{56}. Although T_{44} is not as effective as T_{32}, we can still keep it as it will also slightly improve the mesh quality; however, we can consider applying it less frequently relative to swap T_{32}. From the optimisation of the Delaunay meshes, we also observed that local topological operations converge very rapidly in general, and much fewer operations are done in the second iteration (mesh k.2, k = 1,5 of Tables 6.16 and 6.17), and virtually there is not much improvement even for the second iteration.

As for the CPU time required for the transformations T_{32} and T_{44}, Delaunay meshes of 100,000 points are generated for which the number of tetrahedral elements varies from 657,947 to 659,327. The average CPU times for two cycles of iterations are calculated by dropping the largest and smallest values and taking the average of the remaining three values, as shown in Table 6.18. In the two cycles of iterations, the number of tetrahedral elements processed roughly equals to 657,947 + 659,327 = 1,317,274, from which we can calculate that T_{32} and T_{44} can process about 400,000 and 500,000 tetrahedra, respectively, per second on PC i7 CPU 870 at 2.93 GHz running on XP mode, which is about five times slower than running on the same machine with Windows 7, Visual Studio 2010 on 64 bits.

However, the actual number of swaps is much fewer for T_{44} compared to T_{32} within a second (hence, the efficiency of the two transformations), as a large amount of CPU time is spent on scanning for T_{44}, which includes the evaluation for any possible gain due to the swap. As the swapping operations are local processes, the time complexity for optimisation is linear provided that relevant topological relationships are prepared beforehand, which can be done also in linear time. In summing up, an efficient topological optimisation scheme

Table 6.17 Optimisation of Delaunay triangulation by swaps T_{23}, T_{32}, T_{44} and T_{56}

Mesh	$Scan_{23}$	T_{23}	$Scan_{32}$	T_{32}	$Scan_{44}$	T_{44}	$Scan_{56}$	T_{56}	γ_{min}	γ_{mean}
1.1	166,127	2	7501	7208	31,481	3413	41,804	22	0.00602	0.49449
1.2	114,142	0	500	231	21,558	267	34,216	5	0.00602	0.49693
2.1	164,575	2	7448	7130	31,056	3219	41,392	24	0.01584	0.49386
2.2	114,354	0	571	250	21,611	281	34,200	7	0.01813	0.49650
3.1	163,403	0	7194	6912	30,731	3234	41,123	32	0.01996	0.49681
3.2	114,310	0	518	253	21,355	275	34,321	6	0.01996	0.49955
4.1	165,216	0	7517	7193	31,205	3326	41,460	28	0.01257	0.49284
4.2	114,226	0	542	241	21,600	278	34,136	9	0.01257	0.49542
5.1	165,125	1	7524	7209	31,204	3286	41,456	29	0.00914	0.49375
5.2	114,168	0	506	220	21,602	277	34,236	5	0.00914	0.49619
Aver	279,129	1	7964	7369	52,681	3571	75,669	33	0.01316	0.49692

Table 6.18 Optimisation of Delaunay triangulation by swaps T_{32} and T_{44}

Mesh	$10\gamma_{min}$ T_{32}	γ_{mean} T_{44}	$Scan_{32}$	T_{32}	$10\gamma_{min}$	γ_{mean}	$Scan_{44}$	T_{44}	$10\gamma_{min}$	γ_{mean}
1.1	0.00613	0.4366	76,517	73,961	0.04243	0.5049	220,420	18,727	0.00613	0.4402
CPU	3.30 s	2.65 s	3381	903	0.04243	0.5057	193,068	653	0.00613	0.4408
2.1	0.01157	0.4348	77,052	75,568	0.01979	0.5033	220,276	18,802	0.01157	0.4384
CPU	3.25 s	2.60 s	3310	953	0.01979	0.5042	192,837	610	0.01157	0.4386
3.1	0.00419	0.4352	77,333	74,784	0.00419	0.5036	220,051	18,856	0.00419	0.4388
CPU	3.62 s	2.55 s	3402	945	0.00419	0.5044	192,623	604	0.00419	0.4390
4.1	0.01168	0.4353	77,332	74,707	0.03945	0.5041	220,761	18,903	0.01168	0.4390
CPU	3.18 s	2.77 s	3485	947	0.03945	0.5050	192,903	573	0.01261	0.4391
5.1	0.00506	0.4345	76,877	74,336	0.06084	0.5029	220,622	18,879	0.00506	0.4381
CPU	3.29 s	2.62 s	3441	981	0.06084	0.5038	193,042	613	0.00506	0.4383
Aver	3.28 s	2.62 s	80,426	75,617	0.03334	0.5046	413,321	19,444	0.01545	0.4392

can be formulated by starting with one round of T_{32} iterations, followed by another round of T_{44} swaps and finally further improve the mesh with another round of T_{32} swaps.

6.4.5 Optimisation by both geometrical and topological operations

Mesh optimisation can be achieved by means of either geometrical consideration of node shifting or topological operations of local edge/face swaps. However, there is no reason why topological and geometrical operations cannot be put together to form even more efficient and well-balanced optimisation schemes. Node-shifting methods such as QL, GETMe and LO will be merged with T_{32} and T_{44} element swap operations. There are a number of ways in putting these operations together; hence, four possible scenarios are conceived and tested with Delaunay meshes of randomly generated points. Scheme I, $4 \times QL + 3 \times GETMe + 3 \times LO$, consists of only geometrical operations of four cycles of QL, followed by three rounds of GETMe and then three rounds of LO to serve as a control of mesh optimisation without topological operations; in scheme II, $T_{32} + T_{44} + T_{32} + 3 \times QL + 2 \times GETMe + 2 \times LO$, topological operations T_{32} and T_{44} are applied followed by geometrical operations; in scheme III, $3 \times QL + 2 \times GETMe + 2 \times LO + T_{32} + T_{44} + T_{32}$, geometrical operations are first applied followed by topological operations; and in scheme IV, $T_{32} + 3 \times QL + T_{44} + 2 \times GETMe + T_{32} + 2 \times LO$, topological and geometrical operations are thoroughly mixed up.

The four mesh-optimisation schemes I, II, III and IV have been applied to Delaunay meshes of 100,000 randomly generated points, and the results are shown in Table 6.19. Without optimisation by topological means, scheme I can only achieve a mean γ-quality of 0.5865, whereas schemes II, III and IV achieve γ-quality of 0.7243, 0.7000 and 0.7616, respectively. From the results of schemes II and III, it seems that better results could be obtained by applying the topological operations before geometrical operations. However, as indicated by scheme IV, the best result in terms of geometric mean γ-quality of 0.7616 for Delaunay triangulation of random points could only be achieved by a thorough mixing up of the topological and geometrical operations, as shown in Figure 6.54. Moreover, topological operations also have a great impact in raising the minimum shape quality of a mesh, which can hardly be achieved merely by geometric means. The major improvement in mesh shape quality is due to swap T_{32}, for which three tetrahedra are converted into two, and as a result, there is a large reduction in the number of tetrahedral elements for meshes optimised by schemes II, III and IV, as shown in Table 6.19. As for the CPU time, schemes II, III and IV

Table 6.19 Optimisation schemes combining topological and geometrical operations

Scheme		Mesh1	Mesh2	Mesh3	Mesh4	Mesh5	Average
Initial	NE	659,393	658,734	659,112	658,669	658,981	658,978
	γ_{min}	0.001203	0.00027	0.00096	0.001239	0.000057	0.000746
	γ_{mean}	0.43504	0.43473	0.43545	0.43561	0.4362	0.435406
I	NE	659,393	658,734	659,112	658,669	658,981	658,978
	γ_{min}	0.001977	0.002707	0.003757	0.002186	0.005446	0.003215
	γ_{mean}	0.58542	0.58606	0.5873	0.58669	0.58697	0.58649
	CPU time	30.49	30.40	30.81	30.32	30.95	30.60
II	NE	578,597	578,106	578,599	578,101	578,408	578,362
	γ_{min}	0.00945	0.007087	0.011816	0.017751	0.011402	0.01150
	γ_{mean}	0.72343	0.72375	0.72491	0.72454	0.72468	0.72426
	CPU time	22.47	23.01	22.7	22.72	22.82	22.94
III	NE	576,519	575,996	576,414	575,878	576,427	576,247
	γ_{min}	0.00945	0.007087	0.011816	0.017751	0.011402	0.01150
	γ_{mean}	0.69892	0.69976	0.70071	0.70036	0.70005	0.69996
	CPU time	25.67	25.54	25.47	25.32	25.38	25.48
IV	NE	576,900	576,395	576,823	576,470	576,822	576,682
	γ_{min}	0.00945	0.007087	0.011816	0.016204	0.011402	0.01119
	γ_{mean}	0.76107	0.76147	0.76255	0.76154	0.76116	0.76156
	CPU Time	23.21	23.3	23.31	23.59	23.2	23.32

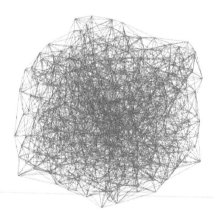

Figure 6.54 Delaunay mesh of 1000 points optimised by topological and geometric operations.

take less computation time compared to scheme I, as topological optimisation takes less CPU time than the geometrical optimisation. The CPU time reported here can only be taken as a reference on a relative basis as the tests were conducted on a notebook i5-2520M CPU at 2.5 GHz, which is rather slow compared to a powerful PC.

From the development so far, it is not hard to realise that shape quality measure plays a crucial role in mesh optimisation. A question naturally arises: can other shape measures such as η, ρ and σ rather than γ be used in the geometrical and topological optimisation? The answer is yes; however, γ is more stable in the optimisation process and could give slightly better results overall (Lo 1997), which can assume positive and negative values in such a way that invalid inverted elements can also be identified in the mesh optimisation. γ-quality is also cheaper in computation than other shape measures, which is a merit worth considering in the optimisation of large-scale meshes as element shape qualities have to be computed repeatedly in the optimisation cycles whether by topological or geometrical means.

Apart from tetrahedral meshes, in three dimensions, optimisation of non-simplicial meshes by means of topological operations has not yet been formulated. Pure hexahedral meshes are very often generated by mapping and sweeping methods, which are structures meshed by nature of their generation, and hence, there is no need to rebuild their topology. As for the mixed meshes of polyhedral elements, pentahedral and pyramid elements are introduced mainly for the liaison between the hexahedral and tetrahedral elements. A solution for the element connections, in general, is difficult enough to leave any room for topological optimisation by a change in element types for which other possible solutions would have already been evaluated right at the spot when the problem first arises. In other words, topological optimisation of polyhedral meshes might be possible; however, it is unlikely to be cost-effective as local element swaps can only be applied to very limited locations, if any, even after time-consuming rigorous analysis.

Chapter 7

Mesh generation by parallel processing

More hands working together in an orderly manner can finish a job in a shorter time.

7.1 INTRODUCTION

With a rapid increase in the problem size from thousands of points to millions of points, it is imperative to devise ever more efficient schemes for the FE mesh generation (MG). With a steady but rapid pace of advancement and upgrading since the emergence of modern computers, today's micro-computers, PCs, are all equipped with more than one processor, and a standard machine with four cores and 16-GB memory is fairly common. Using one single processor for MG represents a pitiful 25% usage of the total capacity of the machine, and efficient parallel meshing schemes making full use of all the processors will simply boost the speed by more than four times, cutting down the generation time to one quarter of that by a serial process.

There are two quite different ways in parallelising the MG process: (i) domain partition – the problem domain is first subdivided into a number of zones, each of which is handled by one processor for MG; and (ii) the problem domain is taken as a whole in MG by several processors operating simultaneously. The strategy by domain decomposition is more restrictive as we have to deal with the boundary between zones and a load balance among all the zones for higher efficiency, but it comes with an advantage that existing MG techniques can all be used directly for parallel meshing. A parallelisation of MG on the entire domain is more general, as no artificial barrier is imposed to limit the shape and the structural connections of the elements. Attempts for parallel meshing have been made on the two most popular methods, namely, the DT and the AFT, for the generation of unstructured meshes. By the nature of the evolution of the generation front that the resulting mesh will depend on the process of how elements are created, it is in general much more difficult to formulate a robust parallelisation algorithm for ADF meshing, especially for large-scale problems (Shostko and Lohner 1995; Wilson and Topping 1998; Lohner 2001). Alternatively, in view of the boundary integrity inherent with ADF, a parallel meshing scheme was developed by Ito et al. (2007) based on a domain decomposition into simpler regions. On the other hand, DT is unique if all the points are in the general positions, and the resulting mesh only depends on the points but not on their order of insertion. By the lemma of Delaunay, DT by point insertion can be made a localised process, which lends itself to many possibilities for parallelisation.

The merits of DT are its robustness with a sound mathematical basis and the availability of an efficient generic insertion algorithm, which can triangulate large practical point sets in linear time complexity. Owing to all these attractive properties of DT, there have been continuous efforts since as early as 1999 in the parallelisation of DT in 2D and 3D with various strategies. Blelloch et al. (1999) presented a dividing path of Delaunay edges by projecting points on a paraboloid surface for parallel triangulation of 2D points using a *divide-and-conquer* algorithm. There was approximately 50% or four times speed-up running in parallel with eight processors for uniformly or non-uniformly distributed points. Kolingerova and Kohout (2002) introduced the *optimistic method* based on the idea that the probability of collision of threads on the same triangle is relatively low for a large data set. 2D points were divided into sub-sets, and concurrent insertions by several processors were handled by synchronisation.

Parallel mesh refinement to triangular and hexahedral meshes is a direct application of the parallel DT to MG. Chrisochoides and Nave (2003) presented a parallel Bowyer–Watson insertion for MG by synchronisation of processors in case the cavities associated with two or more concurrent inserted points intersect. It was reported that code complexity might cause issues of stability near the domain boundary. Based on the *divide-and-conquer* algorithm, Chen et al. (2004) presented a parallel procedure for the *near* DT of 2D points. The main challenge of the method was to merge isolated triangulated patches into one coherent piece. Nave et al. (2004) proposed a parallel Delaunay refinement algorithm by a synchronised point insertion with guaranteed quality provided that certain boundary constraints are fulfilled. A parallel *divide-and-conquer* scheme for 2D DT was proposed by Chen et al. (2006) in which the *affected zone* was introduced to combine *sub-Delaunay* triangulations. A comprehensive account on the parallelisation of DT was given by Chrisochoides et al. (2009). A template for developing parallel Delaunay refinement was presented by Chernikov and Chrisochoides (2010) in which rigorous analysis on how to avoid conflicting Delaunay insertion in mesh refinement was discussed. Simonovski and Cizelj (2011) created partition surfaces by the voxel method, and material zones were processed in parallel by the standard triangulation package. Wu et al. (2011) introduced *ParaStream*, a parallel streaming DT algorithm for LiDAR points on multi-core architectures, in which kd-tree was applied to distribute workload among processors.

Research works have only just been started in recent years in the parallelisation of DT over three or higher dimensions. deCougny and Shephard (1999a,b) proposed parallel refinement and coarsening of tetrahedral meshes by means of a domain partition, and Freitag et al. (1999) presented a parallel mesh smoothing procedure. Kohout and Kolingerova (2003) put forward a parallel DT based on a randomised incremental insertion with edge and face swaps in 3D space in which synchronisation for multiple point insertion was applied at various stages of point insertion. A survey of parallel DT algorithms was presented by Kohout et al. (2005) in which a parallel insertion algorithm was also proposed based on a synchronisation scheme using thread priorities. Beyer et al. (2005) presented a procedure for parallel dynamic and kinetic regular triangulation in 3D based on an incremental construction with parallel flipping of tetrahedra arising from the idea of accessible and non-accessible simplices.

Batista et al. (2010) studied and implemented several parallel geometric algorithms for multi-core computers. Ma (2008) presented a surface extraction algorithm for a large binary image data set based on parallel 3D Delaunay subdivision strategy. Points and surface data were partitioned into zones in which DT was carried out over each zone in parallel, which could then be connected to form the skeleton of the required surface. However, no details were given in the algorithm or in the implementation of the parallel 3D DT. Recently, Camata and Coutinho (2013) proposed a parallel meshing scheme based on Octree partition. Summing up the research work done so far on parallel meshing without domain partitioning, there are many strategies proposed for DT by multiple insertion of points simultaneously with several processors, which can be roughly classified into three main categories as follows:

i. *Creation of Delaunay boundary edges.* Delaunay cutting lines are introduced by local DT or by some more sophisticated method such as the projection onto a paraboloid (Blelloch et al. 1999) to partition the global domain into two roughly equal portions. Further division can be done by introducing more cut lines to produce as many regions as necessary for triangulation by point insertion or any other techniques within each isolated region in parallel.

ii. *Triangulation of points within zones in parallel* and the patches of triangulation are connected by filling up gaps between zones. The given points for triangulation are first allocated to various zones, the ensemble of which is a partition of space covering all the points. DT can now be carried out using points within the zones in parallel. The result is a collection of isolated patches of triangles, which have to be connected properly to form the final DT (Lemaire and Moreau 2000).

iii. *Synchronisation.* Parallel point insertion controlled by some verified sequence/location of insertion, or by means of synchronised insertion for several processors by blocking the access of a particular point or triangles from other processors to avoid conflicts (Kohout and Kolingerova 2003; Chrisochoides and Nave 2003).

In view of the weaknesses mentioned above, a parallel DT algorithm for MG in two, three or possibly higher dimensions based on a partition of the point set into zones will be presented in Sections 7.2, 7.3 and 7.4 (Lo 2012a,b). Processors can work to their full capacity in the simultaneous zonal point insertion in an absolute independent manner in the generation and the elimination of redundant Delaunay triangles and tetrahedra. Artificial Delaunay boundary construction between zones before point insertion is not required; there is no need to connect the Delaunay tetrahedra generated within zones by filling up gaps between zones, nor is there any synchronisation control or access blocking for the concurrent point insertion. The parallel zonal insertion scheme will be a powerful engine for large-scale unstructured MG by a PC or other machines running on a shared memory system for its robustness, efficacy and generality.

MG over surfaces is an important and interesting area in MG, which also defines the boundary of solid objects in a CAD environment. As Delaunay empty sphere criterion cannot be directly applied to MG over surfaces and due to a hierarchical data structure and non-manifold surface modelling, domain partition into zones of well-defined boundary coupled with ADF meshing for individual surface pieces seems to be more convenient for parallel meshing over complicated surfaces or discretised non-manifolds. A brief account on the development of surface partition for parallel meshing will be given in Section 7.5, and a new scheme for surface decomposition based on topological operations guided by specific geometrical criteria will also be introduced.

7.2 FUNDAMENTALS AND STRATEGIES

The idea and the implementation of the parallel zonal insertion scheme in 2D and 3D are given in Sections 7.3 and 7.4. Before a detailed discussion on the algorithm for the parallelisation of Delaunay point insertion, some fundamental concepts and general strategies in parallelising DT will first be explored in this section.

7.2.1 Partition of points and insertion algorithm

The basic concepts of DT and its implementation have been presented in Chapters 3, 4 and 5. Before we could formulate any idea for the parallelisation of DT, we have to review the fundamentals, strategies and developments so far in FE MG by DT with one single processor. In the point insertion algorithm, there are two basic steps: (i) locating the base triangle containing the inserted point and (ii) verifying the circumcircle criterion for all the triangles connected to the inserted point. If points are inserted by means of adjacent cells, as shown in Figure 7.1, the searching path in determining the BASE is a constant depending on the number of points in a cell, and the verification of the circumcircle criterion is a local process if we follow the adjacency relationship of the triangles, thanks to the lemma of Delaunay. From this observation, it can be seen that DT by point insertion is one of the most efficient algorithms due to its simplicity and linearity provided that points are inserted in clusters in a contiguous manner like the growth of crystals.

As for non-uniform point distributions, spatial partition of points can be achieved by the Quadtree/Octree (Shaffer and Samet 1987; Zhou et al. 2011) and kd-trees (Liu et al. 2013) for 2D and 3D points. Nevertheless, the time complexity for the construction of Quadtree/Octree or kd-trees is of order $O(n\log n)$ for a set of n points, which indicates that the time complexity of DT by point insertion cannot be better than $O(n\log n)$ unless the barrier of spatial point partition could be overcome by some more intelligent schemes of lower order. The *multi-grid* insertion proposed by Lo (2013c) to be presented in Section 8.2 is virtually a node partition scheme of linear time characteristics, which has a great potential in the DT of highly non-uniform point distributions by single-processor or multiple-processor insertion.

Point insertion following an ordered cell sequence for DT of points in clusters is a very efficient scheme as the number of operations is almost optimal. Take for instance the 2D point insertion; it takes n scans of triangles on average to locate the BASE triangle, where n is the number of points in a cell, and on average, a point is connected to six triangles for which the circumcircle criterion has to be verified. A simple count illustrates that the number of operations for each point insertion is n + 6 = 10 if each cell contains roughly four points. Numerical tests on the insertion of 10 million randomly generated points by PC show that more than 1 million points can be inserted per second, which is one of the fastest rates ever published known to the author on a PC.

A robust and efficient parallel DT algorithm has to be based on a sound, reliable and fast sequential scheme, and hence, parallelisation of point insertion is considered for the possibility of multiple point insertions by several processors simultaneously. Theoretically, a

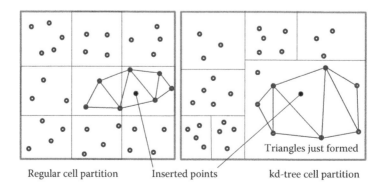

Regular cell partition   Inserted points   kd-tree cell partition

Triangles just formed

Figure 7.1 **Points sorted into cells.**

processor can be assigned to each point for the maximum speed-up. Nevertheless, machines with so many processors are not yet available, and even if they exist, it may not be economical to have so many processors working together as conflicts are bound to occur for the insertion of neighbouring points.

7.2.2 The zonal insertion scheme

Realising that the point insertion scheme is a very efficient algorithm for well-organised points as only a few operations are involved for each point insertion, any additional tests, control or single-processor operations will substantially slow down the overall rate of the parallel insertion process. For example, in a parallel insertion process, if the number of operations in locating the base triangle and in the verification of the circumsphere criterion has to be doubled, the efficiency will drop by 50%. Hence, the basic operations for point insertion should remain more or less the same when there are several processors working together.

Let's start from the ideal situation in which one processor is assigned to each point for simultaneous insertion. We have to ensure that each triangle connected to the inserted point **p** is Delaunay, and the simplest check to assure this is to apply the circumcircle criterion that each triangle connected to **p** contains no point in its interior, as shown in Figure 7.2. An effective means to achieve this is to make use of some spatial control on the points, and a partition of points into zones could well serve this purpose. As shown in Figure 7.2, the patch of triangles associated with point **p** is Delaunay if no point exists in the union of circumcircles passing through point **p**, or if there is no point in the box bounding all the circumcircles of **p**. For an efficient insertion of a large set of points, points are already sorted into cells, and to check whether the bounding box contains any other points will just be a simple task.

It is impractical and unnecessary to assign one processor to each point as there is a lot of redundancy in doing so, and an optimised scheme is to group several cells together into a zone for a single-processor insertion. As a result, cells are grouped into a number of zones depending on the number of processors available. There are two major issues to be addressed in such a scenario: (i) how to make sure that Delaunay triangles are generated between zones and (ii) how to get rid of the redundant elements efficiently on the boundary between zones.

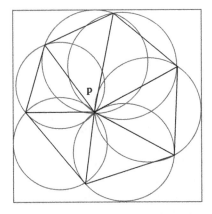

Figure 7.2 Patch of Delaunay triangles.

7.3 PARALLEL DELAUNAY TRIANGULATION IN 2D

Following the idea of the zonal point insertion, points are first partitioned into cells by means of a regular grid, which are then grouped into zones according to the number of processors available. Points are inserted cell by cell within a zone simultaneously with one processor operating in each zone.

7.3.1 Points partitioned into cells

Let N be the number of points in a 2D Delaunay triangulation, and n be the average number of point desirable in a cell. Then

$$nN_xN_y = N \tag{7.1}$$

where
N_x = number of cell divisions along the x-axis
N_y = number of cell divisions along the y-axis

and the number of cells, $N_c = N_xN_y$.

Let $x_{min}, y_{min}, x_{max}, y_{max}$ be the bounds of the (x,y) co-ordinates of the point set. Compute the range in x, $R_x = x_{max} - x_{min}$, and the range in y, $R_y = y_{max} - y_{min}$. N_x and N_y can be determined by substituting $\lambda R_x = N_x$ and $\lambda R_y = N_y$ into Equation 7.1.

The most important requirement in a spatial partition of points into cells is to ensure that each point belongs to one and only one cell, and the sum of points in all the cells equal to the total number of points, i.e.

$$N = \sum_{i=1}^{N_c} n_i \quad \text{where } n_i = \text{number of points in cell i}$$

7.3.2 Grouping cells into zones

Let $N_p = Z_xZ_y$ be the number of processors available for parallel insertion, where Z_x and Z_y are the zonal divisions along the x- and y-axes, respectively, as shown in Figure 7.3. Cells are grouped into zones such that the number of cells in each zone is given by

$$N_z = \text{NINT}\left(\frac{N_x}{Z_x}\right)\text{NINT}\left(\frac{N_y}{Z_y}\right) \quad \text{where NINT (.) returns the nearest integer}$$

For the best performance of parallel insertion, the number of zones ought to be an integral multiple of the number of processors available; for instance, division into 12 zones for three or four processors is a sound division. However, division into 12 zones for eight processors may not be that appropriate, because 12 = 8 + 4, and not all processors will be working to their full capacity all the time. However, since we are working with millions of points, the partition into cells and the division into zones, which are multiples of the number of processors, will never be a problem. The number of cells in some zones near the boundary may not be the same if N_x/Z_x or N_y/Z_y is not a whole number. Such a variation would not cause

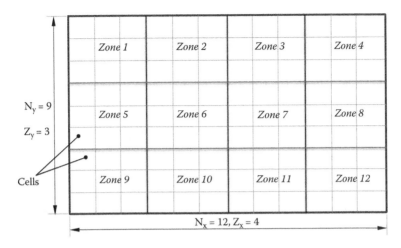

Figure 7.3 12 × 9 = 108 cells partitioned into 4 × 3 = 12 zones.

any trouble in the subsequent operations, as a zone will be identified by the bounding row and the bounding columns, i.e. the starting and ending rows and the starting and ending columns.

A typical computer run for 1000 randomly generated points divided into 3 × 3 = 9 zones is shown in Figure 7.4, in which the average number of points in a cell n = 4, R_x and R_y are in a ratio of 3:2, and hence $N_x = 19$, $N_y = 13$ and $nN_xN_y = 988 \approx 1000$. There are 24 cells in the zones on the left, but there are 28 cells or even 35 cells in the zones on the right-hand side of the domain. The size of the cells may vary slightly in such a way that the number of points in the cells is as close to n = 4 as possible.

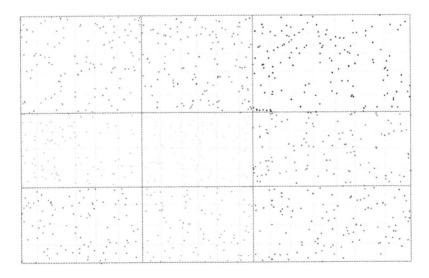

Figure 7.4 1000 random points partitioned into nine zones.

7.3.3 Simultaneous insertion within zones

Similar to a single-processor insertion, the initial triangulation is a rectangular bounding box of two triangles large enough to contain all the points. Point insertion in each zone will be handled by one single processor in a completely independent manner. All the points within each zone (say, zone I, I = 1,N_p) will be first processed by inserting points in those cells belonging to the zone under consideration. This will generate all the Delaunay triangles within the zones; however, Delaunay triangles cutting the zonal boundary are missing, as shown in Figure 7.5. The second step is to construct all the Delaunay triangles at the boundary between zones. To do so, boundary cells are added around the zone, which are the cells contained in two columns on the right and on the left and in two rows on the top and on the bottom of the zone, as shown in Figure 7.6. Boundary triangles are defined as those triangles supported on vertices from the current zone I and vertex or vertices from neighbouring zone(s), as shown in Figure 7.6. This simple definition is a topological one, which works on zones of any geometry and no matter how they are partitioned, i.e. it is also applicable to Quadtree/Octree and kd-tree partitions.

By the lemma of Delaunay, if all the boundary triangles are Delaunay, the triangles within the zone will also be Delaunay. Hence, the next step is to ensure that all the boundary triangles are Delaunay. Compute the circumcircles of all the boundary triangles, and check if any circle crosses the boundary of the augmented zone with a layer of cells added to the boundary of the original zone I. In case a circle cut across the vertical boundary of the rectangular zone on the right, a column of cells will be added to the right of the region. Not all the points in this column have to be inserted, and only those that fall into one of the circumcircles cutting the vertical boundary need to be considered. This operation can be applied to the boundary edge of the zone on the left and the boundary edges at the bottom and at the top wherever necessary until all the circumcircles of the boundary triangles are bounded by the ever-expanding zone of additional columns and rows of cells. As cells contain a roughly equal number of points, the process converges fairly rapidly and evenly on all the zonal boundary edges in one or two layers of cells, as shown in Figure 7.6. After adding one layer of cells around the middle zone I, almost no circumcircles of boundary triangles cut across

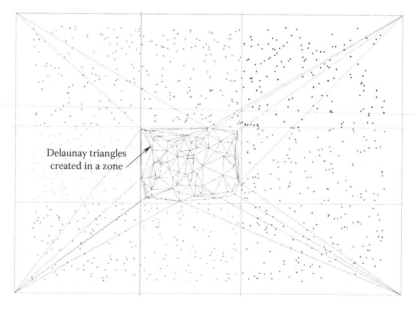

Figure 7.5 Points inserted within a zone.

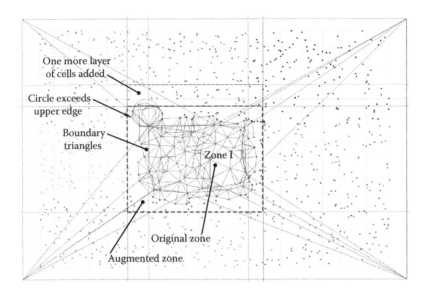

Figure 7.6 Augmented zone with additional columns and rows of cells.

the augmented zonal boundary depicted by the dotted line in Figure 7.6, except that two or three circles exceed the top edge; with one more row of cells added on the top edge as depicted by the dotted line in the rectangular region, all the circumcircles are now bounded.

However, for the concurrent point insertion of a corner zone, as shown in Figure 7.7, no additional column of cells is added to the left of the zone as the boundary of the global domain has already been reached (i.e. column 1 has already been inserted to the zone), and no row of cells is added to the bottom edge either. Similarly, in the concurrent insertion of a zone at the bottom boundary of the global domain, as shown in Figure 7.8, there is no need to add one more row of cells at the bottom edge as we have already arrived at the bottom (i.e. row 1 has already been included to the zone).

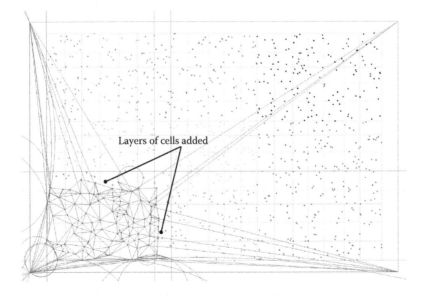

Figure 7.7 Concurrent insertion at a corner zone.

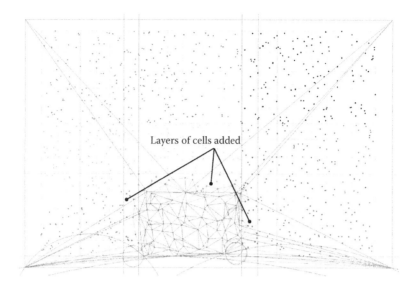

Figure 7.8 Concurrent insertion of a zone at the bottom edge.

Due to a nice property of circumcircles, points considered before need not be reconsidered after further point insertions, i.e. the union of circumcircles will ever become smaller for the introduction of a new point. It suffices to show this with a patch of Delaunay triangles around a point P, as shown in Figure 7.9a. First of all, such a patch of Delaunay triangles always exists for any point distribution and insertion. If a point Q is outside the union of circumcircles, nothing has to be done, and there is no change to the union of circumcircles. Now, suppose point Q is found inside circumcircle PBC; then the circumcircle BQP is bounded by circumcircles BCP and PAB, and the circumcircle PQC is bounded by circumcircles PBC and PCD. Hence, the union of circumcircles shrinks in size with the introduction of a new point Q. On the other hand, if Q is found at the intersection of two circumcircles, as shown in Figure 7.9b, then circumcircle AQP is bounded by circumcircles ABP and PFA, and circumcircle PQC is bounded by circumcircles PBC and CDP. The union of the circumcircles of all the boundary triangles will also shrink with the introduction of a new point, as the union is composed of patches of circumcircles centred at a point.

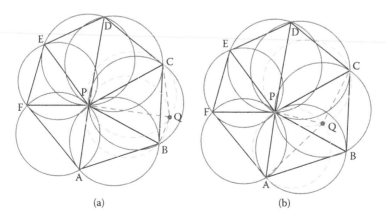

(a)   (b)

Figure 7.9 A patch of circumcircles around a point always shrinks in size: (a) Q found in one circle; (b) Q found in two circles.

7.3.4 Elimination of redundant triangles

We have to realise that the Delaunay triangulation by zonal insertion is complete in the sense that it contains all the Delaunay triangles of the point set as Delaunay triangles in a patch around each point are generated in the insertion process for all the points partitioned into zones. As each zone is processed by a processor independently, all the triangles within a zone are generated by one processor. However, triangles on the boundary between zones may be generated by one or more processors. Let's consider the triangles at a typical junction of four rectangular zones, as shown in Figure 7.10.

Vertices of triangles at the boundary can get zonal labels 1, 2, 3 and 4, and there are 20 combinations, as shown in Figure 7.10. To eliminate redundant triangles, a simple tactic for a partition of triangles into various zones is to assign triangles with vertices {111, 112, 113, 114, 122, 123, 124, 133, 134, 144} to zone 1, triangles with vertices {222, 223, 224, 233, 234, 244} to zone 2, triangles with vertices {333, 334, 344} to zone 3 and triangles with vertices {4,4,4} to zone 4. It is easy to verify that this allocation of triangles will produce a partition of triangles into zones; hence, redundant triangles can only be put in one and only one zone, and the elimination of redundant triangles within each zone can be carried out independently by individual processors concurrently.

In fact, for triangles surrounded by k zones, the number of possible combinations of zonal vertices is given by

$$\text{Possible Combinations} = \sum\left(\sum k\right) = \sum \frac{k(k+1)}{2!} = \frac{k(k+1)(k+2)}{3!}$$

For triangles surrounded by four zones,

$$\text{Possible Combinations} = \frac{4(4+1)(4+2)}{3!} = 20$$

For tetrahedra surrounded by k zones, we have

$$\text{Possible Combinations} = \sum\left(\sum\left(\sum k\right)\right) = \frac{k(k+1)(k+2)(k+3)}{4!}$$

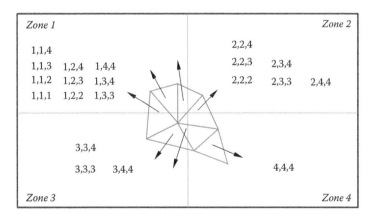

Figure 7.10 Partitioning triangles into zones.

For tetrahedra surrounded by eight zones, we have

$$\text{Possible Combinations} = \frac{8(8+1)(8+2)(8+3)}{4!} = 330$$

7.3.5 Minimum vertex-allocation scheme

The formula for possible zonal label combinations can be easily generalised to higher dimensions by applying more summations to the lower-dimension formulas. However, it is rather tedious to formulate a general scheme for the partitioning of simplices into zones over higher dimensions. With a closer look at the triangles assigned to zones 1, 2, 3 and 4, it is not difficult to find out that zone 1 always receives triangles with vertex label 1, and zone 2 always receives triangles with vertex label 2; the same goes for zones 3 and 4. Hence, a general rule for assigning triangles to zones can be formulated such that triangles with vertex zone labels z_1, z_2 and z_3 will be assigned to zone z given by

$$z = \min(z_1, z_2, z_3)$$

This is a surprisingly simple and elegant scheme by which redundant triangles can be eliminated independently almost without effort. Again, this simple rule is entirely based on topology consideration, so it is also applicable to higher dimensions irrespective of the geometry of the zones and how zones are connected as long as it is a zonal partition of points. That is, d-dimensional simplices are assigned to zone z given by

$$z = \min(z_1, z_2, z_3,..., z_{d+1})$$

Based on this minimum vertex allocation scheme, redundant triangles for the parallel insertion of 1000 points are eliminated and distributed to all the nine zones, as shown in Figure 7.11. Strictly speaking, it is not a convex hull of the point set as the auxiliary corner

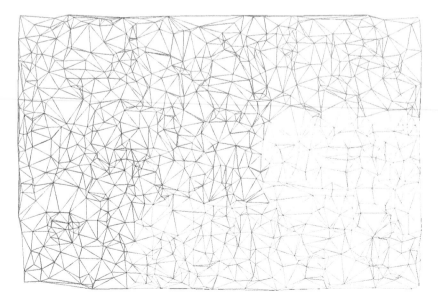

Figure 7.11 Partition of triangles by the minimum vertex scheme.

points for setting up the initial triangulation are not placed to infinity. It is very close to a convex hull, except that some boundary edges of the point set may be missing. However, results generated by parallel point insertion have all been checked against those produced by the sequential point insertion by one processor.

7.3.6 Efficiency analysis

As the zonal point insertion is completely independent, the algorithm is expected to be very efficient and 100% scalable depending only on the characteristics of zonal partition. The extent of zonal boundary depends on how cells are partitioned into zones, and usually, the zonal boundary will increase with the number of zones. Let's consider various schemes of the partition of a unit square into zones for parallel point insertion by four processors, as shown in Figure 7.12. From a simple boundary count, a four-zone partition (2 × 2) will be the most efficient with a boundary of 8 units, whereas a 4 × 1 strip division will have a common boundary of 10 units and the least efficient 2 × 4 division will create a common boundary of 12 units.

For a k × k division into zones, the number of internal boundary cuts is given by

$$\text{cuts} = 2(k - 1)$$

Suppose that m is the average number of layers of cells added to each side of a cut line to achieve convergence for boundary Delaunay triangles; then the increase in the workload μ in triangulating the zones is given by

$$\mu = \frac{N_x N_y + 2(k-1)m(N_x + N_y)}{N_x N_y}$$

where N_x and N_y are, respectively, the number of cell divisions along the x- and y-axes.

For a square domain with equal zonal division on each side ($N_x = N_y$), we have

$$\mu = 1 + \frac{4(k-1)m}{N_x} \Rightarrow m = \frac{(\mu-1)N_x}{4(k-1)}$$

A numerical experiment has been carried out to triangulate 10 million 2D points using various zonal divisions. Setting n = 16, $N_x = N_y = 791$, and the results are shown in Table 7.1.

In terms of redundant triangles generated, as predicted, m is more or less a constant, and it varies only very slightly from 1.10 to 1.16 for zonal divisions 2 × 2, 3 × 3 to 8 × 8.

Boundary = 4 + 2 × 2 = 8   Boundary = 4 + 2 × 3 = 10   Boundary = 4 + 2 × 4 = 12

Figure 7.12 **Partitioning a square into zones.**

Table 7.1 Estimate number of additional layers for convergence

Zones	NT	NE	N_x	μ =NT/NE	m
2 × 2	20,107,826	19,995,274	791	1.00563	1.11
3 × 3	20,216,146	19,995,274	791	1.011	1.10
4 × 4	20,330,252	19,995,274	791	1.017	1.12
5 × 5	20,445,806	19,995,274	791	1.023	1.14
6 × 6	20,564,566	19,995,274	791	1.028	1.11
7 × 7	20,690,300	19,995,274	791	1.035	1.15
8 × 8	20,816,146	19,995,274	791	1.041	1.16

Note: NT and NE are the number of triangles generated and in the convex hull, respectively.

However, in terms of CPU time, the result is quite different, and various zonal divisions almost take the same CPU time for point insertion by a single processor tested on PC, as shown in Figure 7.13. The analysis of point insertion by multiple processors will be deferred to Section 7.3.8 as maximum efficiency can only be attained when the number of zones N_c is an integral multiple of the number of processors employed. The almost constant CPU time for different zonal divisions is a very intriguing phenomenon. As mentioned before, the speed of DT by point insertion is rather sensitive to the order of points inserted, and usually, points inserted in clusters in a contiguous manner would be more efficient than a random insertion following the natural order of the points. A finer zone division will generate more redundant triangles on the common boundary, as shown in Table 7.1. However, point insertion clustered within a smaller zone is likely to be more efficient than the insertion in a larger zone. The extra workload in generating more redundant triangles is just offset by the slightly higher efficiency in DT over a smaller zone.

In fact, in point insertion without zonal division, if points are inserted following the natural order, a complexity of $O(N^2)$ is resulted for a set of N points, and the scalability is well over 100%. Of course, this is not reasonable but is a by-product of points being sorted into cells and inserted following the cell adjacency order. In a point insertion without zonal division, if points are sorted into cells that are then inserted following the cell adjacency order, a linear time complexity is achieved, but it is still slightly slower than a serial insertion with zonal

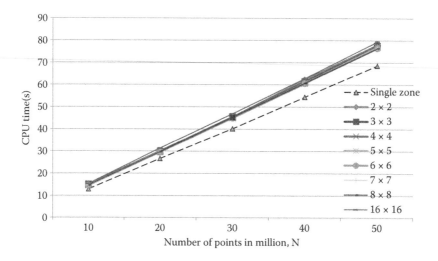

Figure 7.13 Zonal point insertion by a single processor.

Table 7.2 CPU time in seconds for zonal point insertion

N(10^6)	1 × 1	2 × 2	3 × 3	4 × 4	5 × 5	6 × 6	7 × 7	8 × 8	16 × 16
10	12.87	14.92	14.93	14.61	14.48	14.51	14.57	14.67	15.30
20	26.63	29.94	29.72	29.72	29.58	29.52	29.84	29.70	31.36
30	40.17	45.35	45.4	45.29	44.98	44.98	45.19	45.29	46.91
40	54.44	62.17	61.29	61.02	60.53	60.64	60.83	60.73	62.81
50	68.66	77.99	77.55	77.31	77.06	76.58	76.50	76.44	79.42

subdivision, resulting in something like 110% scalability. It is later realised that efficiency computation would not make much sense unless a more efficient point insertion order is devised for point insertion without zonal division. The optimal order is unknown, but point insertion in clusters in a contiguous manner would be very close to optimal. The final scheme employed for single zone insertion is to insert points sorted in cells of the same subdivision and grouping used in parallel insertion, and to follow exactly the same order within zones to minimise the distance between consecutive points. The quasi-optimal insertion scheme has brought down the insertion time to a level about 10% faster than zonal insertion.

The point insertion within a single zone and by means of a subdivision into several zones has been tested thoroughly on a PC, Intel®, Core™, i7 CPU, 870 at 2.93 GHz with 16-GB RAM running on Windows 7 using Intel Fortran VS2010 on 64 bit. The CPU times taken for the insertion of 10 million to 50 million points with various zonal divisions are given in Table 7.2, and the corresponding graph plot is shown in Figure 7.13. From Table 7.2, it is seen that the CPU time for point insertion by zones is not sensitive to the zonal subdivision. For example, the insertion of 50 million points by 2 × 2 division took 78 s, 5 × 5 division took 77 s and 16 × 16 division took 79 s, whereas insertion as a single zone took 69 s, which is about 11% faster than insertion by zonal subdivision. The fact that performance is not tied to how the cells are divided into zones has an important implication that the parallel zonal insertion scheme is almost 100% scalable and is very adaptive to the number of processors employed.

7.3.7 Memory requirement

No extra memory is needed for the parallel zonal insertion scheme compared to the classical sequential insertion process, except a zonal label for each point, which can be easily handled by an array of 2B integers. To minimise the use of memory, the circumcentre and the circumradius of each triangle are not stored, which are recalculated whenever necessary. For each point insertion, about $2 \times (3 + 3) \times 4 + 2 \times 8 = 64$ B memory is required to store the vertices and the neighbours of the two triangles generated and the x- and y-co-ordinates of the point. Thus, a PC with 16 GB, apart from the memory taken up by the operating system, can generate quite comfortably more than 400 million triangles for an insertion of 200 million points without being appreciably slowed down due to lack of memory.

Indeed, given the same amount of memory, a larger point set can be handled by insertion with zonal subdivision, as triangle adjacency information is only needed for one zone at a time, rather than for the entire population of all the triangles as in the case of point insertion within a single zone without subdivision. However, for parallel insertion, only if the zones are not processed all at the same time, some memory saving is possible for the reason stated above.

7.3.8 Test on OpenMP shared memory systems

The first test of 2D DT by parallel zonal insertion is done on PC running on Windows 7 and Intel Fortran VS2010. It is a four-core machine with a maximum of eight pseudo-threads running on OpenMP parallel directives with shared memory architecture. Zonal division would only attain a maximum efficiency if the number of zones is an integral multiple of the number of threads used. Hence, only zonal division into 4 × 4 = 16 and 8 × 8 = 64 zones was tested, and the results of inserting 20 million to 200 million randomly generated points are shown in Table 7.3; the corresponding graphical plot is shown in Figure 7.14. It is seen that the parallel zonal insertion can boost the speed by 4.5 times, up to 7 million Delaunay triangles per second on a PC with a 4 × 4 or 8 × 8 zonal division for the insertion of 10 million points.

The scalability of the parallel zonal insertion was tested by a proper multi-core machine (gridpoint.hku.hk). There are 144 nodes in the machine, but only the IBM HS22 blade server with two 64-bit 6-core Intel Westmere CPU at 2.66 GHz was employed in the test. The IBM HS22 blade server, which supports the OpenMP parallel directives, could provide any number of threads up to 12 to test for scalability. A zonal division into 10 × 12 = 120 zones was tested, which is a common multiple for 2, 4, 6, 8, 10 and 12 processors, so that processors are expected to work to their full capacity for a fair comparison. The results for

Table 7.3 Parallel zonal insertion versus single-processor insertion

$N(10^6)$	1×1	4×4	8×8
10	12.88	2.86	2.85
20	26.63	5.91	5.91
40	54.44	12.26	12.08
60	83.38	18.61	18.56
80	112.5	25.19	24.82
100	142.2	31.71	31.73
120	175.6	38.7	38.14
140	205.1	45.81	45.13
160	234.5	52.77	51.27
180	266.6	59.1	58.53
200	295	66.14	65.73

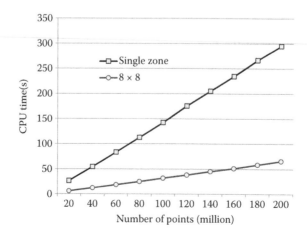

Figure 7.14 Parallel Delaunay triangulation on PC.

Table 7.4 Scalability test using 1 to 12 processors

Processors	CPU time	Speed-up	Efficiency
1	27.2	1.00	100
2	15	1.81	91
4	7.55	3.60	90
6	5.05	5.39	90
8	3.82	7.12	89
10	3.08	8.83	88
12	2.57	10.58	88

Note: CPU time = CPU time recorded for the parallel insertion of 20 million points.

the parallel Delaunay triangulation of 20 million points are shown in Table 7.4, and the plots of the speed-up and the efficiency against the number of processors used are shown in Figure 7.15.

$$\text{Speed-up} = \frac{\text{CPU time of single processor insertion}}{\text{CPU time of parallel insertion}}$$

$$\text{Efficiency} = \frac{\text{Speed-up}}{\text{Number of processors used}}$$

From Figure 7.15, the CPU time required decreases rapidly as more processors are used in the parallel insertion, and the speed-up with the number of processors employed is quite linear with an efficiency of about 90% for 2 to 12 processors; moreover, 15.6 million Delaunay triangles could be generated per second in a parallel insertion by 12 processors. Scalability test for more than 12 processors has not yet been done, as the algorithm has to run on a cluster machine with a parallel environment quite different from the OpenMP system with shared memory.

Finally, processors can run at full capacity only when the number of zones is an integral multiple of the number of processors used. This trivial fact was verified in the insertion of 20 million points with a zonal division of 4 × 4 = 16 zones employing two, four, six and eight processors in turn. The results are shown in Table 7.5, and a plot of CPU time, speed-up

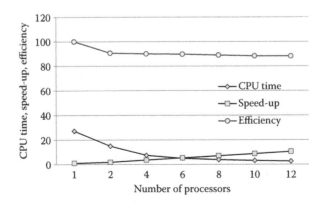

Figure 7.15 Parallel insertion of 20 million points.

Table 7.5 Insertion by 4 × 4 zonal division

Processors	CPU time	Speed-up	Efficiency
1	27.2	1.00	100
2	14.92	1.82	91
4	7.57	3.59	90
6	5.74	4.74	79
8	3.84	7.08	89

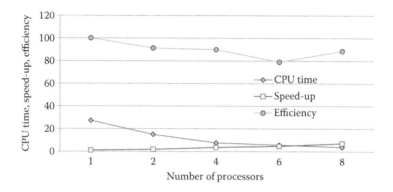

Figure 7.16 Parallel Delaunay triangulation using 4 × 4 zones.

and efficiency is given in Figure 7.16. It can be seen that parallel insertion at high efficiency of 90% was achieved using two, four or eight processors, whereas the efficiency dropped by more than 10% when six processors were used.

7.4 PARALLEL DELAUNAY TRIANGULATION IN 3D

The parallelisation algorithm presented in Section 7.3 for Delaunay point insertion in 2D is a generic scheme, which can be readily extended in higher dimensions. Nothing needs to be changed in the concept or in the procedure moving from a 2D setting to a 3D one, except for some natural modifications such as 2D plane to 3D space, (x, y) co-ordinates to (x, y, z) co-ordinates, triangles to tetrahedra, circles to spheres, etc.

7.4.1 Points partitioned into cells

Let N be the number of points in a 3D Delaunay triangulation, and n be the average number of points desirable in a cell. Then

$$nN_xN_yN_z = N \tag{7.2}$$

where N_x, N_y and N_z are, respectively, the number of cell divisions along the x-, y- and z-axes, and the number of cells $N_c = N_xN_yN_z$.

Let x_{min}, x_{max}, y_{min}, y_{max}, z_{min}, z_{max} be the bounds of the (x, y, z) co-ordinates of the point set; compute $R_x = x_{max} - x_{min}$, $R_y = y_{max} - y_{min}$ and $R_z = z_{max} - z_{min}$, and then N_x, N_y and N_z can be determined by substituting $\lambda R_x = N_x$, $\lambda R_y = N_y$ and $\lambda R_z = N_z$ into Equation 7.2.

Similar to the parallelisation in 2D, the most important requirement in a spatial partition of points into cells is to ensure that each point belongs to one and only one cell, and the sum of points in all the cells is equal to the total number of points, i.e.

$$N = \sum_{i=1}^{N_c} n_i \quad \text{where } n_i = \text{number of points in cell i}$$

7.4.2 Grouping cells into zones

Let $N_p = D_x D_y D_z$ be the number of processors available for parallel insertion, where D_x, D_y and D_z are the zonal divisions along the x-, y- and z-axes, respectively, as shown in Figure 7.17. Then cells are grouped into zones such that each zone will consist of m cells given by

$$m = \text{NINT}\left(\frac{N_x}{D_x}\right)\text{NINT}\left(\frac{N_y}{D_y}\right)\text{NINT}\left(\frac{N_z}{D_z}\right)$$

For best performance of parallel insertion, the number of zones has to be an integral multiple of the number of processors available; for example, division into $2 \times 2 \times 3 = 12$ zones for four or six processors is a sound division. The number of cells in some zones near the boundary may have fewer or more cells if N_x/D_x, N_y/D_y or N_z/D_z is not a whole number. Such a variation would not cause any problem in the subsequent operations, as a zone is identified by the bounding cells in x-, y- and z-directions, i.e. a zone I, I = 1 ~ Np, is specified by Zone I = Zone (N_{x1}, N_{x2}, N_{y1}, N_{y2}, N_{z1}, N_{z2}), where (N_{x1}, N_{x2}), (N_{y1}, N_{y2}) and (N_{z1}, N_{z2}) are, respectively, the starting and ending cell division lines along the x-, y- and z-directions.

7.4.3 Simultaneous insertion in 3D

Analogous to the simultaneous insertion in 2D, the procedure can be applied in exactly the same manner in 3D by replacing the rectangle by a rectangular cuboid, the triangles by tetrahedra, the circle by a sphere, etc. A division into $2 \times 2 \times 2 = 8$ zones for the insertion of 2000 3D points is shown in Figure 7.18, in which the average number of points in a cell

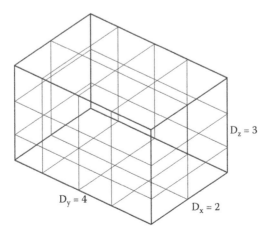

Figure 7.17 **Partitioned into 2 × 4 × 3 = 24 zones.**

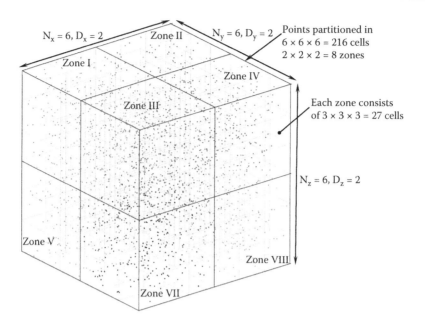

Figure 7.18 2000 random points partitioned into eight zones.

n = 8, and R_x, R_y and R_z are in the ratios 1:1:1; hence, $N_x = N_y = N_z = 6$, and $nN_xN_yN_z = 1728 \approx 2000$. The initial triangulation for each zone is a cuboid of five tetrahedra large enough to contain all the points. Point insertion in each zone will be handled by one single processor in a completely independent manner. All the points within each zone I, I = 1 ~ 8, will be processed concurrently by inserting points in those cells belonging to the zone under consideration. This will generate all the Delaunay tetrahedra within the zone; however,

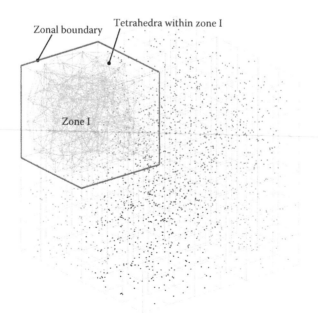

Figure 7.19 Tetrahedra generated within a zone.

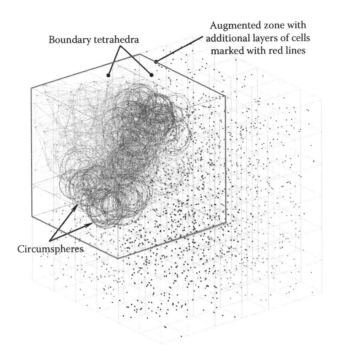

Figure 7.20 **Boundary tetrahedra and circumspheres.**

Delaunay tetrahedra crossing the zonal boundary are missing, as shown in Figure 7.19. The next step is to construct all the Delaunay tetrahedra at the boundary between zones. To do so, boundary cells are added around the zone to all the boundary surfaces of the zone, as shown in Figure 7.20. For easy visualisation, only those boundary tetrahedra with the neighbouring zones and their associated circumspheres are shown, whereas tetrahedra formed with the auxiliary corner points and their associated circumspheres are not shown. Boundary tetrahedra are defined as those tetrahedra supported on vertex or vertices from the current zone and vertex or vertices from the neighbouring zone(s), as shown in Figure 7.20. This simple definition is also applicable to Octree and kd-tree spatial partitions.

Layers of cells can be added to the boundary surfaces of the augmented zone until all the circumspheres of the boundary tetrahedra are bounded to ensure that all boundary tetrahedra are Delaunay, as shown in Figure 7.20. As each cell contains a roughly equal number of points, the process converges fairly rapidly and evenly on all the zonal boundary faces in one or two layers of cells. In this particular example, no additional layer of cells is needed, as all circumspheres of the boundary tetrahedra are already bounded by the augmented zone. Similar to the 2D case, points considered before need not be reconsidered after further point insertions as the union of circumspheres will always shrink with the introduction of a new point.

7.4.4 Elimination of redundant tetrahedra

Redundant tetrahedra on the zonal boundary can be eliminated concurrently by the generic minimum vertex allocation scheme introduced in Section 7.3.5. As each node has been given a zonal label in grouping cells into zones, a tetrahedron with zonal labels z_1, z_2, z_3 and z_4 on its vertices will be assigned to zone z given by

$$z = \min(z_1, z_2, z_3, z_4)$$

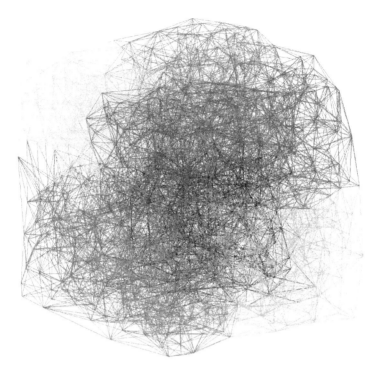

Figure 7.21 Partition of tetrahedra into eight zones by the minimum vertex scheme.

Over each zone, all the tetrahedra generated will be examined and allocated with the minimum vertex allocation scheme simultaneously by all the processors. At the end of a single pass by all the processors, tetrahedra will be partitioned into zones, and redundant tetrahedra are all eliminated as each tetrahedron should belong to one and only one zone. By the minimum vertex allocation, redundant tetrahedra for the parallel insertion of 2000 spatial points are eliminated and distributed into eight zones, as shown in Figure 7.21. Strictly speaking, this is not the convex hull of the given points but a DT of the points with some missing tetrahedra on the boundary surface.

7.4.5 Zonal insertion is Delaunay and complete

We have to assure that the zonal insertion algorithm produces Delaunay triangulation and the Delaunay tetrahedra around each point exist for any point in the zone. Let S be the set of points in zone Z under consideration. Let I be the set of interior tetrahedra in zone Z, where an interior tetrahedron of Z is a tetrahedron with all its four vertices taken from S. Let B be the set of boundary tetrahedra of Z, where a boundary tetrahedron of Z is a tetrahedron with vertex or vertices taken from S and vertex or vertices from the neighbouring zone(s). Hence, the triangulation T constructed in the zonal insertion process consists of tetrahedra from I and B, i.e. $T = I \cup B$

(I) T is Delaunay. Let α and β be two adjacent tetrahedra of T, $\alpha \in T \Rightarrow \alpha \in I \cup B$, $\beta \in T \Rightarrow \beta \in I \cup B$

   i. $\alpha \in I$ and $\beta \in I$. In the insertion process, the vertices of α and β are taken from S. Hence, by construction, α and β are mutually exclusive, which means that the circumsphere of α does not contain any vertex of β in its interior and vice versa.

ii. $\alpha \in I$ and $\beta \in B$. $\beta \in B$, β is Delaunay and the circumsphere of β does not contain any vertex of α in its interior. Since circumsphere inclusion is a reciprocal relationship, the circumsphere of α does not contain any vertex of β in its interior either. That is to say, α and β are mutually exclusive.

iii. $\alpha \in B$ and $\beta \in I$. Similar to (ii) as α is Delaunay.

iv. $\alpha \in B$ and $\beta \in B$. Both α and β are Delaunay; α and β are mutually exclusive.

From (i) to (iv), and by the lemma of Delaunay, T is a Delaunay triangulation.

(II) T consists of all the Delaunay tetrahedra centred at p for each point $p \in S$. Let C be the cluster (ball) of tetrahedra centred at p; we have to show that $C \subset T$, i.e.

$$\forall \text{ tetrahedra } T \in C, \quad T \in I \text{ or } T \in B \Rightarrow T \in I \cup B = T$$

7.4.6 Efficiency analysis

As the zonal point insertion is completely independent, the algorithm is expected to be very efficient and almost 100% scalable depending only on the characteristics of zonal partition. The efficiency of the insertion scheme depends on the amount of redundant tetrahedra between zones, which is, in turn, related to the extent of common zonal boundaries. Depending on how cells are grouped into zones, usually, interior zonal boundaries will increase with the number of zones and how zones are connected together. Let's consider various schemes of the partition of a unit cube into zones for parallel point insertion by eight processors, as shown in Figure 7.22. From a simple boundary count, an eight-zone partition (2 × 2 × 2) is the most efficient with a boundary of 12 units, whereas a 4 × 2 × 1 division will have a common boundary of 14 units, and the least efficient 8 × 1 × 1 division will create a common boundary of 20 units.

Consider the insertion of N points by zonal division of a cubic domain ($N_x = N_y = N_z$); if n is the desirable number of points in a cell, we have

$$nN_xN_yN_z = N \Rightarrow N_x = N_y = N_z = \sqrt[3]{\frac{N}{n}}$$

For a $k \times k \times k$ subdivision into zones, the number of internal boundary faces is given by

Internal faces $= 3(k - 1)$

Suppose that λ is the average number of layers of cells added to each side of an internal face to achieve convergence for boundary Delaunay tetrahedra; then the increase in the workload in triangulating the zones is given by

Increase $= 3(k - 1)(2\lambda)N_x^2$

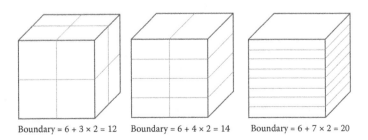

Boundary = 6 + 3 × 2 = 12   Boundary = 6 + 4 × 2 = 14   Boundary = 6 + 7 × 2 = 20

Figure 7.22 **Partition a cube into eight zones.**

Table 7.6 Estimating the number of additional layers for convergence

Division	I	$2 \times 2 \times 2$	$3 \times 3 \times 3$	$4 \times 4 \times 4$	$5 \times 5 \times 5$	$6 \times 6 \times 6$
Zonal insertion of 10 million points, $N_x = 215$						
NE	67,464,342	67,464,342	67,464,342	67,464,342	67,464,342	67,464,342
NB	147,072	284,383	631,810	1,230,369	2,175,883	3,547,541
NT	67,611,414	71,983,723	76,850,496	82,399,338	88,780,183	96,182,524
NR		4,234,998	8,754,344	13,704,627	19,139,958	25,170,641
μ= NR/NE		0.0628	0.1298	0.2031	0.2837	0.3731
λ		1.12	1.16	1.21	1.27	1.34
CPU Time	115.5	126.25	137.41	151.37	167.62	188.99
μ^*		0.0931	0.1897	0.3106	0.4513	0.6363
λ^*		1.67	1.70	1.85	2.02	2.28
Zonal insertion of 20 million points, $N_x = 271$						
NE	135,018,138	135,018,138	135,018,138	135,018,138	135,018,138	135,018,138
NB	236,282	568,667	1,263,026	2,460,783	4,351,733	7,095,280
NT	135,254,420	142,008,870	149,502,380	157,952,381	167,603,326	178,777,499
NR		6,422,065	13,221,216	20,473,460	28,233,455	36,664,081
μ= NR/NE		0.0476	0.0979	0.1516	0.2091	0.2715
λ		1.07	1.11	1.14	1.18	1.23
CPU Time	232.5	255.05	277.27	299.9	328.66	360.69
μ^*		0.0970	0.1926	0.2899	0.4136	0.5514
λ^*		2.19	2.17	2.18	2.34	2.49
Zonal insertion of 30 million points, $N_x = 311$						
NE	202,594,412	202,594,412	202,594,412	202,594,412	202,594,412	202,594,412
NB	303,892	853,194	1,895,403	3,691,170	6,527,964	10,642,326
NT	202,898,304	211,652,978	221,298,077	232,105,002	244,383,071	258,518,694
NR		8,205,372	16,808,262	25,819,420	35,260,695	45,281,956
μ= NR/NE		0.0405	0.0830	0.1274	0.1740	0.2235
λ		1.05	1.08	1.10	1.13	1.16
CPU Time	350.5	379.38	409.74	441.82	481.3	524.08
μ^*		0.0824	0.1690	0.2605	0.3732	0.4952
λ^*		2.14	2.19	2.25	2.42	2.57
Zonal insertion of 40 million points, $N_x = 342$						
NE	270,141,497	270,141,497	270,141,497	270,141,497	270,141,497	270,141,497
NB	337,677	1,137,523	2,527,204	4,921,467	8,703,523	14,190,462
NT	270,479,174	281,053,809	292,652,576	305,504,726	320,174,364	336,983,137
NR		9,774,789	19,983,875	30,441,762	41,329,344	52,651,178
μ= NR/NE		0.0362	0.0740	0.1127	0.1530	0.1949
λ		1.03	1.05	1.07	1.09	1.11
CPU Time	475	510.21	549.51	591.23	638.19	706.74
μ^*		0.0741	0.1569	0.2447	0.3436	0.4879
λ^*		2.11	2.24	2.32	2.45	2.78

Note: NT = total number of tetrahedra generated; NE = number of tetrahedra in the convex hull; NB = number of boundary tetrahedra connected to auxiliary corner points; NR = NT-NB-NE = number of redundant tetrahedra between zones.

The ratio of increase in workload μ is given by

$$\mu = \frac{3(k-1)(2\lambda)N_x^2}{N_x^3} \Rightarrow \lambda = \frac{\mu}{6(k-1)}\sqrt[3]{\frac{N}{n}}$$

The point insertion of a given point set within a single zone and by means of a subdivision into several zones has been tested thoroughly on a PC, Intel®, Core™, i7 CPU, 870 at 2.93 GHz with 16-GB RAM running on Windows 7 using Intel Fortran VS2010 on 64 bit. A series of computer runs by a single processor for the Delaunay triangulation of 10 to 40 million 3D points by means of various zonal subdivisions have been tested. Setting n = 8, $N_x = N_y = N_z = \frac{1}{2}\sqrt[3]{N}$, and the results are tabulated in Table 7.6.

In terms of the redundant tetrahedra created, as predicted, λ is more or less a constant, and it varies from 1.0 to 1.3 for zonal insertion of 10 to 40 million points with zonal divisions 2 × 2 × 2 to 6 × 6 × 6, showing that very little additional work is involved in the insertion with zonal partition, which somewhat increases slightly with the number of zones. There are only 3.6% redundant tetrahedra between zones corresponding to $\lambda = 1.03$ additional layer of cells for the insertion of 40 million points by 2 × 2 × 2 zonal subdivision. Nevertheless, in terms of CPU time, the result is quite different, and the CPU time increases more rapidly with the zonal subdivision. For example, for the insertion of 40 million points, a single processor insertion required 475 s, a 2 × 2 × 2 zonal insertion required 510 s (an increase of 7.4%) and a 6 × 6 × 6 zonal insertion required 707 s (an increase of 48.8%). As a result, the corresponding equivalent additional layer in terms of CPU time, λ^*, increased slightly compared to that obtained based on counting redundant tetrahedra. λ^* varies from 1.67 to 2.78 for the insertion of 10 to 40 million points with various zonal subdivisions from 2 × 2 × 2 to 6 × 6 × 6. The fact that λ^* is larger than λ indicates that there is an overhead in zonal subdivision, and the handling of layers of additional cells to achieve convergence also takes more CPU time compared to ordinary internal point insertion. In the insertion of 40 million points, λ^* increases gradually from 2.11 to 2.78 for various zonal subdivisions, and this range seems to be quite steady for a large number of points.

A plot of the insertion of 10 to 40 million points with various zonal subdivisions is shown in Figures 7.23 and 7.24. It is seen that CPU time increases with the number of zonal

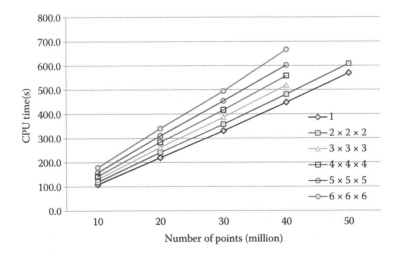

Figure 7.23 CPU time for the insertion of 10 to 40 million points.

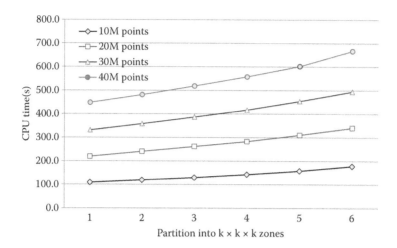

Figure 7.24 **Partition into k × k × k zones.**

subdivision, i.e. more zones in a zonal subdivision, in general, will take more CPU time for the insertion of the same number of points. However, given a zonal subdivision, the CPU time only increases linearly with the total number of points inserted, indicating that the additional work for zonal division is linear. The linear complexity for point insertion by zonal subdivision has an important implication that the parallel zonal insertion is almost 100% scalable for the number of processors available.

7.4.7 Memory requirement

Similar to the parallelisation in 2D, no extra memory is needed for the parallel zonal insertion scheme compared to the classical sequential insertion process, except for a zonal label for each point stored as a 2B integer in a linear array. To minimise the use of memory, the circumcentre and the circumradius of each tetrahedron are not stored, which are recalculated whenever necessary. For each point insertion, about $7 \times (4 + 4) \times 4 + 3 \times 8 = 248 \approx 250$ B memory is required to store the vertices and the neighbours of the seven tetrahedra (rounded up from an average of 6.75) generated and the x-, y- and z-co-ordinates of the point. Thus, a PC with 16 GB, apart from the memory taken up by the operating system, can generate quite comfortably 350 million tetrahedra for an insertion of more than 50 million points without being appreciably slowed down due to lack of memory.

An inherent advantage with zonal insertion in 2D and 3D is that given the same amount of memory, a larger point set can be handled by insertion with zonal subdivision, as information of neighbouring tetrahedra is only needed for one zone at a time, rather than for the entire population of all the tetrahedra as in the case of point insertion within a single zone without subdivision. However, for parallel insertion, if the zones are not processed all at the same time, some memory saving is possible for the reason stated above.

7.4.8 Treatment of degeneracy

Robustness is paramount in parallel point insertion, and all the points have to be kept intact throughout the insertion process. For randomly generated points or natural data points, the situation is not that bad, and there is a case of inconsistency about 1 in 1000 for the generation of 50 million points using double-precision calculations. For artificial objects,

Figure 7.25 **Points on the surface of a sphere are recorded.**

points are created at the corner of a cube, and some control measure has to be done. It might not be the most efficient to rectify the degeneracy within the parallel insertion process, as whether points are in the general positions or not can only be verified most efficiently by the empty sphere test. In view of this situation, in the parallel insertion, points not in the general positions within the machine precision for possible degeneracy will be recorded. In case no machine degeneracy is detected, the DT is valid, and nothing more needs to be done; otherwise, the DT may not be valid, and in this case, degeneracy points are perturbed and a second parallel insertion is carried out. As shown in Figure 7.25, points lying on the surface of a sphere to within machine precision for possible inconsistent DT are detected and recorded in the parallel insertion process.

7.4.9 Test on OpenMP shared memory systems

The first test of 3D DT by parallel zonal insertion is done on PC running on Windows 7 and Intel Fortran VS2010. It is a four-core machine with a maximum of eight pseudo-threads running on OpenMP parallel directives with shared memory architecture. Zonal divisions into $2 \times 2 \times 2$, $3 \times 3 \times 3$, $4 \times 4 \times 4$, $5 \times 5 \times 5$ and $6 \times 6 \times 6$ were tested, and the results of inserting 10 to 50 million randomly generated 3D points are shown in Table 7.7. A plot of the CPU time taken versus the number of points inserted for various zonal subdivisions is shown in Figure 7.26, whereas a plot of the CPU time versus the number of zonal subdivisions for the insertion of 10 to 50 million points is shown in Figure 7.27. It can be seen that the $2 \times 2 \times 2$ zonal partition is the most efficient, and the efficiency drops slightly with the increase in zonal subdivisions. However, the performance of $3 \times 3 \times 3$ and $4 \times 4 \times 4$ subdivisions are similar as $4 \times 4 \times 4 = 64$ is an integral multiple of 8, which helped to raise the

Table 7.7 **CPU time(s) for point insertion by various zonal subdivisions**

Points (million)	Tetrahedra (million)	1	2	3	4	5	6
10	67.6	109.0	25.8	30.5	30.9	35.4	38.8
20	135.3	219.3	51.9	60.4	61.4	68.3	76.2
30	203	330.7	79.7	90.1	90.0	99.7	109.6
40	270.7	448.1	105.7	120.9	120.8	133.2	146.9
50	338.5	569.2	132.9				

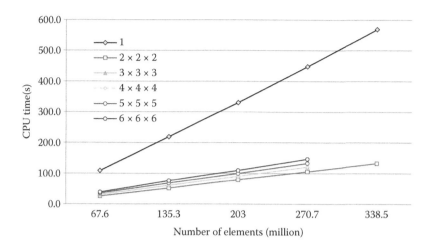

Figure 7.26 CPU time vs. tetrahedra generated by parallel point insertion.

performance of the 4 × 4 × 4 zonal division up to that of the 3 × 3 × 3 zonal division. It is seen that the parallel zonal insertion can boost the speed by 4.5 times, up to 2.6 million Delaunay tetrahedra per second on a PC with the 2 × 2 × 2 zonal subdivision for the insertion of 10 million 3D points.

The scalability of the parallel zonal insertion was tested by a proper multi-core machine (Gridpoint.hku.hk). There are 144 nodes in the machine, but only the IBM HS22 blade server with two 64-bit 6-core Intel Westmere CPU at 2.66 GHz was employed for the test. The IBM HS22 blade server, which supports the OpenMP parallel directives, can provide any number of threads up to 12 to test for scalability. A zonal division 4 × 5 × 6 = 120 zones was adopted, which is a common multiple for 2, 4, 6, 8, 10 and 12 processors, so that processors are expected to work to their full capacity for a fair comparison. The results for the parallel DT of 20 million 3D points are shown in Table 7.8, and the corresponding plots of the speed-up and the efficiency against the number of processors used are shown in Figure 7.28.

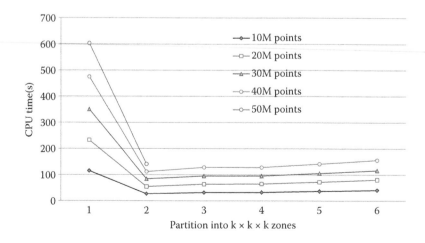

Figure 7.27 CPU time of parallel insertion for various zonal partitions.

Table 7.8   Scalability of test of 1 to 12 processors

No. of processors	CPU time	Speed-up	Efficiency
1	238.8	1.000	100.00
2	153	1.561	78.04
4	76.43	3.124	78.11
6	54.09	4.415	73.58
8	39.57	6.035	75.44
10	32.25	7.405	74.05
12	27.12	8.805	73.38

From Figure 7.28, the CPU time required decreases rapidly as more processors are used in the parallel insertion, and the speed-up with the number of processors employed is quite linear with a 5% difference in efficiency from 78% to 73% for 2 to 12 processors. Although good scalability is observed for parallel zonal insertion, in 3D, the performance of parallel zonal insertion is sensitive to the number of subdivisions, as in higher dimensions, the ratio of boundary to volume increases and hence the amount of work in dealing with redundant simplices increases with the dimensionality. This is only a scalability test in which the zonal subdivision may not be the most efficient for the number of processors employed. Scalability test for more than 12 processors has not yet been done, as the algorithm has to run on a cluster machine with a parallel environment quite different from the OpenMP system with shared memory.

Finally, processors can run at full capacity only when the number of zones is an integral multiple of the number of processors used and the number of zones is kept to a minimum. This trivial fact was verified in the insertion of 20 million points with a zonal division of $2 \times 2 \times 2 = 8$ zones for 2, 4 and 8 processors, $2 \times 2 \times 3 = 12$ zones for 6 and 12 processors and $2 \times 2 \times 5 = 20$ zones for 10 processors. The results are shown in Table 7.9, and a plot of CPU time, speed-up and efficiency is depicted in Figure 7.29. It can be seen that parallel insertion at high efficiency of 90% or more was achieved using 2, 4, 6, 8 or 12 processors, whereas the efficiency dropped to 84% when 10 processors were used. A slightly lower efficiency in the performance of using 10 processors is ascribed to a more complicated zonal subdivision of $2 \times 2 \times 5 = 20$ zones, whereas only $2 \times 2 \times 2 = 8$ or $2 \times 2 \times 3 = 12$ zones were used in the other cases. The efficiency of using six processors dropped drastically from 96% to only

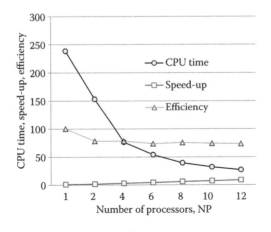

Figure 7.28 Insertion of 20 million points.

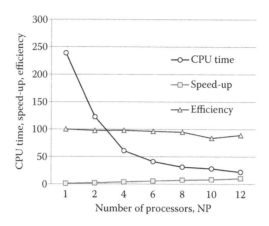

Figure 7.29 Insertion with optimal zonal divisions.

65% for parallel insertion with a zonal division of $2 \times 2 \times 2 = 8$ zones, as shown in Table 7.9; as 8 is not divisible by 6, not all the processors were working at the full capacity throughout the insertion process. From these tests, it can be seen that for parallel insertion over three or higher dimensions, it is better to use the minimum zonal subdivision, which is a multiple of the number of processors, to attain the best possible performance.

The $2 \times 2 \times 2$ zonal parallel insertion has been applied to three non-uniform point distributions on a PC in which 5 million points were randomly generated super-imposed with another 5 million points generated, respectively, along the diagonals, over the surface of a sphere and on a spiral curve (within 1% of the data spread), as shown in Figures 7.30–7.32 for the first 2000 points. These point distributions are designed to simulate the realistic case of adaptive refinement meshing, local concentration in solid and fluid mechanics problems. The CPU time in seconds of single-processor insertion and multi-processor is shown in Table 7.10.

For uniform random points, the speed-up is 4.3 times, which is reduced to approximately 3.5 times for the non-uniform point distributions, showing that the parallel insertion process is not very sensitive to the point distributions as long as loading is fairly balanced. It is interesting to note that the CPU time taken for point insertion of the spherical surface distribution is slightly less than that of the uniform distribution. This may be due to the fact that the average number of points within a cell is more optimised for the spherical distribution tested, which is locally similar to a uniform distribution. The CPU time quoted in Table 7.10 is faster than those given in Table 7.7 for the insertion of the same number of points. It is because the non-uniform distribution tests were done using a compiled Fortran program

Table 7.9 Parallel insertion with optimal zonal divisions

NP	Zones	CPU	SU	Eff
1	1	238.8	1.000	100.00
2	$2 \times 2 \times 2$	122.4	1.951	97.55
4	$2 \times 2 \times 2$	60.96	3.917	97.93
6	$2 \times 2 \times 3$	41.32	5.779	96.32
8	$2 \times 2 \times 2$	31.42	7.600	95.00
10	$2 \times 2 \times 5$	28.43	8.40	84.00
12	$2 \times 2 \times 3$	22.08	10.82	90.13
6	$2 \times 2 \times 2$	60.95	3.92	65.30

Note: CPU = CPU time in seconds, SU = speed-up, Eff = efficiency.

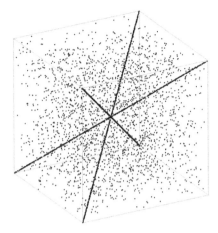

Figure 7.30 Points concentrated along diagonals.

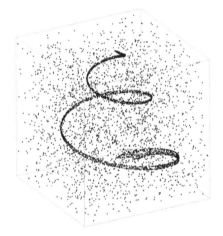

Figure 7.31 Points distributed over a spherical curve.

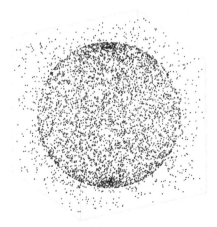

Figure 7.32 Points concentrated on a spiral.

Table 7.10 CPU time(s) for non-uniform distributions of 10 million points

	Uniform	Diagonal	Sphere	Spiral
Single processor	81.7	100.1	79.1	102.5
Multi-processor	18.9	29.5	22.7	29.4
Speed-up	4.3	3.4	3.5	3.5

with full optimisation, whereas the earlier tests were done using a compiled program with debugging mode being turned on.

Remarks: The characteristics of the parallel zonal insertion scheme presented in this chapter are as follows:

i. Processors can work to their full capacity by the simultaneous zonal point insertion in an absolute independent manner for both the triangulation and the elimination of redundant Delaunay tetrahedra.

ii. Artificial Delaunay boundary construction between zones before point insertion is not required.

iii. There is no need to connect the Delaunay tetrahedra generated within zones by filling up voids between zones, nor is there any synchronisation control or access blocking for the concurrent point insertion.

The scalability of the parallel zonal insertion algorithm was tested on a proper multi-core machine running on OpenMP parallel directives with shared memory. Provided the number of zones was an integral multiple of the number of processors used, almost 100% scalability at 90% efficiency was observed for parallel insertion. From the theoretical analysis and the numerical tests conducted, for parallel point insertion over 3D, it is better to use a minimum zonal subdivision, which is a multiple of the number of processors to attain the best possible performance. The challenge that lies ahead is to extend the parallel zonal insertion from the shared memory systems to cluster machines with many nodes. One possibility is to partition the data points into zones, each of which is sent to a multi-core machine (computation node) for parallel triangulation just described in this chapter. As for the inter-zonal boundaries, layers of points can be included in each zone to ensure that boundary simplices generated are all Delaunay.

The proposed algorithm is a generic one that could be readily extended to higher dimensions, as only the Delaunay circumsphere criterion is used in the construction of simplices, and redundant simplices can all be eliminated by the generic minimum vertex allocation scheme. Parallel DT for non-uniform distributed points is another interesting problem, which could possibly be handled also by the zonal insertion algorithm. The *multi-grid* insertion introduced in Section 8.2 could be employed for the parallelisation insertion of highly non-uniform point distributions as it is more efficient than the kd-tree partition in single-processor insertion. At locations of point concentrations, *multi-grid* is just a recursive application of the uniform grid, which has already been demonstrated to be effective in parallel zonal insertion.

7.5 PARTITION OF DISCRETISED SURFACE FOR PARALLEL PROCESSING

7.5.1 Introduction

3D models represented as polygonal meshes or point clouds are extensively used in engineering design, FE analysis, graphics visualisation (Labatut and Keriven 2009), terrain digitisation

(Meyer et al. 2007), etc. Algorithms are imperative to decompose complex triangulated objects consisting of thousands to millions of nodal points into simpler patches of smaller size for parameterisation and parallel processing. Non-manifold discretised surfaces with internal cells and diaphragms can be analysed and decomposed into simple open or closed manifold surfaces based purely on topological operations (Lo 1998b). In line with the development of CAD and surface digitisation, many algorithms have been proposed to decompose complicated triangulated manifold surfaces into patches for various applications (Lohner et al. 1992; Williams 1992; Ozturan et al. 1994; Shephard et al. 1997; Flaherty et al. 1998). Feature recognition separators defined over surfaces form a pool of choices for fitting cutting surfaces to produce separate volumes for hex meshing (Lu et al. 2001). A surface composed of patches is stitched together as a unique bi-parametric patch, thus providing a global parameterisation for free-form surfaces (Noel 2002). The original discretised surface is simplified by local remeshing in such a way that the global error is confined to a specified amount (Balmelli et al. 2002). By means of *blowing bubbles*, surface features can be identified, and curvature at a vertex could be estimated (Mortara et al. 2004). Boundaries of surface patches could be detected by discontinuous tangents of locally fitted surfaces, which provide a means for segmentation of 3D triangulated data points (Meyer and Marin 2004). A parallel ADF surface meshing scheme on B-rep CAD data with object-oriented approach was proposed by Deister et al. (2004). Tremel et al. (2004) presented a meshing procedure in which surface characteristics such as element size, angles and curvatures were computed directly from CAD data. Based on *the minima rule and part salient theory*, candidate contours are defined, which could form loops around mesh specifications (Lee et al. 2005). By solving a maximum hemispherical partitioning problem raised from a weighted Gaussian image, an optimisation algorithm is proposed to decompose a free-form surface into two sub-patches (Tang and Liu 2005). Framework for automatic extraction and annotation of anthropometric features from human-body models with shape measure based on multi-scale geometric and structural analysis was proposed (Mortara et al. 2006). Parallel adaptive FE analysis with a data structure supporting non-manifold geometries was studied by Seol and Shephard (2006).

By means of the grid-based technique, an algorithm for automatic generation of all hex meshes capable of representing the deformed geometry for updated Lagrangian FE calculations was presented (Fernandes and Martins 2007). A brain surface parameterisation method that invokes the Riemann surface structure to generate conformal grids on surfaces of arbitrary complexity including branching topologies was introduced (Wang et al. 2007b). A spherical harmonic decomposition for spherical functions defining 3D-triangulated objects of any genus number into star-shaped surface patches was proposed (Mousa et al. 2008). A *mesh constraint topology* model with automatic adaptation operators was put forward to transform a CAD boundary description model into an FE model (Foucault et al. 2008). The discretisation of curved surfaces by *geodesic Bezier curves* derived from the Euclidean Bezier curves was studied by Morera et al. (2008). Automatic segmentation of 3D models by means of random walks from user-defined seed faces was presented (Lai et al. 2009). A consistent *pant* decomposition framework for mapping surfaces with arbitrary topology was proposed (Li et al. 2009). A local refinement/coarsening algorithm was proposed for nested triangulation with untangling and smoothing procedure over domains with boundary faces projectable on a *meccano* boundary (Montenegro et al. 2009a,b). Simultaneous optimisation of multiple heterogeneous objectives that capture application-specific segmentation criteria was presented (Simari et al. 2009). Patch layouts are generated based on a topologically consistent feature graph, which separates the surface along feature lines into functional and geometric building blocks (Nieser et al. 2010). Moreover, a large-scale parallel adaptive analysis by means of dynamic load balance and domain repartitioning was developed by Zhou et al. (2012).

The method we are going to present may not be the most sophisticated to produce the *best* result. It is, however, one of the simplest schemes with robustness derived on a topological consideration of a cutting zone, which consists of all the triangular facets intersected by a cut plane passing through the centroid of the object and normal to one of the principal directions. As the cut plane is passing through the centroid of the object, the object will be roughly divided into two equal portions on the two sides of the cut plane. A surface will be intersected by cut planes corresponding to the three principal directions, and the one that gives the most balanced portions and is simpler in topology will be selected. There is no need for any global analysis of the structure and other geometrical features for the given surface; instead if any geometrical quantities other than the surface area are needed to be balanced, it can be done by redefining or shifting the boundary between two cut portions. Guided by a reduction formula for the number of cut pieces, the procedure can be repeated in a recursive manner to dissect a complicated surface open or closed into as many pieces as required.

7.5.2 Problem definition and preliminaries

The aim of surface decomposition is to partition a general surface discretised into triangular facets into n 'equal' pieces following a well-defined geometrical criterion. The triangular facets on the original surface are kept intact such that each triangular facet will be contained in one and only one decomposed surface part, and the entire surface can be fully recovered when every surface piece is put back together.

7.5.2.1 Triangulated surfaces

Complicated surfaces open or closed can be conveniently represented as a discretised model in terms of triangular facets. A discretised surface S consisting of N_S triangular facets is given by $S = \{S_i = (v_1, v_2, v_3)_i, i = 1, N_S\}$, where $(v_1, v_2, v_3)_i$ are the vertices of triangular facet S_i. Geometrical checks and topological analysis of surface S can be carried out before it is partitioned into simpler surface parts, as described in Section 8.1. There is no restriction on the topology of the surfaces being handled, and the surface could be simply connected resembling an open surface or a closed surface like a sphere or torus. Multi-connected non-manifold surfaces with internal cells and diaphragms are first decomposed into simpler manifold surfaces by pure topological considerations.

7.5.3 How the surface is cut into n pieces

An efficient robust partition procedure into two pieces is fundamental in cutting a complicated surface into n pieces, as shown in Figure 7.33. An additional piece can be generated by cutting one of the bisected surfaces into two pieces, as shown in Figure 7.33a and b. A given surface can be divided into n pieces by means of the following reduction sequence:

$$n = \begin{cases} \dfrac{n}{2} + \dfrac{n}{2} & \text{if n is even} \\[2mm] \dfrac{n-1}{2} + \dfrac{n+1}{2} & \text{if n is odd} \end{cases}$$

Any number n can be expressed in terms of addition and multiplication of 2 and 1, for instance, $10 = 2 \times 5 = 2 \times (2 \times 2 + 1)$, $17 = 2 \times 2 \times 2 \times 2 + 1$, etc. The partition is not just

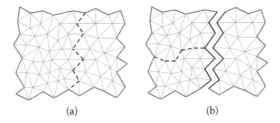

Figure 7.33 **Cut a surface into pieces: (a) two pieces; (b) three pieces.**

indicative but can be exact. Let's say that we would like to divide a surface into exactly 17 pieces of roughly equal area. The surface is first divided into two pieces, which are then adjusted on the common boundary such that the ratio of the area between the two pieces is about 8:9. Similarly the procedure is repeated on the two surface pieces to divide the large one into nine pieces and the smaller one into eight pieces, as shown in Figure 7.34.

In case there is no clue as where is the optimal cut, the discretised surface will be bisected by a plane passing through its centre of gravity with an orientation normal to one of its principal axes of inertia. Cutting along the centre of gravity can give us roughly two equal pieces without much calculation, which can be adjusted along the cut boundary for further balancing of any desirable geometrical quantities. If the centre of gravity is not inside the object, it will still work well by selecting the best cut along one of the principal directions.

$$\text{Centre of gravity } \mathbf{c} = (x_c, y_c, z_c) = \frac{1}{\Delta} \sum_{i=1,N_f} \Delta_i (x_i, y_i, z_i) \quad \text{where } \Delta = \sum_{i=1,N_f} \Delta_i$$

$$\text{Moment of inertia } \mathbf{I} = \sum_{i=1,N_f} \Delta_i \begin{bmatrix} \bar{y}_i^2 + \bar{z}_i^2 & -\bar{x}_i \bar{y}_i & -\bar{z}_i \bar{x}_i \\ -\bar{x}_i \bar{y}_i & \bar{z}_i^2 + \bar{x}_i^2 & -\bar{y}_i \bar{z}_i \\ -\bar{z}_i \bar{x}_i & -\bar{y}_i \bar{z}_i & \bar{x}_i^2 + \bar{y}_i^2 \end{bmatrix}$$

where Δ_i and (x_i, y_i, z_i) are, respectively, the area and the centroid of triangular facet S_i, and $(\bar{x}_i, \bar{y}_i, \bar{z}_i) = (x_i, y_i, z_i) - (x_c, y_c, z_c)$ are the co-ordinates of the centroid relative to the centre of gravity. The principal axes of inertia are given by the eigenvectors of $\mathbf{I}$.

Based on pure geometrical consideration on the shape of the resulting cut pieces, the cutting plane normal to the minor principal axis is preferable to the cutting plane normal to the major principal axis, as shown in Figure 7.35. However, when there is a change of topology, the cut that leads to a simple topology (for instance, from multi-connected to simply connected) is preferred, as shown in Figure 7.36. The cut along the longitudinal axis, normal to

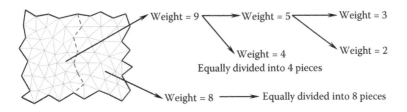

Figure 7.34 **How a surface is divided into n = 17 pieces.**

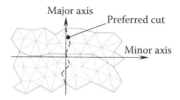

Figure 7.35 Cutting along principal axes.

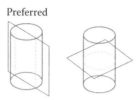

Figure 7.36 Cutting a cylinder into two pieces.

the major principal axis, produces two open surfaces, whereas the cut along the transverse direction, normal to the minor principal axis, produces two cylindrical surfaces.

7.5.4 Euler–Poincare characteristics

The topology of an open surface and that of a cylinder can be easily differentiated by considering the Euler–Poincare characteristics of the objects:

$$\chi = v + f - e$$

where v = Number of vertices, f = Number of faces and e = Number of edges.

For closed surfaces without holes, $\chi = 2$. An open surface is topologically equivalent to a closed surface upon removing one of its faces; hence, $\chi = 1$ for open surfaces. Take for instance an open surface shown in Figure 7.37a; there are 9 vertices, 8 faces and 16 edges, and $\chi = 9 + 8 - 16 = 1$. As for the open cylindrical surface shown in Figure 7.37b, there are 8 vertices, 8 faces and 16 edges, and $\chi = 8 + 8 - 16 = 0$. $\chi = 2$ for closed surfaces can be restored by adding two faces to the open cylinder, one at the top and one at the bottom. In fact, the cylindrical surface is topologically equivalent to an open surface with an interior opening, as shown in Figure 7.37c. If only one face is added to the bottom of the cylinder, effectively we are filling up the interior opening, and an open surface will be resulted.

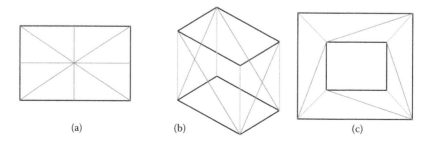

(a)	(b)	(c)

Figure 7.37 Topology of some simple objects: (a) open surface; (b) cylinder with open ends; (c) open surface with a hole.

7.5.5 Procedure for surface decomposition

7.5.5.1 Read in the surface S and carry out some basic topological computation

Input $S = \{S_i = (v_1, v_2, v_3)_i, i = 1, N_S\}, \{p_i = (x_i, y_i, z_i), i = 1, N_p\}$

where N_S and N_p are, respectively, the number of triangles and the number of points in S. Find all the edges on the surface S and the two triangles connected to each edge, $E = \{(E, \Delta_1, \Delta_2)_i, i = 1, N_E, \Delta_1, \Delta_2 \in S\}$. Find the neighbours of each triangular facet S_i, $T = \{(T_1, T_2, T_3)_i, i = 1, N_S; T_1, T_2, T_3 \in S\}$. T_i is set to zero when there is no neighbour at a boundary edge. The boundary of S is given by the collection of edges supported by only one triangular facet. If no such edges exist, then S is a closed surface.

7.5.5.2 Determination of the cutting zone

The cutting zone is composed of a band of triangular facets intersected by the cutting plane. Once cut plane P is determined, the set of triangular facets intersecting with the cut plane P is given by $I = \{S_i \in S, S_i \cap P \neq \varnothing\}$. A line segment AB is intersected by a cut plane P with origin O normal to unit vector $\mathbf{u}$ if points A and B are on the opposite sides of the plane, as shown in Figure 7.38a, i.e. $(OA \cdot \mathbf{u})(OB \cdot \mathbf{u}) \leq 0$. By adjacency relationship, in general, the cutting zone is composed of strips and loops of intersected triangular facets, as shown in Figure 7.38b and Section 4.5.4. Using the cutting zone as the natural separation line, the surface will be automatically partitioned into a number of smaller pieces with the cutting zone as boundaries between surface parts, as shown in Figure 7.39.

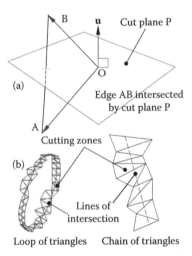

Figure 7.38 Determination of cutting zone: (a) edge AB intersected by cut plane P; (b) triangles intersected form a loop or a chain.

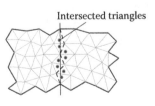

Figure 7.39 Cut line is formed by edges.

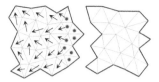

Figure 7.40 Grouping by element adjacency.

7.5.5.3 Subdivide S by the cutting zone

Surface S can be conveniently divided into two pieces by taking all the triangles on one side of the cutting zone following the element adjacency relationship, as shown in Figure 7.40. The remaining triangular facets could be put into another group to form the second piece. A clear cut boundary composed of edges can be defined between the two surfaces if the intersected triangles in the cutting zone are absorbed by the smaller surface piece. In case there are more than two surfaces in the partitioning process, smaller surface parts will be merged to form two major surfaces, as shown in Figure 7.41.

7.5.5.4 Balancing the two resulting surface parts

This is an important step as the two resulting surfaces may not be equally divided for arbitrarily oriented surface with general topology, or sometimes two unequal pieces are needed for different subsequent treatments. The geometrical quantity α of the two pieces of surface to be balanced depends on the application of the actual problem at hand. The geometrical quantity α is usually an increasing function of surface S, i.e. $\alpha(S) \le \alpha(S \cup S_i)$ for arbitrary triangular facet S_i. For instance, geometric quantities commonly considered are the total area of the surface, the sum of dihedral angles of the surface, the sum of Gaussian curvature (Tang and Liu 2005) of the surface, etc. It is noted that dihedral angles between two triangular faces can be conveniently computed from the normals to the faces.

7.5.5.5 Distance from the cut line

The distance of a point on the surface to the cut line is defined as the shortest distance without leaving the surface from the cut line (Martinez et al. 2005). While it is tedious to

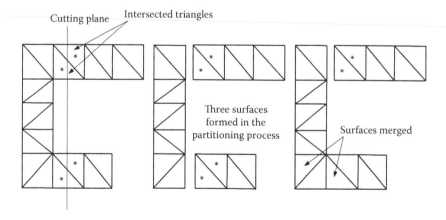

Figure 7.41 Smaller surface parts merged to form one single piece.

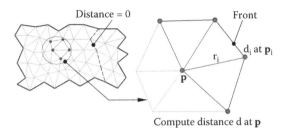

Figure 7.42 **Distance map.**

determine the geodesic distance point by point, it is much easier to set up a distance map for all the points by means of a recursive formula. The distance at a point **p** can be computed from its neighbouring points, as shown in Figure 7.42, such that

$$d = \min_{i \in \text{neighbour}} (d_i + r_i) \tag{7.3}$$

where d_i is the distance of a neighbouring point $\mathbf{p}_i$, and r_i is the distance between **p** and $\mathbf{p}_i$.

As the adjacency relationship of the triangular facets has already been determined, the distance of all the points can be computed by means of a frontal process starting from the cut line, where points are set a datum of zero value. Take a point **p** next to the front where the distance from the datum line is known. Form the patch of triangles around **p**, as shown in Figure 7.42, and the distance at **p** is calculated using formula 7.3 on all its neighbouring frontal points with known distance from the datum. Point **p** is absorbed to the front as its distance from the datum is now known. The front of points of known distance continues to propagate until all the points are processed. This method does not produce the absolute distance from the cut line but the pseudo-distance along the edges of the discretised surface. However, the distance map so computed is accurate enough to differentiate the relative distances of two points from the cut line.

7.5.5.6 Marching on the surface

The original cut line can march forward (shift in one direction) by absorbing one or more triangular facets (elements) nearest to the cut line. In the more general situation where the marching process is based on some user-defined geometrical/topological criterion α, the smaller surface expands by taking triangles from the large surface at the boundary between the two surfaces. The balance criterion α of the surfaces has to be evaluated from time to time when more triangles are taken up by one of the surfaces in the marching process. The final position of the cut line is determined when the two surface parts are more or less balanced measured in α quantity, as shown in Figure 7.43. In the marching process, the boundary will be smoother

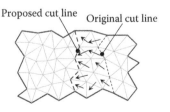

Figure 7.43 **Marching on surface.**

if a layer of triangles are added at a time. However, if the entire layer of triangles has to be split up to take in individual triangles at the final stage of balancing, priority will be given to those triangles touching the cutting line on two edges with a good balance of the α quantity.

7.5.5.7 Optimisation to improve the quality of the cut

After balancing the two surfaces, the cut line can be optimised in terms of its length and the sum of dihedral angles or any other geometric quantities. The proposed simple scheme does not require a tedious global analysis of any geometrical features of the surface so as to identify patterns and structures for the intended purpose. A few layers of triangles are included on both sides of the cut line to form a band (the cut line may also be a closed loop). A distance map is evaluated from one boundary to another over the band of triangles as described in Section 7.5.5.5. As shown in Figure 7.44, find a point on boundary B whose distance to boundary A is a minimum. The new cut line is the path tracing back from this point towards boundary A following the rule of the greatest descent from one point to the next. Apart from the consideration of the minimum distance, the geometrical quantities such as dihedral angle or angle of deviation, etc., could also be taken into account in establishing a path from one boundary to another in an overall optimised manner, as shown in Figure 7.45. The scheme is based on the principle of minimisation of a potential pertinent to the geometrical quantities under consideration. A cut path can be automatically traced out with optimisation on several geometrical parameters just following the weighted distance map with a dihedral angle, a deviation angle, etc. Of course, the shortest cut on the nearly balanced surfaces could be determined by imposing the minimum distance between points A and B as the sole criterion for optimisation.

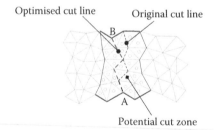

Figure 7.44 **Cut line optimisation.**

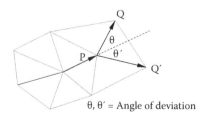

Figure 7.45 **Selection between point Q and Q′.**

7.5.6 Examples

The first example 'Voltaire' is an open surface consisting of 1124 triangular facets and 966 nodal points, as shown in Figure 7.46a. The surface is to be divided into six pieces of equal area. The surface is first divided into two halves by a vertical cut plane. Marching process is applied on the two sides of the cut line so that a band of approximately 1/3 of the total surface area is formed. The original surface is effectively cut into three vertical strips of equal area, as shown in Figure 7.46b. Each vertical strip is then divided into two pieces by a cut plane normal to the minor principal axis of inertia, as shown in Figure 7.46c. The second example, which consists of 452 triangular facets and 228 nodal points, is a closed surface in the form of an ashtray, as shown in Figure 7.47a. The surface is first cut by a vertical plane into two equal halves, as shown in Figure 7.47b. However, when the closed surface is intersected by a horizontal plane, two closed loops are resulted. An optimised cutting path is determined by selecting a loop between the two closed loops produced by the cut plane. If apart from the surface area, the dihedral angles are also taken into consideration, a cutting path indicated by the arrows following the sharp edges between the horizontal and vertical planes of the ashtray is automatically traced out, as shown in Figure 7.47c.

The third example is 'Hugo,' which is a cylindrical surface opened at the top and at the bottom. The surface consists of 4000 triangular facets and 2026 nodal points. The surfaces cutting along three principal axes of inertia are shown, respectively, in

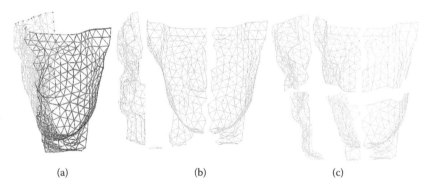

(a)   (b)   (c)

Figure 7.46 Cutting a surface into six pieces: (a) initial surface; (b) surface cut into 3 pieces; (c) surface cut into 6 pieces.

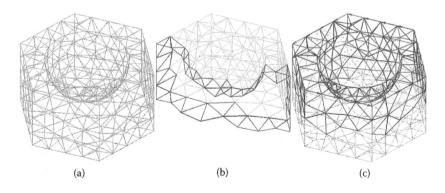

(a)   (b)   (c)

Figure 7.47 A closed surface cut by (i) a vertical and (ii) a horizontal plane: (a) ash tray model; (b) vertical cut plane; (c) horizontal cut plane.

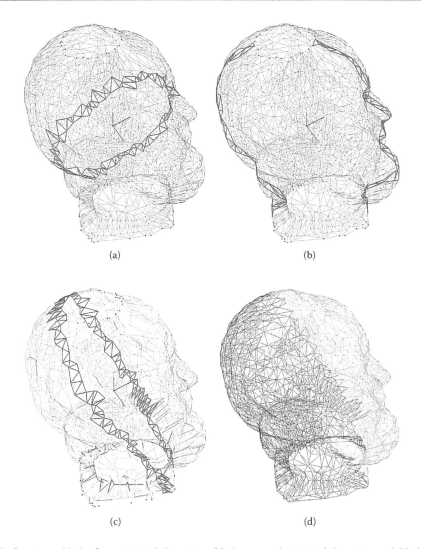

Figure 7.48 Cut along (a) the first principal direction, (b) the second principal direction and (c) third principal direction. (d) Two surfaces of equal area based on a cut along the third principal.

Figure 7.48a–c. Surfaces of equal area after balancing based on the cutting zone associated with the third principal direction are shown in Figure 7.48d. The fourth example is the shuttle plane Columbia modelled by 12,834 triangular facets and 6419 nodal points. The cutting of the shuttle plane along three principal directions of inertia is shown in Figure 7.49. The last example is a finite element model of part of a wheel consisting of 1929 nodes, 3870 triangles and 5805 edges. The wheel is cut along two principal directions, as shown in Figure 7.50. There are two intersection loops for the cut along the second principal direction, as shown in Figure 7.50b; however, the second intersection, which is too small and does not divide the object into two substantial parts, has to be ignored. The remaining two intersection loops practically cut the wheel into four roughly equal portions, which can be further adjusted on the cut boundaries following the procedure described in Section 7.5.5.

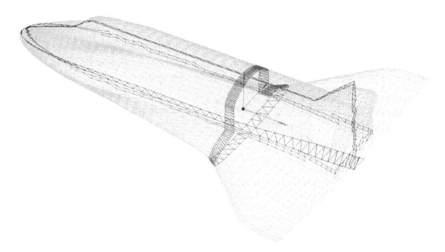

Figure 7.49 Cutting zones of a shuttle plane along three principal directions.

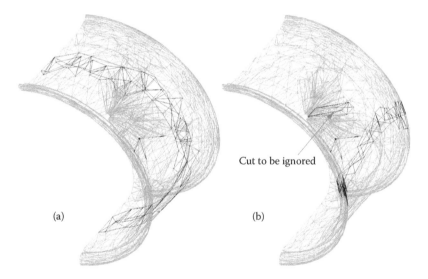

Cut to be ignored

(a)   (b)

Figure 7.50 Part of a wheel cut along two principal directions of inertia: (a) vertical cut; (b) horizontal cut.

7.5.7 Conclusion

A generic surface decomposition algorithm is presented in this chapter, which could subdivide a general open or closed discretised surface into n pieces following any specified geometrical criteria. The procedure has been designed based on the minimum information of a triangulated surface such that only the nodal points of each triangle are needed for surface decomposition. The surface is automatically partitioned without ambiguity into smaller surface parts by cut zones of intersected triangular facets. As only topological operations are involved, individual surfaces with clearly defined boundary could be processed independently before being put back together to recover the original surface. A surface marching process has been proposed to balance the specified geometrical quantities of the cut surfaces following a distance map readily determined from the element adjacency relationship. The cut line can be further refined by a marching procedure within a potential cutting zone, so that other geometrical quantities such as sharp dihedral angles and deviation angles can also be taken into account. The time complexity of the surface decomposition process is basically linear as only the element adjacency relationship and the system of edges are referred to in all the steps for surface decomposition. Once formulated, the surface bisection algorithm can be applied repeatedly on the subdivided surface parts. As the subdivided surface parts are distinct and independent, further subdivision of surfaces could be done in parallel by the same algorithm to speed up the process. For complicated non-manifold objects and surfaces with many connections, they have to be first decomposed into open or closed surfaces by topological operation, as described in Section 8.2.

Chapter 8

Auxiliary meshing techniques

No algorithm is perfect, and there are always alternatives, complements and supplements.

There are many techniques that are not directly applied to generate finite elements but are extremely useful in serving various aspects to facilitate finite element (FE) mesh generation (MG). The surface verification and preparation presented in Section 8.1 aims at analysing whether a given surface is closed and constitutes a well-defined boundary for MG within a volume. The multi-grid insertion introduced in Section 8.2 is a recursive application of a simple regular grid for Delaunay triangulation (DT) of highly non-uniform point distributions. Multi-grid insertion is a generic concept that can be readily applied also in 3D, and such a procedure is presented in detail in Section 8.3. Mesh refinement, especially in 3D, is a rapid and reliable method in producing valid FE mesh in compliance with an element size specification. Guided by a shape quality measure along with optimisation by geometrical and topological means, high-quality FE meshes can be generated for adaptive refinement analysis. Mesh refinement algorithms over surface and within volume according to an element size function are presented in Section 8.4. Following mesh refinement in Section 8.4, volume bounded by analytical surfaces can be easily meshed into tetrahedral elements in two steps: in the first step, an adaptive refinement is applied with respect to the curvature of the boundary surface, and in the second step, nodal points close to the boundary are snapped on to the boundary surface to obtain the final mesh as detailed in Section 8.5.

Similar to the merging of surface meshes introduced in Section 4.5, many objects can only be conveniently defined through an intersection process, and meshes of various characteristics can be created simply by putting one or several solid FE meshes together. A generic mesh merging scheme for tetrahedral elements will be discussed in Section 8.6, in which arbitrary tetrahedral meshes can be automatically combined into one single FE mesh with full compatibility for numerical analysis. Based on the merging of tetrahedral elements, merging of hexahedral elements is also possible by first converting each hexahedral element into five or six tetrahedral elements, most of which can be recovered after mesh merging as shown in Section 8.7. A brief note on the generation of higher-order curvilinear FE meshes is given in Section 8.8. A simple way is to first generate a p1 (linear) mesh, which can then be converted to a p2 mesh by adding mid-side nodes; however, a more sophisticated scheme is to generate the p2 mesh directly following the curved boundary of the object. However, there are a lot of difficulties in the direct generation of the curved edges for p2 elements. As an alternative, the associated p1 mesh is further refined, and by means of the optimisation techniques developed in Chapter 6, the refined p1 mesh can be optimised to produce the curved edges and surfaces of a curvilinear mesh quite automatically in a natural formation process. Adaptive refinement is an effective automatic procedure to control the error of an FE analysis in which element sizes have to adapt to the error of the FE solutions so as to achieve the optimal rate of convergence. Based on the current FE solution, how the size of the FEs can be estimated in practical applications will be given in Section 8.9.

8.1 SURFACE VERIFICATION AND PREPARATION

8.1.1 Introduction

High-quality FE MG for a complex 3D object is a time-consuming process, and the boundary surfaces of the object have to be checked very carefully to ensure they are error-free. With the increasing use of computer graphics in the creation of complicated object shapes, such as aircraft, space vessels, automobile, machine parts and physical models, visual inspection alone is not good enough to affirm the validity and the quality of the FE model, and a thorough verification of the output from a CAD system is mandatory to ensure that the object is well defined and complies with the input requirement of the mesh generator.

With the rapid advent of computing machines and the fast development in parallel computations, we are becoming increasingly ambitious and confident in handling large-scale 3D objects of arbitrary shapes and complexity with unstructured meshes (Coupez et al. 1991; Johnson and Tezduyar 1994, 1997; Frey and Alauzet 2005; Loseille et al. 2010). Baker (1989a,b) gives a summary of the work on 3D MG and a discussion on the future development in the fluid flow problem around an aircraft. The boundary surface modelling technique (Boender et al. 1994; Lo 1995; Armstrong 2000; Alleaume 2009; Quadros and Owen 2012) working under a comprehensive CAD environment has now developed to a point where distinct functions of the mechanical design and the structural analysis of 3D objects of complex geometry and topology are integrated intimately. The practical construction of such a formal link between a solid modeller and FE analysis software for general 3D structures is by no means straightforward. FE MG over curved surfaces and within volumes represent the largest bottleneck in establishing such a link.

Surprisingly there is not much research directly reporting on this topic, partly due to the diverse nature of the subject and the fact that related geometrical and topological operations have all been published and well known under different research topics, and partly due to the difficulty of having little recognised academic value but tremendous practical significance in adapting, modifying and grouping the available geometrical/topological operations in such a way for the purpose of boundary surface verification and automation in MG. Nevertheless, let's take a look at the research works that are somewhat relevant to this topic. Lo (1991b, 1998b) published papers on the analysis and verification of the boundary surface for solid objects, and the materials presented in this section will be mainly based on the ideas of these papers with proper updating and reediting of the formulas, figures and examples. Mackie (1999) worked on the data modelling of object-oriented FE programming. Fine et al. (2000) described an idealisation process of an FE model from a polyhedral domain. Tautges (2001a) proposed CGM, which is a geometry interface for MG and analysis in which facet-based geometry application was discussed. Ribo et al. (2002) proposed algorithms to repair and improve the geometric models of a CAD system for FE MG. Owen and White (2003) discussed mesh-based geometry in which the underlying geometry of an FE mesh was extracted. Automatic mesh-repairing techniques were introduced by Chong et al. (2007), which were directly applied to fix incompatibilities in an FE mesh rather than on the CAD geometry. Foucault et al. (2008) presented the adaptation of CAD model topology for FE analysis, in which geometrical features of a B-rep model were automatically identified to form natural boundaries of an FE mesh. A survey of CAD model simplification techniques for physical-based simulations was conducted by Thakur et al. (2009) in which CAD models were classified and compared. Geuzaine and Remacle (2009) presented GMSH, which is an FE mesh generator with built-in pre-processing and post-processing facilities. Procedures such as edge swapping/hammering and face lifting were proposed to repair CAD surfaces (Yamakawa and Shimada 2009b). Cuillière et al. (2011) discussed a comparison and remeshing scheme for CAD models, in which objects of similar geometrical and topological features were identified in an automatic assembly process. Karamete et al. (2013) presented an algorithm for discrete Booleans with applications to FE modelling of complex systems in which triangulated surfaces are combined by means of Boolean operations derived from the intersection of the triangular elements. Foucault et al. (2013) applied the ADF to composite surfaces in which the geometric features and the constraints of the objects were well respected. Cuillière et al. (2013) discussed the MG and transformation for topology optimisation of objects with heterogeneous geometry.

In this section, a simple data-verification scheme for discretised curved surfaces and objects defined by the boundary surface modelling specifications will be presented. The quality of individual elements, overall topological structures and geometrical correctness in terms of intersections, close touches and sharp angles will all be examined and verified. This will only provide a basic routine surface-verification process for MG and FE analysis; however, much more in-house checking and correcting procedures can, and will definitely, be implemented following the same concept of validation in automation adapted to specific applications under various working environments.

8.1.1.1 Boundary surface of solid objects

The boundary surface of an object to be meshed can be conveniently represented by a collection of triangular facets. The following is a typical data structure for such a representation in which the boundary surface $\mathbf{B}$ is given by

$$\mathbf{B} = \{\Delta_i = (P_1, P_2, P_3)_i, i = 1, N_B\}$$

The spatial points $\mathbf{P} = \{(x_i, y_i, z_i), i = 1, N_P\}$, where N_P and N_B are, respectively, the number of points and the number of boundary triangular facets in $\mathbf{B}$. This data structure is the simplest possible, but is broad enough to include solid objects based on B-rep models in which $\mathbf{B}$ consists of all the triangles on the discretised boundary surfaces.

8.1.2 Preliminary checks and preparations

Simple routine checks are done on the boundary surface $\mathbf{B}$ to find out the limits of the co-ordinates, the range of the node numbers, the number of active nodes, the spacing between nodes, the quality of each boundary triangle, etc.

8.1.2.1 Limits of points

The limits of the co-ordinates x_{min}, x_{max}, y_{min}, y_{max}, z_{min} and z_{max} are calculated:

$$x_{min} = \min_{i=1,Np} x_i, \quad x_{max} = \max_{i=1,Np} x_i, \text{ similarly for } y_{min}, y_{max}, z_{min} \text{ and } z_{max}$$

8.1.2.2 Normalisation of co-ordinates

To have a better idea about the size of the elements, the co-ordinates are mapped onto the interval between 0 and 100 through shifting and the multiplication of a scaling factor λ:

$$x_i \mapsto \lambda(x_i - x_{min}), \quad y_i \mapsto \lambda(y_i - y_{min}), \quad z_i \mapsto \lambda(z_i - z_{min})$$

where λ is a scaling factor such that $x_i, y_i, z_i \in [0,100]$, $i = 1, N_P$. Of course, $[0,100]$ is rather arbitrary, and other ranges could have been chosen for convenience in particular applications.

8.1.2.3 Check if any node is outside the range [1, N_P]

$$1 \leq P_{ij} \leq N_P \quad \forall i = 1, N_B, \quad \forall j = 1, 2, 3$$

8.1.2.4 Find out all the connected node points

Point $P \in \mathbf{P}$ is connected if P is a nodal point of some triangle(s) in $\mathbf{B}$:

$$\exists P_{ij} = P \quad \text{for some } i \in [1, N_B], j = 1, 2, 3$$

8.1.2.5 Check the spacing between nodes

$$\|P_i P_j\| > \delta \quad i, j \in [1, N_P], \quad i \neq j$$

where $\delta = \varepsilon L$, in which L is the diagonal of the containing box, and ε is a tolerance factor.

8.1.2.6 Verification of individual elements

Several parameters ought to be computed to judge the quality and the size distribution of the elements.

i. Area of the triangles Δi, $A_i = \dfrac{1}{2} \|P_1 P_2 \times P_1 P_3\|_i \quad i = 1, N_B$

ii. Total surface area, $S = \displaystyle\sum_{i=1,N_B} A_i$; average element size, $A = \dfrac{S}{N_B}$

iii. The smallest and the largest elements, $A_{\min} = \min\limits_{i=1,N_B} A_i$, $A_{\max} = \max\limits_{i=1,N_B} A_i$

iv. Element shape quality in which three criteria are available:

Smallest angle measure, $\theta = \min (\angle P_1 P_2 P_3, \angle P_2 P_3 P_1, \angle P_3 P_1 P_2)$

Ratio of in-radius(r) to circum radius(R), $\rho = r/R$

$$\alpha - \text{quality} = \frac{\text{signed area}}{\text{sum of edges squared}}, \quad \alpha_i = \frac{4\sqrt{3} A_i}{\|P_1 P_2\|_i^2 + \|P_2 P_3\|_i^2 + \|P_3 P_1\|_i^2}$$

α-quality, which also gives the orientation of the triangle, will be used to measure the shape quality of the boundary surface.

$$\alpha_{\min} = \min_{i=1,N_B} \alpha_i, \text{ geometric mean } \alpha = \left[\prod_{i=1,N_B} \alpha_i \right]^{\frac{1}{NB}} \text{ and report any } \alpha_i < \alpha_o \text{ with threshold}$$

value α_o set arbitrarily at 0.0001.

8.1.3 Analysis of topology

Numbering of triangular facets in $\mathbf{B}$ is completely arbitrary, as shown in Figure 8.1, and this allows the maximum flexibility in the definition of boundary surfaces for solid objects.

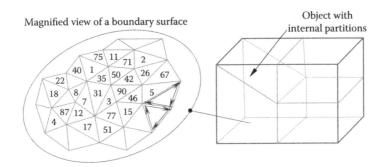

Figure 8.1 Numbering boundary triangles and non-manifold object.

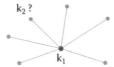

Figure 8.2 Check if k_2 is already connected to k_1.

8.1.3.1 Search for all the edges on boundary surface B

The set of edges is the collection of all the distinct line segments on the boundary surface **B**. Edges of the triangles are examined in turn. Let k_1 and k_2 be the node numbers of an edge, with $k_1 < k_2$. Before recording $k_1 k_2$ as a new edge, a check is made on the nodes already connected to k_1, as shown in Figure 8.2. Let C_{k1} be the set of nodes connected to k_1. Then if $k_2 \in C_{k1}$, edge $k_1 k_2$ is ignored; otherwise, edge $k_1 k_2$ is recorded and C_{k1} is updated to $C_{k1} \mapsto C_{k1} \cup \{k_2\}$. For a given triangular mesh, the average number of nodes connected to a particular node is constant and independent of the total number of nodes in the system. Hence, the efficiency of the scheme for the retrieval of edges is of order N_P. With little additional care during the edge-retrieval process, the three edges of each triangle can be identified. This information is also stored and will be useful later in many topological operations on the elements in **B**. The algorithm presented in Section 2.5.4 for the retrieval of edges in triangular meshes on planar domains and over curved surfaces is even simpler; however, the algorithms just described can work with unoriented triangles, non-manifold objects and non-orientable surfaces as well, which is an important attribute of the algorithm in the decomposition of non-manifold objects into closed surfaces for MG.

8.1.3.2 Elements connected to each edge

For each edge L, L = $1, N_L$, where N_L is the number of edges in **B**, find the list of elements connected to it $\{E_1, E_2, ..., E_n\}$. Since the three edges of each triangle are known, the list of elements connected to each edge can be readily determined by examining each triangular element in turn. Let L_1, L_2 and L_3 be the three edges of triangle Δ_i; then on processing Δ_i, L_1, L_2 and L_3 will include Δ_i in the list of elements connected to them. By scanning through all the triangular elements Δ_i, i = $1, N_B$ sequentially, the list of elements connected to an edge can be established. In fact, the list of elements connected to an edge is the inverse relationship of edges connected to an element, as shown in Figure 8.3.

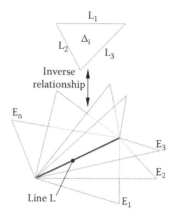

Figure 8.3 Find the elements connected to an edge.

8.1.3.3 Elimination of redundant triangles

Elimination of redundant elements in **B** can be conveniently done by examining the elements around each edge. Let Δ_i and Δ_j be two triangles connected to a particular edge; if $\Delta_i = \Delta_j$ or $(P_{i1}, P_{i2}, P_{i3}) = (P_{j1}, P_{j2}, P_{j3})$ in any order, then Δ_k will be eliminated, where $k = \max(i, j)$.

8.1.3.4 Surface construction

Surfaces can be constructed one at a time starting from any triangular element in **B**. The following are the procedures of surface construction:

1. An open surface of one single triangle (seed) with a boundary consisting of three edges of the triangle is initiated, as shown in Figure 8.4a.
2. The triangles connected to the edges of the construction front are identified and added to the current unclosed surface of triangular facets, as shown in Figure 8.4b.
3. The boundary edges are those edges without neighbours (free edges) or edges shared by more than two triangles, as shown in Figure 8.4c. A surface is obtained if the construction front is reduced to zero, as shown in Figure 8.5.

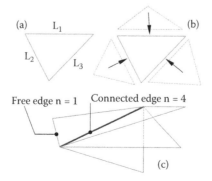

Figure 8.4 Connected edges and free edges: (a) seed triangle; (b) neighbouring triangles; (c) boundary edges.

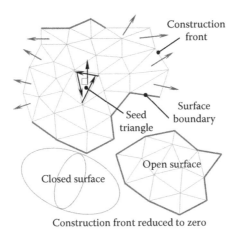

Figure 8.5 **Surface construction.**

Let n(L) be the number of triangles connected to edge L; then L is a boundary edge if n(L) ≠ 2. No search is needed since the elements connected to each edge are known and well prepared beforehand. Efficiency of the scheme is of order N_B. The procedure guarantees that each surface generated is orientable (non-orientable surfaces such as Mobius strip can be detected when there is an inconsistency in the orientation between neighbouring triangles). All the elements belonging to the same surface will be given a unique surface label and oriented with respect to the first (seed) element of the surface. Elements used in the surface construction will be deleted from set **B**, and the surface-construction process terminates if there are no more triangles in **B**. Each triangle in **B** belongs to one and only one surface, i.e. **B** is effectively partitioned into surface parts by the surface-construction procedure. If N_S is the number of surfaces so created, then $N_S \le N_B$.

8.1.3.5 Flagging unused surface parts

All hanging surface parts with free edges, n(L) = 1, are flagged, and they will not be used in the next step of region identification. In 3D MG, these surface parts may represent additional constraints in triangulation.

8.1.4 Region identification

8.1.4.1 Boundary edges

Find the boundary edges of each surface. Boundary edges exist for open surfaces, whereas for closed surfaces such as sphere and torus, there is no boundary edge, as shown in Figure 8.6.

8.1.4.2 Formation of regions

1. Start with any open surface; take note of the two orientations (sides) of the surface, as shown in Figure 8.7a.
2. Pick up any free edge L on the unclosed surface; join it with the other surface (such a surface exists as n(L) > 2) at the free edges for which the dihedral angle θ between the surfaces is the smallest, as shown in Figure 8.7b. The 2D analogy is shown in Figure 8.8; regardless of the complexity of the internal region, the boundary loop can always

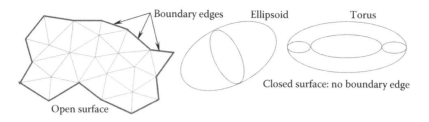

Figure 8.6 **Open and closed surfaces.**

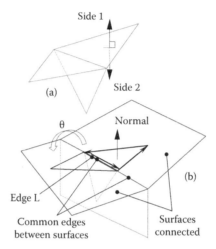

Figure 8.7 **Couple with neighbouring surface: (a) surface orientations; (b) surfaces connected by the least dihedral angle.**

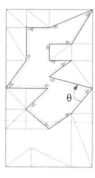

Figure 8.8 **Trace a region in 2D.**

be traced out consistently following the smallest internal angles going in either clockwise or counter-clockwise direction.

3. In merging two surfaces, the region boundary is updated. All the common edges will cancel each other, and the unclosed surface grows in size, as shown in Figure 8.9. Go to step 2 if there are still free boundary edges.

4. No more free edges imply that a closed region has been formed. Each side of a surface can be a wall of only one single region; hence, the region formation process finishes if both sides of the surfaces have been used in the construction of regions. A

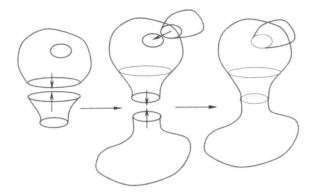

Figure 8.9 Surfaces combined to form a region.

non-manifold object with many internal partitions will be decomposed into regions, each of which is bounded by individual surfaces so collected based on the smallest dihedral angles between surfaces.

8.1.4.3 Validity check of the formation of regions

1. Topology correctness
   The correctness in topology can be verified by the following three conditions:
   a. Every region is closed.
   b. Each surface is traversed two and only two times, one on each side.
   c. For each region, check with the Euler–Poincare formula,

   $$v - e + f = 2(s - h)$$

   where v = number of vertices, e = number of edges, f = number of faces, s = number of surfaces (in this case, it is always equal to 1) and h = number of holes (genus), e.g. a torus has one hole.
2. Geometrical Correctness
   One of the geometrical checks that can be performed is to sum up the volumes of individual regions $\mathbf{R}_i$ shown in Figure 8.10. Then $\displaystyle\sum_{i=1,N_R} \text{vol}(\mathbf{R}_i) = 0$, where N_R is the number of regions so formed in the region construction process. If we define that the

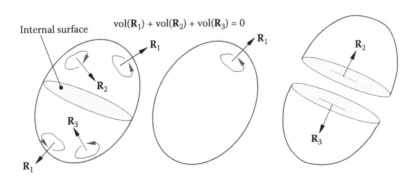

Figure 8.10 Sums of volumes of regions equal to zero.

volumes of regions with outward normal orientations are positive, then regions with inward normal orientations will have negative volumes, such that vol($\mathbf{R}_1$) is positive and vol($\mathbf{R}_2$) and vol($\mathbf{R}_3$) are both negative.

8.1.4.4 Convergence

The process always converges as only a finite number of regions will be formed. The number of facets in $\mathbf{B} = N_B$, the number of surfaces constructed $N_S \leq N_B$ and the number of regions formed $N_R \leq N_B$, as at least four triangular facets are taken to form a region. The process terminates when both sides of all the surfaces are used in region construction. The time complexity of the region formation is of order N_S.

8.1.5 Geometrical aspects

The boundary surfaces may have self-intersections, two triangular facets may be very close to each other and line segments may make sharp angles with a facet, as shown in Figure 8.11. To guarantee that a surface is tractable numerically (within machine precision), all intersections, close touches and sharp angles have to be reported and corrected, if necessary, before MG. Of course, the tolerance for close touches and sharp angles depends on the machine precision and the permissible limits of the mesh generator. Intersections, touches and sharp angles can be detected by examining the geometrical relationship between the boundary triangular facets and the edges.

8.1.5.1 Intersection

How the intersection is determined between a line segment (edge) with a triangular facet is given in Section 2.4.4, and for the sake of completeness, the process will be briefly described here. Given the triangle $P_1P_2P_3$ and line segment Q_1Q_2, calculate the distance h_1 and h_2 of Q_1 and Q_2 from the plane $P_1P_2P_3$. If $h_1h_2 \leq 0$, calculate $\xi = h_1/(h_1 - h_2)$. Intersection point Q on the plane $P_1P_2P_3$ is given by $Q = Q_1 + \xi(Q_2 - Q_1)$. Compute the barycentre co-ordinates of Q, (e, f, g). If $e \geq 0$, $f \geq 0$ and $g \geq 0$, then Q_1Q_2 intersects $P_1P_2P_3$, as shown in Figure 8.12.

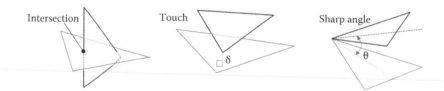

Figure 8.11 **Geometrical verifications.**

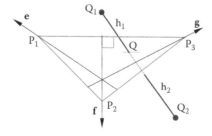

Figure 8.12 **Intersection between triangle $P_1P_2P_3$ and edge Q_1Q_2.**

8.1.5.2 Touch

Edge Q_1Q_2 is said to touch a triangular facet $P_1P_2P_3$ if it comes to within a very close distance with the triangle. Detailed formulas for the determination of the distance between a line segment and a triangular facet are given in Section 2.4.8, and the following are the steps in the determination of touches:

1. Calculate the barycentre co-ordinates of Q_1 and Q_2 w.r.t. triangle $P_1P_2P_3$, (e_1, f_1, g_1) and (e_2, f_2, g_2)

$$e_1 = e \cdot P_2Q_1, \qquad f_1 = f \cdot P_3Q_1, \qquad g_1 = g \cdot P_1Q_1,$$
$$e_2 = e \cdot P_2Q_2, \qquad f_2 = f \cdot P_3Q_2, \qquad g_2 = g \cdot P_1Q_2,$$

where e, f and g are unit vectors along co-ordinate axes.

The minimum distance is given by the distance from an end point of the line segment to the triangular facet, as shown in Figure 8.13a.

$$e_1 \geq 0, \, f_1 \geq 0, \, g_1 \geq 0 \quad \text{and} \quad h_1(h_2 - h_1) \geq 0, \quad \text{then} \quad d_{min} = h_1$$
$$e_2 \geq 0, \, f_2 \geq 0, \, g_2 \geq 0 \quad \text{and} \quad h_2(h_1 - h_2) \geq 0, \quad \text{then} \quad d_{min} = h_2$$

The minimum distance is given by the distance from an interior point of the line segment to the triangular facet, as shown in Figure 8.13b.

$$Q_1Q_2 \text{ is on the side of } P_2P_3 \text{ if } e_1 \leq 0 \text{ or } e_2 \leq 0, \, d_{min} = d(Q_1Q_2, P_2P_3)$$
$$Q_1Q_2 \text{ is on the side of } P_3P_1 \text{ if } f_1 \leq 0 \text{ or } f_2 \leq 0, \, d_{min} = d(Q_1Q_2, P_3P_1)$$
$$Q_1Q_2 \text{ is on the side of } P_1P_2 \text{ if } g_1 \leq 0 \text{ or } g_2 \leq 0, \, d_{min} = d(Q_1Q_2, P_1P_2)$$

Segment Q_1Q_2 is said to be in touch with triangle $P_1P_2P_3$ if

$$d_{min} < \varepsilon \times \min\left(\|P_1P_2\|, \|P_2P_3\|, \|P_3P_1\|, \|Q_1Q_2\|\right) \text{ with } \varepsilon \text{ set arbitrarily to } 1\%$$

The distance between two line segments $d(Q_1Q_2, P_1P_2)$, etc., can be found in Section 2.4.3.

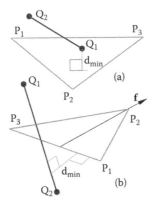

Figure 8.13 Determination of touches: (a) nearest point found at the interior; (b) nearest point found on an edge.

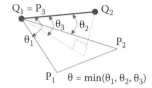

Figure 8.14 Angle θ between edge Q_1Q_2 and triangle $P_1P_2P_3$.

Figure 8.15 Background grid.

8.1.5.3 Sharp angle

If either Q_1 or Q_2 is connected to a vertex of triangle $P_1P_2P_3$, then calculate the angle between the line segment and the triangle, as shown in Figure 8.14. The angle will be called a *sharp angle* if it is smaller than, say, 1°.

8.1.5.4 Use of background grid

The geometrical check is the most time-consuming step. Suppose we have a boundary surface consisting of N_B facets and N_L edges; the number of tests required $n = N_B N_L$, or order N^2. Usually, $N_L \approx 1.5 N_B$; hence, in a system of 10^6 facets, $n = 10^6 \times 1.5 \times 10^6 = 1.5 \times 10^{12}$. Further, suppose that there are 100 operations per test on the average; a machine of 1GIP $\approx$ 250MFLOPs may take roughly 6000 s for checking intersections. This estimate shows that an order $O(N^2)$ process is just too slow. To reduce the amount of computations in geometric calculations, a background grid can be used to localise the searching process, as shown in Figure 8.15, in which the cells intersected by the shaded triangle are shaded in light grey, and the edges in the vicinity of the shaded triangle crossed by the cells are marked in red. For an even distribution of elements, the time complexity for the geometrical check process can be made linear, i.e. order N_B. Background grids for the partition of nodes have been presented in Sections 7.3.1 and 7.4.1 for 2D and 3D. Edges and triangles can be assigned to grid cells according to the cell labels of their vertices.

8.1.6 Examples

The boundary surfaces of a number of engineering objects are analysed in this section. The input data for these examples are randomly numbered and arbitrarily oriented triangular

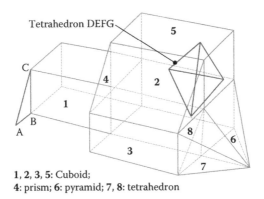

Tetrahedron DEFG

1, 2, 3, 5: Cuboid;
4: prism; **6:** pyramid; **7, 8:** tetrahedron

Figure 8.16 Object composed of eight chambers.

facets. The tasks to be performed are to verify the validity of the data, to construct the boundary and interior partition surfaces, to give each triangular facet a consistent orientation according to the surface to which it belongs, to identify the volume bounded by surfaces and to carry out a geometrical check for intersections, touches and sharp angles.

The first object tested is made of an artificial assembly of planar surfaces in which eight compartments can be identified, as shown in Figure 8.16. A triangular facet ABC and a tetrahedron DEFG were added deliberately to test the ability of the computer program in detecting free edges and intersections. Several objects were downloaded from the repertoire of the GAMMA project of INRIA in France and were put to tests. The second object composed of 5226 triangular facets in the form of a twisted cable is shown in Figure 8.17. In spite of the strange form, only one closed surface and one volume could be identified, and no intersection, sharp angle or touch was found. The aeroplane shown in Figure 8.18 is the third object studied. The analysis indicates that some of the elements on the surface are of very bad shape, and there is also a significant change of element size over the surface. Two touches and 106 sharp angles less than 2° were found, and the aeroplane turned out to be one of the most difficult objects for boundary recovery discussed in Section 5.3. The fourth example is a mould whose surface is represented by 10,098 triangular facets, as shown in Figure 8.19; 1396 redundant triangles were found and removed before they could be triangulated into tetrahedra. The fifth example is taken from a model of a bust of Victor Hugo, as shown in Figure 8.20. Fifty-two free edges were found; it was revealed that the facets did not form a closed

Figure 8.17 A string of 5226 triangles.

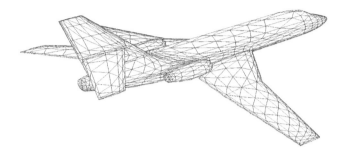

Figure 8.18 Avion falcon of 3762 triangles.

Figure 8.19 A mould of 10,098 triangles.

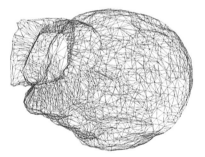

Figure 8.20 Surface model of Victor Hugo.

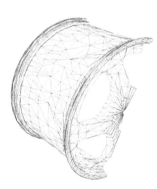

Figure 8.21 Part of a wheel.

Table 8.1 Statistics of boundary surfaces tested

Object	Partition	String	Avion	Mould	Hugo	Wheel
NP	21	2613	1879	4343	2028	1929
NB	53	5226	3762	10,098	4000	3870
NL	65	7893	5643	13,053	6062	5805
λ	1	1	0.001	100	1	1
S	1087	235	237	3205	470	10,814
A_{min}	12.5	0.0287	0.00011	0.00134	0.0109	0.00021
A	20.5	0.045	0.0629	0.3683	0.118	2.79
α_{min}	0.693	0.645	0.0188	0.281	0.135	0.00939
α	0.77	0.772	0.498	0.798	0.699	0.271
Angles <1°	0	0	8	0	0	264
Touches	0	0	2	0	0	280
θ_{min}	27.8	24.1	0.624	9.73	4.5	0.32
δ_{min}	41.7%	36.4%	0.498%	6.31%	5.8%	0.158%
NS	18	1	1	1	1	1
NR	8	1	1	1	0	1
V	1187.5	56.9	54.4	1636	0	10884.8
Grid	3 × 6 × 3	51 × 51 × 15	67 × 67 × 22	48 × 48 × 42	28 × 41 × 36	23 × 21 × 34
Test/facet	10	15	34	84	12	47
Efficiency	6.5	526	111	155	505	123
CPUT (s)	0.00	0.03	0.04	0.11	0.00	0.07
Remark	Two free edges and 12 intersections found	Good-quality smooth surface	Poorly shaped elements with very small size encountered	Big change in element size, redundant elements found	Elements had to be added at the top and at the bottom	Many touches, sharp angles, an object challenging to mesh

Note: NP = Number of points; NB = number of triangles; NL = number of edges; λ = co-ordinate scaling factor; S = total surface area; A_{min} = minimum size of a triangle; A = average size of triangles; α_{min} = minimum quality of a triangle; α = average quality of triangles; θ_{min} = minimum angle in degree, δ_{min} = minimum touch; NS = number of surfaces identified; NR = number of region formed; V = volume of the object; Grid = space partition for geometrical operations; Test/facet = average tests performed for each triangle; Efficiency = ratio of CPU time between no grid and with grid; CPUT = CPU time of PC i7 CPU 870 at 2.93 Ghz running on XP mode.

surface, and no volume could be defined. It was similar to an open cylinder, and triangular facets had to be added at the top and at the bottom to close it up. The last example is a wheel of 1929 nodes and 3870 triangles, as shown in Figure 8.21. This is a very poor boundary surface of 264 angles less than 1° and 280 touches of distance less than 1% of the length of the edge under consideration. The characteristics and the results of analysis of all these examples are summarised in Table 8.1. It is found that the CPU time for checking intersections, touches and sharp angles can be much reduced with the aid of a background grid.

8.2 MULTI-GRID INSERTION OF NON-UNIFORM POINT DISTRIBUTIONS (2D)

8.2.1 Introduction

Uniform or mildly non-uniform distribution point sets can be efficiently handled by the insertion scheme with the aid of a regular grid. However, for highly non-uniform point distributions, the triangulation time may be many times more than what is required by

Figure 8.22 Elongated triangles are removed and reconstructed for each point inserted.

a uniform point distribution. For the DT of severe non-uniform point distributions, the complexity is non-linear such that the triangulation time may be excessively long for large point sets of millions of points. In this section, the difficulties of DT of non-uniform point distributions by the insertion algorithm are discussed, and a versatile multi-grid insertion scheme is proposed and tested with large sets of highly non-uniformly distributed points.

For non-uniform point distributions, spatial partition of points can be achieved by the Quadtree/Octree (Shaffer and Samet 1987; Zhou et al. 2011) or kd-trees (Lemaire and Moreau 2000) for 2D and 3D points. Nevertheless, the time complexity for the construction of Quadtree/Octree or kd-trees is of the order O(nlogn) for a set of n points, which indicates that the time complexity of DT of non-uniform point distributions by the point insertion algorithm is at least O(nlogn) unless the barrier of spatial point partition could be overcome by some more efficient schemes of lower order.

As for the identification of the insertion cavity, the number of triangles checked and removed depends on the insertion sequence as well as the intrinsic characteristics and pattern of the point distributions. No theoretical results have been published so far as to how to devise an optimised order so as to minimise the number of conflicting triangles in the entire triangulation process. While this is not an issue in uniform point distribution, for non-uniform point distributions, the total number of triangles removed varies significantly with the order of point insertion. For highly non-uniform point distributions, over many parts, elongated triangles with large circumscribing circles exist. With the introduction of a newly inserted point near a fan of elongated triangles, a large number of triangles have to be removed and reconstructed, as shown in Figure 8.22. If points are to be inserted in a contiguous manner close to each other, the removal and reconstruction of elongated triangles have to be repeated for each point insertion. Obviously, this is rather time-consuming and unnecessary. The dilemma is that in order to reduce the time in searching for the BASE triangle, points have to be inserted in a contiguous manner; however, in order not to remove and reconstruct the elongated triangles, points have to be inserted with sufficient separation. An effective algorithm ought to optimise both the search for the BASE and the conflicting triangles in the creation of the CORE.

8.2.2 Review on insertion schemes

8.2.2.1 *Random order*

In the earlier development, points were taken by the natural order (the order as their input) or sorted by a lexical axis order (Joe 1989). This scheme has a tendency to produce intermediate triangulations of higher complexity than the final triangulation. The widely used random order was devised to overcome the shortcomings of natural or lexical axis order; it simply states to shuffle the points and insert them one by one (Edelsbrunner and Shah 1996a,b). There are several robust and efficient implementations of the incremental DT construction based on random-order point insertion, such as Clarkson's Hull (Clarkson and Shor 1989) and that contained in the α-shape software (Edelsbrunner and Mucke 1994; Delfinado and Edelbrunner 1995).

8.2.2.2 Biased randomised insertion order

Amenta et al. (2003) presented a biased randomised insertion order (BRIO). Their argument was as follows: Since modern memory architecture is hierarchical and the paging policies favour programs that observe locality of reference, a major concern is cache coherence – a sequence of recent memory references should be clustered locally rather than randomly in the address space. A program implementing a randomised algorithm does not observe this rule and can be drastically slowed down when its address space no longer fits in the main memory. The BRIO preserves enough randomness in the input points so that the performance of a randomised incremental algorithm is unchanged but orders the points by spatial locality to improve cache coherence. However, from the reports of other researchers (Liu and Snoeyink 2005; Zhou and Jones 2005), the practical performance of BRIO is not promising. Indeed, Amenta et al. (2003) considered BRIO a concept rather than a specific order, and BRIO could be implemented by merging with various insertion orders in a number of ways.

8.2.2.3 Hilbert curve

Recently, the pendulum seems to be swinging back to the deterministic order. Due to the work of Liu and Snoeyink (2005), Zhou and Jones (2005), Buchin (2009), Boissonnat et al. (2009) and Alauzet and Loseille (2009), the space-filling curve orders are widely used for constructing DTs. Among them, the Hilbert curve order is considered to be the most efficient because of its locality-preserving behaviour. Liu and Snoeyink (2005) used the Hilbert curve order in their program tess3 for 3D Delaunay tessellation and compared it with qhull, CGAL2.4, pyramid and hull. In their empirical comparisons, tess3 was the fastest for both uniform and non-uniform point distributions. Now the Hilbert curve order is also employed in the latest version of CGAL – CGAL3.7 through mixing with the idea of BRIO in its implementation.

As a variant of the space-filling curve developed by Giuseppe Peano, the Hilbert curve is a fractal continuous space-filling curve first described by Hilbert (1891). Its construction rule in the 2D case is shown in Figure 8.23. A square with an arrow is subdivided into four sub-squares. The ordering of the sub-squares is indicated by a green arrow, which

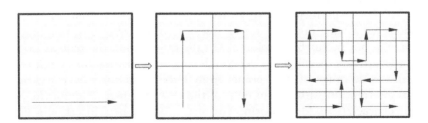

Figure 8.23 **Construction of Hilbert curve.**

connects the centres of neighbouring sub-squares. By repeat subdivision of each sub-square, the curve produced by linking up all the arrows becomes longer and longer, and it is called a Hilbert curve. The division into smaller squares or cells is just like a Quadtree subdivision applied uniformly over the entire domain to produce sub-squares of equal size, which can be controlled by specifying the maximum number of points in a sub-square. The Hilbert curve is designed to provide a path passing through all the sub-squares, so that points will be inserted cell by cell in a reasonably contiguous manner. However, extra time has to be allowed in the cell subdivision and management, and in formulating the path of insertion, which may go through many empty cells.

8.2.2.4 Space partition (background grid)

In triangulation, a background grid is a partition of space to facilitate searching or to establish a neighbourhood relationship for various quantities. Depending on the distribution of the points or objects under consideration and other specific requirements, grids of different characteristics can be constructed. By the use of a background grid, the space will be partitioned into cells, and the ideal scenario is to divide the space such that each cell will hold roughly an equal number of points. According to different point distributions, regular grids of equal and unequal spacings, Quadtree/Octree and kd-tree partitions, etc., have been devised.

Regular grids with equal and unequal spacing are the simplest and easiest to create with linear time complexity; hence, their construction is almost without any cost. Nevertheless, they are very effective in dealing with objects with normal or random distributions, and they are also very useful as a space control even for mildly non-uniform distributions. The basic properties, procedure of construction, general characteristics and memory requirement of a regular grid will be elaborated in Section 8.2.3. Point insertion by means of a regular grid for uniform and non-uniform point distributions will be fully investigated in Section 8.2.5.

Kd-tree construction is an effective space control for highly non-uniformly distributed data. Points are partitioned into two equal portions (which differ at most by one point) by taking the median of the point set. This procedure is repeated as necessary until there is no more than one point in each subdivided region. By the very construction of the kd-tree taking the median to divide a point set, at an intermediate stage, the number of points in a region (cell) is always equal no matter how points are distributed. The price to pay is to determine the median of the points in a cell for each subdivision by means of a sorting process or more effectively by means of the *bin sort* or *address sort*. The points in the cells along with the sequence of pivots of subdivision points form a sorted insertion order for the original point set. In a paper 'A New Insertion Sequence for Incremental Delaunay Triangulation' by Liu et al. (2013), it is reported that kd-tree insertion is superior to random, BRIO and Hilbert curve insertions up to 3 million 3D non-uniformly distributed points in various patterns such as cluster, disc, curved surface, line, spiral, etc.

In view of the better performance of kd-tree insertion over random, BRIO and Hilbert curve insertions for a wide range of non-uniform point distributions, random, BRIO and Hilbert curve will not be further explored. On the other hand, the regular grid insertion, the kd-tree insertion and its enhanced version developed in Section 8.2.3 will be rigorously tested against a wide range of data points from 1 to 100 million over a variety of non-homogeneous distribution patterns. In light of the ease of construction of the regular grid and its linear characteristic for uniform distributed data, a new insertion scheme is proposed by recursive application of regular grid insertion to non-uniform distributed points partitioned into cells by regular grid or kd-tree. Exploiting the local homogeneity of non-uniformly distributed points, regular grid insertion is especially effective over uniform or non-uniform point distributions partitioned into tiny cells of a large number of points (Lo 2013a). The procedure of multi-grid insertion will be developed in Section 8.2.4, and the idea will be fully tested in Section 8.2.5.

8.2.3 Kd-tree insertion scheme

8.2.3.1 Kd-tree construction

Based on the idea of space partition, a kd-tree (short form for k-dimensional tree) is a data structure in the form of a branching tree for organising points in the k-dimensional space. Kd-tree provide a hierarchic spatial relationship between data points, which can be used to find the nearest neighbours and locate points within a zone and for other searching operations, such as construction and point insertion of DT (Devroye et al. 2000, 2004).

1. *Construction of 2d-tree*
   A 2d-tree is generated by selecting a pivot point Q1 from the given point set to separate the remaining points into two regions by passing a vertical line through the chosen point. Points with x-values smaller than that of Q1 will be put to the region on the left, whereas points with x-values greater than that of Q1 will be put to the region on the right, as shown in Figure 8.24; points with equal x-values can be put on either side. In the second step, Q2 and Q3 are selected, respectively, from the regions to further divide them each into two sub-regions. This time, the regions are divided into smaller regions by horizontal lines passing points Q2 and Q3. Obviously, this subdivision process can be repeated by selecting one point from each region as the pivot to create more sub-regions. As one point, the pivot, is taken away for each subdivision, for a set of N points, there are N divisions, and the space will be partitioned into N + 1 regions.
   A balanced tree will be resulted if the median is selected as the pivot for each region sub-division, and in this case, each region will contain approximately an equal number of points (which differ at most by one point). For a balanced tree construction, the level of subdivision can be calculated as the number of points in a region is reduced by half in each subdivision, and all the leaf nodes (nodes at the bottom of the tree) are

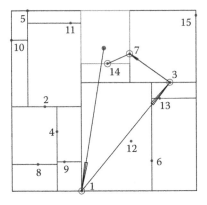

Figure 8.24 **2d-tree of 15 points.**

about the same distance from the root. To construct a balanced kd-tree for a set of N points, the level of subdivision, L, is given by

$$2^L \geq N \Rightarrow L \geq \log_2(N) \Rightarrow L = \text{INT}(\log_2(N)) + 1$$

Hence,

$L = 1, \quad N = 1$
$L = 2, \quad N = 2,3$
$L = 3, \quad N = 4,5,6,7$
$L = 4, \quad N = 8,9,10,11,12,13,14,15,$ and so on.

For a more symmetrical partition of space, the line of subdivision, vertical or horizontal, is rotated for each change of level of subdivision, i.e. L = 1, vertical, L = 2, horizontal, L = 3, vertical and so forth.

2. *Sequence of points*

A kd-tree can be conveniently defined by a proper sequencing of the given set of N points, $P = \{P_i, i = 1, N\}$. The following describes how a region P_i is divided into two sub-regions P_k and P_{k+1}. Find the median from P_i as a pivot along the x- or y-axis depending on the level of subdivision. Set Q_i = median, which divides P_i into P_k containing points on the left of Q_i and P_{k+1} containing points on the right of Q_i. As Q_i is the median in region P_i, the number of points in P_k and the number of points in P_{k+1} are given by

$$N_k = \begin{cases} \dfrac{N_i - 1}{2} & \text{if } N_i \text{ is odd} \\ \dfrac{N_i}{2} - 1 & \text{if } N_i \text{ is even} \end{cases} \qquad N_{k+1} = \begin{cases} \dfrac{N_i - 1}{2} & \text{if } N_i \text{ is odd} \\ \dfrac{N_i}{2} & \text{if } N_i \text{ is even} \end{cases}$$

where N_i = number of points in P_i. The subdivision will be carried out following the order of construction of the regions. As one point, Q_i, will be taken away for the subdivision of region P_i, the sequence $\{Q_i, i = 1, N\}$ will be established after exactly N subdivisions.

Set $P_1 = P = \{P_i, i = 1, N_1\}$, where $N_1 = N$. P_1 will give rise to P_2 and P_3, and P_2 will generate P_4 and P_5, whereas P_3 will generate P_6 and P_7. In turn, P_4 will generate P_8 and P_9, and P_5 will generate P_{10} and P_{11} and so on, as shown in Figure 8.25.

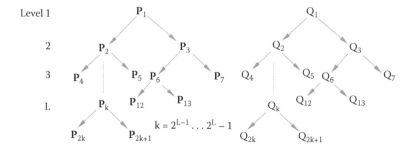

Figure 8.25 Sequence of sub-regions and pivots.

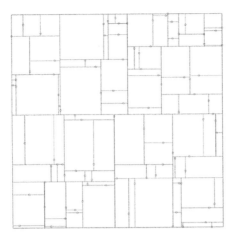

Figure 8.26 A *squarish* 2d-tree.

Let's consider the relationship of the points in the sequence $\{Q_k, k = 1, N^+\}$ associated with region subdivisions, $P_1, P_2, ..., P_N$, where $N^+ = 2^L - 1 \geq N$. Q_1 is the root node, which has child nodes Q_2 and Q_3 and in general, Q_k generates Q_{2k} and Q_{2k+1}. Hence, Q_{2k} and Q_{2k+1} are the children of Q_k, and Q_k is the father of Q_{2k} and Q_{2k+1}, as shown in Figure 8.25. Hence, $Q_{k/2}$ is the father node of Q_k, and in case $k/2 = 0$ or $k=1$, Q_k is already at the top of the tree. Generalising the previous results, the grandfather (two levels up) of Q_k is given by $Q_{k/4}$, etc.

Let's now consider the construction of the 2d-tree for 15 points, as shown in Figure 8.24. The median in the x-value of the given 15 points is determined and is assigned to Q_1. Points sorted to the left of the median are put to region P_2, and points sorted to the right of the median are put to region P_3. The process continues with the subdivision of P_2 and P_3 along the y-axis to give rise to points Q_2 and Q_3 and regions P_4, P_5, P_6 and P_7. Subdivision of regions P_4, P_5, P_6 and P_7 will generate points Q_4, Q_5, Q_6 and Q_7 and regions $P_8, P_9, P_{10}, P_{11}, P_{12}, P_{13}, P_{14}$ and P_{15}. At this stage, there is exactly one point in each region P_8 to P_{15}, which are assigned to Q_8 to Q_{15}, and the construction of the 2d-tree is completed.

The concept and the procedure for the construction of 2d-tree can be generalised naturally to higher dimensions. In three dimensions, $k = 3$, the space/regions are partitioned by a plane perpendicular to x-, y- or z-axis in turn according to the level of subdivision. If Q_k is on the x-aligned plane, child nodes Q_{2k} and Q_{2k+1} are on the y-aligned plane, and grandchild nodes $Q_{4k}, Q_{4k+1}, Q_{4k+2}$ and Q_{4k+3} are on the z-aligned plane and so forth. The construction of a static k-d tree of n points takes $O(n\log^2(n))$ time if an $O(n\log(n))$ sort is employed to compute the median for each region subdivision (see Section 2.7.8). The complexity becomes $O(n\log(n))$ if a linear median finding algorithm is used (Lemaire and Moreau 2000). In the partition of regions, if the regions are cut along their longest side instead of rotating through the axes, the pattern and the performance of the kd-tree are quite different. The behaviour of the kd-trees is much better as points are more closely grouped together, and they are called *squarish* kd-tree, as shown in Figure 8.26.

8.2.3.2 Kd-tree partition of points

A nice feature of kd-tree partition is that at any level of subdivision, each region contains roughly an equal number of points irrespective of the point distribution. In triangulation, it is not necessary to subdivide each region down to the lowest level, such that each region contains no point or only one point. A more general approach for point insertion by kd-tree partition is to allow each region to contain a specified number of points. Let N be the number of points in the set and n be the desirable number of points in a cell; the level of subdivision is given by

$$2^L n = N \Rightarrow L = \log_2\left(\frac{N}{n}\right)$$

The number of cells in the partition $N_c = 2^L$.

Cells are labelled in a sequential order, and for a level L subdivision, cells are labelled as

$$\{2^L, 2^L + 1,..., 2^{L+1} - 1\}.$$

8.2.3.3 Sequence of cell insertion

Squarish 2d-tree is constructed for point insertion cell by cell, and the entire set of points will be inserted when all the cells and pivot points are processed. Unlike the regular grid scheme, there is no natural contiguous sequence for cells generated by a *squarish* kd-tree partition. However, the region splitting pivot points can serve as a link between two regions at a given level of subdivision. As shown in Figure 8.27b, for partition level L = 2, the insertion sequence for the cells and pivot points is given by

$$\{\text{cell } 4 - \text{pivot } 2 - \text{cell } 5 - \text{pivot } 1 - \text{cell } 6 - \text{pivot } 3 - \text{cell } 7\}$$

In general, for a set of points partitioned into 2^L cells, the insertion sequence is also given by a cell–pivot–cell sandwiched scheme. The cells for a level L partition are given by

$$\{2^L, 2^L + 1,..., 2^{L+1} - 1\}$$

and the pivot points are given by $\{1, 2, ..., 2^L - 1\}$.

In the cell–pivot–cell sandwiched scheme, cells of points are inserted sequentially starting with cell 2^L, and the pivot point k following cell j is given by k = j/2, if j is divisible by 2; otherwise, set $j_1 = j/2$, and repeat the division process until j_m is divisible by 2; then pivot k = $j_m/2$. By the kd-tree construction, such a pivot point always exists and is unique for all the cells between 2^L and $2^{L+1} - 2$.

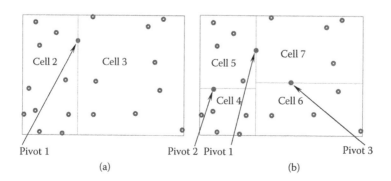

Figure 8.27 Kd-tree partition into cells: (a) partition into two cells; (b) partition into four cells.

For example, a set of 20 points is divided into two regions (cells), in which the cell on the left contains 10 points and the cell on the right contains 9 points, as shown in Figure 8.27a. L = 1, regions = $\{2^1, 2^2 - 1\}$ = {2, 3} and pivot = {1}. For L = 2, the point set is partitioned into four regions, as shown in Figure 8.27b, in which regions = $\{2^2, ..., 2^3 - 1\}$ = {4, 5, 6, 7} and pivots = {1, 2, 3}. Take for instance the level L = 2 partition of 20 points; the cells are {4, 5, 6, 7} and pivots = {1, 2, 3}. The insertion sequence is given as follows:

1. Five points in cell 4 are inserted, and 4 is divided by 2 hence, pivot = 4/2 = 2.
2. Pivot 2 is inserted.
3. Four points in cell 5 are inserted, and 5 is not divisible by 2, 5/2 = 2, and 2 is divisible by 2; hence, pivot = 2/2 = 1.
4. Pivot 1 is inserted.
5. Four points in cell 6 are inserted, and 6 is divisible by 2; hence, pivot = 6/2 = 3.
6. Pivot 3 is inserted.
7. Four points in cell 7 are inserted.
8. All the 20 points in the set have been processed.

By means of the simple cell–pivot–cell sandwiched scheme, points partitioned into kd-tree cells can be inserted in a reasonable contiguous manner without much computation.

8.2.3.4 Kd-tree grid insertion

As the number of points in a cell of kd-tree is roughly equal (differ at most by one), within a kd-tree cell, more sophisticated insertion scheme, say the regular grid insertion, can be applied to enhance the overall efficiency as shown in Figure 8.28. In the numerical tests of insertion of 1 million to 100 million points for various non-uniform distribution patterns, the most efficient scheme is to have about 16 points in each cell. However, if regular grid insertion is available, a more efficient insertion scheme can be devised by putting 1000 to 10,000 points in a cell. Whichever the case, the number of points in a cell is not very sensitive to the overall performance as long as it is within the right order, as it is only a statistical average depending also on the characteristics of the distribution of the points.

8.2.4 Multi-grid insertion

A multi-grid insertion is the result of applying the regular grid insertion to each cell of a spatial partition in a recursive manner. The idea is based on the observation that the regular

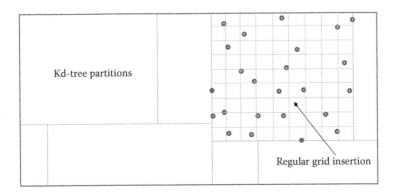

Figure 8.28 Regular grid insertion for kd-tree partitions.

grid insertion scheme is very effective for uniform point distributions with a quasi-linear complexity for very large point sets. Applying regular grid partition to non-uniform point distributions, points are not evenly distributed into cells, such that there are many points in some cells and none in other cells. However, within each cell, the points are more or less evenly distributed, and regular grid insertion can be applied recursively to each cell resulting in a very simple efficient scheme with almost linear complexity for locally evenly distributed points, such as line, circle and cluster distributions. Absolute linear complexity for the non-uniform point distributions with more complicated patterns or topology, like the spiral and the cross distributions, might be more difficult or even impossible as in the DT, points are connected to a large number of distant points through narrow elongated triangles. A quasi-linear insertion scheme is only possible unless the number of triangles removed for each point insertion can be controlled in such a way that it does not increase with the number of points in the set.

8.2.4.1 Regular grid insertion

1. *Construction of Regular Grid*

   Regular (uniform) grid is a simple efficient scheme for uniformly and mild non-uniformly distributed objects, for instance, points generated in a random process. Let $\{P_i = (x_i, y_i), i = 1, N\}$ be a set of N points on a 2D space. A procedure for the construction of a regular grid is given as follows. Let n be the average number of points desirable in a cell; then

$$nN_xN_y = N \tag{8.1}$$

where N_x = number of cell divisions along the x-axis, N_y = number of cell divisions along the y-axis and the number of cells, $N_c = N_x N_y$.

Let $x_{min}, y_{min}, x_{max}, y_{max}$ be the bounds of the (x,y) co-ordinates of the point set; compute the range in x, $R_x = x_{max} - x_{min}$, and the range in y, $R_y = y_{max} - y_{min}$.

N_x and N_y can be determined by substituting $\lambda R_x = N_x$ and $\lambda R_y = N_y$ into Equation 8.1.

The most important requirement in a spatial partition of points into cells is to ensure that each point belongs to one and only one cell, and the sum of points in all the cells is equal to the total number of points, i.e.

$$N = \sum_{k=1,N_c} n_k \quad \text{where } n_k = \text{number of points in cell k}$$

Compute $d_x = \dfrac{R_x}{N_x}$, $d_y = \dfrac{R_y}{N_y}$.

For a set of N points $\{P_i = (x_i, y_i), i = 1, N\}$, P_i is assigned to cell k given by

$$k = N_x(I_y - 1) + I_x \quad \text{with} \quad I_x = INT\left(\frac{x_i - x_{min}}{d_x}\right) + 1, \quad I_y = INT\left(\frac{y_i - y_{min}}{d_y}\right) + 1$$

By this allocation scheme, P_i will be assigned to one and only one cell $k \in \{1, 2, ..., N_c\}$.

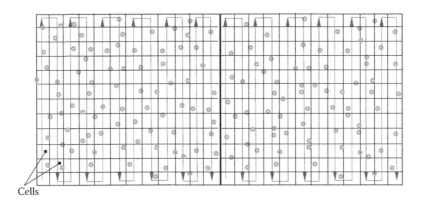

Cells

Figure 8.29 Path of cell by cell insertion in a regular grid.

2. Point Insertion by Regular Grid

The given set of points are partitioned into $N_c = N_x N_y$ cells, which are inserted one by one in a contiguous manner to reduce the searching time in locating the BASE triangle, as shown in Figure 8.29. Given a set of N points, the number of cells is approximately given by

$$N_c \approx \frac{N}{n} \quad \text{where n is the average number of points in a cell.}$$

There is no theoretical optimal value for n; however, from practical experience, the performance of a regular grid is not very sensitive to n, which can produce satisfactory results with values ranging from 4 to 16 for various point distributions. The memory requirement for the construction of a regular grid is relatively low, and all that is needed is an integer array of size given by

Memory required by regular grid = Number of cells + Number of points = N_c + N

8.2.4.2 Multi-grid as a repeated application of the regular grid

Applying regular grid to non-uniform point distribution, the number of points within a cell can have a large variation, as shown in Figure 8.30. In some cells, there are more than 1000 points, whereas in other cells, not even a single point is present. For evenly distributed point sets, the threshold for the application of regular grid is very low at about 100 points, i.e. the CPU time will be reduced for the insertion of any reasonable point distributions more than 100 points by applying a regular grid compared to a direct insertion point by point in one zone. In view of the very low threshold of about 100 points for regular insertion, the efficiency of point insertion can be significantly enhanced by applying the regular grid in a recursive manner to cells containing, say, more than 200 points, as shown in Figure 8.30. The searching time for the location of the BASE triangle can be tightly controlled such that there are about 20 points in each further divided cell. However, the number of conflicting triangles for each point inserted is still governed by the non-uniform distribution characterised with excessive connections by elongated triangles.

The presence of elongated triangles is an intrinsic feature of highly non-uniform point distributions, which poses a significant difficulty over uniform point distributions. For the

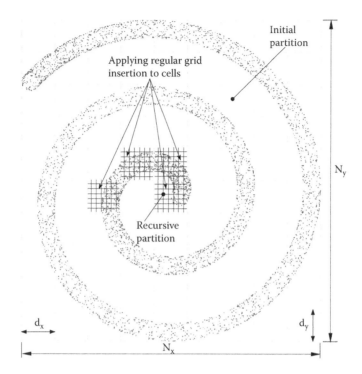

Figure 8.30 Regular/multi-grid insertion.

same number of points, much more CPU time will be needed for the triangulation of severe non-uniformly distributed point sets compared to a uniform distribution or a mildly non-uniform distribution. Nevertheless, using appropriate point-insertion schemes, the time for the determination of the BASE triangle would not be much more than the case of uniform distribution. To avoid excessive removal of closely packed elongated triangles in a point insertion with points too close to each other, some separation control can be exercised by taking one point from a cell at a time instead of inserting all the points within a cell all at the same time. By doing so, as shown in many examples, the average number of conflicting triangles removed for each point insertion can be substantially reduced, resulting in a more stable insertion scheme with higher overall performance.

8.2.4.3 Pseudo-code for the recursive insertion algorithm by multi-grid

Multi-grid Insertion (P)

```
// Input: A set of N point P; Output: Triangles generated by inserting P.
Create regular grid of cells C_k, k=1,Nc for P following Section 8.2.4.1.
Threshold=200
Loop: k=1,Nc
If (n(C_k) > Threshold) Multi-grid Insertion (C_k) // n(C_k) = number of
points in C_k
else
Insert points in C_k by standard point insertion kernel
End If
End Loop k
```

The threshold of 200 is arbitrary, and the algorithm will also work pretty well if the threshold is set to 150 or 250. The process will always converge as a subdivided cell cannot have more than 200 points after a number of subdivisions; for instance, in three subdivisions, a cell may contain more than $200^3 = 8,000,000$ for locally uniform distributions. Even for the most non-uniform point distributions, the level of subdivision seldom exceeds four, unless points are distributed in an exponential manner, which will, however, easily exceed the finite precisions adopted in floating point computations. As for the memory requirement, the memory needed for each subdivision is $Nc + n(C_k)$, where Nc is the number of cell divisions for C_k. The memory for recursive grid subdivision will only grow in a linear manner, i.e. memory required in m subdivisions is given by $M = \sum_{i=1,m} N_c + n(C_{k_i})$, and this is roughly equal to the total number of points processed in the recursive insertion process. As the points in a cell is a sub-set of **P**, we must have $M < N$.

8.2.5 Tests on non-uniform point distributions

Six different distributions: random, line, cross, circle, spiral and cluster of 1 to 100 million points, as shown in Figures 8.31 through 8.36, were triangulated using the regular

Figure 8.31 **Random distribution.**

Figure 8.32 **Line distribution.**

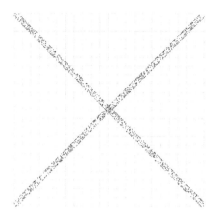

Figure 8.33 **Cross distribution.**

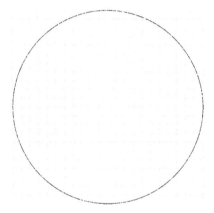

Figure 8.34 **Circle distribution.**

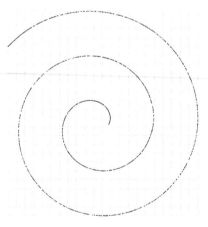

Figure 8.35 **Spiral distribution.**

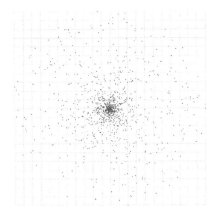

Figure 8.36 Cluster distribution.

grid insertion, the enhanced kd-tree insertion and the multi-grid insertion on a PC, Intel®, Core™, i7 CPU, 870 at 2.93 GHz with 16-GB RAM running on Windows 7 using Intel Fortran VS2010 on 64 bit. Mild non-uniform point distributions of line and cross patterns are shown in Figures 8.32 and 8.33, in which the width is about 1% of the larger dimension, and the cluster distribution is generated with random points biased towards the centre with radius r shortened by

$$r = r_0 \times random\_number^{3-5} \qquad (8.2)$$

The random point distribution and the other five mild non-uniform point distributions of 1 million points were triangulated by the regular grid insertion scheme.

It is found that the CPU times taken for the mild non-uniform distributions were not much higher than that of the uniform distribution, as shown in Table 8.2 and Figure 8.37. In view of this situation, in the tests for the three insertion schemes, more severe non-uniform point distributions were adopted in which the width of point spread was narrowed down to 0.01%, as shown in Figures 8.34 and 8.35, and the cluster distribution was generated using a power of 5 instead of 3 in Equation 8.2.

Table 8.2 Triangulation of 1 million mild non-uniformly distributed points

Distribution	CPU time(s)
Uniform	0.967
Line	1.436
Cross	1.248
Circle	1.389
Spiral	1.498
Cluster	1.076

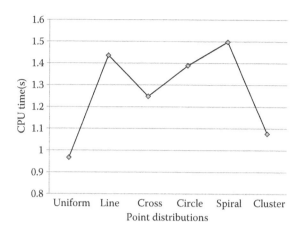

Figure 8.37 Regular grid insertion for mild non-uniform distributions of 1 million points.

The results for the insertion of the highly non-uniform point distributions are shown in Tables 8.3 through 8.8 for various distribution patterns, whereas the corresponding plots of CPU time against the number of points inserted are depicted in Figures 8.38 through 8.43.

For a uniform point distribution, the number of points in a cell is roughly equal to the number of points specified in a cell, and the multi-grid insertion was not invoked as the threshold of 200 points was never reached. Thus, the results for regular grid and multi-grid are identical for the insertion of 1 to 100 million points. The average number of triangles visited for the location of the BASE, NP, varied from 7 to 8, and the average number of conflicting triangles removed in the creation of the insertion cavity, NR, varied from 5.5 to 6.2 showing that local characteristics of randomly generated points do not change with the total number of points in the set. For the insertion of 1 million points, 0.988 s was required, which increased to 116.97 s for the insertion of 100 million points, a mere 18% more than a strict linear time relationship for a 100-fold increase in points.

As points are well sorted in the kd-tree insertion, the BASE searching time as indicated by NP ≈ 5 is slightly lower compared to regular grid insertion. However, the number of triangles removed, NR, is similar to that of the regular insertion, resulting in a slightly faster insertion scheme relative to the regular grid insertion. Nevertheless, if the grid construction time is also taken into account, the total CPU time of kd-tree insertion is always higher than that of the regular grid insertion scheme. This gap widened with the number of points increased, as the time complexity of kd-tree is of order $O(n\log n)$. For the insertion of 100 million points, the construction of kd-tree alone took 242.72 s, which is already more than the total CPU time of 129.96 s for the regular grid insertion.

There is a great difference between regular grid insertion and multi-grid insertion for points concentrated along a straight line, as shown in Table 8.3. The main cause for such a large difference is due to the time taken for the location of the BASE triangle; NP for regular grid was more than 100, and NP for multiple grid varied from 7 to 8, resulting in a six-time difference in the insertion time. NP = 5 for kd-tree insertion is the smallest for the three insertion schemes. However, the NR value of the kd-tree is slightly larger than that of the regular grid insertion or the multi-grid insertion, resulting in a slightly less overall insertion time. Again, if the grid construction time is also taken into consideration, the total CPU time of the kd-tree scheme is still higher than that of multi-grid insertion scheme.

The cross distribution is the first major challenge for various insertion schemes. With a similar NR value, NP = 78 of the regular grid insertion is much higher than NP = 9 of the

Table 8.3 Uniform distribution

Uniform	Regular grid insertion					Multiple grid insertion					Kd-tree insertion				
			CPU time					CPU time					CPU time		
N (million)	NP	NR	Grid	Insert	Total	NP	NR	Grid	Insert	Total	NP	NR	Grid	Insert	Total
1	6.88	5.52	0.063	0.988	1.051	6.88	5.52	0.063	0.988	1.051	5.07	5.55	0.562	1.014	1.576
2	7.03	5.63	0.125	1.966	2.091	7.03	5.63	0.125	1.966	2.091	5.1	5.61	1.513	1.872	3.385
4	7.18	5.73	0.25	4.009	4.259	7.18	5.73	0.25	4.009	4.259	5.08	5.6	3.978	3.791	7.769
6	7.27	5.79	0.39	6.037	6.427	7.27	5.79	0.39	6.037	6.427	5.13	5.64	6.661	5.834	12.495
8	7.33	5.83	0.596	8.159	8.755	7.33	5.83	0.596	8.159	8.755	5.09	5.62	10.156	7.862	18.018
10	7.38	5.86	0.733	10.296	11.029	7.38	5.86	0.733	10.296	11.029	5.12	5.64	13.51	9.922	23.432
20	7.53	5.96	1.9	21.29	23.19	7.53	5.96	1.9	21.29	23.19	5.12	5.65	31.67	20.16	51.83
40	7.69	6.06	4.6	43.79	48.39	7.69	6.06	4.6	43.79	48.39	5.13	5.68	76.99	41.11	118.1
60	7.77	6.11	7.21	68.33	75.54	7.77	6.11	7.21	68.33	75.54	5.08	5.66	128.49	63.62	192.11
80	7.83	6.15	9.95	92.45	102.4	7.83	6.15	9.95	92.45	102.4	5.13	5.68	179.21	84.4	263.61
100	7.87	6.18	12.99	116.97	129.96	7.87	6.18	12.99	116.97	129.96	5.15	5.71	242.72	109.88	352.6

Table 8.4 Line distribution

Line	Regular grid insertion					Multiple grid insertion					Kd-tree insertion				
				CPU time					CPU time					CPU time	
N (million)	NP	NR	Grid	Insert	Total	NP	NR	Grid	Insert	Total	NP	NR	Grid	Insert	Total
1	112.3	3.98	0.047	7.441	7.488	8.86	3.98	0.047	1.233	1.28	5.95	4.92	0.434	0.874	1.308
2	112	4	0.11	16.19	16.3	7.95	3.99	0.11	2.425	2.535	5.59	4.86	1.264	1.716	2.98
4	110.3	4	0.202	35.6	35.802	7.3	4	0.202	5.187	5.389	5.41	4.89	3.385	3.541	6.926
6	109.2	4	0.297	56.176	56.473	7.11	4.01	0.297	7.792	8.089	5.49	4.97	5.506	5.492	10.998
8	108.2	4	0.405	78.203	78.608	7.02	4.01	0.405	10.59	10.995	5.32	5	8.549	7.363	15.912
10	107.4	4.01	0.515	99.544	100.06	7.13	4.01	0.515	13.62	14.135	5.39	5.04	11.232	9.375	20.607
20	104.8	4.02	1.12	208.2	209.32	7.05	4.02	1.12	29.31	30.43	5.38	5.17	26.32	19.3	45.62
40	103.5	4.02	2.31	431.84	434.15	7.37	4.02	2.31	64.74	67.05	5.39	5.27	61.31	39.47	100.78
60	104.1	4.02	3.75	672.64	676.39	7.6	4.02	3.75	101.09	104.84	5.31	5.3	101.59	59.94	161.53
80	105.4	4.02	5.04	913.94	918.98	7.78	4.02	5.04	139.89	144.93	5.37	5.34	140.81	81.54	222.35
100	107.4	4.02	6.58	1191.2	1197.8	8.05	4.02	6.58	183.11	189.69	5.42	5.36	183.77	104.15	287.92

Table 8.5 Cross distribution

Cross N (million)	Regular grid insertion					Multiple grid insertion					Kd-tree insertion				
	NP	NR	CPU time			NP	NR	CPU time			NP	NR	CPU time		
			Grid	Insert	Total			Grid	Insert	Total			Grid	Insert	Total
1	79.6	5.27	0.047	5.273	5.32	9.2	5.3	0.047	1.54	1.587	10.2	12.18	0.437	2.839	3.276
2	80	5.42	0.109	11.653	11.762	9.62	5.4	0.109	3.448	3.557	10.8	13.5	1.279	7.395	8.674
4	79.3	5.52	0.219	26.005	26.224	9.18	5.54	0.219	7.86	8.079	11.2	15.1	3.432	20.062	23.494
6	78.7	5.59	0.312	41.637	41.949	8.95	5.62	0.312	12.71	13.022	12.4	16.4	5.616	36.785	42.401
8	78.2	5.68	0.437	58.906	59.343	8.98	5.65	0.437	18.44	18.877	12.7	18.9	8.658	62.728	71.386
10	77.8	5.74	0.561	76.674	77.235	8.64	5.69	0.561	23.45	24.011	12.9	18.8	11.357	81.713	93.07
20	76.2	5.81	1.185	166.84	168.03	9.15	5.83	1.185	56.89	58.075	14.3	22.3	26.46	242.64	269.1
40	75.4	5.94	2.43	375.3	377.73	9.52	5.93	2.43	139.04	141.47	16.2	26.5	62.34	745.04	807.38
60	75.9	6	3.84	577.28	581.12	9.71	5.99	3.84	229.33	233.17	17.3	29.1	102.27	1382.2	1484.5
80	77	6.03	5.26	813	818.26	10.1	6.04	5.26	338.08	343.34	16.9	28.7	141.35	2006.7	2148.1
100	78.3	6.03	6.69	1068.4	1075.1	9.91	5.89	6.69	431.78	438.47	16	27.1	186.19	2499.6	2685.8

Table 8.6 Circle distribution

Circle	Regular grid insertion					Multiple grid insertion					Kd-tree insertion				
			CPU time					CPU time					CPU time		
N (million)	NP	NR	Grid	Insert	Total	NP	NR	Grid	Insert	Total	NP	NR	Grid	Insert	Total
1	84.4	3.98	0.047	4.992	5.039	8.63	3.98	0.047	1.06	1.107	5.5	4.77	0.453	0.858	1.311
2	87.8	3.99	0.109	10.92	11.029	8.32	3.99	0.109	2.161	2.27	5.67	4.82	1.248	1.732	2.98
4	90	4	0.202	24.446	24.648	8.02	4	0.202	4.62	4.822	5.72	4.91	3.354	3.744	7.098
6	91.7	4	0.312	39.561	39.873	7.83	4	0.312	7.527	7.839	5.86	5	5.584	5.617	11.201
8	92.1	4	0.421	56.238	56.659	7.7	4	0.421	9.938	10.359	5.57	4.96	8.564	7.488	16.052
10	93.6	4.01	0.546	74.366	74.912	7.64	4	0.546	12.79	13.336	5.53	4.98	11.216	9.469	20.685
20	96.6	4.01	1.16	169.26	170.42	7.47	4.01	1.16	28.48	29.64	5.5	5.09	26.32	19.34	45.66
40	99	4.01	2.45	384.67	387.12	7.4	4.01	2.45	61.73	64.18	5.46	5.22	62.86	40.8	103.66
60	101.7	4.02	3.76	602.9	606.66	7.42	4.01	3.76	94.43	98.19	5.44	5.31	102.71	61.12	163.83
80	103.8	4.02	5.16	836.61	841.77	7.47	4.02	5.16	129.92	135.08	5.49	5.35	141.3	81.96	223.26
100	105.8	4.02	6.57	1098.1	1104.7	7.52	4.02	6.57	170.87	177.44	5.51	5.39	185.73	105.63	291.36

Table 8.7 Spiral distribution

Spiral	Regular grid insertion					Multiple grid insertion					Kd-tree insertion				
N			CPU time					CPU time					CPU time		
(million)	NP	NR	Grid	Insert	Total	NP	NR	Grid	Insert	Total	NP	NR	Grid	Insert	Total
1	67.1	7.48	0.047	4.961	5.008	12.1	7.41	0.047	2.153	2.2	8.81	11.6	0.437	2.184	2.621
2	71.9	7.5	0.109	11.373	11.482	12	7.5	0.109	4.929	5.038	9.31	11.4	1.263	5.086	6.349
4	73.6	7.69	0.218	25.99	26.208	12.1	7.68	0.218	11.75	11.968	11.5	14.4	3.416	16.177	19.593
6	74.3	7.78	0.327	41.871	42.198	12.3	7.8	0.327	19.71	20.037	11.5	14.2	5.601	26.005	31.606
8	75.6	7.75	0.452	60.139	60.591	12.3	7.8	0.452	27.91	28.362	12.5	16.7	8.626	44.008	52.634
10	76.4	7.89	0.561	79.857	80.418	12.4	7.88	0.561	37.55	38.111	12.6	16.2	11.357	56.238	67.595
20	79.7	7.96	1.17	186.9	188.07	12.8	7.98	1.17	90.31	91.48	15	19.9	26.47	164.19	190.66
40	83.2	8.08	2.54	449.89	452.43	13.2	8.08	2.54	221.3	223.84	16.4	23.5	61.82	458.2	520.02
60	85.1	8.12	3.81	733	736.81	13.4	8.1	3.81	372.42	376.23	19.2	27.5	107.65	927.24	1034.9
80	86.9	8.16	5.16	1042.8	1048	13.6	8.12	5.16	547.8	552.96	19.1	28	141.52	1300.3	1441.8
100	88.9	8.19	6.68	1406	1412.7	13.9	8.23	6.68	755.17	761.85	19.3	28.2	195.24	1801.4	1996.6

Table 8.8 Cluster distribution

| Cluster | Regular grid insertion | | | | | Multiple grid insertion | | | | | Kd-tree insertion | | | | |
| | | | CPU time | | | | | CPU time | | | | | CPU time | | |
N (million)	NP	NR	Grid	Insert	Total	NP	NR	Grid	Insert	Total	NP	NR	Grid	Insert	Total
1	47.2	4.69	0.062	4.04	4.102	9.56	4.73	0.062	1.287	1.349	5.4	5.55	0.468	0.921	1.389
2	54.2	4.75	0.109	10.452	10.561	10	4.75	0.109	2.683	2.792	5.35	5.55	1.341	1.857	3.198
4	62.7	4.81	0.219	28.345	28.564	10.1	4.8	0.219	5.928	6.147	5.4	5.56	3.619	3.838	7.457
6	81.8	4.83	0.328	72.524	72.852	10.1	4.83	0.328	9.235	9.563	5.46	5.6	5.575	5.913	11.488
8	72.8	4.85	0.468	78.08	78.548	10.3	4.85	0.468	12.84	13.308	5.42	5.61	9.22	7.971	17.191
10	87.3	4.87	0.593	138.34	138.93	10.5	4.87	0.593	16.47	17.063	5.43	5.63	12.168	10.093	22.261
20	88.9	4.92	1.26	276.03	277.29	10.9	4.92	1.26	36.72	37.98	5.5	5.69	28.86	20.66	49.52
40	97.5	4.96	2.76	678.43	681.19	11.1	4.96	2.76	78.08	80.84	5.48	5.72	69.13	42.76	111.89
60	94.6	4.99	4.43	1062.8	1067.2	11	4.99	4.43	118.11	122.54	5.32	5.7	115.45	63.73	179.18
80	102.9	5.01	6.05	1486.4	1492.5	11.4	5	6.05	158.69	164.74	5.31	5.71	159.68	85.21	244.89
100	104	5.03	7.97	1928.1	1936.1	11.5	5.02	7.97	209.69	217.66	5.36	5.74	215.92	111.79	327.71

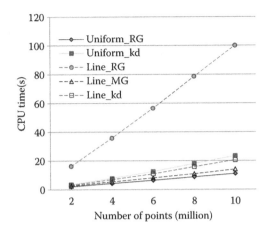

Figure 8.38 CPU time for uniform and line distributions.

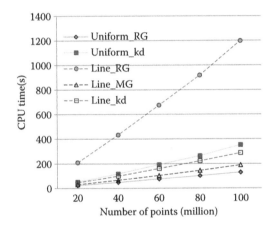

Figure 8.39 CPU time for uniform and line distributions.

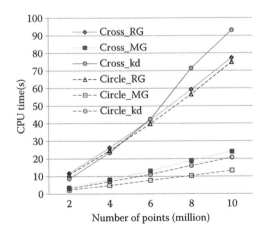

Figure 8.40 CPU time for cross and circle distributions.

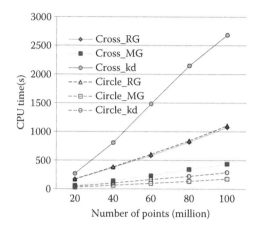

Figure 8.41 CPU time for cross and circle distributions.

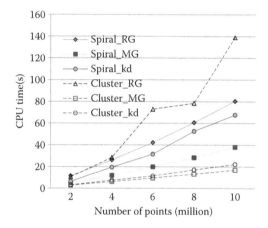

Figure 8.42 CPU time for spiral and cluster distributions.

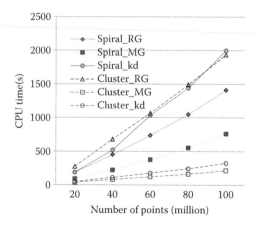

Figure 8.43 CPU time for spiral and cluster distributions.

multi-grid insertion, leading to about four-time difference in the overall performance. As for the kd-tree insertion, surprisingly, the NP value of 16 as well as the NR value of 27 are both much higher than those of the multi-grid insertion, resulting in an insertion time about six times of that of the multi-grid insertion for the insertion of 100 million points.

In spite of a change in the pattern shape, the results for the circle distribution are very similar to those for the straight line distribution, such that the performance of the multi-grid insertion is a couple of times better than that of the regular grid insertion. Considering insertion time alone, kd-tree is the most efficient, but the multi-grid insertion is the best overall if grid construction time is also taken into account.

The spiral distribution is perhaps the most difficult case for the various point insertion schemes. The NP and NR values are considerably higher for the three insertion schemes compared to other distributions, in which the multi-grid insertion is still the best overall. It is noted that NP and NR values increased with the number of points leading to a non-linear increase in CPU time with the number of points inserted. In particular, the NP and NR values of the kd-tree insertion increased rather rapidly with the points in the set, and the high NR value of 28 makes the kd-tree insertion the least efficient among the three insertion schemes for the insertion of 100 million points. The high NR value for kd-tree insertion was probably due to the over-sorting of the points for highly non-uniform distributions such that many elongated triangles had to be removed for each point insertion. As shown in Figure 8.44 for the triangulation of a spiral distribution of 1000 points, elongated triangles are produced joining points at different levels of the spiral, and much narrower triangles are expected when the number of points increases to 100 million. As shown in Table 8.7, the NR value for regular insertion was very stable at 7 or 8, but the NP value was rather high at 76. A good balance can be found with the multi-grid insertion, with an NR value similar to that of the regular insertion and an NP value smaller than that of the kd-tree insertion. This is really a challenging example for the design of the insertion scheme as well as the numerical stability of the algorithm.

A cluster distribution is locally very similar to the uniform distribution, and not many elongated triangles are expected in the triangulation. As points are drastically unevenly distributed, the NP value for regular grid insertion was pretty high, which increased steadily with the

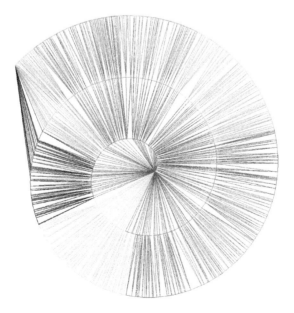

Figure 8.44 Delaunay triangulation of spiral distribution of 1000 points.

number of points inserted up to more than 100 for a large point set of 100 million points, and as a result, the performance of the regular grid insertion was even worse than the case of spiral distribution. On the other hand, similar to uniform point distribution, kd-tree insertion recorded the best performance with low NP and NR values as the number of points in a kd-tree partition is always equal irrespective of the point distribution. Nevertheless, the performance of the multi-grid insertion is still the best overall if grid construction is taken into consideration. Apart from the number of points inserted, the impact of various non-uniform distributions for the three insertion schemes has also been studied. The CPU times for various distribution patterns by the three insertion schemes for the insertion of 1 million, 10 million and 100 million points are plotted, respectively, in Figures 8.45 through 8.47. As shown in Figure 8.45, for the insertion of 1 million points, regular grid insertion took much more CPU time (six times on average) for non-uniform point distributions compared to uniform distributions, whereas kd-tree insertion and multi-grid insertion could substantially reduce the CPU time taken except for the spiral distribution for the multi-grid insertion and the cross and the spiral for the kd-tree insertion.

For the insertion of 10 million points, a similar trend is observed, as shown in Figure 8.46. However, the CPU time taken for the regular grid insertion increased more rapidly for the

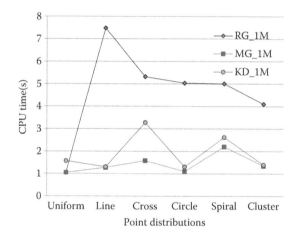

Figure 8.45 CPU time for various distributions of 1 million points.

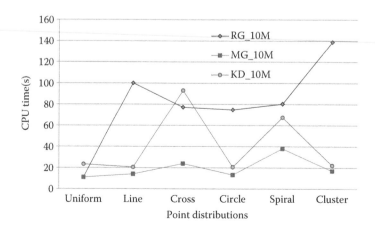

Figure 8.46 CPU time for various distributions of 10 million points.

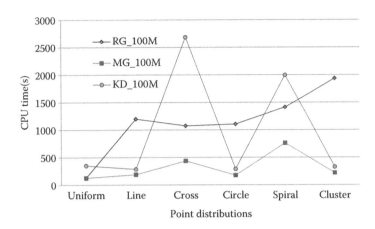

Figure 8.47 CPU time for various distributions of 100 million points.

cluster distribution than that for other distributions. Kd-tree insertion was not very stable either, as rapid increase in the CPU time was observed for the insertion of the cross and the spiral distributions. For the insertion of 100 million points, similar changes from 1 million points to 10 million points are observed, except that the effects are more pronounced. For the case of large point sets of 100 million points, the cluster distribution is the worst distribution pattern for regular grid insertion, whereas cross and spiral distributions are tough cases for the kd-tree insertion. The multi-grid insertion scheme is the most stable and efficient out of the three insertion methods over the range of points inserted and the distribution patterns tested.

One more example of point distribution with mixed patterns to simulate practical MG problems is shown in Figure 8.48, in which there is a more gradual change in nodal spacing between points. The number of points has been increased progressively from 2 million points to 10 million points in order to check the performance of various insertion schemes as the number of points increases in the system. As shown in Table 8.9, for this mildly non-uniform point distribution pattern, the performance of the insertion schemes is very similar as in the case of uniform distribution of randomly generated points. Compared to the

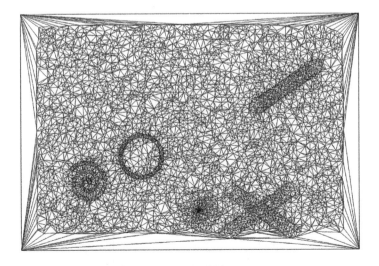

Figure 8.48 Point distribution with mixed patterns of 10000.

Table 8.9 Point distribution with mixed patterns

| Mixed | Regular grid insertion | | | | | Multiple grid insertion | | | | | Kd-tree insertion | | | | |
| | | | CPU time | | | | | CPU time | | | | | CPU time | | |
N (million)	NP	NR	Grid	Insert	Total	NP	NR	Grid	Insert	Total	NP	NR	Grid	Insert	Total
2	12.7	5.29	0.11	2.98	3.09	9.11	5.29	0.109	2.23	2.339	5.2	5.56	1.2	1.81	3.01
4	13.4	5.38	0.23	5.88	6.11	9.15	5.39	0.23	4.36	4.59	5.2	5.6	3.32	3.68	7
6	14.6	5.42	0.34	9.72	10.06	9.19	5.43	0.34	6.61	6.95	5.26	5.64	5.68	5.68	11.36
8	14.2	5.45	0.45	12.67	13.12	9.29	5.46	0.45	8.86	9.31	5.15	5.57	8.38	7.49	15.87
10	14.5	5.49	0.56	16.15	16.71	9.33	5.49	0.56	11.19	11.75	5.19	5.59	11.12	9.49	20.61

uniform point distribution, the regular grid insertion takes slightly more CPU time in the insertion of 10 million points, 16.71 s vs. 11.03 s, whereas the other two insertion schemes take almost the same CPU time for the insertion of the same number of points. Surprisingly, the kd-tree insertion scheme takes even slightly less CPU time as the kd-tree partition of evenly distributed points takes slightly more CPU time than a non-uniformly point distribution. Again, the most efficient scheme is the multiple grid insertion if the grid construction time is also taken into consideration.

8.2.6 Closure

The fundamentals for the DT of non-uniformly distributed points by the insertion method are discussed in this section. In the light of the simplicity and linearity of regular grid insertion, a multi-grid insertion scheme is presented for the triangulation of uniform and non-uniform point distributions by recursive application of the regular grid insertion to an arbitrary subset of the original point set. The regular grid insertion is very sensitive to point distributions, as the time taken in searching for the base triangle increases rapidly with highly non-uniformly distributed points. The kd-tree may have a rather poor performance for triangulations characterised with a large amount of elongated triangles as conflicting triangles have to be removed and reconstructed for the insertion of closely spaced points. The multi-grid insertion is the most efficient and stable for all the distribution patterns tested among the three insertion schemes if the grid construction time is also taken into account. A quasi-linear complexity can be observed for the triangulation of point distribution patterns such as line, circle and cluster, with local features similar to those of a uniform point distribution. However, substantially more CPU time is needed for the triangulation of the more difficult distribution patterns, such as cross and spiral, in which there are many elongated triangles in the triangulation, which leads to a non-linear complexity. Nevertheless, with the multi-grid insertion, the triangulation time of the worst non-uniform distribution of 100 million points tested is about only five times more than that of a uniform distribution.

8.3 MULTI-GRID INSERTION OF NON-UNIFORM POINT DISTRIBUTIONS (3D)

8.3.1 Introduction

Similar to 2D point insertion, relative to a uniform point distribution, non-uniform point sets are significantly more difficult due to the following intrinsic properties of non-uniformly distributed points. (i) More efficient space control or partition has to be applied to highly non-uniform distributed points so that each partitioned cell contains a roughly equal number of points. (ii) In the location of the base tetrahedron by the point insertion algorithm, the CPU time taken or the number of tetrahedra examined before the base tetrahedron could be determined is much longer than the case of uniform distribution as there are many pointed tetrahedra in the triangulation. (iii) In the creation of the insertion cavity (CORE) upon the introduction of a new point, many tetrahedra have to be removed and reconstructed as long thin tetrahedra have large circumspheres, and a point can be connected to many distant points through long stretching tetrahedra. (iv) There are problems of numerical stability both in the determination of the BASE tetrahedron and in the verification of the empty circumsphere criterion, and more CPU time has to be allowed to exercise additional consistency checks and precautions.

As for the identification of the insertion cavity, the number of tetrahedra checked and removed depends on the insertion sequence as well as the intrinsic characteristics and pattern

Figure 8.49 Elongated tetrahedra over a spiral point distribution.

of the point distributions. No theoretical results have been reported so far as to how to devise an optimised order to minimise the number of conflicting tetrahedra in the entire triangulation process. While this is not an issue in uniform point distribution, for non-uniform point distributions, the total number of tetrahedra removed varies significantly with the order of point insertion. For highly non-uniform point distributions, over many parts, long thin tetrahedra with large circumscribing spheres exist. Upon the introduction of a newly inserted point near a cluster of elongated tetrahedra, a large number of tetrahedra have to be removed and reconstructed, as shown in Figure 8.49. If points are to be inserted in a contiguous manner close to each other, the removal and reconstruction of elongated tetrahedra have to be repeated for each point insertion. Obviously, this is rather time-consuming and perhaps unnecessary. The dilemma is that in order to reduce the time in searching for the base tetrahedron, the points have to be inserted in a contiguous manner; however, in order not to remove and reconstruct elongated triangles in an excessive manner, the points have to be inserted with sufficient separation. An effective algorithm ought to optimise both the search for the BASE and the conflicting tetrahedra in the creation of the CORE.

Similar to the triangulation of non-uniform point distributions in 2D, the kd-tree insertion and its enhanced version developed in Section 8.3.2 will be rigorously tested against a wide range of data points from 0.4 to 20 million over a variety of non-uniform distribution patterns. The idea of multi-grid insertion is generic, which is expected to perform well in 3D. A 3D multi-grid insertion scheme is developed by recursive application of regular grid insertion to non-uniform distributed points partitioned into cells by regular grid or kd-tree. Exploiting the local homogeneity of non-uniformly distributed points, regular grid insertion is especially effective over uniform or non-uniform point distributions partitioned into tiny cells of a large number of points. The procedure of multi-grid insertion will be presented in Section 8.3.3, and the multi-grid insertion will be fully tested in Section 8.3.4.

8.3.2 Kd-tree insertion (3D)

8.3.2.1 3d-tree partition of space and points

The concept and the procedure for the construction of 2d-tree can be generalised naturally to higher dimensions, as discussed in Section 2.7.8. In three dimensions, k = 3, the space/

regions are partitioned by a plane perpendicular to x-, y- or z-axis in turn according to the level of subdivision. Using notations of Section 8.2.3.1, if Q_k is on the x-aligned plane, child nodes Q_{2k} and Q_{2k+1} are on the y-aligned plane, and grandchild nodes Q_{4k}, Q_{4k+1}, Q_{4k+2} and Q_{4k+3} are on the z-aligned plane and so forth. The construction of a static kd-tree of n points takes $O(n\log^2(n))$ time if an $O(n\log(n))$ sort is employed to compute the median for each region subdivision. The complexity becomes $O(n\log(n))$ if a linear median-finding algorithm is used. In the partition of regions, if the regions are cut along their longest side instead of rotating through the axes, the pattern and the performance of the kd-tree are quite different. The behaviour of the kd-trees are much better as points are more closely grouped together, and they are called *squarish* kd-trees. It is noted that the construction of 2d-tree and 3d-tree for the same number of points takes a roughly equal amount of CPU time as the procedure is similar, and the number of operations involved is exactly the same. A *squarish* 3d-tree of 1000 points distributed along a straight line is shown in Figure 8.50 in which small cells are found near the diagonal.

The only difference between kd-tree insertions in 2D and 3D is how the space is partitioned into cells according to the given point set. As for the sequence of point insertion, *exactly the same sandwich sequence for 2D can be applied in the 3D case*. For the sake of completeness and easy referencing, some of the more important features of kd-tree insertion are also highlighted in this section. A nice feature of kd-tree partition is that at any level of subdivision, each region contains a roughly equal number of points irrespective of the point distribution. In triangulation, it is not necessary to subdivide each region down to the lowest level, such that each region contains no point or only one point. A more general approach for point insertion by kd-tree partition is to allow each region to contain a specified number of points. Let N be the number of points in the set and n be the desirable number of points in a cell; the level of subdivision is given by

$$2^L n = N \Rightarrow L = \log_2\left(\frac{N}{n}\right)$$

The number of cells in the partition $N_c = 2^L$.

For a level L subdivision, N_c cells are labelled as $\{2^L, 2^L + 1,..., 2^{L+1} - 1\}$ and the pivot points between cells are $\{1, 2,..., 2^L - 1\}$.

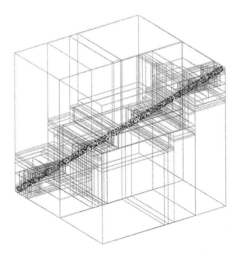

Figure 8.50 **Squarish** 3d-tree of 1000 points.

8.3.2.2 Insertion by a sandwich sequence

Again, a 2d-tree is taken to illustrate the sequence of point insertion cell by cell, and the entire set of points will be inserted when all the cells and pivot points are processed. Unlike the regular grid scheme, there is no natural contiguous sequence for cells generated by a *squarish* 3d-tree partition. However, the region splitting pivot points can serve as a link between two regions at a given level of subdivision. As shown in Figure 8.27, for partition level L = 2, the insertion sequence for the cells and pivot points are given by

{cell 4 – pivot 2 – cell 5 – pivot 1 – cell 6 – pivot 3 – cell 7}

In 3D insertion, for a set of points partitioned into 2^L cells, the insertion sequence is also given by the cell–pivot–cell sandwiched scheme. The cells for a level L partition are given by

$$\{2^L, 2^L + 1,..., 2^{L+1} - 1\}$$

and the pivot points are given by $\{1, 2,..., 2^L - 1\}$.

In the cell–pivot–cell sandwiched scheme, cells of points are inserted sequentially starting with cell 2^L, and the pivot point k following cell j is given by k = j/2, if j is divisible by 2; otherwise, set $j_1 = j/2$, and repeat the division process until j_m is divisible by 2; then pivot k = $j_m/2$. By the kd-tree construction, such a pivot point always exists and is unique for all the cells between 2^L and $2^{L+1} - 2$.

8.3.2.3 Enhanced kd-tree insertion

Kd-tree insertion can be improved following the same idea as in 2D; within a kd-tree cell, a more sophisticated insertion scheme, say the regular grid insertion, can be applied to enhance the overall efficiency, as shown in Figure 8.28. For a classical kd-tree partition, the space is partitioned until there are no more points left in each cell. There are two disadvantages in doing a full kd-tree partition: (i) more CPU time is required for the partition down to the last point; and (ii) points are overly sorted in a full kd-tree partition so that, on average, more non-Delaunay tetrahedra have to be removed in the insertion process. In the numerical tests of insertion of 0.4 to 20 million 3D points for various non-uniform distribution patterns, the most efficient scheme is to have about 20 points in each 3D cell. However, if a regular grid insertion is available, a more efficient insertion scheme can be devised by putting more than 1000 points in a cell. Similar to 2d-tree insertion, the number of points in a cell is not very sensitive to the overall performance as long as it is within the right order, as it is only a statistical average depending also on the characteristics of point distributions.

8.3.3 Multi-grid insertion

Following exactly what is done in 2D, a 3D multi-grid insertion is the result of applying the regular grid insertion to each cell of a spatial partition in a recursive manner. Within each cell, the points are more or less evenly distributed, and regular grid insertion can be applied recursively to each cell resulting in a very simple efficient scheme with almost linear complexity for locally evenly distributed points, such as line, ellipse and cluster distributions. Absolute linear complexity for the non-uniform point distributions with more complicated patterns or topology, like the spiral and the diagonal distributions, might be more difficult as points are connected to a large number of distant points through narrow elongated tetrahedra. A quasi-linear insertion scheme is only possible unless the number of tetrahedra removed for each point insertion can be controlled in such a way that it does not increase with the number of points in the set, which is a daunting task, and so far, not much theoretical results are available. Nevertheless, the situation can be somehow alleviated by keeping a suitable distance between successive insertion points, say, by a distance of one cell. For completeness, the partition of points into spatial cells is given again in the following section.

8.3.3.1 Regular grid (3D)

A 3D uniform grid for a set of N points, $P = \{P_i = (x_i, y_i, z_i), i = 1,N\}$, can be constructed following the same procedure of the 2D uniform grid. As shown in Figure 8.51, a regular grid of $N_x \times N_y \times N_z$ cells (boxes, zones, bins) subdivides uniformly the space of interest into a number of regions such that for any given point, the cell containing it can be rapidly determined without much calculation.

$$\text{Compute } \lambda = \sqrt[3]{\frac{N}{nR_xR_yR_z}}$$

where n = average number of points in a cell, $R_x = x_{max} - x_{min}$, $R_y = y_{max} - y_{min}$, $R_z = z_{max} - z_{min}$.

The number of divisions along each direction is given by

$N_x = \text{NINT}(\lambda R_x)$, $N_y = \text{NINT}(\lambda R_y)$ and $N_z = \text{NINT}(\lambda R_z)$; NINT returns the nearest integer.

The number of cells $N_c = N_xN_yN_z$; compute the size of a cell along each direction

$d_x = R_x/N_x$, $d_y = R_y/N_y$, $d_z = R_z/N_z$.

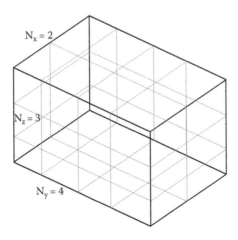

Figure 8.51 Partitioned into 2 × 4 × 3 = 24 cells.

Determine cell k to which point $P_i = (x_i, y_i, z_i)$ belongs:

$$I_x = [(x_i - x_{min})/d_x] + 1, \quad I_y = [(y_i - y_{min})/d_y] + 1, \quad I_z = [(z_i - z_{min})/d_z] + 1, \quad [.] = \text{Integral part}$$

$$k = (I_z - 1)N_{xy} + (I_y - 1)N_x + I_x$$

where $N_{xy} = N_x N_y$.

Points in the cells, or the regular grid of point set **P**, can be fully specified by a single array **A**, such that points assigned to cell k are given by $\{A_m, m = A_k + 1, A_{k+1}\}$. Whether it is a 2D grid or a 3D grid, by means of the pointer approach, the memory required for the grid construction is always a linear array **A** of size equal to $N + N_c$.

8.3.3.2 Point insertion by regular grid

The given set of points are partitioned into $N_c = N_x N_y N_z$ cells, which are inserted cell by cell in a contiguous manner to reduce the searching time in locating the BASE tetrahedron, as shown in Figure 8.29, for the insertion of a layer of cells. Given a set of N points, the number of cells is approximately given by

$$N_c \approx \frac{N}{n} \quad \text{where n is the average number of points in a cell.}$$

There is no theoretical optimal value for n; however, from practical experience, the performance of a regular grid is not very sensitive to n, which can produce satisfactory results with values ranging from 15 to 30 for various point distributions. The memory requirement for the construction of a regular grid is relatively low, and all that is needed is an integer array of size given by

Memory required by regular grid = number of cells + number of points = $N_c + N$

8.3.3.3 Multi-grid as a repeated application of the regular grid

Applying regular grid to non-uniform point distribution, the number of points within a cell can have a large variation. In some cells, there are more than 1000 points, whereas in other cells, not even a single point is present. Similar to 2D point insertion, for evenly distributed point sets, the threshold for the application of regular grid is very low at about 100 points, i.e. the CPU time will be reduced for the insertion of any reasonable point distributions more than 100 points by applying a regular grid compared to a direct insertion point by point in one zone. In view of the very low threshold of about 100 points for regular insertion, efficiency of point insertion can be significantly enhanced by applying the regular grid in a recursive manner to cells containing, say, more than 200 points. The searching time for the location of the BASE tetrahedron can be tightly controlled such that there are about 20 points in each cell that is further divided. However, the number of conflicting tetrahedra for each point inserted is still governed by the non-uniform distribution characterised with excessive connections by elongated tetrahedra.

Presence of elongated tetrahedra is an intrinsic feature of highly non-uniform point distributions, which poses a significant difficulty over uniform point distributions. For the same number of points, much more CPU time will be needed for the triangulation of severe non-uniformly distributed point sets compared to a uniform distribution or a mildly non-uniform distribution. Nevertheless, using appropriate point insertion schemes, such as the kd-tree and the multi-grid insertions, the time for the determination of the BASE tetrahedron would not be much more than the case of uniform distribution. To avoid excessive removal of closely packed elongated tetrahedra in a point insertion with points too close to each other, some separation control can be exercised by taking one point from a cell at a time instead of inserting all the points within a cell all at the same time. From the experience of various point distributions, the number of conflicting tetrahedra removed for each point insertion can be substantially reduced, resulting in a more stable insertion scheme with higher overall performance.

8.3.4 Tests on non-uniform point distributions

Six different distributions, namely, random, line, diagonal, ellipse, spiral and cluster, of 0.4 to 20 million points, as shown in Figures 8.52 through 8.57, were triangulated using

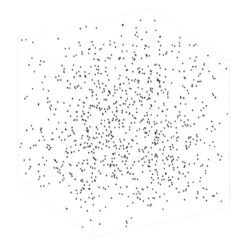

Figure 8.52 **Random distribution.**

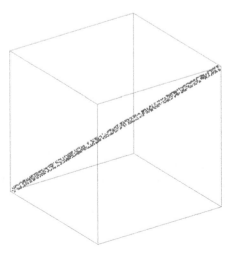

Figure 8.53 **Line distribution.**

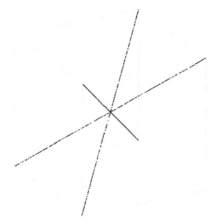

Figure 8.54 Diagonal distribution.

Figure 8.55 Ellipsoid distribution.

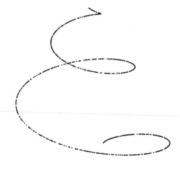

Figure 8.56 Spiral distribution.

the regular grid insertion, enhanced kd-tree insertion and multi-grid insertion. Mild non-uniform point distributions of line pattern are shown in Figure 8.52, in which the width is about 1% of the larger dimension, and the cluster distribution is generated with random points biased towards the centre with radius r shortened by

$$r = r_0 \times random\_number^{3-5} \tag{8.3}$$

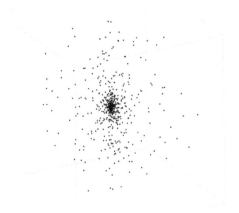

Figure 8.57 **Spiral distribution.**

The random point distribution and the other five mild non-uniform point distributions of 1 million points were triangulated by the regular grid insertion scheme.

Again, it is found that the CPU times taken for the insertion of 1 million mild non-uniform distributions were not much higher than that of the uniform distribution, as shown in Table 8.10 and Figure 8.58. In view of this situation, in the tests for the three insertion schemes, more severe non-uniform point distributions were adopted, in which the width of point spread was narrowed down to 0.01% as shown in Figures 8.54 and 8.56, and the cluster distribution was generated using a power of 5 instead of 3 in Equation 8.3.

Table 8.10 **Triangulation of mildly non-uniform point distributions**

Distribution	CPU time(s)
Uniform	7.83
Line	8.33
Cross	7.74
Ellipse	6.94
Spiral	8.71
Cluster	14.0

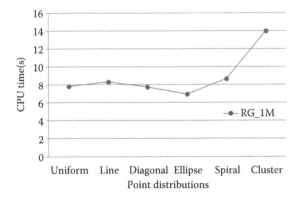

Figure 8.58 **CPU time for the regular insertion of mildly non-uniformly distributed points.**

Table 8.11 Uniform distribution

Uniform	Regular grid insertion					Multiple grid insertion					Kd-tree insertion				
				CPU time					CPU time					CPU time	
N (million)	NP	NR	Grid	Insert	Total	NP	NR	Grid	Insert	Total	NP	NR	Grid	Insert	Total
0.4	7.4	26.4	0.04	3.28	3.32	7.4	26.4	0.04	3.28	3.32	7.92	28.5	0.485	3.23	3.715
0.8	7.54	26.9	0.08	6.69	6.77	7.54	26.9	0.08	6.69	6.77	7.53	28.8	1.35	6.59	7.94
1.2	7.57	27.2	0.12	9.99	10.11	7.57	27.2	0.12	9.99	10.11	7.73	29.1	2.2	9.82	12.02
1.6	7.63	27.3	0.17	13.6	13.77	7.63	27.3	0.17	13.6	13.77	8.09	29.4	3.14	13.2	16.34
2	7.64	27.5	0.22	17	17.22	7.64	27.5	0.22	17	17.22	7.53	28.9	4.21	16.6	20.81
4						8.63	26.5	0.37	33.4	33.77	7.23	28.8	9.55	31	40.55
8						8.66	26.9	0.76	66.5	67.26	7.64	29.3	23.4	69.2	92.6
12						8.74	27.1	1.26	100.2	101.46	8.1	29.8	40.6	102.4	143
16						8.75	27.3	1.85	136.3	138.15	7.62	29.3	62.5	135.5	198
20						8.78	27.4	2.39	169.7	172.09	7.86	29.5	84.2	183.1	267.3

Table 8.12 Line distribution

Line N (million)	Regular grid insertion					Multiple grid insertion					Kd-tree insertion				
			CPU time					CPU time					CPU time		
	NP	NR	Grid	Insert	Total	NP	NR	Grid	Insert	Total	NP	NR	Grid	Insert	Total
0.4	211.7	18.8	0.03	11.3	11.33	30.6	18.8	0.03	3.85	3.88	8.91	38.2	0.36	3.89	4.25
0.8	226.7	19	0.06	25.7	25.76	28.9	19	0.06	8.3	8.36	9.06	38.3	0.92	7.75	8.67
1.2	231.9	19.1	0.08	41.5	41.58	26.8	19.1	0.08	12.9	12.98	8.72	38	1.58	11.6	13.18
1.6	237.1	19.2	0.1	59.3	59.4	25.3	19.2	0.1	17.7	17.8	9.06	38.4	2.35	15.9	18.25
2	238.4	19.3	0.12	77.9	78.02	25.1	19.3	0.12	23	23.12	8.5	37.9	3.16	19.3	22.46
4						23.5	19.4	0.26	48.2	48.46	8.51	38.1	7.32	37.5	44.82
8						22.6	19.6	0.49	102.5	102.99	9.44	38.4	18.5	79	97.5
12						21.5	19.6	0.76	156.6	157.36	9.01	38.6	32.5	120.1	152.6
16						21.6	19.7	1.03	217.6	218.63	9.46	38.5	46.5	164.6	211.1
20						21.2	19.7	1.28	277.2	278.48	9.37	38.7	66	207.5	273.5

Table 8.13 Diagonal distribution

Diagonal	Regular grid insertion					Multiple grid insertion					Kd-tree insertion				
				CPU time					CPU time					CPU time	
N (million)	NP	NR	Grid	Insert	Total	NP	NR	Grid	Insert	Total	NP	NR	Grid	Insert	Total
0.4	130.4	18.5	0.03	7.12	7.15	28.7	18.5	0.03	3.45	3.48	9.43	35.5	0.36	3.75	4.11
0.8	144.9	19	0.05	15.9	15.95	27.3	19	0.05	7.11	7.16	9.79	35.2	0.89	7.65	8.54
1.2	150.4	19.1	0.08	25.4	25.48	25.3	19.1	0.08	10.8	10.88	9.6	35.8	1.47	11.3	12.77
1.6	155	19.3	0.1	35.8	35.9	25.3	19.3	0.1	14.7	14.8	9.5	35.4	2.25	14.9	17.15
2	156.8	19.3	0.13	45.8	45.93	23.5	19.3	0.13	18.4	18.53	9.16	35	3.12	18.8	21.92
4						21.9	19.5	0.27	37.2	37.47	9.15	35.1	8.05	36.4	44.45
8						20.8	19.7	0.49	80.5	80.99	8.77	35.2	18.8	75.2	94
12						20.4	19.8	0.76	127.2	127.96	9.36	35.5	32.6	115.3	147.9
16						20	19.9	1.03	178	179.03	8.71	35.2	45.9	150.3	196.2
20						21.3	19.9	1.3	231.7	233	8.94	35.5	60.8	190.6	251.4

Table 8.14 Ellipsoid distribution

Ellipse	Regular grid insertion					Multiple grid insertion					Kd-tree insertion				
N			CPU time					CPU time					CPU time		
(million)	NP	NR	Grid	Insert	Total	NP	NR	Grid	Insert	Total	NP	NR	Grid	Insert	Total
0.4	12.6	17.7	0.03	2.56	2.59	9.81	17.7	0.03	2.38	2.41	6.93	23.3	0.48	2.55	3.03
0.8	13.4	17.2	0.06	5	5.06	9.66	17.2	0.06	4.76	4.82	6.97	23.1	1.19	5.14	6.33
1.2	13.8	17.2	0.08	7.61	7.69	9.41	17.2	0.08	7.26	7.34	6.81	23.1	2.09	7.53	9.62
1.6	14.4	17.2	0.1	10.1	10.2	9.3	17.2	0.1	9.55	9.65	7.06	23.6	2.93	10.3	13.23
2	14.7	17.3	0.12	12.8	12.92	9.24	17.3	0.12	12.3	12.42	6.82	23.5	3.94	12.8	16.74
4						9.33	18	0.26	25.2	25.46	6.98	24.5	8.91	25.4	34.31
8						9.69	18.8	0.48	51.7	52.18	7.21	25.7	23.4	54	77.4
12						9.92	19.2	0.76	80.5	81.26	7.64	26.7	39.1	88	127.1
16						10.1	19.5	1.03	109.9	110.93	7.44	26.9	58.9	118.3	177.2
20						10.2	19.8	1.29	140.8	142.09	7.66	27.4	79.3	150.6	229.9

Table 8.15 Spiral distribution

Spiral	Regular grid insertion					Multiple grid insertion					Kd-tree insertion				
				CPU time					CPU time					CPU time	
N (million)	NP	NR	Grid	Insert	Total	NP	NR	Grid	Insert	Total	NP	NR	Grid	Insert	Total
0.4	109.3	22.6	0.03	6.82	6.85	18.8	22.6	0.03	3.43	3.46	10.1	48.3	0.38	4.8	5.18
0.8	141.4	21.6	0.05	16.9	16.95	20	21.6	0.05	7.15	7.2	10.2	44.8	0.98	9.12	10.1
1.2	157.3	21.4	0.08	27.7	27.78	20.4	21.4	0.08	11	11.08	10	43.2	1.7	13.6	15.3
1.6	167.8	21	0.1	39.7	39.8	20.7	21	0.1	14.8	14.9	10.2	42.4	2.53	18	20.53
2	177.3	20.9	0.13	53	53.13	20.9	20.9	0.13	18.8	18.93	10.1	41.5	3.43	21.7	25.13
4						21	20.5	0.26	39	39.26	9.74	40.5	7.93	41.4	49.33
8						21.5	20.4	0.48	82	82.48	9.5	39.5	19.9	86.2	106.1
12						22.2	20.3	0.76	131.8	132.56	9.98	39.2	35.7	132.7	168.4
16						21.7	20.3	1.03	188.1	189.13	9.55	38.9	52.8	173.4	226.2
20						22.2	20.3	1.29	236.7	237.99	9.61	38.8	70.9	217.5	288.4

Table 8.16 Cluster distribution

Cluster	Regular grid insertion					Multiple grid insertion					Kd-tree insertion				
N	NP	NR	CPU time			NP	NR	CPU time			NP	NR	CPU time		
(million)			Grid	Insert	Total			Grid	Insert	Total			Grid	Insert	Total
0.4	43	23.4	0.03	5.25	5.28	14.5	23.4	0.03	4.06	4.09	11.4	33	0.46	3.77	4.23
0.8	43.8	24	0.05	10.8	10.85	14.3	24	0.05	7.99	8.04	12.1	33.5	1.15	7.59	8.74
1.2	53	23.9	0.08	17.3	17.38	14.3	23.9	0.08	11.5	11.58	12.2	33.7	1.93	11.4	13.33
1.6	53.3	24	0.11	23.7	23.81	14.8	24	0.11	15.6	15.71	12.8	34	2.65	15.2	17.85
2	50.1	24.4	0.13	28.5	28.63	14.8	24.4	0.13	19.8	19.93	13	34.2	3.6	19.2	22.8
4						15.1	24.5	0.27	46.7	46.97	14.4	34.7	9.24	41.2	50.44
8						15.4	24.8	0.49	92.5	92.99	16.1	35.3	22.8	89.1	111.9
12						15.4	24.9	0.76	140.3	141.06	17.8	35.7	39.8	140	179.8
16						15.5	25.1	1.03	187.8	188.83	18.5	35.9	58.5	187.2	245.7
20						15.4	25.2	1.3	238.5	239.8	19.3	36	78.9	250	328.9

The results for the insertion of the highly non-uniform point distributions are shown in Tables 8.11 through 8.16 for various distribution patterns, whereas the corresponding plots of CPU time against the number of points inserted are depicted in Figures 8.59 through 8.64. For uniform point distribution, the number of points in a cell is roughly equal to the number of points specified in a cell, and the multi-grid insertion was not invoked as the threshold of 200 points was never exceeded. Thus, the results for regular grid and multi-grid are identical for the insertion of 0.4 to 20 million points. The average number of tetrahedra visited for the location of the BASE, NP, was very steady at 7.5, and the average number of conflicting tetrahedra removed in the creation of the CORE, NR, was about 27 showing that local characteristics of randomly generated points do not change with the total number of points in the set. For the insertion of 2 million points, 17 s was required, which increased to 170 s for the insertion of 20 million points, a strict linear time relationship for a 10-fold increase of points.

As points are well sorted in the kd-tree insertion, the BASE searching time, as indicated by NP ≈ 7.7, is similar to that of the regular grid insertion. However, the number of tetrahedra removed, NR = 29, is slightly larger than that of the regular insertion, resulting in an equally fast insertion scheme relative to the regular grid insertion. Nevertheless, if the grid construction time is also taken into account, the total CPU time of kd-tree insertion is always higher than that of the regular grid-insertion scheme. This gap widened with the

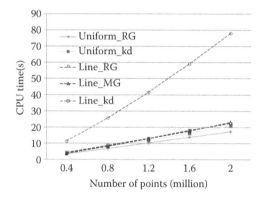

Figure 8.59 **CPU time for uniform and line distributions (2 million points).**

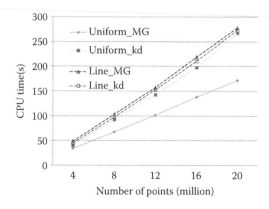

Figure 8.60 **CPU time for uniform and line distributions (20 million points).**

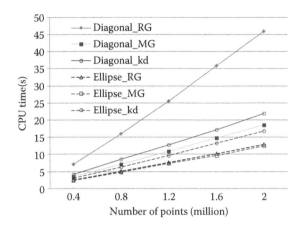

Figure 8.61 CPU time for diagonal and ellipse distributions (2 million points).

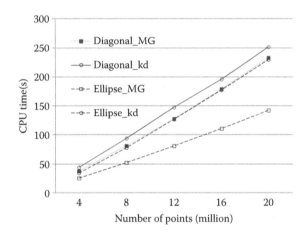

Figure 8.62 CPU time for diagonal and ellipse distributions (20 million points).

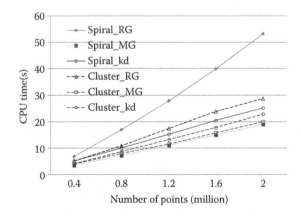

Figure 8.63 CPU time for spiral and cluster distributions (2 million points).

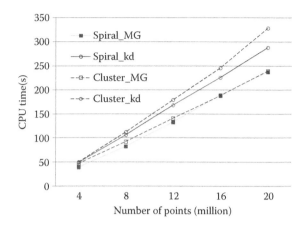

Figure 8.64 **CPU time for spiral and cluster distributions (20 million points).**

number of points increased, as the time complexity of the kd-tree is of the order $O(n\log n)$. For the insertion of 20 million points, the construction of the kd-tree alone took 84.2 s, which is already half of the total CPU time of 171.1 s for the regular grid insertion.

There is a great difference between regular grid insertion and multi-grid insertion for points concentrated along a straight line, as shown in Table 8.11. The main cause for such a large discrepancy is due to the time taken for the location of the BASE tetrahedron; NP for regular grid was more than 200, and NP for multiple grid varied from 20 to 30, resulting in a four-time difference in the insertion time. NP = 9 for kd-tree insertion is the smallest for the three insertion schemes. However, the NR value of the kd-tree is double of that of the regular grid insertion or the multi-grid insertion, resulting in only a slightly more efficient insertion scheme overall. However, if the grid construction time is also taken into consideration, the total CPU time of the kd-tree scheme is just about the same compared to the multi-grid insertion scheme.

The diagonal distribution was the first major challenge for the various insertion schemes. With a similar NR value, NP = 150 of the regular grid insertion is much higher than NP = 25 of the multi-grid insertion, leading to about a three-time difference in the overall performance. As for the kd-tree insertion, with an NP value of 9.5 and an NR value of 35, excluding the grid construction time, the insertion time is slightly less than that of the multi-grid insertion for the insertion of 20 million points.

As points are fairly evenly spread over the ellipsoidal surface, the performance of the three insertion schemes is roughly the same for the ellipse distribution. It is interesting to note that the CPU time taken for the ellipse distribution is even less than that of the uniform random distribution for the insertion of 0.4 to 20 million points. A possible cause for this to happen is that perhaps the average number of points in a cell for the ellipse distribution is closer to the optimal number of insertion (15–30) compared to the case of the uniform distribution in which the specified number of points in a cell had been set equal to 8.

The spiral distribution was another difficult case for various point insertion schemes. The NP value of 160 for the regular insertion is much higher compared to that of the uniform distribution, resulting in a three-time increase in the CPU time for the insertion of 2 million points. With reference to the uniform distribution, the NP values for multi-grid insertion increased only slightly from 7.5 to 20, resulting in a slight increase in the CPU time from 170 to 237 s for the insertion of 20 million points. Although there is not much difference in the overall performance of the kd-tree insertion with a small NP value of 10 and a large

NR value of 43, it is appreciably slower than the multi-grid insertion if the grid construction time is also taken into consideration.

A cluster distribution is locally very similar to the uniform distribution, and not many elongated tetrahedra are expected in the triangulation. As points are drastically unevenly distributed, the NP value for regular grid insertion was pretty high, which increased steadily with the number of points inserted up to more than 50 for a cluster of 2 million points; as a result, the regular grid insertion took 50% more CPU time compared to uniform distribution. On the other hand, the multi-grid and kd-tree insertions both recorded an excellent performance of a marginal increase in the CPU time relative to what is needed for the uniform distribution. Nevertheless, the performance of the multi-grid insertion is still the best overall if grid construction is taken into account.

Apart from the number of points inserted, the impact of various non-uniform point distributions for the three insertion schemes has also been studied. The CPU times for various distribution patterns by the insertion schemes for the insertion of 2 million and 20 million points are plotted, respectively, in Figures 8.65 and 8.66. For the insertion of 2 million points, regular grid insertion took much more CPU time (four times more in the worst case of line distribution) for non-uniform point distributions compared to uniform distributions,

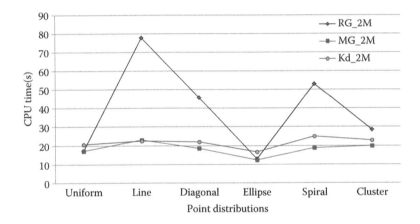

Figure 8.65 CPU time for various distributions of 2 million points.

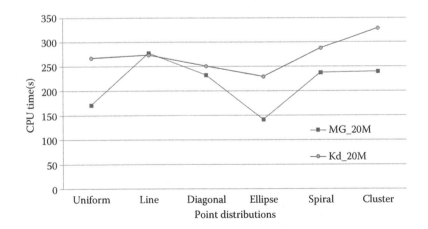

Figure 8.66 CPU time for various distributions of 20 million points.

Figure 8.67 Galaxy distribution of 6000 points.

whereas both the kd-tree insertion and the multi-grid insertion could substantially reduce the CPU time taken for point sets of non-uniform distributions as large as 20 million points.

One more difficult point distribution in the pattern of galaxy in which there is vast empty space between clusters is considered, as shown in Figure 8.67. Two tests with a total of 3 million and 6 million points, respectively, were triangulated by the regular grid insertion, the multi-grid insertion and the enhanced kd-tree insertion. As shown in Table 8.17, more or less the same results are obtained as the other non-uniform point distributions, in which the regular grid insertion took three times more CPU time compared to the multi-grid insertion, as the path for the search of the base tetrahedron was substantially longer. Again, the multi-grid insertion is the most efficient among the three insertion schemes if the grid generation time is also taken into account.

As mentioned in Section 8.3.1, the performance of the point insertion algorithm is related to how fast the BASE tetrahedron is located and how many non-Delaunay tetrahedra have to be removed for each point insertion. The time complexity of the point insertion algorithm depends, therefore, not only on the insertion method employed but also on the distribution and the characteristics of the point set. The primary purpose of the multi-grid insertion is to allocate an optimal number of points within a cell (controlled space) so as to speed up the search for the BASE tetrahedron, while the random order of the points within a cell is maintained to avoid excessive removal of non-Delaunay tetrahedra for the insertion of neighbouring points. In this respect, the multi-grid insertion is very effective in speeding up the search for the BASE tetrahedra without increasing the number of non-Delaunay tetrahedra to be removed, as in the case of kd-tree partition due to over sorting.

Table 8.17 Galaxy distribution of 3 and 6 million points

Galaxy	Regular grid insertion					Multiple grid insertion					Kd-tree insertion				
N				CPU time					CPU time					CPU time	
(million)	NP	NR	Grid	Insert	Total	NP	NR	Grid	Insert	Total	NP	NR	Grid	Insert	Total
3	261.2	24.1	0.2	100.2	100.4	31.3	24.1	0.2	32.1	32.3	10.5	54.7	4	38.4	42.4
6	167.2	26.1	0.4	153.9	154.3	21.2	26.1	0.4	62.1	62.5	10.3	43.2	10.9	68.2	79.1

In the light of the numerical examples done, it can be seen that the path for the search of the BASE tetrahedron in the regular grid insertion increases rapidly with the non-uniformity of the point set from 7.5 for uniformly distributed points to more than 200 for a line distribution, as pointed tetrahedra connecting points far apart are created, as shown in Figure 8.49. With the introduction of the multi-grid insertion scheme, the path for the line distribution is cut down to 30 or less. The path can be further reduced by means of the kd-tree partition; however, the price to pay is an increase in the number of non-Delaunay tetrahedra to be removed for each point insertion as points are so close to each other in a kd-tree partition, resulting in a more or less equally efficient insertion scheme without counting the grid construction time. It is an open question whether the path could be reduced to the level of that of the kd-tree without increasing the non-Delaunay tetrahedra to be removed for each point insertion. Nevertheless, the spiral and galaxy point distributions may be among the worst cases as tetrahedra are connected to boundary points; but then even in these difficult cases, the CPU time needed is less than twice for that of a uniform point distribution by the multi-grid insertion algorithm.

The enhanced kd-tree and multi-grid insertion have been applied to two practical examples of adaptive mesh refinement, as shown in Figures 8.68 and 8.69. The first practical example is a machine part of 256,403 tetrahedral elements and 54,924 nodes, which was

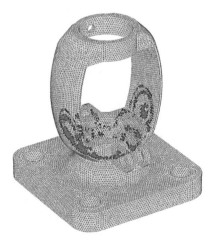

Figure 8.68 **Machine part.**

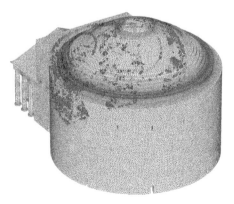

Figure 8.69 **A model of a church.**

Table 8.18 Performance of insertion schemes for practical examples

Mesh	Regular grid insertion					Multiple grid insertion					Kd-tree insertion				
			CPU time(s)					CPU time(s)					CPU time(s)		
	NP	NR	Grid	Insert	Total	NP	NR	Grid	Insert	Total	NP	NR	Grid	Insert	Total
Machine	7.92	38.2	0.21	32.97	33.18	7.34	38.2	0.21	29.21	29.42	17.4	31.0	3.9	29.2	33.1
Church	8.35	34.1	0.29	44.2	44.49	7.66	34.1	0.29	42.1	42.39	39.2	38.0	4.74	61.3	65.04

refined by adding more nodes within a relatively small part of the object, and as a result, a total of 3,318,541 nodes were inserted and 19,653,929 tetrahedral elements were constructed, as shown in Table 8.18. The second practical example is an FE model of a church consisting of 192,196 nodes and 772,140 tetrahedral elements. The mesh was refined by adding more nodes to the interior of the mesh rather locally to within a small region, and the final mesh consists of 4,824,332 nodes and 28,631,864 tetrahedral elements.

As expected, the gradual nodal spacing variation within an adaptive FE mesh is very similar to the case of mild non-uniform point distributions, and about 1 million points are inserted in 10 s. The performances of the three insertion schemes, namely, the regular grid insertion, multiple grid insertion and kd-tree insertion are more or less the same. As the number of cells is of the same order of the points, for a smooth variation of nodal spacing over refinement meshes, some cells may contain up to only a hundred times of the average number of points in a cell. This slight variation in the number of points within cells, less than 1000, would not have much impact in the time for triangulation. Moreover, the nodal distribution of an FE mesh is less drastic compared to the benchmark non-uniform point distributions such as the cross and spiral with a narrow spread of 0.01%, in such a way that much fewer pointed and narrow tetrahedra are formed, which indeed locally resemble a mesh of uniform point distribution.

8.3.5 Possibility for parallelisation

By means of a regular grid, the parallel DT by zonal insertion can be applied to a mild non-uniform point distribution or a point distribution with a gradual change of nodal spacing arising from adaptive meshing, as already discussed in Section 7.4.10. The 2 × 2 × 2 zonal parallel insertion has been applied to non-uniform point distributions in which 5 million points were randomly generated super-imposed with another 5 million points generated,

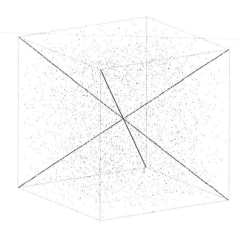

Figure 8.70 Diagonals.

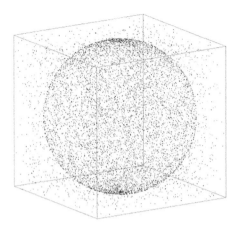

Figure 8.71 Sphere.

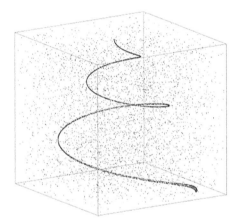

Figure 8.72 Spiral.

respectively, along the diagonals, over the surface of a sphere and on a spiral curve, as shown in Figures 8.70 through 8.72. These point distributions have been designed to simulate a practical case of adaptive refinement meshing of a large number of points, with local concentration in solid and fluid-mechanic problems. The CPU times in seconds of single-processor insertion and multi-processor insertion are shown in Table 8.19.

For highly non-uniform point distributions with point concentration within a region of 0.01% of the larger dimension, the multi-grid insertion could be applied. As the multi-grid is just a regular grid on top of another regular grid, the parallel insertion designed for regular grid can also be readily extended to multi-grid insertion with very little modifications. However, it is more difficult for the parallelisation of kd-tree insertion, as the cells of a kd-tree partition is not in alignment going from one cell to the adjacent cells.

Table 8.19 CPU time(s) for non-uniform distributions of 10 million points

	Uniform	Diagonal	Sphere	Spiral
Single processor	81.7	100.1	79.1	102.5
Multi-processor	18.9	29.5	22.7	29.4
Speed up	4.3	3.4	3.5	3.5

8.3.6 Closure

The regular grid insertion, enhanced kd-tree insertion and multi-grid insertion have all been extended to 3D in this section. Excellent performance of the multi-grid and enhanced kd-tree insertions is observed for large highly non-uniformly distributed point sets, such that the CPU time taken is no more than double of that of a uniform distribution for the insertion of 20 million points. Compared to the multi-grid insertion, the NP value of kd-tree insertion is usually smaller, but this is often offset by a larger NR value for triangulations characterised with a large amount of elongated tetrahedra, resulting in a similar overall performance for the two schemes. The multi-grid insertion is the most efficient and stable for all the distribution patterns tested among the three insertion schemes if the grid construction time is also taken into account. A quasi-linear complexity can be observed for the triangulation of distribution patterns with local features similar to those of a uniform point distribution, such as line, ellipse and cluster distributions. However, a bit more CPU time is needed for the triangulation of more difficult distribution patterns, such as diagonal and spiral, in which there are many elongated tetrahedra in the triangulation leading to a slightly non-linear time complexity. Nevertheless, with the multi-grid insertion, the triangulation time of the worst non-uniform distribution of 20 million points tested is about only 60% more than that of a uniform distribution, which is already a very promising result so far achieved.

8.4 MESH GENERATION AND ADAPTATION BY EDGE REFINEMENT

8.4.1 Introduction

The importance of the MG by refinement deserves perhaps an entire chapter rather than just a section in the chapter of auxiliary techniques. However, in spite of the diverse applications (Ortiz and Quigley 1991; Field and Smith 1991; Ruppert 1994; Muthukrishnan et al. 1995; Traxler 1997; Hertel and Kronmuller 1998; Muller-Hannemann and Weihe 2000; Bechet et al. 2002; Wang et al. 2006), the basic idea is pretty simple that it could be well covered within this section. Only mesh refinement will be discussed even though mesh *de-refinement* or *coarsening* (Benzley et al. 2005; Staten et al. 2008; Anderson et al. 2009) is equally important in many situations especially in the adaptive analysis involving anisotropic meshes and in reducing the data points yet maintaining the main features of a surface mesh. Mesh refinement is carried out by dividing edges of a mesh in order to produce elements of size and shape in compliance with a specified metric field, or the edges are refined in such a way that it will be reproduced in the DT, as discussed in Section 5.4. Since boundary recovery cannot be done in a systematic manner to ensure the quality of the elements produced, mesh refinement is particularly useful in 3D as boundary conformity will not be violated in the refinement process, and even full boundary integrity with topological constraints can also be achieved quite easily with additional care in the element connections. As mesh refinement is a local process, virtually a linear time complexity can be achieved, which is a crucial attribute for parallelisation and large-scale problems. Along with coarsening, complicated meshes in compliance with general anisotropic metric fields for fluid dynamic computations can also be created. Mesh refinement is a deterministic robust process such that a solution always exists, and a valid FE mesh is always maintained throughout the refinement process. The resulting mesh is well under control and verified in each refinement step before it is accepted to ensure the quality of the mesh produced.

Refinement of triangular meshes on a planar domain or on a curved surface can be carried out by the method of bisection of the longest edge (LE). It has been proved that the bisection process always terminates in a finite number of steps, and the quality of the resulting triangles can be guaranteed (Rivara 1984). However, the idea of bisection of the LE cannot be extended to higher dimensions as the quality of the resulting tetrahedra is unbounded, i.e. very poor tetrahedral elements may be created in the bisection process. The local bisection refinement of simplicial grids was also studied by Maubach (1995). Alternatively, a local refinement of tetrahedral meshes was proposed in which a finite number of classes of similar tetrahedra were created in the refinement such that the quality of the tetrahedral mesh was guaranteed (Liu and Joe 1995). A year later, a local refinement of eight-tetrahedron subdivision was formally presented in which a finite number of sub-classes of similar tetrahedra could be created with guaranteed quality. The procedure could be extended to the neighbouring tetrahedral elements to maintain a conforming mesh (Liu and Joe 1996).

Lee and Lo (1997a,b) presented an adaptive analysis by refining T4 and T10 tetrahedral meshes according to the error of the FE solution. Lo (1998a) introduced an edge-sorting scheme for 3D mesh refinement in compliance with a specified node spacing function. A 3D anisotropic mesh refinement method in compliance with a general metric specification was also presented (Lo 2001). Lohner and Cebral (2000) generated non-isotropic meshes based on directional refinement. Arnold et al. (2000) proposed a bisection algorithm, which was compared with those of Bansch (1991), Liu and Joe (1995) and Maubach (1995). Mesh conformity of the bisection procedure was also rigorously verified. Baker (2002) proposed a mesh-enhancing scheme in which coarsening and enrichment were combined with an r-refinement to produce a flexible approach for mesh adaptation of time-evolving domains. Plaza et al. (2004) showed that in the repeated partition process of tetrahedra by the eight-tetrahedron subdivision and the LE bisection, only three dissimilar types are produced. Hence, the quality of the resulting mesh can be guaranteed. Plaza et al. (2005) discussed the non-degeneracy property of the eight-tetrahedra LE partition and gave an estimate on the asymptotic value of the refined elements based on numerical experiments. Li et al. (2005) presented mesh optimisation and adaptation involving refinement, coarsening, projecting boundary vertices, shape correction and other techniques according to the given metric field. Gruau and Coupez (2005) proposed a mesh refinement scheme based on an anisotropic metric by specifying the number of layers across the thickness direction and subdividing elements within a topological neighbourhood around a node. Remacle et al. (2005) proposed a series of local modifications to an FE mesh in compliance with the prescribed anisotropic metric field. Dai and Schmidt (2005) presented a refinement scheme of triangular meshes on surfaces for large deformation problems. Si and Gaertner (2005) presented a semi-constrained DT in which tetrahedral mesh for a solid object bounded by a discretised surface is created by introducing Steiner points at strategic positions on the boundary edges.

Alauzet et al. (2006) presented a mesh-coarsening and refinement scheme with shape optimisation for the fluid flow problem. The problem domain was also decomposed for parallel processing for which the load balancing and the inter-partition communication issues were addressed. Wessner et al. (2006) discussed the anisotropic mesh refinement issues for semiconductor manufacturing processes. A refinement tree-partitioning method related to Octree and space-filling curves for adaptive parallel processing was proposed by Mitchell (2007). An algorithm was presented to generate surface and volume meshes for modelling molecules of arbitrary sizes and shapes by a series of surface modifications and enhancement techniques (Yu et al. 2008a). Yu et al. (2008b) put forward a geometric model and MG with adaptation targeted to biomedical systems by means of mesh refinement and coarsening with feature preserving. Loseille and Alauzet (2009) discussed in detail an anisotropic mesh adaptation for 3D-steady Euler equations based on the optimisation of a functional. Dey et al. (2012) discussed a restricted Delaunay refinement to generate triangular meshes at the interface surfaces of objects obtained by labelling images from various modalities.

The three common bisection schemes, namely, those of Rivara and Levin (1992), Liu and Joe (1995, 1996) and Plaza and Carey (2000), were compared by Padron et al. (2005).

1. The longest-edge (LE) bisection (Rivara and Levin 1992): Single LE bisection consists of dividing the tetrahedron into two tetrahedra by the midpoint of the longest edge.
2. Liu–Joe eight-tetrahedron (8T) partition: Tetrahedron T is first mapped onto a special tetrahedron S with vertices $\{(0, \frac{1}{\sqrt{2}}, 0), (-1,0,0), (1,0,0), (0,0,1)\}$, which is partitioned into eight tetrahedra following a particular sequence based on the edge lengths of S, which has one long edge and three short edges, and the partition of T is obtained by an inverse mapping from S to T.
3. Eight-tetrahedron LE (8T-LE) partition (Plaza et al. 2005): The 8T-LE partition can be achieved by performing a sequence of bisections of the edges of the original tetrahedron, as shown in Figure 8.73, by the following steps:
   a. LE bisection of tetrahedron T to produce tetrahedra T1 and T2
   b. Bisection of T1 and T2 by the LE with the original tetrahedron T producing four tetrahedra $T_{ij} = \{T11, T12, T21, T22\}$
   c. Bisection of each tetrahedron T_{ij} at the edge shared with the original tetrahedron T

Based on n repeated subdivisions of each typical degenerate shape of tetrahedron, such as needle, wedge, sliver and cap, into about 2 million tetrahedra, it was found that the η ratios between the initial tetrahedron and the final tetrahedra, $\rho = \eta(T_n)/\eta(T_1)$, for the worse case of the three bisection schemes are, respectively,

$$\rho(\text{8T-LE}) = 0.3155, \quad \rho(\text{8T}) = 0.1417 \quad \text{and} \quad \rho(\text{LE}) = 0.01783$$

It was also concluded that the three partitioning schemes are equivalent in some particular cases, and the 8T and 8T-LE partitions are stable in the number of tetrahedra generated, but LE partition produces a variable number of tetrahedra.

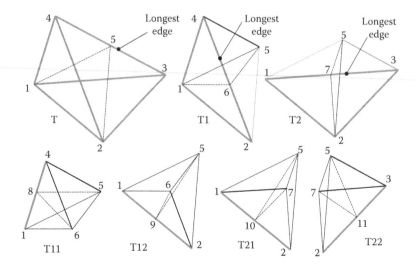

Figure 8.73 A tetrahedron is divided into eight tetrahedra by the 8T-LE partition.

8.4.2 Refinement of discretised surfaces

Very often, mesh refinements are required for curved surfaces to capture the field variable in an FE analysis, in an adaptive refinement analysis, and to refine the boundary of a 3D object for MG. If the curved surface is well defined in analytical form or by means of a parametric representation, direct MG on the surface, as described in Section 4.4, or MG on a parametric planar domain, as discussed in Section 4.2, can be applied to generate FE meshes of the required characteristics. However, in case the curved surface is only available in a discretised form of a triangular FE mesh, a mesh refinement rather than a complete mesh regeneration is a very attractive option in producing a refined mesh in compliance with the specified node spacing function without violating the original geometry represented by triangular facets.

With a local topology of a planar domain, triangulated curved surfaces can be refined by the LE bisection and trisection algorithms (Marquez et al. 2008; Plaza et al. 2010, 2012; Suarez et al. 2012). The LE bisection possesses a number of promising features, and the two most important characteristics allowing it for efficient mesh refinement are as follows: (i) the non-degeneracy property, i.e. the minimum angle of the mesh is bounded; and (ii) the LE bisection always terminates in a finite number of steps. The LE propagation path (LEPP) algorithm (Suarez et al. 2005, 2008) is one of the most popular techniques to refine a triangular mesh, and the following procedure is an adaptation of the algorithm for the refinement of a triangulated curved surface with respect to a given node spacing function ρ.

8.4.2.1 Statement of the problem

Given a mesh of triangular elements $S = \{\Delta_i, i = 1, N_s\}$ and node spacing function $\rho(x)$, $x \in S$, refine mesh S such that $E_{AB}^2 \leq 2\rho(A)\rho(B)$ for all edges in S joining node points A and B.

8.4.2.2 Algorithm: Refinement of triangular mesh

1. Build the initial list of edges to be refined, $E = \left\{ E_{AB} \in S: \ E_{AB}^2 \leq 2\rho(A)\rho(B) \right\}$
2. Take the LE L in **E**; identify the two triangles T_1 and T_2 connected to L
3. If L is the LE shared by triangles T_1 and T_2, then
   a. Delete L from the list **E**
   b. Bisect edge L at the mid-point M as shown in Figure 8.74
   c. Verify if edges AM, MB, DM and MC should be added to **E**

   Otherwise
   a. Search for the LE in a chain of connected triangles by the adjacency relationship, as shown in Figure 8.75, and such a search will terminate, say, in triangle T_k.

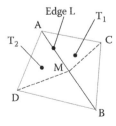

Figure 8.74 **Bisection of edge L.**

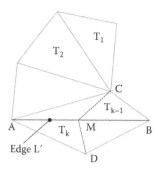

Figure 8.75 Bisection of the longest edge in a chain of connected triangles.

    b. Bisect the LE L′ between triangles T_k and T_{k-1} (L′ $\notin$ **E** as L′ is longer than L).
    c. Verify if edges AM, MB, DM and MC should be added to **E** (this step may be skipped if we do not start with a very coarse mesh to interpolate a rapidly changing spacing function)
  4. Update mesh **S** and its adjacency relationship, and if **E** is not empty, go to step (2)

As we can see by the design of the LEPP algorithm, the mesh will be slightly over-refined to guarantee the quality of the triangles as some edges not in **E** are also bisected. If the adjacency relationship is well computed or already available, the time complexity of the refinement algorithm is virtually linear as in case of a search, T_k can be located quite rapidly by adjacency in scanning a few triangles. Take for example the boundary surface mesh of 450 triangular facets of a solid object, which is to be refined to as small as 1% in linear scale, as shown in Figure 8.76. The refined mesh of 72,252 triangles with smaller elements being placed along three orthogonal planes is shown in Figure 8.77. The minimum α-value and the mean α-value of the initial boundary surface are 0.051 and 0.576, respectively. For the refined mesh, upon smoothing by node shifting, the minimum α-value improved from 0.045 to 0.152, and the mean α-value improved from 0.825 to 0.840.

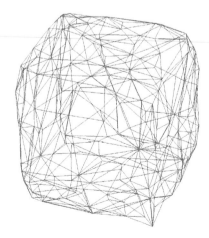

Figure 8.76 Initial boundary surface.

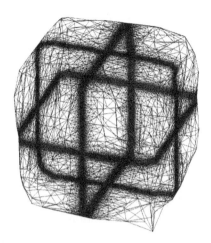

Figure 8.77 **Boundary surface refined along three orthogonal planes.**

8.4.3 3D refinement in compliance with a specified node-spacing function

Adaptive unstructured refinement is basically applied only to simplicial meshes, though hexahedral meshes can also be refined using transition elements, as discussed in Section 5.8.15. As the 8T subdivision is less flexible, the quality of the resulting mesh is very often inferior to that of the initial mesh. A simple and efficient refinement procedure for tetrahedral meshes based on successive bisection of edges is presented in this section (Laug and Borouchaki 2003a). The quality of the elements produced can be optimised if the subdivision is performed in the sequence according to the length of the edges to be divided. Such an order of priority can be established by a simple sorting process on all the edges for which refinement are needed. This list of ordered segments has to be updated from time to time to take into account the new edges created during the subdivision process. As the ring of tetrahedra around the LE are all refined at the same time, mesh conformity is always maintained at each refinement step, which, in general, is the most crucial part for a refinement algorithm. The shape quality of the mesh can be further improved by the standard optimisation procedures discussed in Chapter 6 during and at the end of mesh refinement.

8.4.3.1 The algorithm

Mesh refinement can be achieved by introducing a new node in a tetrahedron's interior, face or edge to divide it, respectively, into four, three and two tetrahedra. Insertion of a new node on an edge is a better choice because the new tetrahedra produced are of better quality, and the nodes can be placed everywhere in the mesh, both in the interior and on the external faces. In this section, an algorithm is presented in which edges of the mesh and the associated tetrahedral elements are bisected until the specified refinement is reached.

The input for the procedure is a coarse mesh **M** of tetrahedral elements, which are defined by the element vertices and the nodal co-ordinates.

$$\text{Tetrahedral mesh } \mathbf{M} = \{T_i, i = 1, N_T\}, \quad T_i = \{P_{ij}, j = 1, 4\}, \quad \{(x_i, y_i, z_i), i = 1, N_P\}$$

where N_P and N_T are, respectively, the number of points and of tetrahedral elements in mesh **M**. The algorithm will refine the initial mesh **M** according to the given node spacing function by adding extra nodes at the edges of the elements. It is proposed that the refined mesh would have the best result, and the quality of the elements could be guaranteed if nodes are inserted at the mid-point of the LE of the elements. However, it would be computationally prohibitive if nodes are always added at the mid-point of the absolutely LE of the entire mesh. As a result, a scheme is proposed such that nodes are inserted at the mid-point of one of the LEs or the quasi-LEs. It is expected that the quality of the resulting mesh would only differ slightly from the ideal mesh in which the absolutely LE is used for every node insertion.

The mesh refinement procedure is summarised into the following steps:

1. Extract the list of edges from the tetrahedral mesh **M**.
2. Determine the set of edges **L** that need refinement.
3. Arrange the set of edges $\mathbf{L} = \{L_i, i = 1,n\}$ in order of their length, such that $L_1 \leq L_2 \ldots \leq L_n$.
4. Starting from the last edge L_n, for each edge in **L**, bisect all the elements connected to it. The new edges so generated in the bisection of edges that need further refinement will be collected in the set **E**.
5. The mesh refinement process by bisecting edges in **L** is suspended if there is an edge in **E** that is longer than the LE in **L**, i.e. edge L_n currently under consideration.
6. Set **L** is updated by properly merging the edges in the two sets **L** and **E** such that the edges in **L** will always be arranged in ascending order of their lengths. Go to step (4) and the refinement process continues.
7. The refinement process terminated if there are no more edges left in the two sets **L** and **E**.

8.4.3.1.1 Extraction of edges

Extraction of edges from a mesh has been discussed in Section 2.5.6. The set of edges **L** is the collection of all the distinct line segments of tetrahedral mesh **M**. What has to be done is to examine the six edges of each tetrahedral element in turn.

8.4.3.1.2 Determine the set of edges for refinement, **L**

Let $L \in \mathbf{L}$ be an edge joining nodes P_1 and P_2, and h_1 and h_2 be the node spacings at nodes P_1 and P_2, respectively. Then edge L with length r will be refined if

$$r^2 > 2h_1h_2 \tag{8.4}$$

8.4.3.1.3 Ordering of edges in **L**

Let L be a unique line and $\mathbf{R}_L$ be the ring of tetrahedral elements connected to this edge L. To ensure that all the tetrahedral elements are to be divided by the generalised bisection process, L should be the longest among all the edges in $\mathbf{R}_L$, i.e.

$$L \geq L_{ij} \quad \text{for i = 1, m and j = 1, 6}$$

where L_{ij} is the jth edge of the ith tetrahedron in $\mathbf{R}_L$, and m is the number of tetrahedral elements in $\mathbf{R}_L$.

Goliaz and Tsiboukis (1994) proposed to carry out the refinement process on an element-by-element basis. Following the given labelling sequence of the tetrahedral elements, subdivision of a tetrahedron is done by adding a new node at its LE. Subdividing an edge of a tetrahedron results in the refinement of all tetrahedra connected to this edge in order to preserve the conformity of the mesh, for which the shape of the elements can then be improved through cycles of mesh optimisation based on the Delaunay transformations and node repositioning. However, this is not a generalised bisection process as the LE of a tetrahedron may not be the LE in the ring of tetrahedra connected to this edge. To ensure that elements are generated by the generalised bisection, an edge should not be divided unless it is also the longest in the ring of tetrahedra. Working on an element-by-element basis is not as efficient as the checking for the LE in a ring of tetrahedra could be time consuming. Indeed, much of the checking work is repeated when we pass from one tetrahedron to its neighbours.

An alternative for mesh refinement by the generalised bisection is to proceed on a line-by-line basis. A scheme ensuring that each subdivision is a generalised bisection of a ring of tetrahedra is to cut always the LE of the mesh. This implies that all the edges satisfying refinement criterion 8.4 have to be sorted in order of their length. A thorough sorting could be time-consuming and runs at a time complexity of nlog(n) for a system of n objects. Assuming that the generalised bisection is not very sensitive to a slight variation in length and minor misplacement of edges in the order list, the first-level *address sort* (*bin sort*), which has a linear time characteristic, is the most suitable to provide the priority sequence for the bisection of edges.

Let $L = \{L_i, i = 1,n\}$ be the set of edges to be refined. Let r_i be the length of edge L_i; compute $r_{min} = min_{i=1,n} r_i$ and $r_{max} = max_{i=1,n} r_i$. A new order, k, for line L_i is given by

$$k = NINT\left[(n-1)\left(\frac{r_1 - r_{min}}{r_{max} - r_{min}}\right)\right] + 1$$

Effectively, the edges are sorted in the ascending order of magnitude, i.e. shorter lines will be placed in the front part of the list, and longer lines are placed towards the end of the list. To assess the error of this sorting process, suppose that two edges L_i and L_j are evaluated to occupy the same position k:

$$NINT\left[(n-1)\left(\frac{r_i - r_{min}}{r_{max} - r_{min}}\right)\right] + 1 = NINT\left[(n-1)\left(\frac{r_j - r_{min}}{r_{max} - r_{min}}\right)\right] + 1 = k$$

$$\Rightarrow |r_i - r_j| < \frac{r_{max} - r_{min}}{n-1}$$

Hence, if there are 1000 edges that need subdivision, then the error would be at most $r_{max}/1000$. In a coarse mesh where there are not many edges that need subdivision, more space than necessary can be allowed for sorting (say, 1000 irrespective of the number of edges) to ensure that the error between two consecutive edges can be limited to 0.1% of r_{max}.

8.4.3.1.4 Bisection of a ring of tetrahedra

Bisection of an edge IJ results in the refinement of the ring of tetrahedra connected to this edge. The ring of tetrahedra associated to an edge can be obtained by a simple algorithm 2.5.9 without searching. As the four neighbours of each tetrahedron are known, starting from any element that has edge IJ as one of its edges, the algorithm identifies successively all the elements around line segment IJ when it comes back to the first element where the journey started. The bisection of a ring of n tetrahedra associated with edge IJ, $\{IJP_1P_2, ..., IJP_{n-1}P_n, IJP_nP_1\}$, will produce 2n tetrahedra $\{IKP_1P_2, ..., IKP_{n-1}P_n, IKP_nP_1, KJP_1P_2, ..., KJP_{n-1}P_n, KJP_nP_1\}$ in which K is a new node added to the mid-point of edge IJ, as shown in Figure 8.78. The new edges

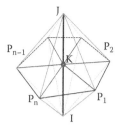

Figure 8.78 Bisecting a ring of tetrahedra.

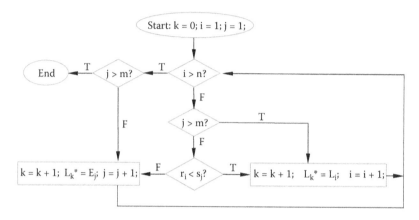

Figure 8.79 Flow chart for merging of sets **L** and **E**.

generated are IK, KJ, KP_1, ..., KP_n. Those new edges satisfying criterion 8.4, which need further refinement, are stored in the set E.

8.4.3.1.5 *Merging of the two sets **L** and **E***

The number of edges in L decreases as refinement is carried out by bisecting lines in L. Suppose that there are n edges remaining in L; then L can be expressed as $L = \{L_i, i = 1,n\}$. Subdivision of edges in L is suspended if there is a line segment in E that is longer than the edge L_n currently under consideration. Before merging, the edges in E are ordered in the same way as described in Section 8.4.3.1.3. Let m be the number of line segments in E; then E can be written as $E = \{E_j, j = 1,m\}$. These two ordered sets can be merged easily to form a new set $L^* = \{L_k^*, k = 1,m + n\}$ consisting of edges from the sets L and E. Replacing L by L^*, go back to Section 8.4.3.1.4, and the refinement process continues. Let r_i and s_j be, respectively, the lengths of edges $L_i \in L$ and $E_j \in E$. A simple algorithm for the merging of the two ordered sets L and E is given by the flow chart shown in Figure 8.79.

8.4.3.2 *Optimisation of element shape*

The quality of the refined mesh can be significantly improved through cycles of element shape optimisation. As discussed in Chapter 6, mesh optimisation procedures can be classified into two main categories: (i) those that do not involve a change in mesh topology (node–element connectivity relationship), for instance, node repositioning; and (ii) those that involve a structural modification of the mesh such as face/edge-swapping operations.

8.4.3.3 Examples

The refinement algorithm is tested with meshed objects of various shapes and subject to different node spacing distribution requirements. The data for the initial mesh are the vertices of the tetrahedral elements and the co-ordinates of the nodal points. A node spacing function has to be specified, which allows the element size of the mesh to be evaluated at any point within the meshed domains. The algorithm is coded in Visual Fortran on PC i7 CPU 870 at 2.93 GHz running on XP mode with QuickWin graphic supports. The algorithm is to refine the initial mesh by repeatedly bisecting the edges of the mesh following the priority sequence of the edge length until the specified node spacing is achieved.

The first example is the refinement of a cube along a vertical edge. The cube shown in Figure 8.80 is divided into five tetrahedral elements, which are refined into 34,616 node points and 179,553 elements, as shown in Figure 8.81. The size of the elements at the refinement edge is 0.05 unit, whereas those at the far end are of size 10 units, which is also the length of the cube. Owing to the rapid change in the element size, this is the only example where the γ-quality has decreased slightly from a relatively high value of 0.7791 to 0.7374. The same cube is used in the second example in which refinement is carried out following one octant of a spherical surface. The initial mesh consists of six tetrahedral elements with a γ-value of 0.6606, which is improved to 0.7244 in the refined mesh of 186,524 node points and 1,024,650 elements, as shown in Figure 8.82.

In the third example, a more complicated density distribution function was employed. It is intended to generate elements of relatively smaller size on a spherical surface inside a cube.

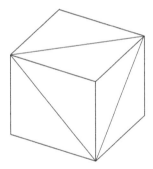

Figure 8.80 **A cube of five tetrahedra.**

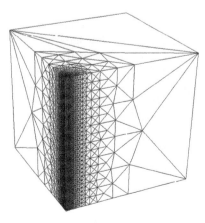

Figure 8.81 **Refined along an edge.**

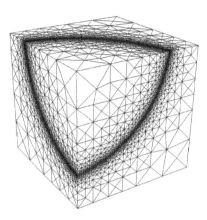

Figure 8.82 Refined over an octant of a spherical surface.

Figure 8.83 Refined over an interior spherical surface.

The centre of the spherical surface whose radius is of 2 units is positioned at the centroid of the cube. The size of the elements on the surface is to be refined to 0.05 unit, which increases linearly away from the surface to a maximum value of 5 units at the corners of the cube, a change of 100 times in linear scale or 10^6 in volume. A cross section of the final mesh consisting of 67,326 node points and 364,778 elements is shown in Figure 8.83. The mean γ-quality of the mesh is 0.771 indicating a very high quality of the refined mesh in spite of the complicated pattern of the density distribution and a rapid change in element size.

A more realistic physical object made up of 24 elements for adaptive refinement analysis is taken as example 4, as shown in Figure 8.84. The lengths of the external edge and the internal edges are 10 and 4 units, respectively, whereas the height of the object is 5 units. Small elements of size 0.05 unit are to be generated at the internal sharp corners, and elements

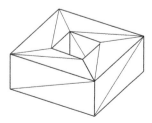

Figure 8.84 Object with internal opening.

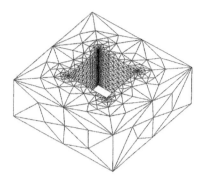

Figure 8.85 **Refined mesh: 35,843 tetrahedra.**

of a relatively large size are to be generated away from the reflex edges. The refined mesh consists of 7720 node points and 35,843 elements, as shown in Figure 8.85. The quality of the refined mesh is much better than that of the initial mesh with the mean γ-coefficient increased from 0.384 to 0.721.

The fifth example is the refinement of the same cube divided into six tetrahedra over a spherical surface, which is larger than the cube. The entire refined mesh, which is a bit unclear with so many edges, is shown in Figure 8.86, and rings of refinement on the boundary surface are shown in Figure 8.87. The largest elements are of size about 2 units, whereas

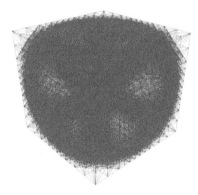

Figure 8.86 **Mesh of 1,071,207 tetrahedra.**

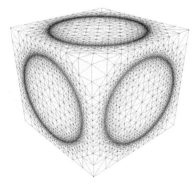

Figure 8.87 **Boundary surface of example 5.**

the smallest elements are of size 0.05 unit in linear scale. Based on these specifications, the cube is refined into 202,810 node points and 1,071,207 elements. In spite of the rapid change in the element size from 0.05 to 2 units in about five layers, the refined mesh attains a mean γ-quality of 0.7243, which is quite consistent with the other refined meshes.

The sixth example is about the refinement of a sphere. This example demonstrates the ability of the algorithm in generating adaptive refined meshes for engineering objects with curved boundaries. To recover the curved surface of the object, new nodes on the boundary as soon as they are generated due to refinement are projected onto the boundary surface of the object in each refinement cycle. As shown in Figure 8.88, the initial mesh that consists of 96 fairly good-quality tetrahedral elements with a mean γ-value equal to 0.7698 is refined to a mesh of 39,123 elements with an improved mean γ-quality equal to 0.7878, as shown in Figure 8.89. The polyhedron is refined progressively into elements of specified size in compliance with the requirement that elements have to be concentrated at three specific locations. It is also noted that the curved surface of the sphere is recovered when more elements are generated on the boundary of the refined mesh.

The last example is designed to test the time complexity of the algorithm. Elements are to be refined to a size of 0.01 unit at the re-entrant corner of an indented object whose sides are of lengths equal to 10 units, a change in linear scale of 1000 times. The initial mesh of 15 tetrahedra is refined into 216,275 node points and 1,110,181 tetrahedral elements, as shown in Figure 8.90. The characteristics of initial and final meshes together with some important

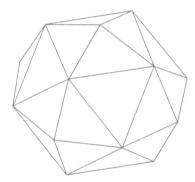

Figure 8.88 **Example 6: 96 tetrahedra.**

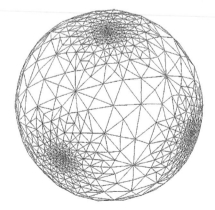

Figure 8.89 **Refined mesh: 39,123 tetrahedra.**

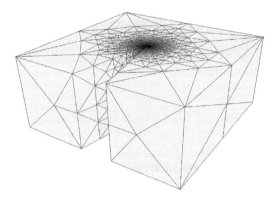

Figure 8.90 Mesh refinement around a reflective edge.

Table 8.20 Characteristics of initial and refined meshes

Example		1	2	3	4	5	6	7
Initial mesh	NP	8	8	8	16	8	33	14
	NE	5	6	6	24	6	96	15
	γ	0.7791	0.6606	0.6606	0.3837	0.6606	0.7698	6.472
Refined	NP*	34,616	186,524	67,326	7720	202,810	7654	216,275
mesh	NE*	179,553	1,024,650	364,778	35,843	1,071,207	39,123	1,110,181
	γ*	0.7374	0.7244	0.7713	0.7211	0.7342	0.7878	0.7545
Refinement	T1	0.05	0.28	0.1	0.01	0.29	0.01	0.29
Optimisation	T2	1.44	8.11	3.01	0.28	8.78	0.32	9.55
Efficiency	N1 = NE*/T1	3.59	3.66	3.65	3.58	3.69	3.91	3.83
(in millions)	N2 = NE*/T2	0.125	1.26	0.121	0.128	0.122	0.122	0.116

Note: NP = number of nodal points in the initial mesh; NE = number of elements in the initial mesh; γ = geometric mean γ-quality of the initial mesh; NP* = number of nodal points in the refined mesh; NE* = number of elements in the refined mesh; γ* = geometric mean γ-quality of the refined mesh; T1,T2 = CPU time in seconds on PC.

statistics of all the examples are summarised in Table 8.20. As this refinement algorithm is simple as well as direct, it is robust and could reproduce exactly all the symmetrical patterns of node distributions or even element connections implicitly specified by the node spacing functions. The quality of the refined mesh in general will be improved after optimisation, except for example 1, in which the initial γ-quality is fairly high at 0.7791, and the gain in mesh quality will be higher, in general, for initial meshes of relatively poor shape quality. As expected, the CPU time for refinement bears a linear relationship with the number of elements generated, at a very promising rate of about 4 million tetrahedra per second on a PC with XP mode. Much more CPU time is required for shape optimisation as it is an iterative process, and many cycles of iterations are needed to achieve the optimal results.

8.4.3.4 Refinement according to an anisotropic metric field

Nothing needs to be changed in the mesh refinement in compliance with an anisotropic metric field, except in the determination of the set of refinement edges L in Section 8.4.3.1.2. For the anisotropic refinement, the element requirement is specified by a metric tensor M defined over the entire problem domain Ω. A metric tensor at a point of the domain Ω can be represented by a symmetric positive-definite matrix. In two dimensions, a 2×2 matrix is needed, whereas in three dimensions, a 3 × 3 matrix is required. In a Riemann space defined by M on Ω, the length of edge PQ with respect to M in Ω is given by

$$\|PQ\|_M = \int_0^1 \sqrt{PQ^T \cdot M(P + tQ) \cdot PQ} \; dt \tag{8.5}$$

where M(P + tQ) is the metric at point P + tQ, t ∈ [0,1]. $\|PQ\|_M$ can be evaluated by numerical integration using two or more Gaussian points. If r is greater than $\sqrt{2}\,\|PQ\|_M$, where r is the Euclidean distance between P and Q, then edge PQ is bisected. Alternatively, the element size along a particular direction represented by unit vector **v** can also be given by

$$h = \lambda_1 (\mathbf{v} \cdot \mathbf{u}_1)^2 + \lambda_2 (\mathbf{v} \cdot \mathbf{u}_2)^2 + \lambda_3 (\mathbf{v} \cdot \mathbf{u}_3)^2 \tag{8.6}$$

where unit vectors $\mathbf{u}_1$, $\mathbf{u}_2$ and $\mathbf{u}_3$ are the principal stretch directions (eigenvectors) of metric tensor M, and λ_1, λ_2 and λ_3 are the required element sizes along the three principal directions, respectively. In the actual implementation, edge PQ will be divided if $r^2 > 2h_1 h_2$, where h_1 and h_2 are the desirable sizes computed using Equation 8.5 or 8.6 at the two Gaussian points on edge PQ. The metric M can also be designed to produce the so-called *unit mesh* (Section 4.2.6) such that edges measured by $\|PQ\|_M$ should have an ideal lengths of 1 unit, and edges having length evaluated with respect to the unit metric M greater than $\sqrt{2}$ ought to be refined. The anisotropic refinement is tested with the refinement of a standard cube under different node spacing distributions. A homogeneous metric field is used as the first test, in which the principal stretch directions are along the principal axis of the cube. Edge lengths of 3, 1 and 0.2 units are required along the three principal directions. The resulting mesh depicted in Figure 8.91 consisting of 26,199 node points and 133,766 elements shows that element sizes are quite different along various directions as specified by the metric tensor.

   The same metric definition is used in the second test. However, in this case, the principal directions of the metric tensor have been rotated so that one of the principal directions is now along the diagonal of the cube. The resulting mesh shown in Figure 8.92 consisting of 42,555 nodes and 218,062 elements shows that the principal directions are along the diagonal of the faces, which is the result of the projection of the main diagonal onto the three orthonormal faces. A non-homogeneous size map is employed in the third test. Here,

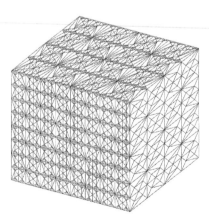

Figure 8.91 Homogeneous anisotropy.

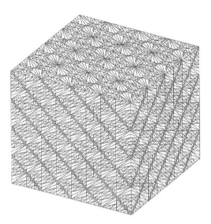

Figure 8.92 **Refinement with rotated axis.**

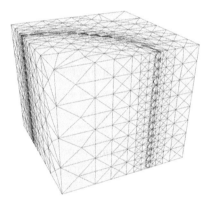

Figure 8.93 **Cylindrical anisotropy.**

we would like to define a cylindrical size distribution inside a cube. A very small size of 0.01 is required normal to the cylindrical surface, which is placed at a distance of 7 units from a vertical edge of the cube. Along the other two principal directions, relatively larger sizes between 1 and 2 units are prescribed. The resulting mesh consisting of 7264 nodal points and 35,871 elements is shown in Figure 8.93, in which elongated elements are generated along the circumferential direction of the cylindrical surface. The cluster of elements and their alignment are fairly obvious that a cylindrical surface in sandwich layers is clearly seen inside the cube.

In the last test, the same metric of the previous test is applied again. This time, the principal direction along the vertical axis of the cube has been rotated so that it is along one of the diagonals of the cube. The resulting mesh consisting of 8683 nodal points and 45,437 elements is shown in Figure 8.94. It is interesting to note that a cylindrical surface can be identified inside the cube with its axis along the diagonal of the cube. Anisotropic refinement here presented based on edge refinements just suggests a possibility of one of the basic components to produce anisotropy in an FE mesh. While promising results are readily obtained for isotropic refinement, the result of anisotropic refinement is less satisfactory probably due to the fact that there is a discrepancy in the geometry of the initial mesh and the required anisotropy. Hence, unless coarsening or de-refinement is also used along with

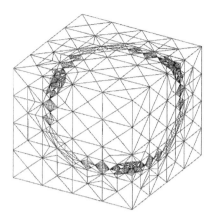

Figure 8.94 **Axis rotated along a diagonal.**

some local topological operations to provide a compatible environment for the creation of anisotropic elements, there is a limitation in the generation of anisotropic meshes merely based on edge refinement alone. The works of Gruau and Coupez (2005), Remacle et al. (2005), Alauzet and Frey (2005), Sahni et al. (2006), Loseille and Alauzet (2009), Loseille and Lohner (2009) and Loseille et al. (2010) are referred to for details of those treatments in anisotropic refinement and adaptation.

8.4.4 Refinement of non-simplicial elements

Non-simplicial meshes in 2D and 3D such as quad and hex meshes can be refined uniformly across all the elements in the mesh along the principal directions, as shown in Figure 8.95. In order to be compatible with adjacent elements, there is a severe restriction that each element has to be subdivided in exactly the same manner. As for non-uniform refinement, it seems that the only practical way is perhaps to refine quads (Tchon et al. 2004: Garimella 2009; Liang et al. 2009) and hex (Sun et al. 2011) elements around a point, as shown in Figure 8.96. A quad is divided into three quads, and a hex is divided into four hex, a smaller one close to the point and three bigger ones on the opposite faces. However, for the non-uniform subdivision of quads and hex, distorted elements will inevitably be generated from the original regular elements. Refer to Sections 5.8.14 and 5.8.15; in case we would like to keep all the elements as regular as possible in the refinement process, transition quad and hex elements can be employed, as shown in Figure 8.97.

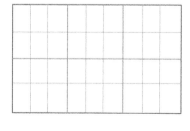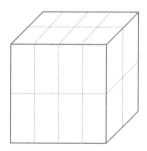

Figure 8.95 **Uniform refinement.**

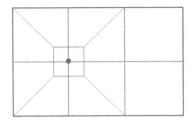

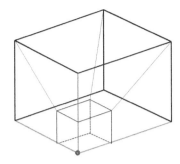

Figure 8.96 **Non-uniform refinement around a point.**

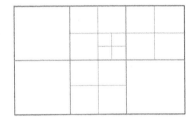

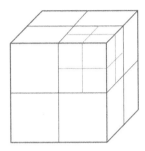

Figure 8.97 **Refinement with transition elements.**

8.5 MESHING VOLUME BOUNDED BY ANALYTICAL CURVED SURFACES

8.5.1 Introduction

There are MG problems for solid objects in which the boundary integrity requirement is less restricted to the geometrical aspect only without any topological constraints. Such meshing problems encompass a large class of objects including those scanned by MRI, biomedical parts or organs, objects bounded by analytical surfaces or functions defined over discrete surface patches in a CAD system. By adapting from an existing FE model, Couteau et al. (2000) introduced the mesh-matching algorithm to construct 3D meshes for anatomical structures, and meshes with guaranteed quality were produced by Ollivier-Gooch and Boivin (2001). While the AFT and the boundary recovery techniques of DT are systematic approaches, which work perfectly well for these objects, it is just unnecessary to introduce a boundary surface mesh to define the object prior to MG. A scheme whereby MG is possible without a boundary mesh or the boundary surface, which will be formed more or less automatically at the same time when the solid meshes are built, would be more efficient in terms of both computation time and manual data preparation.

The Meccano method is a recent attempt towards this goal (Cascon et al. 2009; Montenegro et al. 2009a,b, 2010). To mesh an object, a Meccano approximation is first introduced, which is an assembly of cuboids whose boundary surface provides a one-to-one correspondence (known as *admissible mapping*) to the boundary of the physical object to be meshed. The cuboids of the Meccano approximation are decomposed to hex elements based on the topological structure on the boundary surface, each of which is further divided into six tetrahedra to form an initial coarse tetrahedral mesh. The tetrahedral mesh is then refined mainly based on the boundary characteristics and the geometrical features of the physical object, though other mesh requirements can also be taken into consideration in the refinement. Nodes on the boundary of the Meccano approximation are mapped onto the boundary of the solid object by means of the admissible mapping defined earlier. By this mapping, the refined tetrahedral mesh will undergo a severe deformation such that many tetrahedral elements will be flattened, elongated or even inverted. Interior nodes have to be relocated to untangle the invalid elements and to open up the flattened or elongated elements. The final mesh has to be optimised by refinement and de-refinement along with some other smoothing procedures involving geometrical and topological operations.

The major difficulty with the Meccano method is the setting up of the admissible mapping for which manual intervention is necessary for objects with general geometry and/ or specific requirements. Poor quality elements will be generated by forcing the Meccano approximation onto the physical boundary of the solid object. Inverted elements have to be untangled, which is not always possible, and even though this could be done in a majority of cases, inevitably, elements of lower quality have to be accepted. The method relies heavily on the ability of the refinement and de-refinement techniques developed and the available optimisation procedures to improve the overall quality of the final mesh. The other restriction is that the method is limited to objects with the same topology of a cube in the current implementation, though objects with more complicated topological structures can be envisaged by the introduction of interior boundary surfaces; however, the mapping process has not been formally studied, and how elements will be deformed and corrected is still an open question.

With the introduction of a generic refinement algorithm in Section 8.4.3, high-quality refinement meshes can be readily generated in compliance with a specified node spacing function. Accordingly, a tetrahedral mesh can be generated with appropriate refinement following the boundary of the given solid object. The boundary of the object can be recovered simply by projecting nodes onto the boundary surface. Compared to the Meccano method, by projecting nodes close to object boundary, elements will suffer much less distortion, and meshes of much higher quality can be resulted. Based on the idea of refinement over object boundaries, Sullivan et al. (1997) employed the DT to mesh volumes of multiple material types, and Yu et al. (2008a,b) applied the mesh refinement and projection techniques for MG and adaptation in modelling molecular shapes and biomedical parts. By means of the Octree refinement techniques, Su et al. (2004), Qian et al. (2009) and Zhang et al. (2010) presented an MG scheme to produce hex and tetrahedral meshes over a domain with multiple material types. Using a Delaunay-based surface meshing algorithm, Oudot et al. (2010) generated tetrahedral meshes over volume bounded by curved surfaces. Qian and Zhang (2012) proposed a scheme for the generation of unstructured hex meshes for an object defined by CAD B-Rep surfaces. Following a bottom-up algorithm, Gosselin and Ollivier-Gooch (2011) triangulated volume bounded by piecewise smooth surfaces based on a controlled Delaunay point insertion on the boundary surfaces. Sazonov and Nithiarasu (2012) generated surface and volume meshes for biomedical objects by means of the marching cube method.

8.5.2 MG algorithm by refinement and boundary fitting

As tetrahedral mesh is more adaptive to curved surfaces, MG based on the refinement of a tetrahedral mesh rather than a hex mesh for objects bounded by analytical surfaces will be presented. The essential steps are given as follows.

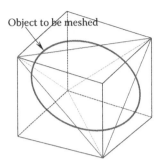

Figure 8.98 Object inside a cube.

8.5.2.1 Initial embedding mesh

The object to be meshed defined by a collection of boundary surfaces is put in a cube, which is divided into five or six tetrahedra, as shown in Figure 8.98.

8.5.2.2 Mesh refinement over object boundary

Mesh refinement algorithm presented in Section 8.4.3 is applied to the initial tetrahedral mesh following the object boundary surfaces with a mesh refinement compatible with the curvatures of the boundary surface and the size specifications, as shown in Figure 8.99.

8.5.2.3 Projection of nodes close to boundary surface

To recover the object boundary, nodes near the boundary have to be projected onto the boundary surface. The main issue here is to devise an objective measure to judge whether a point is close to the object boundary or not. Obviously, a distance measure will be problem-dependent and subject to the geometrical characteristics of the object under consideration. Instead of prescribing a specific threshold or tolerance, a more generic approach is to judge based on the shape quality of the elements associated with the projection. To this end, a node close to the object boundary will be projected if the minimum γ-quality of the patch of elements incident to the node is not inferior to that prior to the nearest point projection.

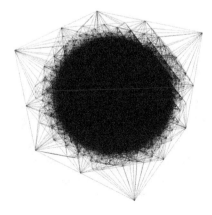

Figure 8.99 Mesh refined along object boundary.

8.5.2.4 Cutting of intersecting edges

Partly due to nodes not all being projected onto the boundary surface and partly due to the fact that the boundary surfaces in general may not be contained in the refined mesh (i.e. boundary surfaces are covered with triangular facets of the refined mesh), there are edges of the refinement mesh cutting across the boundary surface. To create a conforming mesh by getting rid of these intersecting edges, the edge bisection procedure described in Section 8.4.3.1 can be applied. To this end, the intersection point on the penetrating edge is determined, and the ring of tetrahedra around this edge will be divided by the introduction of an extra node at the intersection point. Effectively, an additional point is created at the boundary surface with faces of tetrahedra in good alignment with the boundary surface. Of course, intersection point too close to an end point of the edge will not be introduced, which indeed would rarely happen due to close point projection in Section 8.5.2.3. Anyway, a point too close to an existing point will not be created, and this situation will be dealt with in the phase of boundary point projection in Section 8.5.2.6. In the current formulation, intersection points within 5% to an end point of the cutting edge will not be created so as to prevent elements of poor quality being formed near the object boundary.

8.5.2.5 Elimination of elements not belonging to the object

By the procedures in Sections 8.5.2.3 and 8.5.2.4, the object boundary is virtually in place to within a few percentages depending on the actual situation. Tetrahedral elements outside the object boundary can be readily eliminated by checking their centres of gravity relative to the object boundary, as shown in Figure 8.100. The boundary surface of the same object would be much more irregular with an indentation of half an element size on the average if boundary preparation procedures in Sections 8.5.2.3 and 8.5.2.4 have not been applied, as shown in Figure 8.101. It is noted that further mesh refinement will only reduce the size but not the pattern of boundary irregularity.

8.5.2.6 Boundary point projection

With the elimination of elements outside the object, the boundary of the meshed object is formally defined as a result. The boundary nodes can now be identified, which are vertices belonging to one of the boundary triangular facets. As a final step, the boundary nodes are all projected onto the analytical boundary surfaces to further match the geometry between

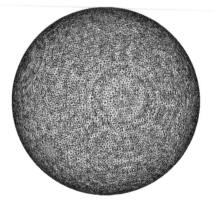

Figure 8.100 Elements within boundary retained.

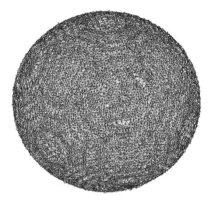

Figure 8.101 **Mesh with rugged boundary.**

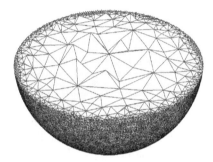

Figure 8.102 **A cross-section view.**

the tetrahedral mesh and the object. Shape optimisation operations of Chapter 6 can then be applied to the resulting mesh with fixed boundary nodes not subject to any nodal movement imposed by the smoothing procedures. A cut section of the FE mesh after optimisation is shown in Figure 8.102 in which elements of relatively small size are found only at the boundary of the object. Some more examples are given in Section 8.5.3 to illustrate the performance and the characteristics of the MG algorithm by refinement.

8.5.3 Example of mesh adaptation by refinement

The first example is to mesh a sphere of a radius of 5 units (Table 8.21). The spherical surface is embedded in a cube of five tetrahedra, which is refined into 398,494 elements with the smallest size of about 0.1 unit, as shown in Figure 8.99. According to the formulas of geometrical control given in Section 4.2.5, the gap-to-radius ratio $\varepsilon = \delta/\rho$ is less than 0.005%. A total of 11,734 nodes close to the boundary of the sphere are projected onto the spherical surface, and 47,923 penetrating edges along with their associated tetrahedral elements are divided at the intersection points. Elements outside the spherical surface are eliminated, and it turns out that 329,455 tetrahedra are retained, of which the mean γ-quality is 0.5713, as shown in a cross-section view in Figure 8.102.

The MG time by refinement, node projection and cutting of intersecting edges excluding mesh optimisation only takes 0.8 s on a PC i7 CPU 870 at 2.93 GHz running on XP mode for a mesh of about 400,000 elements. As small elements are formed around the boundary surface of the sphere coupled with boundary node projection and division of

Table 8.21 Statistics of MG by refinement

Example	Sphere	Ellipsoid	Two ellipsoids	Internal void
Nodes in the refined mesh	74,254	65,203	76,425	73,517
Elements in the refined mesh	398,494	351,877	412,703	396,461
Nodes projected on surface	11,734	10,502	12,539	18,198
Intersecting edges divided	47,923	24,307	29,011	12,515
Boundary node projection	40,186	22,959	26,755	21,324
Elements in the object	329,455	250,158	278,860	227,769
γ-quality of the object	0.5713	0.6318	0.6224	0.6458
CPU time for MG	0.80 s	0.60 s	0.79 s	0.68 s
Discrepancy in volume	0.018%	0.02%	Not computed	0.003%

intersecting edges, the difference in volume between the FE mesh and the sphere is pretty small at 0.018%.

The second example is an ellipsoid with semi-principal axes of 5, 4 and 3 units along x, y and z directions, respectively. The initial tetrahedral mesh is refined such that elements are of different sizes in compliance with the curvatures of the boundary surface, such that the smallest elements are of size 0.025 unit around the apex of the major axis. A total of 65,203 nodes and 351,877 elements are generated in the refined mesh, as shown in Figure 8.103a. After nodal projections and division of penetrating edges, tetrahedral elements not belonging to the ellipsoid are removed leaving 250,158 tetrahedral elements within the oval surface, as shown in Figure 8.103b and c. The γ-quality of the final mesh is 0.6318, and the CPU time take for MG is 0.6 s. There is a discrepancy of 0.02% in the volume between the FE mesh and the physical object.

The third example is the MG of the same ellipsoid, but this time, a spherical opening of radius 2 units is introduced by including an interior boundary surface within the ellipsoid. The initial mesh is refined over both the exterior and interior boundary surfaces into 412,703 tetrahedral elements. Exactly the same node projection and edge division procedures are applied

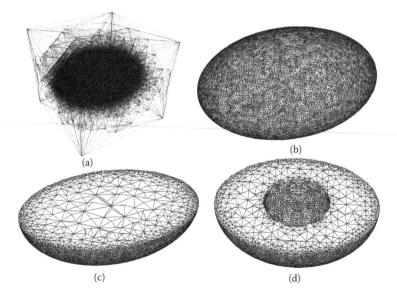

(a)

(b)

(c)

(d)

Figure 8.103 Meshing ellipsoids with and without internal opening: (a) refinement over boundary surfaces; (b) remove elements outside surfaces; (c) section view of ellipsoid; (d) ellipsoid with an internal opening.

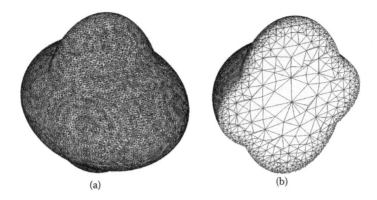

(a)   (b)

Figure 8.104 Mesh of volume bounded by two ellipsoids: (a) entire mesh; (b) half section.

to the two boundary surfaces. Upon removal of elements outside the object, 278,860 tetrahedra are left in the volume within the boundary surfaces, as shown in Figure 8.103d. The last example is a volume defined by two analytical surfaces in which another ellipsoid with semi-principal axes of 2, 3 and 5 units along x, y and z directions is added to the ellipsoid employed in the second example. The initial tetrahedral mesh is refined on the ellipsoidal surfaces into 396,461 elements. On removal of elements outside the object, an FE mesh of 227,769 tetrahedral elements is generated with a γ-quality of 0.646, as shown in Figure 8.104a. A half mesh cross section revealing the internal structure of the mesh is shown in Figure 8.104b.

MG by refinement is a versatile and efficient scheme in meshing geometrical objects bounded by analytical surfaces. As FEs only need to fit the geometrical boundary of the object, the algorithm is applicable to objects of general topological shapes, simply or multi-connected, with or without internal openings, bounded by one or several surfaces, and the object may even be disjointed into a number of pieces. Exactly the same boundary rectification steps of node projection and edge division can be applied to recover objects of the most general shapes and topology. The computation time mainly depends on the number of elements in the mesh more or less linearly but not on the geometry or the topology of the object. As shown by the examples, meshing a sphere using the same number of elements takes roughly the same CPU time as meshing a volume bounded by two surfaces or with an internal void. Though more evidence and an in-depth study are required, the quality of the mesh seems to depend on the number of penetrating edges on the boundary surfaces as there is little difference in the quality of the initial refined mesh. Anyway, though the proposed boundary treatment is simple enough, there still is much room for improvement of the quality of the elements on the boundary through the use of some more sophisticated techniques. In theory, volumes of different material types with internal partition surfaces can also be meshed as long as the regions along with the internal surfaces within the object are well defined and represented.

8.6 MERGING OF TETRAHEDRAL MESHES

8.6.1 Introduction

Nowadays, advanced industrial production often systematically goes through the cycle of conception, CAD geometrical modelling, data preparation for simulation and meshing, FE analysis and optimisation. For a proper maintenance of the products and possible upgrades for additional features, another round of simulation and refinement has to be carried out to meet the new requirements. However, for the sake of rapid turnaround, it may not be a good idea to go right back to the early conceptual stage, as after cycles of simulation, modification and update, the FE mesh might have a substantial departure from the CAD model, or simply the CAD model is already outdated or not even available at all for one reason or another. Furthermore, very often the new demands would require just some minor adjustments from the original design, and a few modifications of the existing FE mesh will produce a new model for engineering simulation addressing all the changes required.

Lou et al. (2010) put forward an FE merging scheme to enrich FE triangular meshes for rapid prototyping of alternate solutions for industrial maintenance. On the introduction of a feature to a triangulated surface mesh, a cavity at the vicinity of the new feature will be created in the existing surface mesh, where triangular elements could be generated to fill up the gap linking the two parts together. Another possibility to combine two surface parts is to consider the intersection of the given surfaces (Lo 1995). The interaction of curved surfaces discretised into triangles and/or quadrilaterals will produce intersection segments, which could always be grouped into structural loops for analysis and further manipulation. Triangles intersected by the intersection loops are removed to produce gaps between the intersection loops and the curved surfaces. By filling up the void with triangles, all intersection loops will be incorporated into the given surfaces. The surfaces will be automatically merged together as all intersection segments exist on both of the given surfaces to ensure compatibility. This method has been proved to be versatile and efficient in generating new surface meshes from existing discretised objects (Shostko and Lohner 1999; Lo and Wang 2003, 2004).

In response to the demands for maintenance and update of products in service similar to those of the triangular surface meshes, there is an urgent need to develop automatic algorithms to merge solid tetrahedral meshes. Apart from the aforementioned needs for a versatile solid mesh merger, such a tool could also offer the following applications.

1. Handling minor modifications and mesh update of existing meshed objects
2. Working as a supplementary MG tool for objects that could only be conveniently defined by a combination of some fundamental shapes put together
3. Incorporating foreign parts into an existing meshed object (Mouton et al. 2010)
4. Putting existing meshes together in a variety of ways to create new models
5. Opening up a way to merge other solid meshes that can be converted into tetrahedral meshes

Not much research work has been reported in the literature so far in the development of such an automatic scheme in merging arbitrary tetrahedral meshes (Yamakawa and Shimada 2009), although there are robust algorithms proposed for the merging of triangulated surfaces (Cebral et al. 2002; Lo and Wang 2005a) and the insertion of surface mesh into a tetrahedral mesh (Ebeida et al. 2009; Cuillière et al. 2010). The major difficulty lies in the finding of a reliable way for the determination of intersections and in formulating a systematic plan in filling up any cavity to accommodate all the intersection parts. For a consistent treatment of intersections between two solid objects, we have to concentrate on their patterns and structural form. In the algorithm to be presented in this section, rather than focusing on the intersection points and segments as individual components, we aim at finding distinct non-overlapping intersection loops as a single entity on the boundary of the 3D objects under consideration. Instead of micro-management in linking up broken intersected tetrahedral elements one by one, the intersection parts will be recorded and treated as volumes of intersection. In terms of regions of intersection, we could exercise a tight control and have a much better understanding on the geometry and the topology of the intersection parts. Apart from the determination of intersection loops, which is an integral part for the interaction of two meshed objects, the operations involved for the identification and the definition of the regions of the intersection are purely topological, which could be crucial to enhancing the stability of the entire merging process.

A generic algorithm will be presented in Section 8.6.2 to merge arbitrary solid tetrahedral meshes automatically into one single valid finite element mesh (Lo 2013c). The intersection segments in the form of distinct non-overlapping loops between the boundary surfaces of the given solid objects are determined by the neighbour-tracing technique discussed in Section 4.5.4. Each intersected triangle on the boundary surface will be triangulated to incorporate the intersection segments onto the boundary surface of the objects. Tetrahedra on the boundary surface with intersected triangular facets as one of the faces are each divided into as many tetrahedra as the number of sub-triangles on the triangulated face. There is a natural partition of the boundary surfaces of the solid objects by the intersection loops into a number of zones. Volumes of intersection can now be identified by the collected bounding surfaces from the patches of the surface partition. While mesh compatibility has already been established on the boundary of the solid objects, mesh compatibility has yet to be restored on the bounding surfaces of the regions of intersection. Tetrahedral elements intersected by the cut surfaces are removed, and new tetrahedra can be generated to fill the resulting cavity to restore mesh compatibility at the cut surfaces. Upon restoring mesh compatibility over volumes of intersection, the objects are ready to be combined as all regions of intersection can now be detached freely from the objects. All operations, besides the determination of intersection loops, are entirely topological, and no parameter and tolerance is needed in the entire merging process. The examples of various characteristics are presented to show the steps and the performance of the mesh-merging procedure.

8.6.2 Algorithm: Merging tetrahedral mesh

Problem statement: Given two 3D objects Ω and $\bar{\Omega}$, each of which is a valid finite element mesh of tetrahedral elements such that $\Omega = \{T_i, i = 1, N_T\}$ and $\bar{\Omega} = \{\bar{T}_i, i = 1, N_{\bar{T}}\}$, where N_T and $N_{\bar{T}}$ are, respectively, the number of tetrahedral elements in Ω and $\bar{\Omega}$. The merging problem is to find the union of the two given meshes into a single valid finite element mesh $\hat{\Omega}$ consisting of only tetrahedral elements, i.e. $\hat{\Omega} = \Omega \cup \bar{\Omega} = \{\hat{T}_i, i = 1, N_{\hat{T}}\}$, such that points found in Ω or $\bar{\Omega}$ are found in $\hat{\Omega}$ as well.

$$\forall i \in \{1,2,...N_T\} \text{ and } \forall j \in \{1,2,...N_{\bar{T}}\}, \; x \in T_i \text{ or } x \in \bar{T}_j \Rightarrow x \in \hat{T}_k \quad \exists k \in \{1,2,...N_{\hat{T}}\}$$

The objects to be dealt with are very general as long as they are valid finite element meshes, and there is no restriction on the size, shape and the topology of the objects, which could be convex, concave, simply or multi-connected, etc. The algorithm proposed can even handle objects that are disjoint; however, without loss of generality, we simply assume that objects are connected pieces, since Boolean operations can be applied sequentially between different intersecting objects to form ever more complex objects. To merge two objects together, it is necessary to consider their intersections, which by definition are the parts common to both objects. Hence, the union of the two objects/meshes is given by the objects putting together subtracting the parts in common from either one of the objects. For a robust treatment of the merging of solid objects, the nature and the fundamentals of intersection have to be carefully reviewed. The boundary of a meshed 3D object is a closed discretised surface of triangular facets. The intersections between two objects are regions common to the two objects bounded by boundary surface parts from the two objects. In other words, the intersection between two solid objects can be fully determined by considering solely the boundary surfaces of the two objects. This idea is crucial to the mesh-merging algorithm to be presented.

Intersections between two meshed objects or their boundary surfaces are intrinsic to their geometry and topology, even though their determination is based entirely on geometrical computations between line segments and facets. The topological features of intersection thus would not be affected by the method of its determination, or by a change of co-ordinates, or by any continuous mapping such as bending, twisting and scaling. As a result, in the determination of intersection between two objects, it may not be a good idea to take intersection as discrete information of intersection points and line segments, as finite precision arithmetic will often produce inconsistent results. Instead, we have to focus on the pattern and form of intersection as a whole with reference to the geometrical characteristics for their determination. Hence, for the intersection between discretised curved surfaces, we aim at getting surface patches bounded by loops of intersection segments, and for the intersection between two solid objects, we would like to see volumes or regions of intersections bounded by surface patches derived from the boundary of the two objects under consideration. In summary, the boundary of the regions of intersection is composed of surface patches of intersection, whose boundary, in turn, is given by the intersection loops, and that's why intersection loops are fundamental to the mesh-merging process.

The following are the steps of the mesh-merging algorithm:

1. Determine all the intersection loops between the boundary surfaces of the given meshed solid objects by means of the neighbour-tracing method introduced in Section 4.5.4.
2. Perform 2D triangulation on all the intersected triangular facets on the boundary so that intersection segments are incorporated on the boundary surfaces of the given objects.
3. Each tetrahedral element with a triangulated boundary facet is divided into as many tetrahedra as the number of triangles on the triangulated face. As all the sub-triangles on the triangulated face are visible to the opposite node, this subdivision exists and is simple to carry out.
4. As each intersection segment is present as an edge of a tetrahedron on the boundary surfaces of the given objects, the intersection loops will partition the boundary surface of the objects into a number of zones.
5. Regions or volumes of intersection could be identified and defined by collecting bounding surfaces from intersection loops and surface patches of the boundary surface partition.
6. Incompatible tetrahedra intersected by the bounding surfaces of all the regions of intersection are removed, and mesh compatibility is restored by filling up the void with new tetrahedra between the bounding surfaces and the cavities so created.
7. All the regions of intersection common to both objects can now be detached freely from the solid meshes, and the two solid meshes are now ready to be merged with full compatibility at all intersecting surfaces.

8.6.2.1 Intersection of boundary surfaces

Let Ω and $\bar{\Omega}$ be the given 3D solid objects discretised into tetrahedral meshes, simply or multi-connected with or without internal openings. In a broader sense and within the scope of the meshing algorithm, Ω and $\bar{\Omega}$ could be extended to include a collection of solid objects, even though it is general enough to consider the special case of a single object in the two sets Ω and $\bar{\Omega}$. The boundaries of Ω and $\bar{\Omega}$ are closed surfaces denoted, respectively, by $\partial\Omega$ and

$\partial\bar{\Omega}$. To determine the intersection between Ω and $\bar{\Omega}$, it is much simpler to concentrate on their boundary, and it turns out that it is also sufficient to determine the regions or volumes of intersection by considering solely the intersection of their boundaries, i.e. $\partial\Omega$ and $\partial\bar{\Omega}$.

When $\partial\Omega$ interacts with $\partial\bar{\Omega}$, the result is a finite number of closed loops of line segments, as shown in Figure 8.105. The loops are non-intersecting and distinct as boundary surfaces are smooth and closed, such that valid intersections on the boundary surface can always be represented by loops of intersection segments. Even in the worst case where loops sharing a common point as in the case of the intersection of cylindrical surfaces of equal diameter, the situation could be easily resolved using the non-intersecting principle by accepting an intersection loop composed of two half circles as shown in Figure 8.106.

The intersection loops due to the intersection of the two boundary surfaces can be conveniently determined in a robust manner by the method of neighbour tracing making the best use of the property of continuity of intersection lines, as discussed in Section 4.5.4. Along each intersection line segment, the intersecting triangles on each surface are neighbours to each other to form a closed chain, as shown in Figure 8.107. From the chain of triangles that contain all the intersection segments, by means of the neighbouring relationship, an intersection loop can always be constructed by tracing neighbouring triangles one after the other. The reliability of the method is greatly enhanced as intersection can be locally controlled and guided by the adjacency relationship, and a clear direction could be specified as how an intersection loop should progress from one triangular facet onto the next. Most importantly, the neighbour-tracing method ensures that a closed loop of intersection could always be defined on a closed surface starting from any intersection segment between a pair of intersecting triangles. As the neighbour-tracing process is carried out

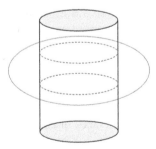

Figure 8.105 Intersection between a sphere and a cylinder.

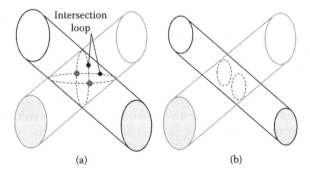

Figure 8.106 Intersection of cylindrical surfaces: (a) equal diameter; (b) unequal diameter.

Figure 8.107 Intersection segments form a closed loop, and intersected triangles are neighbours to each other.

simultaneously on the two boundary surfaces, the resulting intersection loop could be recorded on the two boundary surfaces without ambiguity for later cross-referencing and identification.

8.6.2.2 Incorporating intersection loops into meshes Ω and $\bar{\Omega}$

The intersection segments have to be incorporated into meshes Ω and $\bar{\Omega}$ so that compatibility could be established on the boundary of Ω and $\bar{\Omega}$. Boundary surface compatibility is a crucial step for mesh compatibility between tetrahedral elements in Ω and $\bar{\Omega}$, which will be further elaborated in Sections 8.6.2.2 to 8.6.2.5. Intersection segments could be readily introduced into a tetrahedral mesh in a consistent manner if we take a magnified view on the face of each tetrahedral element being intersected. Tetrahedron ABCD is intersected on the boundary facet ABC, as shown in Figure 8.108, where intersection segments are marked with thicker lines. The 2D triangular domain ABC can be easily triangulated into a triangular mesh with all the intersection segments kept intact in the process. Once all the intersected triangular facets are triangulated, all the intersection segments will be present as edges of the triangulated boundary surfaces of the objects. As the meshing of triangular facets with a couple of interior line segments is a pretty simple task, in the current implementation, the ADF method is adopted. What we have to do is to take the triangulation of the facets as a mini-scale meshing problem on a planar domain. The boundary of the triangular facet can be entered segment by segment following the counter-clockwise convention for exterior boundary, and as for the interior intersection line segments, they could be entered as internal constraints, i.e. for line segment PQ, it is presented as two directed line segments PQ and QP. As shown in Figure 8.108, the exterior boundary segments along with the interior line segments automatically partition the triangular facets into three separate regions for meshing.

Each triangle on the boundary face ABC is visible to node D, and a natural division exists for each triangle to form a tetrahedron with node D. As a result, tetrahedron ABCD is divided into as many tetrahedra as the number of triangles on the triangular facet ABC, with intersection segments on the face ABC being kept intact as part of the tetrahedral mesh. Such a process could be carried out independently and successively for each intersected tetrahedral element in turn. When such a 2D triangulation process is done on all the intersected tetrahedral elements in Ω and $\bar{\Omega}$, the intersection segments (loops) will be

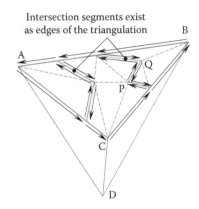

Intersection segments exist
as edges of the triangulation

Figure 8.108 Triangulating facet ABC.

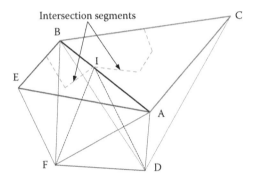

Intersection segments

Figure 8.109 Tetrahedra ABCD and ABFE are not direct neighbours.

incorporated into the tetrahedral meshes, and boundary surface compatibility is established as each intersection segment is present in the resulting modified tetrahedral meshes Ω and $\bar{\Omega}$.

While compatibility is established on the boundary surfaces, compatibility between intersected tetrahedral elements has to be checked. As shown in Figure 8.109, the adjacent intersected tetrahedral elements on the boundary surface may not be direct neighbours of each other. Referring to Figure 8.109, ABCD and ABFE are adjacent intersected tetrahedral elements on the boundary surface. However, tetrahedral elements ABCD and ABFE are not direct neighbours to each other, and compatibility could not be maintained in tetrahedral element ABDF, which also shares the common line segment AB. To restore compatibility, simply divide each tetrahedral element between tetrahedra ABCD and ABFE sharing the common line segment AB into two tetrahedral elements, in which ABDF is divided by intersection point I at edge AB into tetrahedra AIDF and IBDF. Such a process is a standard procedure in the optimisation and refinement of tetrahedral meshes, as shown in Figure 8.110.

8.6.2.3 Volume (region) of intersection

Based on the method of neighbour tracing in the determination of the intersection between two boundary surfaces, as described in Section 8.6.2.1, intersections are represented as a

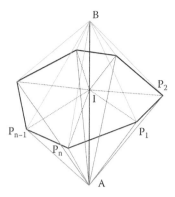

Figure 8.110 Dividing edge AB at I.

finite number of intersection loops. Let L be the set of loops of intersection between $\partial\Omega$ and $\partial\bar{\Omega}$, i.e. $L = \{\ell_1, \ell_2, ..., \ell_{N_L}\}$ such that $\ell_i \in \partial\Omega$ and $\ell_i \in \partial\bar{\Omega}$ for $1 \leq i \leq N_L$, where N_L is the number of intersection loops between $\partial\Omega$ and $\partial\bar{\Omega}$. It is remarked that Ω and $\bar{\Omega}$ are the modified tetrahedral meshes, which have undergone the surface triangulation, subdivision of intersected tetrahedral elements and restoration of compatibility for adjacent triangulated tetrahedral elements on the boundary, as described in Section 8.6.2.2.

The set of intersection loop L will partition naturally each of the boundary surfaces into a number of zones of surface patches, as shown in Figure 8.111, such that $\partial\Omega = \bigcup\limits_{i=1}^{N_\Omega} S_i$ and $\partial\bar{\Omega} = \bigcup\limits_{i=1}^{N_{\bar\Omega}} \bar{S}_i$ where S_i and $\bar{S}_i$ are, respectively, surface partitions in $\partial\Omega$ and $\partial\bar{\Omega}$ bounded by intersection loop(s); N_Ω and $N_{\bar\Omega}$ are, respectively, the number of such surface patches in the partition of $\partial\Omega$ and $\partial\bar{\Omega}$.

Based on the surface-neighbouring relationship, the surface partition by intersection loops can be easily carried out by a pure topological process. Once we have the intersection segments incorporated into the tetrahedral meshes Ω and $\bar{\Omega}$, take any triangular facet on the boundary surface, the patch will grow from this seed triangle by attaching more triangles to it (by means of the adjacency relationship) until the entire patch is bounded by intersection segments. Such a patch of triangles will be given a label and recorded as an individual

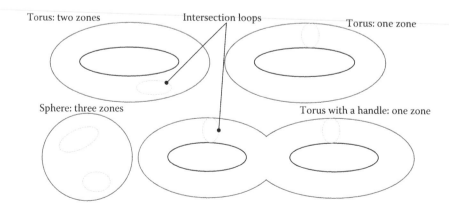

Figure 8.111 Boundary surfaces partitioned into zones by intersection loops.

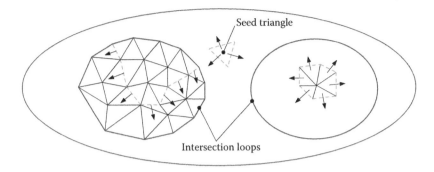

Figure 8.112 Partition into zones by marching towards the bounding loops.

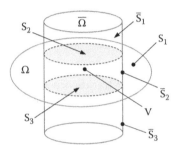

Figure 8.113 Region of intersection bounded by surface patches.

surface patch, and each triangle in this patch will be given the same surface label. The same procedure can be repeated when there are triangular elements remaining on the boundary surface unlabelled. Again, any remaining triangular element could be taken as the seed element to grow into another surface patch. When all the triangular facets on the boundary surfaces $\partial\Omega$ and $\partial\bar{\Omega}$ have been processed, each of the boundary surface will be partitioned into a number of surface patches by the intersection loops, as shown in Figure 8.112.

With the introduction of boundary-surface partition by intersection loops into patches S_i and $\bar{S}_i$, the volume of intersection between Ω and $\bar{\Omega}$ can be conveniently identified and defined. Let V be one of the regions of intersection between Ω and $\bar{\Omega}$; V is bounded by surface patches S_i and $\bar{S}_i$. As shown in Figure 8.113, V is bounded by surface patches S_2, S_3 and $\bar{S}_2$ from $\partial\Omega$ and $\partial\bar{\Omega}$. Objects Ω and $\bar{\Omega}$ can be merged together by ensuring that each intersection region V is compatible at the boundary-surface patches.

8.6.2.4 Identification of intersection volumes (regions)

The identification of regions of intersection between two tetrahedral meshes is based on the observation that each intersection region is bounded by surface patches S_i and $\bar{S}_i$ from $\partial\Omega$ and $\partial\bar{\Omega}$ separated by intersection loops. Starting from intersection loop, $\ell \in L$, determine the two surface patches S_i and $S_j \subset \partial\Omega$ attached to ℓ. As intersection loops are boundary lines of penetration from one volume into another, identify between S_i and S_j the cutting surface into mesh $\bar{\Omega}$. Let S_i and $S_j \subset \partial\Omega$, $\bar{S}_i$ and $\bar{S}_j \subset \partial\bar{\Omega}$ be the surface patches connected to intersection loop ℓ, as shown in Figure 8.114, where arrows are pointing towards the volume interior. S_i is a cut surface if it penetrates into $\bar{\Omega}$, for which dihedral angle θ between triangles BAD and ABE following the rotation axis BA is smaller than angle ϕ between triangles BAD and ABC following the same rotation axis BA.

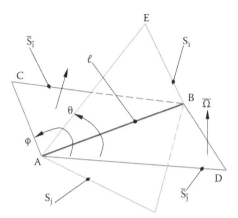

Figure 8.114 Determination of cutting surfaces.

Whether a triangular facet connected to an edge is inside a volume or not can be determined by considering the volumes of the tetrahedra so formed, as shown in Figure 8.114, where arrows are pointing towards the interior of the object. Let V_1 and V_2 be the volumes of tetrahedra ABCE and BADE, i.e. $V_1 = V(A, B, C, E)$ and $V_2 = V(B, A, D, E)$. If both V_1 and V_2 are positive, then face ABE is between faces ABC and BAD, or E is inside the volume bounded by faces ABC and BAD. On the contrary, if V_1 and V_2 are both negative, then point E is outside of the volume bounded by faces ABC and BAD. However, if either V_1 or V_2 is positive and the other negative, then the volume $V = V(A, B, C, D)$ has to be checked. If V is positive, E is outside, and on the other hand, if V is negative, E is inside the bounded volume.

As each loop connects to exactly two cutting surfaces, one from $\partial\Omega$ and the other from $\partial\overline{\Omega}$, between S_i and $S_j \subset \partial\Omega$ connected to loop ℓ, suppose S_k is the cutting surface into mesh $\overline{\Omega}$, and similarly, $\overline{S}_{\overline{k}}$ is the cutting surface in $\partial\overline{\Omega}$ connected to ℓ cutting into volume Ω. Check if there are any more intersection loops connected to surface patches S_k and $\overline{S}_{\overline{k}}$, and include all these loops and surface patches to this region until no more surface patches and intersection loops could be collected for this region. This loop and surface patch collection process is well defined and will be completed rapidly as each intersection loop and each cutting surface belong to one and only one volume of intersection. The collection process is repeated when there are intersection loops remaining untreated without being assigned to any region of

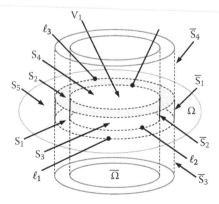

Figure 8.115 Intersection between hollow cylinder and ellipsoid.

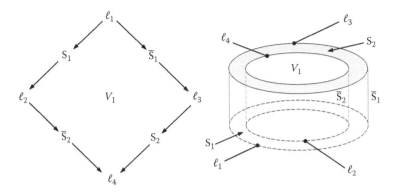

Figure 8.116 **Topology between intersection loops and cut surfaces.**

intersection. The collection process terminates when all the intersection loops are dealt with and being given a region label.

For example, consider the intersection of an ellipsoid and a hollow cylinder, as shown in Figure 8.115. There are four intersection loops $\{\ell_1, \ell_2, \ell_3, \ell_4\}$, whereas the boundary of the sphere $\partial\Omega$ is partitioned into five patches $\{S_1, S_2, S_3, S_4, S_5\}$, and the boundary of the hollow cylinder $\partial\bar{\Omega}$ is partitioned into four zones $\{\bar{S}_1, \bar{S}_2, \bar{S}_3, \bar{S}_4\}$. Starting with intersection loop ℓ_1, cut surfaces S_1 and $\bar{S}_1$ are identified. ℓ_2 connected to S_1 and ℓ_3 connected to $\bar{S}_1$ are now included. Following ℓ_2, $\bar{S}_2$ is collected, and as for ℓ_3, we arrive at S_2. Finally, S_2 and $\bar{S}_2$ close the boundary of the region meeting at loop ℓ_4, and the entire region formation process is depicted in Figure 8.116. As all the intersection loops have been considered and dealt with, the intersection loops and the cutting surfaces constitute only one single region of intersection, and the process of region identification terminates. From Figure 8.116, it is obvious to see that apart from loop ℓ_1, the recovery of region V_1 might as well be done starting with any other intersection loop belonging to the region.

8.6.2.5 Mesh compatibility

The final step in the merging process is to ensure mesh compatibility at the bounding surfaces of all the regions of intersection. By the region construction process, mesh compatibility is already achieved at the intersection loops and on some bounding surface patches depending on which object is taken as the master object and which is taken as the slave object. Of course, any of the two objects Ω and $\bar{\Omega}$ could be taken as the master or the slave, and the choice is rather arbitrary as master and slave are just names to denote two interacting objects. In case there is no particular preference to make such a choice, usually it is more advantageous to select the more complicated object as the master and the less complicated object as the slave, as more mesh modifications would be applied to the slave object so as to enable matching with the master object at the intersection interfaces. To obtain the best results at the expense of extra computations, a second mesh merging can as well be carried out by interchanging the master object with the slave object, and from these two possibilities, the one that gives a more promising result overall will be selected as the final merged mesh, as shown in Example 1. As a convention, Ω represents the master object, and $\bar{\Omega}$ represents the slave object. As shown in Figure 8.117, there is only one region of intersection, which is bounded by S_1 and S_2 from the master object Ω and $\bar{S}_1$ from the slave object $\bar{\Omega}$. Taking the master object Ω as the underlying mesh, mesh compatibility is already achieved

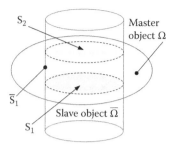

Figure 8.117 **Master object and slave object.**

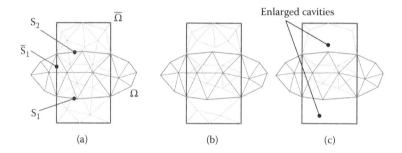

Figure 8.118 **Restoring mesh compatibility by modifying the slave object: (a) cross-sectional view; (b) tetrahedra intersect with S_1 and S_2 are removed; (c) cavity enlarged by deleting more tetrahedra.**

on $\bar{S}_1$, and the slave object can be merged with the master object if mesh compatibility is also restored on S_1 and S_2, which cut across the slave object $\bar{\Omega}$. The situation can be visualised from a cross section of a longitudinal cut of the two objects, as shown in Figure 8.118a. To restore mesh compatibility on cut surfaces while keeping the master object intact, it is necessary to modify the slave object.

The slave object $\bar{\Omega}$ can be modified according to the following steps:

1. All tetrahedral elements in $\bar{\Omega}$ that intersect with S_1 and S_2 are removed. This will leave a cavity, which may be connected or disconnected in the vicinity of cut surfaces S_1 and S_2, as shown in Figure 8.118b.
2. The cavity is enlarged by further removing tetrahedra in $\bar{\Omega}$, which are connected to the initial cavity.
3. The boundary of the cavity in $\bar{\Omega}$ along with cut surfaces S_1 and S_2 will define tiny volumes between Ω and $\bar{\Omega}$, as shown in Figure 8.118c.
4. Mesh compatibility between Ω and $\bar{\Omega}$ can be restored by filling in the volumes so created between Ω and $\bar{\Omega}$ with valid tetrahedral elements, as shown in Figure 8.119.

On the contrary, we can as well take the cylinder $\bar{\Omega}$ as the master object and the ellipsoid Ω as the slave object. In this case, mesh compatibility is already achieved on S_1 and S_2. To restore mesh compatibility on surface $\bar{S}_1$, we have to modify the ellipsoid by removing all the tetrahedral elements in the ellipsoid intersected with $\bar{S}_1$. Again, to create a larger cavity, we further remove the layer of tetrahedra that are connected to the intersected tetrahedral elements. The resulting cavity is of the shape of a cylindrical surface similar to $\bar{S}_1$, as shown in Figure 8.119a. Advanced MG algorithms such as the ADF method, boundary recovery for DT with the minimisation of Steiner points and meshing software in the public domain (Si

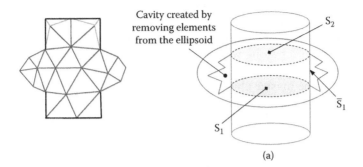

Figure 8.119 Mesh compatibility restored. (a) Cylinder = master and sphere = slave.

2007; Geuzaine and Remacle 2009) are available for the generation of tetrahedral meshes conformable with specified triangulated boundary surfaces.

8.6.2.6 Merging of tetrahedral meshes

A region of intersection is bounded by cut surfaces, which are patches on the boundary surface bounded by intersection loops. Each surface patch is given a unique surface label, and all the triangular facets belonging to this patch will be given the same surface label or zone number. All the triangular facets from $\partial\Omega$ and $\partial\overline{\Omega}$ bearing the zone number corresponding to the bounding surfaces of the region are collected. As shown in Figure 8.120a,

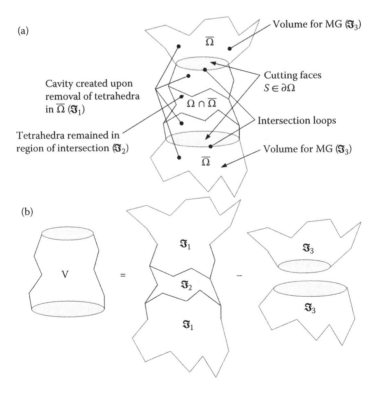

Figure 8.120 Restore compatibility by local MG: (a) compatibility restored by local MG; (b) region of intersection is detached.

each intersection loop, which is also the boundary of a cut surface, will divide the boundary surface of the cavity into two parts. The part of the boundary surface that is consistent in orientation with the cut surface defining a volume will be selected. The triangles on the cut surfaces of the master object along with triangles on the boundary of the cavity created in the slave object will form tiny volume(s) for the generation of tetrahedral elements, so that compatibility between the master object and the slave object could be restored.

In the current implementation, the ADF approach is adopted for the generation of tetrahedral elements over the bounded cavity. Following a consistent orientation (surface normals all pointing outwards), the triangles so collected from the master cut surfaces and the cavity created in the slave object are input to the 3D mesh generator. On the completion of local MG to fill up the voids, the region of intersection V can be expressed by $V = \mathfrak{I}_1 + \mathfrak{I}_2 - \mathfrak{I}_3$, as shown in Figure 8.120b, where $\mathfrak{I}_1$ is the set of tetrahedra in the slave object intersected by the master cut surfaces and those that are further removed to enlarge the cavity, $\mathfrak{I}_2$ is the set of tetrahedra in the slave object that are remaining inside the region of intersection (those tetrahedra can be easily identified by means of the neighbouring relationship) and $\mathfrak{I}_3$ is the set of tetrahedra generated to fill the cavity created in the slave object.

Region V is common to both Ω and $\bar{\Omega}$, and it is also compatible to $\bar{\Omega}$ on the cut surfaces after MG within the slave object $\bar{\Omega}$ on the side of right orientation with reference to cut surfaces from Ω. Hence, region V can be readily subtracted from $\bar{\Omega}$. To form a merged object from Ω and $\bar{\Omega}$, simply subtract V from the union of the two objects, i.e.

$$\Omega \cup \bar{\Omega} = \Omega + (\bar{\Omega} - V)$$

In the event that we have more than one region (volume) of intersection, simply subtract all those volumes of intersection from the slave object $\bar{\Omega}$, i.e.

$$\Omega \cup \bar{\Omega} = \Omega + (\bar{\Omega} - V_1 - V_2 - ... - V_n)$$

where n is the number of regions of intersection between Ω and $\bar{\Omega}$.

Remarks on numerical stability: The only numerical (geometrical) computation in the entire mesh-merging process is the determination of the intersections between two boundary surfaces. In merging realistic objects, minute intersection segments, which are true results of the actual geometrical feature, as shown in Figure 8.121, rarely occur, which should be dealt with and handled by a separate process. On the other hand, if the short intersection segments are due to the meshes used such that nodes accidentally (unnecessarily) come close to each other, then a simple reallocation of the relevant nodes on the boundary surface could

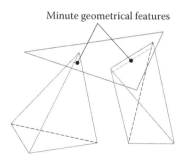

Figure 8.121 **Short intersection segments due to minute geometrical features.**

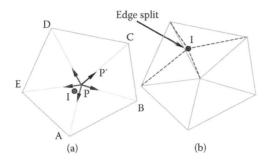

Figure 8.122 Avoiding short intersection segments by relocating nodal point P and edge splitting: (a) node relocation; (b) edge splitting.

much alleviate the situation. As shown in Figure 8.122, node P is shifted away from the intersection point I at which short intersection segments are found. Usually, short intersection segments could be removed by shifting mesh nodal points on the boundary surface away from intersection points, while keeping the geometry of the object intact.

As shown in Figure 8.122a, P is the centre node of a patch of surface triangles (polygon ABCDE). Node P is allowed to shift by, say, 10% along the edge PA, PB, PC, PD and PE in turn. The position of P' is taken as the point farthest away from all the intersection points on the patch. In case no improvement could be made, the shift is reduced progressively from 10% to 8% and 6% and so forth, until a satisfactory position could be found. Another possibility is that the intersection point I is away from all the nodal points but falls close to an edge, which can be easily dealt with by splitting the edge concerned, as shown in Figure 8.122b.

8.6.3 Examples

In this section, there are five examples of various characteristics and complexity, from a simple academic example to illustrate the basic theory and the implementation to the more complicated case of the interaction of industrial objects downloaded from the public domain. The mesh-merging examples were done on a PC, Intel®, Core™, i7 CPU, 870 at 2.93 GHz with 16-GB RAM running on Windows 7 using Intel Fortran VS2010 on 64 bits. The summary of the essential features of the examples is shown in Table 8.22, where the data presented are mostly numbers as the merging process is based, to a great extent, on topological operations. As shown in Table 8.22, excluding time for graphics display, the CPU time for the determination of the intersection loops in all the examples is only a tiny fraction of the total time taken for which the most complex example of the machine part is still less than 1 s. The shape quality of the tetrahedral elements has not been calculated, as this is not a main concern or restriction of the merging algorithm. As far as finite element applications are concerned, the resulting meshes could be further enhanced in connectivity and shape by means of standard optimisation procedures. In the examples, however, no enhancement has been applied to the raw meshes, which are the direct output of the merging process, in order to show clearly the steps of mesh merging and to enable features of intersection to be identified in the merged object.

Example 1 is an academic example to show the basic ideas of the merging process. We are interested to find out a hollow cube being cut across transversely by a solid plate. Both objects are first meshed into tetrahedral elements in which the hollow cube consists of 125 nodes and 288 tetrahedra, and the solid plate consists of 192 nodes and 588 tetrahedra. In the intersection between these two objects, there are 192 intersection segments making

Table 8.22 Summary of the statistics of the examples 1 to 5

Example	1		2		3		4		5	
Objects	Plate	Cube	Plate	String	String	String	Hand	Plate	Cuboid	Machine
NN	192	125	2255	3366	3366	3366	6222	1953	60	21,241
NE	588	288	9600	11,248	11,248	11,248	22,391	7200	144	71,222
NF	308	240	2400	5226	5226	5226	12,440	2800	104	39,200
NI	192		421		613		744		624	
NL	4		7		11		5		3	
NFT	92	96	202	258	336	334	513	280	35	548
NZ	5	4	8	7	12	12	6	6	4	2
NR	1		4		11		2		1	
NT	1570		22,624		24,948		25,099		73,931	
NB	788		7896		11,120		13,928		38,696	
TI	0.01		0.01		0.01		0.02		0.02	
T2	0.10		0.26		0.29		0.48		0.82	
Figures	8.123a–f		8.124a–i		8.125a–d		8.126a–e		8.127a–d	

Note: NN = number of nodes in the object; NE = number of tetrahedral elements in the object; NF = number of triangles on the boundary surface; NI = number of intersection points/segments; NL = number of intersection loops; NFT = number of faces triangulated; NZ = number of zones (patches) on the boundary surface; NR = number of regions (volumes) of intersection; NB = number of faces on the boundary of the merged object; TI = CPU time in seconds for intersection; T2 = total CPU time for mesh merging.

up four intersection loops, which divide the boundary of the hollow cube into four zones (patches) and that of the plate into five zones, as shown in Figure 8.123a and b. One region in the form of a shortened hollow cube bounded by 528 triangles from four cut surfaces and four intersection loops is recovered, as shown in Figure 8.123c. The hollow cube, in which there are fewer elements, is taken as the slave object. When compatibility is restored on the top and the bottom surfaces of the region of intersection, the volume of intersection could be taken out from the hollow cube, as shown in Figure 8.123d. On the other hand, if the solid plate is taken as the slave object, the volume of intersection can be taken from the plate to divide it into two regions, as shown in Figure 8.123e. The hollow cube and the solid plate could be merged either by putting objects in Figure 8.123a and d or objects in Figure 8.123b and e together, and the resulting object is shown in Figure 8.123f.

The penetration of a string into a wooden board is considered in Example 2. Full details for all the steps of the merging process will be given in this interesting example as many regions of intersection are found in the intersection of the objects. The string in the form of a curved solid object is downloaded from the public domain (INRIA GAMMA 2007), which consists of 3366 nodes, 11,248 tetrahedral elements and 5226 triangular facets on the boundary surface. The board is first decomposed into hexahedral elements, each of which is further divided into six tetrahedral elements, and the resulting tetrahedral mesh consists of 2255 nodes, 9600 tetrahedra and 2400 triangles on the boundary surface. In the interaction of these two objects, there are 421 intersection points (segments) from which seven closed loops can be retrieved, as shown in Figure 8.124a and b. The intersection segments cut across 202 triangles on the surface of the wooded board, each of which is triangulated in turn, and new tetrahedral elements are formed. The modified mesh consists of 10,615 tetrahedra and 3242 triangles on the surface, as shown in Figure 8.124c. As for the string, 258 triangles are intersected by the loops, and the intersected triangles are triangulated to incorporate the intersection segment into the tetrahedral mesh. The triangular facets on the boundary surface increases from 5226 to 6068, as shown in Figure 8.124d. The boundary

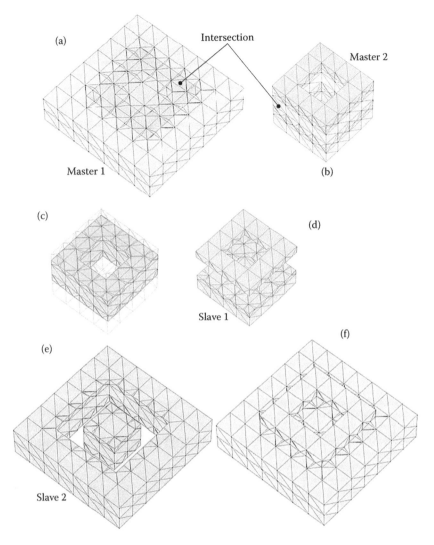

Figure 8.123 Merging of a solid plate and a hollow cube: (a) plate; (b) hollow cube; (c) region of intersection; (d) subtracting the region of intersection from the hollow cube; (e) subtracting the region of intersection from the plate; (f) merging of the hollow cube and the plate.

surface of the board is partitioned by the intersection loops into eight zones, whereas seven patches could be identified on the boundary surface of the string, as shown, respectively, in Figure 8.124e and f.

The relationships between intersection loops, surface patches and regions recovered, which are automatically built up by the computer by means of searching, matching and topological operations, are shown in Table 8.23. For instance, in the construction of region R_1, we start with loop L_1, which is connected to surfaces S_1 and S_2 of the master object (board) and $\bar{S}_3$ and $\bar{S}_4$ of the slave object (string). It is found that S_2 and $\bar{S}_3$ are the cutting surfaces associated with L_1. However, $\bar{S}_3$ is also connected to L_5, for which the other cutting surface from the master object is S_6. As no more loops or surfaces could be included, the collection process terminates, and R_1 is bounded by 368 triangles from S_2, S_6 and $\bar{S}_3$, as shown in Figure 8.124e and f. The construction of other regions of intersection

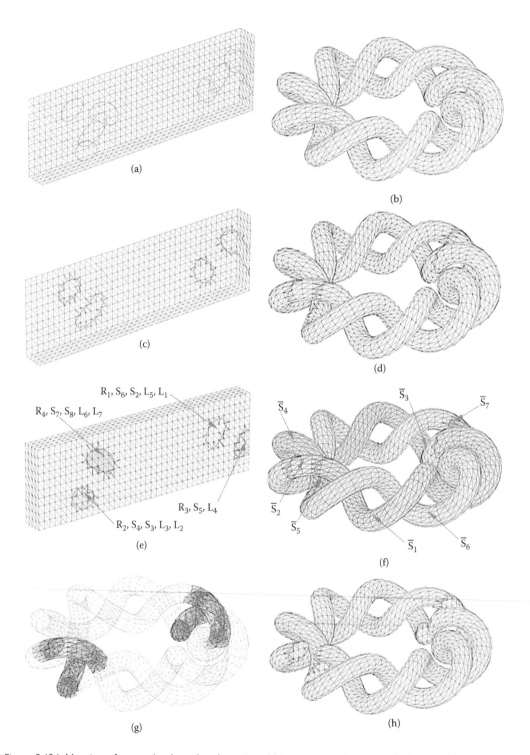

Figure 8.124 Merging of a wooden board and a string: (a) intersection loops on the board; (b) intersection loops on the string; (c) intersection segments incorporated; (d) intersection incorporated in the string; (e, f) boundary surfaces of the board and the string are partitioned into zones; (g) regions identified and local meshing; (h) intersection removed from the string.

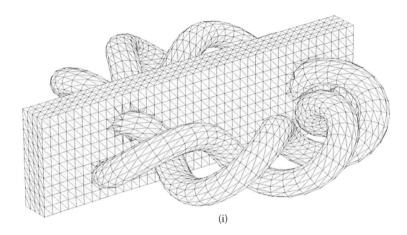

(i)

Figure 8.124 (Continued) Merging of a wooden board and a string. (i) Final mesh of merging the board and the string.

Table 8.23 Relationship between intersection loops, surfaces and regions for example 2

Loop	Master surfaces		Slave surfaces		Cut surfaces		Region
L_1	S_1	S_2	$\bar{S}_3$	$\bar{S}_4$	S_2	$\bar{S}_3$	R_1
L_2	S_1	S_3	$\bar{S}_4$	$\bar{S}_5$	S_3	$\bar{S}_5$	R_2
L_3	S_1	S_4	$\bar{S}_5$	$\bar{S}_6$	S_4	$\bar{S}_5$	R_2
L_4	S_1	S_5	$\bar{S}_6$	$\bar{S}_7$	S_5	$\bar{S}_7$	R_3
L_5	S_1	S_6	$\bar{S}_1$	$\bar{S}_3$	S_6	$\bar{S}_3$	R_1
L_6	S_1	S_7	$\bar{S}_1$	$\bar{S}_2$	S_7	$\bar{S}_2$	R_4
L_7	S_1	S_8	$\bar{S}_2$	$\bar{S}_6$	S_8	$\bar{S}_2$	R_4

follows a similar path based on the relationship between intersection loops and cut surfaces, as depicted in Table 8.23. Finally, all four regions of intersection are recovered such that R_2 is bounded by 356 triangles from S_3, S_4 and $\bar{S}_5$; R_3 is bounded by 314 triangles from S_5 and $\bar{S}_7$; and R_4 is bounded by 376 triangles from S_7, S_8 and $\bar{S}_2$. Cavities are created at the cutting surfaces of the regions by taking tetrahedra away from the slave object (string), and MG is carried out at the cavities so as to restore compatibility at the cutting surfaces S_2, S_3, S_4, S_5, S_6, S_7 and S_8 of the master object, as shown in Figure 8.124g. A new object or mesh is resulted by subtracting all the regions of intersection from the slave object, as shown in Figure 8.124h. The reduced slave object can now be readily combined with the master object with full compatibility at the cutting surfaces, as shown in Figure 8.124i.

Example 3 demonstrates the intersection of two identical strings put rather arbitrarily together. Owing to the twisting of the curved boundary of the two strings, there are many intersection segments that could be grouped into 11 loops. The boundaries of the master object and the slave object are each partitioned into 12 zones, as shown in Figure 8.125a and b. Eleven regions are identified, as shown in Figure 8.125c, and the merging of the two strings is shown in Figure 8.125d.

A hand penetrating a flat board is considered in Example 4. The hand, which consists of 6222 nodes, 22,391 tetrahedra and 12,440 triangles on the boundary surface, is downloaded from the public domain. The board is created by first generating hexahedral elements,

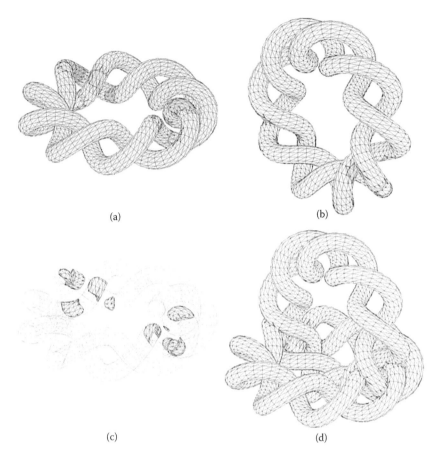

Figure 8.125 Merging of two strings into a single object: (a) first string partitioned into 12 zones; (b) second string divided into 12 zones; (c) eleven regions identified; (d) mesh of merging of strings.

followed by a subdivision of the hexahedra into tetrahedra. The resulting tetrahedral mesh consists of 1953 nodes, 7200 tetrahedra and 2800 triangles on the boundary surface. The intersection as a result of hand penetration produces 744 intersection segments (points), which could be grouped into five loops. The boundaries of the hand and that of the board are each partitioned into six zones, as shown in Figure 8.126a and b. Two regions of intersection could be recovered from the intersection loops and surface patches, as shown in Figure 8.126c. Two holes will be perforated in the board by subtracting the regions of intersection from the slave object, as shown in Figure 8.126d. When the hand is inserted into the perforated board, the two objects merge with a tight fit at the contact surfaces, as shown in Figure 8.126e.

Example 5 is about the intersection of a cuboid with a machine part. The machine part, which consists of 21,241 nodes, 71,222 tetrahedra and 39,200 triangles on the boundary surface, is a realistic engineering object provided by Prof. Z.Q. Guan of Dalian University of Technology, China. The cuboid is created by first generating hexahedral elements, followed by a subdivision of the hexahedra into tetrahedra. The large difference in the element size between the objects is adopted to test the performance of the merging algorithm. As shown in Figure 8.127a, elongated triangles of large aspect ratios are created on the surface of the cuboid. The 104 triangles on the surface of the cuboid are divided into 1312 triangular facets, and the number of tetrahedral elements increases from 144 to 1395. There are 624

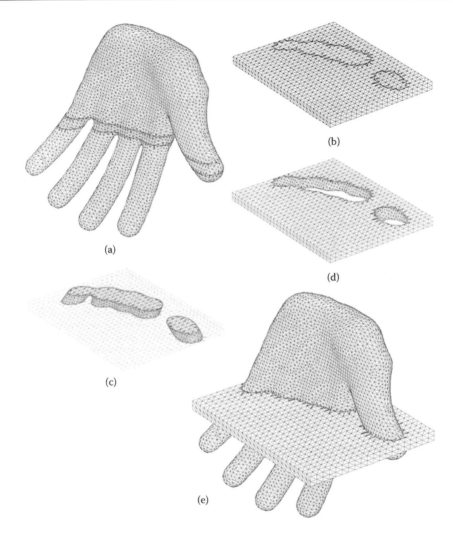

Figure 8.126 Tetrahedral mesh of a hand penetrated into a board: (a) five loops divide the hand into six zones; (b) the board is also divided into six zones; (c) two regions of intersection identified; (d) intersection parts are removed from the board; (e) hand inserted into the board.

intersection segments, which could be grouped into three intersection loops. The cuboid is partitioned into four zones, whereas the machine part is partitioned into two zones, as shown in Figure 8.127a and b. One region of intersection is identified, as shown in Figure 8.127c, and the merging of the two objects is shown in Figure 8.127d.

8.6.4 Closure

Apart from the determination of surface intersections in the form of intersection loops, all operations for mesh merging are topological in nature, such as the adjacency relationship, surface partition from a seed triangle, boundary operations to obtain bounding surfaces from a volume and loops from surface patches, etc. However, the quality of the resulting mesh depends on the geometry of the objects as well as how they are meshed; besides optimisation, what else could be done to improve mesh quality will be of interest for further research.

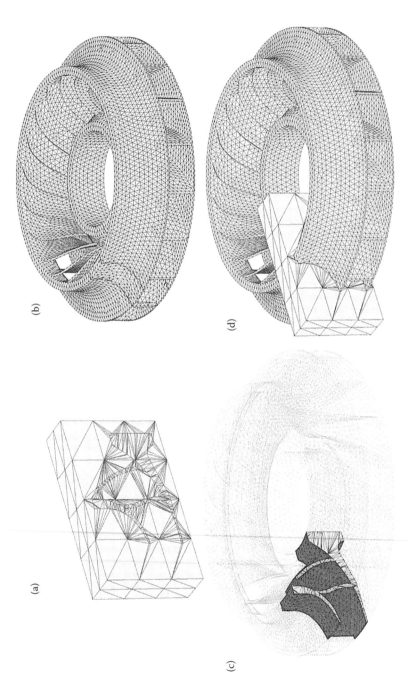

(a)

(b)

(c)

(d)

Figure 8.127 Merging objects of different element sizes: (a) boundary surface of cuboid is divided into four zones; (b) machine part divided into two zones; (c) region of intersection; (d) machine part merged with cuboid.

8.7 MERGING OF HEXAHEDRAL MESHES

8.7.1 Introduction

While advanced meshing algorithm are available in the generation of isotropic and aniso-tropic tetrahedral meshes for complex 3D industrial and bio-mechanical objects, a versatile mesh generator capable of generating high-quality unstructured hexahedral elements for FE analysis of complex industrial objects is yet to be developed. Nevertheless, hexahedral elements are superior to tetrahedral elements in the following aspects: (i) high-quality structured regular hexahedral elements can be generated rapidly by means of mapping, extrusion and sweeping techniques; (ii) hexahedral elements could be generated following the domain boundary in layers and in alignment with important geometrical features for easy visualisation and inspection, whereas the validity of tetrahedral meshes can only be verified by computer; and (iii) low-order high-performance hexahedral elements are available, which are considered to be more computationally efficient than the tetrahedral counterparts. Hexahedral meshes of various characteristics can be generated by different approaches as proposed by researchers working in diverse scientific and engineering fields, and a comprehensive account of the current status in hex meshing has been given in Section 5.8.

In view of the success of merging arbitrary solid tetrahedral FE meshes in addressing the industrial need for modification, update and manipulation of meshed objects, it is natural to ask whether we can also merge hexahedral meshes. Before we answer this question, we have to know what we expect of the resulting mesh by merging. Inevitably, tetrahedra will be created as a result of intersection, which can only be connected to hexahedral elements by means of a pyramid element. Hence, without loss of generality, merging of hexahedral meshes is possible by first dividing each hexahedral element into five or six tetrahedral elements. Non-intersected hexahedral elements can be retrieved from the merged tetrahedral mesh as the constituent tetrahedra as a subdivision of the original hexahedral elements are intact and present in the mesh. The mesh-merging algorithm provides a means to combine, modify and insert new features to existing hexahedral and tetrahedral meshes. It is also a useful tool in a general MG package to create new meshes from existing hexahedral and tetrahedral meshes through Boolean operations. High-quality regular hexahedral elements of the original mesh will be preserved, which is important for FE analysis as hexahedral elements are rather sensitive to shape distortions. Generic mesh merger will definitely be a good companion to many existing hexahedral mesh generators in creating ever complex meshes, which could only be conveniently defined and created by putting different meshes together.

A generic algorithm will be developed in this section to merge structured and unstructured hexahedral meshes automatically into one single valid FE mesh of hexahedral, tetrahedral and pyramid elements. Like the merging of tetrahedral meshes, the procedure is robust and efficient as all operations such as loops of intersection, incorporation of intersection segments, partition of boundary surfaces and identification of regions of intersection are deterministic and topological. Examples with details for each step of the mesh-merging process are presented to elucidate the main ideas of the algorithm.

8.7.2 Algorithm: Merging hexahedral mesh

Problem statement: Given two 3D objects Ω and $\bar{\Omega}$, each of which is a valid finite element mesh of hexahedral elements such that $\Omega = \{H_i, i = 1, N_H\}$ and $\bar{\Omega} = \{\bar{H}_j, j = 1, N_{\bar{H}}\}$, where N_H and $N_{\bar{H}}$ are, respectively, the number of hexahedral elements in Ω and $\bar{\Omega}$, the merging problem is to find the union of the two given meshes into a single valid finite element mesh

$\hat{\Omega}$ consisting of hexahedral elements $\hat{H}_k$, tetrahedral elements $\hat{T}_m$ and pyramid elements $\hat{Y}_n$, i.e. $\hat{\Omega} = \Omega \cup \bar{\Omega} = \{\hat{H}_k, k = 1, N_{\hat{H}}; \hat{T}_m, m = 1, N_{\hat{T}}; \hat{Y}_n, n = 1, N_{\hat{Y}}\}$, where

$N_{\hat{H}}$ = Number of non-intersected hexahedra
$N_{\hat{T}}$ = Number of tetrahedra in the merged mesh to ensure compatibility
$N_{\hat{Y}}$ = Number of pyramid elements at the interface of hexahedra and tetrahedra

such that points in Ω or $\bar{\Omega}$ are found in $\hat{\Omega}$ as well.

$\forall i \in \{1, 2, ..N_H\}$ and $\forall j \in \{1, 2, ..N_{\bar{H}}\}$, $\mathbf{x} \in H_i$ or $\mathbf{x} \in \bar{H}_j$

$\Rightarrow \mathbf{x} \in \hat{H}_k$ or $\mathbf{x} \in \hat{T}_m$ or $\mathbf{x} \in \hat{Y}_n$ for some $k \in \{1, 2, ..N_{\hat{H}}\}$, $m \in \{1, 2, ..N_{\hat{T}}\}$, $n \in \{1, 2, ..N_{\hat{Y}}\}$

Similar to the merging of tetrahedral meshes, the objects to be dealt with are very general as long as they are valid hexahedral FE meshes, and there is no restriction on the size, shape and topology of the objects, which could be convex, concave, simply or multi-connected, etc. The algorithm proposed can even handle objects that are disjoint; however, without loss of generality, we simply assume that the objects are connected pieces, since Boolean operations can be applied sequentially between different intersecting objects in turn to form ever more complex objects. The robustness of the algorithm is attributed to the determination of intersection loops between the given objects as topological intrinsic features rather than a set of independent intersection points and segments and a consistent topologic treatment in the intersection loops, boundary surface partition and construction of the regions of intersection. To ensure FE mesh compatibility, pyramid elements have to be created at the interface between tetrahedral and hex elements. Irregular hex elements could also be retrieved from the remaining tetrahedral elements by some exhaustive topology searching and shape identification procedures (Meshkat and Talmor 2000; Fernandes and Martins 2007).

8.7.2.1 Hexahedron decomposed into tetrahedra

A hexahedron can be conveniently decomposed into five or six tetrahedra. As shown in Figure 8.128, hexahedron H(1,2,3,4,5,6,7,8) is divided into five tetrahedra, namely, T(8,6,5,1), T(8,7,6,3), T(8,6,1,3), T(1,3,4,8) and T(1,2,3,6).

Hexahedron H(1,2,3,4,5,6,7,8) can also be divided into two wedge elements (pentahedra), each of which is further divided, respectively, into three tetrahedra T(2,3,4,7), T(2,6,7,8), T(4,7,8,2) and T(8,6,5,1), T(6,2,1,8), T(1,4,8,2), as shown in Figure 8.129. In the merging of hexahedral meshes, however, each hexahedron will be divided into six tetrahedra, as compatibility can be ensured between hexahedral elements on all quadrilateral faces for such a subdivision, as shown in Figure 8.130a.

Of course, as shown in Figure 8.130b, the two symmetrical subdivisions of a hexahedron into six tetrahedra, named type 6A and type 6B, are similar. There is no preference for one subdivision over the other except for skewed hexahedral meshes where a certain division will produce elements of better geometrical shape, or there is a need to fit some specified boundary requirements. However, hexahedral elements have to be all divided into either type 6A or type 6B, but not arbitrarily into types 6A and 6B as incompatibility arises between the two subdivisions, as shown in Figure 8.130b. Another possibility exists such that hexahedra of a hexahedral mesh can also be divided into five tetrahedra for full compatibility. First of

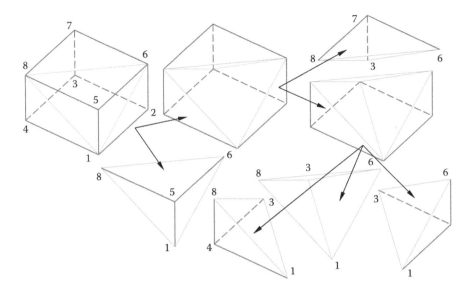

Figure 8.128 Dividing a hexahedron into five tetrahedra.

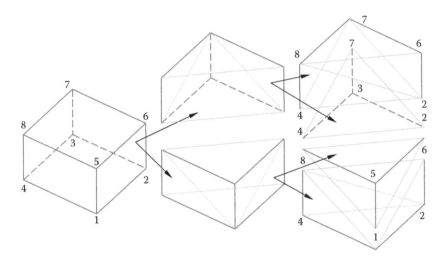

Figure 8.129 Dividing a hexahedron into six tetrahedra.

all, we have to note that there are two symmetrical ways in dividing a hexahedron into five tetrahedra, which are named type 5A and type 5B. As shown in Figure 8.130c, each subdivision type 5A or type 5B is created by detaching four tetrahedra from two pairs of opposite corners as indicated by the arrows, leaving another tetrahedron (the fifth one) at the centre of the cube sharing a common face with each corner tetrahedron. It is further realised that hexahedral subdivision type 5A can be connected harmoniously to hexahedral subdivision type 5B in any three principal directions and vice versa. Hence, hexahedral elements can each be divided into five tetrahedra for full compatibility, provided that neighbouring elements are of subdivisions type 5A and type 5B in an alternative manner, like the arrangement of some molecular structures, as shown in Figure 8.131.

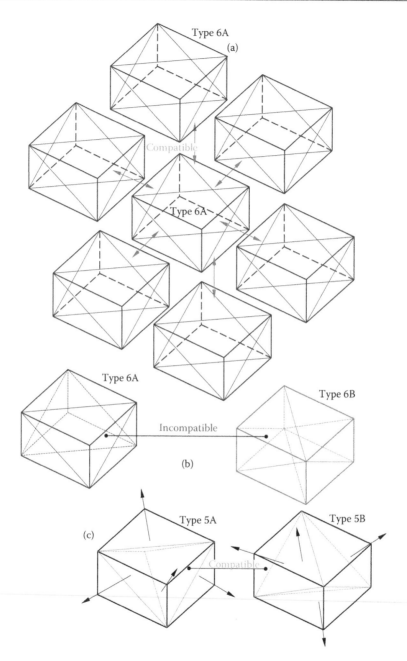

Figure 8.130 **Compatibility between triangulated hexahedra: (a) compatibility between same type of sub-division; (b) incompatible between two types of subdivisions; (c) compatibility between sub-divisions of hexahedron into five tetrahedra.**

For unstructured hexahedral meshes, neighbouring hexahedral elements can be con-nected to form chords and sheets by means of connections –6A–6A–6A–, –6B–6B–6B– or –5A–5B–5A–5B–. As for isolated hexahedral elements, identify the face orientations to check if any one of the four types 5A, 5B, 6A and 6B could be used. In case not a fit can be found, as a last resort, the hexahedral element is divided into 12 tetrahedra by inserting a point at its centre.

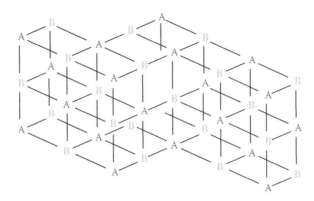

Figure 8.131 Connection of type 5A and type 5B hexahedra.

8.7.2.2 Merging of hexahedral meshes

The tetrahedral meshes converted from the two given hexahedral meshes are merged by the tetrahedral mesh-merging algorithm presented in Section 8.6.2 into one single tetrahedral finite element mesh. Let's consider the merging of a square board and a rectangular prism.

A square board is discretised into $2 \times 10 \times 10 = 200$ hexahedral elements, each of which is further divided into six tetrahedral elements, as shown in Figure 8.132a. The rectangular prism is first discretised into 36 hexahedral elements, each of which is then divided into six tetrahedral elements, as shown in Figure 8.132b. All the intersection loops between the boundary surfaces of the two solid objects are determined by the neighbour-tracing method, as shown in Figure 8.132c. Intersection segments are incorporated onto the boundary surface of the given objects by applying the 2D triangulation on all the intersected triangular facets on the boundary, as shown in Figure 8.132d. As each intersection segment is present as an edge of a tetrahedron on the boundary surfaces of the solid objects, the intersection loops will partition the boundary surface of the object into a number of zones, as shown in Figure 8.132e. One region of intersection can be identified and defined by collecting bounding surfaces from intersection loops and boundary surface patches, as shown in Figure 8.132f. Incompatible tetrahedra intersected by the bounding surfaces of all the regions of intersection are removed, and mesh compatibility is restored by filling up the gap with new tetrahedra between the bounding surfaces and the cavities so created. All the regions of intersection common to both objects can now be detached freely from the solid meshes, and the two solid meshes are now ready to be merged with full compatibility at all intersecting surface parts, as shown in Figure 8.132g.

8.7.2.3 Recovery of hexahedral elements from tetrahedral elements

As the original hexahedral elements before subdivision into tetrahedra are known, their recovery from the intersected tetrahedral mesh is straightforward. For each hexahedral element in the original mesh, collect all the tetrahedra bounded by the hexahedral domain. Let T be the set of tetrahedral elements bounded by hexahedron $H(1,2,3,4,5,6,7,8)$, as shown in Figure 8.133. Retrieve the boundary of T, ∂T. If the number of triangular facets on ∂T equals 12, verify if the boundary facets of T are those of the hexahedron, i.e.

$$\partial T = \partial H = \{2\text{-}6\text{-}7, 2\text{-}7\text{-}3, 7\text{-}4\text{-}3, 7\text{-}8\text{-}4, 1\text{-}4\text{-}8, 1\text{-}8\text{-}5, 1\text{-}5\text{-}6, 1\text{-}6\text{-}2, 6\text{-}5\text{-}8, 6\text{-}8\text{-}7, 1\text{-}2\text{-}4, 2\text{-}3\text{-}4\}$$

If $\partial T = \partial H$, restore hexahedron $H(1,2,3,4,5,6,7,8)$ and delete all the tetrahedra in T.

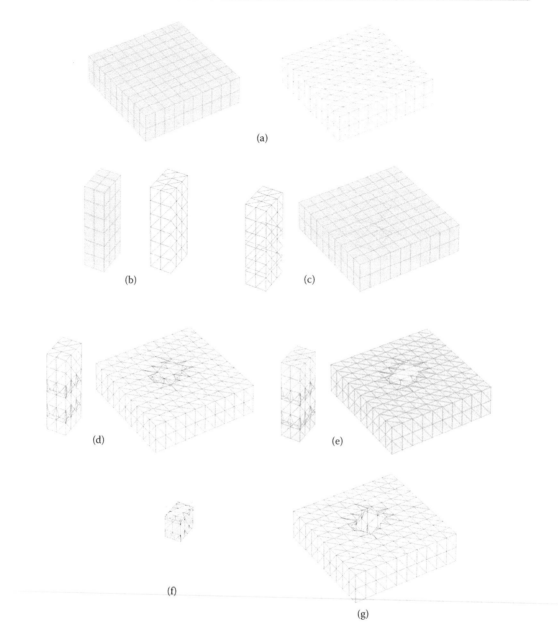

Figure 8.132 Merging of two hexahedral meshes: (a) board divided into 200 hexahedral elements and 1200 tetrahedral elements; (b) prism divided into 36 hexahedral and 216 tetrahedral elements; (c) intersection loops found on the prism and the board; (d) intersection loops incorporated onto the surfaces of the objects; (e) boundary surfaces partitioned into zones; (f) region of intersection; (g) region of intersection taken away leaving a hole at the middle of the board.

8.7.2.4 Compatibility between hexahedral and tetrahedral elements

Upon the completion of the recovery of the hexahedral elements from the tetrahedral meshes, the finite element meshes are meshes of a mixture of hexahedral and tetrahedral elements. Mesh compatibility cannot be ensured at the interfaces between hexahedral and tetrahedral elements (Owen and Saigal 2001). To restore compatibility between hexahedral

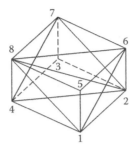

Figure 8.133 **Hexahedron H(1, 2, 3, 4, 5, 6, 7, 8).**

and tetrahedral elements, a pyramid element has to be introduced at the interface between hexahedral and tetrahedral elements, as shown in Figure 8.134. As hexahedral elements are retrieved from a valid tetrahedral mesh, a quadrilateral face of a hexahedron can only be connected to two tetrahedra, as shown in Figure 8.135. There are two cases to consider.

1. If the two tetrahedra share a common opposite node E, as shown in Figure 8.135a, then the two tetrahedra can be merged to form pyramid element ABCDE.
2. In case the two tetrahedra do not share a common node, as shown in Figure 8.135b, node I has to be inserted at the mid-point of edge BD. All the tetrahedra connected to line BD are each divided into two tetrahedra by the standard edge-subdivision process. Node I is then lifted slightly away from the quadrilateral face ABCD to create pyramid element ABCDI.

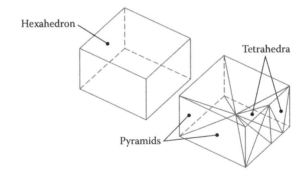

Figure 8.134 **Pyramid as an interface between hexahedral and tetrahedral elements.**

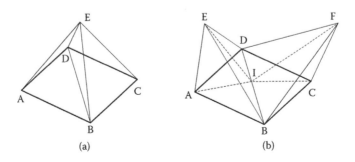

Figure 8.135 **Connection between tetrahedral elements and a quadrilateral face: (a) two tetrahedra form a pyramid at face ABCD; (b) tetrahedral elements cannot merge into a pyramid at face ABCD.**

From the tetrahedral mesh of the prism shown in Figure 8.132d, 24 hexahedral elements are recovered, 28 pyramid elements are created at the interfaces and 292 tetrahedral elements remain in the mesh, as shown in Figure 8.136. From the tetrahedral mesh of the board shown in Figure 8.132g, 128 hexahedral elements are recovered, 69 pyramid elements are created at the interfaces and 560 tetrahedral elements remain in the mesh, as shown in Figure 8.137. The merged finite element mesh of hexahedral, tetrahedral and pyramid elements of the prism and the board is shown in Figure 8.138.

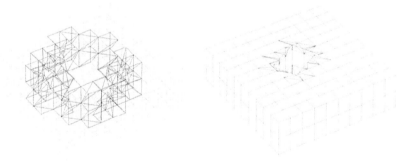

Figure 8.136 Prism: NH = 24, NT = 292, NY = 28, NF = 220, NQ = 52.

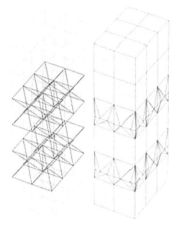

Figure 8.137 Board: NH = 128, NT = 560, NY = 69, NF = 240, NQ = 258, NF = number of triangular faces, NQ = number of quadrilateral faces.

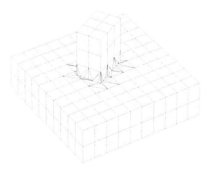

Figure 8.138 Board and prism merged.

Table 8.24 Statistics for examples 1, 2 and 3

Example	Objects	NH	NT	NL	NZ	NR	NH*	NT*	NY*	Figures
1	Cylinder1	1200	7200	8	9	2	860	2913	209	8.139 a–g
	Cylinder2	640	3840		8		468	2546	512	
2	Curveplate1	2400	14,400	8	9	2	2165	3319	170	8.140 a–f
	Handle	1200	7200		8		1008	2600	576	
3	Curveplate1	2400	14,400	3	4	1	1454	4474	462	8.141 a–e
	Curveplate2	2400	14,400		4		2208	2800	576	

Note: NH = number of hexahedral elements in the mesh; NT = number of tetrahedral elements in the mesh; NL = number of boundary intersection loops; NZ = number of zones (patches) on the boundary surface; NR = number of regions (volumes) of intersection; NH* = number of hexahedral elements recovered; NT* = number of tetrahedral elements remained in the mesh; NY* = number of pyramid elements created at interfaces.

8.7.3 Examples

In this section, three examples of merging hexahedral meshes of various characteristics and complexity are presented. The summary of the essential features of the examples is shown in Table 8.24 in which the number of hexahedral elements before and after merging is given. The shape quality of the tetrahedral elements has neither been optimised nor calculated, as this is not a main concern or restriction to the merging algorithm. As far as finite element applications are concerned, the tetrahedral and pyramid elements could be further enhanced in connectivity and shape by means of standard optimisation procedures. As for the hexahedral elements, only those regular hexahedral elements of the original mesh are retrieved from the tetrahedral mesh.

The intersection of two hollow cylinders is considered in example 1. The larger cylinder is discretised into 1200 hexahedral elements, each of which is further divided into six tetrahedral elements, as shown in Figure 8.139a. The smaller cylinder is discretised into 640 hexahedral elements, which is further divided into 3840 tetrahedral elements, as shown in Figure 8.139b. Eight intersection loops are found in the intersection of the cylinders, which divide the larger cylinder into nine zones and the smaller cylinder into eight zones, as shown in Figure 8.139c. From the surface partition of the boundary surfaces, two regions of intersection can be identified, as shown in Figure 8.139d. A total of 468 hexahedral elements (green in colour) are recovered, 512 pyramid elements (yellow in colour) are formed and 2546 tetrahedral elements (blue in colour) remain in the intersected tetrahedral mesh of the smaller cylinder, as shown in Figure 8.139e; 860 hexahedral elements are recovered, 209 pyramid elements are formed and 2913 tetrahedral elements remain in the intersected tetrahedral mesh of the larger cylinder, as shown in Figure 8.139f. The finite element mesh as a result of merging the two cylinders is shown in Figure 8.139g. The intersection of a curved hollow cylinder and a curved plate is considered in example 2, as shown in Figure 8.140a–f. The intersection of two curved plates is studied in example 3, as shown in Figure 8.141a–e.

8.7.4 Closing remarks

In view of the success of merging arbitrary solid tetrahedral finite element meshes, the algorithm is extended to merge hexahedral meshes by first dividing each hexahedral element into five or six tetrahedral elements. Non-intersected regular hexahedral elements are readily recovered from the intersected tetrahedral mesh as the constituent tetrahedra as a subdivision of the original hexahedral elements are intact and present in the mesh. If necessary,

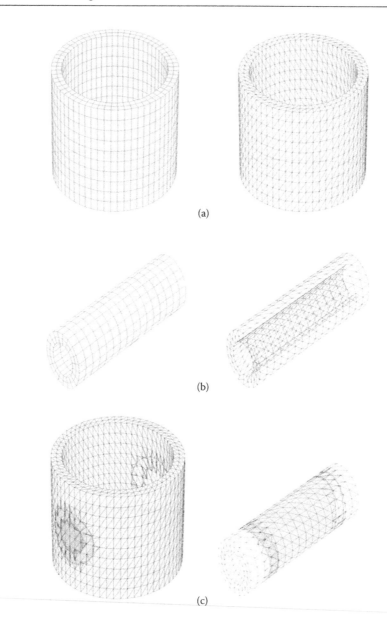

Figure 8.139 Merging of two cylinders of different size. (a) Larger cylinder: NH = 1200, NT = 7200, (b) smaller cylinder: NH = 640, NT = 3840 and (c) larger cylinder divided into nine zones and smaller cylinder divided into eight zones.

non-regular hexahedral elements could be retrieved from the tetrahedral mesh by some shape identification and topological searching algorithms. The resulting mesh consists of a mixture of hexahedral and tetrahedral elements. To restore mesh compatibility between hexahedral and tetrahedral elements, pyramid elements are introduced at the interfaces between hexahedral and tetrahedral elements.

The mesh-merging algorithm provides a means to combine, modify and insert additional features to existing hexahedral meshes. It is also a powerful tool to create new meshes from existing hexahedral and tetrahedral meshes. The merging algorithm is reliable and efficient,

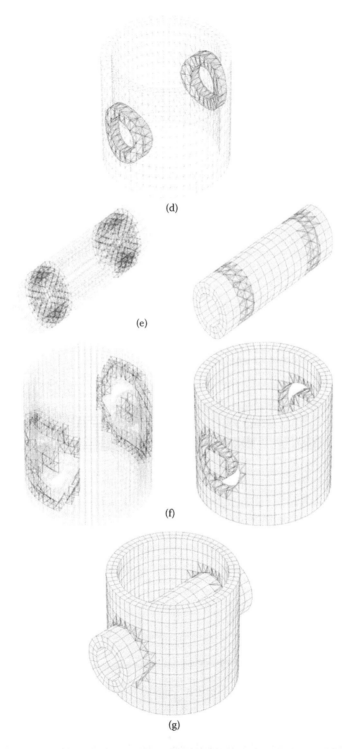

(d)

(e)

(f)

(g)

Figure 8.139 (Continued) Merging of two cylinders of different size. (d) Two regions of intersection, (e) smaller cylinder: NH = 468, NT = 2546, NY = 512, (f) larger cylinder: NH = 860, NT = 2913, NY = 209 and (g) cylinders merged.

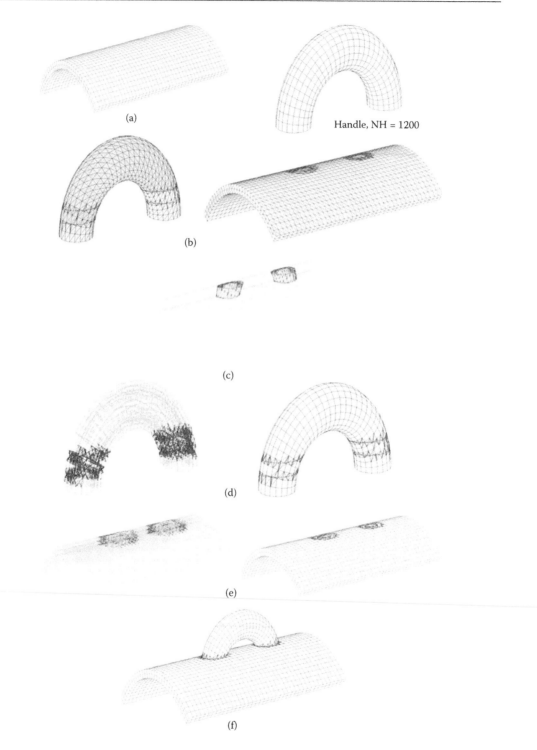

Figure 8.140 Merging a curved plate and a hollowed cylinder. (a) Hexahedral meshes: curved plate, NH = 2400; (b) handle partitioned into eight zones and curved plate partitioned into nine zones, (c) two regions of intersection, (d) hollow cylinder: NH = 1008, NT = 2600, NY = 576, (e) curved plate: NH = 2165, NT = 3319, NY = 170 and (f) Hollow cylinder and curved plate merged.

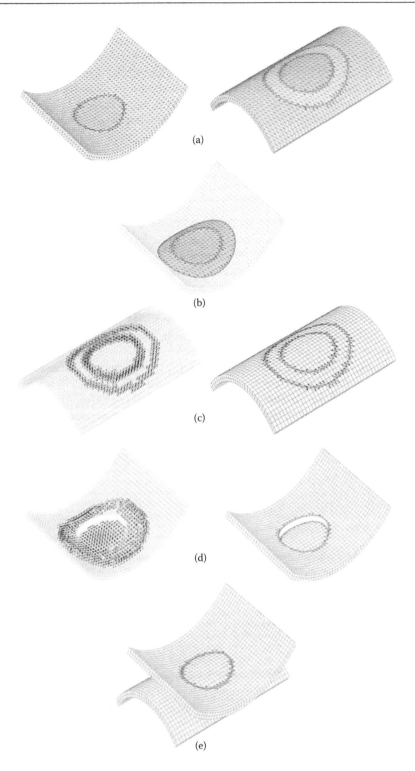

Figure 8.141 Merging of two curved plates. (a) Curved plates partitioned into four zones, (b) region of intersection, (c) Lower curved plate: NH = 2208, NT = 2800, NY = 576, (d) upper curved plate: NH = 1454, NT = 4474, NY = 462 and (e) curved plates merged.

as all operations from the subdivision of hexahedral elements into tetrahedra, merging of tetrahedral meshes, to the recovery of hexahedral elements are a deterministic process without iteration and exhaustive searching. The quality of the hexahedral elements could be well preserved as the recovered hexahedral elements are those of the original mesh. Although the mesh quality improvement is not the main focus of the mesh-merging algorithm, for finite element applications, the shape quality of hexahedral and tetrahedral elements can all be enhanced by the optimisation procedures described in Chapter 6.

8.8 CURVILINEAR FINITE ELEMENT MESH

8.8.1 Introduction

Higher-order curvilinear finite elements (p2 or p3 elements, etc.), as shown in Figure 8.142, refer to elements with a curved boundary, i.e. curved edges and faces, in contrast to linear p1 elements whose edges are straight line segments and faces are planar facets. Structured curvilinear non-simplicial meshes such as hexahedral and pentahedral meshes can be generated by a mapping process, whereas there are basically two ways in producing unstructured curvilinear meshes of simplicial elements for an object bounded by curved surfaces, namely, (i) directly generating of curvilinear mesh from the CAD geometry, as shown in Figure 8.143a, and (ii) indirectly generating a linear p1 mesh first and then converting it to a curvilinear mesh, as shown in Figure 8.143b.

Curvilinear elements are much more flexible in representing smooth curved boundaries than linear elements such that fewer elements are needed in the FE model. Higher-order p-version adaptive analysis may have an exponential rate of convergence over smooth regions without singularity. Shephard and his group have done a great deal of work in the generation of

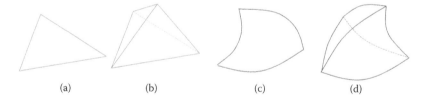

(a)   (b)   (c)   (d)

Figure 8.142 Linear and curvilinear elements in two and three dimensions: (a) linear triangle; (b) linear tetrahedron; (c) curved triangle; (d) curved tetrahedron.

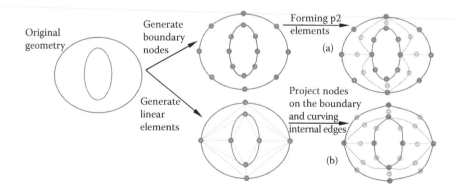

Figure 8.143 Generation of p2 meshes by direct and indirect methods: (a) direct approach; (b) indirect approach.

curvilinear meshes on 2D, 3D and over surfaces following an isotropic metric or an anisotropic metric requirement. Dey et al. (2001, 2006) discussed issues related to the generation of curvilinear mesh and presented an iterative algorithm based on node relocations, edge/face swaps, splits and collapses for curving a p1 mesh to a p2 mesh. Curved triangular elements over smooth surfaces were generated on a parametric domain for molecular surface modelling (Laug and Borouchaki 2002). Li et al. (2003) proposed a two-step boundary node projection method along with local remeshing around curved geometry as an indirect approach for the generation of curvilinear tetrahedral elements. By the standard mesh modification procedures, curvilinear tetrahedral meshes were generated by Luo et al. (2004) with particular emphasis in dealing with singular features. Luo et al. (2010) generated curvilinear tetrahedral meshes by capturing the geometry of a thin section through an element collection procedure to mark the thin section with an empty polygon. Sahni et al. (2010) presented a procedure to generate an anisotropic curved boundary layer for adaptive viscous flow simulations. Making use of the properties of Bezier tetrahedral elements to determine the determinant of the Jacobian, adaptive meshes moving across a curved domain were generated by Luo et al. (2011). Again by the coefficients related to the determinant of the Jacobian of Bezier tetrahedra, George and Borouchaki (2012) proposed a global iterative algorithm to optimise the determinants of the Jacobian of p2 tetrahedral elements. Based on the results of a linear elasticity to reposition nodes, Xie et al. (2013) presented a procedure to generate curvilinear tetrahedral elements of any order. Curvilinear meshes were untangled by the optimisation using the conjugate gradient approach in conjunction with the *barrier method* on an objective function consisting of two parts to take care of the Jacobians of the elements and the other constraints (Toulorge et al. 2013).

8.8.2 Generation of curvilinear meshes

In this section, a generic algorithm for the generation of curvilinear FE meshes based on the optimisation of p1 elements is presented. The boundary of the object to be meshed is bounded by a collection of curved edges for 2D domain and curved surfaces for 3D domain. The algorithm can be classified as an indirect approach to generate linear elements first, which are then converted to the required curvilinear elements by the following steps.

8.8.2.1 Generation of a linear element mesh

A linear element mesh is generated using the ending (corner) nodes of the curved edges (faces) on the object boundary, which can be considered as a more restricted boundary setting for MG as, besides geometry, the curvilinear mesh has to conform with the topology of the discretised curved boundary edges (faces) as well. As shown in Figure 8.144b, a triangular mesh is generated using the curved edges of the object shown in Figure 8.144a. When mid-side nodes are added to all the edges retrieved by *Algorithm Edges_T3* in Section 2.5.4, an approximation to the curvilinear element mesh is created.

8.8.2.2 Snap of boundary node and mesh subdivision

Mid-side nodes on the boundary edges of the linear mesh have to be snapped onto the boundary. If a valid p2 mesh is produced, the MG can be stopped at this point. However, in general, the quality of some curved elements may be so poor that their Jacobian (here, Jacobian means the determinant of Jacobian) is negative, and at times, some elements may even be inverted due to excessive nodal movements in the boundary node-snapping process. In order to rectify this situation and to improve the quality of the curved elements, the linear triangular mesh formed in the step discussed in Section 8.8.2.1 is refined uniformly such

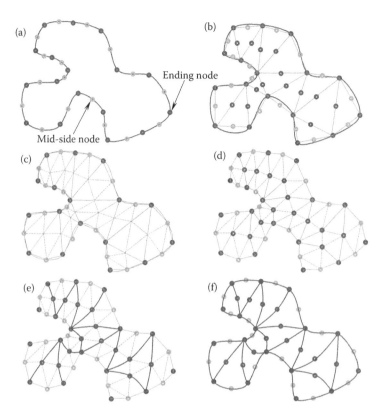

Figure 8.144 Generation of p2 curvilinear mesh by the indirect approach: (a) boundary composed of curved edges; (b) linear element mesh with boundary nodes; (c) snap nodes on the boundary and subdivide the p1 mesh; (d) mesh optimisation to improve the quality of the refined mesh; (e) internal edges of p2 elements become curved; (f) p2 mesh formed with the internal and the boundary edges.

that each triangle is divided into four smaller triangles by joining up the mid-side nodes, as shown in Figure 8.144c.

8.8.2.3 Quality improving by mesh optimisation

Mesh optimisation techniques can now be applied to the subdivided triangular mesh to untangle any inverted triangles and improve the general quality of the triangular elements by widening their interior angles, as shown in Figure 8.144d. Internal edges of p1 mesh in Figure 8.144b turn naturally into curves by the optimisation procedures, which can be traced out quite clearly in Figure 8.144e. As shown in Figure 8.144f, a curvilinear p2 mesh is resulted by just updating the nodal co-ordinates of the p1 mesh (p2 approximation) of Figure 8.144b.

Remarks: In case inverted elements or elements with negative Jacobian still exist in the mesh after one cycle of mesh refinement and optimisation, the mesh subdivision followed by mesh optimisation can be repeated until all the inverted elements are removed, and all the elements with negative Jacobian are cleared (Lopez et al. 2008). The iteration cycle of refinement and optimisation will converge as inverted elements can be untangled by the mesh optimisation and the quality of p1 mesh can be guaranteed subject to the boundary conditions of the domain, such that the Jacobians of all the elements in the refined mesh are positive. Further refinement and optimisation in general can further improve the quality of a valid p2 mesh.

In the current implementation, only node shifting is applied; however, mesh of better quality could be achieved if topological operations are also employed. Adaptive refinement can also be applied at locations of difficult boundary conditions characterised with large curvatures.

The quality of curvilinear elements can be further improved by node shifting optimising the Jacobian of the curved elements connected to an interior node; in 2D, an edge node is shared by only two elements, and a corner node is shared by three or more elements. In this regard, it is necessary to devise an efficient scheme to determine the minimum Jacobian of a curvilinear element. In the event that higher-order curvilinear elements are required, the p1 mesh is simply subdivided according to the order of the element, for instance, 2 × 2 subdivision for p2 mesh, 3 × 3 subdivision for p3 mesh and so on. Even though no concrete results have been obtained as yet for curvilinear meshes in 3D, in theory, the refinement and optimisation should work out pretty well as the two key features for success are still there with a tetrahedral mesh – (i) inverted tetrahedral elements can be untangled by mesh optimisation; and (ii) all levels of refined meshes of tetrahedral elements with non-negative Jacobian can always be generated.

8.8.3 Examples in 2D

In this example, the domain boundary is created by means of B3 splines on a sequence of arbitrarily generated nodal points, as shown in Figure 8.145a. C^2 continuous B3 spline curve S is defined in terms of curve segments. For a B3 spline curve loop defined by n control

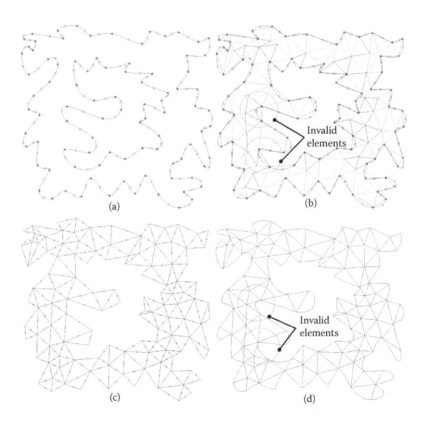

Figure 8.145 Generation of p2 mesh by refinement and optimisation of p1 meshes: (a) domain bounded by two closed loops; (b) p1 mesh based on boundary nodes; (c) mid-side nodes added to the p1 mesh; (d) curvilinear mesh approximated by p1 mesh.

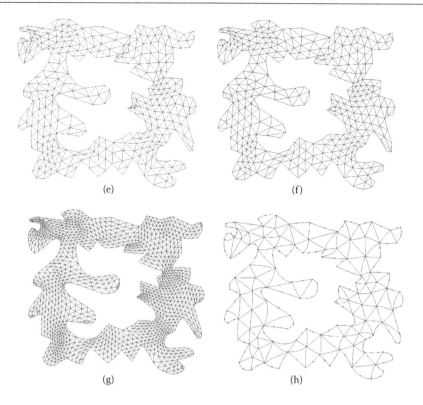

(e)

(f)

(g)

(h)

Figure 8.145 (Continued) Generation of p2 mesh by refinement and optimisation of p1 meshes: (e) p2 elements subdivided into p1 triangles; (f) optimised p1 mesh; (g) further refined and optimised p1 mesh; (h) completed p2 mesh of 148 elements.

points $\{P_1, P_2, \ldots, P_n\}$, there are n curve segments $S_k = S_k(P_k, P_{k+1}, P_{k+2}, P_{k+3})$, k = 1, n, in which if k + j > n, k + j $\mapsto$ k + j – n. A point P(t) on curve segment S_k is given by

$$S = \bigcup_{k=1,n} S_k \text{ with } S_k: P(t) = [t^3 \quad t^2 \quad t \quad 1]\frac{1}{6}\begin{bmatrix} -1 & 3 & -3 & 1 \\ 3 & -6 & 3 & 0 \\ -3 & 0 & 3 & 0 \\ 1 & 4 & 1 & 0 \end{bmatrix}\begin{bmatrix} P_k \\ P_{k+1} \\ P_{k+2} \\ P_{k+3} \end{bmatrix}$$

Fifty-nine and 31 control points are put in to define, respectively, the exterior and the interior boundary loops, on which 59 and 31 p2 (parabolic) curved boundary segments are generated, as shown in Figure 8.145a. The control points are placed in such a way that curved segments of unequal lengths and large curvatures are created. At this moment onwards, the MG problem is fully defined by a domain boundary of 90 parabolic curved segments, and the control points and the B3 spline will not be referred to anymore. Based on the ending points of the boundary segments, a p1 triangular mesh of 148 elements is generated, as shown in Figure 8.145b, in which there are potential invalid elements where the boundary segments cut into the p1 mesh. As shown in Figure 8.145c, the p1 mesh is converted to a p2 mesh by adding 264 mid-side nodes to the edges, out of which 90 mid-side nodes have to be snapped onto the boundary to create curved p2 triangles. The first p2 mesh of 148 curved T6 triangles is thus generated, as shown in Figure 8.145d; the MG can be stopped at this point if there are no invalid elements in the mesh, and the quality of the elements are all up to an acceptable level.

As there are invalid elements in the first p2 mesh, as shown in Figure 8.145d, each element of the p2 mesh is subdivided into four p1 triangles, as shown in Figure 8.145e. Standard mesh smoothing by node shifting is carried out, and the optimised p1 mesh is shown in Figure 8.145f. In order to further improve the quality of the p2 mesh, the p1 mesh is refined again such that each p1 triangle is again divided into four smaller triangles and the optimised mesh is shown in Figure 8.145g. In the refinement process, mid-side nodes on the boundary triangles have to be projected onto the curved boundary segments to produce p2 meshes of better quality. Finally, a third p2 mesh of the original 148 T6 curved triangles with optimised mid-side nodes is generated, as shown in Figure 8.145h.

8.9 ADAPTIVE REFINEMENT ANALYSIS

The idea of an adaptive refinement analysis is to control the discretisation error by increasing the number of degrees of freedom in areas where the previous FE model is inadequate (Hinton et al. 1991; Lee and Lo 1995, 1997, 1999; Lo and Lee 1998; Lee and Hobbs 1999; Lo 2002; Huang 2005a,b). In the adaptive refinement FE analysis, the node spacing or the element size specification is determined based on the error in the FE solution (Babushka and Rheinboldt 1978); the adaptive meshing can be extended to as far as moving mesh strategy in gradient flow equations (Huang and Russell 1999; Cao et al. 1999) and solutions exhibiting strong anisotropy (Farrell et al. 2011). Since the element size distribution is the main concern in FE MG, only the h-refinement will be considered but not the hp-refinement (Zeinkiewicz et al. 1989; Ainsworth and Senior 1997a,b) or the hr-refinement (Mosler and Ortiz 2007; Li et al. 2001). The basic ideas and the essential steps in adaptive meshing will be briefly described in this section. Taking the 3D linear elasticity problem as an example, how the element size is determined for an optimal rate of convergence is discussed.

8.9.1 Fundamentals in solid mechanics and error in FE solution

Let $\hat{\mathbf{u}}$ be the FE solution on the meshed problem domain Ω, and our task is to set up an element size function over the FE mesh for the next FE analysis based on an error estimate on the displacement field $\hat{\mathbf{u}}$. The FE strain and stress are computed from $\hat{\mathbf{u}}$ as follows.

$$\hat{\boldsymbol{\varepsilon}} = \frac{1}{2}(\nabla\hat{\mathbf{u}} + \hat{\mathbf{u}}\nabla), \quad \hat{\boldsymbol{\sigma}} = \mathbf{C} : \hat{\boldsymbol{\varepsilon}}$$

where ∇ is the gradient operator on the problem domain Ω, and C is the four-order material constitutive tensor. Error in displacement $\mathbf{e}_u$ and error in stress $\mathbf{e}_\sigma$ are defined as

$$\mathbf{e}_u = \hat{\mathbf{u}} - \mathbf{u}, \quad \mathbf{e}_\sigma = \hat{\boldsymbol{\sigma}} - \boldsymbol{\sigma}$$

where $\mathbf{u}$ and $\boldsymbol{\sigma}$ are, respectively, the exact solution of displacement and stress. The energy norm and the error in energy norm are given by

$$\|\mathbf{u}\|_\Omega = \left[\int_\Omega \boldsymbol{\sigma} : \boldsymbol{\varepsilon} \; d\Omega\right]^{\frac{1}{2}} \text{ and } \|\mathbf{e}_u\|_\Omega = \left[\int_\Omega \mathbf{e}_\sigma : \mathbf{e}_\varepsilon \; d\Omega\right]^{\frac{1}{2}} \quad \text{where } \mathbf{e}_\varepsilon = \mathbf{C}^{-1} : \mathbf{e}_\sigma$$

For isotropic material, stress and strain are related by

$$\boldsymbol{\sigma} = C : \boldsymbol{\varepsilon} = \lambda \mathrm{tr}(\boldsymbol{\varepsilon})\mathbf{I} + 2\mu\boldsymbol{\varepsilon}, \quad \boldsymbol{\varepsilon} = C^{-1} : \boldsymbol{\sigma} = \frac{1}{E}\left[(1+\nu)\boldsymbol{\sigma} - \nu\mathrm{tr}(\boldsymbol{\sigma})\mathbf{I}\right] \quad \mathbf{I} = \text{Identity tensor}$$

where λ and μ are Lamé constants, E is the Young's modulus and ν is the Poisson's ratio, which are related as follows:

$$\lambda = \frac{\nu E}{(1+\nu)(1-2\nu)}, \quad \mu = \frac{E}{2(1+\nu)}, \quad E = \frac{\mu(3\lambda+2\mu)}{\lambda+\mu}, \quad \nu = \frac{\lambda}{2(\lambda+\mu)}.$$

Since the absolute error is problem-dependent, a dimensionless relative error $\eta = \dfrac{\|e_u\|_\Omega}{\|u\|_\Omega}$ is more useful as the criterion for convergence of the adaptive refinement process. A specific value η_0, say, 1% or 2%, can also be prescribed to set an acceptable level of accuracy for the FE solution, such that $\eta < \eta_0$.

8.9.2 *A priori* and *a posteriori* error estimates

A posteriori error estimate is needed to assess the error of the FE solution, and *a priori* error estimate provides a guide as to how elements should be refined to achieve an optimal rate of convergence (Rheinboldt 1985). Displacement **u** is the only independent variable employed in the displacement formulation. If the order of FE interpolation p is kept constant, the error of the FE solution and the element size h are related by *a priori* estimate (Babuska and Szabo 1982)

$$\|e_u\|_\Omega \le \alpha h^{\min(p,\phi)} \tag{8.7}$$

where ϕ is the strength of singularity present, which characterises the smoothness of the solution, and α is a constant dependent on the problem but not on the element size h. As element size h and the total number of degrees of freedom of the FE mesh N is related by $h \propto N^{-\frac{1}{\dim}}$, with dim = 3, Equation 8.7 can be rewritten in the form

$$\|e_u\|_\Omega \le \tilde{\alpha} N^{-\frac{\min(p,\phi)}{3}} \quad \text{in which } \tilde{\alpha} \text{ is independent of N.}$$

For the displacement FE formulation, p must be greater than or equal to 1 to ensure convergence of the FE solution. However, for linear elasticity problem, the value of ϕ is always less than 1. Hence, in general, we have

$$\phi \le p$$

This means that whenever there are singularities within the problem domain Ω such as a re-entrant corner or a sudden change in boundary conditions, the rate of convergence of the FE solution will be reduced. Nevertheless, the rate of convergence can be improved by the construction of a proper *optimal* mesh in such a way that the error of the solution is equally distributed among all the elements in the mesh. The effect of the singularities will then be suppressed, and the convergence rate will become

$$\left\| e_u \right\|_\Omega \le \tilde{\alpha} N^{-\frac{p}{3}} \tag{8.8}$$

The main objective of the adaptive analysis is to achieve the above optimal rate of convergence and to reach the desired accuracy with a minimum number of degrees of freedom. It should be noted that the presence of singularity will significantly affect the size distribution of an optimal mesh. Small elements have to be placed at these singularities, and the resulting optimal mesh will be rapidly graded towards these positions (Kelly et al. 1983).

The *a posteriori* error estimate here adopted is based on the Zienkiewicz and Zhu (1987) (Z^2) error estimator. The central idea of this error estimate is to obtain a better approximation of the exact stress from the FE solution following a post-processing procedure. There are many ways in obtaining an improved stress field from the FE solution (Lee and Lo 1997a,b); however, in terms of simplicity and efficiency, the super-convergent patch recovery (SPR) technique proposed in Zienkiewicz and Zhu (1992) will be employed for stress recovery, though recovery by equilibrium patches can also be used for triangular and tetrahedral elements (Boroomand and Zienkiewicz 1997).

8.9.3 Super-convergence and optimal sampling points

It is noted that on many occasions, the displacements or the field variables are more accurately sampled at the nodes defining an element and that the gradients or stresses are best sampled at some interior points within the element.

8.9.3.1 One-dimensional example

Consider a heat-conduction problem of a second-order equation

$$\frac{d}{dx}\left(k \frac{du}{dx} \right) + \beta u + Q = 0$$

A typical finite element analysis using linear interpolation is shown in Figure 8.146. The field variable u is, in general, more accurate at nodal points; however, for the gradient, we observe

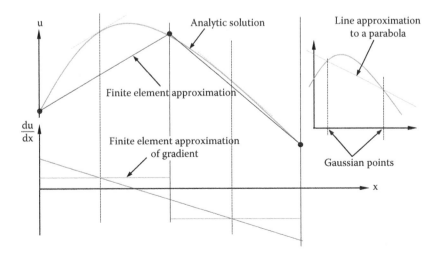

Figure 8.146 Finite element approximation to the heat conduction problem.

large discrepancies at nodal points, but somewhere within the element, the results are nearly exact. Summarising from the one-dimensional example, the strategic points are as follows:

1. Displacements are best sampled at the nodes.
2. Gradients' (or stresses) best accuracy are at Gaussian points corresponding to the polynomial of the field variable used in the approximation.

At such points, the order of convergence of the function or its gradient is one order higher than that would be anticipated, and thus, these points are referred to as *super-convergent*. The same concept can be readily extended to elements of higher dimensions. For instance, the super-convergent points for stress evaluation over a Q8 or L9 element are the 2 × 2 Gaussian points and those for the H20 hexahedral elements are the 2 × 2 × 2 Gaussian points. However, the super-convergent points for triangular and tetrahedral elements are less conspicuous and are more difficult to identify.

8.9.3.2 Super-convergent patch recovery

Attempts are generally made to recover the nodal values of stresses from those super-convergent sampling points. Within the element, the recovered (smoothed) stress σ^* are defined by FE interpolation from the smoothed stress (σ_k^*) at nodal points, as shown in Figure 8.147. The SPR procedure for stress recovery will be elucidated by means of 2D examples, as shown in Figure 8.148. To recover stress component by component, we assume for each stress component a complete second-order polynomial over a patch of Q8 quadratic quadrilateral elements, i.e.

$$\sigma_i^* = \underset{1 \times m}{P} \underset{m \times 1}{a} \qquad i = 1, 2, 3 \text{ for 2D elasticity}$$

$$Polynomial\ P = \begin{bmatrix} 1 & x & y & x^2 & xy & y^2 \end{bmatrix}$$

$$Coefficients\ a = \begin{bmatrix} a_1 & a_2 & a_3 & a_4 & a_5 & a_6 \end{bmatrix}^T$$

By means of the least square fit (LSF) over strategic super-convergent points, we have

$$LSF = \sum_{\alpha=1}^{N} \left(\sigma_{\alpha i}^s - P_\alpha a \right)^2 \quad \text{where N = number of superconvergent points}$$

$$P_\alpha = P(x_\alpha, y_\alpha), \ \sigma_\alpha^s = \text{stress at super-convergent point } \alpha$$

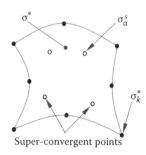

Super-convergent points

Figure 8.147 SPR over a Q8 isoparametric element.

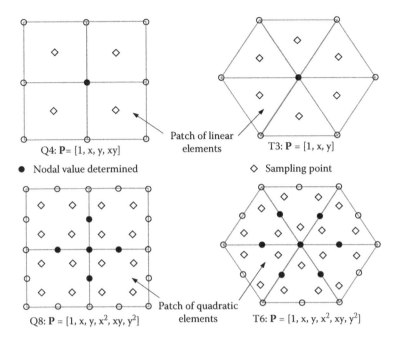

Figure 8.148 SPR sampling strategy for common 2D elements.

$$\text{LSF}(\mathbf{a}+\eta\mathbf{v}) = \sum_{\alpha=1}^{N}\left[\sigma_{\alpha i}^{s} - \mathbf{P}_{\alpha}(\mathbf{a}+\eta\mathbf{v})\right]^{2} \Rightarrow \frac{d}{d\eta}\text{LSF}(\mathbf{a}+\eta\mathbf{v}) = \sum_{\alpha=1}^{N}\left[\sigma_{\alpha i}^{s} - \mathbf{P}_{\alpha}(\mathbf{a}+\eta\mathbf{v})\right]^{T}(\mathbf{P}_{\alpha}\mathbf{v})$$

$$\frac{d}{d\eta}\text{LSF}(\mathbf{a}+\eta\mathbf{v})\bigg|_{\eta=0} = \sum_{\alpha=1}^{N}\left[\sigma_{\alpha i}^{s} - \mathbf{P}_{\alpha}\mathbf{a}\right]^{T}(\mathbf{P}_{\alpha}\mathbf{v}) = 0 \Rightarrow \sum_{\alpha=1}^{N}\left[\sigma_{\alpha i}^{s} - \mathbf{P}_{\alpha}\mathbf{a}\right]^{T}\mathbf{P}_{\alpha} = 0$$

$$\sum_{\alpha=1}^{N}\left[\sigma_{\alpha i}^{s} - \mathbf{P}_{\alpha}\mathbf{a}\right]^{T}\mathbf{P}_{\alpha} = 0 \Rightarrow \sum_{\alpha=1}^{N}\mathbf{P}_{\alpha}^{T}\left[\sigma_{\alpha i}^{s} - \mathbf{P}_{\alpha}\mathbf{a}\right] = 0 \Rightarrow \sum_{\alpha=1}^{N}\mathbf{P}_{\alpha}^{T}\mathbf{P}_{\alpha}\mathbf{a} = \sum_{\alpha=1}^{N}\sigma_{\alpha i}^{s}\mathbf{P}_{\alpha}^{T}$$

$$\text{or } \mathbf{a} = \mathbf{A}^{-1}\mathbf{b} \quad \text{where} \underset{m\times m}{\mathbf{A}} = \sum_{\alpha=1}^{N}\underset{m\times 1}{\mathbf{P}_{\alpha}}^{T}\underset{1\times m}{\mathbf{P}_{\alpha}} \text{ and } \underset{m\times 1}{\mathbf{b}} = \sum_{\alpha=1}^{N}\sigma_{\alpha i}^{s}\underset{m\times 1}{\mathbf{P}_{\alpha}^{T}}$$

Components of smoothed stress $\boldsymbol{\sigma}^{*}$ are computed in turn from the components of stresses at super-convergent points, and enhanced stresses at nodal points k = 1, n (n = number of nodes in the element) are evaluated. Thus, a smoothed high-quality stress field can be set up over the entire mesh by the standard FE interpolation (H_k) such that $\boldsymbol{\sigma}^{*} = H_k\boldsymbol{\sigma}_k^{*}$.

$$\text{SPR: } \boldsymbol{\sigma}_{\alpha}^{s} \rightarrow \boldsymbol{\sigma}_{\alpha}^{*} \rightarrow H_k\boldsymbol{\sigma}_{\alpha}^{*} = \boldsymbol{\sigma}^{*}, \ \boldsymbol{\sigma}^{*} = \begin{bmatrix} \sigma_1^{*} \\ \sigma_2^{*} \\ \sigma_3^{*} \end{bmatrix} = \begin{bmatrix} \mathbf{Pa}_1 \\ \mathbf{Pa}_2 \\ \mathbf{Pa}_3 \end{bmatrix} \text{ or } \boldsymbol{\sigma}_k^{*} = \begin{bmatrix} \mathbf{P}_k\mathbf{a}_1 \\ \mathbf{P}_k\mathbf{a}_2 \\ \mathbf{P}_k\mathbf{a}_3 \end{bmatrix} k = 1, n$$

8.9.3.3 The Herrmann theorem and optimal sampling points

The LSF has additional justification in self-adjoint problems, for which the minimisation of the total potential energy (TPE) is equivalent to the LSF of the approximation stresses to the exact solution.

$$\text{TPE}(\mathbf{u}) = \frac{1}{2}\int_{\Omega}(\mathbf{Lu})^{\mathrm{T}}\mathbf{A}(\mathbf{Lu}) - \mathbf{f}^{\mathrm{T}}\mathbf{u}\,\mathrm{d}\Omega$$

$$\text{and LSF}(\mathbf{u}) = \frac{1}{2}\int_{\Omega}\big(\mathbf{L}(\mathbf{u}-\tilde{\mathbf{u}})\big)^{\mathrm{T}}\mathbf{A}\big(\mathbf{L}(\mathbf{u}-\tilde{\mathbf{u}})\big)\mathrm{d}\Omega$$

where $\mathbf{L}$ = linear operator, matrix $\mathbf{A}$ is symmetric, $\mathbf{f}^{\mathrm{T}}\mathbf{u}$ = external work and $\tilde{\mathbf{u}}$ = exact solution.

Herrmann Theorem

LSF = TPE + *constant*

Minimisation of the TPE and the LSF projection onto exact solutions are equivalent problems. Noting that $\delta\text{TPE}(\tilde{\mathbf{u}}) = 0$, a proof can be established by variation, i.e. $\delta\text{TPE} = \delta\text{LSF}$.

In 3D, the number and the positions of the sampling points in terms of barycentre (volume) co-ordinates for tetrahedral elements T4, T10 and T20 for stress recovery are as follows.

T4 : 1 sampling point at the centre $\left(\dfrac{1}{4}, \dfrac{1}{4}, \dfrac{1}{4}, \dfrac{1}{4}\right)$

T10 : 4 sampling points at (b, a, a, a), (a, b, a, a), (a, a, b, a),

(a, a, a, b), $a = \dfrac{5-\sqrt{5}}{20}$, $b = \dfrac{5+3\sqrt{5}}{20}$

T20 : 5 sampling points at $\left(\dfrac{1}{4}, \dfrac{1}{4}, \dfrac{1}{4}, \dfrac{1}{4}\right)$, $\left(\dfrac{1}{2}, \dfrac{1}{6}, \dfrac{1}{6}, \dfrac{1}{6}\right)$,

$\left(\dfrac{1}{6}, \dfrac{1}{2}, \dfrac{1}{6}, \dfrac{1}{6}\right)$, $\left(\dfrac{1}{6}, \dfrac{1}{6}, \dfrac{1}{2}, \dfrac{1}{6}\right)$, $\left(\dfrac{1}{6}, \dfrac{1}{6}, \dfrac{1}{6}, \dfrac{1}{2}\right)$

The basis polynomials adopted in the construction of the smoothed stress field are

T4 : $\mathbf{P} = [1, x, y, z]$

T10: $\mathbf{P} = [1, x, y, z, xy, yz, xz, x^2, y^2, z^2]$

T20: $\mathbf{P} = [1, x, y, z, xy, yz, xz, x^2, y^2, z^2, x^3, y^3, z^3, x^2y, y^2z, z^2x, zx^2, xy^2, yz^2, xyz]$

The rate of convergence of the recovery stress is affected by the smoothness of the solution as well as the FE mesh. If the exact solution is smooth without any singularity, the recovery stress will be super-convergent independent of the design of the FE mesh. Both uniformly refined and adaptively refined meshes will give a super-convergent recovered stress field. However, in the presence of singularities, adaptive refinement is required to ensure the super-convergence of the recovered stresses.

8.9.4 Adaptive refinement strategy

As the exact stress σ is not available except in some problems where analytical solutions exist, the exact error of the FE solution cannot be computed. However, we can obtain an estimate of the error in the FE solution by replacing the exact stress σ with the more accurate recovered stress σ^*, i.e.

$$\overline{\|e_u\|}_\Omega = \left[\int_\Omega e_\sigma^* : e_\varepsilon^* \, d\Omega \right]^{\frac{1}{2}} \quad \text{where } e_\sigma^* = \hat{\sigma} - \sigma^* \text{ and } e_\varepsilon^* = C^{-1} : e_\sigma^*$$

As FE solution is known on an element-by-element basis, in practice, estimated error norm is computed in terms of individual elements, and the error over the entire problem domain Ω is evaluated as a summation of errors in the elements.

$$\overline{\|e_u\|}_\Omega = \left[\sum_{i=1}^{N_e} \overline{\|e_u\|}_{\Omega_i}^2 \right]^{\frac{1}{2}} \quad \text{where } \overline{\|e_u\|}_{\Omega_i}^2 = \int_{\Omega_i} e_\sigma^* : e_\varepsilon^* \, d\Omega$$

$\overline{\|e_u\|}_{\Omega_i}$ is the estimate error norm of the ith element, N_e is the number of elements in the FE mesh and Ω_i represents the domain of element i. The accuracy of the error estimator can be assessed by the effectivity index θ, which is defined as the ratio between the predicted error and the actual error, i.e.

$$\theta = \frac{\overline{\|e_u\|}_\Omega}{\|e_u\|_\Omega}$$

The error estimate is said to be asymptotically exact if θ converges to 1 as the mesh is refined. The effectiveness of the Z^2 error estimate depends, obviously, on the quality of the smoothed stress field σ^*. Zienkiewicz and Zhu (1992) showed that whenever the recovery stress is converging at a rate higher than that of the FE stress, the error estimator will be asymptotically exact. The aim of an adaptive refinement procedure is to take into account the information derived from the error estimate to design an optimal mesh such that the discretisation error is equally distributed within all the elements. Let $\|\hat{u}\|_\Omega$ be the energy norm of the FE solution; the estimated energy norm $\|u\|_\Omega$ is given by

$$\overline{\|u\|}_\Omega = \left[\|\hat{u}\|_\Omega^2 + \overline{\|e_u\|}_\Omega^2 \right]^{\frac{1}{2}}$$

Similarly, the estimated relative error of the FE solution $\bar{\eta}$ is given by

$$\bar{\eta} = \frac{\overline{\|e_u\|}_\Omega}{\|u\|_\Omega}$$

The expected number of elements needed to achieve the target relative error (η_0), NE′, is calculated by assuming that the effect of singularity can be eliminated by an even error distribution, and the convergence rate implied by Equation 8.8 holds. We have

$$NE' = NE \left[\frac{\bar{\eta}}{\eta_0} \right]^{\frac{3}{p}}$$

where NE is the number of elements in the current FE mesh. As the refinement will only be optimal if the error of the mesh is uniformly distributed among all the elements, the local allowable error norm (e_a) is defined for each element.

$$e_a = \eta_0 \frac{\overline{\|u\|}_\Omega}{\sqrt{NE'}}$$

The local refinement indicator for the ith element, ρ_i, is thus given by

$$\rho_i = \frac{\overline{\|e_u\|}_{\Omega_i}}{e_a}$$

The error norm for a particular finite element converges in a rate (Onate and Bugeda 1993)

$$\|e_u\|_{\Omega_i} = \left[\int_{\Omega_i} O\left(h_i^\beta\right) C^{-1} O\left(h_i^\beta\right) d\Omega \right]^{\frac{1}{2}} = O\left(h_i^\beta\right)\Omega_i^{\frac{1}{2}} = O\left(h_i^{\beta+\frac{dim}{2}}\right) = O\left(h_i^{\beta+\frac{3}{2}}\right) \tag{8.9}$$

where h_i is the size of the ith element. β is the convergence rate of the element error norm, which is equal either to the strength of singularity ϕ if the element is located near a singularity (i.e. connected to a singularity in the actual implementation) or to the order of interpolation p of the finite element. Following Equation 8.9, the refined element size h_i' is given by

$$h_i' = h_i \rho_i^{\frac{-2}{2\beta+3}} \tag{8.10}$$

The FE mesh will be refined in cycles of adaptive analysis until the FE solution falls below the target error, i.e. $\bar{\eta} \le \eta_0$. If the error estimator is asymptotically exact, the estimated relative error will be pretty close to the actual error.

The refined element size h'_i given by Equation 8.10 is defined only over the ith element. However, the input required by an automatic mesh generator are the element sizes specified at the nodal points of the mesh. Such a nodal value of the element size can be readily calculated by averaging the values of all the elements incident to a given node. Usually, the required element size at the nodal points of the FE mesh are computed and stored for MG, and within a tetrahedral element (T4, T10 or T20), nodal spacing is evaluated by FE interpolation. As the problem domain Ω is discretised into FE elements, a piecewise continuous map of the node spacing function is thus established over the entire problem domain. A typical module in the determination of the element size in an adaptive refinement analysis will take as input the FE solution of displacements (field variables) at the nodal points and output the values of the required element size at the corresponding nodal points. The element size can be smoothed out based on an element area gradient (Howlett and Zundel 2009).

8.9.5 Examples

Two examples are included in this section to illustrate the adaptive refinement procedure. The FE sizes of adaptive refinement meshes are computed by the procedure described in Sections 8.9.1–8.9.4, and the MG is done by the generic refinement algorithm given in Section 8.4.3. The first example is about an L-shaped object under tension on one of its faces, and the dimensions, support conditions and loading are shown in Figure 8.149. The only singularity is along the re-entrant corner line AB, which is of singularity strength of 0.5445. The final adaptively refined FE mesh of 31,607 T10 elements is shown in Figure 8.150. A graphic plot of the adaptive refinement analysis using T4, T10 and T20 elements is depicted in Figure 8.153.

A notch under uniform tensile force is considered as the second example, as shown in Figure 8.151a. Owing to symmetry, only one-eighth of the notch is modelled, and the geometry along with the boundary and loading conditions are shown in Figure 8.151b. Again, only one line (line AB) of singularity is present due to a sudden change in boundary condition, and it is expected that highly graded elements are needed around this singular line. An adaptive refinement FE mesh of 30,425 T10 elements for this problem is shown in Figure 8.152. A graphic plot of the adaptive refinement analysis using different orders of tetrahedral elements is given in Figure 8.154.

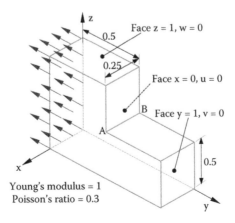

Figure 8.149 L-shaped domain subject to a horizontal load.

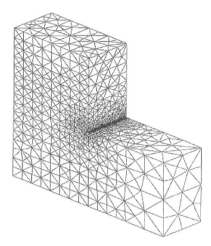

Figure 8.150 Adaptive refinement mesh of 48,214 nodes and 31,607 T10 elements.

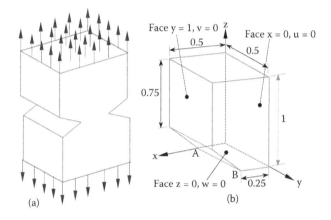

Figure 8.151 Notch under uniform tensile force. (a) Notch under tensile load and (b) model of one-eighth of the notch.

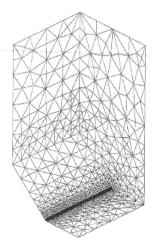

Figure 8.152 Refinement mesh of 49,512 nodes and 30,425 T10 elements.

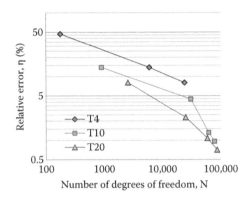

Figure 8.153 **Rate of convergence for example 1.**

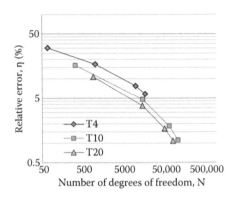

Figure 8.154 **Rate of convergence for example 2.**

From the graphs shown in Figures 8.153 and 8.154, it is observed that in spite of line singularities in Examples 1 and 2, the FE solution converges with the optimal rate of convergence as predicted by Equation 8.7 for the adaptive refinement analysis based on the element size computed using Equation 8.9. Although T20 can achieve a higher convergence rate, for the same target relative error η_o of 1%, T10 element is the most efficient in terms of both memory requirement and computational cost.

References

Aftosmis MJ, Berger MJ, Melton JE (1998) Robust and efficient Cartesian mesh generation for component-based geometry, *AIAA Journal*, **36**, Issue 6, 952–960.

Aiffa M, Flaherty JE (2003) A geometrical approach to mesh smoothing, *Computer Methods in Applied Mechanics and Engineering*, **192**, 4497–4514.

Ainsworth M, Senior B (1997a) An adaptive refinement strategy for hp-finite element computations, *Applied Numerical Mathematics*, **26**, Issues 1–2, 165–178.

Ainsworth M, Senior B (1997b) Aspects of an adaptive hp-finite element method: Adaptive strategy, conforming approximation and efficient solvers, *Computer Methods in Applied Mechanics and Engineering*, **150**, Issues 1–4, 65–87.

Akhras G, Dhatt G (1976) An automatic code relabeling scheme for minimizing a matrix or network bandwidth, *International Journal for Numerical Methods in Engineering*, **10**, 787–797.

Akkiraju N, Edelsbrunner H (1996) Triangulating the surface of a molecule, *Discrete Applied Mathematics*, **71**, Issues 1–3, 5–22.

Alauzet F (2010) Size gradation control of anisotropic meshes, *Finite Elements in Analysis and Design*, **46**, 181–202.

Alauzet F, Frey PJ (2005) Anisotropic mesh adaptation for CFD computation, *Computer Methods in Applied Mechanics and Engineering*, **194**, 5068–5082.

Alauzet F, George PL, Mohammadi B, Frey P, Borouchaki H (2003) Transient fixed point-based unstructured mesh adaptation, *International Journal for Numerical Methods in Engineering*, **43**, 729–745.

Alauzet F, Li X, Seol S, Shephard MS (2006) Parallel anisotropic 3D mesh adaptation by mesh modification, *Engineering with Computers*, **21**, 247–258.

Alauzet F, Loseille A (2009) On the use of space filling curves for parallel anisotropic mesh adaptation, *Proceedings of the 18th International Meshing Roundtable*, Springer, 337–357.

Alleaume A (2009) Automatic non-manifold topology recovery and geometry noise removal, *Proceedings of the 18th International Meshing Roundtable*, Springer, 266–279.

Alliez P, Cohen-Steiner D, Yvinec M, Desbrun M (2005) Variational tetrahedral meshing, *ACM Transactions on Graphics*, **24**, Issue 3, 617–625.

Amenta N, Choi S, Rote G (2003) Incremental constructions con BRIO, *Proceedings of the 19th Annual ACM Symposium on Computational Geometry*, ACM Press, New York, 211–219.

Amezua E, Hormaza MV, Hernandez A, Ajuria MBG (1995) A method for the improvement of 3D solid finite-element meshes, *Advances in Engineering Software*, **22**, Issue 1, 45–53.

Anderson BD, Benzley SE, Owen SJ (2009) Automatic all quadrilateral mesh adaptation through refinement and coarsening, *Proceedings of the 18th International Meshing Roundtable*, Springer, 556–574.

Antiga L, Ene-Lordache B, Remuzzi A (2003) Computational geometry for patient-specific reconstruction and meshing of blood vessels from MR and CT angiography, *IEEE Transactions on Medical Imaging*, **22**, Issue 5, 674–684.

Armstrong CG (2000) Modeling requirements for finite-element analysis, *Computer-Aided Design*, **26**, Issue 7, 573–578.

Arnold DN, Mukherjee A, Pouly L (2000) Locally adapted tetrahedral meshes using bisection, *SIAM Journal on Scientific Computing*, **22**, Issue 2, 431–448.

Attali D, Boissonnat JD (2004) A linear bound on the complexity of the Delaunay triangulation of points on polyhedral surfaces, *Discrete & Computational Geometry*, **31**, 369–384.

Aubry R, Houzeaux G, Vazquez M (2011) A surface remeshing approach, *International Journal for Numerical Methods in Engineering*, **85**, 1475–1498.

Aurenhammer F (1991) Voronoi diagrams – A survey of a fundamental geometric data structure, *ACM Computing Surveys*, **23**, 345–405.

Avis D, Bhattacharya BK (1983) Algorithm for computing d-dimensional Voronoi diagrams and their duals, in *Advances in Computing Research*, **1**, Ed. FP Preparata, JAI Press, Greenwich, CT, 159–188.

Babuska I, Rheinboldt WC (1978) Error estimates for adaptive finite element computations, *SIAM Journal on Numerical Analysis*, **15**, 736–754.

Babuska I, Szabo B (1982) On the rate of convergence of the finite element method, *International Journal for Numerical Methods in Engineering*, **18**, 323–341.

Baehmann PL, Wittchen SL, Shephard MS, Grice KR, Yerry MA (1987) Robust geometrically-based automatic two-dimensional mesh generation, *International Journal for Numerical Methods in Engineering*, **24**, 1043–1078.

Bajaj C, Schaefer S, Warren J, Xu GL (2002) A subdivision scheme for hexahedral meshes, *Visual Computer*, **18**, Issues 5–6, 343–356.

Bajaj CL, Coyle EJ, Lin KN (1999) Tetrahedral meshes from planar cross-sections, *Computer Methods in Applied Mechanics and Engineering*, **179**, Issues 1–2, 31–52.

Baker TJ (1989a) Developments and trends in three-dimensional mesh generation, *Applied Numerical Mathematics*, **5**, 275–309.

Baker TJ (1989b) Automatic mesh generation for complex three-dimensional regions using a constrained Delaunay triangulation, *Engineering with Computers*, **5**, 161–175.

Baker TJ (1992) Tetrahedral mesh generation by a constrained Delaunay triangulation, in *Artificial Intelligence, Expert System and Symbolic Computing*, Eds. EN Houtis and JR Rice, Elsevier Science Publishers, BV, North-Holland.

Baker TJ (1997) Mesh adaptation strategies for problems in fluid dynamics, *Finite Elements in Analysis and Design*, **25**, Issues 3–4, 243–273.

Baker TJ (2002) Mesh movement and metamorphosis, *Engineering with Computers*, **18**, 188–198.

Baker TJ (2005) Mesh generation: Art or science? *Progress in Aerospace Science*, **41**, Issue 1, 29–63.

Balanis CA (1989) *Advanced Engineering Electromagnetics*, Wiley, New York.

Balmelli L, Vetterli M, Liebling TM (2002) Mesh optimization using global error with application to geometry simplification, *Graphical Models*, **64**, 230–257.

Bansch E (1991) Local mesh refinement in 2 and 3 dimensions, *Impact Computer Science Engineering*, **3**, 181–191.

Batista VHF, Millman DL, Pion S, Singler J (2010) Parallel geometric algorithms for multi-core computers, *Computational Geometry – Theory and Applications*, **43**, Issue 8, 663–677.

Bechet E, Cuilliere JC, Trochu F (2002) Generation of a finite element mesh from stereolithography (STL) files, *Computer-Aided Design*, **34**, Issue 1, 1–17.

Benabbou A, Borouchaki H, Laug P, Lu J (2009) Geometrical modeling of granular structures in two and three dimensions. Applications to nanostructures, *International Journal for Numerical Methods in Engineering*, **80**, 425–454.

Benabbou A, Borouchaki H, Laug P, Lu J (2010) Numerical modeling of nanostructured materials, *Finite Elements in Analysis and Design*, **46**, Issues 1–2, 165–180.

Bentley JL (1975) Multidimensional binary search trees used for associative searching, *Communications of the ACM*, **18**, Issue 9, 509–517.

Bentley JL, Ottmann TA (1979) Algorithms for reporting and counting geometric intersections, *IEEE Transaction on Computers*, **28**, Issue 9, 643–647.

Benzley SE, Harris NJ, Scott M, Borden M, Owen SJ (2005) Conformal refinement and coarsening of unstructured hexahedral meshes, *Journal of Computing and Information Science in Engineering*, **5**, Issue 4, 330–333.

Benzley SE, Perry E, Merkley K, Clark B, Sjaardema G (1995) A comparison of all-hexahedral and all-tetrahedral finite element meshes for elasto-plastic analysis, *Proceedings of the 4th International Meshing Roundtable*, Sandia National Laboratories, 179–191, October 1995.

Bern M, Eppstein D (2000) Quadrilateral meshing by circle packing, *International Journal of Computational Geometry & Applications*, **10**, Issue 4, 347–360.

Bern M, Eppstein D, Gilbert J (1994) Provably good mesh generation, *Journal of Computer and System Science*, **48**, Issue 3, 384–409.

Bern M, Eppstein D, Teng SH (1999) Parallel construction of quadtrees and quality triangulation, *International Journal of Computational Geometry & Applications*, **9**, Issue 6, 517–532.

Bern M, Eppstein D, Yao F (1991) The expected extremes in Delaunay triangulation, *International Journal of Computational Geometry & Applications*, **1**, Issue 1, 79–91.

Bern M, Mitchell S, Ruppert J (1995) Linear-size nonobtuse triangulation of polygons, *Discrete & Computational Geometry*, **14**, Issue 4, 411–428.

Beyer T, Schaller G, Deutsch A, Meyer-Hermann M (2005) Parallel dynamic and kinetic regular triangulation in three dimensions, *Computer Physics Communications*, **172**, 86–108.

Blacker TD, Stephenson MB (1991) Paving – A new approach to automated quadrilateral mesh generation, *International Journal for Numerical Methods in Engineering*, **32**, Issue 4, 811–847.

Blacker TD, Stephenson MB, Canann S (1991) Analysis automation with paving – A new quadrilateral meshing technique, *Advances in Engineering Software and Workstations*, **13**, Issues 5–6, 332–337.

Blelloch GE, Hardwick JC, Miller GL, Talmor D (1999) Design and implementation of a practical parallel Delaunay algorithm, *Algorithmica*, **24**, Issues 3–4, 243–269.

Boender E, Bronsvoort WF, Post FH (1994) Finite-element mesh generation from construction solid geometrical models, *Computer-Aided Design*, **26**, Issue 5, 379–392.

Boissonnat JD, Devillers O, Hornus S (2009) Incremental construction of the Delaunay triangulation and the Delaunay graph in medium dimension, *Proceedings of the 25th Annual Symposium on Computational Geometry*, Aarhus, Denmark, June 8–10, 2009.

Boissonnat JD, Sharir M, Tagansky B, Yvinec M (1998) Voronoi diagram in higher dimensions under certain polyhedral distance functions, *Discrete & Computational Geometry*, **19**, 485–519.

Bonet J, Peraire J (1991) An alternating digital tree (ADT) algorithm for geometric searching and intersection problems, *International Journal for Numerical Methods in Engineering*, **31**, 1–17.

Boroomand B, Zienkiewicz OC (1997) Recovery by equilibrium in patches (REP), *International Journal for Numerical Methods in Engineering*, **40**, 137–164.

Borouchaki H, Frey PJ (1998) Adaptive triangular-quadrilateral mesh generation, *International Journal for Numerical Methods in Engineering*, **41**, Issue 5, 915–934.

Borouchaki H, George PL (1996) Delaunay mesh with anisotropic specification, *Comptes Rendus de Academie des Sciences Series I – Mathematique*, **323**, Issue 10, 1141–1146.

Borouchaki H, George PL (1997) Aspects of 2D Delaunay mesh generation, *International Journal for Numerical Methods in Engineering*, **40**, 1957–1975.

Borouchaki H, George PL, Hecht F, Laug P, Saltel E (1997a) Delaunay mesh generation governed by metric specification: 1. Algorithms, *Finite Elements in Analysis and Design*, **25**, 61–83.

Borouchaki H, George PL, Lo SH (1996) Optimal Delaunay point insertion, *International Journal for Numerical Methods in Engineering*, **39**, 3407–3437.

Borouchaki H, George PL, Lo SH (2000b) Boundary enforcement by face splits in Delaunay based mesh generation, *Numerical Grid Generation in Computational Field Simulations*, The International Society of Grid Generation, 203–221, September 2000.

Borouchaki H, George PL, Mohammadi B (1997b) Delaunay mesh generation governed by metric specification: 2. Applications, *Finite Elements in Analysis and Design*, **25**, 85–109.

Borouchaki H, Laug P (2004) Simplification of composite parametric surface meshes, *Engineering with Computers*, **20**, Issue 3, 176–183.

Borouchaki H, Laug P, George PL (1999) About parametric surface meshing, *2nd Symposium on Trends in Unstructured Mesh Generation*, USNCCM'99, University of Colorado, Boulder, CO.

Borouchaki H, Laug P, George PL (2000a) Parametric surface meshing using combined advancing-front generalized Delaunay approach, *International Journal for Numerical Methods in Engineering*, **49**, Issues 1–2, 233–259.

Borouchaki H, Lo SH (1995) Fast Delaunay triangulation in three dimensions, *Computer Methods in Applied Mechanics and Engineering*, **128**, Issues 1–2, 153–167.

Borouchaki H, Villard J, Laug P, George PL (2005) Surface mesh enhancement with geometric singularities identification, *Computer Methods in Applied Mechanics and Engineering*, **194**, Issues 48–49, 4885–4894.

Bossen FJ, Heckbert PS (1996) A pliant method for anisotropic mesh generation, *5th International Meshing Roundtable*, Sandia National Laboratories, 63–76, October 1996.

Boutora Y, Takorabet N, Ibtiouen R, Mezani S (2007) A new method for minimizing the bandwidth and profile of square matrices for triangular finite elements mesh, *IEEE Transactions on Magnetics*, **43**, Issue 4, 1513–1516.

Bowyer A (1981) Computing Dirichlet tessellations, *Computer Journal*, **24**, 162–166.

Boyd SK, Muller R (2006) Smooth surface meshing for automated finite element model generation from 3D image data, *Journal of Biomechanics*, **39**, Issue 7, 1287–1295.

Branets L, Carey GF (2010) Condition number bounds and mesh quality, *Numerical Linear Algebra with Applications*, **17**, 855–869.

Branets LV, Garanzha VA (2002) Distortion measure of trilinear mapping. Application to 3D grid generation, *Numerical Linear Algebra with Applications*, **9**, 511–526.

Bruyns CD, Senger S (2001) Interactive cutting of 3D surface meshes, *Computers & Graphics*, **25**, Issue 4, 635–642.

Buchin K (2009) Constructing Delaunay triangulations along space-filling curves, *Algorithms – ESA 2009 Proceedings, Lecture Notes in Computer Science*, **5757**, 119–130.

Buscaglia GC, Dari EA (1997) Anisotropic mesh optimization and its application in adaptivity, *International Journal for Numerical Methods in Engineering*, **40**, Issue 22, 4119–4136.

Bykat A (1976) Automatic generation of triangular grid: I. Subdivision of general polygon into convex subregions; II. Triangulation of convex polygons, *International Journal for Numerical Methods in Engineering*, **3**, 1329–1342.

Calvo NA, Idelsohn SR (2000) All-hexahedral element meshing: Generation of the dual mesh by recurrent subdivision, *Computer Methods in Applied Mechanics and Engineering*, **182**, Issues 3–4, 371–378.

Camacho DLA, Hopper RH, Lin GM, Myers BS (1997) An improved method for finite element mesh generation of geometrically complex structures with application to the skullbase, *Journal of Biomechanics*, **25**, Issue 6, 1067–1070.

Camata JJ, Coutinho ALGA (2013) Parallel implementation and performance analysis of a linear octree finite element mesh generation scheme, *Concurrency and Computation – Practice and Experience*, **182**, Issues 3–4, 826–842.

Canann SA (1991) *Plastering and optismoothing: New approach to automated, 3D hexahedral mesh generation and mesh smoothing*, Ph.D. Dissertation, Brigham Young University, Provo, UT.

Canann SA, Liu YC, Mobley AV (1997) Automatic 3D surface meshing to address today's industrial needs, *Finite Elements in Analysis and Design*, **25**, Issues 1–2, 185–198.

Canann SA, Muthukrishnan SN, Phillips RK (1998) Topological improvement procedures for quadrilateral finite element meshes, *Engineering with Computers*, **14**, Issue 2, 168–177.

Cao WM, Huang WZ, Russell RD (1999) An r-adaptive finite element method based upon moving mesh PDEs, *Journal of Computational Physics*, **149**, Issue 2, 221–244.

Cascon JM, Montenegro R, Escobar JM, Rodriguez E, Montero G (2009) The Meccano method for automatic tetrahedral mesh generation of complex genus-zero solids, *Proceedings of the 18th International Meshing Roundtable*, Springer, 462–480.

Cass RJ, Benzley SE, Meyers RJ, Blacker TD (1996) Generalized 3-D Paving: An automatic quadrilateral surface mesh generation algorithm, *International Journal for Numerical Methods in Engineering*, **39**, 1475–1489.

Cavendish JC (1974) Automatic triangulation of arbitrary planar domains for the finite element method, *International Journal for Numerical Methods in Engineering*, **8**, 679–696.

Cavendish JC, Field DA, Frey WH (1985) An approach to automatic three-dimensional finite element mesh generation, *International Journal for Numerical Methods in Engineering*, **21**, 329–347.

Cebral JR, Camelli FE, Lohner R (2002) A feature-preserving volumetric technique to merge surface triangulations, *International Journal for Numerical Methods in Engineering*, **55**, 177–190.

Cebral JR, Lohner R, Choyke PL, Yim PJ (2001) Merging of intersecting triangulations for finite element modeling, *Journal of Biomechanics*, **34**, Issue 6, 815–819.

Chabanas M, Luboz V, Payan Y (2003) Patient specific finite element model of the face soft tissues for computer-assisted maxillofacial surgery, *Medical Image Analysis*, **7**, Issue 2, 131–151.

Chae SW, Jeong JH (1997) Unstructured surface meshing using operators, *Proceedings of the 6th International Meshing Roundtables*, 281–291.

Chalasani S, Thompson D (2004) Quality improvement in extruded meshes using topologically adaptive generalized elements, *International Journal for Numerical Methods in Engineering*, **60**, Issue 6, 1139–1159.

Chand KK (2005) Component-based hybrid mesh generation, *International Journal for Numerical Methods in Engineering*, **62**, 747–773.

Chanzy P, Devroye L, Zamora-Cura C (2001) Analysis of range search for random k-d trees, *Acta Informatica*, **37**, Issues 4–5, 355–383.

Chazelle B (1984) Convex partitions of polyhedral – A lower bound and worst case optimal algorithm, *SIAM Journal on Computing*, **13**, Issue 3, 488–507.

Chen J, Zhao D, Huang Z, Zheng Y, Gao S (2011) Three-dimensional constrained boundary recovery with an enhanced Steiner point suppression procedure, *Computers & Structures*, **89**, 455–466.

Chen JC (2006) Efficient sample sort and the average analysis of PE sort, *Theoretical Computer Science*, **369**, 44–66.

Chen MB, Chuang TR, Wu JJ (2004) Efficient parallel implementations of near Delaunay triangulation with high performance Fortran, *Concurrency Computation: Practice and Experience*, **16**, 1143–1159.

Chen MB, Chuang TR, Wu JJ (2006) Parallel divide-and-conquer scheme for 2D Delaunay triangulation, *Concurrency and Computation: Practice and Experience*, **18**, 1595–1612.

Chernikov AN, Chrisochoides NP (2010) A template for developing next generation parallel Delaunay refinement methods, *Finite Elements in Analysis and Design*, **46**, 96–113.

Cherouat A, Borouchaki H, Giraud-Moreau L (2010) Mechanical and geometrical approaches applied to composite fabric forming, *International Journal of Material Forming*, **46**, S1189–S1204.

Chew LP (1989) Constrained Delaunay triangulations, *Algorithmica*, **4**, Issue 1, 97–108.

Chong CS, Kumar AS, Lee HP (2007) Automatic mesh-healing technique for model repair and finite element model generation, *Finite Elements in Analysis and Design*, **43**, 1109–1119.

Chrisochoides N, Chernikov A, Fedorov A, Kot A, Linardakis L, Foteinos P (2009) Towards exascale parallel Delaunay mesh generation, *Proceedings of the 18th International Meshing Roundtable*, Springer, 319–336.

Chrisochoides N, Nave D (2003) Parallel Delaunay mesh generation kernel, *International Journal in Numerical Methods and Engineering*, **58**, 161–176.

Chung SW, Kim SJ (2003) A remeshing algorithm based on bubble packing method and its application to large deformation problems, *Finite Elements in Analysis and Design*, **39**, Issue 4, 301–324.

Clarkson KL, Shor PW (1989) Application of random sampling in computational geometry, *Discrete & Computational Geometry*, **4**, Issue 5, 387–421.

Clemencon B, Borouchaki H, Laug P (2006) Ridge extraction and its application to surface meshing, *Engineering with Computers*, **24**, Issue 3, 287–304.

Cockayne E, Mihalkovic M (1999) Stable quasicrystalline sphere packing, *Philosophical Magazine Letters*, **79**, Issue 7, 441–448.

Coelho LCG, Gattass M, De Figueiredo LH (2000) Intersecting and trimming parametric meshes on finite element shells, *International Journal for Numerical Methods in Engineering*, **47**, Issue 4, 777–800.

Collins RJ (1973) Bandwidth reduction by automatic renumbering, *International Journal for Numerical Methods in Engineering*, **6**, 345–356.

Cook WA (1974) Body oriented (natural) coordinates for generating three-dimensional meshes, *International Journal for Numerical Methods in Engineering*, **8**, 27–43.

Cortis CM, Friesner RA (1997) An automatic three-dimensional finite element mesh generation system for the Poisson–Boltzmann equation, *Journal of Computational Chemistry*, **18**, Issue 13, 1570–1590.

Coupez T, Soyris N, Chenot JL (1991) 3-D finite-element modeling of the forging process with automatic remeshing, *Journal of Materials Processing Technology*, **27**, Issues 1–3, 119–133.

Couteau B, Payan Y, Lavallee S (2000) The mesh-matching algorithm: An 3D mesh generator for finite element structures, *Journal of Biomechanics*, **33**, Issue 8, 1005–1009.

Cuillière JC (1998) An adaptive method for the automatic triangulation of 3D parametric surfaces, *Computer-Aided Design*, **30**, Issue 2, 95–161.

Cuillière JC, Bournical S, Francois V (2010) A mesh-geometry-based solution to mixed-dimensional coupling, *Computer-Aided Design*, **42**, Issue 6, 509–522.

Cuillière JC, Francois V, Drouet JM (2013) Automatic mesh generation and transformation for topology optimization methods, *Computer-Aided Design*, **45**, 1489–1506.

Cuillière JC, Francois V, Souaissa K, Benamara A, BelHadjSalah H (2009) Automatic CAD models comparison and re-meshing in the context of mechanical design optimization, *Proceedings of the 18th International Meshing Roundtable*, Springer, 231–245.

Cuillière JC, Francois V, Souaissa K, Benamara A, BelHadjSalah H (2011) Automatic comparison and remeshing applied to CAD model modification, *Computer-Aided Design*, **43**, 1545–1560.

Cuthill EH (1972) Several strategies for reducing the bandwidth of sparse symmetric matrices, in *Sparse Matrices and Their Applications*, Eds. DJ Rose and RA Willoughby, Plenum Press, New York, 157–160.

Dai M, Schmidt DP (2005) Adaptive tetrahedral meshing in free-surface flow, *Journal of Computational Physics*, **208**, 228–252.

deCougny HL, Shephard MS (1999a) Parallel volume meshing using face removals and hierarchical repartitioning, *Computer Methods in Applied Mechanics and Engineering*, **174**, Issues 3–4, 275–298.

deCougny HL, Shephard MS (1999b) Parallel refinement and coarsening of tetrahedral meshes, *International Journal for Numerical Methods in Engineering*, **46**, Issue 7, 1101–1125.

Deister F, Tremel U, Hassan O, Weatherill N (2004) Fully automatic and fast mesh size specification for unstructured mesh generation, *Engineering with Computers*, **20**, 237–248.

Delaunay B (1934) Sur la sphere vide, *Bulletin de l'Académie des Sciences de l'URSS. Classe des sciences mathématiques et na*, Issue 6, 793–800 (Mi izv4937).

Delfinado CJA, Edelbrunner H (1995) An incremental algorithm for Betti numbers of simplicial complexes on the 3-sphere, *Computer-Aided Geometrical Design*, **12**, 771–784.

Devillers O, Preparata FP (1999) Further results on arithmetic filters for geometric predicates, *Computational Geometry*, **13**, 141–148.

Devillers O, Teillaud M (2011) Perturbations for Delaunay and weighted Delaunay 3D triangulations, *Computational Geometry*, **44**, 160–168.

Devroye L (1981) On the average complexity of some bucketing algorithms, *Computer & Mathematics with Applications*, **7**, Issue 5, 407–412.

Devroye L (1985) The expected length of the longest probe sequence for bucket searching when the distribution is not uniform, *Journal of Algorithms*, **7**, Issue 1, 1–9.

Devroye L, Jabbour J, Zamora-Cura C (2000) Squarish k-d trees, *SIAM Journal of Computing*, **30**, Issue 5, 1678–1700.

Devroye L, Klincsek T (1981) Average time behavior of distributive sorting algorithms, *Computing*, **26**, Issue 1, 1–7.

Devroye L, Lemaire C, Moreau JM (2004) Expected time analysis for Delaunay point location, *Computational Geometry*, **29**, 61–89.

Dey S, Flaherty JE, Ohsumi TK, Shephard MS (2006) Integration by table look-up for p-version finite elements on curved tetrahedra, *Computer Methods in Applied Mechanics and Engineering*, **195**, 4532–4543.

Dey S, Janoos F, Levine JA (2012) Meshing interfaces of multi-label data with Delaunay refinement, *Engineering with Computers*, **28**, 71–82.

Dey S, O'Bara RM, Shephard MS (2001) Towards curvilinear meshing in 3D: The case of quadratic simplices, *Computer-Aided Design*, **33**, 199–209.

Dhondt GD (2001) A new automatic hexahedral mesher based on cutting, *International Journal for Numerical Methods in Engineering*, **50**, Issue 9, 2109–2126.

Dietrich N (1997) Quadrilateral mesh generation via geometrically optimized domain decomposition, *Proceedings of the 6th International Meshing Roundtables*, 281–291.

Dillard SE, Bingert JF, Thoma D, Hamann B (2007) Construction of simplified boundary surfaces from Seriel-sectioned metal micrographs, *IEEE Transactions on Visualization and Computer Graphics*, **13**, Issue 6, 1528–1535.

Dirichlet GL (1850) Über die reduction der positiven quadratischen formen mit drei understimmten ganzen zahlen, *Zeitschrift für Angewandte Mathematik und Mechanik*, **40**, Issue 3, 209–227.

Dompierre J, Vallet MG, Bourgault Y, Fortin M, Habashi WG (2002) Anisotropic mesh adaptation: Towards user-independent and solver-independent CFD. Part III: Unstructured meshes, *International Journal for Numerical Methods in Engineering*, **39**, Issue 8, 675–702.

Dompierre J, Vallet MG, Labbe P, Guibault F (2005) An analysis of simplex shape measures for anisotropic meshes, *Computation Methods in Applied Mechanics and Engineering*, **194**, 4895–4914.

Donev A, Torquato S, Stillinger FH, Connelly R (2004) A linear programming algorithm to test for jamming in hard-sphere packings, *Journal of Computational Physics*, **197**, 139–166.

Du Q, Gunzburger M (2002) Grid generation and optimization based on centroidal Voronoi tessellations, *Applied Mathematics and Computation*, **133**, Issues 2–3, 591–607.

Du Q, Gunzburger M, Ju L (2010) Advances in studied and applications of centroidal Voronoi tessellation, *Numerical Mathematics – Theory Methods and Applications*, **3**, Issue 2, 119–142.

Du Q, Wang D (2004) Constrained boundary recovery for three dimensional Delaunay triangulations, *International Journal for Numerical Methods in Engineering*, **61**, 1471–1500.

Du Q, Wang D (2006) Recent progress in robust and quality Delaunay mesh generation, *Journal of Computational and Applied Mathematics*, **195**, 8–23.

Dwyer RA (1987) A faster divide-and-conquer algorithm for constructing Delaunay triangulations, *Algorithmica*, **2**, 137–151.

Ebeida MS, Davis RL, Freund RW (2010) A new fast hybrid adaptive grid generation technique for arbitrary two-dimensional domains, *International Journal for Numerical Methods in Engineering*, **84**, 305–329.

Ebeida MS, Mestreau E, Zhang Y, Dey S (2009) Mesh insertion of hybrid meshes, *Proceedings of the 18th International Meshing Roundtable*, Springer, 358–375.

Edelsbrunner H (1987) *Algorithm in Combinational Geometry*, Springer-Verlag, New York.

Edelsbrunner H, Mucke EP (1994) 3-dimensional alpha-shapes, *ACM Transactions on Graphics*, **13**, Issue 1, 43–72.

Edelsbrunner H, Preparata FP, West DB (1990) Tetrahedrizing points sets in 3 dimensions, *Journal of Symbolic Computation*, **10**, Issues 3–4, 335–347.

Edelsbrunner H, Shah NR (1996a) Triangulating topological spaces, *International Journal of Computational Geometry & Application*, **7**, Issue 4, 365–378.

Edelsbrunner H, Shah NR (1996b) Incremental topological flipping works for regular triangulations, *Algorithmica*, **15**, Issue 3, 223–241.

Edwards MG (2002) Unstructured, control-volume distributed, full-tensor finite-volume schemes with flow based grids, *Computational Geosciences*, **6**, Issues 3–4, 433–452.

Egidi N, Maponi P (2008) Block decomposition technique in the generation of adaptive grids, *Mathematics and Computers in Simulation*, **78**, Issues 5–6, 593–604.

Eppstein D (1997) Faster circles packing with application to nonobtuse triangulation, *International Journal of Computational Geometry & Applications*, **7**, Issue 5, 485–491.

Erten H, Ungor A, Zhao C (2009) Mesh smoothing algorithm for complex geometric domains, *Proceedings of the 18th International Meshing Roundtable*, Springer, 173–193.

Escobar JM, Montenegro R (1996) Several aspects of three-dimensional Delaunay triangulation, *Advances in Engineering Software*, **27**, Issues 1–2, 27–39.

Escobar JM, Rodriguez E, Montenegro R, Montero G, Gonzalez-Yuste JM (2003) Simultaneous untangling and smoothing of tetrahedral meshes, *Computer Methods in Applied Mechanics and Engineering*, **192**, Issue 25, 2775–2787.

Esposito A, Catalano MSF, Malucelli F, Tarricone L (1998) A new matrix bandwidth reduction algorithm, *Operations Research Letters*, **23**, 99–107.

Farrell PE, Micheletti S, Perotto S (2011) An anisotropic Zienkiewicz–Zhu type error estimate for 3D applications, *International Journal for Numerical Methods in Engineering*, 85, Issue 6, 671–692.

Feng YT, Han K, Owen DRJ (2003) Filling domains with disks: An advancing front approach, *International Journal for Numerical Methods in Engineering*, 56, Issue 5, 699–713.

Fernandes JLM, Martins PAF (2007) All hexahedral remeshing for the finite element analysis of metal forming processes, *Finite Elements in Analysis and Design*, 43, 666–679.

Fernandez JW, Mithraratne P, Thrupp SF, Tawhai MH, Hunter PJ (2004) Anatomically based geometric modelling of the musculo-skeletal system and other organs, *Biomechanics and Modeling in Mechanobiology*, 2, Issue 3, 139–155.

Ferrant M, Nabavi A, Macq B, Jolesz FA, Kikinis R, Warfield SK (2001) Registration of 3-D intraoperative MR images of the brain using a finite-element biomechanical model, *IEEE Transactions on Medical Imaging*, 20, Issue 12, 1384–1397.

Field DA (2000) Qualitative measures for initial meshes, *International Journal for Numerical Methods in Engineering*, 47, Issue 4, 887–906.

Field DA, Smith WD (1991) Graded tetrahedral finite-element meshes, *International Journal for Numerical Methods in Engineering*, 31, Issue 3, 413–425.

Filip D, Magedson R, Markot R (1986) Surface algorithm using bounds on derivatives, *Computer-Aided Geometric Design*, 3, 295–311.

Fine L, Remondina L, Leon JC (2000) Automatic generation of FEA models through idealization operators, *International Journal for Numerical Methods in Engineering*, 49, Issues 1–2, 83–108.

Finkel R, Bentley JL (1974) Quad trees: A data structure for retrieval on composite keys, *Acta Informatica*, 4, Issue 1, 1–9.

Finnigan PM, Kela A, Davis JE (1989) Geometry as a basis for finite-element automation, *Engineering with Computers*, 5, Issues 3–4, 147–160.

Flaherty JE, Loy RM, Ozturan C, Shephard MS, Szymanski BK, Teresco JD, Ziantz LH (1998) Parallel structures and dynamic load balancing for adaptive finite element computation, *Applied Numerical Mathematics*, 26, 241–263.

Folwell NT, Mitchell SA (1999) Reliable whisker weaving via curve contraction, *Engineering with Computers*, 15, Issue 3, 292–302.

Formaggia L, Perotto S (2000) Anisotropic error estimate for finite element method, *31st Computational Fluid Dynamics Lecture Series 2000–05*, von Karman Institute for Fluid Dynamics.

Foucault G, Cuilliere JC, Francois V, Leon JC, Maranzana R (2008) Adaptation of CAD model topology for finite element analysis, *Computer-Aided Design*, 40, 176–196.

Foucault G, Cuilliere JC, Francois V, Leon JC, Maranzana R (2013) Generalizing the advancing front method to composite surfaces in the context of meshing constraints topology, *Computer-Aided Design*, 45, 1408–1425.

Fournier A, Montuno DY (1984) Triangulating simple polygons and equivalent problems, *ACM Transactions on Graphics*, 3, Issue 2, 153–174.

Freitag LA, Jones M, Plassmann P (1999) A parallel algorithm for mesh smoothing, *SIAM Journal on Scientific Computing*, 20, Issue 6, 2023–2040.

Freitag LA, Knupp PM (2002) Tetrahedral mesh improvement via optimization of the element condition number, *International Journal for Numerical Methods in Engineering*, 53, 1377–1391.

Freitag LA, Ollivier-Gooch C (1996) A comparison of tetrahedral mesh improvement techniques, *Proceedings of the Fifth International Meshing Roundtable*, Sandia National Laboratories, 87–108.

Freitag LA, Ollivier-Gooch C (1997) Tetrahedral mesh improvement using swapping and smoothing, *International Journal for Numerical Methods in Engineering*, 40, 3979–4002.

Freitag LA, Plassmann P (2000) Local optimization-based simplicial mesh untangling and improvement, *International Journal for Numerical Methods in Engineering*, 49, 109–125.

Frey P, Sarter B, Gautherie M (1994) Fully automatic mesh generation for 3-D domains based upon voxel sets, *International Journal for Numerical Methods in Engineering*, 37, Issue 16, 2735–2753.

Frey PJ, Alauzet F (2005) Anisotropic mesh adaptation for CFD computations, *Computer Methods in Applied Mechanics and Engineering*, 194, 5068–5082.

Frey PJ, Borouchaki H, George PL (1998) 3D Delaunay mesh generation coupled with an advancing-front approach, *Computer Methods in Applied Mechanics and Engineering*, 157, Issues 1–2, 115–131.

Frey PJ, George PL (2000) *Mesh Generation: Application to Finite Elements*, HERMES Science Publishing, Oxford, UK.

Frey WH (1987) Selective refinement: A new strategy for automatic node placement in graded triangular meshes, *International Journal for Numerical Methods in Engineering*, 24, 2183–2200.

Fujisawa T, Inaba M, Yagawa G (2003) Parallel computing of high-speed compressible flows using a node-based finite-element method, *International Journal for Numerical Methods in Engineering*, 58, Issue 3, 481–511.

Fukuda J, Suhara J (1972) Automatic mesh generation for finite element mesh generation scheme, in *Advance in Computational Methods in Structural Mechanics and Design*, Eds. JT Oden, RW Clough and Y Yamamoto, UAH Press, Huntsville, Alabama.

Garimella R (2009) Conformal refinement of unstructured quadrilateral meshes, *Proceedings of the 18th International Meshing Roundtable*, Springer, 30–44.

Garimella RV, Shashkov MJ, Knupp PM (2004) Triangular and quadrilateral surface mesh quality optimization using local parametrization, *Computation Methods in Applied Mechanics and Engineering*, 193, 913–928.

George PL (1997) Improvement on Delaunay based three-dimensional automatic mesh generator, *Finite Elements in Analysis and Design*, 25, Issues 3–4, 297–317.

George PL, Borouchaki H (1998) *Delaunay Triangulation and Meshing, Application to Finite Elements*, HERMES, Paris, ISBN 2-86601-692-0.

George PL, Borouchaki H (2012) Construction of tetrahedral meshes of degree two, *International Journal for Numerical Methods in Engineering*, 90, 1156–1182.

George PL, Borouchaki H, Laug P (2002) An efficient algorithm for 3D adaptive meshing, *Advances in Engineering Software*, 33, Issues 7–10, 377–387.

George PL, Borouchaki H, Saltel E (2003) Ultimate robustness in meshing an arbitrary polyhedron, *International Journal for Numerical Methods in Engineering*, 58, 1061–1089.

George PL, Hecht F, Saltel E (1990) Automatic 3D-mesh generation with specified meshed boundaries, *IEEE Transactions on Magnetics*, 26, Issue 2, 771–774.

George PL, Hecht F, Saltel E (1991) Automatic mesh generator with specified boundary, *Computer Methods in Applied Mechanics and Engineering*, 92, 269–288.

George PL, Hermeline F (1992) Delaunay's mesh of a polyhedron in dimension d. Application to arbitrary polyhedral, *International Journal for Numerical Methods in Engineering*, 33, 975–995.

George PL, Seveno E (1994) The advancing-front mesh generation method revisited, *International Journal for Numerical Methods in Engineering*, 37, Issue 21, 3605–3619.

Geuzaine C, Remacle JF (2009) GMSH: A 3-D finite element mesh generator with built-in pre- and post-processing facilities, *International Journal for Numerical Methods in Engineering*, 79, Issue 11, 1309–1331.

Ghadyani H, Sullivan J, Wu Z (2010) Boundary recovery for Delaunay tetrahedral meshes using local topological transformations, *Finite Elements in Analysis and Design*, 46, Issues 1–2, 74–83.

Ghosh S, Mallett RL (1994) Voronoi cell finite-elements, *Computers & Structures*, 50, Issue 1, 33–46.

Goliaz NA, Dutton RW (1997) Delaunay triangulation and 3D adaptive mesh generation, *Finite Elements in Analysis and Design*, 25, Issues 3–4, 331–341.

Goliaz NA, Tsiboukis TD (1994) An approach to refining three-dimensional tetrahedral meshes based on Delaunay triangulations using local transformations, *Computer-Aided Geometric Design*, 8, 123–142.

Gordon WJ, Hall CA (1973a) Construction of curvilinear co-ordinates systems and applications to mesh generation, *International Journal for Numerical Methods in Engineering*, 7, Issue 4, 461–477.

Gordon WJ, Hall CA (1973b) Transfinite element methods – Blending function interpolation over arbitrary curved element domain, *Numerische Mathematik*, 21, Issue 2, 109–129.

Gordon WJ, Hall CA (1973c) Discretization error bound for transfinite elements, *SIAM Review*, 15, Issue 1, 254+.

Gosselin S, Ollivier-Gooch C (2011) Constructing constrained Delaunay tetrahedralization of volumes bounded by piecewise smooth surfaces, *International Journal of Computational Geometry & Applications*, 21, Issue 5, 571–594.

Greaves DM, Borthwick AGL (1999) Hierarchical tree-based finite element mesh generation, *International Journal for Numerical Methods in Engineering*, **45**, Issue 4, 447–471.

GRIDPOINT (2010) Available at http://www.itservices.hku.hk/ccsystem/gridpoint/hardwaregrid point.html.

Gruau C, Coupez T (2005) 3D tetrahedral unstructured and anisotropic mesh generation with adaptation to natural and multidomain metric, *Computer Methods in Applied Mechanics and Engineering*, **194**, 4951–4976.

Guan Z, Song C, Gu Y (2006) The boundary recovery and sliver elimination algorithms of three-dimensional constrained Delaunay triangulation, *International Journal for Numerical Methods in Engineering*, **68**, 192–209.

Gursoy HN, Patrikalakis NM (1992) An automatic coarse and fine surface mesh generation scheme based on medial axis transform. 1. Algorithms, *Engineering with Computers*, **8**, Issue 3, 121–137.

Hansen G, Zardecki A, Greening D, Bos R (2005) A finite element method for three-dimensional unstructured grid smoothing, *Journal of Computational Physics*, **202**, Issue 1, 281–297.

Hartmann E (1998) A marching method for the triangulation of surfaces, *The Visual Computer*, **14**, 95–108.

Hermeline F (1980) *Une methode automatique de maillage en dimension n*, Thesis, Université Paris VI, Paris, France.

Herrmann LR (1976) Laplacian-isoparametric grid generation scheme, *Journal of the Engineering Mechanics Division ASCE*, **12**, 749–759.

Hertel R, Kronmuller H (1998) Adaptive finite element mesh refinement techniques in three-dimensional micromagnetic modeling, *IEEE Transactions on Magnetics*, **34**, Issue 6, 3922–3930.

Hilbert D (1891) Über die stetige Abbildung einer Linie auf ein Flächenstück, *Mathematische Annalen*, **38**, 459–460.

Hinton E, Rao NVR, Ozakca M (1991) Mesh generation with adaptive finite-element analysis, *Advances in Engineering Software and Workstations*, **13**, Issues 5–6, 238–262.

Ho-Le K (1988) Finite element mesh generation methods: A review and classification, *Computer-Aided Design*, **20**, Issue 1, 27–38.

Holmes DG, Snyder DD (1988) The generation of unstructured triangular meshes using Delaunay triangulation, *Numerical Grid Generation in Computational Fluid Mechanics*, Miami, FL, 643–652.

Holroyd FC, Mason DC (1990) Efficient linear Quadtree construction algorithm, *Image and Vision Computing*, **8**, Issue 3, 218–224.

Horgan TJ, Gilchrist MD (2003) The creation of three-dimensional finite element models for simulating head impact biomechanics, *International Journal of Crashworthiness*, **8**, Issue 4, 353–366.

Horgan TJ, Gilchrist MD (2004) Influence of FE model variability in predicting brain motion and intracranial pressure changes in head impact simulation, *International Journal of Crashworthiness*, **9**, Issue 4, 401–418.

Howlett J, Zundel A (2009) Size function smoothing using an element area gradient, *Proceedings of the 18th International Meshing Roundtable*, Springer, 1–12.

Huang WZ (2005a) Measuring mesh qualities and application to variational mesh adaptation, *SIAM Journal on Scientific Computing*, **26**, Issue 5, 1643–1666.

Huang WZ (2005b) Metric tensors for anisotropic mesh generation, *Journal of Computational Physics*, **204**, Issue 2, 633–665.

Huang WZ, Russell RD (1999) Moving mesh strategy based on a gradient flow equation for two-dimensional problems, *SIAM Journal on Scientific Computing*, **20**, Issue 3, 998–1015.

Inoue K, Itoh T, Yamada A, Furuhata T, Shimada K (2001) Face clustering of a large scale CAD model for surface mesh generation, *Computer-Aided Design*, **33**, Issue 3, 251–261.

INRIA, GAMMA (2007) Automatic mesh generation and adaptation methods, Available at http://www-c.inria.fr/gamma.

Ito Y, Nakahashi K (2002) Surface triangulation for polygonal models based on CAD data, *International Journal for Numerical Methods in Fluids*, **39**, Issue 1, 75–96.

Ito Y, Shih AM, Erukala AK, Soni BK, Chernikov A, Chrisochoides NP, Nakahashi K (2007) Parallel unstructured mesh generation by an advancing front method, *Mathematics and Computers in Simulations*, **75**, 200–209.

Ito Y, Shih AM, Soni BK (2009) Octree-based reasonable-quality hexahedral mesh generation using a new set of refinement templates, *International Journal for Numerical Methods in Engineering*, **77**, Issue 13, 1809–1833.

Jerier JF, Richefeu V, Imbault D, Donze FV (2010) Packing spherical discrete elements for large scale simulations, *Computer Methods in Applied Mechanics and Engineering*, **199**, Issues 25–28, 1668–1678.

Ji SB, Ford JC, Greenwald RM, Beckwith JG, Paulsen KD, Flashman LA, McAllister TW (2011) Automatic subject-specific, hexahedral mesh generation via image registration, *Finite Elements in Analysis and Design*, **47**, Issue 10, 1178–1185.

Jin H, Tanner RI (1993) Generation of unstructured tetrahedral meshes by advancing front technique, *International Journal for Numerical Methods in Engineering*, **36**, Issue 11, 1805–1823.

Joe B (1989) Three-dimensional triangulations from local transformations, *SIAM Journal on Scientific and Statistical Computing*, **10**, Issue 4, 718–741.

Joe B (1991a) Delaunay versus max min solid angle triangulation for 3-dimensional mesh generation, *International Journal for Numerical Methods in Engineering*, **31**, Issue 5, 987–997.

Joe B (1991b) Construction of three-dimensional Delaunay triangulations using local transformations, *Computer-Aided Geometric Design*, **8**, 123–142.

Joe B (1991c) Geompack – A software package for the generation of meshes using geometric algorithms, *Advances in Engineering Software and Workstations*, **13**, Issues 5–6, 325–331.

Joe B (1992) Three-dimensional boundary-constrained triangulations, in *Artificial Intelligence, Expert System and Symbolic Computing*, Eds. EN Houstis and JR Rice, Elsevier Science Publishers BV, North-Holland IMACS.

Joe B (1993) Construction of k-dimensional Delaunay triangulations using local transformations, *SIAM Journal on Scientific Computing*, **14**, Issue 6, 1415–1436.

Joe B (1994) Tetrahedral mesh generation in polyhedral regions based on convex polyhedron decompositions, *International Journal for Numerical Methods in Engineering*, **37**, 268–287.

Joe B (1995a) Quadrilateral mesh generation in polygonal regions, *Computer-Aided Design*, **27**, 209–222.

Joe B (1995b) Construction of 3-dimensional improved-quality triangulations using local transformations, *SIAM Journal on Scientific Computing*, **16**, Issue 6, 1292–1307.

Johnson AA, Tezduyar TE (1994) Mesh update strategies in parallel finite-element computations of flow problems with moving boundaries and interfaces, *Computer Methods in Applied Mechanics and Engineering*, **119**, Issues 1–2, 73–94.

Johnson AA, Tezduyar TE (1997) Parallel computation of incompressible flows with complex geometries, *International Journal for Numerical Methods in Engineering*, **24**, Issue 12, 1321–1340.

Johnson AA, Tezduyar TE (1999) Advanced mesh generation and update methods for 3D flow simulations, *Computational Mechanics*, **23**, Issue 2, 130–143.

Johnson E, Zhang Y, Shimada K (2009) Using parameterization and spring to determine aneurysm wall thickness, *Proceedings of the 18th International Meshing Roundtable*, Springer, 397–414.

Johnston BP, Sullivan JM (1992) Fully automatic two-dimensional mesh generation using normal offsetting, *International Journal for Numerical Methods in Engineering*, **33**, 425–442.

Johnston BP, Sullivan JM (1993) A normal offsetting technique for automatic mesh generation in 3 dimensions, *International Journal for Numerical Methods in Engineering*, **36**, Issue 10, 1717–1734.

Johnston BP, Sullivan JM, Kwasnik A (1991) Automatic conversion of triangular finite element meshes to quadrilateral elements, *International Journal for Numerical Methods in Engineering*, **31**, 67–84.

Jones MT, Plassmann PE (1997) Adaptive refinement of unstructured finite-element meshes, *Finite Elements in Analysis and Design*, **25**, 41–60.

Joun MS, Lee MC (1997) Quadrilateral finite element generation and mesh quality control for metal forming simulations, *International Journal for Numerical Methods in Engineering*, **40**, Issue 21, 4059–4075.

Jung YH, Lee K (1993) Tetrahedron based Octree encoding for automatic mesh generation, *Computer-Aided Design*, **25**, Issue 3, 141–153.

Kaminsky J, Rodt T, Gharabaghi A, Fordter J, Brand G, Samii M (2005) A universal algorithm for an improved finite element mesh generation mesh quality assessment in comparison to former automated mesh-generator and an analytic model, *Medical Engineering & Physics*, **27**, Issue 5, 383–394.

Kanungo T, Mount DM, Netanyahu NS, Piatko CD, Silverman R, Wu AY (2002) An efficient k-means clustering algorithm: Analysis and Implementation, *IEEE Transactions on Pattern Analysis and Machine Intelligence*, **24**, Issue 7, 881–892.

Karamete BK, Beall MW, Shephard MS (2000) Triangulation of arbitrary polyhedra to support automatic mesh generation, *International Journal for Numerical Methods in Engineering*, **49**, Issues 1–2, 167–191.

Karamete BK, Dey S, Mestreau EL, Aubry R, Bulat-Jara FA (2013) An algorithm for discrete Booleans with applications to finite element modeling of complex systems, *Finite Elements in Analysis and Design*, **68**, 10–27.

Kaveh A, Bondarabady HAR (2002) A multi-level finite element nodal ordering using algebraic graph theory, *Finite Elements in Analysis and Design*, **38**, 245–261.

Kawamura Y, Islam MS, Sumi Y (2008) A strategy of automatic hexahedral mesh generation by using an improved whisker-weaving method with a surface mesh modification procedure, *Engineering with Computers*, **24**, Issue 3, 215–229.

Kelly DW, Gago LPSR, Zienkiewicz OC (1983) A-posteriori error analysis and adaptive processes in the finite-element method, *International Journal for Numerical Methods in Engineering*, **19**, 1593–1619.

Kim HJ, Swan CC (2003a) Voxel-based meshing and unit-cell analysis of textile composites, *International Journal for Numerical Methods in Engineering*, **56**, Issue 7, 977–1006.

Kim HJ, Swan CC (2003b) Algorithm for automated meshing and unit cell analysis of periodic composites with hierarchical tri-quadratic tetrahedral elements, *International Journal for Numerical Methods in Engineering*, **58**, Issue 11, 1683–1711.

Kim JH, Kim HG, Lee BC, Im S (2003) Adaptive mesh generation by bubble packing method, *Structural Engineering and Mechanics*, **15**, Issue 1, 135–149.

Klingner B, Shewchuk J (2008) Aggressive tetrahedral mesh improvement, *Proceedings of the 16th International Meshing Roundtable*, 3–23.

Knupp PM (2000a) Achieving finite element mesh quality via optimization of the Jacobian matrix norm and associated quantities. Part II – A framework for surface mesh optimization, *International Journal for Numerical Methods in Engineering*, **48**, Issue 3, 401–420.

Knupp PM (2000b) Achieving finite element mesh quality via optimization of the Jacobian matrix norm and associated quantities. Part II – A framework for volume mesh optimization and the condition number of the Jacobian matrix, *International Journal for Numerical Methods in Engineering*, **48**, 1165–1185.

Knupp PM (2001) Algebraic mesh quality metric, *SIAM Journal on Scientific Computing*, **23**, Issue 1, 193–218.

Knupp PM (2003a) Algebraic mesh quality metrics for unstructured initial meshes, *Finite Elements and Design*, **39**, 217–241.

Knupp PM (2003b) A method for hexahedral mesh shape optimization, *International Journal for Numerical Methods in Engineering*, **58**, 319–332.

Knupp PM (2006) Mesh quality improvement for SciDAC applications, *Journal of Physics, Conference Series*, **46**, 458–462.

Knupp PM (2012) Introducing the target-matrix paradigm for mesh optimization via node-movement, *Engineering with Computers*, **28**, 419–429.

Knuth DE (1975) *The Art of Computing*, 2nd Edition, Addison-Wesley, Reading, MA.

Kobbelt L, Hesse T, Prautzsch H, Schweizerhof K (1997) Iterative mesh generation for FE computations on free form surfaces, *Engineering Computations*, **14**, Issues 6–7, 806–820.

Kohout J, Kolingerova I (2003) Parallel Delaunay triangulation in E^3: Make it simple, *Visual Computer*, **19**, 532–548.

Kohout J, Kolingerova I, Zara J (2005) Parallel Delaunay triangulation in E^2 and E^3 for computers with shared memory, *Parallel Computing*, **31**, 491–522.

Kolingerova I, Kohout J (2002) Optimistic parallel Delaunay triangulation, *Visual Computer*, **18**, 511–529.

Kraft P (1999) Automatic remeshing with hexahedral elements: Problems, solutions and applications, *Proceedings of the 8th International Meshing Roundtable*, South Lake Tahoe, CA, 357–367, October 1999.

Krysl P, Ortiz M (2001) Variational Delaunay approach to the generation of tetrahedral finite element meshes, *International Journal for Numerical Methods in Engineering*, 50, Issue 7, 1681–1700.

Kwok W, Haghighi K, Kang E (1995) An efficient data structure for the advancing front triangular mesh generation technique, *Communication in Numerical Methods in Engineering*, 11, Issue 5, 465–473.

Kwon GH, Chae SW, Lee KJ (2003) Automatic generation of tetrahedral meshes from medical images, *Computers & Structures*, 81, Issues 8–11, 765–775.

Labatut P, Pons JP, Keriven R (2009) Robust and efficient surface reconstruction from range data, *Computer Graphics Forum*, 28, Issue 8, 2275–2290.

Labelle F, Shewchuk JR (2006) Isosurface stuffing: Fast tetrahedral meshes with good dihedral angles, *ACM Transactions on Graphics*, 26, Issue 3, Article 57, 1–10.

Lacroix D, Prendergast PJ (2002) A mechano-regulation model for tissue differentiation during fracture healing: Analysis of gap size and loading, *Journal of Biomechanics*, 35, Issue 9, 1163–1171.

Lai MW, Bensley S, White D (2000) Automatic hexahedral mesh generation by generalized multiple source to multiple target, *International Journal for Numerical Methods in Engineering*, 49, Issues 1–2, 261–275.

Lai YC (1998) A three-step renumbering procedure for high-order finite element analysis, *International Journal for Numerical Methods in Engineering*, 41, 127–135.

Lai YC, Weingarten VI, Eshraghi H (1996) Matrix profile and wavefront reduction based on the graph theory and wavefront minimization, *International Journal for Numerical Methods in Engineering*, 39, 1137–1159.

Lai YK, Hu SM, Martin RR, Rosin PL (2009) Rapid and effective segmentation of 3D models using random walks, *Computer-Aided Geometric Design*, 26, 665–679.

Lapeer RJ, Prager RW (2001) Fetal head moulding: Finite element analysis of a fetal skull subjected to uterine pressures during the first stage of labour, *Journal of Biomechanics*, 34, Issue 9, 1125–1133.

Lau TS, Lo SH (1996) Finite element mesh generation over analytical curved surfaces, *Computers & Structures*, 59, Issue 2, 301–309.

Lau TS, Lo SH, Lee CK (1997) Generation of quadrilateral mesh over analytical curved surfaces, *Finite Elements in Analysis and Design*, 27, 251–272.

Laug P (2010) Some aspects of parametric surface meshing, *Finite Elements in Analysis and Design*, 46, Issues 1–2, 216–226.

Laug P, Borouchaki H (2000) Automatic generation of finite element meshes for molecular surfaces, *European Congress on Computational Methods in Applied Sciences and Engineering, ECCOMAS 2000*, Barcelona, Spain.

Laug P, Borouchaki H (2001) Molecular surface modelling and meshing, *Proceedings, 10th International Meshing Roundtable*, Sandia National Lab, 31–41.

Laug P, Borouchaki H (2002) Molecular surface modeling and meshing, *Engineering with Computers*, 18, Issue 3, 199–210.

Laug P, Borouchaki H (2003a) Interpolating and meshing 3D surface grids, *International Journal for Numerical Methods in Engineering*, 58, Issue 2, 209–225.

Laug P, Borouchaki H (2003b) Generation of finite element meshes on molecular surfaces, *International Journal of Quantum Chemistry*, 93, Issue 2, 131–138.

Laug P, Borouchaki H (2004) Curve linearization and discretization for meshing composite parametric surfaces, *Communications in Numerical Methods in Engineering*, 20, Issue 11, 869–876.

Lawson CL (1977) Software for C1 surface interpolation, in *Mathematical Software III*, Ed. J Rice, Academic Press, New York, 161–194.

Lee CK (1999) Automatic metric advancing front triangulation over curved surfaces, *Engineering Computations*, 16, 230–263.

Lee CK (2003a) Automatic metric 3D surface mesh generation using subdivision surface geometrical model. Part I: Construction of underlying geometrical model, *International Journal for Numerical Methods in Engineering*, 56, Issue 11, 1593–1614.

Lee CK (2003b) Automatic metric 3D surface mesh generation using subdivision surface geometrical model. Part II: Mesh generation algorithm and examples, *International Journal for Numerical Methods in Engineering*, 56, Issue 11, 1615–1646.

Lee CK, Chiew SP, Lie ST, Nguyen TBN (2010) Adaptive mesh generation procedures for thin-walled tubular structures, *Finite Elements in Analysis and Design*, **46**, Issues 1–2, 114–131.

Lee CK, Hobbs RE (1998) Automatic adaptive finite element mesh generation over rational B-spline surfaces, *Computers & Structures*, **69**, Issue 5, 577–608.

Lee CK, Hobbs RE (1999) Automatic adaptive finite element mesh generation over arbitrary two-dimensional domain using advancing front technique, *Computers & Structures*, **71**, 9–34.

Lee CK, Lo SH (1994) A new scheme for the generation of graded quadrilateral mesh, *Computers & Structures*, **52**, Issue 5, 847–857.

Lee CK, Lo SH (1995) An automatic adaptive refinement procedure using triangular and quadrilateral meshes, *Engineering Fracture Mechanics*, **50**, Issues 5–6, 671–686.

Lee CK, Lo SH (1997a) Automatic adaptive refinement finite element procedure for 3D stress analysis, *Finite Element in Analysis and Design*, **25**, Issues 1–2, 135–166.

Lee CK, Lo SH (1997b) Automatic adaptive 3D finite element refinement using different order tetrahedral elements, *International Journal for Numerical Methods in Engineering*, **40**, 2195–2226.

Lee CK, Lo SH (1999) A full 3D finite element analysis using adaptive refinement and PCG solver with back interpolation, *Computer Methods in Applied Mechanics and Engineering*, **160**, 175–191.

Lee KY, Kim II, Cho DY, Kim TW (2003) An algorithm for automatic 2D quadrilateral mesh generation with line constraints, *Computer-Aided Design*, **35**, Issue 12, 1055–1068.

Lee M, Samet H (2000) Navigating through triangular meshes implemented as linear quadtree, *ACM Transactions on Graphics*, **19**, Issue 2, 79–121.

Lee Y, Lee S, Shamir A, Cohen-Or D, Seidel HP (2005) Mesh scissoring with minima rule and part salience, *Computer-Aided Geometric Design*, **22**, 444–465.

Lee YK, Lee CK (2002) Automatic generation of anisotropic quadrilateral meshes on three-dimensional surfaces using metric specifications, *International Journal for Numerical Methods in Engineering*, **53**, 2673–2700.

Lee YK, Lee CK (2003) A new anisotropic quadrilateral mesh generation scheme with enhanced local mesh smoothing procedures, *International Journal for Numerical Methods in Engineering*, **58**, 277–300.

Lee YK, Yang DY (2000) A grid-based approach to non-regular mesh generation for automatic remeshing with metal forming analysis, *Communication in Numerical Methods in Engineering*, **16**, Issue 9, 625–635.

Leland RW, Melander DJ, Meyers RW, Mitchell SA, Tautges TJ (1998) The Geode algorithm: Combining hex/tet plastering, dicing and transition elements for automatic all-hex mesh generation, *Proceedings of the 7th International Meshing Roundtable*, Sandia National Laboratories, 515–521, October 1998.

Lelong-Ferrand J, Arnaudiès JM (1977) *Cours de Mathématiques, Tome 3, Géométrie et cinématique*, Dunod Université, Bordas.

Lemaire C, Moreau JM (2000) A probabilistic result on multi-dimensional Delaunay triangulations, and its application to the 2D case, *Computational Geometry*, **17**, 69–96.

Lewis RW, Zheng Y, Usmani AS (1995) Aspects of adaptive mesh generation based on domain decomposition and Delaunay triangulation, *Finite Elements in Analysis and Design*, **20**, Issue 1, 47–70.

Li H, Cheng GD (1998) New method for graded mesh generation of all hexahedral finite elements, *Numerical Grid Generation in Computational Field Simulations, Proceedings of the 6th International Conference*, University of Greenwich, 1031–1038.

Li H, Cheng GD (2000) New method for graded mesh generation of all hexahedral finite elements, *Computers & Structures*, **76**, Issue 6, 729–740.

Li R, Tang T, Zhang PW (2001) Moving mesh methods in multiple dimensions based on harmonic maps, *Journal of Computational Physics*, **170**, Issue 2, 562–588.

Li SP, Ng KL (2003a) Monte Carlo study of the sphere packing problem, *Physica A*, **321**, 359–363.

Li SP, Ng KL (2003b) Study of the unequal spheres packing problem: An application to radio-surgery treatment, *International Journal of Modern Physics C*, **14**, Issue 6, 815–823.

Li TS, Mckeag RM, Armstrong CG (1995) Hexahedral meshing using midpoint and integer programming, *Computer Method in Applied Mechanics and Engineering*, **124**, Issues 1–2, 171–193.

Li X, Gu X, Qin H (2009) Surface mapping using consistent pants decomposition, *IEEE Transactions on Visualization and Computer Graphics*, **15**, Issue 4, 558–571.

Li X, Shephard MS, Beall MW (2003) Accounting for curved domains in mesh adaptation, *International Journal for Numerical Methods in Engineering*, **58**, 247–276.

Li X, Shephard MS, Beall MW (2005) 3D anisotropic mesh adaptation by mesh modification, *Computer Method in Applied Mechanics and Engineering*, **124**, Issues 1–2, 171–193.

Li XY, Teng SH, Ungor A (2000) Biting: Advancing front meets sphere packing, *International Journal for Numerical Methods in Engineering*, **194**, 4915–4950.

Liang X, Ebeida MS, Zhang Y (2009) Guaranteed-quality all quadrilateral mesh generation with feature preservation, *Computer Methods in Applied Mechanics and Engineering*, **199**, 2072–2083.

Lie ST, Lee CK, Wong SM (2001) Modelling and mesh generation of weld profile in tubular Y-joint, *Journal of Constructional Steel Research*, **57**, 547–567.

Lim A, Lin J, Rodrigues B, Xiao F (2006) Ant colony optimization with hill climbing for bandwidth minimization problem, *Applied Soft Computing*, **6**, 180–188.

Lim A, Lin J, Xiao F (2007) Particle swarm optimization and hill climbing for the bandwidth minimization problem, *Applied Intelligence*, **26**, 175–182.

Lin TJ, Guan ZQ, Chang JH, Lo SH (2014) Vertex-ball spring smoothing: An efficient method for unstructured dynamic hybrid meshes, *Computers and Structures*, **136**, 24–33.

Liu A, Joe B (1994a) Relationship between tetrahedron shape measures, *Biosystems and Information Technology*, **34**, Issue 2, 268–287.

Liu A, Joe B (1994b) On the shape of tetrahedra from bisection, *Mathematics of Computation*, **63**, Issue 207, 141–154.

Liu A, Joe B (1995) Quality local refinement of tetrahedral meshes based on bisection, *SIAM Journal on Scientific Computing*, **16**, Issue 6, 1269–1291.

Liu A, Joe B (1996) Quality local refinement of tetrahedral meshes based on 8-subtetrahedron subdivision, *Mathematics of Computation*, **65**, Issue 215, 1183–1200.

Liu J, Chen B, Chen Y (2007) Boundary recovery after 3D Delaunay tetrahedralization without adding extra nodes, *International Journal for Numerical Methods in Engineering*, **72**, 744–756.

Liu J, Yan J, Lo SH (2013) A new insertion sequence for incremental Delaunay triangulation, *Acta Mechanica Sinica*, **29**, Issue 1, 99–109.

Liu JF (2003) Automatic mesh generation of 3-D geometric models, *Acta Mechanica Sinica*, **19**, Issue 3, 285–288.

Liu SS, Rajit G (1997) Automatic hexahedral mesh generation by recursive convex and sweep volume decomposition, *Proceedings of the 6th International Meshing Roundtable*, Sandia National Laboratories, 217–231.

Liu Y, Lo SH, Guan ZQ, Zhang HW (2014) Boundary recovery for 3D Delaunay triangulation, *Finite Elements in Analysis and Design*, **84**, 32–43.

Liu Y, Snoeyink J (2005) A comparison of five implementations of 3d Delaunay tessellations, *Combinatorial and Computational Geometry*, **52**, 439–458, MERI Publications.

Lo SH (1985) A new mesh generation scheme for arbitrary planar domains, *International Journal for Numerical Methods in Engineering*, **21**, 1403–1426.

Lo SH (1988a) Perspective projection of non-convex polyhedra, *International Journal for Numerical Methods in Engineering*, **26**, 1485–1506.

Lo SH (1988b) A hidden-line algorithm using picture subdivision technique, *Computers & Structures*, **28**, Issue 1, 37–45.

Lo SH (1988c) Finite element mesh generation over curved surfaces, *Computers & Structures*, **29**, 731–742.

Lo SH (1989a) Delaunay triangulation of non-convex planar domains, *International Journal for Numerical Methods in Engineering*, **28**, 2695–2707.

Lo SH (1989b) Generating quadrilateral elements on plane and over curved surfaces, *Computers & Structures*, **31**, Issue 3, 421–426.

Lo SH (1991a) Automatic mesh generation and adaptation by using contours, *International Journal for Numerical Methods in Engineering*, **31**, 689–707.

Lo SH (1991b) Volume discretization into tetrahedra – I: Verification and orientation of boundary surfaces, *Computers & Structures*, **39**, Issue 5, 501–511.

Lo SH (1991c) Volume discretization into tetrahedra – II: 3D triangulation by advancing-front approach, *Computers & Structures*, **39**, Issue 5, 501–511.

Lo SH (1992) Generation of high-quality gradation finite element mesh, *Engineering Fracture Mechanics*, **41**, Issue 2, 191–202.

Lo SH (1995) Automatic mesh generation over intersecting surfaces, *International Journal for Numerical Methods in Engineering*, **38**, 943–954.

Lo SH (1997) Optimization of tetrahedral meshes based on element shape measures, *Computers & Structures*, **63**, Issue 5, 951–961.

Lo SH (1998a) 3D mesh refinement in compliance with a specified node spacing function, *Computational Mechanics*, **21**, 11–19.

Lo SH (1998b) Analysis and verification of boundary surfaces of solid objects, *Engineering with Computers*, **14**, 36–47.

Lo SH (2001) 3D anisotropic mesh refinement in compliance with a general metric specification, *Finite Elements in Analysis and Design*, **38**, 3–19.

Lo SH (2002) Finite element mesh generation and adaptive meshing, *Progress in Structural Engineering and Materials*, **4**, Issue 4, 381–399.

Lo SH (2012a) Parallel Delaunay triangulation in three dimensions, *Computer Methods in Applied Mechanics and Engineering*, **237–240**, 88–106.

Lo SH (2012b) Parallel Delaunay triangulation – Application to two dimensions, *Finite Elements in Analysis and Design*, **62**, 37–48.

Lo SH (2012c) Automatic merging of hexahedral meshes, *Finite Elements in Analysis and Design*, **55**, 7–22.

Lo SH (2013a) Delaunay triangulation of non-uniform point distribution by means of multi-grid insertion, *Finite Elements in Analysis and Design*, **63**, 8–22.

Lo SH (2013b) Dynamic grid for mesh generation by the advancing front method, *Computers & Structures*, **123**, 15–27.

Lo SH (2013c) Automatic merging of tetrahedral meshes, *International Journal for Numerical Methods in Engineering*, **93**, Issue 11, 1191–1215.

Lo SH, Cheung YK, Leung YT (1982) Automatic finite element mesh generation, *Proceedings of the International Conference on Finite Element Method*, Beijing, China, 931–937.

Lo SH, Lau TS (1992) Generation of hybrid finite element mesh, *Microcomputers in Civil Engineering*, **7**, 235–241.

Lo SH, Lau TS (1998) Mesh generation over curved surfaces with explicit control on discretization error, *Engineering Computations*, **15**, Issue 3, 357–373.

Lo SH, Lee CK (1992) On using meshes of mixed element types in adaptive finite element analysis, *Finite Elements in Analysis and Design*, **11**, 307–336.

Lo SH, Lee CK (1994) Generation of gradation meshes by the background grid technique, *Computers & Structures*, **50**, Issue 1, 21–32.

Lo SH, Lee CK (1998) On constructing accurate recovered stress fields for the finite element solution of Reissner–Mindlin plate bending problems, *Computer Methods in Applied Mechanics and Engineering*, **160**, 175–191.

Lo SH, Liu JF (2002) Automatic mesh generation on a regular background grid, *Journal of Computer Science and Technology*, **17**, Issue 6, 882–887.

Lo SH, Wan KH, Sze KY (2005) Adaptive refinement analysis using hybrid-stress transition elements, *Computers & Structures*, **84**, Issues 31–32, 2213–2230.

Lo SH, Wang WX (2003) An algorithm for the intersection of quadrilateral surfaces by tracing of neighbours, *Computer Methods in Applied Mechanics and Engineering*, **192**, 2319–2338.

Lo SH, Wang WX (2004) A fast robust algorithm for the intersection of triangulated surfaces, *Engineering with Computers*, **20**, Issue 1, 11–21.

Lo SH, Wang WX (2005a) Finite element mesh generation over intersecting curved surfaces by tracing of neighbours, *Finite Elements in Analysis and Design*, **41**, 351–370.

Lo SH, Wang WX (2005b) Generation of finite element mesh with variable size over an unbounded 2D domain, *Computer Methods in Applied Mechanics and Engineering*, **194**, Issues 45–47, 4668–4884.

Lo SH, Wang WX (2005c) Generation of anisotropic mesh by ellipse packing over an unbounded domain, *Engineering with Computers*, **20**, Issue 4, 372–383.

Lo SH, Wang WX (2005d) Generation of tetrahedral mesh of variable element size by sphere packing over an unbounded 3D domain, *Computer Methods in Applied Mechanics and Engineering*, **194**, Issues 48–49, 5002–5018.

Lo SH, Wu D, Sze KY (2010) Adaptive meshing and analysis using transitional quadrilateral and hexahedral elements, *Finite Elements in Analysis and Design*, **46**, 2–16.

Lohner R (1994) Progress in grid generation via the advancing front technique, *3rd International Meshing Roundtable*, Albuquerque, NM, 24–25 October 1994.

Lohner R (1995) Mesh adaptation in fluid-mechanics, *Engineering Fracture Mechanics*, **50**, Issues 5–6, 819–847.

Lohner R (1996a) Regridding surface triangulations, *Journal of Computational Physics*, **126**, Issue 1, 1–10.

Lohner R (1996b) Extensions and improvements of the advancing front grid generation technique, *Communications in Numerical Methods in Engineering*, **12**, Issue 10, 683–702.

Lohner R (1996c) Progress in grid generation via the advancing front technique, *Engineering with Computers*, **12**, Issues 3–4, 186–210.

Lohner R (1997) Automatic unstructured grid generator, *Finite Elements in Analysis and Design*, **25**, Issues 1–2, 111–134.

Lohner R (2001) A parallel advancing front grid generation scheme, *International Journal for Numerical Methods in Engineering*, **51**, Issue 6, 663–678.

Lohner R, Camberos J, Merriam M (1992) Parallel unstructured grid generation, *Engineering with Computers*, **12**, Issues 3–4, 168–177.

Lohner R, Cebral J (2000) Generation of non-isotropic unstructured grids via directional enrichment, *International Journal for Numerical Methods in Engineering*, **49**, Issues 1–2, 219–232.

Lohner R, Onate E (1998) An advancing front point generation technique, *Communications in Numerical Methods in Engineering*, **14**, Issue 12, 1097–1108.

Lohner R, Onate E (2004) A general advancing front techniques for filling space with arbitrary objects, *International Journal for Numerical Methods in Engineering*, **61**, Issue 12, 1977–1991.

Lohner R, Onate E (2010) Advancing front techniques for filling space with arbitrary separated objects, *Finite Elements in Analysis and Design*, **46**, Issues 1–2, 140–151.

Lohner R, Parikh P (1988) Three-dimensional grid generation by the advancing front method, *International Journal for Numerical Methods in Fluids*, **8**, 1135–1149.

Lopez EJ, Nigro NM, Storti MA (2008) Simultaneous untangling and smoothing of moving grid, *International Journal for Numerical Methods in Engineering*, **76**, Issue 7, 994–1019.

Loseille A, Alauzet F (2009) Optimal 3D highly anisotropic mesh adaptation based on continuous mesh framework, *Proceedings of the 18th International Meshing Roundtable*, Springer, 575–594.

Loseille A, Dervieux A, Alauzet F (2010) Fully anisotropic goal-oriented mesh adaptation for 3D steady Euler equations, *Journal of Computational Physics*, **229**, 2866–2897.

Loseille A, Lohner R (2009) On 3D anisotropic local remeshing for surface, volume and boundary layer, *Proceedings of the 18th International Meshing Roundtable*, Springer, 611–630.

Lou R, Pernot JP, Mikchevitch A, Veron P (2010) Merging enriched finite element triangular meshes for fast prototyping of alternate solutions in the context of industrial maintenance, *Computer-Aided Design*, **42**, 670–681.

Lu Y, Gadh R, Tautges TJ (2001) Feature based hex meshing methodology: Feature recognition and volume decomposition, *Computer-Aided Design*, **33**, 221–232.

Luo H, Spiegel S, Lohner R (2010) Hybrid grid generation method for complex geometries, *AIAA Journal*, **48**, Issue 11, 2639–2647.

Luo XJ, Shephard MS, Lee LQ, Ge LX, Ng C (2011) Moving curved mesh adaptation for higher-order finite element simulations, *Engineering with Computers*, **27**, 41–50.

Luo XJ, Shephard MS, O'Bara RM (2004) Automatic p-version mesh generation for curved domains, *Computer-Aided Design*, **20**, 273–285.

Ma YL (2008) A parallel surface extraction algorithm for large binary image data sets based on an adaptive 3D Delaunay subdivision strategy, *IEEE Transaction and Visualization and Computer Graphics*, **14**, Issue 1, 160–172.

Mackerle J (2001) 2D and 3D finite element meshing and remeshing – A bibliography, *Engineering Computations*, **18**, Issues 7–8, 1108–1197.

Mackie RI (1999) Object-oriented finite element programming – The importance of data modeling, *Advances in Engineering Software*, **30**, Issues 9–11, 775–782.

Marechal L (2009) Advances in Octree-based all-hexahedral mesh generation: Handling sharp features, *Proceedings of the 18th International Meshing Roundtable*, Springer, 64–86.

Marquez A, Moreno-Gonzalez A, Plaza A, Suarez JP (2008) The seven-triangle longest-side partition of triangles and mesh quality improvement, *Finite Elements in Analysis and Design*, **44**, 748–758.

Marsden JE, Hughes TJR (1983) *Mathematical Foundation of Elasticity*, Prentice-Hall, Englewood Cliffs, NJ.

Martinez D, Velho L, Carvalho PC (2005) Computing geodesics on triangular meshes, *Computers & Graphics*, **29**, 667–675.

Masa S, Noel F, Leon JC (1999) Generation of quadrilateral meshes on free-form surfaces, *Computers & Structures*, **71**, Issue 5, 505–524.

Maubach JM (1995) Local bisection refinement for N-simplicial grids generated by reflection, *SIAM Journal on Scientific Computing*, **16**, 210–227.

Mavriplis DJ (1997) Unstructured grid techniques, *Annual Review of Fluid Mechanics*, **29**, 473–514.

McMorris H, Kallinderis Y (1997) Octree-advancing front method for generation of unstructured surface and volume meshes, *AIAA Journal*, **35**, Issue 6, 976–984.

Melander DJ, Tautges JT, Benzley SE (1997) Generation of multi-million element meshes for solid model-based geometries: The Dicer Algorithm, *AMD – Trends in Unstructured Mesh Generation*, ASME, **220**, 131–135.

Merhof D, Grosso R, Tremel U, Greiner G (2007) Anisotropic quadrilateral mesh generation: An indirect approach, *Advances in Engineering Software*, **38**, 860–867.

Meshkat S, Talmor D (2000) Generating a mixed mesh of hexahedra, pentahedra and tetrahedra from an underlying tetrahedral mesh, *International Journal for Numerical Methods in Engineering*, **49**, Issues 1–2, 17–30.

Meyer A, Marin P (2004) Segmentation of 3D triangulated data points using edge constructed with a C1 discontinuous surface fitting, *Computer-Aided Design*, **36**, 1327–1336.

Meyer M, Kirby RM, Whitaker R (2007) Topology, accuracy, and quality of isosurface meshes using dynamic particles, *IEEE Transactions on Visualization and Computer Graphics*, **13**, Issue 6, 1704–1711.

Meyers RJ, Tautges TJ, Tuchinsky PM (1998) The Hex-Tet hex-dominant meshing algorithm as implemented in CUBIT, *Proceedings of the 7th International Meshing Roundtable*, 151–158.

Min W (1997) Generating hexahedron-dominant mesh based on shrinking-mapping method, *Proceedings of the 6th International Meshing Roundtable*, 171–182.

Miranda ACO, Martha LF, Wawrzynek PA, Ingraffea AR (2009) Surface mesh regeneration considering curvatures, *Engineering with Computers*, **25**, Issue 2, 207–219.

Misztal MK, Baerentzen JA, Anton F, Erleben K (2009) Tetrahedral mesh improvement using multi-face retriangulation, *Proceedings of the 18th International Meshing Roundtable*, Springer, 539–555.

Mitchell SA (1998) The all-hex geode – Template for conforming a diced tetrahedral mesh to any diced hexahedral mesh, *Proceedings of the 7th International Meshing Roundtable*, Sandia National Laboratories, 295–305.

Mitchell SA, Vavasis SA (2000) Quality mesh generation in higher dimensions, *SIAM Journal on Computing*, **19**, Issue 4, 1334–1370.

Mitchell WF (2007) A refinement-tree based partitioning method for dynamic load balancing with adaptively refined grids, *Journal of Parallel and Distributed Computing*, **67**, 417–429.

Moller P, Hansbo P (1995) On advancing front mesh generation in 3 dimensions, *International Journal for Numerical Methods in Engineering*, **38**, Issue 21, 3551–3569.

Montenegro R, Cascon JM, Escobar JM, Rodriguez E, Montero G (2009a) An automatic strategy for adaptive tetrahedral mesh generation, *Applied Numerical Mathematics*, **59**, 2203–2217.

Montenegro R, Cascon JM, Escobar JM, Rodriguez E, Montero G (2009b) The automatic Meccano method to mesh complex solids, *WCECS 2009: World Congress on Engineering and Computer Science*, Vols. I and II, 955–960, San Francisco, October 20–22, 2009.

Montenegro R, Cascon JM, Escobar JM, Rodriguez E, Montero G (2010) The Meccano method for simultaneous volume parametrization and mesh generation of complex solids, *9th World Congress on Computational Mechanics and 4th Asian Pacific Congress on Computational Mechanics*, Sydney, Australia, July 19–23, 2010.

Morera DM, Carvalho PC, Velho L (2008) Modeling on triangulations with geodesic curves, *The Visual Computer*, 24, 1025–1037.

Morgan K, Peraire J (1998) Unstructured grid finite element methods for fluid mechanics, *Reports on Progress in Physics*, 61, Issue 6, 569–638.

Mortara M, Patane G, Spagnuolo M (2006) From geometric to semantic human body models, *Computers and Graphics*, 30, 185–196.

Mortara M, Patane G, Spagnuolo M, Falcidieno B, Rossignac J (2004) Blowing bubbles for multi-scale analysis and decomposition of triangle meshes, *Algorithmica*, 38, 227–248.

Mosler J, Ortiz M (2007) Variational h-adaptation in finite deformation elasticity and plasticity, *International Journal for Numerical Methods in Engineering*, 72, 505–523.

Mousa MH, Chaine R, Akkouche S, Galin E (2008) Towards an efficient triangle-based spherical harmonics representation of 3D objects, *Computer Aided Geometrical Design*, 25, 561–575.

Mouton T, Borouchaki H, Bennis C (2010) Hybrid mesh generation for reservoir flow simulation: Extension to highly deformed corner point geometry grids, *Finite Elements in Analysis and Design*, 46, 152–164.

Muller-Hannemann M (1998) Hexahedral mesh generation by successive dual cycle elimination, *Proceedings of the 7th International Meshing Roundtable*, Sandia National Laboratories, 365–378, October 1998.

Muller-Hannemann M, Weihe K (2000) Quadrangular refinements of convex polygon with an application to finite element meshes, *International Journal of Computational Geometry & Applications*, 10, Issue 1, 1–40.

Murdoch P, Benzley SE (1995) The spatial twist continuum, *Proceedings of the 4th International Meshing Roundtable*, Sandia National Laboratories, 243–251.

Murdoch P, Benzley SE, Blacker T, Mitchell SA (1997) The spatial twist continuum: A connectivity based method for representing all-hexahedral finite element meshes, *Finite Elements in Analysis and Design*, 28, Issue 2, 137–149.

Muses C (1997a) The dimensional family approach in (hyper) sphere packing: A topological study of new patterns, structures, and interdimensional functions, *Applied Mathematics and Communication*, 88, 1–26.

Muses C (1997b) A sphere-packing breakthrough via dimensional families, *Applied Mathematics and Communication*, 83, 1–2.

Muthukrishnan SN, Shiakolas PS, Nambiar RV, Lawrence KL (1995) Simple algorithm for adaptive refinement of 3-dimensional finite-element tetrahedral meshes, *AIAA Journal*, 33, Issue 5, 928–932.

Nardelli E, Proietti G (2006) Efficient unbalanced merge-sort, *Information Science*, 176, 1321–1337.

Nave D, Chrisochoides N, Chew LP (2004) Guaranteed-quality parallel Delaunay refinement for restricted polygonal domains, *Computational Geometry*, 28, 191–215.

Neto JBC, Wawrzynek PA, Carvalho MTM, Martha LF, Ingraffea AR (2001) An algorithm for three-dimensional mesh generation for arbitrary regions with cracks, *Engineering with Computers*, 17, Issue 1, 75–91.

Nieser M, Schulz C, Polthier K (2010) Patch layout from feature graphics, *Computer-Aided Design*, 42, 213–220.

Nishioka T, Tokudome H, Kinoshita M (2001) Dynamic fracture-path predication in impact fracture phenomena using moving finite element method based on Delaunay automatic mesh generation, *International Journal of Solid and Structures*, 38, Issues 30–31, 5273–5301.

Noel F (2002) Global parameterization of a topological surface defined as a collection of trimmed bi-parametric patches: Application to automatic mesh construction, *International Journal for Numerical Methods in Engineering*, 54, 965–986.

Ollivier-Gooch C, Boivin C (2001) Guaranteed-quality simplicial mesh generation with cell size and grading control, *Engineering with Computers*, 17, Issue 3, 269–286.

Onate E, Bugeda G (1993) A study of mesh optimality criteria in adaptive finite element analysis, *Engineering Computations*, **10**, Issue 4, 307–321.

Orenstein JA, Merrett JA, Devroye L (1983) Linear sorting with O(LOG-N) processors, *BIT*, **23**, Issue 2, 170–180.

Ortiz M, Quigley JJ (1991) Adaptive mesh refinement in strain localization problems, *Computer Methods in Applied Mechanics and Engineering*, **90**, Issues 1–3, 781–804.

Oudot S, Rineau L, Yvinec M (2010) Meshing volumes with curved boundaries, *Engineering with Computers*, **26**, 265–279.

Owen SJ (1998) A survey of unstructured mesh generation technology, *Proceedings, 7th International Meshing Roundtable*, Sandia National Lab, 239–267.

Owen SJ (1999) Constrained triangulation: Application to hex-dominated mesh generation, *Proceedings of the 8th International Meshing Roundtable*, South Lake Tahoe, CA, 31–41, October 1999.

Owen SJ (2001) Hex-dominant mesh generation using 3D constrained triangulation, *Computer-Aided Design*, **33**, 211–220.

Owen SJ, Canann SA, Saigal S (1997) Pyramid elements for maintaining tetrahedra and hexahedra conformability, *AMD – Trends in Unstructured Mesh Generation, ASME*, **220**, 123–129.

Owen SJ, Saigal S (2000) H-Morph: An indirect approach to advancing front hex meshing, *International Journal for Numerical Methods in Engineering*, **49**, 289–312.

Owen SJ, Saigal S (2001) Formation of pyramid elements for hexahedra to tetrahedra transition, *Computer Methods in Applied Mechanics and Engineering*, **190**, Issue 34, 4505–4518.

Owen SJ, Shepherd JF (2009) Embedding features in a Cartesian grid, *Proceedings of the 18th International Meshing Roundtable*, Springer, 116–138.

Owen SJ, Staten ML, Canann SA, Saigal S (1998) Advancing front quad meshing using local triangles transformations, *Proceedings of the 7th International Meshing Roundtable*.

Owen SJ, Staten ML, Canann SA, Saigal S (1999) Q-morph: An indirect approach to advancing-front quad meshing, *International Journal for Numerical Methods in Engineering*, **44**, 1317–1340.

Owen SJ, White DR (2003) Mesh-based geometry, *International Journal for Numerical Methods in Engineering*, **58**, 375–395.

Ozturan C, De Cougny HL, Shephard MS, Flaherty JE (1994) Parallel adaptive mesh refinement and redistribution on distributed memory computers, *Computer Methods in Applied Mechanics and Engineering*, **119**, Issues 1–2, 123–137.

Padron MA, Suarez JP, Plaza A (2005) A comparative study between some bisection based partitions in 3D, *Applied Numerical Mathematics*, **55**, 357–367.

Pain CC, Umpleby AP, De Oliveira CRE, Goddard AJH (2001) Tetrahedral mesh optimization and adaptivity for steady-state and transient finite element calculations, *Computer Methods in Applied Mechanics and Engineering*, **190**, 3771–3796.

Park S, Washam CJ (1979) Drag method as a finite element mesh generation scheme, *Computers & Structures*, **10**, Issues 1–2, 343–346.

Parthasarathy VN, Kodiyalam S (1991) A constrained optimization approach to finite element mesh smoothing, *Finite Elements in Analysis and Design*, **9**, Issue 4, 309–320.

Pavlakos CJ, Jones JS, Mitchell SA (1997) An immersive environment for exploration of CUBIT meshes, *Proceedings of the 6th International Meshing Roundtable*, Sandia National Laboratories, 47–48, October 1997.

Peraire J, Peiro J, Formaggia L, Morgan K, Zienkiewicz OC (1988) Finite element Euler computations in three dimensions, *International Journal for Numerical Methods in Engineering*, **26**, 2135–2159.

Perucchio R, Saxena M, Kela A (1989) Automatic mesh generation from solid models based on recursive spatial decomposition, *International Journal for Numerical Methods in Engineering*, **28**, 2469–2501.

Peterson RC, Jimack PK, Kelmanson MA (1999) The solution of two-dimensional free-surface problems using automatic mesh generation, *International Journal for Numerical Methods in Engineering*, **31**, Issue 6, 937–960.

Plaza A, Carey GF (2000) Refinement of simple grids based on the skeleton, *Applied Numerical Mathematics*, **2**, Issue 32, 195–218.

Plaza A, Falcon S, Suarez JP (2010) On the non-degeneracy property of the longest-edge trisection of triangles, *Applied Mathematics and Computation*, **216**, 862–869.

Plaza A, Falcon S, Suarez JP, Abad P (2012) A local refinement algorithm for the longest-edge trisection of triangle meshes, *Mathematics and Computers in Simulation*, **82**, 2971–2981.

Plaza A, Padron MA, Suarez JP (2005) Non-degeneracy study of the 8-tetrahedra longest-edge partition, *Applied Numerical Mathematics*, **55**, 458–472.

Plaza A, Padron MA, Suarez JP, Falcon S (2004) The 8-tetrahedra longest-edge partition of right-type tetrahedra, *Finite Elements in Analysis and Design*, **41**, 253–265.

Prakash S, Ethier CR (2001) Requirements for mesh resolution in 3D computational hemodynamics, *Journal of Biomechanics Engineering – Transactions of the ASME*, **123**, Issue 2, 134–144.

Prassl AJ, Kickinger F, Ahammer H, Grau V, Schneider JE, Hofer E, Vigmond EJ, Trayanova NA, Plank G (2009) Automatically generated anatomically accurate meshes for cardiac electrophysiology problems, *IEEE Transactions on Biomedical Engineering*, **56**, Issue 5, 1318–1330.

Pressburger Y, Perucchio (1995) A self-adaptive FE system based on recursive spatial decomposition and multigrid analysis, *International Journal for Numerical Methods in Engineering*, **38**, Issue 8, 1399–1421.

Price MA, Armstrong CG (1995) Hexahedral mesh generation by medial surface subdivision: 1. Solids with convex edges, *International Journal for Numerical Methods in Engineering*, **38**, 3335–3359.

Price MA, Armstrong CG (1997) Hexahedral mesh generation by medial surface subdivision: 2. Solids with flat and concave edges, *International Journal for Numerical Methods in Engineering*, **40**, 111–136.

Qian J, Zhang Y (2012) Automatic unstructured all-hexahedral mesh generation from B-Reps for non-manifold CAD Assemblies, *Engineering with Computers*, **28**, 345–359.

Qian J, Zhang Y, Wang W, Lewis AC, Qidwai MAS, Geltmacher AB (2009) Quality improvement of non-manifold hexahedral meshes for critical feature determination of microstructure materials, *Proceedings of the 18th International Meshing Roundtable*, Springer, 211–230.

Qian J, Zhang Y, Wang W, Lewis AC, Qidwai MAS, Geltmacher AB (2010) Quality improvement of non-manifold hexahedral meshes for critical feature determination of microstructure material, *International Journal for Numerical Methods in Engineering*, **82**, 1406–1423.

Quadros WR, Shimada K, Owen SJ (2004) Skeleton-based computational method for the generation of a 3D finite element mesh sizing function, *Engineering with Computers*, **20**, Issue 3, 249–264.

Quadros WR, Owen SJ (2012) Defeaturing CAD models using a geometry-based size field and facet-based reduction operations, *Engineering with Computers*, **28**, Issue 3, 211–224.

Quey R, Dawson PR, Barbe F (2011) Large-scale 3D random polycrystals for the finite element method: Generation, meshing and remeshing, *Computer Methods in Applied Mechanics and Engineering*, **200**, Issues 17–20, 1729–1745.

Radin C (2004) Orbit of orbs: Sphere packing meets Penrose Tilings, *The American Mathematical Monthly; February 2004*, **111**, 2; Academic Research Library, 137.

Rajan VT (1994) Optimality of the Delaunay triangulation in $\mathbb{R}^d$, *Discrete & Computational Geometry*, **12**, 189–202.

Ramos A, Simoes JA (2006) Tetrahedral versus hexahedral finite elements in numerical modelling of the proximal femur, *Medical Engineering & Physics*, **28**, 916–924.

Ran L, Borouchaki H, Benali A, Bennis C (2012) Hex-dominant mesh generation for subterranean formation modeling, *Engineering with Computers*, **28**, Issue 3, 255–268.

Rand A, Walkington N (2009) Collars and intestines: Practical conforming Delaunay refinement, *Proceedings of the 18th International Meshing Roundtable*, Springer, 481–497.

Rashid MM, Selimotic M (2006) A three-dimensional finite element method with arbitrary polyhedral elements, *International Journal for Numerical Methods in Engineering*, **28**, Issue 9, 916–924.

Rassineux A (1997) 3D mesh adaptation, optimization of tetrahedral meshes by advancing front technique, *Computer Methods in Applied Mechanics and Engineering*, **141**, 335–354.

Rassineux A (1998) Generation and optimization of tetrahedral meshes by advancing front technique, *International Journal for Numerical Methods in Engineering*, **41**, Issue 4, 651–674.

Remacle JF, Geuzaine C, Compere G, Marchandise E (2010) High-quality surface remeshing using harmonic maps, *International Journal for Numerical Methods in Engineering*, **83**, 403–425.

Remacle JF, Henrotte F, Carrier-Baudouin T, Bechet E, Marchandise E, Geuzaine C, Mouton T (2013) A frontal Delaunay quad mesh generator using the L^∞ norm, *International Journal for Numerical Methods in Engineering*, **94**, 494–512.

Remacle JF, Li X, Shephard MS, Flaherty JE (2005) Anisotropic adaptive simulation of transient flows using discontinuous Galerkin methods, *International Journal for Numerical Methods in Engineering*, **62**, 899–923.

Rheinboldt WC (1985) Error estimates for nonlinear finite element computations, *Computers & Structures*, **20**, 91–98.

Ribo R, Bugeda G, Onate E (2002) Some algorithm to correct a geometry in order to create a finite element mesh, *Computers & Structures*, **80**, Issues 16–17, 1399–1408.

Rivara MC (1984) Mesh refinement based on the generalized bisection of simplices, *SIAM Journal on Numerical Analysis*, **2**, 604–613.

Rivara MC (1997) New longest edge algorithms for the refinement and/or improvement of unstructured triangulations, *International Journal for Numerical Methods in Engineering*, **40**, 3313–3324.

Rivara MC, Inostroza P (1997) Using longest-side bisection techniques for the automatic refinement of Delaunay triangulations, *International Journal for Numerical Methods in Engineering*, **40**, Issue 4, 581–597.

Rivara MC, Levin C (1992) A 3-d refinement algorithm suitable for adaptive and multi-grid techniques, *Journal of Computational and Applied Mathematics*, **8**, 281–290.

Roca X, Sarrate J (2010a) Local dual contributions: Representing dual surfaces for block meshing, *International Journal for Numerical Methods in Engineering*, **83**, Issue 6, 709–740.

Roca X, Sarrate J (2010b) An automatic and general least-square projection procedure for sweep meshing, *Engineering with Computers*, **26**, Issue 4, 391–406.

Ruiz-Girones E, Roca X, Sarrate J (2009) A new procedure to compute imprints in multi-sweeping algorithms, *Proceedings of the 18th International Meshing Roundtable*, Springer, 280–299.

Ruiz-Girones E, Roca X, Sarrate J (2012) The receding front method applied to hexahedral mesh generation of exterior domains, *Engineering with Computers*, **28**, Issue 4, 391–408.

Ruiz-Girones E, Sarrate J (2010) Generation of structural hexahedral meshes in volumes with holes, *Finite Elements in Analysis and Design*, **46**, 280–299.

Ruppert J (1994) A Delaunay refinement algorithm for quality 2-dimensional mesh generation, *Journal of Algorithms*, **18**, Issue 3, 548–585.

Ruppert J, Seidel R (1992) On the difficulty of triangulating 3-dimensional nonconvex polyhedral, *Discrete & Computational Geometry*, **7**, Issue 3, 227–253.

Rypl D (2010) Sweeping of unstructured meshes over generalized extruded volumes, *Finite Elements in Analysis and Design*, **46**, Issues 1–2, 203–215.

Sahimi M, Darvishi R, Haghighi M, Rasaei MR (2010) Upscaled unstructural computational grids for efficient simulation of flow in fractured porous media, *Transport in Porous Media*, **83**, 195–218.

Sahni O, Luo XJ, Jansen KE, Shephard MS (2010) Curved boundary layer meshing for adaptive viscous flow simulations, *Finite Elements in Analysis and Design*, **46**, 132–139.

Sahni O, Muller J, Jansen KE, Shephard MS, Taylor CA (2006) Efficient anisotropic adaptive discretization of the cardiovascular system, *Computer Methods in Applied Mechanics and Engineering*, **195**, Issues 41–43, 5634–5655.

Samani A, Bishop J, Yaffe MJ, Plewes DB (2001) Biomechanical 3D finite element modeling of the human breast using MRI data, *IEEE Transactions on Medical Imaging*, **20**, Issue 4, 271–279.

Sarrate J, Huerta A (2000) Efficient unstructured quadrilateral mesh generation, *International Journal for Numerical Methods in Engineering*, **49**, Issue 10, 1327–1350.

Sarrate J, Huerta A (2001) An improved algorithm to smooth graded quadrilateral meshes preserving the prescribed element size, *Communication in Numerical Methods in Engineering*, **17**, Issue 2, 89–99.

Sastry SP, Shontz SM (2009) A comparison of gradient- and hessian-based optimization methods for tetrahedral mesh quality improvement, *Proceedings of the 18th International Meshing Roundtable*, Springer, 631–648.

Sazonov I, Nithiarasu P (2012) Semi-automatic surface and volume mesh generation for subject-specific biomedical geometries, *International Journal for Numerical Methods in Biomedical Engineering*, **28**, 133–157.

Schneiders R (1996) A grid-based algorithm for the generation of hexahedral element meshes, *Engineering with Computers*, **12**, Issues 3–4, 168–177.

Schneiders R (1997) An algorithm for the generation of hexahedral element meshes based on an Octree technique, *Proceedings of the 6th International Meshing Roundtable*, 183–194.

Schneiders R (2000) Octree-based hexahedral mesh generation, *International Journal of Computational Geometry & Applications*, **10**, Issue 4, 383–398.

Schneiders R, Bunten R (1995) Automatic-generation of hexahedral finite-element meshes, *Computer Aided Geometric Design*, **12**, Issue 7, 693–707.

Schroeder WJ, Shephard MS (1990) A combined Octree Delaunay method for fully automatic 3D mesh generation, *International Journal for Numerical Methods in Biomedical Engineering*, **29**, Issue 1, 37–55.

Scott MA, Benzley SE, Owen SJ (2006) Improved many-to-one sweeping, *International Journal for Numerical Methods in Engineering*, **65**, Issue 3, 332–348.

Secchi S, Simoni L (2003) An improved procedure for 2D unstructured Delaunay mesh generation, *Acta Mechanica Sinica*, **19**, Issue 2, 162–171.

Seol ES, Shephard MS (2006) Efficient distributed mesh data structure for parallel automated adaptive analysis, *Engineering with Computers*, **22**, 197–213.

Sezer L, Zeid I (1991) Automatic quadrilateral triangular free-form mesh generation for planar regions, *International Journal for Numerical Methods in Engineering*, **32**, Issue 7, 1441–1483.

Shaffer CA, Samet H (1987) Optimal Quadtree construction algorithms, *Computer Vision Graphics and Image Processing*, **37**, Issue 3, 402–419.

Shamos MI, Hoey D (1991) Closest point problems, *Proceedings of the 16th Annual Symposium on Computational Geometry*, 357–363.

Shaw RD, Pitchen RG (1978) Modification to the Suhara–Fukuda method of network generation, *International Journal for Numerical Methods in Engineering*, **12**, 93–99.

Sheehy DJ, Armstrong CG, Robinson DJ (1996) Shape description by medial surface construction, *IEEE Transactions on Visualization and Computer Graphics*, **2**, Issue 1, 62–72.

Sheffer A, Etzion M, Rappoport A, Bercovier M (1999) Hexahedral mesh generation using the embedded Voronoi graph, *Engineering with Computers*, **15**, Issue 3, 248–262.

Shenton DN, Cendes ZJ (1985) 3-dimensional finite-element mesh generation using Delaunay tessellation, *IEEE Transactions on Magnetics*, **21**, Issue 6, 2535–2538.

Shephard MS (1988) Approach to the automatic generation and control of finite element meshes, *Applied Mechanics Reviews*, **41**, 169–185.

Shephard MS, Flaherty JE, Bottasso CL, De Cougny HL, Ozturan C, Simone ML (1997) Parallel automatic adaptive analysis, *Parallel Computing*, **23**, 1327–1347.

Shephard MS, Georges MK (1991) Automatic 3-dimensional mesh generation by the finite Octree technique, *International Journal for Numerical Methods in Engineering*, **32**, Issue 4, 709–749.

Shephard MS, Georges MK (1992) Reliability of automatic 3D mesh generation, *Computer Methods in Applied Mechanics and Engineering*, **101**, Issues 1–3, 443–462.

Shepherd J, Benzley S, Mitchell S (1999) Interval assignment for volume and holes, *Second Symposium on Trends in Unstructured Mesh Generation*, University of Colorado, Boulder, CO, August 1999.

Shepherd JF (2009) Conforming hexahedral mesh generation via geometric capture methods, *Proceedings of the 18th International Meshing Roundtable*, Springer, 85–102.

Shepherd JF, Johnson CR (2008) Hexahedral mesh generation constraints, *Engineering with Computers*, **24**, Issue 3, 195–213.

Sherbrooke EC, Patrikalakis NM, Brisson E (1996a) An algorithm for the medial axis transformation of 3D polyhedral solids, *IEEE Transactions on Visualization and Computer Graphics*, **2**, Issue 1, 44–61.

Sherbrooke EC, Patrikalakis NM, Wolter FE (1996b) Differential and topological properties of medial axis transform, *Graphical Models and Image Processing*, **58**, Issue 6, 574–592.

Sherwin SJ, Peiro J (2002) Mesh generation in curvilinear domain using high-order elements, *International Journal for Numerical Methods in Engineering*, **53**, Issue 6, 207–223.

Shewchuk JR (1997) Adaptive precision floating-point arithmetic and fast robust geometric predicates, *Discrete & Computational Geometry*, **18**, 305–363.

Shewchuk JR (2002) Delaunay refinement algorithm for triangular mesh generation, *Computational Geometry – Theory and Applications*, **22**, Issues 1–3, 21–74.

Shewchuk JR (2008) General-dimensional constrained Delaunay and constrained regular triangulation, I: Combinational properties, *Discrete & Computational Geometry*, **39**, 580–637.

Shimada K, Gossard DC (1998) Automatic triangular mesh generation of trimmed parametric surfaces for finite element analysis, *Computer-Aided Geometric Design*, **15**, Issue 3, 199–222.

Shimada K, Yamada A, Itoh T (2000) Anisotropic triangulation of parametric surfaces via close packing of ellipsoids, *International Journal of Computational Geometry & Applications*, **10**, Issue 4, 417–440.

Shivanna KH, Tadepalli SC, Grosland N (2010) Feature-based multiblock finite element mesh generations, *Computer-Aided Design*, **42**, Issue 12, 1108–1116.

Shostko A, Lohner R (1995) 3-Dimensional parallel unstructured grid generation, *International Journal for Numerical Methods in Engineering*, **38**, Issue 6, 905–925.

Shostko AA, Lohner R, Sandberg WC (1999) Surface triangulation over intersecting geometries, *International Journal for Numerical Methods in Engineering*, **44**, 1359–1376.

Si H (2007) TetGen. Available at http://tetgen.berlios.de.

Si H (2008) Adaptive tetrahedral mesh generation by constrained Delaunay refinement, *International Journal for Numerical Methods in Engineering*, **75**, Issue 7, 856–880.

Si H (2009) An analysis of Shewchuk's Delaunay refinement algorithm, *Proceedings of the 18th International Meshing Roundtable*, Springer, 498–518.

Si H (2010) Constrained Delaunay tetrahedral mesh generation and refinement, *Finite Elements in Analysis and Design*, **46**, 33–46.

Si H, Gaertner K (2005) Meshing piecewise linear complexes by constrained Delaunay tetrahedralizations, *Proceedings of the 14th International Meshing Roundtable*, Springer-Verlag New York Inc., 147–163.

Si H, Gaertner K (2010) Boundary conforming Delaunay mesh generation, *Computational Mathematics and Mathematical Physics*, **50**, Issue 1, 38–53.

Si H, Gaertner K (2011) 3D boundary recovery by constrained Delaunay tetrahedralization, *International Journal for Numerical Methods in Engineering*, **85**, Issue 11, 1341–1364.

Simari P, Nowrouzezahrai D, Kalogerakis E, Singh K (2009) Multi-objective shape segmentation and labeling, *Eurographics Symposium on Geometry Processing*, Blackwell Publishing, Oxford, **28**, Number 5.

Simonovski I, Cizelj L (2011) Automatic parallel generation of finite elements for complex spatial structures, *Computational Materials Science*, **50**, Issue 5, 1606–1618.

Sirois Y, Dompierre J, Vallet MG, Guibault F (2005) Measuring the conformity of non-simplicial elements to an anisotropic metric field, *International Journal for Numerical Methods in Engineering*, **64**, 1944–1958.

Smith NP, Pullan AJ, Hunter PJ (2000) Generation of an anatomically based geometric coronary model, *Annals of Biomedical Engineering*, **28**, Issue 1, 14–25.

Smith TS, Farouki RT (2001) Gauss map computation for free-form surfaces, *Computer Aided Geometric Design*, **18**, Issue 9, 831–850.

Sohn D, Cho YS, Im S (2012) A novel scheme to generate meshes with hexahedral elements and poly-pyramid elements: The carving technique, *Computer Methods in Applied Mechanics and Engineering*, **201**, 208–227.

Sowa H, Koch E (2004) Hexagonal and trigonal sphere packings. II. Bivariant lattice complexes, *Acta Crystallographica. Section A*, **60**, 158–166.

Srinivasan V, Nackman LR, Tang JM, Meshkat SN (1992) Automatic mesh generation using the symmetrical axis transformation of polygonal domain, *Proceedings of the IEEE*, **80**, Issue 9, 1485–1501.

Stadler M, Holzapfel GA (2004) Subdivision schemes for smooth contact surfaces of arbitrary mesh topology in 3D, *International Journal for Numerical Methods in Engineering*, **60**, Issue 7, 1161–1195.

Staten ML, Benzley S, Scott M (2008) A methodology for quadrilateral finite element mesh coarsening, *Engineering with Computers*, **24**, Issue 3, 241–251.

Staten ML, Canann SA, Owen SJ (1999) BMSweep: Locating interior nodes during sweeping, *Engineering with Computers*, **15**, Issue 3, 212–218.

Staten ML, Kerr RA, Owen SJ, Blacker TD, Stupazzini M, Shimada K (2010a) Unconstrained plastering – Hexahedral mesh generation via advancing-front geometry decomposition, *International Journal for Numerical Methods in Engineering*, **81**, 135–171.

Staten ML, Shepherd JF, Ledoux F, Shimada K (2010b) Hexahedral mesh matching: Converting non-conforming hexahedral-to-hexahedral interfaces into conforming interfaces, *International Journal for Numerical Methods in Engineering*, **82**, Issue 12, 1475–1509.

Su P, Drysdale RLS (1997) A comparison of sequential Delaunay triangulation algorithms, *Computational Geometry*, **7**, 361–385.

Su Y, Lee KH, Kumar AS (2004) Automatic hexahedral mesh generation for multi-domain composite models using a hybrid projective grid-based method, *Computer-Aided Design*, **36**, Issue 3, 203–215.

Su Y, Lee KH, Kumar AS (2005) Automatic hexahedral mesh generation using a new grid-based method with geometry and mesh transformation, *Computer Method in Applied Mechanics and Engineering*, **194**, Issues 39–41, 4071–4096.

Suarez JP, Moreno T, Abad P, Plaza A (2012) Properties of the longest-edge n-section refinement scheme for triangular meshes, *Applied Mathematics Letters*, **25**, 2037–2039.

Suarez JP, Plaza A, Carey GF (2005) The propagation problem in longest-edge refinement, *Finite Elements in Analysis and Design*, **42**, 130–151.

Suarez JP, Plaza A, Carey GF (2008) Propagation of longest-edge mesh pattern in local adaptive refinement, *Communications in Numerical Methods in Engineering*, **24**, 543–553.

Sukumar N, Tabarraei A (2004) Conforming polygonal finite elements, *International Journal for Numerical Methods in Engineering*, **61**, Issue 12, 2045–2066.

Sullivan GJ, Baker RL (1994) Efficient Quadtree coding of images and video, *IEEE Transactions on Image Processing*, **3**, Issue 3, 327–331.

Sullivan JM, Charron G, Paulsen KD (1997) A three-dimensional mesh generator for arbitrary multiple material domains, *Finite Elements in Analysis and Design*, **25**, Issues 3–4, 219–241.

Sun L, Zhao G, Ma X (2011) Adaptive generation and local refinement of three-dimensional hexahedral element mesh, *Finite Elements in Analysis and Design*, **50**, Issue 1, 184–200.

Sußner G, Greiner G (2009) Hexagonal Delaunay triangulation, *Proceedings of the 18th International Meshing Roundtable*, Springer, 519–538.

Sutou A, Dai Y (2002) Global optimization approach to unequal sphere packing problems in 3D, *Journal of Optimization and Applications*, **114**, Issue 3, 671–684.

Sze KY (1992) Efficient formulation of robust hybrid elements using orthogonal stress–strain interpolants and admissible matrix formulation, *International Journal for Numerical Methods in Engineering*, **35**, Issue 1, 1–20.

Sze KY, Ghali A (1993a) Hybrid hexahedral element for solids, plates shells and beams by selective scaling, *International Journal for Numerical Methods in Engineering*, **36**, Issue 9, 1519–1540.

Sze KY, Ghali A (1993b) Hybrid plane quadrilateral element with corner rotations, *Journal of Structural Engineering – ASCE*, **119**, Issue 9, 2552–2572.

Sze KY, Lo SH (1999) A twelve-node hybrid stress brick element for beam column analysis, *Engineering Computations*, **16**, Issue 6/7, 752–766.

Sze KY, Lo SH, Yao LQ (2002) Hybrid-stress solid elements for shell structures based upon a modified variational functional, *International Journal for Numerical Methods in Engineering*, **53**, 2617–2642.

Tabarraei A, Sukumar N (2005) Adaptive computations on conforming Quadtree meshes, *Finite Elements in Analysis and Design*, **41**, 686–702.

Tabarraei A, Sukumar N (2007) Adaptive computations using material forces and residual-based error estimates on quadtree meshes, *Computer Methods in Applied Mechanics and Engineering*, **196**, Issues 25–28, 2657–2680.

Tabarraei A, Sukumar N (2008) Extended finite element method on polygonal and quadtree meshes, *Computer Methods in Applied Mechanics and Engineering*, **197**, 425–438.

Tacher L, Parriaux (1996) Automatic node generation in N-dimensional space, *Communications in Numerical Methods in Engineering*, **12**, 243–248.

Taddei F, Pancanti A, Viceconti M (2004) An improved method for the automatic mapping of computed tomography numbers onto finite element models, *Medical Engineering & Physics*, **26**, Issue 1, 61–69.

Taddei F, Viceconti M, Manfrini M, Toni A (2003) Mechanical strength of a femoral reconstruction in paediatric oncology: A finite element study, *Proceedings of the Institution of Mechanical Engineers Part H – Journal of Engineering in Medicine*, **217**, Issue H2, 111–119.

Talbert JA, Parkinson AR (1991) Development of an automatic two-dimensional finite element mesh generator using quadrilateral elements and Bezier curve boundary definition, *International Journal for Numerical Methods in Engineering*, **29**, 1551–1567.

Tam TKH, Armstrong CG (1991) Finite element mesh generation by medial axis subdivision, *Advances in Engineering Software*, **13**, 313–324.

Tam TKH, Armstrong CG (1993) Finite-element mesh control by integer programming, *International Journal for Numerical Methods in Engineering*, **36**, Issue 15, 2581–2605.

Tang K, Liu YJ (2005) An optimization algorithm for free-form surface partitioning based on weighted Gaussian image, *Graphical Models*, **67**, 17–42.

Tautges TJ (1999) On automating the generation of hex meshes for assembly geometries, *Second Symposium on Trends in Unstructured Mesh Generation*, University of Colorado, Boulder, CO, August 1999.

Tautges TJ (2001a) CGM: A geometry interface for mesh generation, analysis and other applications, *Engineering with Computers*, **17**, 299–314.

Tautges TJ (2001b) The generation of hexahedral meshes for assembly geometry: Survey and progress, *International Journal for Numerical Methods in Engineering*, **50**, Issue 12, 2617–2642.

Tautges TJ, Blacker T, Mitchell S (1996) The whisker-weaving algorithm: A connectivity based method for constructing all-hexahedral finite element meshes, *International Journal for Numerical Methods in Engineering*, **39**, 3327–3349.

Tautges TJ, Liu SS, Liu YL, Kraftcheck J, Rajit G (1997) Feature recognition applications in mesh generation, *AMD – Trends in Unstructured Mesh Generation, ASME*, **220**, 117–121.

Tawhai MH, Hunter P, Tschirren J, Reinhardt J, McLennan G, Hoffman EA (2004) CT-based geometry analysis and finite element models of the human and ovine bronchial tree, *Journal of Applied Physiology*, **97**, Issue 6, 2310–2321.

Tchon KF, Dompierre J, Camarero (2004) Automatic refinement of conforming quadrilateral and hexahedral meshes, *International Journal for Numerical Methods in Engineering*, **59**, 1539–1562.

Teran J, Molino N, Fedkiw R, Bridson R (2005) Adaptive physics based tetrahedral mesh generation using level sets, *Engineering with Computers*, **21**, Issue 1, 2–18.

Thacker WC (1980) A brief review of techniques for generating irregular computational grids, *International Journal for Numerical Methods in Engineering*, **15**, 1335–1341.

Thakur A, Banerjee AG, Gupta SK (2009) A survey of CAD model simplification techniques for physics-based simulation applications, *Computer-Aided Design*, **41**, 65–80.

Thompson DS, Soni BK (1999) Generation of quad and hex dominant semi-structured meshes using advancing layer scheme, *Proceedings of the 8th International Meshing Roundtable*, South Lake Tahoe, CA, 171–178, October 1999.

Thompson JF (1982) Elliptic grid generation, *Applied Mathematics and Computation*, **10**, 79–105.

Toulorge T, Geuzaine C, Remacle JF, Lambrechts J (2013) Robust untangling of curvilinear meshes, *Journal of Computational Physics*, **254**, 8–26.

Tournois J, Srinivasan R, Alliez P (2009) Perturbing slivers in 3D Delaunay meshes, *Proceedings of the 18th International Meshing Roundtable*, Springer, 156–173.

Traxler CT (1997) An algorithm for adaptive mesh refinement in n dimensions, *Computing*, **59**, Issue 2, 115–137.

Trease HE, Barrett GL (1999) Three-dimensional, parallel, hybrid grid generation for ASCI program, *Second Symposium on Trends in Unstructured Mesh Generation*, University of Colorado, Boulder, CO, August 1999.

Tremel U, Deister F, Hassan O, Weatherill NP (2004) Automatic unstructured surface mesh generation for complex configurations, *International Journal for Numerical Methods in Fluid*, **45**, 341–364.

Tuchinsky P, Clark BW (1997) The hex-tet, hex-dominant automesher. An interim progress report, *Proceedings of the 6th International Meshing Roundtable*, 183–193.

Turkiyyah GM, Storti DW, Ganter M, Chen H, Vimawala M (1997) An accelerated triangulation method for computing the skeletons of free form solid models, *Computer-Aided Design*, **29**, Issue 1, 5–19.

Valette S, Chassery JM, Prost R (2008) Generic Remeshing of 3D triangular meshes with metric-dependent discrete Voronoi diagrams, *IEEE Transaction on Visualization and Computer Graphics*, **14**, Issue 2, 369–381.

Van Oosterom A, Strackee J (1983) The solid angle of a plane triangle, *IEEE Transaction on Biomedical Engineering*, **BME-30**, Issue 2, 125–126.

Vartziotis D, Athanasiadis T, Goudas I, Wipper J (2008) Mesh smoothing using the geometric element transformation method, *Computation Methods in Applied Mechanics and Engineering*, **197**, 3760–3767.

Vartziotis D, Wipper J (2009) The geometric element transformation method for mixed mesh smoothing, *Engineering with Computers*, **25**, 287–301.

Vartziotis D, Wipper J (2011) A dual element based geometric element transformation method for all-hexahedral mesh smoothing, *Computation Methods in Applied Mechanics and Engineering*, **200**, 1186–1203.

Vartziotis D, Wipper J (2012) Fast smoothing of mixed meshes based on the effective geometric element transformation method, *Computation Methods in Applied Mechanics and Engineering*, **201**, 65–68.

Vartziotis D, Wipper J, Papadrakakis M (2013) Improving mesh quality and finite element solution accuracy by GETMe smoothing in solving the Poisson equation, *Finite Elements in Analysis and Design*, **66**, 36–52.

Vartziotis D, Wipper J, Schwald B (2009) The geometric element transformation method for tetrahedral mesh smoothing, *Computation Methods in Applied Mechanics and Engineering*, **199**, 169–182.

Vassberg JC (1999) Multi-block mesh extrusion driven by a globally elliptic system, *Second Symposium on Trends in Unstructured Mesh Generation*, University of Colorado, Boulder, CO, August 1999.

Vemuri BC, Cao Y, Chen L (1998a) Fast collision algorithms with applications to particle flow, *Computer Graphics*, **17**, Issue 2, 121–134.

Vemuri BC, Chen L, Vu-Quoc L, Zhang X, Walton O (1998b) Efficient and accurate collision detection for granular flow simulation, *Graphical Models and Image Processing*, **60**, Issue 6, 403–422.

Verdonschot N, Fennis WMM, Kuijs RH, Stolk J, Kreulen CM, Creugers NHJ (2001) Generation of 3-D finite element models of restored human teeth using micro-CT techniques, *International Journal of Prosthodontics*, **14**, Issue 4, 310–315.

Viceconti M, Bellingeri L, Cristofolini L, Toni A (1998) A comparative study on different methods of automatic mesh generation of human femurs, *Medical Engineering & Physics*, **20**, Issue 1, 1–10.

Viceconti M, Davinelli M, Taddei F, Cappello A (2004) Automatic generation of accurate subject-specific bone finite element models to be used in clinical studies, *Journal of Biomechanics*, **37**, Issue 10, 1597–1605.

Voronoi G (1908) Nouvelles applications des parametres continus à la theorie des forms quadratiques. Recherches sue les parallelloedres primitifs, *Journal für die reine und angewandte Mathematik*, **134**, 198–287.

Vyas V, Shimada K (2009) Tensor-guided hex-dominant mesh generation with targeted all-hex regions, *Proceedings of the 18th International Meshing Roundtable*, Springer, 376–396.

Wada Y, Yoshimura S, Yagawa G (1999) Intelligent local approach for automatic hexahedral mesh generation, *Second Symposium on Trends in Unstructured Mesh Generation*, University of Colorado, Boulder, CO, August 1999.

Wang D, Hassan O, Morgan K, Weatherill N (2006) EQSM: An efficient high quality surface grid generation method based on remeshing, *Computer Methods in Applied Mechanics and Engineering*, **195**, Issues 41–43, 5621–5633.

Wang J, Yu Z (2009) A novel method for surface mesh smoothing: Application in biomedical modeling, *Proceedings of the 18th International Meshing Roundtable*, Springer, 194–210.

Wang Q, Shi XW, Guo C, Guo YC (2012) An improved GPS method with a new pseudo-peripheral nodes finder in finite element analysis, *Finite Elements in Analysis and Design*, **48**, 1409–1415.

Wang WX, Ming CY, Lo SH (2007a) Generation of triangular mesh with specified size by circle packing, *Advances in Engineering Software*, **38**, Issue 2, 133–142.

Wang Y, Lui LM, Gu X, Hayashi KM, Chan TF, Toga AW, Thompson PM, Yau ST (2007b) Brain surface conformal parameterization using Riemann surface structure, *IEEE Transactions on Medical Imaging*, **26**, Issue 6, 853–865.

Wang ZL, Teo JCM, Chui CK, Ong SH, Yan CH, Wang SC, Wong HK, Teoh SH (2005) Computational biomechanical modelling of the lumbar spine using marching-cubes surface smoothened finite element voxel meshing, *Computer Method and Programs in Biomedicine*, **80**, Issue 1, 25–35.

Wang ZM, Kwan AKH, Chan HC (1999) Mesoscopic study of concrete I: Generation of random aggregate structure and finite element mesh, *Computers & Structures*, **70**, Issue 5, 533–544.

Watson DF (1981) Computing the n-dimensional Delaunay tessellation with application to Voronoi polytopes, *Computer Journal*, **24**, 167–172.

Weatherill NP (1988) A method for generating irregular computational grids in multiply connected planar domains, *International Journal for Numerical Methods in Fluids*, **8**, Issue 2, 181–197.

Weatherill NP (1990) The integrity of geometrical boundaries in the 2-dimensional Delaunay triangulation, *Communication in Numerical Methods in Engineering*, **6**, 101–109.

Weatherill NP, Hassan O (1994) Efficient 3-dimensional Delaunay triangulation with automatic point creation and imposed boundary constraints, *International Journal for Numerical Methods in Engineering*, **37**, 2005–2039.

Weiler F, Schneiders R (1996) Automatic geometry adaptive generation of quadrilateral and hexahedral element meshes for the FEM, *Proceedings of the 5th International Conference on Numerical Grid Generation in Computational Field Simulations*, Mississippi State University, 689–697.

Wessner W, Cervenka J, Heitzinger C, Hossinger A, Selberherr S (2006) Anisotropic mesh refinement for the simulation of three-dimensional semiconductor manufacturing processes, *IEEE Transactions on Computer-Aided Design of Integrated Circuit and Systems*, **25**, Issue 10, 2129–2139.

Weyer S, Frohlich A, Riesch-Oppermann H, Cizelj L, Kovac M (2002) Automatic finite element meshing of planar Voronoi tessellations, *Engineering Fracture Mechanics*, **69**, Issue 8, 945–958.

White DR, Kinney P (1997) Redesign of the Paving algorithm: Robustness enhancement through element by element meshing, *Proceedings of the 6th International Meshing Roundtable*, Sandia National Laboratories, 323–335.

White DR, Saigal S, Owen SJ (2004) CCSweep: Automatic decomposition of multi-sweep volumes, *Engineering with Computers*, **20**, 222–236.

White DR, Tautges TJ (1999) Automatic scheme selection for Toolkit hex meshing, *Second Symposium on Trends in Unstructured Mesh Generation*, University of Colorado, Boulder, CO, August 1999.

White DR, Tautges TJ (2000) Automatic scheme selection for toolkit hex meshing, *International Journal for Numerical Methods in Engineering*, **49**, 127–144.

Williams PL (1992) Visibility ordering meshed polyhedra, *ACM Transactions on Graphics*, **11**, Issue 2, 103–126.

Wilson JK, Topping BHV (1998) Parallel adaptive tetrahedral mesh generation by the advancing front technique, *Computers & Structures*, **68**, 57–78.

Wordenweber B (1984) Finite-element mesh generation, *Computer-Aided Design*, **16**, Issue 5, 285–291.

Wright JP, Jack AG (1994) Aspects of 3-dimensional constrained Delaunay meshing, *International Journal for Numerical Methods in Engineering*, **37**, 1841–1861.

Wu D, Lo SH, Sheng N, Sze KY (2010) Universal three-dimensional connection hexahedral elements based on hybrid-stress theory for solid structures, *International Journal for Numerical Methods in Engineering*, **81**, Issue 3, 307–334.

Wu D, Sze KY, Lo SH (2009) Two- and three-dimensional transition elements for adaptive refinement analysis of elasticity problems, *International Journal for Numerical Methods in Engineering*, **78**, 587–630.

Wu H, Guan X, Gong J (2011) ParaStream: A parallel Delaunay triangulation algorithm for LiDAR points on multicore architectures, *Computers & Geosciences*, **37**, 1355–1363.

Xia H, Tucker PG (2009) Distance solutions for medial axis transformation, *Proceedings of the 18th International Meshing Roundtable*, Springer, 246–265.

Xie ZQ, Sevilla R, Hassan O, Morgan K (2013) The generation of arbitrary order curved meshes for 3D finite element analysis, *Computational Mechanics*, **51**, Issue 3, 361–374.

Xu H, Newman S (2006) An angle-based optimization approach for 2D finite element smoothing, *Finite Elements in Analysis and Design*, **42**, 1150–1164.

Yamada A, Shimada K, Itoh T (1999) Meshing curved wire-frame models through energy minimization and packing of ellipses, *International Journal for Numerical Methods in Engineering*, **46**, Issue 8, 1221–1236.

Yamakawa S, Shimada K (2003) Fully-automated hex-dominant mesh generation with directionality control via packing rectangular solid cells, *International Journal for Numerical Methods in Engineering*, **57**, Issue 15, 2099–2129.

Yamakawa S, Shimada K (2009a) Converting a tetrahedral mesh to a prism-tetrahedral hybrid mesh for FEM accuracy and efficiency, *International Journal for Numerical Methods in Engineering*, **80**, Issue 1, 74–102.

Yamakawa S, Shimada K (2009b) Removing self intersection of a triangular mesh by edge swapping, edge hammering and face lifting, *Proceedings of the 18th International Meshing Roundtable*, Springer, 13–29.

Yang YJ, Yong JH, Sun JG (2005) An algorithm for tetrahedral mesh generation based on conforming constrained Delaunay tetrahedralization, *Computer & Graphics – UK*, **29**, Issue 4, 606–615.

Yerry MA, Shephard MS (1983) A modified Quadtree approach to finite element mesh generation, *IEEE Computer Graphics*, **3**, Number 1, 39–46.

Yerry MA, Shephard MS (1984) Automatic three-dimensional mesh generation by the modified-octree technique, *International Journal for Numerical Methods in Engineering*, **20**, 1965–1990.

Yiu KFC, Greaves DM, Cruz S, Saalehi A, Borthwick AGL (1996) Quadtree grid generation: Information handling, boundary fitting and CFD applications, *Computers and Fluids*, **25**, Issue 8, 759–769.

Young PG, Beresford-West TBH, Coward SRL, Notarberardino B, Walker B, Abdul-Aziz A (2008) An efficient approach to converting three-dimensional image data into high accurate computational models, *Philosophical Transactions of the Royal Society A – Mathematical Physical and Engineering Sciences*, **366**, Issue 1878, 3155–3173.

Yu Z, Holst MJ, Cheng Y, McCammon JA (2008a) Feature-preserving adaptive mesh generation for molecular shape modeling and simulation, *Journal of Molecular Graphics and Modelling*, **26**, Issue 8, 1370–1380.

Yu Z, Holst MJ, McCammon JA (2008b) High-fidelity geometric modeling for biomedical applications, *Finite Elements in Analysis and Design*, **44**, 715–723.

Zannoni C, Mantovani R, Viceconti M (1998) Material properties assignment to finite element models of bone structures: A new method, *Medical Engineering & Physics*, **20**, Issue 10, 735–740.

Zavattieri PD, Dari EA, Buscaglia GC (1996) Optimization strategies in unstructured mesh generation, *International Journal for Numerical Methods in Engineering*, **39**, Issue 12, 2055–2071.

Zhang H, Zhao G, Ma X (2007) Adaptive generation of hexahedral element mesh using an improved grid-based method, *Computer-Aided Design*, **39**, Issue 10, 914–928.

Zhang Y, Bajaj C (2006) Adaptive and quality quadrilateral/hexahedral meshing from volumetric data, *Computer Methods in Applied Mechanics and Engineering*, **195**, 942–960.

Zhang Y, Bajaj C, Sohn BS (2005) 3D finite element meshing from image data, *Computer Methods in Applied Mechanics and Engineering*, **194**, Issues 48–49, 5083–5106.

Zhang Y, Hughes TJR, Bajaj CL (2010) An automatic 3D mesh generation method for domains with multiple materials, *Computer Methods in Applied Mechanics and Engineering*, **199**, Issues 5–8, 405–415.

Zhang Y, Xu G, Bajaj C (2006) Quality meshing of implicit solvation models of biomolecular structures, *Computer-Aided Geometric Design*, **23**, Issue 6, 510–530.

Zhao G, Zhang H, Cheng L (2008) Geometry-adaptive generation algorithm and boundary match method for initial hexahedral element mesh, *Engineering with Computers*, **24**, 321–339.

Zhou K, Gong M, Huang X, Guo B (2011) Data-parallel Octrees for surface reconstruction, *IEEE Transactions on Visualization and Computer Graphics*, **17**, Issue 5, 669–681.

Zhou M, Xie T, Seol S, Shephard MS, Sahni O, Jansen KE (2012) Tools to support mesh adaptation on massively parallel computers, *Engineering with Computers*, **28**, 287–301.

Zhou S, Jones CB (2005) HCPO: An efficient insertion order for incremental Delaunay triangulation, *Information Processing Letters*, **93**, Issue 1, 37–42.

Zhou T, Shimada K (2000) An angle-based approach to two-dimensional mesh smoothing, *Proceedings of the 9th International Roundtable*, Sandia National Laboratories, 373–384.

Zhu JZ, Hinton E, Zienkiewicz OC (1991a) Adaptive finite-element analysis with quadrilaterals, *Computers & Structures*, **40**, Issue 5, 1097–1104.

Zhu JZ, Zienkiewicz OC, Hinton E, Wu J (1991b) A new approach to the development of automatic quadrilateral mesh generation, *International Journal for Numerical Methods in Engineering*, **32**, 849–866.

Zienkiewicz OC, Phillips DV (1971) An automatic mesh generation scheme for plane and curved surfaces by isoparametric coordinates, *International Journal for Numerical Methods in Engineering*, **3**, 519–528.

Zienkiewicz OC, Zhu JZ (1987) A simple error estimator and adaptive procedure for practical engineering analysis, *International Journal for Numerical Methods in Engineering*, **24**, 337–357.

Zienkiewicz OC, Zhu JZ (1991) Adaptivity and mesh generation, *International Journal for Numerical Methods in Engineering*, **32**, Issue 4, 783–810.

Zienkiewicz OC, Zhu JZ (1992) The superconvergent patch recovery and a posteriori error estimates. Part I: The recovery technique, *International Journal for Numerical Methods in Engineering*, **33**, 1331–1364.

Zienkiewicz OC, Zhu JZ, Gong NG (1989) Effective and practical h-p-version adaptive analysis procedure for the finite element method, *International Journal for Numerical Methods in Engineering*, **28**, 879–891.

Zuo JZ, Deng XM, Sutton MA (2005) Advances in tetrahedral mesh generation for modelling of three-dimensional regions with complex curvilinear crack shapes, *International Journal for Numerical Methods in Engineering*, **63**, Issue 2, 256–275.

Appendix

Though nobody knows, the appendix may have some useful functions after all.

A.1 AREA OF TRIANGLE

$$\text{Area of triangle } ABC = \frac{1}{2}\|AB \times AC\| = \frac{1}{2}\|\mathbf{b} \times \mathbf{c}\|$$

$$AB \times AC = (b_2 c_3 - c_2 b_3,\ b_3 c_1 - c_3 b_1,\ b_1 c_2 - c_1 b_2)$$

where $\mathbf{b} = AB = (b_1, b_2, b_3)$ and $\mathbf{c} = AC = (c_1, c_2, c_3)$.

On a 2D plane, the vector product is reduced to

$$AB \times AC = (0,\ 0,\ b_1 c_2 - c_1 b_2)$$

Hence, the signed area of triangle ABC is given by (Figure A.1)

$$\text{Area of triangle} = \frac{1}{2}(b_1 c_2 - c_1 b_2)$$

A.2 VOLUME OF TETRAHEDRON

$$\text{Volume of tetrahedron } ABCD = \frac{1}{6}\|AB \times AC \cdot AD\| = \frac{1}{6}\|\mathbf{b} \times \mathbf{c} \cdot \mathbf{d}\|$$

$$\text{Signed volume} = \frac{1}{6}\begin{vmatrix} b_1 & b_2 & b_3 \\ c_1 & c_2 & c_3 \\ d_1 & d_2 & d_3 \end{vmatrix}$$

where $\mathbf{b} = AB = (b_1, b_2, b_3)$, $\mathbf{c} = AC = (c_1, c_2, c_3)$, $\mathbf{d} = AD = (d_1, d_2, d_3)$ and $|\cdot| = $ determinant.

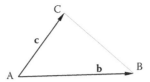

Figure A.1 Triangle ABC.

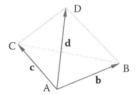

Figure A.2 Tetrahedron ABCD.

A.3 IN-RADIUS AND CIRCUMRADIUS OF TRIANGLE

As the in-circle is tangent to the sides of the triangle, as shown in Figure A.3, the distance between the in-centre and the sides of the triangles is equal to r. Hence,

$$A = \frac{1}{2}rs_1 + \frac{1}{2}rs_2 + \frac{1}{2}rs_3 \Rightarrow 2A = r(s_1 + s_2 + s_3)$$

$$\Rightarrow r = \frac{A}{s}$$

The area of the triangle, A, is given by the Heron's formula

$$A = \sqrt{s(s - s_1)(s - s_2)(s - s_3)}$$

and half perimeter $s = \frac{1}{2}(s_1 + s_2 + s_3)$

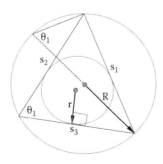

Figure A.3 In-radius and circumradius of a triangle.

The circumradius R of a triangle can also be determined by area consideration:

$$A = \frac{1}{2}s_2s_3\sin(\theta_1) \Rightarrow 4AR = s_2s_3[2R\sin(\theta_1)] = s_2s_3s_1 \Rightarrow R = \frac{s_1s_2s_3}{4A}$$

A.4 CIRCUMCENTRE AND CIRCUMSPHERE OF A TETRAHEDRON

In Delaunay triangulation, the circumsphere inclusion test is applied to determine the non-Delaunay tetrahedra with respect to the insertion point. Given a tetrahedron ABCD, by definition, the circumcentre O can be determined by equating the distance from O to the vertices, as shown in Figure A.4.

$$\|OB\|^2 = \|OA\|^2 \Rightarrow (OB + OA) \cdot (OB - OA) = 0$$

$$\Rightarrow 2AB \cdot O = b^2 - a^2$$

Similarly, $2AC \cdot O = c^2 - a^2$, $2AD \cdot O = d^2 - a^2$
Rewriting the three equations in matrix form, we have

$$2\begin{bmatrix} AB \\ AC \\ AD \end{bmatrix} \cdot O = 2\begin{bmatrix} b_1 - a_1 & b_2 - a_2 & b_3 - a_3 \\ c_1 - a_1 & c_2 - a_2 & c_3 - a_3 \\ d_1 - a_1 & d_2 - a_2 & d_3 - a_3 \end{bmatrix}\begin{bmatrix} x \\ y \\ z \end{bmatrix} = \begin{bmatrix} b^2 - a^2 \\ c^2 - a^2 \\ d^2 - a^2 \end{bmatrix}$$

where

$$A = (a_1, a_2, a_3), B = (b_1, b_2, b_3), C = (c_1, c_2, c_3), D = (d_1, d_2, d_3), O = (x, y, z)$$

and

$$a^2 = A \cdot A = a_1a_1 + a_2a_2 + a_3a_3, b^2 = B \cdot B, c^2 = C \cdot C, d^2 = D \cdot D$$

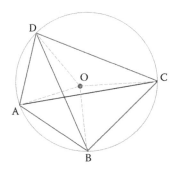

Figure A.4 **Circumsphere of tetrahedron ABCD.**

To determine the circumcentre O, a 3×3 matrix has to be solved. Knowing the circumcentre O, the circumradius r is given by the distance from any one of the vertices, i.e.

$$r = \|OA\| = \|OB\| = \|OC\| = \|OD\|$$

A.5 INVERSE OF MATRIX

i. 2×2 matrix

$$\text{Matrix } A = \begin{bmatrix} a_{11} & a_{12} \\ a_{21} & a_{22} \end{bmatrix}, \quad A^{-1} = \frac{1}{\det(A)} \begin{bmatrix} a_{22} & -a_{12} \\ -a_{21} & a_{11} \end{bmatrix} \quad \text{where } \det(A) = a_{11}a_{22} - a_{12}a_{21}$$

ii. 3×3 matrix

$$\text{Matrix } A = \begin{bmatrix} a_{11} & a_{12} & a_{13} \\ a_{21} & a_{22} & a_{23} \\ a_{31} & a_{32} & a_{33} \end{bmatrix}, \quad A^{-1} = \frac{1}{\det(A)} \begin{bmatrix} c_{11} & c_{12} & c_{13} \\ c_{21} & c_{22} & c_{23} \\ c_{31} & c_{32} & c_{33} \end{bmatrix}$$

where $\det(A) = a_{11}(a_{22}a_{33} - a_{23}a_{32}) + a_{12}(a_{23}a_{31} - a_{21}a_{33}) + a_{13}(a_{21}a_{32} - a_{22}a_{31})$

$$c_{11} = a_{22}a_{33} - a_{32}a_{23}, \quad c_{12} = a_{32}a_{13} - a_{12}a_{33}, \quad c_{13} = a_{12}a_{23} - a_{22}a_{13}$$

$$c_{21} = a_{23}a_{31} - a_{33}a_{21}, \quad c_{22} = a_{33}a_{11} - a_{13}a_{31}, \quad c_{23} = a_{13}a_{21} - a_{23}a_{11}$$

$$c_{31} = a_{21}a_{32} - a_{31}a_{22}, \quad c_{32} = a_{31}a_{12} - a_{11}a_{32}, \quad c_{33} = a_{11}a_{22} - a_{21}a_{12}$$

A.6 NEWTON–RAPHSON METHOD

In mesh generation, very often, an approximation has to be sought to a non-linear system of equations and/or constraints, and the Newton–Raphson method is one of the most effective means in finding a solution.

i. Single variable scalar function $f(x) = 0$ for real number function f on real number x.

As $f(x)$ may have many zeros, one has to start with an initial guess x_0, which is reasonably close to the expected solution. At an intermediate step k, one has $f(x_k) \neq 0$, and one would like to improve the solution by adding a small value Δx to x_k, such that

$$f(x_k + \Delta x) = 0 \Rightarrow f(x_k) + f'(x_k)\Delta x \approx 0 \quad \text{or} \quad \Delta x \approx -\frac{f(x_k)}{f'(x_k)}$$

Hence,

$$x_{k+1} = x_k + \Delta x = x_k - \frac{f(x_k)}{f'(x_k)}$$

The procedure can be repeated until $\|f(x_{k+1})\| < \varepsilon$ for some pre-determined tolerance ε. In case $f'(x) = \dfrac{d}{dx} f(x)$ cannot be determined analytically, numerical differentiation can be used, such that

$$\frac{d}{dx} f(x) = \lim_{h \to 0} \frac{f(x+h) - f(x)}{h} \approx \frac{f(x+h) - f(x)}{h} \text{ for some small value h}$$

ii. Vector functions $\mathbf{F}(\mathbf{x}) = 0$ of multiple variables

$$\mathbf{F} = \{f_1, f_2, \dots f_n\} = \{f_i, i = 1, n\} \in \mathbb{R}^n; \quad \mathbf{x} = \{x_1, x_2, \dots x_n\} = \{x_i, i = 1, n\} \in \mathbb{R}^n$$

$$\mathbf{F}(\mathbf{x}_k + \Delta\mathbf{x}) = 0 \Rightarrow \mathbf{F}(\mathbf{x}_k) + \nabla\mathbf{F}(\mathbf{x}_k)\Delta\mathbf{x} \approx 0 \quad \text{or} \quad \Delta\mathbf{x} \approx -(\nabla\mathbf{F}(\mathbf{x}_k))^{-1}\mathbf{F}(\mathbf{x}_k)$$

$$\text{Gradient matrix } \nabla\mathbf{F} = \begin{bmatrix} \dfrac{\partial f_1}{\partial x_1} & \cdots & \dfrac{\partial f_1}{\partial x_n} \\ \vdots & \ddots & \vdots \\ \dfrac{\partial f_n}{\partial x_1} & \cdots & \dfrac{\partial f_n}{\partial x_n} \end{bmatrix}, \quad \Delta\mathbf{x} = \begin{bmatrix} \Delta x_1 \\ \vdots \\ \Delta x_n \end{bmatrix} \quad \text{and } \mathbf{x}_{k+1} = \mathbf{x}_k + \Delta\mathbf{x}$$

The procedure can be repeated until $\|f(x_{k+1})\| < \varepsilon$ for some pre-determined tolerance ε. In case $\dfrac{\partial f_i}{\partial x_j}$ cannot be determined analytically, numerical differentiation can be used, such that

$$\frac{\partial f_i}{\partial x_j} = \lim_{h \to 0} \frac{f_i(x_j + h) - f(x_j)}{h} \approx \frac{f_i(x_j + h) - f(x_j)}{h} \text{ for some small value h}$$

A.7 NUMERICAL INTEGRATION

Numerical integration is needed to evaluate the distance between two points, the area or the volume of an element created in a parametric domain governed by a non-Euclidean metric. Gauss–Legendre integration is the most efficient to evaluate an integral defined on the interval [–1, 1], and the numerical integration using n Gaussian points is given by

$$I = \int_{-1}^{+1} f(x)\,dx \approx \sum_{k=1,n} w_k f(x_k)$$

The co-ordinates of sampling points x_k and the corresponding weights w_k for one-point to four-point integration are given in Table A.1. Instead of using higher-order integration formulas, complicated integrals can also be evaluated by using subdivided intervals.

Table A.1 Gauss–Legendre integration formulas for one-to-four integration points

n	x_k	w_k	Highest polynomial exactly integrated
1	0	2	1
2	$\pm 1/\sqrt{3}$	1	3
3	0	8/9	
	$\pm\sqrt{3/5}$	5/9	5
4	$\pm\sqrt{\dfrac{3-2\sqrt{6/5}}{7}}$	$\dfrac{1}{2}+\dfrac{1}{6\sqrt{6/5}}$	7
	$\pm\sqrt{\dfrac{3+2\sqrt{6/5}}{7}}$	$\dfrac{1}{2}-\dfrac{1}{6\sqrt{6/5}}$	

A.8 JACOBIAN MATRIX – TRANSFORMATION OF DERIVATIVES

Let (ξ, η, ζ) be a set of reference element co-ordinates and (x, y, z) be the corresponding set of global co-ordinates referring to the same point related through FE interpolation function H.

$$x = H_a(\xi, \eta, \zeta)x_a, \quad y = H_a(\xi, \eta, \zeta)y_a, \quad z = H_a(\xi, \eta, \zeta)z_a, \quad a = 1, n$$

where n is the number of nodes in the finite element and repeated indices represent a summation. By chain rule of partial differentiation on function $\phi = \phi(\xi, \eta, \zeta)$, we can write

$$\frac{\partial\phi}{\partial\xi} = \frac{\partial\phi}{\partial x}\frac{\partial x}{\partial\xi} + \frac{\partial\phi}{\partial y}\frac{\partial y}{\partial\xi} + \frac{\partial\phi}{\partial z}\frac{\partial z}{\partial\xi} \qquad \frac{\partial\phi}{\partial\eta} = \frac{\partial\phi}{\partial x}\frac{\partial x}{\partial\eta} + \frac{\partial\phi}{\partial y}\frac{\partial y}{\partial\eta} + \frac{\partial\phi}{\partial z}\frac{\partial z}{\partial\eta}$$

$$\frac{\partial\phi}{\partial\zeta} = \frac{\partial\phi}{\partial x}\frac{\partial x}{\partial\zeta} + \frac{\partial\phi}{\partial y}\frac{\partial y}{\partial\zeta} + \frac{\partial\phi}{\partial z}\frac{\partial z}{\partial\zeta}$$

$$\begin{bmatrix} \dfrac{\partial\phi}{\partial\xi} \\ \dfrac{\partial\phi}{\partial\eta} \\ \dfrac{\partial\phi}{\partial\zeta} \end{bmatrix} = \begin{bmatrix} \dfrac{\partial x}{\partial\xi} & \dfrac{\partial y}{\partial\xi} & \dfrac{\partial z}{\partial\xi} \\ \dfrac{\partial x}{\partial\eta} & \dfrac{\partial y}{\partial\eta} & \dfrac{\partial z}{\partial\eta} \\ \dfrac{\partial x}{\partial\zeta} & \dfrac{\partial y}{\partial\zeta} & \dfrac{\partial z}{\partial\zeta} \end{bmatrix} \begin{bmatrix} \dfrac{\partial\phi}{\partial x} \\ \dfrac{\partial\phi}{\partial y} \\ \dfrac{\partial\phi}{\partial z} \end{bmatrix} = \mathbf{J} \begin{bmatrix} \dfrac{\partial\phi}{\partial x} \\ \dfrac{\partial\phi}{\partial y} \\ \dfrac{\partial\phi}{\partial z} \end{bmatrix} \quad \text{or} \quad \begin{bmatrix} \dfrac{\partial\phi}{\partial x} \\ \dfrac{\partial\phi}{\partial y} \\ \dfrac{\partial\phi}{\partial z} \end{bmatrix} = \mathbf{J}^{-1} \begin{bmatrix} \dfrac{\partial\phi}{\partial\xi} \\ \dfrac{\partial\phi}{\partial\eta} \\ \dfrac{\partial\phi}{\partial\zeta} \end{bmatrix}$$

$$\text{where } \mathbf{J} = \begin{bmatrix} \dfrac{\partial x}{\partial\xi} & \dfrac{\partial y}{\partial\xi} & \dfrac{\partial z}{\partial\xi} \\ \dfrac{\partial x}{\partial\eta} & \dfrac{\partial y}{\partial\eta} & \dfrac{\partial z}{\partial\eta} \\ \dfrac{\partial x}{\partial\zeta} & \dfrac{\partial y}{\partial\zeta} & \dfrac{\partial z}{\partial\zeta} \end{bmatrix}$$

in which $\dfrac{\partial x}{\partial \xi} = \dfrac{\partial H_a}{\partial \xi} x_a, \quad \dfrac{\partial y}{\partial \xi} = \dfrac{\partial H_a}{\partial \xi} y_a, \quad \dfrac{\partial z}{\partial \xi} = \dfrac{\partial H_a}{\partial \xi} z_a, \quad$ etc.

The determinant of Jacobian (det($\mathbf{J}$)) can be computed using the formulas given in Appendix A.5.

A.9 LAGRANGIAN POLYNOMIAL

The polynomial P(x) passing through n distinct points (x_k, y_k) k = 1,n is given by

$$P(x) = \sum_{k=1}^{n} L_n^k(x) y_k$$

where Lagrangian polynomials $L_n^k(x) = \dfrac{(x - x_1)\ldots(x - x_{k-1})(x - x_{k+1})\ldots(x - x_n)}{(x_k - x_1)\ldots(x_k - x_{k-1})(x_k - x_{k+1})\ldots(x_k - x_n)}$

A.10 BEZIER CURVE

Given n + 1 points $\{P_0, P_1, \ldots, P_n\}$ on a plane, a point corresponding to parameter t on Bezier curve $\mathbf{B}$ is given by

$$P(t) = \sum_{i=0}^{n} P_i B_{i,n}(t), \quad B_{i,n}(t) = \binom{n}{i} t^i (1-t)^{n-i}, \quad \binom{n}{i} = \frac{n!}{i!(n-i)!}, \quad t \in [0,1]$$

For $n = 3$, $\quad P(t) = \begin{bmatrix} t^3 & t^2 & t & 1 \end{bmatrix} \begin{bmatrix} -1 & 3 & -3 & 1 \\ 3 & -6 & 3 & 0 \\ -3 & 3 & 0 & 0 \\ 1 & 0 & 0 & 0 \end{bmatrix} \begin{bmatrix} P_0 \\ P_1 \\ P_2 \\ P_3 \end{bmatrix} \quad t \in [0,1]$

Some of the properties of Bezier curve are

i. End points: $P(0) = P_0$, $P(1) = P_n$.
ii. Tangent: $P'(0) = P_0 P_1$, $P'(1) = P_{n-1} P_n$.
iii. Symmetry: $\mathbf{B}(P_0, P_1, \ldots, P_n) = \mathbf{B}(P_n, P_{n-1}, \ldots, P_0)$.
iv. Bezier curve only depends on the given points but not on the co-ordinate system.
v. Global behaviour: P(t) is a function of all the points $\{P_0, P_1, \ldots, P_n\}$.

Bezier surface can be constructed on a grid of spatial points using direct cross products.

A.11 TRANSFORMATION OF CO-ORDINATES

i. Polar co-ordinates and cylindrical co-ordinates
In Figure A.5a, the Cartesian co-ordinates (x, y) for polar co-ordinates (r, θ) are given by

$$x = r\cos(\theta), \quad y = r\sin(\theta)$$

ii. Spherical co-ordinates
As shown in Figure A.5b, the Cartesian co-ordinates (x, y, z) corresponding to spherical co-ordinates (r, θ, φ) are given by

$$x = r\cos(\phi)\cos(\theta), \quad y = r\cos(\phi)\sin(\theta), \quad z = r\sin(\phi)$$

A.12 LIST OF FORTRAN PROGRAM ADF2D

ADF2D is a mesh generation program of 2D triangular meshes based on the ADF approach (Lo 1992). The input is a list of boundary segments {MA(I), MB(I), I = 1,NB} following the convention, as depicted in Section 3.6.2. This is the final boundary for mesh generation by ADF2D, and in case boundary nodes need to be added by some boundary discretisation process, it should be done by a separate procedure earlier beforehand. The output is a mesh of triangular elements {ME(I), I = 1,3*NE}, where NE = number of elements. The element size is computed based on the line segments on the boundary, as described in Section 3.5.6.3. To cope with a general node spacing function, the part related to the generation of interior nodes ought to be modified such that the height of the triangle to be created is no longer related to the boundary segments but computed from the specified node spacing function.

An example of mesh generation is shown in Figure A.6. In this example, the four edges of a 100 × 100 square are each divided into 10 segments. There are nine circular openings of a diameter of 1 unit: four of them are located 30 units from the edges, another four are located at 20 units normal to the mid-points of the edges and the last one is at the centre of the square. The circular openings are each divided into 20 segments, making up a total of 220 boundary segments for mesh generation. The triangular mesh generated by ADF2D is shown in Figure A.6a in which there are 3634 nodal points and 7064 triangular elements with an α-quality equal to 0.93526. After two cycles of Laplace smoothing, the α-quality of

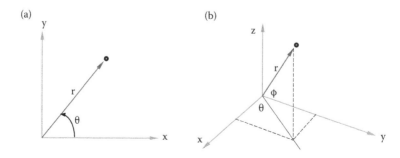

Figure A.5 Transformation of co-ordinates. (a) Polar co-ordinates. (b) Spherical co-ordinates.

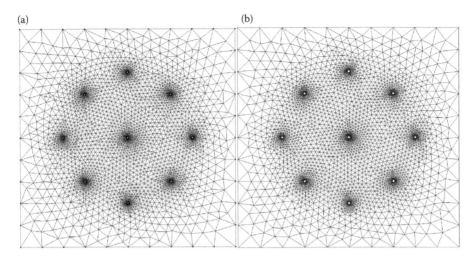

Figure A.6 Triangular mesh generated by ADF2D. (a) Initial mesh, α = 0.93526. (b) Improved mesh, α = 0.96120.

the triangular mesh is improved to 0.96120, as shown in Figure A.6b. **Exercise:** *Following Section 3.6.3, introduce the background grid to speed up the mesh generation process.*

```
      SUBROUTINE AFT2D (NN,X,Y,NB,MA,MB,NE,ME)
!     ADF MESHING ON PLANAR DOMAINS
!     WITH ELEMENT SIZE BASED ON BOUNDARY SEGMENTS
!     INPUT: COORDINATES OF NODAL POINTS, {X(I), Y(I), I=1,NN}
!                    BOUNDARY SEGMENTS, {MA(I),MB(I),I=1,NB}
!     OUTPUT: TRIANGULAR ELEMENTS, {ME(I), I=1,3*NE}
!     WORKING ARRAY: {XP(I),YP(I),DP(I), I=1,5000}
!     MID-POINTS AND LENGTHS OF SEGMENTS
!     {MA(I),MB(I),I=1,1000} CANDIDATE SEGMENTS ON THE FRONT

      IMPLICIT DOUBLE PRECISION (A-H,O-Z)
      DIMENSION X(*),Y(*),ME(*),MA(*),MB(*)
      DIMENSION MS(1000),MT(1000),XP(5000),YP(5000),DP(5000)
!     COMPUTE THE LENGTHS OF THE BOUNDARY SEGMENTS AND THEIR MID-POINTS
      NP=NB
      DO I=1,NB
      IA=MA(I)
      IB=MB(I)
      XP(I)=(X(IA)+X(IB))/2
      YP(I)=(Y(IA)+Y(IB))/2
      DP(I)=(X(IB)-X(IA))**2+(Y(IB)-Y(IA))**2
      ENDDO

!     PREPARATION WORKS FOR THE BASE SEGMENT,
!     J1-J2 = LAST SEGMENT ON THE FRONT
    5 J3=0
      J1=MA(NB)
      J2=MB(NB)
      NB=NB-1
      X1=X(J1)
      Y1=Y(J1)
      X2=X(J2)
```

```
       Y2=Y(J2)
       A=Y1-Y2
       B=X2-X1
       DD=A*A+B*B
       TOR=DD/100
       XM=(X1+X2)/2
       YM=(Y1+Y2)/2
       RR=1.25*DD+TOR
       XC=XM+A
       YC=YM+B
       C=X2*Y1-X1*Y2+TOR

!      FILTER OFF SEGMENTS TOO FAR AWAY FROM THE BASE SEGMENT
    9  NS=0
       DO 11 I=1,NB
       IA=MA(I)
       IB=MB(I)
       IF (DPL(X(IA),Y(IA),X(IB),Y(IB),XC,YC).GT.RR) GOTO 11
       NS=NS+1
       MS(NS)=IA
       MT(NS)=IB
   11  CONTINUE

!      DETERMINE CANDIDATE NODES ON THE GENERATION FRONT
       DO 22 I=1,NS
       J=MS(I)
       P=X(J)
       Q=Y(J)
       IF ((P-XC)**2+(Q-YC)**2.GT.RR.OR.A*P+B*Q.LT.C) GOTO 22
       CALL CHKINT (J1,J2,J,X1,Y1,X2,Y2,P,Q,NS,MS,MT,X,Y,*22)
       CALL CIRCLE (X1,Y1,X2,Y2,P,Q,XC,YC,RR)
       J3=J
   22  CONTINUE

       IF (J3.EQ.0) THEN
       H=SQRT(RR-TOR-DD/4)
       R=SQRT(RR-TOR)
       AREA=SQRT(DD)*(R+H)
       ALPHA=AREA/((R+H)**2+0.75*DD)
       ELSE
       AREA=A*X(J3)+B*Y(J3)+X1*Y2-X2*Y1
       S=DD+(X(J3)-X1)**2+(Y(J3)-Y1)**2+(X(J3)-X2)**2+(Y(J3)-Y2)**2
       ALPHA=SQRT(12.0)*AREA/S
       ENDIF

!      CREATE INTERIOR NODES,
!      CHECK THEIR QUALITIES AND COMPARE WITH FRONTAL NODE J3
       XX=XM+A/2
       YY=YM+B/2
       S1=0
       S2=0
       DO 44 I=1,NP
       S=(XP(I)-XX)**2+(YP(I)-YY)**2+TOR
       S1=S1+DP(I)/S
```

```
   44  S2=S2+1/S
       F=SQRT(0.75*S1/(S2*DD))
       F1=F
       DO 111 I=1,5
  111  F1=(2*F1**3+3*F)/(3*F1*F1+2.25)
       S=F*DD/AREA
       IF (S.GT.1) S=1/S
       BETA=S*(2-S)*ALPHA
       T=1/ALPHA-SQRT(ABS(1/ALPHA**2-1))

       DO 66 I=1,9
       S=(11-I)*F1/10
       GAMMA=SQRT(3.0)*S*S*(2-S/F)/(S*S*F+0.75*F)
       IF (GAMMA.LT.BETA) GOTO 1
       P=XM+A*S
       Q=YM+B*S
       IF ((P-XC)**2+(Q-YC)**2.GT.RR) GOTO 66
       CALL CHKINT (J1,J2,0,X1,Y1,X2,Y2,P,Q,NS,MS,MT,X,Y,*66)
       D=(X(MT(1))-X(MS(1)))**2+(Y(MT(1))-Y(MS(1)))**2
       H=DPL(X(MS(1)),Y(MS(1)),X(MT(1)),Y(MT(1)),P,Q)
       DO 99 J=2,NS
       S=DPL(X(MS(J)),Y(MS(J)),X(MT(J)),Y(MT(J)),P,Q)
       IF (S.GE.H) GOTO 99
       H=S
       D=(X(MT(J))-X(MS(J)))**2+(Y(MT(J))-Y(MS(J)))**2
   99  CONTINUE
       IF (H.GT.D*T**2) GOTO 3
   66  CONTINUE
    1  II=3*NE

!      IF NO NODE CAN BE FOUND TO FORM A VALID ELEMENT
!      WITH THE BASE SEGMENT, ENLARGE THE SEARCH RADIUS
!      ENLARGE THE SEARCH RADIUS
       IF (J3.NE.0) GOTO 2
       IF (RR.GT.100*DD) THEN
       WRITE (*,*) '*** Mesh generation failed! ***'
       RETURN
       ENDIF
       XC=XC+XC-XM
       YC=YC+YC-YM
       RR=(XC-X1)**2+(YC-Y1)**2+TOR
       GOTO 9

!      NODE J3 IS FOUND TO FORM VALID ELEMENT WITH BASE SEGMENT J1-J2
!      UPDATE GENERATION FRONT WITH FRONTAL NODE J3
    2  NE=NE+1
       ME(II+1)=J1
       ME(II+2)=J2
       ME(II+3)=J3
       DO 77 I=1,NB
       IF (MA(I).NE.J3.OR.MB(I).NE.J1) GOTO 77
       MA(I)=MA(NB)
       MB(I)=MB(NB)
       NB=NB-1
       GOTO 7
```

```
 77  CONTINUE
     NB=NB+1
     MA(NB)=J1
     MB(NB)=J3
  7  DO 88 I=1,NB
     IF (MA(I).NE.J2.OR.MB(I).NE.J3) GOTO 88
     IF (NB.EQ.1) RETURN
     MA(I)=MA(NB)
     MB(I)=MB(NB)
     NB=NB-1
     GOTO 5
 88  CONTINUE
     NB=NB+1
     MA(NB)=J3
     MB(NB)=J2
     GOTO 5

!    INTERIOR NODE NN CREATED,
!    UPDATE GENERATION FRONT WITH INTERIOR NODE NN
  3  NN=NN+1
     X(NN)=P
     Y(NN)=Q
     II=3*NE
     NE=NE+1
     ME(II+1)=J1
     ME(II+2)=J2
     ME(II+3)=NN
     NB=NB+1
     MA(NB)=J1
     MB(NB)=NN
     NB=NB+1
     MA(NB)=NN
     MB(NB)=J2
     GOTO 5
     END

!    CALCULATE THE DISTANCE BETWEEN POINT (X3,Y3)
!    TO LINE SEGMENT (X1,Y1)-(X2,Y2)
     FUNCTION DPL(X1,Y1,X2,Y2,X3,Y3)
     IMPLICIT DOUBLE PRECISION (A-H,O-Z)
     R=(X2-X1)**2+(Y2-Y1)**2
     S=(X2-X1)*(X3-X1)+(Y2-Y1)*(Y3-Y1)
     T=(X3-X1)**2+(Y3-Y1)**2
     DPL=T-S*S/R
     IF (S.GT.R) DPL=(X3-X2)**2+(Y3-Y2)**2
     IF (S.LT.0) DPL=T
     END

!    CALCULATE THE CIRCUMCIRCLE OF TRIANGLE (X1,Y1),(X2,Y2),(P,Q)
     SUBROUTINE CIRCLE (X1,Y1,X2,Y2,P,Q,XC,YC,RR)
     IMPLICIT DOUBLE PRECISION (A-H,O-Z)
     A1=X2-X1
     A2=Y2-Y1
     B1=P-X1
     B2=Q-Y1
```

```
      AA=A1*A1+A2*A2
      BB=B1*B1+B2*B2
      AB=A1*B1+A2*B2
      DET=AA*BB-AB*AB
      C1=0.5*BB*(AA-AB)/DET
      C2=0.5*AA*(BB-AB)/DET
      XX=C1*A1+C2*B1
      YY=C1*A2+C2*B2
      RR=1.000001*(XX*XX+YY*YY)
      XC=X1+XX
      YC=Y1+YY
      END


      SUBROUTINE CHKINT (J1,J2,J,X1,Y1,X2,Y2,P,Q,NB,MA,MB,X,Y,*)
      IMPLICIT DOUBLE PRECISION (A-H,O-Z)
      DIMENSION MA(*),MB(*),X(*),Y(*)
      DATA TOL/0.000001/
!     Check if there are any intersections between line segment
!     (P,Q)-(X1,Y1) and the non-Delaunay segments MA(i)-MB(i), i=1,NB
      C1=Q-Y1
      C2=P-X1
      C=Q*X1-P*Y1
      CC=C1*C1+C2*C2
      TOR=-TOL*CC*CC
      DO 11 I=1,NB
      IA=MA(I)
      IB=MB(I)
      IF (J.EQ.IA.OR.J.EQ.IB.OR.J1.EQ.IA.OR.J1.EQ.IB) GOTO 11
      XA=X(IA)
      YA=Y(IA)
      XB=X(IB)
      YB=Y(IB)
      IF ((C2*YA-C1*XA+C)*(C2*YB-C1*XB+C).GT.TOR) GOTO 11
      H1=YB-YA
      H2=XB-XA
      H=XA*YB-XB*YA
      IF ((H2*Y1-H1*X1+H)*(H2*Q-H1*P+H).LT.TOR) RETURN 1
   11 CONTINUE
!     Check if there are any intersections between line segment
!     (P,Q)-(X2,Y2) and the non-Delaunay segments MA(i)-MB(i), i=1,NB
      C1=Q-Y2
      C2=P-X2
      C=Q*X2-P*Y2
      CC=C1*C1+C2*C2
      TOR=-TOL*CC*CC
      DO 22 I=1,NB
      IA=MA(I)
      IB=MB(I)
      IF (J.EQ.IA.OR.J.EQ.IB.OR.J2.EQ.IA.OR.J2.EQ.IB) GOTO 22
      XA=X(IA)
      YA=Y(IA)
      XB=X(IB)
      YB=Y(IB)
      IF ((C2*YA-C1*XA+C)*(C2*YB-C1*XB+C).GT.TOR) GOTO 22
      H1=YB-YA
```

```
      H2=XB-XA
      H=XA*YB-XB*YA
      IF ((H2*Y2-H1*X2+H)*(H2*Q-H1*P+H).LT.TOR) RETURN 1
   22 CONTINUE
      END
```

A.13 LIST OF FORTRAN PROGRAM INSERT

INSERT is a Delaunay triangulation program for three-dimensional spatial points. The input to the insertion kernel INSERT is the point to be inserted IN and the current Delaunay triangulation {ME(i), MF(i), i = 1,4*NE}, where NE is the number of tetrahedra. The output is the updated (modified) Delaunay triangulation with the point to be inserted IN included in the mesh. To construct the Delaunay triangulation of NN randomly generated points, what has to be done is to insert the points 1 to NN in a loop one by one by means of the insertion kernel INSERT. INSERT is the simplest version for Delaunay point insertion in which the circumspheres and normals to the faces of the tetrahedra are not stored; the visibility check in forming the CORE and the random walk or a change of the seed element in the search for the BASE have not been installed either. *As an exercise, readers are encouraged to implement all these precautions with reference to Section 5.5.2.*

An example of Delaunay triangulation of randomly generated points is shown in Figure A.7. In this example, 1000 points are randomly generated, which are then inserted sequentially in an initial Delaunay triangulation of a cube divided into five tetrahedra, as shown in Figure A.7a. The final mesh, as shown in Figure A.7b, consists of 8 + 1000 = 1008 points and 6520 elements, which is the result of the compression from all the 24,255 tetrahedra created throughout the point insertion process. The γ-qualities of the tetrahedral elements of randomly generated points are quite low with $\gamma_{min} = 0.002045$ and $\gamma_{mean} = 0.3436$.

(a)   (b)

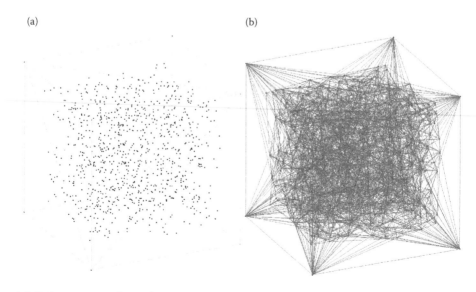

Figure A.7 Delaunay triangulation by point insertion. (a) 1000 randomly generated points. (b) Delaunay triangulation of 1000 points.

```
!       LOOP TO INSERT POINTS 1 TO NN
!       NEM = MAXIMUM NUMBER OF ELEMENTS ALLOWED IN ARRAYS ME, MF AND MT
        DO IN=1,NN
        CALL INSERT (IN,NE,ME,MF,MT,X,Y,Z)
        IF (NE+100.GT.NEM) CALL COMPRESS (NE,ME,MF,MT)
        ENDDO
        CALL COMPRESS (NE,ME,MF,MT)

!       DELAUNAY TRIANGULATION: INSERTION KERNEL
!       INPUT: POINT TO BE INSERTED = IN,
!       DELAUNAY TRIANGULATION {NE,ME,MF}
!       OUTPUT: DELAUNAY TRIANGULATION {NE,ME,MF}
!       INCLUDING INSERTED POINT IN
!       ELEMENT NODE NUMBERS: ME(I), I=1,4*NE
!       NEIGHBORING ELEMENTS: MF(I), I=1,4*NE
!       NUMBER OF ELEMENTS = NE
!       NODAL COORDINATES: X(I),Y(I),Z(I), I=1,NN
!       MT(I), I=1,NE : ARRAY TO FLAG DELETED ELEMENTS,
!       MT(I)=1:DELETE, MT(I)=0:KEEP

        SUBROUTINE INSERT (IN,NE,ME,MF,MT,X,Y,Z)
        IMPLICIT DOUBLE PRECISION (A-H,O-Z)
        DIMENSION ME(*),MF(*),X(*),Y(*),Z(*)
        LOGICAL(1) MT(*)
        DIMENSION MK(1000),ML(1000),M1(1000),M2(1000),MM(3,4)
        DATA MM/2,4,3,1,3,4,1,4,2,1,2,3/
!       MK(*),ML(*),M1(*),M2(*):
!       WORKING ARRAYS FOR THE CONSTRUCTION OF THE CORE

!       PREPARATION AND SEARCH FOR THE BASE TET KB
!       CONTAINING INSERTION POINT IN
        XI=X(IN)
        YI=Y(IN)
        ZI=Z(IN)
        CALL BASE (NE,ME,MF,X,Y,Z,IN,KB)
        MT(KB)=1
        KK=4*KB-4
        NB=0
        NF=0
        DO I=1,4
        KI=MF(KK+I)
        IF (KI.EQ.0) THEN
        NB=NB+1
        M1(NB)=KB
        M2(NB)=I
        ELSE
        NF=NF+1
        MK(NF)=KB
        ML(NF)=I
        ENDIF
        ENDDO

!       DETERMINE THE BOUNDARY FACETS OF THE CORE:
!       FACETS SEPARATING DELAUNAY AND NON-DELAUNAY TETS
!       NB = NUMBER OF BOUNDARY FACETS OF THE CORE
```

```
!      M1(I) = TET SUPPORTING THE FACE I,
!      M2(I) = WHICH FACE OF TET M1(I), I=1,NB
   1   J=ML(NF)
       K=MK(NF)
       NF=NF-1
       L=MF(4*K-4+J)
       IF (MT(L).NE.0) GOTO 3
       LL=4*L-4
       I1=ME(LL+1)
       I2=ME(LL+2)
       I3=ME(LL+3)
       I4=ME(LL+4)
       CALL SPHERE (I1,I2,I3,I4,X,Y,Z,XX,YY,ZZ,RR)
       IF ((XI-XX)**2+(YI-YY)**2+(ZI-ZZ)**2.LT.RR) THEN
       MT(L)=1
       DO 11 I=1,4
       KI=MF(LL+I)
       IF (KI.EQ.K) GOTO 11
       IF (KI.EQ.0) THEN
       NB=NB+1
       M1(NB)=L
       M2(NB)=I
       ELSE
       NF=NF+1
       MK(NF)=L
       ML(NF)=I
       ENDIF
  11   CONTINUE
       ELSE
       NB=NB+1
       M1(NB)=K
       M2(NB)=J
       ENDIF
   3   IF (NF.GT.0) GOTO 1

!      CORE VERIFICATION AND CORRECTION: TO BE INSTALLED

!      FORMING ELEMENTS BY JOINING THE INSERT NODE
!      AND THE BOUNDARY FACETS
       NA=NE
       DO I=4*NE+1,4*(NE+NB)
       MF(I)=-1
       ENDDO
       DO I=1,NB
       K=M1(I)
       J=M2(I)
       KK=4*K-4
       I1=ME(KK+MM(1,J))
       I2=ME(KK+MM(2,J))
       I3=ME(KK+MM(3,J))
       NE4=NE*4
       ME(NE4+1)=I1
       ME(NE4+2)=I2
       ME(NE4+3)=I3
       ME(NE4+4)=IN
```

```
      L=MF(KK+J)
      MF(NE4+4)=L
      NE=NE+1
      MT(NE)=0
      IF (L.NE.0) MF(4*L-4+NEIB(L,MF,K))=NE
      MF(KK+J)=NE
      ENDDO

!     ESTABLISH THE ADJACENCY RELATIONSHIP
      DO I=1,NB
      JJ=M2(I)
      IT=NA+I
      DO 22 II=1,4
      IF (II.EQ.JJ) GOTO 22
      K=M1(I)
      KK=4*K-4
      J3=ME(KK+II)
      J4=ME(KK+JJ)
      IS=4*IT-4+NODE(IT,ME,J3)
      IF (MF(IS).GE.0) GOTO 22
      J=II

    2 L=MF(4*K-4+J)
      IF (MT(L).NE.0) THEN
      J=NODE(L,ME,J4)
      N=NEIB(L,MF,K)
      J4=ME(4*L-4+N)
      K=L
      GOTO 2
      ENDIF
      MF(IS)=L
      N=NOP(L,IT,ME,J3)
      MF(4*L-4+N)=IT
   22 CONTINUE
      ENDDO
      END



      FUNCTION NODE(K,ME,I)
      DIMENSION ME(*)
!     FIND FROM ELEMENT K THE POSITION OF NODE I
      KK=4*K-4
      DO NODE=1,4
      IF (ME(KK+NODE).EQ.I) RETURN
      ENDDO
      STOP "ERROR IN NODE"
      END



      FUNCTION NEIB(L,MF,K)
      DIMENSION MF(*)
!     FIND FROM ELEMENT L THE POSITION OF NEIGHBOURING ELEMENT K
      LL=4*L-4
      DO NEIB=1,4
      IF (MF(LL+NEIB).EQ.K) RETURN
```

```
      ENDDO
      STOP "ERROR IN NEIB"
      END


      FUNCTION NOP(L,K,ME,J)
      DIMENSION ME(*)
!     FIND FROM ELEMENT L THE POSITION OF THE NODE NOT SHARING WITH
!     NEIGHBOURING ELEMENT K (OPPOSITE NODE = J)
      KK=4*K-4
      N=ME(KK+1)+ME(KK+2)+ME(KK+3)+ME(KK+4)-J
      LL=4*L-4
      N=ME(LL+1)+ME(LL+2)+ME(LL+3)+ME(LL+4)-N
      DO NOP=1,4
      IF (ME(LL+NOP).EQ.N) RETURN
      ENDDO
      STOP "ERROR IN NOP"
      END



      SUBROUTINE COMPRESS (NE,ME,MF,MT)
!     TO CONSOLIDATE ELEMENT STORAGE ME, MF
!     TAKING INTO ACCOUNT OF THE DELETED ELEMENTS MT(*)
      ELEMENTS MT(*)
      DIMENSION ME(*),MF(*)
      LOGICAL(1) MT(*)
      N=NE
      DO 11 I=1,NE
      IF (MT(I).EQ.0) GOTO 11
      II=4*I-4
      N=N+1
    1 N=N-1
      IF (N.LT.I) GOTO 2
      IF (MT(N).NE.0) GOTO 1
      NN=N*4-4
      DO J=1,4
      ME(II+J)=ME(NN+J)
      K=MF(NN+J)
      MF(II+J)=K
      IF (K.GT.0) THEN
      L=NEIB(K,MF,N)
      MF(4*K-4+L)=I
      ENDIF
      ENDDO
      MT(I)=0
      MT(N)=1
      N=N-1
   11 CONTINUE
    2 WRITE (*,*) "ELEMENTS BEFORE AND AFTER COMPRESSION =",NE,N
      NE=N
      END


!     LOCATE THE BASE TETRAHEDRON K WHICH CONTAINS INSERTED POINT I
      SUBROUTINE BASE (NE,ME,MF,X,Y,Z,I,K)
      IMPLICIT DOUBLE PRECISION (A-H,O-Z)
```

```
      DIMENSION ME(*),MF(*),X(*),Y(*),Z(*)
      K=NE
    1 KK=4*K-4
      I1=ME(KK+1)
      I2=ME(KK+2)
      I3=ME(KK+3)
      I4=ME(KK+4)
      K1=MF(KK+1)
      K2=MF(KK+2)
      K3=MF(KK+3)
      K4=MF(KK+4)
      V1=VOL(I,I2,I3,I4,X,Y,Z)
      V2=VOL(I1,I,I3,I4,X,Y,Z)
      V3=VOL(I1,I2,I,I4,X,Y,Z)
      V4=VOL(I1,I2,I3,I,X,Y,Z)
      V=VOL(I1,I2,I3,I4,X,Y,Z)
      IF (V1.GE.0.AND.V2.GE.0.AND.V3.GE.0.AND.V4.GE.0) RETURN
      IF (K1.NE.0.AND.V1.LT.V) THEN
      K=K1
      V=V1
      ENDIF
      IF (K2.NE.0.AND.V2.LT.V) THEN
      K=K2
      V=V2
      ENDIF
      IF (K3.NE.0.AND.V3.LT.V) THEN
      K=K3
      V=V3
      ENDIF
      IF (K4.NE.0.AND.V4.LT.V) K=K4
      GOTO 1
      END
      FUNCTION VOL(I1,I2,I3,I4,X,Y,Z)
!     COMPUTE THE VOLUME OF TETRAHEDRON I1-I2-I3-I4
      IMPLICIT DOUBLE PRECISION (A-H,O-Z)
      DIMENSION X(*),Y(*),Z(*)
      X1=X(I1)
      Y1=Y(I1)
      Z1=Z(I1)
      A1=X(I2)-X1
      A2=Y(I2)-Y1
      A3=Z(I2)-Z1
      B1=X(I3)-X1
      B2=Y(I3)-Y1
      B3=Z(I3)-Z1
      C1=X(I4)-X1
      C2=Y(I4)-Y1
      C3=Z(I4)-Z1
      VOL=C1*(A2*B3-A3*B2)+C2*(A3*B1-A1*B3)+C3*(A1*B2-A2*B1)
      END

      SUBROUTINE SPHERE (I1,I2,I3,I4,X,Y,Z,XX,YY,ZZ,RR)
!     COMPUTE THE CIRCUMSPHERE OF TETRAHEDRON I1-I2-I3-I4
      IMPLICIT DOUBLE PRECISION (A-H,O-Z)
      DIMENSION X(*),Y(*),Z(*)
```

```
X1=X(I1)
Y1=Y(I1)
Z1=Z(I1)
X21=X(I2)-X1
X31=X(I3)-X1
X41=X(I4)-X1
Y21=Y(I2)-Y1
Y31=Y(I3)-Y1
Y41=Y(I4)-Y1
Z21=Z(I2)-Z1
Z31=Z(I3)-Z1
Z41=Z(I4)-Z1
C11=Y31*Z41-Y41*Z31
C12=Y41*Z21-Y21*Z41
C13=Y21*Z31-Y31*Z21
C21=Z31*X41-Z41*X31
C22=Z41*X21-Z21*X41
C23=Z21*X31-Z31*X21
C31=X31*Y41-X41*Y31
C32=X41*Y21-X21*Y41
C33=X21*Y31-X31*Y21
D=2*(X21*C11+Y21*C21+Z21*C31)
A1=X21**2+Y21**2+Z21**2
A2=X31**2+Y31**2+Z31**2
A3=X41**2+Y41**2+Z41**2
XX=X1+(A1*C11+A2*C12+A3*C13)/D
YY=Y1+(A1*C21+A2*C22+A3*C23)/D
ZZ=Z1+(A1*C31+A2*C32+A3*C33)/D
RR=(XX-X1)**2+(YY-Y1)**2+(ZZ-Z1)**2
END
```

Index

Page numbers followed by f and t indicate figures and tables, respectively.